D1754897

Handbook of Small Modular Nuclear Reactors

Woodhead Publishing Series in Energy

Handbook of Small Modular Nuclear Reactors

Second Edition

Edited by

Daniel T. Ingersoll
Former Director of Research Collaborations at NuScale Power LLC, Oak Ridge, TN, United States

Mario D. Carelli
Former Chief Scientist for Research & Technology at Westinghouse Electric Co., Pittsburgh, PA, United States

WP
WOODHEAD
PUBLISHING

ELSEVIER An imprint of Elsevier

Woodhead Publishing is an imprint of Elsevier
The Officers' Mess Business Centre, Royston Road, Duxford, CB22 4QH, United Kingdom
50 Hampshire Street, 5th Floor, Cambridge, MA 02139, United States
The Boulevard, Langford Lane, Kidlington, OX5 1GB, United Kingdom

Copyright © 2021 Elsevier Ltd. All rights reserved.

No part of this publication may be reproduced or transmitted in any form or by any means, electronic or mechanical, including photocopying, recording, or any information storage and retrieval system, without permission in writing from the publisher. Details on how to seek permission, further information about the Publisher's permissions policies and our arrangements with organizations such as the Copyright Clearance Center and the Copyright Licensing Agency, can be found at our website: www.elsevier.com/permissions.

This book and the individual contributions contained in it are protected under copyright by the Publisher (other than as may be noted herein).

Notices
Knowledge and best practice in this field are constantly changing. As new research and experience broaden our understanding, changes in research methods, professional practices, or medical treatment may become necessary.

Practitioners and researchers must always rely on their own experience and knowledge in evaluating and using any information, methods, compounds, or experiments described herein. In using such information or methods they should be mindful of their own safety and the safety of others, including parties for whom they have a professional responsibility.

To the fullest extent of the law, neither the Publisher nor the authors, contributors, or editors, assume any liability for any injury and/or damage to persons or property as a matter of products liability, negligence or otherwise, or from any use or operation of any methods, products, instructions, or ideas contained in the material herein.

Library of Congress Cataloging-in-Publication Data
A catalog record for this book is available from the Library of Congress

British Library Cataloguing-in-Publication Data
A catalogue record for this book is available from the British Library

ISBN: 978-0-12-823916-2 (print)

ISBN: 978-0-12-823917-9 (online)

For information on all Woodhead publications
visit our website at https://www.elsevier.com/books-and-journals

Publisher: Brian Romer
Acquisitions Editor: Maria Convey
Editorial Project Manager: Chiara Giglio
Production Project Manager: Anitha Sivaraj
Cover Designer: Mark Rogers

Typeset by SPi Global, India

Dedication

This book is dedicated to all pioneers, practitioners, and first adopters of small modular reactors who are collaborating to create the future of nuclear energy.

Contents

Contributors	xv
Preface	xvii
Introduction	xix

Part One Fundamentals of small modular nuclear reactors (SMRs) — 1

1 Small modular reactors (SMRs) for producing nuclear energy: An introduction — 3
Neil Todreas
- 1.1 Introduction — 3
- 1.2 Incentives and challenges for achieving commercial deployment success — 7
- 1.3 Overview of different types of SMRs — 10
- 1.4 Public health and safety — 19
- 1.5 The current status of SMRs — 23
- 1.6 Future trends — 24
- 1.7 Conclusion — 24
- 1.8 Sources of further information and advice — 24
- Appendix: Nomenclature — 25
- References — 26

2 Small modular reactors (SMRs) for producing nuclear energy: International developments — 29
Daniel T. Ingersoll
- 2.1 Introduction — 29
- 2.2 Water-cooled reactors — 31
- 2.3 Gas-cooled reactors — 38
- 2.4 Liquid metal-cooled reactors — 41
- 2.5 Molten-salt-cooled reactors — 44
- 2.6 Future trends — 47
- 2.7 Sources of further information — 49
- References — 50

3	Integral pressurized-water reactors (iPWRs) for producing nuclear energy: A new paradigm	51
	M.D. Carelli	
	3.1 Introduction	51
	3.2 The imperatives for nuclear power	52
	3.3 The integral pressurized-water reactor (iPWR)	54
	3.4 Addressing the safety imperative	56
	3.5 Satisfying the economic competitiveness imperative	61
	3.6 Future trends	63
	3.7 Conclusion	64
	3.8 Sources of further information and advice	65
	References	65

Part Two Small modular nuclear reactor (SMR) technologies   67

4	Core and fuel technologies in integral pressurized water reactors (iPWRs)	69
	Andrew Worrall	
	4.1 Introduction	69
	4.2 Safety design criteria	70
	4.3 Design features to achieve the criteria	75
	4.4 Integral pressurized water reactor (iPWR) design specifics	82
	4.5 Conclusion	91
	References	93

5	Key reactor system components in integral pressurized water reactors (iPWRs)	95
	Randall J. Belles	
	5.1 Introduction	95
	5.2 Integral components	96
	5.3 Connected system components	108
	5.4 Future trends	112
	5.5 Sources of further information and advice	113
	References	114

6	Instrumentation and control technologies for small modular reactors (SMRs)	117
	Dara Cummins and Edward (Ted) Quinn	
	6.1 Introduction	117
	6.2 Safety system instrumentation and controls	119
	6.3 NSSS control systems instrumentation	127
	6.4 BOP instrumentation	131
	6.5 Diagnostics and prognostics	131
	6.6 Processing electronics	132

6.7	Cabling	135
6.8	Future trends and challenges	136
6.9	Conclusion	143
	References	143

7 Human-system interfaces in small modular reactors (SMRs) — 147
Jacques Hugo

7.1	Introduction	147
7.2	Human-system interfaces for small modular reactors	149
7.3	The state of HSI technology in existing nuclear power plants	151
7.4	Advanced HSIs and the human factors challenges	152
7.5	Differences in the treatment of HSIs in the nuclear industry	155
7.6	How to identify and select advanced HSIs: Five dimensions	157
7.7	Operational domains of HSIs	162
7.8	HSI technology classification	167
7.9	HSI architecture and functions	173
7.10	Implementation and design strategies	175
7.11	Future trends	178
7.12	Conclusion	182
	References	183

8 Safety of integral pressurized water reactors (iPWRs) — 187
Bojan Petrovic

8.1	Introduction	187
8.2	Approaches to safety: Active, passive, inherent safety and safety by design	189
8.3	Testing of SMR components and systems	196
8.4	Probabilistic risk assessment (PRA)/probabilistic safety assessment (PSA)	204
8.5	Security as it relates to safety	210
8.6	Future trends	211
	References	213

9 Proliferation resistance and physical protection (PR&PP) in small modular reactors (SMRs) — 217
Lap-Yan Cheng and Robert A. Bari

9.1	Introduction	217
9.2	Methods of analysis	222
9.3	System response and outcomes	223
9.4	Steps in the Generation IV International Forum (GIF) evaluation process	226

9.5	Lessons learned from performing proliferation resistance and physical protection (PR&PP)	229
9.6	Physical security	232
9.7	Future trends	234
9.8	Sources of further information and advice	236
	References	237

Part Three Implementation and applications   239

10 Economics and financing of small modular reactors (SMRs)   241
S. Boarin, M. Mancini, M. Ricotti, and G. Locatelli

10.1	Introduction	241
10.2	Investment and risk factors	245
10.3	Capital costs and economy of scale	252
10.4	Capital costs and multiple units	255
10.5	Capital costs and size-specific factors	260
10.6	Competitiveness of multiple small modular reactors (SMRs) versus large reactors	263
10.7	Competitiveness of SMRs versus other generation technologies	269
10.8	External factors	271
10.9	Future trends	273
10.10	Sources of further information and advice	274
	References	275

11 Licensing of small modular reactors (SMRs)   279
Richard L. Black

11.1	Introduction	279
11.2	US Nuclear Regulatory Commission (NRC) licensing of small modular reactors (SMRs): An example	280
11.3	Non-LWR advanced reactor SMR licensing	288
11.4	Industry codes and standards to support SMR licensing	290
11.5	International strategy and framework for SMR licensing	291
11.6	Conclusion	297
	References	297

12 Construction methods for small modular reactors (SMRs)   299
N. Town and S. Lawler

12.1	Introduction	299
12.2	Options for manufacturing	302
12.3	Component fabrication	309
12.4	Advanced joining techniques	317
12.5	Supply chain implications	318
12.6	Conclusion	322
	Reference	322

13	**Hybrid energy systems using small modular nuclear reactors (SMRs)**	323
	Shannon M. Bragg-Sitton	
	13.1 Introduction	323
	13.2 Principles of HESs	328
	13.3 Evaluating the merit of proposed hybrid system architectures	330
	13.4 The when, why, and how of SMR hybridization	336
	13.5 Coupling reactor thermal output to nonelectric applications	342
	13.6 Future trends	349
	Acknowledgments	353
	References	353

Part Four International R&D and deployment 357

14	**Small modular reactors (SMRs): The case of Argentina**	359
	Dario F. Delmastro	
	14.1 Introduction	359
	14.2 Small modular reactor (SMR) research and development in Argentina	359
	14.3 Integrated pressurized water reactor: CAREM	362
	14.4 Deployment of SMRs in Argentina	370
	14.5 Future trends	370
	14.6 Sources of further information and advice	372
	References	372
15	**Small modular reactors (SMRs): The case of Canada**	375
	Metin Yetisir	
	15.1 Introduction	375
	15.2 Canada's SMR strategy	376
	15.3 SMR markets and potential applications in Canada	378
	15.4 Canadian regulatory framework	384
	15.5 Support for development and deployment	386
	15.6 Future trends	388
	15.7 Conclusion	390
	Acknowledgments	391
	References	391
16	**Small modular reactors (SMRs): The case of China**	395
	Danrong Song	
	16.1 Introduction	395
	16.2 SMRs in the People's Republic (PR) of China: HTR-200	396
	16.3 SMRs in PR of China: ACP100	399

	16.4	Deployment of SMRs in PR of China	406
	16.5	Future trends	407
	Acknowledgments		408
	References		408
17	**Small modular reactors (SMRs): The case of Japan**		**409**
	Tsutomo Okubo		
	17.1	Introduction	409
	17.2	Small modular nuclear reactor (SMR) R&D in Japan	410
	17.3	SMR technologies in Japan	412
	17.4	Deployment of SMRs in Japan	422
	17.5	Future trends	423
	17.6	Sources of further information and advice	423
	References		423
18	**Small modular reactors (SMRs): The case of the Republic of Korea**		**425**
	Suhn Choi		
	18.1	Introduction	425
	18.2	Korean integral pressurized-water reactor: System-integrated Modular Advanced ReacTor	428
	18.3	Development of other small modular nuclear reactor (SMR) programs in the Republic of Korea	443
	Acknowledgment		463
	References		463
	Further reading		464
19	**Small modular reactors (SMRs): The case of Russia**		**467**
	Vladimir Kuznetsov		
	19.1	Introduction	467
	19.2	OKBM Afrikantov small modular reactor (SMR) projects being deployed and developed in Russia	469
	19.3	SMRs being developed by Joint Stock Company (JSC) NIKIET in Russia	480
	19.4	SMR projects developed by JSC AKME Engineering in Russia	489
	19.5	Deployment of SMRs in Russia	494
	19.6	Future trends	496
	19.7	Conclusion	497
	19.8	Sources of further information	498
	References		499
20	**Small modular reactors (SMRs): The case of the United Kingdom**		**503**
	Kevin W. Hesketh and Nicholas J. Barron		
	20.1	Introduction	503

20.2	History of nuclear power development in the United Kingdom	503
20.3	Strategic requirements and background to UK interest in modular reactors	505
20.4	UK R&D activities to support modular reactor development	507
20.5	Future role of SMRs/AMRs in low-carbon energy generation	515
20.6	Conclusions	517
Appendix 20.1		518
Appendix 20.2		518
References		519

21 Small modular reactors (SMRs): The case of the United States of America   521
Gary Mays

21.1	Introduction	521
21.2	Near-term SMR activities in United States	522
21.3	Longer-term activities: US Department of Energy Office of Nuclear Energy (DOE-NE) small modular reactor (SMR) R&D program	530
21.4	A-SMR concept evaluations	534
21.5	DOE-NE GAIN program and A-SMRs	541
21.6	DOE-NE Nuclear Energy University Program and A-SMRs	546
21.7	DOE-NE National Reactor Innovation Center	546
21.8	DOE-NE R&D efforts related to development of microreactors	548
21.9	DOE-ARPA-E R&D for modeling and simulation of innovative technologies for advanced reactors	549
21.10	Future trends	550
References		552

Part Five   Global perspectives   555

22 Small modular reactor (SMR) adoption: Opportunities and challenges for emerging markets   557
Geoffrey Black, David Shropshire, and Kathleen Araújo

22.1	Introduction	557
22.2	SMR market deployment potential	559
22.3	Recent climate goals and initiatives	566
22.4	Disruptive change: A closer look at global shifts and SMR options	570
22.5	Challenges and opportunities	574
22.6	Conclusion	586

22.7	Sources of further information and advice	**587**
References		**588**

23 Small modular reactors (SMRs): The case of developing countries — **595**

D. Goodman

23.1	Introduction	**595**
23.2	Measuring development	**596**
23.3	Trade-offs of small modular reactors (SMRs) in developing countries	**597**
23.4	Characteristics of developing countries that make deployment of SMRs viable	**598**
23.5	SMR choices in developing countries	**601**
23.6	Obstacles and innovations	**603**
23.7	Conclusion	**606**
Acknowledgments		**606**
References		**606**

Index — **611**

Contributors

Kathleen Araújo Energy Policy Institute, Boise State University, Boise, ID, United States

Robert A. Bari Brookhaven National Laboratory, Upton, NY, United States

Nicholas J. Barron Reactor Core Technology, National Nuclear Laboratory, Sellafield, United Kingdom

Randall J. Belles Oak Ridge National Laboratory, Oak Ridge, TN, United States

Geoffrey Black Department of Economics, College of Business and Economics, Boise State University, Boise, ID, United States

Richard L. Black Consultant, McLean, VA, United States

S. Boarin Politecnico di Milano, Milan, Italy

Shannon M. Bragg-Sitton Idaho National Laboratory, Idaho Falls, ID, United States

M.D. Carelli Formerly of Westinghouse Electric Co., Pittsburgh, PA, USA

Lap-Yan Cheng Brookhaven National Laboratory, Upton, NY, United States

Suhn Choi Korea Atomic Energy Research Institute, Daejeon, Republic of Korea

Dara Cummins Independent Contractor, Loudon, TN, United States

Dario F. Delmastro National Atomic Energy Commission and Universidad Nacional de Cuyo, San Carlos de Bariloche, Río Negro, Argentina

D. Goodman Consultant, USA

Kevin W. Hesketh Fuel and Core, National Nuclear Laboratory, Preston, United Kingdom

Jacques Hugo Jacques Hugo Associates, Pretoria, South Africa

Daniel T. Ingersoll NuScale Power LLC (retired), Oak Ridge, TN, United States

Vladimir Kuznetsov Consultant, Austria

S. Lawler Rolls-Royce plc, Derby, UK

G. Locatelli University of Lincoln, Lincoln, UK

M. Mancini Politecnico di Milano, Milan, Italy

Gary Mays Oak Ridge National Laboratory, Oak Ridge, TN, United States

Tsutomo Okubo Japan Atomic Energy Agency (retired), Oarai-Machi, Japan

Bojan Petrovic Georgia Institute of Technology, Atlanta, GA, United States

Edward (Ted) Quinn Technology Resources, Dana Point, CA, United States

M. Ricotti Politecnico di Milano, Milan, Italy

David Shropshire Nuclear Science and Technology Directorate, Idaho National Laboratory, Idaho Falls, ID, United States

Danrong Song Nuclear Power Institute of China, Chengdu, People's Republic of China

Neil Todreas Massachusetts Institute of Technology, Cambridge, MA, United States

N. Town Rolls-Royce plc, Derby, UK

Andrew Worrall Oak Ridge National Laboratory, Oak Ridge, TN, United States

Metin Yetisir Advanced Reactor Technologies, Canadian Nuclear Laboratories, Chalk River, ON, Canada

Preface

The modern era of small modular nuclear reactors began roughly at the turn of the millennium. Smaller sized nuclear reactors have been a part of the nuclear heritage from its beginning in the 1950s. However, designing of new nuclear plants, small and large, was largely suspended during the 1990s when countries such as the United States experienced a low demand for new generating capacity and other countries such as France, Japan, and the Republic of Korea proceeded to deploy standardized plant designs. Near the conclusion of the 1990s, the US government initiated a research program to stimulate the idled nuclear industry and specifically targeted the development of smaller, more robust nuclear plant designs.

One of the designs that emerged from the US research initiative was the International Reactor Innovative and Secure (IRIS) design, which was developed by an international consortium of partners that eventually spanned more than 20 organizations in 10 countries. The team was led by Dr. Mario Carelli, Chief Scientist at Westinghouse Electric Company (WEC), and consisted of a diverse set of academic, research, and industrial partners. One of those partners was Oak Ridge National Laboratory (ORNL). Dr. Daniel Ingersoll, a Senior Program Manager at ORNL, led ORNL's participation in the IRIS project until 2008 when he was assigned technical leadership of a new program initiated by the US Department of Energy focused on research and development of small modular reactors.

By the end of 2010, several new SMR designs had emerged in the United States and globally, and customer interest in a new generation of designs was expanding rapidly. In late 2011 Carelli was approached by Woodhead Publishing to serve as Editor for a new project to publish a handbook on small modular nuclear reactors. Given the size of the project and the dynamic nature of the topic, Carelli invited Ingersoll to join as coeditor. At the same time, Ingersoll left ORNL and moved to NuScale Power, a relatively new company dedicated to the design of an especially innovative SMR. Carelli and Ingersoll collaborated to develop the scope and organization of the Handbook and amassed a collection of 20 experts to contribute specific chapters focused on many different aspects of SMRs. The first edition of the *Handbook of Small Modular Nuclear Reactors* was released at the end of 2014.

A second edition of the Handbook was requested by the publisher in 2019. The result is this newly updated and expanded *Handbook of Small Modular Nuclear Reactors*. Significant changes incorporated into this second edition include the following:

- The creation of a final Part V devoted to broad international markets and perspectives.
- The updating of most chapters to include new developments occurring during the 5 years since release of the first edition.

- The addition of three new chapters: two in Part IV (R&D activities in Canada and the United Kingdom) and one in Part V (global market assessment).

The energy landscape and especially the nuclear energy landscape continues to evolve in a very dynamic and somewhat unpredictable fashion as energy demand changes, old technologies are abandoned, new technologies are introduced, and sociopolitical policies fluctuate. It is our hope that this updated Handbook will provide the reader with the best, most accurate understanding of the state of the art in small modular nuclear reactors.

The Editors

Introduction

This handbook provides a thorough and authoritative introduction to today's hottest new development in nuclear plant design and deployment: small modular reactors (SMRs). Building on the global success of large nuclear plants, SMRs offer the potential to expand the use of clean, reliable nuclear energy to a broader range of customers and energy applications.

The early commercial nuclear power reactors designed and built from the 1950s and 1960s were low-power plants (up to a few hundred megawatts) and were built to demonstrate the commercial viability of nuclear energy. These plants were comparable with their fossil-fueled counterparts, both in output and construction time (a few years). They were moderately successful; however, their unit capital costs ($/kW) were substantially higher than for comparable fossil plants. As the nuclear plant cost kept increasing to improve performance and safety, it became necessary to also increase the output power to maintain competitive energy prices; thus the plant size increased rapidly from a few hundreds of megawatts to nearly 2000 MW today. Such a drastic increase had several effects: Only a few manufacturers, either large conglomerates or state-owned enterprises, remained in operation worldwide; plant costs became stratospheric, creeping into tens of billions dollars; and the time from contract signing to initiating power production exceeded a decade.

Starting in the 1990s new SMR designs emerged worldwide and have gained increasing momentum in the new millennium with the intent to complement large plants and offer a more diverse option to potential customers. The new small plants have several traits in common with earlier designs, such as size (from tens to a few hundreds of megawatts), relative simplicity, and a shorter construction time owing to increased factory-based fabrication. Also, SMRs can cover a wide range of applications and deployment times. Those proposed for power-producing applications in the short term are designs of the light water reactor (LWR) type, while SMRs best suited for other applications such as fuel breeding and waste burning employ different coolants and are deployable over the long term.

Many developers of SMRs, even the near-term LWRs, are quite different from the large LWR manufacturers. They include smaller-sized manufacturers and new enterprises. For example, the two LWR-based SMR vendors in the United States are currently NuScale Power, an entirely new enterprise dedicated to a specific SMR design, and Holtec International, a recognized leader in fuel storage facilities but new to reactor design. The international SMR development community reflects a similar demographic with many designs evolving from the research community or nontraditional suppliers.

This handbook is composed of 23 chapters structured into five parts, each chapter being authored by a recognized expert in the field.

- Part I (Fundamentals of SMRs) provides a comprehensive introduction to SMR technologies, existing commercial designs, and fundamental design strategies. The three authors contributing to this section have been eminent proponents of SMRs since the 1990s and have led the development of integral pressurized water reactor (iPWR) designs, which are the prevailing design strategy for SMRs and the focus of this handbook. Part I is articulated over three chapters: overview of SMR features and technologies, review of several current SMRs being developed and deployed worldwide, and an introduction to iPWRs as a specific SMR design category.
- Part II (SMR technologies) reviews the key technologies that are fundamental to the iPWR design, focusing on what is new and different while also providing insight on potential opportunities and challenges. Six chapters, written by internationally recognized authorities in their field, address several basic SMR technologies: the reactor core and fuel, reactor system components, performance monitoring and control, human-system interfaces, safety, and proliferation resistance and physical protection.
- Part III (Implementation and applications) addresses four key areas critical to successful deployment of SMRs: economics and financing, hybrid energy systems using SMRs, licensing, and manufacturing methods. As with Part II the four authors of Part III are recognized authorities in their field.
- Part IV (International R&D and deployment) reviews state-of-the-art projects and programs for SMR development and deployment. Eight chapters focus on countries that are most active in the development and deployment of SMRs, presented in alphabetical order: Argentina, Canada, China, Japan, Republic of Korea, Russia, the United Kingdom, and the United States. The authors are accomplished researchers and directly involved in their country's SMR activities.
- Part V (Global perspectives), composed of just two chapters, provides a timely snapshot of the global market for SMRs and also offers a timeless perspective on how SMR deployment might impact economic growth and environmental conditions in developing countries. It is a reminder that SMRs promise not only to be a better and more economical solution for new energy but also promote improved opportunities and quality of life in emerging economies.

This handbook is intended to be useful to those with a general interest in SMRs, as well as to those looking for detailed technical information. It is further intended that this handbook serves as a guide, through its copious references, to further learning on the subject.

Part One

Fundamentals of small modular nuclear reactors (SMRs)

Small modular reactors (SMRs) for producing nuclear energy: An introduction

Neil Todreas
Massachusetts Institute of Technology, Cambridge, MA, United States

1.1 Introduction

Just what are small modular reactors (SMRs)? This question is first answered simply along with a brief history of the evolution of this class of reactors. Subsequent sections detail the incentives and challenges to achieving successful commercial deployments, the different types of SMRs based on coolants employed, and, finally, the current status and future trends in the worldwide effort to develop and deploy this reactor type.

1.1.1 Defining SMRs

"Small" refers to the reactor power rating. While no definitive range exists, a power rating from approximately 10–300 MWe has generally been adopted. The minimum rating assures that the reactor delivers power suitable for the practical industrial application of interest. The maximum rating constrains these designs to power levels at which the expected advantages of serial production and incremental deployment as well as the match to electric grid siting opportunities and constraints can be realized. In addition to a growing interest in SMRs, there has been a recent surge of interest in nuclear reactors with output below 10 MWe, which are commonly referred to as microreactors.

"Modular" refers to the unit assembly of the nuclear steam supply system (NSSS), which, when coupled to a power conversion system or process heat supply system, delivers the desired energy product. The unit assembly can be assembled from one or several submodules. The desired power plant can then be created from one or several modules as necessary to deliver the desired power rating. Importantly the deployment of modules can also be sequenced over time both to match regional load growth and to levelize the timing of capital spending over a prescribed time horizon. Construction of the plant by assembly of factory-built elements or modules is the technique of modular construction. Although it is an integral part of the construction strategy envisioned for all SMRs, this technique is not uniquely applied to SMRs. Rather, it is now being employed for relevant construction elements of nuclear power plants of all power ratings, although the modules for large plants are considerably different in size, not typically amenable to rapid assembly as is being proposed for SMRs.

"Reactor" is a term more broadly applied to vessels in which all manner of chemical processes are conducted. However, in our case, reactor refers to a system in which a controlled nuclear fission process is conducted.

1.1.2 Strategy for development of SMRs

Small reactors and the modular construction of reactors are not new. Historically, early reactors for commercial production of electricity were of small size, a consequence of the prudent engineering process of constructing plants starting at small ratings to gain the needed construction and operating experience necessary to move confidently to larger ratings. Now, after a half-century of experience, commercial civil reactors are being deployed with ratings up to 1660 MWe. Additionally, small units were built for terrestrial deployment to provide electric power for remote, vulnerable military sites; for ocean deployment; for propulsion of submarines, naval, and commercial ships; and for aircraft propulsion. Modular construction techniques historically have also been used for serial production of selected products. However, what is new is the vision of small rated power reactors composed of a single or multiple modules sized to markets of small- or large-sized electric grids, thereby creating new nuclear generating sites, which require significantly reduced capital investments and capital investment rates. The further economic premise is that electric generation cost can be made sufficiently comparable to that of existing large-sized plants by employing a strategy of economy of numbers (manufacture of multiple identical modules) and simplification of design versus the traditional economy of scale.

1.1.3 Evolution of SMRs

Commercial electric power began with small reactors of light water-cooled design. Key examples are the Shippingport, 60-MWe reactor designed by the Westinghouse-operated Bettis Naval Atomic Power Laboratory, which started operation in 1958; the Yankee Rowe reactor, 185 MWe (Westinghouse) in 1960; the Indian Point One reactor, 275 MWe (B&W) in 1962 (all pressurized water reactor [PWR] designs); and Dresden 210 MWe (General Electric) in 1960 (a boiling-water reactor (BWR) design).

The eight military reactors for terrestrial application developed by the US Army Nuclear Power Program included (1) the stationary plants operated at Fort Belvoir, Virginia, which started operation in April 1957, 7 months before Shippingport and 5 years before criticality of the Ft. Greely, Alaska reactor; (2) the portable reactor operated at McMurdo Sound at the South Pole in 1962; and (3) a barge-mounted reactor operated off the coast of Panama City, Panama, in 1967. These plants ranged from 1.75 to 10 MWe and performed either a heating or desalinization function in addition to the generation of electricity. Another example of a portable reactor is the Russian PAMIR reactor designed primarily to power remote military radar outposts. The first was the TES-3, a 2 MWe nuclear plant completed in 1961. The design was modified in the 1980s to a smaller, more mobile 630 kW reactor.

The much larger US naval program, which pioneered the application of nuclear power for the propulsion of submarines and surface ships, has produced multiple pressurized water reactors and one sodium-cooled reactor of small ratings. Additionally, several countries have followed suit with naval propulsion—most notably Russia, which expanded its development of water-cooled submarine reactors to submarines using lead-bismuth coolant and has also built nuclear powered naval surface ships and ice breakers.

Commercial (merchant marine) propulsion has also been exploited through the development of ocean freighters and ice breakers. Four freighters, all with reactors of light water design, have been built and operated albeit without commercial success: (1) the *US Savannah*, 74 MWt, in effective service starting 1962; (2) the German *Otto Hahn*, 38 MWt, 1968; (3) the Japanese *Mutsu*, 36 MWt, 1972; and (4) the only vessel still in operation under nuclear power, the Russian *Sevmorput*, 135 MWt, delivered in 1988, which also has ice-breaking capability.

The *Otto Hahn* reactor design is of special interest since its integral design characteristic is the typical configuration being exploited by several modern PWR SMR vendors. As extensively elaborated in Chapter 3, the term integral design means the colocation of all components and piping of the primary coolant system in the single pressure vessel. By contrast the typical large-rated PWRs are loop systems with the primary system components, for example, the steam generators, primary coolant pumps and pressurizer connected by piping to each other, and the pressure vessel, which houses the reactor core and the control elements.

To date Russia alone has constructed and operated nine nuclear-powered ice breakers, starting in 1959 with the *Lenin*. Two vessel classes have been built: the Arktika class, each vessel with two OK-900A reactors each of 171 MWt, and the Taymy class, each vessel with a single KLT-40M reactor of 135 MWt. (NB: All reactors of the ocean vessels noted previously drive propulsion shafts; thus their ratings are only in MWt.) Also, Russia has constructed a nonself-propelled floating nuclear power station, the Akademik Lomonosov, to provide power supply to remote coastal towns. The reactor station which achieved commercial operation in May 2020 consists of two modified ice-breaker reactors, each a KLT-40S reactor of 35 MWe. With these reactors the station can provide either 70 MWe of power, 300 MWt of district heating, or 240,000 m^3/day of fresh water.

The development of a nuclear propulsion system for military aircraft was initiated in 1946 as the US Nuclear Energy for the Propulsion of Aircraft (NEPA) project and continued under the name of the Aircraft Nuclear Propulsion (ANP) program. Two different systems for nuclear-powered jet engines were pursued—a direct air cycle concept developed by General Electric and an indirect air cycle by Pratt and Whitney. Only the direct air cycle program advanced sufficiently to produce reactors. The first product of the GE program was the Aircraft Reactor Experiment (ARE), which operated for 1000 h in 1954. It was a 2.5-MWt nuclear reactor experiment using molten fluoride salt (NaF-ZrF_4-UF_4) as fuel, a beryllium oxide (BeO) moderator, and liquid sodium as a secondary coolant. In 1955 this program produced the successful X-39 engine with heat supplied by the Heat Transfer Reactor Experiment-1 (HTRE-1). The HTRE-1 was replaced by the HTRE-2 and eventually the HTRE-3 unit powering the two jet turbines. Additionally, an operating reactor named the aircraft shield test reactor (ASTR) was flown aboard a modified B-36 bomber to test shielding rather than powering the plane. The HTRE-3 used a shield system of flight-type design but was not taken to power before the program was canceled in 1961.

Experience with these earlier reactors has led to the current interest in reduced size modular power plants. Table 1.1 lists several current SMRs under development, which encompass all coolant technologies being exploited for larger nuclear reactors.

Additional reactor designs not included in Table 1.1 are under development by national research institutions but have not yet reached the commercialization stage. For example, the fluoride-salt-cooled high-temperature reactor (FHR) (Forsberg

Table 1.1 Examples of current small (>10 MWe) modular reactors proposed by commercial industries.

Reactor design	Power rating (MWe)	Country	Vendor/AE
Light water-cooled (PWR)			
ACP100	100	China	CNNC/Guodian
CAREM	27	Argentina	CNEA/INVAP
KLT-40S	35	Russia	OKBM
NuScale	60	United States	NuScale Power/Fluor
RITM-200	50	Russia	OKBM
SMART	100	S. Korea	KAERI
SMR-160	160	United States	Holtec
Light water-cooled (BWR)			
BWRX-300	300	United States/Japan	GE-Hitachi
VK-300	250	Russia	NIKIET
Gas-cooled			
EM2	265	United States	General Atomics
GT-MHR	288	Russia	OKBM
HTR-PM	105	China	INET Tsinghua University
Xe-100	75	United States	X Energy, LLC
Sodium-cooled			
4S	10/50	Japan	Toshiba
PRISM	311	United States	GE-Hitachi
Lead-cooled			
SVBR-100	100	Russia	JSC EDB
BREST	300	Russia	AKME-engineering
Molten salt-cooled			
IMSR	190	Canada	Terrestrial Energy, Inc.
LFTR	250	United States	Flibe Energy, Inc.

et al., 2013) is a 180-MWe reactor with 700°C peak operating temperature coupled to an air-Brayton combined cycle system.

1.2 Incentives and challenges for achieving commercial deployment success

The question arises why interest in SMRs has reemerged and burgeoned over the last decade. The reason is that SMRs offer an attractive vehicle to surmount the current barriers to deployment of the current generation of large-rated advanced light water plants (the Generation III+ designs) and alternative coolant (Generation IV) plants. Principal among these barriers is the large initial investment required to construct a reactor, the attendant significant financial risk to the investor, and the mismatch of reactor size to the electric power grid serviced by many electricity-generating entities.

Given the incentives for SMR deployment, what are the challenges? The major uncertainties are the ability to reduce the financial risk sufficiently to attract investors, the ability to reduce the projected levelized unit electricity cost (LUEC) differential between that of SMRs and the competition offered by lower-cost natural gas power plants and large nuclear plants, and compatibility of fuel cycles with existing facilities. These incentives and challenges are elaborated next.

1.2.1 Incentives

The two major incentives for SMR deployment are as follows.

1.2.1.1 Reduction of initial investment and associated financial risk

The modular concept allows the investor to achieve the level of total power supply desired by time-sequenced construction increments. Each module increment not only does cost less than that of the large monolithic competitor plant, but also the time profile of capital investments can be somewhat offset by revenues from the earliest module deployments as they achieve commercial operation. However, when module construction is staggered, great care must be taken to insure that construction does not adversely impact the safety of the operating SMR.

1.2.1.2 Improved match to smaller electric power grids

A significant number of potential nuclear power plant customers have constraints on the size of allowable and needed increments of power capacity additions, which are smaller than the 1000 MWe and larger ratings of currently offered advanced reactors. The allowable size of additions reflects the somewhat contorted grid layout and interconnections in several US regions. Needed size increments reflect anticipated growth in load demand and incentives to replace older, small generating stations, mostly coal burning, with those using other fuels. Also, since the smaller SMRs should take less time to build than 1000 MWe units, demand forecasts need to be projected for fewer years out than are presently needed. Further markets for small nuclear units are emerging in smaller developing countries, which have not previously embarked on nuclear power utilization. In developed countries with well-established nuclear power programs, remote regions and sites vital for national security exist, which have power needs that can ideally be supplied by SMRs. Additionally, SMRs in these countries can supply process heat on the scale appropriate to commercial chemical processing plant needs.

These major incentives for SMRs are buttressed by several other desirable factors deriving from the small SMR characteristics:

- effective protection of plant investment from the potential to achieve a reactor design with enhanced safety characteristics;
- possible reduction of the current 10-mile emergency planning zone by virtue of the smaller core inventory and potential for added safety design features;
- reduction of transmission requirements and a more robust, more reliable grid;
- use of components that do not require the ultraheavy forgings of today's gigawatt-scale nuclear power plants and are rail shippable, which could be supplied by a reinvigorated US heavy industry; and
- suitability for the district heating mission.

1.2.2 Challenges

The three major challenges for SMR deployment are as follows.

1.2.2.1 Sufficient reduction of financial risk

The investor-perceived financial risk arises from three key factors:

- NRC licensing requirements, which could affect the capital and operating cost of these SMRs regarding plant staffing, security requirements, insurance and licensing fees, and decommissioning funding;
- the validity of the expected learning curve to reduce capital costs through factory manufacture;
- the more typical nuclear construction concerns, such as the following:
  - construction and commercial operation schedule delay due to regulatory related delays,
  - construction cost overrun due to constructor inexperience such as the recent EPR Finnish and French construction activities and unforeseen mandated design enhancements such as those arising from the Fukushima accident,
  - loss of investment due to operational and maintenance cost escalation or occurrence of a severe reactor accident.

All reactors are equally designed to a top level set of regulatory requirements, which however are not fully harmonized internationally. In the United States, these requirements have been made much more explicit for water-cooled reactors, since among the other coolants, only the Fort St. Vrain helium-cooled reactor received a US Nuclear Regulatory Commission (NRC) (commercial) operating license. The explicit existing definition of water-cooled reactor regulatory requirements is a major benefit to light water reactor (LWR) SMRs in comparatively assessing the licensability of other SMR coolant types. However, even for LWR SMRs, the following factors significant to potential cost reductions compared with current GWe stations and regulatory acceptance will need to be resolved:

- the reactor control strategy leading to reduction in the number of required operators,
- the reactivity control issues related to the desired long duration of the irradiation cycle to be accomplished by some designs without the use of soluble poisons,
- the definition of the mechanistic source term for fission product release in a severe accident,
- the potential for and consequences of multimodule interactions,
- establishment of emergency planning and preparedness consistent with the reduced power rating and footprint of SMRs.

Finally, LWR plant vendors are assuming that their designs will be accepted in a timely manner by the regulator. They base their optimism on the contention that their designs employ proven, current licensed concepts using proven components and systems configurations at power levels sufficiently low to allow the enhanced use of passive safety features, which have already been reviewed and approved for the larger Generation III+ advanced light water reactor (ALWRs). This assumption, even if proven correct, needs to reflect regulatory acceptance of at least some of the factors noted previously cast in a manner yielding economic benefit to the SMR.

For SMRs using nontraditional coolants such as helium, sodium, lead-bismuth, or molten salts, the regulatory challenge is more difficult since the NRC staff lack familiarity with these reactor designs. Additionally, given the still largely prescriptive nature of light water-based regulations in the United States, the licensing process is not amenable to the newer more innovative designs. There have been calls for using a technology-neutral licensing process to license these new reactor concepts such that the inherent design features can be recognized by the regulator. The development of such a process is underway but is proceeding slowly.

1.2.2.2 Projected LUEC

The impact of the concept of modularity in reducing the cost of small, replicated, and mostly factory-built units is paramount. Proponents refer to this as the competition between the traditional economy of scale, which has led to GWe-sized plants and the new economy of numbers, which characterizes the construction of SMRs.

Further the dramatically reduced power rating of SMRs provides significant potential for passive safety systems, which simplify or eliminate active safety systems compared with those of current-generation reactors. Also the SMRs can eliminate their reliance upon support systems as compared with the current LWRs' need for such systems. The American Nuclear Society's report on SMR generic licensing issues (see Tables 1.3 and 1.4 in American Nuclear Society, 2010) identifies specific candidate safety and support systems for such simplifications and eliminations. However, projections among analysts vary as to whether SMRs can achieve lower LUECs than traditional large plants. For example, the OECD has reported (OECD, 2011) that the investment component of LUEC from an SMR would probably be higher than that of a large plant, even taking into account the SMR reduced construction schedule, shop fabrication, and learning curve. Further the OECD concluded that SMRs, including twin-unit and multimodule plants, generally have higher values of LUEC than nuclear power plants with larger reactors. Thus achievement of a competitive SMR LUEC will be very difficult to accomplish: reference to independently validated projections is essential for developing realistic cost estimates.

1.2.2.3 Fuel cycle compatibility with facilities and strategy

The SMRs of different coolant types employ very different fuel types. The water-cooled and the lead-/bismuth-cooled SMRs use uranium dioxide (UO_2) ceramic fuel; the gas-cooled SMRs use graphite and silicon carbide-coated UO_2 particles in graphite compacts or pebbles; the sodium-cooled reactor uses metallic UZr with minor actinides; and the lead-cooled SMR uses mononitride-mixed fuel (UN-PuN). The water-cooled SMR fuel is the same as that of the operating plants and of the GEN III+ plants currently being deployed. All the liquid metal-cooled reactor fuels will have an enrichment significantly more than the 5% of current water-cooled fuel.

Although a US national repository is not yet identified, this water-cooled SMR fuel will be handled consistent with the anticipated US policy yet to be finalized. The gas-cooled SMR fuel, the same as that used in the Fort St. Vrain reactor, has significantly more volume per unit energy generation but lower heat load per unit volume than LWR UO_2 fuel. The characteristic of this fuel will require a different overall disposal strategy, although it would likely be compatible with the strategy of the national repository for ceramic UO_2-zircaloy clad fuel since the tristructural isotropic-type (TRISO) fuel particles form good barriers that provide excellent fission product retention.

The fuel of sodium- and lead-cooled SMR reactors exploits the inherent incentive of these fast neutron spectrum reactors to undergo reprocessing and recycling. This fuel cycle will entail construction and operation of reprocessing and fuel fabrication facilities, while most likely it would also be integrated with reprocessing of some light water fleet fuels as feedstock for the plutonium needed for initial loading of a growing fleet of fast reactors. The spent fuel constituents ultimately requiring disposal will be predominantly fission products of much less volume than the spent fuel bundles of thermal spectrum water reactors per equivalent unit of energy generated. However, the deployment of fast spectrum SFRs based on the closed fuel cycle would require significant expansion of reprocessing and fuel fabrication facilities compared with the needs for the existing LWR fleet and LWR SMRs operating on the once-through fuel cycle.

1.3 Overview of different types of SMRs

As with the current large-rated reactors, SMR coolants can be light water, gas, or liquid metal. Key SMR examples of these primary system coolant types with their principal design parameters are presented in Table 1.2. The coolant properties that dictate the different design characteristics of these SMRs are presented in Table 1.3. Principal among them are as follows:

- the very high outlet temperature (750–950°C) of the high-temperature gas reactor (HTGR) possible with the use of helium as coolant and graphite as the principal core material,

Table 1.2 Reactor characteristics by coolant.

Coolant	PWR[a] Light water	BWR[b] Light water	HTGR Helium[d]	HTGR Helium[e]	SFR[c] Sodium	LFR Lead[f]	LFR Lead-bismuth[g]
Power (MWt/MWe)	530/180	750/250	250/100	625/283	840/311	700/300	280/101.5
Power density (kW$_t$/l core)	69	39.5	3.2	6.8	215	116	160
Specific power (kW$_t$/kg HM)	26.8	11.6	89.7	~120	83.6	14.5	30.8
Fuel geometry	Rods	Rods	Pebbles	Prismatic graphite blocks	Rods	Rods	Rods
Fuel material/cladding	UO_2/Zr-4	UO_2/Zr	UO_2/TRISO	UCO/TRISO	(U + Pu)/SS	(U + Pu)N/SS	$UO_2^{1,h}$
Primary system temperature inlet/outlet (°C)	295/319	190/285	250/750	325/750	360/499	420/540	340/490
Primary operating pressure (MPa)	14.2	6.9	7.0	6.0	0.1	0.1	0.1
Secondary operating pressure (MPa)	5.7	NA	13.3	16.7	14.7	18	6.7
Plant thermal efficiency (%)	34	33.3	42	45	37	43^2	36.3

NA, not applicable since the BWR only has a primary system.
Numerical values of characteristics are rounded.
[a] Pers. Comm, D. Langley (mPower) to N. Todreas (MIT), January 2013.
[b] VK-300—Gabaraev et al. (2004); Kuznetsov et al. (2001).
[c] PRISM—Triplett et al. (2012).
[d] HTR-PM—Zhang et al. (2009); Zhang (2012).
[e] SC-HTGR—AREVA (2012).
[f] BREST—Smirnov (2012); Glazov et al. (2007)^2.
[g] SVBR-100—Toshinsky and Petrochenko (2012); MOX and N fuel options proposed^1.
[h] Likely EP823 or EP450.

Table 1.3 Reactor coolant properties of significance[a].

Coolant	Water[b] PWR	Water[b] BWR	Helium[c]	Sodium[d]	Lead[d]	Lead-bismuth[d] (0.445 Pb–0.555 Bi)
Atomic weight	18		4	23	207	208
Phase change at 1 atm						
Melting point (°C)	0		NA	98	327	124
Boiling point (°C)	100		−267	892	1737	1670
Density, ρ (kg/m^3)	704.9	754.7	3.54	880	10,536	10,180
Specific heat, c_p (J/kg K)	5739	5235	5191	1272	147	146
Heat capacity, ρc_p (MJ/m^3 K)	4.05	3.95	0.018	1.07	1.55	1.49
Heat transfer capability						
Thermal conductivity, k (W/mK)	0.543	0.585	0.31	66	15	15
Heat transfer coefficient ($\infty 10^{-4}$), h (W/m^2 K)	3.80	1.90	0.65	18.1	2.81	2.75
Dynamic viscosity ($\infty 10^4$), μ (kg/ms)	0.846	0.945	4.0	2.6	20	15
Kinematic Viscosity ($\infty 10^7$), $\nu = \mu/\rho$ (m^2/s)	1.20	1.26	1.13	2.95	1.91	1.47
Thermal expansion coefficient ($\infty 10^5$), α (1/°C)	326	250	–	29	11	13
Prandtl number, Pr	0.89	0.85	0.66	0.005	0.020	0.015

[a] Typical reactor values.
[b] Property values at PWR average and BWR inlet conditions from Todreas and Kazimi (2012).
[c] Property values at 537°C and 6 MPa from Petersen (1970).
[d] Property values at 450°C from Hejzlar et al. (2009).

yielding a high plant thermal efficiency and supply of reactor heat for processes requiring high temperature heat;
- the low primary operating pressure of the liquid metal reactors permitted by the low vapor pressure of their primary coolant at their high operating temperature;
- the high power density of the sodium-cooled reactor possible because of its operation with a fast neutron spectrum coupled with a very high heat transfer coefficient that allows tight packing of its fuel pins.

The predominant use of light water in both pressurized and boiling-water large-rated reactors currently in use can be readily replicated for SMR application. The smaller primary system components of pressurized water SMRs allows their arrangement within the pressure vessel as is already done even for large power rated BWRs. This PWR configuration, the integral reactor, was pioneered (as discussed in Section 1.1.3) in the commercial merchant vessel, the German *Otto Hahn*, and is a principal configuration of current PWR SMRs as elaborated in Chapter 3.

Helium has been the gas coolant of SMR choice, although carbon dioxide is used in advanced gas reactors (AGRs) operating in the United Kingdom, which are currently slated for retirement. The liquid metal coolants of SMR choice are sodium, lead, and lead-bismuth. Sodium has been exploited significantly for large-rated reactors based on early work with sodium-potassium and sodium, while more exotic coolants such as lithium have been used for electricity-generating space reactors, for example, the Systems for Nuclear Auxiliary Power (SNAP) series. For SMRs, attention is focused on sodium and the variants of lead cooling—both pure lead and lead-bismuth eutectic.

Differentiation among reactor types and specific reactor designs within a coolant-type design is based on their satisfaction of a selected mission and then a set of criteria including operational reliability, protection of public health and safety, and finally economic competitiveness. The salient characteristics of the SMR reactors as they relate to these factors are presented next. Chapter 2 and the chapters in Part Four elaborate the detailed technical features of SMRs covering this range of primary coolants.

1.3.1 Reactor mission

The principal mission adopted for commercial SMRs has been the generation of electricity. All reactor coolant types address this mission. For those plants designed to be deployable to remote locations, whether placed terrestrially or dispatched as barge-mounted reactors, the added cogeneration capabilities for desalinization and district heating exist. Of the water-cooled SMRs, the Russian PWR and BWR systems have been designed for these additional missions. Additionally, propulsion as accomplished by Russian ice-breaker vessels using the KLT-40S reactor and its planned replacement, the RITM-200 reactor, is a further reactor mission.

The heliumgas-cooled reactor can operate at high enough outlet coolant temperature, 750°C in initial designs, to provide a process heat capability. This process heat can be used directly for various industrial processes such as shale oil recovery and the production of hydrogen by relatively high-temperature thermochemical cycles.

Hydrogen production from water by electrolysis can be accomplished at the lower - outlet temperature of the sodium- and lead-cooled reactors, on the order of 500–550° C, but these SMRs have not embraced this mission due to current shrinking US interest.

1.3.2 Operational reliability

Certainly, this criterion is best met by reactor concepts using conventional components and systems operating at coolant temperatures and pressures within the envelope of significant operating experience. Water as a coolant for SMRs has been selected explicitly because of the satisfaction of these conditions. Experience with water reactors using the essential design features selected for water-cooled SMRs goes back to the beginnings of the nuclear electricity generation and propulsion age. The major caveat regarding the achievable reliability of water-cooled SMRs relates to those having selected the integral configuration, the placement of all NSSS components, and piping within a single pressure vessel. While the *Otto Hahn* merchant vessel successfully used this reactor configuration and operated commercially for 9 years, the potential reduction in operational reliability of this configuration due to its limited accessibility for primary system component monitoring, maintenance, and repair can be confidently assessed only through many more years of operating reactor experience.

Sodium-cooled reactors have generally had a mixed, albeit limited, record of operating experience. The US Experimental Breeder Reactor II (EBR-II) and British Dounreay Fast Reactor (DFR) records were exemplary, the Russian BOR-60 and BN-600 and the French Phenix reactor experience was on balance satisfactory, while the Japanese Monju experience has been very troubled, principally due to a sodium leakage event as was the Superphenix experience. Similarly the lead-/bismuth-cooled Russian submarine reactors operated reliably but with the need for careful attention to coolant chemistry control and freeze prevention after the major accident in 1968 before adequate understanding existed of the need for rigorous control of coolant oxygen concentration to prevent lead oxide slag formation (Toshinsky and Petrochenko, 2012). Helium-cooled reactors, for example, the experimental reactors AVR and THTR in Germany and the commercial Fort St. Vrain unit in the Unites States, also have had a mixed operating record.

Hence, it can be concluded that, based on operating experience, the water-cooled SMR class has a significant advantage over the other coolant types with regard to its promise of operational reliability. The operational reliability of nonwater cooled reactors will be uncertain until sufficient demonstration plant operational experience is accumulated.

The principal coolant characteristics influencing this operational experience—for example, coolant toxicity, corrosion effect on bounding surfaces, and coolant freezing and boiling temperatures—are shown in Table 1.4. Coolant toxicity has been expressed in terms of radiological, biological, and chemical factors.

Biological consequences arise from decay of ^{210}Bi, which yields ^{210}Po. The polonium then chemically combines with lead as PbPo(s). Should water enter the primary

Table 1.4 Inherent coolant characteristics affecting operational reliability.

	Water[a]	Helium	Sodium[b]	Lead[b]	Lead-bismuth[b]
Radiological	$^{16}O(n, p)^{16}N$ $^{16}N \rightarrow {}^{16}O$ + 5 to 7 MeVγ ($T_{1/2}$ = 7.1 s)	None but erosion created dust liftoff from sudden depressurization can cause mechanical clogging	$^{23}Na(n, \gamma)^{24}Na$ ($T_{1/2}$ = 15 h) 1.38, 2.76 MeV γs $^{23}Na(n, 2n)^{22}Na$ ($T_{1/2}$ = 2.6 year) 1.28 MeV γ	$^{204}Pb(n, \gamma)^{205}Pb$ ($T_{1/2}$ = 51.5 days) 1.28 MeV γ	Same as lead plus $^{209}Bi(n, \gamma)$ $^{210}Bi(e)\ ^{210}Po$ $^{210}Po\ (\alpha, \gamma$ low prob.) ^{206}Pb ($T_{1/2}$ = 138 days) 5.3 MeV α; 805 keV γ
Toxicity					
Biological	$^{6}Li(n, \alpha)^{3}T$ $^{10}B(n, 2\alpha)^{3}T$ $^{10}B(n, \alpha)^{7}Li(n, n\alpha)^{3}T$ ($T_{1/2}$ = 12.3 years)	None	None	Trace amounts of Po from ^{205}Pb to ^{210}Po by neutron capture and β^- decay	$PbPo(s)$ + H_2O = PbO + $H_2Po(g)$ (volatile alpha-emitting aerosol)
Chemical	None	Asphyxiation hazard	None	Exposure to high levels of lead through inhalation, ingestion or occasionally skin contact can lead to the medical condition known as lead poisoning	Same as for lead

Continued

Table 1.4 Continued

	Water	Helium	Sodium	Lead	Lead-bismuth
Corrosion	Prevention of stress corrosion cracking of stainless steel requires significant attention. Also significant corrosion-induced crud formation potential	None	Sodium is practically noncorrosive with respect to stainless steel. Corrosion is lower than for lead or water	Aggressive corrosion by: $\sum$ direct dissolution by a surface reaction $\sum$ intergranular attack Oxide film formation tends to inhibit the corrosion rates. Need to limit velocity to about 3 m/s to avoid cladding corrosion	Same as for lead
Melting (freezing)/ boiling points (°C)	0/100	NA	98/883	327/1737 High freezing temp—need trace heating	125/1670 Lower freezing temperature advantageous versus lead

[a] Lin (1996).
[b] Todreas et al. (2008).

system due to a failure of the ingress penetration barrier coincident with a steam generator tube leak, it would react with the PbPo(s) to produce $H_2Po(g)$, a volatile alpha-emitting aerosol of biological inhalation concern. The designers of the lead-/bismuth-cooled SVBR-100 reactor (see Table 1.2), who are well versed in Russian submarine experience, cite that operating experience has resulted in the development of measures for providing adequate radiation safety. For water-cooled reactors, water chemistry measures typically include introduction of boron and lithium in the form of boric acid and lithium hydroxide for corrosion control, although some SMRs, for example, the B&W mPower design, have eliminated the use of soluble boron for reactivity control. Neutron activation of ^6Li and ^{10}B produces tritium, ^3T, albeit in small quantities, which nevertheless is a biological hazard if ingested.

Occupational contact hazards of a chemical nature exist for lead through high levels of exposure due to inhalation and occasionally skin contact. Similarly, asphyxiation due to accidental immersion in helium (or in nitrogen typically used to inert BWR containments) is a potential hazard. The more significant, well-recognized chemical oxidation reactions of zirconium cladding and sodium are covered as a safety concern under potential energy release in Section 1.4.1.

Of all the coolants, helium, because it is an inert gas, poses the least corrosion potential, and its activation is minimal as demonstrated by the Fort St. Vrain experience that showed very low activity in the coolant compared to light water reactors. The aggressive attack of lead and lead-bismuth on metal cladding (e.g., in HT-9 and the Russian equivalents EP 823 and EP 450) has forced the limitation of coolant velocity in lead- and lead-/bismuth-cooled core designs to 3 m/s. This in turn has necessitated the provision of a large coolant flow area to bound core coolant temperature rise. Hence, lead and lead-bismuth cores have fuel pins spaced with a large pitch/diameter square lattice array. However, the development (Short and Ballinger, 2012) of a composite material for cladding and structural application may mitigate such limitations.

Finally the operability of liquid metal coolant systems requires trace heaters around piping and components of sodium, lead and lead-bismuth reactors to prevent coolant freezing when insufficient heat is available from power operation or decay heat. The high freezing temperature of lead, 327°C, compared with the modest values for sodium, 98°C, and lead-bismuth, 125°C, renders lead disadvantageous as a reactor coolant in this regard. However, with these high freezing temperatures both lead and lead-bismuth eutectic will solidify in ambient air, providing a means for sealing small leaks in the primary coolant boundary. On the other hand the high boiling points with the attendant low vapor pressures of these liquid metal coolants allow reactor operation at atmospheric pressure without the source of stored energy associated with a high-pressure coolant. Operation at low pressure allows reduction of the required thickness of the pressure vessel and other primary pressure boundary components. Nevertheless, for the heavy lead coolant, the dimensioning of these vessels must be carefully evaluated to satisfy seismic design criteria.

1.3.3 Economic implications of SMR technologies

The economic characteristics of large water power reactors are known from years of construction and operating experience. The cost of sodium-cooled reactors based on deployments of demonstration units in the late 1900s has led to capital cost estimates of 110%–125% that of water-cooled reactors (Waltar et al., 2012). Experience with gas-cooled and certainly lead/lead-/bismuth-cooled reactors has not been sufficient to allow a comparable projection of overnight capital costs compared with water-cooled reactor experience. Hence, while it is accepted that the capital cost of individual SMR units will be far lower than that of the large-rated reactors employing the same coolant, the capital cost per KWe for SMRs compared with large-rated reactors, although likely larger, is as yet not established. We can only project comparative costs of SMRs employing the various coolants on the basis of the previously noted large-rated reactor experience.

Other potential measures of comparative economic characteristics of variously cooled SMRs are the fundamental parameters of core power density and specific power. The power density, kilowatts/liter, reflects the core volume and hence is often a measure of the vessel containment and plant size necessary for a given power rating. Exceptions do exist if the reactor vessel or containment size is dictated by considerations other than core power density. For example, the SPRISM sodium-cooled fast reactor vessel is sized to accommodate decay heat removal through an air-cooled chimney outside the guard vessel: BWR containments by virtue of their use of in-containment coolant pools for pressure suppression are much smaller than those of PWRs, which control pressure by large air-filled containment volume. The power density is thus a relative indication of capital cost, albeit for plants using comparable design strategies and principal materials. The specific power, kW/kg_{IHM}, reflects the mass of initial heavy metal (IHM) or fuel needed for a given power rating. The specific power is thus a relative indication of fuel cycle cost, but for plants using comparable fuels.

However, it is clear that not all SMRs employing the various coolants of interest use comparable materials or fuels. Hence the relative values of power density and specific power presented in Table 1.5 for various coolants do not necessarily forecast the comparative economic character of reactors employing various coolants. Nevertheless, these parameters provide an insight regarding the significant benefit to sodium-cooled reactors from their high relative parametric values, a benefit which likely keeps their costs close to water-cooled designs even though they use an exotic liquid metal coolant requiring considerable costly instrumentation and purification systems, and their enrichment is much higher than that of water-cooled designs.

Table 1.5 Nominal average power density and specific power of SMRs of various coolants.

	PWR	BWR	Helium	Sodium	Lead
Power density (kW/l)	100	51	6	280	110
Specific power (kW/kg$_{IHM}$)	38	27	100	60	45

Furthermore, the low parametric values of the helium-cooled reactor indicate the inherent economic disadvantage of large reactor volume, which this reactor coolant type faces. However, unless one considers all aspects of the design, surrogate parameters of cost can be misleading. What is needed is an integrated cost analysis to include the design of the reactor system, all needed safety systems, the power conversion system considering thermal efficiency, operating staff size, maintenance cost, and fuel costs to evaluate the economic competitiveness of any design as measured in cents/kWh of power produced.

1.4 Public health and safety

All SMRs will be designed to meet the same top level set of regulatory requirements. However, the inherent characteristics of each coolant significantly influence the means by which such requirements are achieved.

Neutronic-based coolant density coefficients of reactivity are of sufficient magnitude as to affect the design of all liquid-cooled reactors. For PWRs and BWRs the moderator coefficient is designed to remain negative for accident conditions, but for sodium-and lead/lead-/bismuth eutectic-cooled fast reactors, the reactivity coefficients that would be present in significant coolant voiding events are unavoidably positive and protected by other design features. Helium-cooled reactors of either the pebble or prismatic type have the unique feature of low power density coupled with a high heat capacity core and reflector that yields a design such that the reactor, upon an increase in temperature, neutronically reduces the power to a very low level, that is, well below 1% of full power.

The inherent means to cope with severe accident conditions and design basis events are key features of the variously cooled SMRs, namely,

- bounding the potential energy release,
- mitigation of release of fission products by scrubbing in primary coolants,
- response to the loss of coolant accident (LOCA) and provision for ultimate removal of decay heat.

1.4.1 Potential energy release

In accident situations, reactor materials can undergo chemical reactions that release stored energy in addition to the generation of decay heat. The primary reactions that occur at operating or modestly low temperatures are listed in Table 1.6 and elaborated in the succeeding text:

- For the sodium reactor, oxidation of the sodium coolant released by steam generator tubing failure by contact with secondary system water that also produces hydrogen of less concern is the sodium reaction with air that causes relatively low heat release but vigorous emission of oxide fumes. The sodium leak in the Monju reactor to air from failure of an instrument penetration in December 1995 caused only modest sodium leakage to the piping compartment. The event forced the shutdown of the reactor for 14 years, even though the overwhelming portion of this period was due to loss of public confidence versus the need for

Table 1.6 Energy release reactions and fission product scrubbing in various coolants.

	Water	Helium	Sodium[a]	Lead/lead-bismuth
Energy release	Zr-water/steam reactions: $Zr(s) + 2H_2O(l) \rightarrow$ $ZrO_2(s) + 2H_2(g) +$ 537.8 kJ/(mol Zr) (500 K) $Zr(s) + 2H_2O(g) \rightarrow$ $ZrO_2(s) + 2H_2(g) +$ 583.6 kJ/(mol Zr) (1477 K) The hydrogen produced can be oxidized as $H_2(g) + \frac{1}{2}O_2(g) \rightarrow H_2O(g) +$ 241.8 kJ/(mol H_2) (298 K)	Air reactions: $C(s) + O_2(g) \rightarrow CO_2(g)$ 393.15 kJ/(mol C) (798 K) $C(s) + CO_2(g) \rightarrow 2 CO(g)$ −171.4 kJ/(mol C) (798 K) $CO(g) + \frac{1}{2}O_2(g) \rightarrow CO_2(g)$ + 282.3 kJ/(mol CO) (798 K)	Water reactions: $Na(l) + H_2O(g) \rightarrow NaOH(l) +$ $\frac{1}{2}H_2(g) + 160.1$ kJ/(mol Na) (798 K) $Na(l) + NaOH(l) \rightarrow$ $Na_2O(s) + \frac{1}{2}H_2(g) +$ 13.3 kJ/(mol Na) (798 K) The hydrogen produced can be oxidized as $Na(l) + \frac{1}{2}H_2(g) \rightarrow NaH(s) +$ 57.3 kJ/(mol Na) (798 K) Burning reaction, zone of small flames at the sodium-air interface Air reaction: Na_2O oxide is produced that upon burning in air forms Na_2O_2. In the molten sodium only Na_2O oxide is stable	Water reactions: Virtually no reaction with cold water or steam Air reaction: results in Pb_2O and then PbO. At the temperature of 450°C the latter is transformed to Pb_2O_3 and then at 450–470°C to Pb_3O_4. All these unstable compositions dissociate into PbO and O_2

Fission product (FP) scrubbing in primary coolants[b]	(1) Volatile FPs belonging to alkali metals such as Cs, K, Rb will form X-OH chemical compounds and will remain in the water. (2) Volatile FPs belonging to halogens such as I, Cl, Br will dissolve in water in ionic form such as I(−1), Cl(−1), and Br(−1). (3) Nonvolatile FPs such as Sr, Ba, Y, La, Zr, Nb, Mo, Tc and Rh do not dissolve significantly in water. (Sr and Ba will react with water to form soluble oxides)	None	(1) Volatile FPs belonging to alkali metals such as Cs, K, Rb have the same electronic structures as sodium (Na) atoms and dissolve in sodium, but they have very high vapor pressures and will evaporate with sodium during long accident times. (2) Volatile FPs belonging to halogens such as I, Cl, Br will form Na-X (NaI, NaCl, and NaBr) chemical compounds with sodium. (3) Nonvolatile FPs such as Sr, Ba, Y, La, Zr, Nb, Mo, Tc, and Rh do not dissolve significantly in sodium	Same as sodium[c]

[a] Endo et al. (1990).
[b] Pers. Comm. H. Endo (JNS Organization) to E. Baglietto (MIT), January 2013.
[c] Pers. Comm. G Toshinsky (SSC IPPE) to N. Todreas (MIT), July 2013.

repairs and refurbishment. The EBR-II had to deal with numerous sodium leaks during its 30-year operating lifetime. These leaks were safely managed and the reactor operated as both a research reactor and a small power demonstrator.
- For graphite-moderated reactors, graphite oxidation from inadvertent air ingress, release of stored energy due to atom displacements in graphite (Wigner energy), can also occur as happened in the UK Windscale reactor, but the elevated operating temperature of modern SMR gas-cooled reactors eliminates this energy storage mechanism.

All other chemical reactions of interest occur at very high temperatures that would be encountered if the reactors suffered conditions of core degradation. These include the following:

- For water-cooled reactors, oxidation of the zircaloy, and steel core cladding and structures by the primary water coolant, this reaction not only is strongly exothermic but also produces hydrogen. The hydrogen when mixed with dry air is flammable in a composition range between 4% and 75% H_2. Typically, containments are sized in PWRs to maintain hydrogen content below 4% by volume; BWRs employ smaller containments by virtue of their pressure suppression design, which then requires either inerting (Mark I and II designs) or employment of hydrogen recombiners and igniters (Mark III design) to prevent hydrogen burning or explosions.
- For all liquid-cooled reactors, oxidation of metals that may exist in molten core material (called corium) by water and carbon dioxide released from thermal decomposition of the concrete containment basemat upon contact with corium; corium contact with the basemat could only occur if the reactor vessel failed.

A major positive characteristic of lead and lead-bismuth coolants is that their reactions with water/steam and air are slight and hence of no reactor safety consequence.

1.4.2 Mitigation of the release of fission products

An important benefit of water and sodium coolants is their ability to scrub or retain fission products that would be released from the fuel and pass through these coolants in the event of a severe accident. This coolant characteristic would reduce the amount of fission products that might otherwise escape to the environment if the containment were to be bypassed. As also detailed in Table 1.6, the various chemical-based fission products behave differently with regard to their retention in water and in sodium.

The conclusions, which can be drawn, are that relative scrubbing capabilities are (1) higher for water for alkali fission products, (2) higher for sodium for halogen fission products, (3) similar for nonvolatile fission products, and (4) indeterminate due to lack of evidence for volatiles such as Sb and Te. Lead coolant behavior is similar to that of sodium. For gas reactors the coolant does not scrub fission products, but scrubbing occurs as plateout on cold surfaces.

1.4.3 LOCA and decay heat removal

The LOCA challenges reactor safety by raising peak cladding and fuel temperatures from stored core energy and decay heat generation. The response to this challenge differs among the SMR types as follows—however, for fully integral SMRs, no large primary coolant diameter piping exists and thus no classic large LOCAs can occur:

- For the water reactors the primary coolant released flashes in the containment, creating steam expansion which pressurizes the containment and can cause mechanical damage to equipment. While equipment can be secured from this threat and containment can be sized both in volume and wall thickness to survive this threat, the threat of exposure of fuel, even after shutdown of the fission process, requires a means to replenish core coolant inventory. Passive, gravity-driven core reflood systems are the current design vehicle. They must be sized both in delivery head and volume sufficient to rewet the cladding if the fuel is uncovered or simply maintain the cladding wet if the reactor system can be designed to prevent core exposure even during a design basis-LOCA as are all integral PWR-type SMRs. For both situations the sufficiency of core coolant inventory reverses the trend of increasing temperature before the zirconium-based cladding reaches the regulatory limit, now 1204°C, at which its ductility and hence its integrity are threatened. Ultimate removal of decay heat is achieved by means of dedicated decay heat removal loops, which transfer heat to the environment or passive conduction heat removal through the reactor containment.
- For gas reactors, timely replacement of coolant inventory at pressure is impractical. However, the use of high conductivity graphite as the core moderating material offers a radial conduction path for core energy to an ex-vessel heat sink. The graphite core material provides a significant heat sink that maintains temperatures at allowable levels until passive heat removal capability can match the decay heat level. For cores of modest dimensions, the length of this path is short enough and the heat storage capacity of the graphite moderator is large enough to allow steady-state power ratings of hundreds of megawatts electric. These ratings are made possible by the use of coated particle fuel with its high 1600°C limit for onset of significant fission product diffusion or leakage through disrupted fuel coatings.
- For liquid metal reactors the very low vapor pressure of coolant even at the high operating temperature allows the NSSS to be housed in a pool of coolant within a thin-walled reactor vessel, which itself is surrounded by a close fitting thin-walled guard vessel. Even upon loss of integrity of the reactor vessel, the coolant inventory is retained in the guard vessel keeping the cladding covered with coolant and the decay heat is removed by a dedicated in-vessel natural circulation coolant loop and/or radial heat flow through the guard vessel to a dedicated air chimney system, both of which discharge heat outside the containment.
- All three design solutions are satisfactory, although they operate on different principles and have configurations of differing passive safety responses.

1.5 The current status of SMRs

The United States, Russia, South Korea, China, Japan, Argentina, and Canada all have concepts under design and component/system testing is underway in several cases. The designs that have progressed the farthest are in the United States, Russian, Chinese, and Canadian programs. The US Department of Energy (DOE) in 2012 launched a two-staged program offering $452 million in government cost-shared grants over 5 years to support the design and certification of SMRs leading to reactor deployment by 2025. The first stage was awarded to Babcock and Wilcox (B&W) in November 2012, while the second stage funding was awarded to NuScale in December 2013. While B&W later terminated their design efforts, NuScale is well advanced in the licensing process for their design, and a domestic utility is proceeding with the deployment of the first US SMR in Idaho. Russian activities are numerous and are centered on the delivery of electricity and cogenerated heat and/or desalinized water to remote locations, through vessel-mounted reactors or terrestrial installations. The fleet of nuclear-powered ice breakers is being expanded as well. China has a two-unit commercial helium-cooled pebble bed plant (100 MWe per unit) under construction and is developing both terrestrial and marine-based SMRs for commercial power. The development and deployment programs in several leading countries are summarized in Chapter 2 and elaborated in Part Four of this handbook (Chapters 14–21).

1.6 Future trends

Developments to monitor whose achievement would indicate the future strength of the SMR program are as follows:

- Russia: sustained construction and deployment of SMRs for specific terrestrial and ocean missions.
- The United States: design certification by the US NRC of an SMR design, continued authorization of SMR research and development funding to the DOE by the US Congress and administration, and firm commitment of a US utility to construct an SMR.
- Canada: continued national support and execution of the planned deployment of a selected SMR concept at the Canadian National Nuclear Laboratory.
- China: successful operation of the pebble bed reactor (HTR-PM) to assess whether cost and performance targets for electricity production based on its experimental HTR-10 steam cycle design have been achieved.
- Worldwide: announcement of firm interest by a developing country in an SMR compared with a large Generation III+ water-cooled unit.

1.7 Conclusion

It is likely that, given the operability and licensing experience of LWRs, the prospects for deployment of these reactors and principally the PWR are the most promising. However, while SMR deployments for specific missions as is occurring in Russia and China will undoubtedly continue, the prospect for large-scale deployment of SMRs for electricity production in the United States and in developing nations is as yet uncertain. The current resolution of this uncertainty awaits achievement of regulatory certification of at least one design and more certainty in the costs, both overnight capital cost and the delivered electricity cost.

1.8 Sources of further information and advice

The references of this chapter provide further information on all items presented. For ongoing information on the status and evolution of the SMR program worldwide, consult the following website of the World Nuclear Association, which is continuously updated to provide such SMR developments: www.world-nuclear.org/info/Nuclear-Fuel-Cycle/Power-Reactors/Small-Nuclear-Power-Reactors/#.UVsR9EL3Cu4.

Appendix: Nomenclature

AGR	advanced gas reactor (CO_2 cooled of UK origin)
ANP	aircraft nuclear propulsion program
ARE	US aircraft reactor experiment
ASTR	aircraft shield test reactor
AVR	German pebble bed (uranium) helium-cooled reactor (Arbeitsgemeinschaft Versuchsreaktor)
B&W	Babcock and Wilcox Company
BOR	Russian sodium-cooled fast reactor
BWR	boiling water reactor
DFR	Dounreay fast reactor
DOE	Department of Energy
EBR-II	experimental breeder reactor II
FHR	fluoride-salt-cooled high-temperature reactor
GEN III+	generation III+ reactor.
GEN IV	generation IV reactor
GWe	gigawatt electricity
HTGR	high-temperature gas reactor
HTRE-1	heat transfer reactor experiment-1
HTRE-2	heat transfer reactor experiment-2
HTRE-3	heat transfer reactor experiment-3
IHM	initial heavy metal
KLT	Russian PWR ice breaker reactor
LOCA	loss-of-coolant accident
LUEC	levelized unit electricity cost
LWR	light water reactor
Monju	Japanese sodium-cooled fast reactor
NEPA	nuclear energy for the propulsion of aircraft
NGNP	next generation nuclear plant
NRC	Nuclear Regulatory Commission
NSSS	nuclear system steam supply
PWR	pressurized water reactor
RITM	Russian PWR ice-breaker reactor (new development)
SMR	small modular reactor
SNAP	systems for nuclear auxiliary power
SVBR	lead-bismuth fast reactor
THTR	thorium high-temperature reactor (German thorium pebble bed fueled, helium-cooled reactor)
TRISO	tristructural isotropic type
UO_2	uranium dioxide
UZr	uranium zirconium

References

American Nuclear Society, 2010. Interim Report of the American Nuclear Society: President's Special Committee on Small and Medium Sized Reactor (SMR) Generic Licensing Issues. July.

AREVA, 2012. A steam cycle HTGR. Nucl. Eng. Int. 57 (699), 29–33.

Endo, H., Sakai, T., Miyaji, N., Haga, K., 1990. Fission product behavior in sodium system based on inpile loop experiment. Session 3In: International Fast Reactor Safety Meeting, Snowbird UTvol. IV.

Forsberg, C., Peterson, P.F., Andreades, H., Dempsay, L., 2013. Fluoride-salt-cooled high temperature reactor (FHR) with natural-gas assist for peak and intermediate electricity loads. In: American Nuclear Society Annual Meeting Transactions, Paper 7436, Atlanta, June 16–20vol. 108.

Gabaraev, B.A., Kuznetzov, Y.N., Romenkov, A.A., Mishanina, Y.A., 2004. Nuclear desalination complex with VK-300 boiling-type reactor facility. In: Proceedings, World Nuclear Association Annual Symposium, London, Sept. 8–10pp. 1–16.

Glazov, A.G., Leonov, V.N., Orlov, V.V., Sila-Novitskii, A.G., Smirnov, V.S., Filin, A.I., Tsikunov, V.S., 2007. BREST reactor and plant-site nuclear fuel cycle. Atomic Energy 103, 15–21.

Hejzlar, P., Todreas, N.E., Shwageraus, E., Nikiforova, A., Petroski, R., Driscoll, M.J., 2009. Cross-comparison of fast reactor concepts with various coolants. Nucl. Eng. Des. 239, 2672–2691.

Kuznetsov, Y.N., Romenkov, A.A., Alekseev, A.I., Lisitsa, F.D., Tokarev, Y.I., Yarmolenko, O.A., 2001. NPP with VK-300 boiling water reactor for power and district heating grids. In: Small and Medium Sized Reactors: Status and Prospects, International Seminar Organized by the International Atomic Energy Agency, Cairo, May 27–31, Paper IAEA-SR-218/32.

Lin, C.C., 1996. Radiochemistry in Nuclear Power Reactors, Commission on Physical Sciences, Mathematics, and Applications. National Academy Press, Washington, DC.

OECD, 2011. Current Status, Technical Feasibility and Economics of Small Nuclear Reactors. Nuclear Energy Agency, Nuclear Development Division June.

Petersen, H., 1970. The properties of helium: density, specific heats, viscosity, and thermal conductivity at pressures from 1 to 100 bar and from room temperature to about 1800 K. Risö Report No. 224Danish Atomic Energy Commission, Denmark.

Short, M.P., Ballinger, R.G., 2012. A functionally graded composite for service in high-temperature lead- and lead-bismuth-cooled nuclear reactors—I: design. Nucl. Technol. 177 (3), 366–381.

Smirnov, V.S., 2012. Lead-cooled fast reactor BREST—project status and prospects. In: International Workshop on Innovative Nuclear Reactors Cooled by Heavy Liquid Metals: Status and Perspectives, Pisa, April 17–20.

Todreas, N.E., Kazimi, M., 2012. Nuclear Systems Volume I: Thermal Hydraulic Fundamentals, second ed. Taylor & Francis, Boca Raton, FL.

Todreas, N.E., Hejzlar, P., Fong, C.J., Nikiforova, A., Petroski, R., Shwageraus, E., Whitman, J., 2008. Flexible conversion ratio fast reactor systems evaluation. (Final Report), MIT CANES Report MIT-NFC-PR-101Massachusetts Institute of Technology, Cambridge, MA.

Toshinsky, G., Petrochenko, V., 2012. Modular lead-bismuth fast reactors in nuclear power. Sustainability 4, 2293–2316.

Triplett, B.S., Loewen, E., Dooies, B., 2012. PRISM: a competitive small modular sodium-cooled reactor. Nucl. Technol. 178 (2), 186–200.

Waltar, A., Todd, D.R., Tsvetkov, P.V., 2012. Fast Spectrum Reactors. Springer.

Zhang, Z., 2012. HTR-PM project status—update. In: 6th International Topical Meeting on High Temperature Reactor Technology, HTR-2012—Nuclear Energy for the Future, Oct. 28–Nov. 1. Miraikan, Tokyo.

Zhang, Z., Wu, Z., Wang, D., Xu, Y., Sun, Y., Li, F., Dong, Y., 2009. Current status and technical description of Chinese 2×250MWth HTR-PM demonstration plant. Nucl. Eng. Des. 239, 1212–1219.

Small modular reactors (SMRs) for producing nuclear energy: International developments

Daniel T. Ingersoll
NuScale Power LLC (retired), Oak Ridge, TN, United States

2.1 Introduction

The pursuit of small modular reactors is both a persistent and global phenomenon with wide-spread interest from developers and customers alike. Some of the earliest concepts emerged in the 1970s for merchant ship propulsion and industrial process heat applications. Small, niche-application reactor designs have traditionally been in the domain of the research community; however, there are now more than 50 designs with commercial involvement ranging from minor evolutions of operational reactors to exotic liquid-fueled and fission-fusion hybrid designs. Chapter 1 of this handbook describes many of the motivations that drive customer interests, including lower upfront capital investment, better match to projected demand, better compatibility with electrical grid infrastructure, and greater flexibility in site locations. Ingersoll (2016) provides in-depth background on SMRs, their history, key attributes, and their motivations.

Reactor developers worldwide are seeking to develop SMR designs to meet a large anticipated market demand (NEA, 2016). Designs are emerging from both traditional reactor vendors and new startup companies and also are being developed in both traditional and new nuclear supplier countries. Summarizing the status of emerging reactor designs is challenging due to the sheer number of designs, diversity of technologies, and rapidly changing list as new designs are introduced and other designs are abandoned. Third Way, a nongovernment organization that has been very active in promoting advanced nuclear technologies, issued a report in June 2015 (Third Way, 2015) that listed 48 advanced nuclear concepts/designs under development. By the end of 2016, the list had grown to 60, and by mid-2019, the list was just shy of 80. These include all reactor types from microreactors through large reactors and both fission and fusion systems. Roughly, half of these designs are in the SMR category. Therefore rather than attempt to provide a comprehensive review of SMRs that are under development or deployment, this chapter will instead provide a "sampler" of prominent SMR designs.

The SMRs discussed in this chapter, listed in Table 2.1, are examples of designs currently being developed by commercial companies, frequently in collaboration with the research community, and most have some level of engagement by a licensing

Table 2.1 Examples of new SMR designs being developed worldwide.

Reactor design	Power rating (MWe)	Country	Vendor
Light water cooled (PWR)			
ACP100	100	China	CNNC
CAREM	27	Argentina	CNEA
Flexblue	160	France	DCNS
KLT-40S	35	Russia	OKBM
NuScale	60	United States	NuScale Power
RITM-200	50	Russia	OKBM
SMART	100	Rep. of Korea	KAERI
SMR-160	160	United States	Holtec (SMR)
Light water cooled (BWR)			
BWRX-300	300	United States/ Japan	GE-Hitachi
VK-300	250	Russia	NIKIET
Gas-cooled			
EM2	265	United States	General Atomics
GT-MHR	288	Russia	OKBM
HTR-PM	105	China	INET Tsinghua University
Xe-100	75	United States	X-Energy
Liquid metal cooled			
4S	10/50	Japan	Toshiba
PRISM	311	United States	GE-Hitachi
SVBR-100	100	Russia	JSC EDB
Molten salt cooled			
IMSR	190	Canada	Terrestrial Energy
LFTR	250	United States	Flibe Energy

authority. As such, these designs are considered to have the potential to be deployed within the next 10–15 years depending on developer commitment and customer interest. An effort was made to highlight designs that span the gamut from traditional technology and engineering to novel technology and highly innovative engineering. Keeping with the widely accepted definition of "small," the designs presented in this chapter all have an electrical output of less than nominally 300 MWe. On the lower end, the designs reviewed here have an output power of at least 35 MWe to distinguish them from a growing interest in very small "microreactors" (typically less than 10 MWe).

As discussed in Chapter 1, SMRs do not represent a unique reactor technology, but rather reflect the same spectrum of technologies considered for large plants. Brief summaries of several commercially developed SMRs designs are provided in the following sections organized by technology type. This is a natural approach for organizing the designs since the different technology classes generally target different energy applications and share many of the same development paths. Within a technology class the designs are presented by vendor country in alphabetical order.

The information presented in this chapter was gleaned from publically available information gathered in reports, papers, presentations, web sites, and personal communications. Many of these sources are listed in the bibliography section. Because all of the designs have commercial interests, detailed data regarding design specifications are often treated as proprietary and are not publically available. Also, most of the designs continue to evolve at a rapid pace and design parameters change equally fast. In some cases, multiple sources for a single design have conflicting information. Every attempt was made to present the latest and most accurate information publically available. The single most valued resource was the latest edition of *Advances in Small Modular Reactor Technology Developments* report issued biannually by the International Atomic Energy Agency (IAEA) (IAEA, 2018). Descriptions and data contained in this report were provided by the developing organizations and are considered the most reliable. However, it is sometimes difficult to glean factual information from marketing-based material; hence, no claims are made regarding the accuracy or currency of the information. In several cases, expanded information about specific designs and their deployment status can be found in the respective country chapters within Part IV of this handbook.

2.2 Water-cooled reactors

Over 80% of commercial power reactors operating worldwide use light (normal) water as the primary reactor coolant. Given the extensive operational experience with light water reactors (LWR), it is the technology of choice for reactor vendors who want to get their product to market quickly and for potential customers who are concerned about investment risk. Heavy water (containing deuterium instead of hydrogen) reactors make up a small fraction of the operating fleet and an even smaller fraction of new SMR designs. Hence, only LWR SMRs are considered here. The 10 LWR SMR designs listed in Table 2.2 and discussed below represent over half of the 19 SMR designs presented in this chapter. Although they all share a common coolant choice, other design features are quite varied. Overall configurations include traditional loop design (steam generator vessels are connected to reactor vessel by large pipes), compact loop design (steam generator vessels are flanged directly to the reactor vessel), and integral system design (most or all primary system components are located within the reactor vessel). Although there are a few SMR designs

Table 2.2 Examples of commercial SMR designs based on light water reactor technology.

Country	SMR	Designer	Configuration	Electrical output (MWe)	Modules/ plant
Argentina	CAREM	CNEA	Integral	27	1
China	ACP-100	CNNC	Integral	100	Up to 8
France	Flexblue	DCNS	Loop	160	Up to 6
Rep. of Korea	SMART	KAERI	Integral	100	1
Russia	KLT-40S	OKBM	Compact loop	38	2
Russia	RITM-200	OKBM	Integral	50	1
Russia	VK-300	NIKIET	Boiling	250	1
United States/ Japan	BWRX-300	GE-Hitachi	Boiling	300	1
United States	NuScale	NuScale Power	Integral	60	Up to 12
United States	SMR-160	Holtec	Compact loop	160	1

being developed based on boiling water reactor technology (BWR), most use pressurized water reactor (PWR) technology. Individual module output capacities range from 35 to 300 MWe and plant-level strategies range from 1 to 12 modules per plant. Also the plants may be sited on land, operated on floating barges or submerged below the ocean surface.

2.2.1 Argentina: Central Argentina de Elementos Modulares design

The Central Argentina de Elementos Modulares (CAREM) design has been under development for a number of years and was first introduced at an IAEA conference in 1984. The National Atomic Energy Commission (CNEA) drew on its experience in the design of research reactors to develop CAREM with the primary goals of enhanced safety and reduced costs. It is a simplified integral design with natural circulation of the primary coolant. The prototype design has a capacity of 27 MWe and commercial sized units are expected to have capacities of 150–300 MWe with forced circulation being employed for units greater than 150 MWe. The design uses self-pressurization, that is, it does not use sprays or heaters to maintain normal system pressure. The core is composed of 61 hexagonal fuel assemblies with 108 fuel pins per assembly and 25 control rods, which are positioned using internal hydraulic control rod drive mechanisms. The primary coolant flows over the shell side of 12 compact helical coil steam generators. On the secondary side the feed water and output steam are collected into a common set of annular pipes that surrounds the reactor pressure vessel.

The CAREM design uses passive safety systems and anticipates a postaccident grace period of 36 h without operator action or power. Several test facilities have been built to test novel components such as the hydraulic control rod drive mechanisms and validate the safety analysis methods. The CNEA received the support of the Argentina government in 2009 to construct the prototype plant adjacent to the existing Atucha nuclear site. Site preparations began in 2012 and construction began in 2014. First criticality of the prototype CAREM SMR is anticipated in 2022. More detail regarding the CAREM design and development effort is given in Chapter 14 of this handbook.

2.2.2 People's Republic of China: ACP-100 design

The ACP-100 is an integral pressurized water reactor being developed by the Nuclear Power Institute of China (NPIC) for the China National Nuclear Corporation (CNNC). The design draws heavily on technology developed by CNNC for the larger LWR plants, including the CNP-600 and the ACP-600/1000. The forced circulation of the primary coolant is driven by externally mounted reactor coolant pumps. Also external to the reactor pressure vessel are the control rod drive mechanisms and the pressurizer. The core is composed of 57 partial-height CF2-type fuel assemblies. Reactivity is controlled using control rods and boron shim dispersed in the primary coolant. The reactor vessel and other primary system components are contained in a traditional large-volume containment structure that is 29 m diameter and 45 m tall. Safety-grade batteries provide backup power for up to 72 h in the case of a station blackout and sufficient water is provided in the spent fuel pool to allow 7 days grace period before fuel is uncovered.

The 100-MWe design is intended to provide electricity and also process steam for water desalination, district heating, and industrial applications. It is primarily intended for inland locations within China as a replacement of or alternative to coal-fired plants. Approval for the design project was given in 2010 and budget for the construction of two ACP-100 modules was approved in 2011. A first-of-a-kind demonstration plant is being constructed in Changjian Count, Hainan Province. A preliminary safety analysis report was completed in 2018 and initial site preparations and long-lead manufacturing was begun in 2019. More detailed information on the ACP-100 is given in Chapter 16 of this handbook.

2.2.3 France: Flexblue design

The Flexblue SMR concept, which is being developed by the French DCNS organization, is one of the more unique entries into the commercial SMR competition. Based substantially on DCNS's submarine design experience, the SMR is intended to be operated on the sea floor at a depth between 50 and 100 m. The 50-m upper limit isolates the plant from surface storm effects, while the 100 m lower limit protects against excessive pressure on the 14 m diameter by 146-m long outer hull. The 160 MWe SMR can be deployed in multiunit "farms" that are remotely operated from a coastal command and control facility. Up to six modules can be operated from a single control room. A heavy lift cargo ship transports the module to and from its operating site. Refueling is expected to occur every 2 or 3 years, at which time it will be returned to a central facility for refueling and maintenance. Later versions of the transport ship may allow for refueling at the site of deployment. Major overhauls of the module are expected every 10 years. At the end of the module's life, expected to be 60 years, the module is transported to a decommissioning facility for final disposal.

The ocean environment greatly simplifies traditional safety aspects of the design. The ocean water provides an assured ultimate heat sink and an additional barrier to fission products in the case of an accident. The passive safety systems ensure that the module can survive an unlimited time without operator action or power following an accident. The initial concept was completed in 2016; however, it is not clear whether further development of the Flexblue design will be continued.

2.2.4 Republic of Korea: SMART design

As with many SMRs the primary goals of the System-Integrated Modular Advanced Reactor (SMART) design are enhanced safety and reliability. The SMART design, developed by the Korea Atomic Energy Research Institute (KAERI), is a 100-MWe integral system with 57 half-height (2 m) fuel assemblies using a standard 17 × 17 pin array design. Reactivity control includes control rods, burnable poisons within the fuel, and soluble boron within the primary coolant. The 25 magnetic jack control rod drive mechanisms are external to the reactor pressure vessel, as are the four reactor coolant pumps that circulate the primary coolant. The integral system includes eight once-through helical coil steam generators and an internal pressurizer. The reactor pressure vessel and passive safety systems are contained within a traditional large-volume containment structure.

The SMART design is supported by an extensive testing program, including safety tests, methods validation tests, and component/system performance tests. Specific components that were tested include the fuel assemblies, steam generators, control rod drive mechanisms, and safety injection systems, as well as overall thermal-hydraulic performance.

The initial concept started in 1997 and achieved Standard Design Approval from the Korean regulator in 2012. Although SMART can be used for electricity only, a cogeneration design option has been developed that provides for 90 MWe of electrical output in addition to 40,000 tons/day of potable water using a four-unit multieffects distillation plant. This cogeneration plant can supply sufficient electricity and water to support approximately 100,000 people. The design is intended primarily for export, and no large-scale deployment is expected within country. In 2016 KAERI signed an agreement with Saudi Arabia to pursue the deployment of SMART plants in Saudi Arabia for cogeneration of electricity and water. Additional details regarding the SMART design and deployment plans are provided in Chapter 18 of this handbook.

2.2.5 Russian Federation: KLT-40S design

The KLT-40S, developed by OKBM Afriantov, is a barge-mounted floating SMR producing nominally 35 MWe per module. The design is a compact loop configuration with most primary system components external to the reactor pressure vessel.

The length of the hot and cold leg pipes connecting the reactor vessel with the two steam generator vessels is kept very short to reduce the likelihood of a large-break loss-of-coolant accident and because of space constraints on the barge. The design is based substantially on the successfully operated KLT-40 reactors that provide propulsion to a fleet of commercial icebreakers.

The 1.2 m diameter by 1.2-m-tall reactor core consists of 121 hexagonal fuel assemblies with uranium enrichment of less than 20%. Control rods and burnable poison rods are used to control reactivity. The four external steam generators are a helical coil type. The four primary reactor coolant pumps are also external to the reactor vessel.

The KLT-40S is normally configured with two units on a 30 m wide by 140-m-long barge and will be delivered to the operating site fully constructed and ready for operation. Initial customers include the isolated communities along Russia's Arctic coast, although the developer anticipates exporting the floating plants, especially to the island countries of Southeast Asia. To extend operations in more remote locations, the barge will be dispatched with multiple core loadings, and the reactors will be refueled on-board after 30–36 months. It is expected that the barge will return to a centralized facility every 10–12 years for major overhaul and fuel replacement.

The first KLT-40S floating power plant began construction in 2007, and the barge was launched in the Baltiysky Zavod shipyards in June 2010 and moved to Saint Petersbug for further construction. The first barge, named the Akademik Lomonosov, received an operational license in June 2019. After undergoing startup testing at its operating site in Pevek, it was connected to the regional grid in December 2019. Additional details of the KLT-40S design and deployment are provided in Chapter 19 of this handbook.

2.2.6 Russian Federation: RITM-200 design

Although potentially deployed as a stationary or floating power plant, the Russian RITM-200 is primarily intended to provide propulsion for the next generation of Russian icebreakers. Developed by OKBM Afriantov, it is a 50-MWe integral system reactor that uses a fuel type and configuration similar to the KLT-40S. The vertically mounted four main reactor coolant pumps are external to the reactor pressure vessel, as the control rod drives mechanisms and a gas pressurizer. Four internal steam generators extract heat from the primary coolant system. Because of the external main coolant pumps and pressurizer, additional water injection systems are provided to mitigate the consequences of a large break loss-of-coolant accident.

By moving the steam generator into the reactor pressure vessel, the reactor system and containment is very compact compared with the KLT-40S. With overall dimensions of $6.4 \times 6.4 \times 15.5$ m and weighing approximately 1200 t, the RITM-200 occupies 45% less volume and is 35% lighter while producing 40% more power compared with the KLT-40S. It is expected to operate continuously for 26,000 h (3.5 years) and refueled after 7 years.

The RITM-200 received design approval in 2012. To date, six modules have been manufactured, and four have been installed in pairs on two icebreaker ships, the

Arktika and the Ural. After startup testing the Arktika is expected to commence service in 2020. Additional details of the RITM-200 design and deployment are provided in Chapter 19 of this handbook.

2.2.7 Russian Federation: VK-300 design

Roughly, 20% of the water-cooled large reactors operating globally are of the boiling water reactor (BWR) type, while the remaining 80% are of the pressurized water reactor (PWR) type. Similarly a relatively small percentage of water-cooled SMR designs being developed use BWR technology, that is, the coolant is allowed to boil within the primary coolant system. The VK-300 being developed by NIKIET in Russia is an example of a BWR SMR. It is based substantially on the VK-50 BWR prototype, which has operated in Russia since 1965 and incorporates component technologies used in the current fleet of VVER-1000 reactors.

The VK-300 can produce up to 250 MWe from a direct Rankine cycle, although it is also suitable for cogeneration of electricity and process steam. It uses natural circulation of the primary coolant and incorporates passive safety systems such as the decay heat removal system. The 4%-enriched uranium oxide fuel is contained in 313 fuel assemblies. Primary containment is provided by a steel-lined reinforced concrete structure.

Although the conceptual design of the VK-300 was initiated in 1998, detailed design work, design validation studies, and commercialization efforts did not begin until 2013. These activities are continuing but no timeline for potential deployment has been provided.

2.2.8 United States and Japan: BWRX-300 design

One of the most recent entries into the water-cooled SMR development community is the BWRX-300 design, which is being developed by a global alliance of General Electric and Hitachi (GE-Hitachi). The first public release of information regarding the design was in 2018 with the announcement that Dominion Company, a US utility, became an investor in the project.

The BWRX-300 is a 300 MWe simplified BWR that draws heavily on the design and engineering of the much larger 1520 MWe ESBWR design that has not yet been constructed but which was certified by the US NRC in 2014. The main focus of the design is to minimize plant capital cost through design simplification. As with the ESBWR the BWRX-300 uses natural circulation of the primary coolant and numerous passive safety features. To date, limited public information has been provided regarding the details of the design.

In 2019 GE-Hitachi announced multiple agreements, especially with Eastern European countries such as Poland, Estonia, and the Czech Republic, for pursuing the development and potential deployment of BWRX-300 plants. The vendor initiated licensing activities with the US Nuclear Regulatory Commission (NRC) in early 2020 and indicates a possible first plant deployment by 2030.

2.2.9 United States: NuScale design

The NuScale concept was initially developed during the 2000–03 timeframe by a research team at Oregon State University (OSU), Idaho National Engineering and Environmental Laboratory, and NEXTANT. After continued development by OSU, NuScale Power LLC was formed in 2007 to commercialize the promising SMR design. Fluor joined NuScale Power in 2011 as its major investor and strategic partner. Initially, each NuScale Power Module produced 50 MWe from a simplified integral system configuration that uses natural circulation of the primary coolant. A power uprate program in 2017 resulted in increasing the module power to nominally 60 MWe. The core is composed of 37 half-height 17×17 pin array fuel assemblies and is controlled using 16 control rods and soluble boron within the primary coolant. Superheated steam is produced in two comingled helical coil steam generators that surround the central hot leg riser. A NuScale plant can be scaled to accommodate up to 12 modules in a single plant, which can be installed incrementally to best match a desired investment or demand profile.

The reactor pressure vessel is contained within a compact, high-pressure, steel containment vessel, which is immersed in a below-grade pool shared by all modules. The shared pool is the ultimate heat sink for residual heat removal. A fail-safe emergency core cooling system can provide an unlimited postaccident grace period with no operator action, no AC or DC power, and no makeup water. Each module has an independent skid-mounted turbine-generator set for power conversion and can continue to operate, while other modules are being refueled.

A scaled integral test facility was built in 2003 as part of the original design development project and has been used throughout the design refinement process to inform design modifications and validate its safety performance. A 12-module control room simulator was commissioned in 2012 to evaluate operator performance for multimodule control rooms. A design certification application was submitted to the US Nuclear Regulatory Commission in early 2016. The safety review is progressing on schedule and design approval is expected to be issued in 2020. Construction of a first-of-a-kind commercial plant is being pursued by a consortium of utilities in the Western United States and is planned to be sited within the Idaho National Laboratory Site near Idaho Falls, Idaho, with first commercial operation expected in 2026.

2.2.10 United States: SMR-160 design

In 2010 Holtec International introduced their entry into the SMR competition: the 140 MWe Small Modular Underground Reactor (HI-SMUR). Holtec, a longtime participant in the nuclear industry for their fuel transport and storage products, assembled a diverse team led by a newly created subsidiary, SMR LLC, to develop the HI-SMUR design. In late 2011 several design changes were made: the external steam generator system was changed substantially from a two-stage horizontal arrangement to a vertical arrangement, the reactor vessel was shortened by approximately 40%, and the power was increased to 160 MWe. The design is now referred to as SMR-160.

The SMR-160 is a compact loop configuration: the separate reactor vessel and steam generator vessel are flanged directly to each other. Hence, all large pipes have been removed from the primary system, although the primary coolant is circulated external to the reactor vessel. As an additional modification to a classic loop-type design, the SMR-160 pressurizer is located at the top of the steam generator vessel. The reactor pressure vessel has a significantly larger height-to-diameter aspect ratio (15:3) compared with other SMRs to enhance the natural circulation of the primary coolant. The core uses full-height traditional 17×17 pin array PWR fuel assemblies and the core is expected to be refueled as a single cartridge that will be discharged to a below-grade spent fuel pool for cooling before being moved to a dry storage facility. Like most contemporary SMR designs, the reactor vessel and spent fuel pool will be entirely below grade level.

Preliminary design of the SMR-160 was completed in 2019. The vendor expects to commence commercial design licensing activities in 2020.

2.3 Gas-cooled reactors

Gas-cooled reactors are the second most common reactor technology used for commercial power application, due largely to several carbon dioxide-cooled reactors deployed in the United Kingdom. However, all gas-cooled SMRs under development today use helium as the primary coolant. Both Germany and the United States previously built and operated helium-cooled test or demonstration reactors, and China and Japan each currently have small helium-cooled test reactors in operation. The key advantage of helium-cooled reactors is that the reactor can operate at much higher temperatures using a single-phase coolant, which is simpler to manage. Typical gas-cooled reactors operate in the range of 700–800°C outlet temperature, compared with 300–325°C for light water reactors. The advantage of the higher temperature is a higher efficiency conversion of the core heat to electricity and the ability to support a much broader range of industrial heat applications. The key drawback to gas-cooled reactors is that gases have a much lower heat capacity than liquids; therefore, the gas must be pumped at high velocity to remove the core heat. A related consequence is that the temperature differential across the core is very large, typically 500°C compared with 25–50°C for a pressurized light water reactor. This temperature differential creates material challenges within the core and on the secondary side of the plant.

The helium-cooled SMR designs under development generally fall into either a pebble bed or a prismatic configuration. In the case of the pebble bed, the fuel is dispersed in spheres, each about the size of a billiard ball. These spherical fuel elements stochastically migrate through the core and are continually removed and reinserted into the core for additional burnup. The prismatic configuration uses a fixed rod geometry for the fuel and the rods are contained within monolithic blocks of graphite that are stacked to form the reactor core. In the case of prismatic configurations, the core is refueled in batch mode similar to fuel assembly based LWRs.

Table 2.3 lists diverse examples of four gas-cooled reactor SMR designs that currently have significant commercial support and are described in the following sections.

Table 2.3 Examples of commercial SMR designs based on gas-cooled reactor technology.

Country	SMR	Designer	Configuration	Electrical output (MWe)	SMRs/ plant
China	HTR-PM	INET	Pebble bed	105	2
Russia	GT-MHR	OKBM	Prismatic	288	1
United States	EM2	General Atomics	Prismatic	265	4
United States	Xe-100	X-Energy	Pebble bed	75	4

2.3.1 People's Republic of China: HTR-PM design

The high-temperature reactor pebble bed module (HTR-PM) is a pebble bed-type high-temperature, helium-cooled reactor. The fuel is UO_2 enriched to 8.5% and contained in TRISO graphite-coated particles that are dispersed in 6-cm-diameter graphite spheres. The 3-m-diameter by 11-m-tall core region represents a tall graphite "hopper" containing 420,000 randomly packed spherical fuel elements. The fuel elements migrate downward through the core as spheres are moved from the central discharge channel in the bottom reflector and optionally reinserted at the top of the core if maximum burnup has not been achieved. The graphite block reflector that defines the core region is contained within a 5.7-m-diameter by 25-m-tall steel pressure vessel. The helium coolant flows upward through the side reflector and then downward through the core region before flowing through a cross-duct to the helium/water steam generator contained in a separate steel pressure vessel. The steam generator is a once-through, counter flow heat exchanger with multiple helical coil modules.

The HTR-PM is a successor to the HTR-10, a 10 MW test reactor operated at the Tsinghua University. The HTR-10 was used to demonstrate the safety response of the HTR-PM, including its response to a loss of off-site power, a main helium blower failure, and a loss of main heat sink. The budget for construction of a commercial demonstration plant, the HTR-200, was approved in 2008; however, construction was delayed after the destruction of the Fukushima Daiichi plant in Japan. Final approval was granted in 2012 and construction of the demonstration plant is underway in Shidao Bay, Shandong Province, China. Major equipment has been installed, and the plant is expected to begin operation in 2020. Additional information regarding the HTR-PM is provided in Chapter 16 of this handbook.

2.3.2 Russian Federation: GT-MHR design

The gas turbine modular high-temperature reactor (GT-MHR) has a long history and got its start in the United States by General Atomics during the mid-1980s as part of a US program to develop smaller reactor designs with assured safety. Since that time, variants of the GT-MHR have been considered for several US national programs, including the New Production Reactor program, the Weapons Material Disposition program and the Next Generation Nuclear Plant program. In 1993 a joint program between the United States and the Russian Federation was formed to build a demonstration GT-MHR for the disposition of weapons grade material. Subsequently the technology was transferred to Russia, where OKBM Afrikantov continues to develop and promote the design.

The GT-MHR design uses TRISO-coated uranium or plutonium oxide particle fuel that is dispersed into cylindrical pellets and stacked into vertical fuel channels within hexagonal graphite moderator blocks that are 0.36 m wide (across flats) by 0.79 m tall. The moderator blocks are stacked 10 rows high in 66 columns forming an annular ring. Additional nonfueled moderator blocks fill the center of the annulus and surround the core as an outer reflector. The heated helium flows to a vertically mounted gas turbine via a concentric cross-duct and electricity is produced at approximately 48% efficiency using a direct Brayton conversion cycle.

As will other gas-cooled reactor systems, the GT-MHR has a strong safety case owing to the very robust TRISO-coated fuel particles, the annular core design to facilitate decay heat removal, and a passive exvessel heat-removal system that dissipates the heat to the atmosphere. Also the graphite moderator provides a strong negative temperature coefficient of reactivity, which limits power excursions in unanticipated transients.

Although the GT-MHR design is relatively mature and has been considered for multiple programs, there is no current commitment to construct a demonstration or commercial unit. However, elements of the technology continue to be demonstrated, and design work continues on a uranium-fueled demonstration project.

2.3.3 United States: EM^2 design

After decades of being focused on the GT-MHR and its variants, General Atomics introduced in 2010 a new gas-cooled SMR called the Energy Multiplier Module (EM^2). Although leveraging some of the GT-MHR technologies, the design is a significant departure from the thermal-spectrum GT-MHR. The EM^2 is a fast-spectrum reactor with a primary purpose of consuming spent nuclear fuel. Specifically, it is a "breed and burn" type reactor that converts fertile elements contained within the special fuel assemblies into fissile elements to sustain the fission process, while also consuming minor actinides produced in precursor once-through reactors such as LWRs. The reactor core uses uranium-carbide fuel clad in silicon-carbon composite material. The reactor is loaded with nearly equal portions of low-enriched uranium fuel (approximately 14.5% ^{235}U average enrichment) and fertile material. Subsequent EM^2 reactors are loaded entirely with previously discharged fuel. An anticipated power conversion efficiency of 53% is achieved by operating with an outlet helium temperature of 850°C and by using a vertically mounted, variable speed direct Brayton cycle power conversion unit coupled to a Rankine bottoming cycle.

It is expected that the reactor will operate for 30 years without refueling or fuel shuffling. After 30 years the core will be replaced with a fresh load of recycled used fuel, and the discharged fuel will be processed using a dry oxidation process to produce feedstock for subsequent reactors. Units are expected to be deployed in four-module plants with independent balance-of-plant equipment for each module. Fuel development and qualification tests are underway but are expected to be the longest lead challenge for eventual deployment. No licensing or deployment timeline has been provided.

2.3.4 United States: Xe-100 design

The newly formed company, X-Energy, is developing a small pebble bed, helium-cooled SMR similar to the pebble bed modular reactor (PBMR) design, which began in 1996 in South Africa. The Xe-100 has a 200 MW cylindrical core design with an outlet temperature of 750°C coupled to an indirect Rankine cycle that produces approximately 75 MWe. Its intended use is for low-cost electricity generation and cogeneration for process heat applications.

The fuel is TRISO-coated uranium oxide-carbide particles, which are dispersed in 6-cm-diameter spherical fuel elements randomly packed in a graphite-reflected core region—220,000 in all. The fuel elements migrate through the core approximately six times before being discharged from the reactor. Operational reactivity is controlled using control rods. Inherent safety characteristics of the design and passive safety features provide for a 7-day grace period in response to accidents.

Several test facilities were constructed to support the underlying technology of the original PMBR design and its derivatives, including a helium test facility, a heat transfer test facility, and numerous component-level test facilities. Also, considerable prelicensing activities with the US NRC were completed for the PBMR project before it was abandoned. Many of these tests remain valid for the Xe-100 design, and the licensing experience remains relevant. The Xe-100 conceptual design was completed in 2019, and the preliminary design is expected to be completed in 2021 with a potential for first plant construction beginning in 2025.

2.4 Liquid metal-cooled reactors

Liquid metal-cooled reactors are third behind water-cooled and gas-cooled reactors in terms of global commercial reactor experience. Test and demonstration reactors using sodium, lead, or lead-bismuth coolant have been constructed in the United States, Russian, France, Japan, the United Kingdom, and most recently in China and India. Interest in liquid metal coolants is driven be the desire to develop fast spectrum reactors, that is, reactors that generate most of the power from fissions resulting from high-energy neutrons compared with water-cooled reactors, which generate power from thermal neutron-induced fissions. The key advantage of fast spectrum reactors is that there are more neutrons produced per fission, and these "excess" neutrons can be used for purposes other than sustaining the basic chain reaction. Initially the extra neutrons were intended to be used to "breed" fuel, that is, produce new fissile fuel faster than it is consumed. As more and more uranium reserves were discovered worldwide, interest in fast spectrum reactors turned away from the breeding function to a resource recovery, that is, producing energy from the unburned fuel discharged from

Table 2.4 Examples of commercial SMR designs based on liquid metal-cooled reactor technology.

Country	SMR	Designer	Coolant	Electrical output (MWe)	SMRs/plant
Japan	4S	Toshiba	Sodium	10 or 50	1
Russia	SVBR-100	AKME	Lead-bismuth	101	1
United States	PRISM	General electric	Sodium	311	2

water-cooled reactors, and waste management, that is, consuming the associated long-lived waste products from partially burned fuel.

Another advantage of liquid metal-cooled reactors is that the metals have high-boiling temperatures, which allow the reactors to operate with a single-phase coolant without pressurization of the primary coolant. Also the coolant can be heated to a moderately high temperature, typically around 500°C. Although higher than the 300–325°C outlet temperature in a water-cooled reactor, it is still lower than the 750–850°C outlet temperature in a gas-cooled reactor. The higher temperature increases the power conversion efficiency relative to water-cooled reactors and can allow for more compact power conversion systems using supercritical Rankine or Brayton cycles.

Described in this section and listed in Table 2.4 are three exemplary liquid metal-cooled reactor designs that have near-term deployment potential by virtue of commercial support and some level of engagement by a licensing authority. Two designs use liquid sodium as the coolant, while the third design uses a lead-bismuth coolant. Many other designs and projects are underway but are either limited to research studies or are experimental/test reactors intended to be precursors for large commercial plants.

2.4.1 Japan: 4S design

The Toshiba Super Safe Small and Simple (4S) reactor design is a sodium-cooled fast-spectrum reactor with an output of either 10 or 50 MWe. The reactor has a compact core design with steel-clad metal-alloy uranium fuel. A unique feature of the core design is that it does not require refueling over the 30-year lifetime of the plant (10 MWe version). This is accomplished by designing for a high conversion of the fertile material in the core and by using a slowly moving reflector to compensate for fuel burnup over the core lifetime. A 50-MWe design option is available with a 10 years refueling cycle. The U-10% Zr metal alloy fuel has an enrichment of less than 20% ^{235}U and is clad in HT-9 alloy. The basic layout of 4S is a pool-type configuration, with the electromagnetic pumps and a single intermediate heat exchanger contained inside the primary vessel. An intermediate sodium loop delivers heat from the primary system to the external steam generator used to generate steam for the

Rankine power conversion system. The reactor containment consists of a lower nitrogen-filled steel guard vessel and an upper steel dome.

The 4S design is supported by extensive testing facilities in Japan, including the Toshiba Sodium Loop Test facility. The design is targeting the diverse and remote energy market where alternative energy sources are very expensive or difficult to sustain. A consortium including the local government in Galena, Alaska, initiated preapplication meetings with the US NRC in 2007 and continued licensing activities until 2013. The Galena project has been effectively terminated; however, Toshiba is exploring alternative first customers. More information on the 4S design is provided in Chapter 17 of this handbook.

2.4.2 Russian Federation: SVBR-100 design

The SVBR-100 builds on the former Russian experience with lead-bismuth reactor technology used for several submarine propulsion units. It is a pool-type fast spectrum reactor with forced circulation of the low-pressure primary coolant using two main circulation pumps. The initial core fuel is UO_2 with ^{235}U enrichment below 20%, although subsequent fuel loads may contain U-Pu mixed oxide fuel or UN-PuN fuel. The core is composed of 61 fuel assemblies; however, a fresh core is loaded as a single cassette with whole-core replacement every 8 years. Internal straight-tube steam generators are used to supply steam to external steam separators and a Rankine power conversion unit. Although material corrosion issues plagued the early Russian submarine experience, corrosion problems were generally resolved by the conclusion of the earlier program.

The 100-MWe SVBR-100 is envisioned to supply both electricity and process heat for nonelectrical applications. A single unit is expected to deliver 580 tons/h of process steam, 70 Gcal/h of district heat, or 200,000 tons/day of desalinated water if configured for these applications.

The SVBR-100 design effort was approved in 2006. A demonstration project was approved in 2011, at which time the reactor designer, OKB Gildropress, teamed with AKME Engineering to construct the first plant. Design and site licensing has been ongoing since 2011 and approval was received in 2015. A license to begin construction is anticipated in 2021 and operation in 2025. Additional information regarding the SVBR-100 design is provided in Chapter 19 of this handbook.

2.4.3 United States: PRISM design

The US Advanced Liquid Metal Reactor program during the 1980s resulted in the design of the 160-MWe Power Reactor Inherently Safe Module (PRISM) sodium-cooled reactor. The PRISM design was one of the first advanced reactor designs to employ significant use of passive safety features and was designed as a power module to be used in multiple three-unit packs to form a large electrical-capacity power plant. The PRISM design was originally intended to be a breeder reactor for improved uranium resource management but more recently has been focused on recouping the unused energy content in discharged LWR fuel and also to consume the very long-lived higher actinide elements that dominate the long-term hazard in a geologic repository.

General Electric, now teamed with Hitachi, has resumed development of the 311-MWe PRISM design, although the name has been changed to Power Reactor Innovative and Small Module. The design has similarities with LWR-based integral designs except that the internal steam generators are replaced with intermediate heat exchanges that transfer heat to two secondary sodium loops. External secondary heat exchangers are coupled to a supercritical Rankine power conversion unit. The design uses U-Pu-Zr metal fuel and can accommodate actinide waste products from LWR spent fuel. Four electromagnetic pumps circulate the sodium coolant in the pool-type primary system, which operates at nearly atmospheric pressure.

The current deployment strategy is to couple two PRISM modules into a single "power block" with a shared turbine generator. One or more power blocks would be colocated with a small electrorefining fuel recycle facility. Although there was significant regulator review of the PRISM design earlier, there are no immediate plans to license PRISM for commercial power production.

2.5 Molten-salt-cooled reactors

In the 1960s and 1970s, a completely different reactor type was developed at Oak Ridge National Laboratory in the United States resulting in the construction of two experimental reactors. This new reactor type used a molten fuel-salt mixture circulated through a graphite moderator block to achieve a very compact, high-power reactor system. Although originally intended for aircraft propulsion applications, which were abandoned in the early 1960s, the technology for molten salt reactors (MSRs) continued to be developed for another 10 years as a candidate "breeder" reactor that could produce more fuel than it consumed. The technology was abandoned in the early 1970s in favor of sodium-cooled reactors for the fuel breeding mission.

Today, several countries are studying MSRs for potential commercial power or high-temperature process heat applications. Salt-cooled reactors offer opportunities to increase primary system temperatures above that available from LWRs while offering benefits of liquid coolants such as good heat capacity and heat transfer characteristics. The high-boiling point of salt (greater than 1000°C) allows the primary system to be operated at low (nearly ambient) pressure. Additionally, liquid salts are transparent, which improves inspection and maintenance operations relative to opaque liquid metals, and the very high temperatures possible with salt-cooled reactors enable them to be used for a broad range of industrial process heat applications.

The earlier MSRs were fueled with uranium or thorium fluorides dissolved in molten lithium and beryllium fluoride salts. Because the fuel was liquid, fission products could be removed from the fuel and fresh fuel could be added while the reactor continued to operate. Suitable containment materials and systems for chemical control of the molten salts were developed; however, this technology has been stagnant for more than 30 years. Recently a new variant of the MSR has emerged that uses unfueled molten fluoride salts to cool a graphite-moderated core containing standard

Table 2.5 Examples of commercial SMR designs based on molten salt-cooled reactor technology.

Country	SMR	Designer	Coolant	Electrical output (MWe)	SMRs/plant
Canada	IMSR	Terrestrial Energy	Fl-Be	190	1
United States	KP-FHR	Kairos Power	Fl-Be	140	1 or mult.
United States	LFTR	Flibe Energy	Fl-Be	250	1

graphite-coated particle fuel similar to gas-cooled reactors. This concept is referred to as a fluoride-salt, high-temperature reactor (FHR) and takes advantage of the excellent thermodynamic characteristics of the liquid salt to overcome many of the engineering challenges of the high-pressure gas-cooled reactors. Also the use of solid fuel in the FHR avoids many chemical and regulatory challenges associated with the circulating fluid fuel in the MSR.

There has been a recent resurgence of interest in both the MSR and the FHR concepts. The MSR is one of the designated Generation IV (GIF, 2002) concepts, and the FHR is a hybrid of the MSR and the VHTR concept (another Generation IV concept). Described in the following sections and listed in Table 2.5 are three examples of molten salt-cooled SMR designs (two MSRs and one FHR) that currently have commercial development support.

2.5.1 Canada: IMSR design

The integral molten salt reactor (IMSR) is being designed by Terrestrial Energy in Canada. It produces a thermal energy of 400 MWe that can yield 190 MWe using a steam Rankine power conversion unit or combined with an external heat exchanger to provide process heat for industrial applications.

Similar to the integral LWRs described earlier, the IMSR design places all primary system components within a sealed reactor vessel. A low-enriched uranium-fluoride salt is circulated through a graphite-moderated core region and then passes through an internal heat-exchanger that transfers the heat from the primary fueled coolant to a clean nitrate salt secondary coolant. Forced circulation of the primary fueled coolant is achieved using pumps located at the top of the sealed reactor vessel. The strong negative temperature coefficient of reactivity that is inherent to liquid-fueled reactors helps to prevent power transients and facilitates load-following operations. A novel feature of the IMSR design is that the integral primary system is completely replaced every 7 years.

The conceptual design of the IMSR was completed in 2015. Prelicensing engagement with the Canadian Nuclear Safety Commission (CNSC) was initiated in 2017. In 2019 the IMSR was selected by the CNSC and the US NRC for a joint technical review. Early evaluation of candidate sites is underway at the Canadian Nuclear Laboratory's Chalk River facility.

2.5.2 United States: KP-FHR design

In the mid-2000s, Oak Ridge National Laboratory (ORNL) and the University of California-Berkeley (UCB) collaborated on the conceptual development of entirely new reactor category that incorporated technologies from both gas-cooled reactors and molten salt-cooled reactors. The original concept, the advanced high-temperature reactor (AHTR), was a large reactor concept that used fixed graphite-coated particle fuel embedded in graphite blocks similar to the GT-MHR. However, instead of using helium as the coolant, the AHTR used molten fluoride salt. Beginning in 2011, UCB teamed with the Massachusetts Institute of Technology and the University of Wisconsin to develop a variant of the AHTR based on the a pebble bed design. The university consortium also initiated several lab-scale tests to demonstrate various separate effects associated with a liquid salt-cooled pebble bed reactor concept.

Kairos Power was formed in 2016 to commercialize the KP-FHR design. The KP-FHR is a small, liquid salt-cooled, pebble bed reactor concept that is targeting low-cost electricity generation and high-temperature process heat applications. With a maximum thermal power of 320 MW, it is expected to deliver up to 140 MWe using a steam Rankine power conversion cycle. The TRISO-coated uranium particle fuel is dispersed in graphite spheres. The primary coolant is a lithium-fluoride/beryllium-fluoride (Flibe) salt, which has a boiling temperature of 1430°C, thus assuring that the coolant will remain in liquid phase at near atmospheric pressure. Because the freezing temperature of the Flibe is 459°C, the reactor system must remain hot during all operations to avoid coolant solidification. Heat from the primary coolant is transferred to a secondary nitrate salt coolant in an external heat exchanger.

Kairos Power is in the conceptual design development phase of the KP-FHR small modular reactor. Headquartered in Alameda, California, Kairos announced in early 2020 the intent to construct a research and development center in Albuquerque, New Mexico to expand their technology testing capabilities. They anticipate potential deployment by 2030.

2.5.3 United States: LFTR design

Flibe Energy was one of the first contemporary companies to promote the resurrection of MSR technology and additionally promote moving the nuclear industry to a thorium-based fuel cycle. Their 250-MWe liquid fluoride thorium reactor (LFTR) design targets the objective of providing low cost electricity and effective fuel management.

The core is divided into two regions: a driver region and a blanket region. The ^{233}U fuel is dissolved within the Flibe salt and circulated in the driver region. The blanket region contains ^{232}Th fertile material that becomes transmuted to ^{233}U, which is processed online and inserted into the driver region. Thus the reactor can run continuously with only the addition of thorium fertile material into the blanket region.

The core operating temperature of 600–700°C can accommodate a supercritical carbon dioxide power conversion unit with improved efficiency.

Flibe Energy continues to develop the materials and chemistry technologies through commercial investments and federal grants. Currently the LFTR design is still in the conceptual development stage.

2.6 Future trends

As stated in the Introduction, the pursuit of SMR designs has been both persistent and global. There is no reason to expect this direction to change. First and foremost, SMRs are intended to provide energy consumers with abundant, safe, clean and affordable energy. The use of renewable energy sources are on the rise, but physical constraints of the variability and energy density of sunlight and wind will constrain their total contribution to meeting the global energy demand. The large and escalating cost of large base-load plants (coal and nuclear) has severely constrained the number of utilities who can consider that option. Continuing concerns for environmental issues, including air quality and climate impacts, will increasingly penalize carbon-based energy sources. Collectively, these conditions will promote the further development and deployment of SMRs.

Nineteen representative SMR designs from seven countries using four different reactor technologies are included in this chapter. Many more designs at various stages of development are also being explored worldwide. Although this would seem to give customers an abundant set of choices, even more designs are likely to emerge. This will be driven by a number of considerations:

- *Market opportunities in emerging countries.* Even modest demand projections predict a staggering amount of energy demand globally, especially for emerging economies. Geographical considerations, population demographics, and electrical grid infrastructures of these countries will tend to favor deployment of SMRs.
- *Market opportunities in established countries.* Established nuclear countries are generally experiencing low demand growth with periodic fluctuations driven by economic cycles. SMRs provide an attractive option for base-load capacity that can be constructed in smaller increments than traditional large plants and therefore allow the utility to be more responsive to demand fluctuations.
- *Broadening energy applications.* With only a few exceptions, existing commercial nuclear plants supply only electricity. Nuclear energy is moving in a direction to support non-electrical uses, and SMRs will accelerate that movement. There will be increased use of nuclear power for water desalination, district heating and cooling, and industrial applications. With this diversity of applications will come a diversity of plant requirements, which will stimulate new designs and perhaps even new technologies. Some of these will be derived from the many existing SMR designs or may include entirely new designs and technologies.
- *Moving away from fossil fuels.* Global warming considerations are prompting many countries to move away from fossil fuel-based energy sources, especially coal. Most existing coal plants are of a modest size, typically producing less than 500 MWe. As these plants are retired, SMRs offer an attractive alternative for replacement—perhaps occupying the same site and using the same site infrastructure as a cost-efficiency.

Balancing the forces that will encourage new SMR designs are counter forces that will cause many current designs to be abandoned. Some of the key challenges that face new SMR development include the following:

- *Significant cost of bringing a new design to market.* The cost to complete a new reactor plant design and to qualify it for construction, including rigorous regulatory review, may exceed $1 billion and can span more than a decade. The simplified designs and larger safety margins that characterize many SMR designs should reduce the amount of detailed design effort and shorten the regulatory review cycle relative to large plant designs. However, these savings are balanced by the need to validate design and technology innovations and address regulatory implications of the innovations. Even allowing for some cost savings due to the simplified designs, it is very difficult for private industry to sustain a significant financial investment for a protracted period of time. Therefore some of the current designs will never make it to market.
- *Low cost alternative energy sources.* The relatively sudden drop in natural gas prices in the United States due to a rapid increase in production rates resulting from new hydraulic fracturing (fracking) technology has created a significant barrier to the construction of new nuclear plants. While many utility leaders are committed to building a diverse portfolio of energy generation capacity, the availability of low-cost natural gas and the low capital investment needed to use it will discourage the development of new nuclear technologies.
- *Lack of policies on greenhouse gas emissions.* The lack of a national policy in several countries, including the United States, regarding greenhouse gas emissions creates a market uncertainty that increases the reluctance of industry and investors to pursue alternative energy sources. Coupled with the sizeable investments and lengthy development schedules for new nuclear, this additional uncertainty will limit the availability of investment funding for new SMR development.

Regarding the potential for new designs and technologies, two classes of SMRs are of special interest: high-temperature reactors and mobile reactors. There has been persistent interest in high-temperature reactor technologies to improve power conversion efficiency and to accommodate a broad range of industrial heat applications. While helium appears to be the coolant of choice for many earlier high-temperature reactor designs, alternative coolants are likely to be pursued, especially liquid coolants that have superior thermodynamic properties relative to helium. An example of this was discussed in Section 2.5. New innovative designs employing this technology are anticipated.

Regarding mobile reactors the Russian Federation is already developing several floating nuclear power plants to supply energy to coastal or river-accessible communities. China is also developing similar technology and designs. If successful, there likely will be increasing interest from countries with significant island-based populations to pursue this option. Also, several potential industrial applications lend well to mobile energy sources, such as advanced oil recovery processes that require the heat source to migrate with the oil recovery equipment as the resource fields are depleted, typically after 10–12 years. There have been a few mixed experiences with developing mobile nuclear plants, but they may again reemerge if the demand is sufficiently compelling. This market will likely drive increased interest in microreactors, which are already beginning to emerge in the developer community.

2.7 Sources of further information

As discussed earlier the information presented in this chapter was compiled from a substantial collection of papers, reports, presentations, web sites, and personal communications—too many to list here. Instead a list of primary resources is provided below. Of particular note are the several resources from the International Atomic Energy Agency (IAEA). The IAEA has been very active for several years in coordinating interests in developing and using SMRs, although in their vernacular, "SMR" originally meant "small- or medium-sized reactor." Hence, earlier IAEA publications and presentations also include designs with capacities between 300 and 700 MWe, which is the nominal range for "medium-sized" reactors. In more recent reports the IAEA has adopted the more accepted definition of SMR used in this chapter. Regardless the IAEA represents a valuable, objective repository of information on global SMR development and application. Additional information on some of the SMR designs is included in the chapters contained in Section IV of this handbook.

Recommended resources in chronological order are listed in the succeeding text:

- International Atomic Energy Agency, Advanced Reactor Information System, http://aris.iaea.org.
- World Nuclear Association, https://www.world-nuclear.org/information-library/nuclear-fuel-cycle/nuclear-power-reactors/small-nuclear-power-reactors.aspx
- International Atomic Energy Agency, *Advances in Small Modular Reactor Technology Developments*, https://aris.iaea.org/Publications/SMR-Book_2018.pdf, 2018.
- Kadak, A.C., *A Comparison of Advanced Nuclear Technologies*, Columbia Center on Global Energy Policy, 2017.
- International Atomic Energy Agency, *Advances in Small Modular Reactor Technology Developments*, https://aris.iaea.org/Publications/SMR-Book_2016.pdf, 2016.
- Ingersoll, D.T., *Small Modular Reactors: Nuclear Power Fad or Future?*, Woodhead Publishing, Cambridge, UK, 2016.
- Third Way, *The Advanced Nuclear Industry*, https://www.thirdway.org/report/the-advanced-nuclear-industry, 2015.
- International Atomic Energy Agency, "Status of Small and Medium Sized Reactor Designs: A Supplement to the IAEA Advanced Reactors Information System (ARIS)," 2012.
- Ingersoll, D.T., Guest Editor, "Special Issue on Small Modular Reactors," *Nuclear Technology*, **178**, No. 2, May 2012.
- International Atomic Energy Agency, "Common User Considerations by Developing Countries for Future Nuclear Energy Systems: Report on Stage 1," NP-T-2.1, 2009.
- Ingersoll, D.T., "Deliberately Small Reactors and the Second Nuclear Era," *Progress in Nuclear Energy*, **51**, 589–603, 2009.
- International Atomic Energy Agency, "Status of Innovative Small and Medium Sized Reactor Designs: Reactors without Onsite Refueling," IAEA-TECDOC-1536, January 2007.
- International Atomic Energy Agency, "Status of Innovative Small and Medium Sized Reactor Designs: Reactors with Conventional Refueling Schemes," IAEA-TECDOC-1485, March 2006.
- International Atomic Energy Agency, "Innovative Small and Medium Sized Reactors: Design Features, Safety Approaches and R&D Trends," IAEA-TECDOC-1451, May 2005.
- Nuclear Energy Agency, "Small and Medium Reactors: Status and Prospects," 1991.

References

Generation IV International Forum, 2002. A Technology Roadmap for Generation IV Nuclear Systems. GIF-002-00.

Ingersoll, D.T., 2016. Small Modular Reactors: Nuclear Power Fad or Future? Woodhead Publishing, Cambridge, UK.

International Atomic Energy Agency, 2018. Advances in Small Modular Reactor Technology Developments, 2018 Edition.

Nuclear Energy Agency, 2016. Small Modular Reactors: Nuclear Energy Market Potential for Near-term Deployment.

Third Way, 2015. The Advanced Nuclear Industry. Available from: https://www.thirdway.org/report/the-advanced-nuclear-industry.

# Integral pressurized-water reactors (iPWRs) for producing nuclear energy: A new paradigm

M.D. Carelli
Formerly of Westinghouse Electric Co., Pittsburgh, PA, USA

3.1 Introduction

Over 60 years ago nuclear power has been ushered into mankind, representing a quantum change which rivals and potentially surpasses those of the combustion engine and electricity. It has advantages and disadvantages, proponents and detractors, just like any other human endeavor. The underlying puzzle is why other innovations have successfully overcome – in a reasonably short time – the inevitable initial distrust, setbacks and potentially suffocating legislation, while nuclear power has not yet flourished.

True, nuclear weapons are not exactly the best introduction to nuclear energy, but that potential connection was overcome in the span of a decade; the advanced world economies accepted nuclear power, development plans were outlined and implemented with a variety of prototypes followed by a variety of progressively improving designs. So, the first couple of decades were not that different for nuclear power than they were for previous endeavors, except that nuclear power progressed at a more sedate pace, as would be expected because of the much higher financial exposure, and of course the drastically different consequences of potential accidents. However, at that time (mid-1970s), instead of becoming a staple of human development, nuclear power became a more and more controversial issue, crowned by the 1979 Three Mile Island accident. Since then it has moved in fits and starts, characterized, on one side, by the need for economical and reliable power and, on the other, by three major accidents, plus a variety of minor or nuisance-type, but highly publicized, ones and a substratum of polarized politics. This resulted in divided public opinion, showing roughly one-third firmly pro, one-third firmly con and the remaining one-third going back and forth depending on the latest happenings.

Consequently, for the past 40 years nuclear power, while it has expanded overall, has been unable to fulfil its potential, nor to maintain its promises; it will remain so for the foreseeable future, unless it goes through a drastic change from its current *modus operandi*. If properly planned and executed, the catalyst for this change could be the deployment of SMRs, starting immediately with the integral PWR designs (iPWRs), whose technology and characteristics are examined in detail in Part II of this Handbook.

3.2 The imperatives for nuclear power

Nuclear power plants must:

- be economically competitive;
- have superior safety; and
- satisfactorily manage their waste.

These are the three imperatives, which are shared by all power plants, but have a particular importance, especially the last two, in the case of nuclear plants.

Economic competitiveness is an obvious given. It is always a necessary condition, but in the case of nuclear power it is not sufficient, as it must be accompanied by the other two. Also, it must be substantially better than the competition and bring in additional considerations to compensate for the mere fact of being nuclear.

Safety must be far superior to conventional power plants because the consequences of a nuclear accident might impact a much larger population because of the dispersion of radioactive effluents. How critical safety is for acceptance of nuclear plants has been underscored by the aftermath of the three major accidents which have occurred. Objectively speaking, Three Mile Island (1979) was actually a minor accident, with very moderate radiation release, but it became the catalyst to halt nuclear growth for more than a decade. In the meantime came Chernobyl (1986) which was eventually overcome because of the atypical conditions of this Soviet reactor. When nuclear was on the rebound, Fukushima (2011) happened and several countries recoiled from nuclear, the most striking example being Germany, which is very willingly accepting the economic penalty of forgoing its nuclear plants.

Disposal of the waste is on a completely different scale for nuclear power than for any other power source. For the latter, handling of the waste is a minor, or at least manageable, issue. For nuclear plants, even though various technical solutions are available to deal with their waste, the mantra is that nuclear waste 'will poison the earth for millennia'.

So, where does nuclear power stand now in fulfilling the three imperatives?

- Economics: green light, mostly. It is a highly competitive field, but the bottom line is that hundreds of nuclear power plants are operating and no utility would be willing to spend billions of dollars to build a plant which is not economically competitive.
- Safety: yellow light. Nuclear plant safety has proven to be fundamentally sound; older plants have been improved or shut down and new designs have much improved safety. The key issue is the eventual occurrence of 'big' accidents.
- Waste disposal: flashing red light. Resolution of the waste problem has not made any substantial progress from the early days of nuclear power and disposal at site. Technical solutions do exist, but the problem is political.

Small modular reactors (SMRs) can satisfactorily address the three imperatives. Moving in reverse, in regards to the waste disposal imperative, fast spectrum reactors run in a burner mode will dispose of the plutonium and minor actinides. Large and smaller fast and thermal reactors using both uranium and thorium cycles can significantly reduce the present waste legacy as well as avoid future additions. To turn this third imperative from red to green requires the political will to move ahead with available technological solutions.

The scenario regarding the safety imperative is completely different. The iPWR is the very type of plant uniquely capable of reaching the ultimate safety in a reasonably short time using extensively proven technology.

The basic tenet of nuclear safety is very simple. The plant must be capable, through its intrinsic design and auxiliary safety systems, to survive any conceivable accident without releasing excessive radiation. Of course the rub is in the definition of 'conceivable accident' and 'excessive radiation', as well as the adopted design and the choice of auxiliary safety systems.

Major accident probabilities are defined through the core damage frequency (CDF), that is the probability that a postulated accident results in core damage, which is automatically considered as cause for radiation release to the environment. The CDF target for the early reactor designs was of the order of E–4/yr (0.0001 events/yr), that is the probability for core damage and radiation release, accounting for all hypothetical accidents, was once in 10 000 years; very small indeed, especially considering that the design lifetime was 30 years. However, the number of plants kept increasing, as well as their operating life. If the probability of core damage remained at E–4/yr for every plant and there are, say, 400 plants around the world with an average lifetime of 30 years, the probability of a major accident at any plant in the world would be 400 E–4 or 0.04 every single year, 1.2 over a 30-year lifetime, which is approximately the Three Mile Island/Chernobyl time frame. After approximately another 30 years, Fukushima. The pattern is there.

It is almost certain that another major accident in the next 15–30 years will be the end of the line for nuclear power. It is therefore necessary to decrease the CDF value. After Three Mile Island, a very significant amount of effort was spent to increase the reliability and redundancy of the safety systems. This improved the safety, but also significantly increased the reactors' price tag. No wonder that no new nuclear plants were ordered for quite some time. A breakthrough occurred with the adoption of passive safety systems in lieu of the previously active systems. The CDF for the new designs dropped to the order of E–6; also, existing designs are being retrofitted to decrease their CDF. However, the lifetime of the new designs as well as of the retrofitted ones is now projected to be of the order of 60 years. Finally, the new passive designs with a CDF of E–6 have a very hefty price tag and thus the market tendency is to prop up the older designs for as long as possible.

A clean breakthrough is necessary, i.e. to introduce reactors having a very low CDF as well as a low price tag to easily replace the old plants.

First, the CDF. If the CDF probability is decreased to E–8, which is the frequency of a severe event known as 'act of God', 1000 plants with a lifetime of 60 years will yield a failure probability over their 60 year lifetime of 6E–4, or 6 major accidents over 10 000 years. Even with 10 000 nuclear plants worldwide, we would have major accidents at an average interval of 170 years. The safety issue has disappeared.

Is it possible to have nuclear plants with such an infinitesimal CDF value? For the current generations of plants operating, in construction or offered now, the answer is no. With a new generation of plants, specifically SMRs of the iPWR type presented in

Section 3.3, the answer is possibly yes. It becomes a definite yes if the iPWR design is safety-driven, as will be discussed in Section 3.4. Finally, this safety driven iPWR must be economically competitive to cover a majority of the market; this will be discussed in Section 3.5.

3.3 The integral pressurized-water reactor (iPWR)

Of the three imperatives (economics, safety, waste disposal) the iPWR does not bring any new approach to the third one, since it is a PWR. Thus, the waste management aspect is not discussed any further here.

Regarding the first imperative, the SMRs in general do not appear at first to be the logical design for achieving competitive economics, since they go against the economy of scale that has prompted larger and larger plants. A new type of design therefore needs to be developed, one that is simpler, requiring fewer components, and that can be built in a shorter time than the present light-water reactors (LWRs). Here the introduction of the iPWR, which capitalizes on the multi-decade development, construction and operating experience of the PWR, is a plant representing the vast majority of nuclear plants built to date. The iPWR designs proposed and investigated do indeed promise economic competitiveness with present nuclear plants as well as conventional energy sources.

However, the *raison d'être* of the iPWR is how it aggressively and innovatively addresses the second imperative, superior safety. The configuration has intrinsically this capability, for example by eliminating the occurrence of large LOCAs (loss of coolant accidents) through locating the steam generators inside the vessel, or of control rod ejection accidents, through also locating the control rod drive mechanisms inside the vessel. Most remarkably, the iPWR has the capability, if properly designed, to address synergistically the first two imperatives, that is an increase in safety is concomitant with a decrease in cost, as will be elaborated later in Section 3.5.

3.3.1 The evolution of iPWR design

Integral reactors have been adopted in nuclear-powered submarines; it is not well known, however, that the first, and so far the only, 'commercial' iPWR was operational as early as 1964. It was the nuclear ship *Otto Hahn*, a German nuclear-powered freighter and research facility, which was launched in 1964 and commissioned in 1968. She sailed 650 000 nautical miles in 10 years without any technical problems, but was eventually docked because of nuclear hysteria, with ports and harbors refusing entry to a nuclear ship. The *Otto Hahn* featured helical steam generators, a solution favored by several current designs.

As far as terrestrial power reactors are concerned, the first iPWR type design to be proposed was the PIUS (Process Inherent Ultimate Safety) reactor by the Swedish ASEA-Atom in the early 1980s. It was conceived in response to the Three Mile Island accident with the objective of replacing the active safety approach with inherent safety. Essentially, the reactor was placed in a large pool of borated water in a concrete pressure vessel located underground. Core cooling was by natural circulation with upper and lower density locks to prevent mixing between the circulated hot reactor coolant and the cold pool water. PIUS got quite a bit of attention, but never real traction. Initially ASEA-Atom had a 500 MWe integral design with steam generators inside the vessel, but later switched to a more conventional 640 MWe with steam generators outside.[1] PIUS was first and foremost a natural circulation reactor, with the integral configuration being a mean of implementation, which was eventually dropped.

A true iPWR was proposed in the mid-1980s by US Combustion Engineering. This was the MAP (Minimum Attention Plant), a 900 MWt self-pressurized, full natural circulation design with multiple once-through steam generators located inside the vessel.[2] MAP was eventually shelved by Combustion Engineering (which had become ABB-CE) in favor of the 320 MWe SIR (Safe Integral Reactor), developed in the late 1980s in collaboration with Rolls-Royce, Stone and Webster and the United Kingdom Atomic Energy Authority (UKAEA).[3] The SIR core design, based on the CE System 80 commercial PWR, had a 55 kW/liter core power density, about half that of traditional PWRs, 24 month refueling cycle, 12 once-through integral steam generators arranged in an annular space above the core and an integral pressurizer in the vessel head. The six wet-winding glandless coolant pumps were mounted around the upper circumference of the vessel.

SIR was typical of iPWR designs which, starting in the early 1990s, were developed, and still continue to be, across the world, most notably in Russia, Argentina, Republic of Korea, Japan, China, as discussed in Part IV of this Handbook. In the USA the first significant effort was the IRIS (International Reactor Innovative and Secure), developed from the late 1990s to the end of 2009, when it was terminated.[4,5] The IRIS project was led by Westinghouse with a major part of the work being performed by the international partners, which included industry, laboratories and academia. While in other countries the current effort on iPWR designs has continued on the concepts developed earlier, the current US effort has basically started anew, spurred by the Department of Energy (DOE) solicitation of SMR designs, which was finalized in 2011.

Going back to SIR, a very significant contribution to advancing the state of the art was a seminal paper by the UKAEA partner investigating the cost benefits of smaller reactors.[6] It showed that the traditional capital cost economy of scale does not hold when other factors typical of smaller plants are taken into consideration. The paper listed and discussed quite a number of these factors: increased factory fabrication, more replication, multiple units at a single site, improved availability, faster progression along the learning curve, bulk ordering, better match to demand, smaller front-end investment, reduced construction time, increased lifetime, design appropriate to site and, elimination/downgrading of some safety systems. Most of these factors will be discussed in Chapter 10. The last factor, about the safety systems, is of momentous importance and will be fully explored in Section 3.4.

The next two sections, dealing with the safety and economics imperatives, are based on work performed during the 10 years of development of the IRIS design. There are many reasons for this choice, aside from the familiarity of this author.

The key one is that IRIS has systematically sought both safety excellence beyond the generally accepted limits and the synergism between safety and economics. Practical reasons are that IRIS is no longer being pursued and its work has been fully documented in over 500 open literature publications which provide a jumping point for iPWR designs, both present and future.

3.4 Addressing the safety imperative

Nuclear reactor safety is achieved through a sound design and the use of safety systems, i.e. protective systems which counteract the accident and/or attenuate its consequences. The ideal scenario would be a design so perfect that no safety systems are necessary, that is, a design where accidents either cannot occur or, if they do, their consequences are acceptable. Obviously this is Utopia, but the integral configuration offers a very good approximation to it. The most immediate, and universally adopted, possibility offered by the integral configuration is the elimination of the large LOCAs, simply because there are no large pipes to be broken. This is only one of the many opportunities offered to the designer. The IRIS project developed a unique approach, articulated over three tiers.

- The first tier, called Safety-by-Design, is a significant step beyond passive safety. The underlying principle is to intrinsically eliminate as many potential accidents as possible by proper design, rather than coping with their consequences through safety systems, either active or passive.
- The second tier is provided by simplified passive safety systems, which protect against the remaining potential accidents and mitigate their consequences.
- The third tier is provided by active systems which are *not* required to perform safety functions (i.e. are not safety grade) and are not accounted for in deterministic safety analyses, but are used as necessary to improve reliability and decrease the CDF. Their use and characteristics are optimized through a design based on probabilisitic safety assessment (PRA).

The iPWR offers the possibility of being able by design to: (1) eliminate some of the accidents (e.g. large LOCAs, control rod ejection); (2) decrease the probability of occurrence for the vast majority of the remaining accidents; and, (3) mitigate the consequences.

In loop-type PWRs there are typically eight accidents classified as Class IV design basis events (DBEs), i.e. accidents which can cause a radiation release to the environment. Thus, the DBEs eventually dictate the necessary safety systems. Table 3.1 (from Petrovic et al.[5]) summarizes the design characteristics; the safety implications of each design characteristic; the impacted accident and events; the related Class IV accident; and, bottom line, how they fare under the Safety-by-Design approach used by the IRIS reactor. As shown in the table, systematic implementation of the Safety-by-Design approach enables elimination of three out of the eight DBEs typically considered for LWRs, while four more are downgraded to a lower severity class (as low as Class 1 for the locked rotor accident) where radiation release will not occur. The only remaining Class IV accident is the fuel-handling accident, because the iPWR, like practically all the power reactors, needs to be periodically refueled.

Table 3.1 Implementation of Safety-by-Design™ in IRIS

IRIS design characteristic	Safety implication	Positively impacted accidents and events	Class IV design basis events	Safety-by-design impact on Class IV events
Integral layout	No large primary piping	Large break LOCAs	Large break LOCA	Eliminated
Large, tall vessel	Increased water inventory	Other LOCAs Decrease in heat removal events		
	Increased natural circulation Accommodates internal control rod drive mechanisms	Control rod ejection Head penetrations failure	Spectrum of control rod ejection accidents	Eliminated
Heat removal from inside the vessel	Depressurizes primary system by condensation and not by loss of mass	Other LOCAs		
	Effective heat removal by steam generator and emergency heat removal system	Other LOCAs All events requiring effective cooldown anticipated transient without scram (ATWS)		
Reduced size, higher design-pressure containment	Reduced driving force through primary opening	Other LOCAs		
Multiple, integral, shaftless coolant pumps	No shaft Decreased importance of single pump failure	Shaft seizure/break	Reactor coolant pump shaft break	Eliminated
		Locked rotor	Reactor coolant pump seizure	Downgraded

Continued

Table 3.1 Continued

IRIS design characteristic	Safety implication	Positively impacted accidents and events	Class IV design basis events	Safety-by-design impact on Class IV events
High design-pressure steam generator system	No steam generator safety valves	Steam generator tube rupture	Steam generator tube rupture	Downgraded
	Primary system cannot over-pressure secondary system	Steam line break	Steam system piping failure	Downgraded
	Feed-water/steam piping designed for full reactor coolant system pressure reduces piping failure probability	Feed line break		
Once-through steam generators	Limited water inventory	Feed line break Steam line break	Feed-water system pipe break	Downgraded
Integral pressurizer	Large pressurizer volume/reactor power	Overheating events, including feed line break ATWS		
Spent fuel pool underground	Security increased	Malicious external acts	Fuel-handling accidents	Unaffected

This is of course very impressive, but in reality DBEs very seldom do occur. Three Mile Island started with a small LOCA and Fukushima was the consequence of an external event (Chernobyl was atypical). So, attention must be paid to have proper, reliable passive safety systems (second tier). This practice is now adopted by all iPWR designs and most of the other LWR designs. The importance of the third tier, the active non-safety grade systems, was originally not recognized by the large LWRs where all the emphasis was placed on the passive systems, but the active systems do play a critical role in improving reliability and decreasing the CDF through an interactive and iterative PRA and safety design procedure.

From the very beginning of the IRIS design process, the PRA was iteratively used to guide and improve the design, as shown in Figure 3.1. The process is conceptually quite straightforward, but also quite time consuming, requiring tens of design iterations. It enabled IRIS to move from an initial CDF value of 2E–6 to a final value of 2E–8, at least a decade less than the most advanced of the other LWRs. As it will be recollected, E–8 is an act of God probability.

In the case of IRIS the cylindrical primary vessel has a 6 m diameter and the spherical containment a 25 m diameter. The containment is housed in the auxiliary building which also houses the remaining components of the nuclear steam supply system (NSSS), such as the spent fuel pool. The IRIS auxiliary building is of a cylindrical shape which is intrinsically more resistant to impact than an angular configuration. The containment is located one-third underground and also underground is the spent fuel pool. IRIS was designed several years before the occurrence of Fukushima, which would have been served very well by the IRIS Safety-by-Design.

Large break LOCAs cannot occur in an integral reactor, but small break LOCAs do. While their consequences are mostly insignificant from a technical standpoint, they do cause a negative financial, regulatory and public acceptance impact. And, of course, there is the possibility of triggering a higher-level accident. One of the

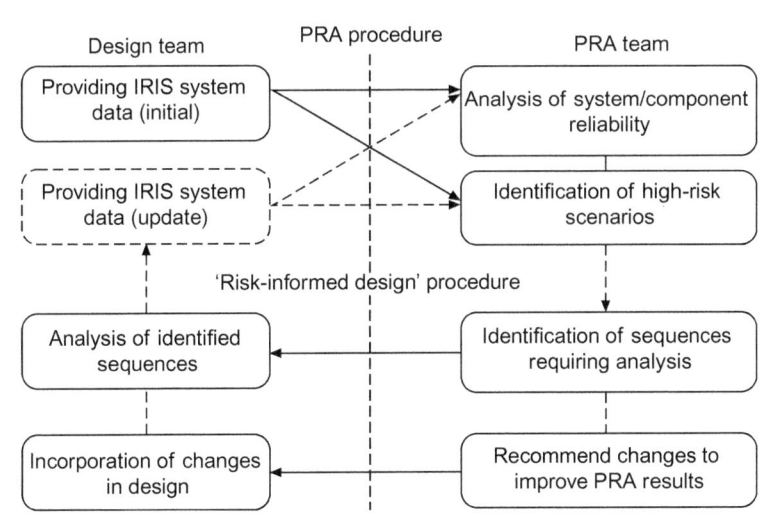

Figure 3.1 IRIS – PRA guided design.

accomplishments of the IRIS Safety-by-Design was to completely neutralize the small LOCAs. First, all lower level vessel penetrations are eliminated up to 2 m above the top of the core and there are no penetrations on the vessel head because of the internal control rods. The grand total is seven penetrations for safety and auxiliary systems. Furthermore, IRIS was designed such that if a penetration fails and a small LOCA does occur, there are no adverse consequences.

Once a break occurs, steam exits from the vessel into the containment, initiating the vessel depressurization; as the internal steam generators remove the decay heat, steam condenses with a further depressurization effect. The IRIS containment was designed such that at the same wall thickness and design strain for traditional loop PWRs it can take an operating pressure approximately four times higher due to its spherical shape and diameter being about half. Because of the simultaneous vessel pressure decrease (due to the steam generators' heat removal) and containment pressure increase that is safely allowed by the improved IRIS containment design, the two pressures equalize quickly and no further steam exits the break. LOCAs of various break sizes and elevations were evaluated and the LOCA duration was about half an hour, with the core remaining safely covered in all cases (for the worst size/elevation combination, about 2 m of water were still left above the top of the core). Once the steam egression phase is over, the vessel and containment are thermo-hydraulically coupled through the break and long-term cooling of the vessel/containment system can be controlled through external cooling of the containment.

The complex coupled behavior of this patented design was extensively analyzed and experimental verification was planned at the SIET test facility in Piacenza, Italy, where tests of the Westinghouse AP 600 passive systems had been performed. The design of the IRIS test facility mockup and the extensive test campaign plans were completed when the IRIS program was terminated.

Historically the emphasis in the safety design and analysis has been on the internal events. However, as more and more improved handling of the internal events is being achieved, the external events, *in primis* seismic, have become determinant in establishing the total CDF. And, of course, Fukushima has been quite a reminder. The very compact iPWR can dispose of the seismic events through the Safety-by-Design approach, since it can be sited on seismic isolators in a partly underground location.

Elimination of the consequences from the seismic events, which are by far the most significant of the external events, will keep the total CDF in the order of E–8/yr as determined by the internal events. On the other hand, seismic CDF for non-isolated plants could be of the order of E–6/yr or more, i.e. of the same order of magnitude as the internal events in the most recent LWR designs, thus significantly affecting the overall safety. Seismic is the most critical of the external events, while others can be kept under E–8/yr through proper application of the Safety by Design. An in-depth discussion can be found in Carelli *et al.*[7] and Alzbutas *et al.*[8]

Another consideration is the plant resistance to terrorist attacks, which can be handled through Security-by-Design, rooted on the same principles as the Safety-by-Design. A typical example is a very low profile above ground (in the IRIS case it was less than 30 m), an uninviting choice to airborne terrorists with many taller targets to choose from. This application of the Security-by-Design would be even more effective for other iPWRs which are smaller than the 335 MWe IRIS.

Integral pressurized-water reactors (iPWRs) for producing nuclear energy: A new paradigm   61

In the same vein as the Safety-by-Design and Security-by-Design, the iPWR offers intrinsic possibilities to improve proliferation resistance and physical protection, as addressed in Chapter 9.

3.5 Satisfying the economic competitiveness imperative

In December 2009 the winner of a competitive tender by the United Arab Emirates for a large LWR was announced. The proponent of a higher priced, and losing, entry was quoted as blaming the loss on the fact that superior safety is expensive. *Au contraire*! In a plant properly designed according to the Safety-by-Design approach the increased level of safety is accompanied by a decrease in cost, as elaborated by Carelli.[9] In fact, such a simpler plant is safer (there are fewer 'things' which can go wrong) and at the same time cheaper (there are fewer 'things').

This inverse connection between safety and cost has not been fully exploited by reactor designers, not going much further than the elimination of large LOCAs and associated piping systems. Rather, relying on proven technology, supply chain and infrastructure of the large PWRs has been taken by traditional LWR vendors as the sure path to economic competitiveness. This conventional, sure-footed approach completely misses the fact that the iPWR is not just a smaller PWR, but it is a completely different design, with its own individual challenges and rewards. If properly identified, addressed and fully exploited the iPWR does indeed yield both increased safety and decreased cost. The systematic approach adopted in the IRIS design is in principle applicable to all iPWRs in general. It starts by eliminating the following major components and systems present in large LWRs:

- all large piping to/from the reactor vessel;
- steam generator pressure vessel;
- canned motors and seals of primary pumps;
- pressurizer vessel and pressurizer spray system;
- vessel head penetrations due to external control rods drive mechanisms (CRDMs);
- vessel bottom penetrations and seals due to in-core instrumentation;
- all active safety systems;
- high-pressure emergency core cooling system;

Also, the following major components have been reduced:

- shielding;
- number and complexity of passive safety systems;
- number of valves;
- size of containment and nuclear building;
- number of NSSS buildings (from two or more to one);
- number of large forged components (from approximately a dozen to one).

The only added major component/system is the seismic isolators.

In most iPWR designs the control rod drive mechanisms are located above the core, with the steam generators arranged outside in an annular configuration against the vessel. This leaves a coolant downcomer between the core and the vessel which is a quite effective shielding of the vessel wall and provides another iPWR intrinsic cost reduction. In fact, in some designs like IRIS the downcomer is large enough to reduce by several orders of magnitude the neutron fluence to the vessel wall, practically eliminating the vessel embrittlement and allowing increased plant lifetime. Also, the routine personnel exposures, and thus the ALARA costs, are reduced because the radiation level outside the iPWRs vessel is significantly decreased in respect to traditional LWRs.

The economy of scale, where the capital cost per unit power increases as the plant size decreases obviously applies and does favor larger plants. But traditional LWRS and iPWRS are on different, roughly parallel, cost versus power curves, with the iPWR being, even significantly, below because of the simpler and more economic design. In addition, SMRs in general enable economy of multiples versus single monolithic plants by relying on bulk/serial components fabrication (e.g. many small serial steam generators versus a few one-of-a-kind), accelerated learning and multiple units savings.

Modular construction and multiple modules deployment yield shorter construction schedule, module deployment tailored to demand with reduced spin reserve (i.e. reduced requirement for purchase power).

Quantification of the various factors is shown in Table 3.2, which compares the IRIS SMR (335 MWe) as part of a four-unit deployment against a single large PWR of 1340 MWe. The SMR starts with a large (1.7) penalty factor due to the economy of scale, but the other factors unique to SMRs in general (factors 2 through 5) and iPWRs in particular (factor 6) bring the estimated penalty for the iPWR down to 1.05, which is within the uncertainty level.

Two other factors have a significant financial impact favoring the SMRs in general: improved cash flow and reduced capital at risk. Construction of large plants takes a long time, with no income until completion. On the other hand, staggered deployment of the smaller SMRs enables 'bootstrapping' with the first unit generating income to support construction of the second unit and so on. As a result, the maximum cash outflow, or capital at risk, is significantly reduced as shown in Figure 3.2, which compares the cash flow of the same four 335 MWe IRIS plants (1340 MWe total) deployed every three years against a single 1340 MWe large PWR. An in-depth

Table 3.2 Comparison of major factors affecting capital cost in SMRs and large plants

Factor	SMR/large capital cost factor ratio	
	Individual	Cumulative
(1) Economy of scale	1.7	1.7
(2) Multiple units	0.86	1.46
(3) Learning	0.92	1.34
(4) (5) Construction schedule and timing	0.94	1.26
(6) Design specific	0.83	1.05

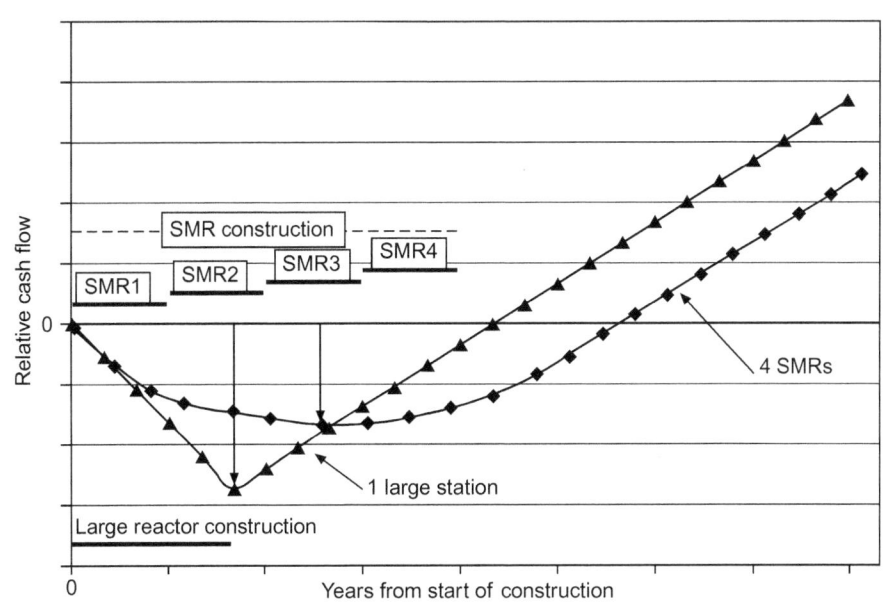

Figure 3.2 Staggered modular build reduces maximum cash outlay and capital at risk.

discussion will be found in Carelli et al.[10] and Boarin and Ricotti[11] iPWR economics is the subject of Chapter 10.

Finally, a critical consideration for iPWRs, which has a very significant impact in terms of both economics and public acceptance: a direct consequence of reducing the probability of catastrophic accidents to the order of E–8 is that the plant can be licensed with a drastically reduced emergency planning zone, in principle to the plant boundary, although in practice to a few kilometers.[12]

3.6 Future trends

There are essentially three components of nuclear power at any time: existing operating plants, plants under order/construction and future plants on the drawing board. While they do not necessarily, and actually should not, move in the same direction, they are strictly interconnected. Logically, as existing plants get old they are replaced by the next generation, while further advanced designs are progressing from the drawing board to eventual entries on the market. However, existing plants (Generation II) have been kept operating through lifetime extensions, while new plant designs (Generation III and III+) have not been deployed in the expected numbers; actually, the cancellations have outnumbered the fulfilled orders. It appears that this situation, be it because of competition by cheap natural gas or because of the Fukushima aftermath, will not change for some time. In the past, such a stagnation pattern has brought down any push to build new generation designs. For example, Generation IV was launched about 15 years ago and has produced only studies, conferences and papers.

Will history be repeated for the SMRs in general and iPWRs in particular? It is possible. However, there is one key difference between now and then. Nuclear power is running out of time and out of chances; a change is needed, with SMRs in general and iPWRs in particular possibly being the vehicle of such change. In fact, in addition to the technical reasons abundantly discussed already, there is a key, fundamental, economic difference. The SMRs in general, and specifically the iPWRs, do not require the massive financial investments typical of the behemoth plants 10–20 or more times their size. This means that government financial investment and support are not as critical as in the past. Private initiatives à la Fluor/NuScale, B&W/mPower or Bill Gates/TerraPower have already shown up.

The possible and auspicable new trend will be for the private initiative to continue the technical development of the iPWR designs and their move to prototype construction, with the governments providing, in addition to seeding financial support, the institutional environment for the effort to succeed.

Another new trend will be the significant role SMRs will play in developing and non-nuclear countries. While SMRs in general, and iPWRs in particular, will be first deployed in present nuclear countries, as in fact is clearly requested by countries considering their first nuclear plant, subsequent significant deployment will be outside the established nuclear states. Deployment of proven SMRs will be of particular significance for developing countries because it will represent a major endeavor not only technical, but also sociological, as discussed in Chapter 20.

3.7 Conclusion

In the early days of nuclear power, on the wings of the enthusiasm for this new and very powerful source of energy, it was stated that electricity produced by nuclear power would be too cheap to meter, a hyperbole which has lived forever in infamy. Every time new nuclear power developments are promoted, it is there for nuclear power opponents to ridicule and for nuclear power proponents to cringe and go on the defensive.

Hyperbole and naivety aside, that statement had indeed a sound physical basis. Human progress is fueled by the utilization of energy sources and it has leapt in bounds with utilization of sources of progressively higher density, from burning wood to hydropower, to fossil fuels. Abundance of power has brought improved products at lower cost. Nuclear fuel represents a several orders of magnitude increase over the power density of fossil fuels, thus the 'too cheap to meter'. However, nuclear power plants, which in principle are quite simple and in the early days became operational after a few years, have become extremely complicated and the timeframe from inception of design to power production has gone from a few to many years and in some cases forever.

The iPWR represents both a clean break and a drastic change. Its simplicity and small size allow a relatively quick construction and deployment in countries going nuclear for the first time. It is a necessary and sufficient solution, by positively addressing both the economic and safety imperatives. It can be designed to be competitive with the larger plants and it is the best solution in particular situations such as developing countries or remote areas. It has a catastrophic failure probability akin to an act of God, i.e. once every few hundred years for a fleet up to 10 000 plants worldwide.

The following chapters of this Handbook will discuss the iPWR technological characteristics (Part II), critical implementation aspects (Part III) and current status across the world (Part IV).

3.8 Sources of further information and advice

For the reader interested in the IRIS design, the *Nuclear Technology* article (B. Petrovic *et al.*, 2012[5]) provides a detailed walk through and a copious list of references.

References

Hannerz, K., 1988. Making Progress on PIUS Design and Verification. Nucl. Eng. Int. 33, Nov.

Turk, R.S., Matzie, R.A., Dec. 1986. The Minimum Attention Plant Inherent Safety through LWR Simplification. ASME 86-WA/NE-15.

Matzie, R.A., et al., 1992. Design of the Safe Integral Reactor. Nuclear Eng. Design 136, 73–83.

Carelli, M.D., 2003. IRIS: A Global Approach to Nuclear Power Renaissance. Nucl. News 46, 32–42.

Petrovic, B., et al., 2012. Pioneering Role of IRIS in the Resurgence of Small Modular Reactors. Nucl. Technol. 178, 126–152.

Haynes, M.R., Shepherd, J., 1985. SIR – Reducing Size Can Reduce Cost. Nucl. Energy. 30(2).

Carelli, M.D., et al., 2004. The Design and Safety Features of the IRIS Reactor. Nucl. Eng. Design 230, 151.

Alzbutas, R., et al., 2005. External Events Analysis and Probabilistic Risk Assessment Application for IRIS Plant Design. In: Proc. 13th Int. Conf. Nuclear Engineering (ICONE-13), Beijing, China, May 16–20.

Carelli, M.D., 2008. IRIS: Safety Increases Yield Cost Decreases. In: Proc. NUTHOS-7: the 7th International Topical Meeting on Nuclear Reactor Thermal-Hydraulics, Operation and Safety, Seoul, Korea, Oct. 5–9.

Carelli, M.D., et al., 2007. Smaller Sized Reactors Can Be Economically Attractive. In: Proceedings of ICAAP 2007, Paper 7569, Nice, France, May 13–18.

S. Boarin and M.E. Ricotti, 'Cost and Probability Analysis of Modular SMRs in Different Deployment Scenarios', Proceedings of ICONE 17, paper 75741, Brussels, Belgium.

Carelli, M.D., Petrovic, B., Ferroni, P., 2008. IRIS Safety by Design and its Implications to Lessen Emergency Planning Requirements. Int. J. Risk Assess. Mgmt. 8, 123.

Part Two

Small modular nuclear reactor (SMR) technologies

Core and fuel technologies in integral pressurized water reactors (iPWRs)

Andrew Worrall
Oak Ridge National Laboratory, Oak Ridge, TN, United States

4.1 Introduction

Small modular reactors (SMRs) share many of the same nuclear design principles as other reactor types, small and large and thermal and fast. In particular, integral pressurized water reactors (iPWRs) build upon the extensive nuclear design experience of the hundreds of larger pressurized water reactors (PWRs) operating around the world today. Nuclear design is the fundamental discipline that involves several major criteria/objectives for the reactor core designer/vendor, including the following:

- Safety—the fuel and the core must be designed to be able to withstand all operational demands and anticipated accidents. Closely related to this is that it must be licensable. In general, this means that constraints on fuel rod power peaking, total reactivity, reactivity coefficients, control rod worths, shutdown margin, and delayed neutron fraction have to be analyzed and demonstrated to be within limits.
- Economics—the fuel and the core must be designed to be able to produce the required energy that the utility demands, over the required time period requested, while minimizing the fuel costs.
- Reliability—the fuel and the core must be designed to operate in a predictable and reliable manner.
- Operations—the fuel and the core must be designed for relative ease of operation but with minimum complexity to the operator. For example, the utility may wish to increase cycle length in subsequent years.
- Strategy—the fuel and the core must be designed to be able to achieve utility strategic requirements, if and when set, such as plutonium management, load follow, and fast ramp up in power.

"Nuclear design" is also often referred to as "core design," "in core fuel management," or "core analysis," but regardless of the term used, there are various design features that the designer can work with to develop an effective and viable design, meeting all

---

[☆] This manuscript has been authored by UT-Battelle, LLC, under Contract No. DE-AC05-00OR22725 with the US Department of Energy. The US Government retains, and the publisher, by accepting the article for publication, acknowledges that the US Government retains a nonexclusive, paid-up, irrevocable, worldwide license to publish or reproduce the published form of this manuscript, or allow others to do so, for US Government purposes. DOE will provide public access to these results of federally sponsored research in accordance with the DOE Public Access Plan (http://energy.gov/downloads/doe-public-access-plan).

of the objectives listed earlier. For iPWRs (and large PWRs), these include, but are not limited to, the following:

- enrichment of the fissile material,
- burnable poison (BP) types,
- BP loadings (location, number, and weight percent),
- locations of fuel (fresh and irradiated) within the core,
- control rod locations and level of insertion,
- frequency and number of fuel assemblies replaced during its maintenance outage.

Therefore the nuclear design of any reactor, including iPWRs, is more specifically focused on the core physics characteristics, the safety and operational performance of the fuel and the core, and the reactivity control systems in the fuel and the core. The designer, wherever possible, has to stay within the licensing window of experience, including fuel irradiation experience, and validated operations for the nuclear design codes, for example, fuel types, reactor types, irradiation time, and energy output. At the same time the designer has to be looking at ways of making the nuclear design as economic as possible, whether that is by minimizing the fuel costs or extracting the most energy from the fuel; all has to be done while staying within the safety envelope of the reactor.

It should be noted that nuclear design is not concerned with the specific design of the fuel or core components nor the thermomechanical performance, or chemical aspects of the fuel, for example, rod internal pressure and fission gas release. This is generally referred to as "fuel design" and is not the subject of this chapter.

However, because of the central role the nuclear design fulfills, there are also a variety of disciplines and interfaces that the designer works with on a regular basis (see Fig. 4.1), including other analysis teams (thermal hydraulics, fuel performance, mechanical design, etc.), licensing, fuel manufacturing and procurement (since the nuclear designer sets the enrichments, number of assemblies, number and type of burnable poisons, etc.), and the broader safety analysis groups (transient analysis, severe accident analysis for source terms, etc.).

For the reader less familiar with the nuclear design process, this chapter begins by introducing and summarizing the nuclear design process for PWRs and then moves into the specifics of iPWRs; the principles and requirements are the same. The chapter describes (i) the important safety design criteria and principles in nuclear design of reactors, and in particular, iPWRs; (ii) what design features are used to achieve a viable and economic nuclear design; and (iii) how the iPWR designers and vendors have addressed the design principles and features in their respective reactors.

4.2 Safety design criteria

As outlined earlier there are a number of design criteria that have to be taken into account when completing the nuclear design of an iPWR. This section highlights the major safety design criteria for PWRs (including iPWRs); the other criteria, including economics, are discussed in later sections.

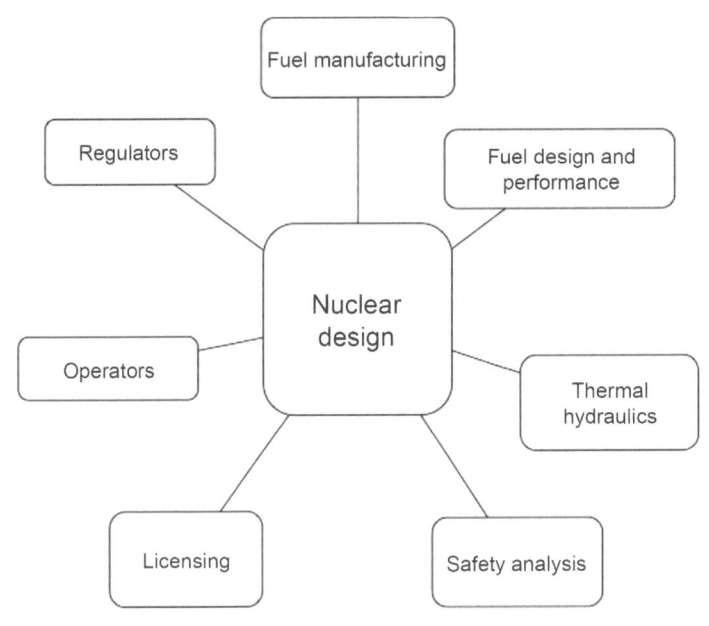

Core and fuel technologies in integral pressurized water reactors 71

Fig. 4.1 An indication of some of the interactions nuclear design has with other stakeholders and teams.

4.2.1 Fuel burnup

Burnup of fuel is the amount of energy extracted per mass of initial fuel loaded, and the units are generally megawatt-days per metric ton of heavy metal loaded (MW d/MTHM). For example, if a 500 megawatt (MW) thermal SMR uses four metric tons (MT) of uranium contained within the fuel and operates at full power for 1 year, the average burnup of the fuel will be:

$$\text{Average burnup} = (500*365)/4 \tag{4.1}$$

which is 45,625 MW d/MTHM, or 45.6 GW d/MTHM.

On this basis, it can be seen that the burnup requirements are derived from the utility needs, both in terms of the energy output of the reactor and the refueling frequency, for example, yearly or every 18 months.

For the fuel to achieve its design burnup, both for cycle of operation when it is first loaded and for its design lifetime, sufficient excess reactivity has to be present in the fuel, and the fuel has to be replenished on an appropriate frequency. For a typical iPWR, this could range anywhere between 1 and 5 years, although it is more typically 12–24 months for modern LWRs.

Although the excess reactivity is not a design limit or requirement directly, it is key to ensuring that the PWR core maintains criticality at full power operating conditions throughout the cycles of operation, including compensation for fuel depletion, buildup of fission product poisons (such as xenon and samarium), and loss of reactivity due to changes in temperature of fuel, moderator, etc. There is no specific limit on the amount of excess reactivity allowed either, but there are other design parameters such as negative reactivity coefficients or shutdown margin (see Sections 4.2.2 and 4.2.4) that are affected by the level of excess reactivity. Since the multiplication of neutrons varies in distribution in the core from one cycle to the next, and during the cycle of operations itself, it is important that the excess in reactivity is controlled both globally within the core and also locally within the fuel assemblies, and the designer has a number of ways in which the core global and local reactivity and hence power can be controlled; these are described in Section 4.3.2.

There are, however, limits on burnups, partially due to not only vendor warrantees associated with the fuel but also licensing limits for the maximum fuel rod average burnup, but this is fuel design and vendor specific in both cases. Fuel rod average burnups of 60–62 GW d/MTHM are typical for large PWRs, and since many iPWR designs rely on the fuel experience gained in large PWRs, a similar limit can be envisaged for all iPWRs, at least within the first few cycles of operation, that is, before greater experience of iPWR operation and fuel irradiation is gained.

4.2.2 Reactivity coefficients

Changes in the temperature of the fuel or core that happen from a change in power can cause changes in reactivity, which in turn result in effects on power. These feedback effects have an important consequence on the safety of reactors. For example, if an increase in power and therefore temperature leads to an increase in reactivity, this will result in a further increase in power (and temperature), and if this is not controlled, the unstable condition could lead to an accident; this is known as a positive reactivity feedback effect. However, if an increase in power and temperature leads to a decrease in reactivity, then the initial power level will be reduced, along with the temperature, and the core will be stable; this is known as a negative reactivity feedback effect. Clearly the latter condition is how the specific iPWR core should be designed.

The term "temperature coefficient" is used to express the effects of changes in temperature on reactivity and is defined as the change in reactivity per degree change in temperature. The two major reactivity temperature coefficients of interest in nuclear design of PWRs and iPWRs are fuel temperature and moderator temperature coefficients. They are considered separately as they are not only caused by different conditions that have to be analyzed but also occur at different rates.

The fuel temperature coefficient is particularly important as it acts with little or no delay in a power rise and as such it is important that is negative. The use of slightly enriched uranium fuels [as currently used in all light-water reactors (LWRs) and planned for iPWRs] ensures that the fuel temperature coefficient is negative due to "Doppler broadening." As the fuel temperature increases, the thermal vibration of the atoms in the fuel also increases, which results in a wider range of neutron energies, and the resonance peaks in the U-238 absorption cross section are broadened, increasing the probability of a neutron capture in the U-238, and not producing a fission; this is "Doppler broadening." In general the fuel temperature coefficient only becomes of concern for those fuels with lower U-238 content that seen in traditional fuels, for example, by increasing the proportion of U-235, or adding another material such as plutonium. Therefore there is relatively little nuclear design in the control of this reactivity coefficient for PWRs, including iPWRs.

The moderator temperature will rise more slowly in the event of a power rise in a PWR core due to the time to transfer the heat from the fuel to the water moderator/coolant. In PWRs (including iPWRs) the increase in temperature results in a reduction in the moderator density and also reduces the moderator-to-fuel ratio. These effects generally reduce the reactivity of the core and therefore result in a negative moderator temperature coefficient.

However, for all PWRs (including iPWRs) that have boron present in the coolant, the moderator temperature coefficient becomes less negative (or more positive) if the boron concentration in the moderator is increased. The reduction in the moderator density caused by the temperature rise reduces the density of the soluble boron that is present, therefore reducing the absorption in the boron, and this effect is clearly increased for higher boron concentrations. Therefore the nuclear design of the fuel and the core addresses this by using burnable poison (BP) rods (see Section 4.3.2 for further details) to limit the amount of soluble boron needed in the coolant to control the excess reactivity and ensure a negative moderator temperature coefficient in the core throughout the cycle of operation.

4.2.3 Power distribution

Many fuel operating limits (performance and safety) for LWRs (including iPWRs) are directly related to the maximum linear power density of the fuel. As a first approximation, for a given fuel rod geometry, the peak fuel temperature, the surface heat flux, the decay heat generation rate, and the stored thermal energy are proportional to the fuel rod linear power density. As such, it is a key role of the nuclear designer to maximize total power output while minimizing power peaking (and hence the linear power density), both for the fuel assembly as a whole in the core and for the individual fuel rods.

The attempt by the nuclear designer to reduce the maximum rod power in the core does not negate the need for full thermomechanical and thermal hydraulic analyses for the reactor, but it does aid the overall design process for the fuel and the core. The key considerations for the nuclear designer are that the power distribution of the core must be such that:

- The fuel will not exceed specified peak linear heat rates under normal operating conditions
  - a value that is determined as part of the safety analysis for the core.
- The fuel does not exceed the design basis set for departure from nucleate boiling (DNB)
  - DNB is the point at which the heat transfer from a fuel rod rapidly decreases due to the insulating effect of a steam blanket that forms on the rod surface when the temperature continues to increase—this value is calculated by thermal hydraulics analysis.
- Under normal and abnormal operating conditions, the maximum linear heat rate will not cause the fuel to melt—a value provided to the nuclear designer.
- The fuel rod power and burnup are consistent with the assumptions and analysis in the fuel rod thermomechanical performance analysis—a subsequent analysis, once the nuclear design is completed and can be iterative if subsequent violations in design criteria for the fuel performance are found.

To demonstrate these requirements the nuclear designer has also to complete the analysis under a range of extreme power shapes and variations, not just under nominal reactor operating conditions. The extreme shapes used reflect experience at operating reactors (in the first instance based on large PWRs), but the shapes are chosen to be deliberately conservative, for example, axial power skewed by control rod insertion/withdrawal and load follow conditions.

For the fuel assembly power in a PWR or in a specific iPWR, there is no design limit or constraint. (It is often said that fuel rods, not fuel assemblies, fail!) But, it is an important design optimization consideration to ensure that those assemblies with continually higher powers do not lead to limitations in the fuel rod performance and its associated integrity (e.g., violations in rod internal pressure due to fission gas release) or that assembly burnups do not vary too significantly within the batch such that the fuel assemblies are used inefficiently (see Section 4.3.3 for details).

4.2.4 Shutdown margin

In general, there are two independent reactivity control systems provided for iPWRs or large LWRs, each relying on different principles to operate. In large, traditional PWRs, the primary reactivity control system is the use of soluble boric acid in the coolant. But as demonstrated in many iPWRs, this does not have to be the case, as a number of iPWRs are designed without the presence of boron in the coolant.

The second control system is the use of control rods, which have to be capable of ensuring subcriticality under normal operations, including anticipated operational occurrences, and with enough margin to tolerate malfunctions such as stuck rods. The control systems have to be able to control the rate of reactivity change during normal operational power maneuvers (such as xenon depletion).

In both cases the systems have to ensure that the fuel design limits are not exceeded, for example, local power variations are less than the limits defined for pellet-clad interaction. In addition, at least one of the systems has to be able to ensure subcritical conditions at cold conditions.

4.2.5 Maximum reactivity insertion rate

To avoid damage to either the fuel or the core components, the designer has to be cognizant of the maximum reactivity insertion rate, due to control rod withdrawal (during normal or accident conditions), or by dilution of the boron in the coolant (if present). Limits on the maximum linear heat rate and margins to departure from nucleate boiling have to be demonstrated. By analyzing the reactivity worth and the potential allowed speed of movement of the control rods, the designer can limit the reactivity insertion rate.

4.2.6 Power stability

For a variety of reasons, the reactor core can become unstable to power oscillations, for example, initiated by instability in the reactivity control systems or turbines. In addition, spatial effects caused by the fission product xenon can also be a trigger across and up and down the core. Instability is particularly an issue for large boiling water reactors (BWRs) due to the large core size, although it can occur in smaller cores but will be least significant in those cases.

Stability in the control systems for the reactor hardware means that total core power oscillations are not usually possible, and nuclear design of the core and control rods can ensure that the xenon effects are self-limiting.

4.3 Design features to achieve the criteria

4.3.1 Setting the enrichment of the fissile material

The first stage in the nuclear design of an iPWR (or any reactor, large or small) is to determine the enrichment requirements for the fuel to provide the energy output over the time period requested by the utility. [As for large PWRs the current design limit, primarily on fuel manufacturing and transportation (both from a criticality control perspective), is 5 wt% U-235.] Once a reactor is at equilibrium conditions (e.g., after several cycles/years of operation), the cycle length and operations are more constant, for example, how many assemblies in a reload of fuel or target burnup. At these conditions the enrichment is unlikely to change a great deal. However, in the early cycles of operation, and as equilibrium operations are approached, the enrichment will need to vary to reflect these changes. Similarly, if the utility changes cycle length, such as going from 12 to 18 months between outages, again the enrichment requirements will change. Even though eventually the enrichments may not need to be changed to achieve the required cycle length, the nuclear design, including the development of the loading pattern (see Section 4.3.3), and assessment of the key safety and performance criteria still have to be completed.

Clearly the designer has to define the enrichment to obtain the required cycle length. But equally important is that the fuel needs to have sufficient enrichment to ensure sufficient reactivity not only for the next cycle but also for its design lifetime, which is typically anywhere between one and four cycles of operation depending on the iPWR and the fuel design (as described later in this section). For the initial estimate and prior to completing the detailed nuclear design using reactor analysis tools (such as CASMO-SIMULATE or PARAGON-ANC), the designer usually relies on either rules of thumb or an experience base that they can call upon for a specific reactor to provide the initial estimate. Linear reactivity models can also be used to assist in the estimation.

The nuclear designer has to work closely with the utility at this point because they will have analyzed the optimum cycle length of operation for their reactor, from an electricity demand perspective, duration of outage, and for planning, for example, if there are several iPWR units on the same site, there will be a master schedule for maintenance and refueling outages throughout the year. The utility will have forecast the practical and economic optimum for operations, including the potential for early shutdown or stretch out in operations.

Since the enrichment of the fuel is governed by the desired cycle length of operation and the burnup of the discharged fuel, the fraction of the core replenished each cycle has to also be considered. Generally in iPWRs (as in all large PWRs), a fraction of the core (referred to as a "batch") is replaced after each cycle of operation. The remainder of the fuel is then reloaded back into the core, albeit in different locations to their previous location—the locations within the core of the fresh and previously irradiated fuels are known as the "core loading pattern," which is described in Section 4.3.3. However, in at least some iPWR designs, there is a plan to discharge all of the fuel after each cycle. The proportion of the core replaced and the frequency, which is replaced, is known as the "fuel management" scheme. Here are some examples to illustrate:

- 4 ∞ 12 *month fuel management scheme:* ¼ of the fuel assemblies are replaced every 12 months;
- 3 ∞ 18 *month fuel management scheme*: ⅓ of the fuel assemblies are replaced every 18 months;
- 2 ∞ 24 *month fuel management scheme*: ½ of the fuel assemblies are replaced every 24 months;
- 1 ∞ 48 *month fuel management scheme*: all of the fuel assemblies are replaced every 48 months.

It should be noted that the cycle length is the time interval between one cycle beginning and the start of the next subsequent cycle. This means that the durations quoted earlier include the time required for the maintenance and refueling outage. As such, they are *not* the length of time that the core is operating at full power. Therefore, in calculating the actual energy produced in that time, capacity factors have to be taken into account.

The capacity (or load) factor is the percentage of electrical power that a reactor actually produced in a given period compared with the electrical power that could be produced if the unit was operated continuously at full power in the same period. For example, if a reactor's nameplate capacity was 1000-MW h of electrical power, but in a given year, it produced 800-MW h, and then the capacity factor would be quoted as 80%.

In the earlier examples the first number not only indicates the proportion of the core replaced but also refers to the number of cycles of irradiation that the fuel will be in the core for 4, 3, 2, and 1, respectively. Therefore, to calculate the discharge burnup of the fuel batch on average, one simply needs to multiply the cycle length (in MW d/MTHM) by the number of cycles.

Therefore, at equilibrium, the discharge burnup of a batch of fuel assemblies is

$$B_{\text{Discharge}} = N \infty S \infty L \infty C \qquad (4.2)$$

where

$N =$ number of batches,
$S =$ specific power of the iPWR (MWth/ton),
$L =$ length of cycle (in days),
$C =$ capacity factor (%).

For example, assume an iPWR has a thermal power of 500 MW, 12 MTHM of fuel, a capacity factor of 90%, and a cycle length of 15 months (∼450 days), for a three-batch scheme, the discharge burnup would be

$$B_{\text{Discharge}} = 3 \infty (500/12) \infty 450 \infty 0.90$$
$$= 50\ 625 \text{ MW d/MTHM} \tag{4.3}$$

It should be noted that this is the *batch average* burnup and not all of the fuel assemblies discharged at the end of that cycle of operation will have achieved the same burnup because the fuel in that batch is positioned in different locations in the core and will be irradiated at slightly different powers may have had control rods inserted into them during operation, experience different operating powers and hence burnups. Similarly, because of their location in the core, in the assembly, next to guide thimbles, BPs, etc., each fuel rod will have a different burnup, as will the pellets in the fuel rods. As explained earlier the assemblies, rods, and pellets will have limits that the designer has to check are not violated.

For example, for a quoted "batch average" burnup of 45 GW d/MTHM, a typical peak assembly burnup within that batch would be 50 GW d/MTHM; a peak *pin* burnup would be typically 55 GW d/MTHM, and a peak *pellet* of the order of 60 GW d/MTHM. This illustrates very clearly why it is important for the designer to minimize the variation of the burnup of pins within the assembly and of assemblies within the batch, because large variations will lead to violations of the limits/warrantees potentially and therefore limit the utilization of the entire batch unnecessarily.

An increase in the number of batches results in an increase in discharged burnup achievable, while requiring lower enrichments, and therefore results in a lower fuel cost overall. This is most easily represented by the equation:

$$B_{\text{Discharge}} = \frac{2n}{(n+1)} \tag{4.4}$$

where n = number of batches and B_1 = burnup of single-batch core (which is equivalent to the cycle length).

For example, a two-batch core design will have an equilibrium discharge burnup ~33% greater than a single batch, and a three-batch core design will be 50% greater than the single batch. The theoretical limit in the improvement is 100% over the single batch. However, increasing the number of batches decreases the cycle length and results in more frequent refueling, which decreases the capacity factor of the iPWR. Even with a small number of assemblies in the core that need refueling and/or reloading, this can have a notable impact on the economics, and so generally a compromise of two to four batches is used.

4.3.2 BPs

Once the enrichment and number of assemblies required has been determined as outlined earlier, the designer needs to evaluate the need for BPs in the fuel assemblies. Since sufficient enrichment (and reactivity) has to be present for the lifetime of the fuel, the excess reactivity has to be controlled, particularly during the first cycle of irradiation. In general for large PWRs and for iPWRs, the core-wide (global) reactivity during the cycle is controlled either by means of boron in the coolant or by use of control rods. The BP content (type of material, number of rods, and content in the rods) is not a design requirement, other than it assists in controlling power peaking *within* (fuel rod powers) and *between* the assemblies (assembly powers), and it assists in reducing core-wide excess reactivity, thereby reducing the soluble boron concentration and resulting in a negative moderator temperature coefficient (MTC).

Materials are chosen for BPs that initially absorbs neutrons (have a high neutron capture cross section), but upon capture, they become an isotope with a low absorption cross section, that is, during irradiation, they are "burnable." Examples include boron, gadolinium, erbium, and dysprosium; the first two are used today on a routine basis in commercial, large PWRs, and these are the most likely candidates for iPWRs also. The absorbing material can either be mixed in with the fuel itself during the manufacture of the fuel pellets (known as "integral BPs") or can be loaded as separate components into the fuel assemblies (into the guide tubes) and so can be removed at the end of a cycle of operation (known as "discrete BPs").

Boron (specifically B-10) burns out quickly because of its high absorption cross section, whereas gadolinium, because of self-shielding effects, tends to burn out more slowly. Generally, this makes boron-based BPs more suitable for short cycles of operation and gadolinium more suitable for longer cycles, but varying the weight percent of the poison material is used to tailor the rate of burnout required.

A comparison of the burnout rates and extent of the reactivity hold-down can be seen in Fig. 4.2, for example, containing gadolinium. It can be seen that the overall reactivity doesn't quite return to the level of the no burnable poison cases. This is due to residual absorption (albeit relatively small) from the remaining gadolinium

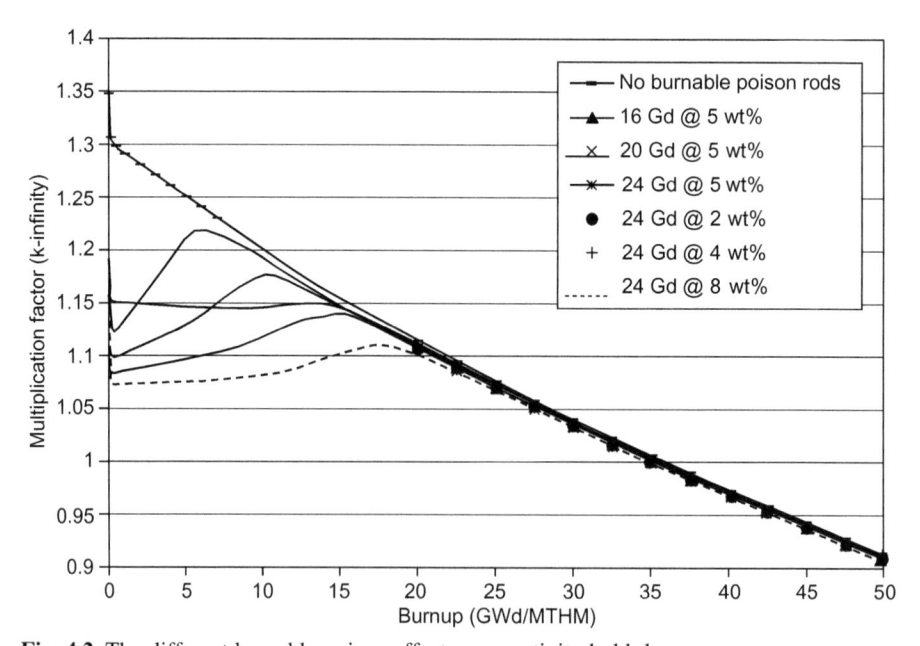

Fig. 4.2 The different burnable poison effects on reactivity hold-down.

in the fuel. For fuels that contain boron as the BP, such as IFBA (which is an "integral fuel burnable absorber" technology where a boron coating is sprayed onto the outside of the fuel pellets), there is no residual absorption in the fuel.

Other performance considerations for the designer may include helium buildup (a result of neutron capture in B-10) or thermal conductivity, both of which can limit the fuel performance. For example, gadolinium has a lower thermal conductivity than uranium, and so the fissile enrichment of the carrier material for the gadolinium poison is deliberately lowered to avoid power peaking concerns.

For those iPWRs that are looking at either very long cycle lengths (of the order of a few years) or those that need a much longer-lasting reactivity hold-down (for example, those that do not use boron in the coolant), erbia is another contender. Erbia has a relatively low absorption cross section compared with boron-10 or gadolinium, which means it depletes slowly. In addition, all of the isotopes of erbia have reasonable absorption cross sections, which means the isotopes produced by capture also have a reactivity hold-down effect.

The designer has to consider the magnitude of the reactivity hold-down, the duration, and the rate of depletion. Combining the number of BP rods, weight percent of the BP in those rods, and the BP material type gives the designer sufficient degrees of freedom to achieve the desire outcome. Examples of indicative gadolinium pin locations are provided in Fig. 4.3 by way of illustration. Note that the BPs are loaded near the water holes (instrument and guide thimbles) as additional thermalization of neutrons improves their effectiveness.

4.3.3 In-core fuel management

Once an initial estimate has been determined of the enrichment, the number of fuel assemblies to be loaded in the cycle, and the types of BPs that are most suited for the iPWR being developed, the next phase is to determine how to load the core and which of the assemblies loaded in the previous cycle of operation are available and suitable to be reloaded. A major part of the nuclear design effort is concerned with this phase, typically referred to as the in-core fuel management, and it brings together many of the key interactions and issues discussed earlier.

The objective is to load the core to provide a flat/smooth distribution of reactivity across the core, that is, high and low k-infinities balancing each other. This distribution of the assemblies in the core is known as the "loading pattern." If reactivity is too high, then assembly powers will be high, and this in turn will result in power peaking in the fuel rods being too high (see Section 4.2.3). The core is loaded in quarter core symmetry to avoid power and performance tilts in the core, for example, one quadrant running at higher power than the others. This means that the fuel assemblies are loaded in groups of four, one in each quadrant. In most iPWRs (and most large PWRs), the fresh fuel is loaded toward the center, and the previously irradiated batches of fuel are loaded toward the edge of the core. This improves the neutron economy by reducing the radial neutron leakage out of the core and is known as a "low leakage loading pattern" (or L^3P). This is particularly important for iPWRs because the smaller cores, with a high surface-to-volume ratio, result in a much higher neutron leakage than large PWRs. This does tend to make the iPWRs more prone to power peaking (with all of the

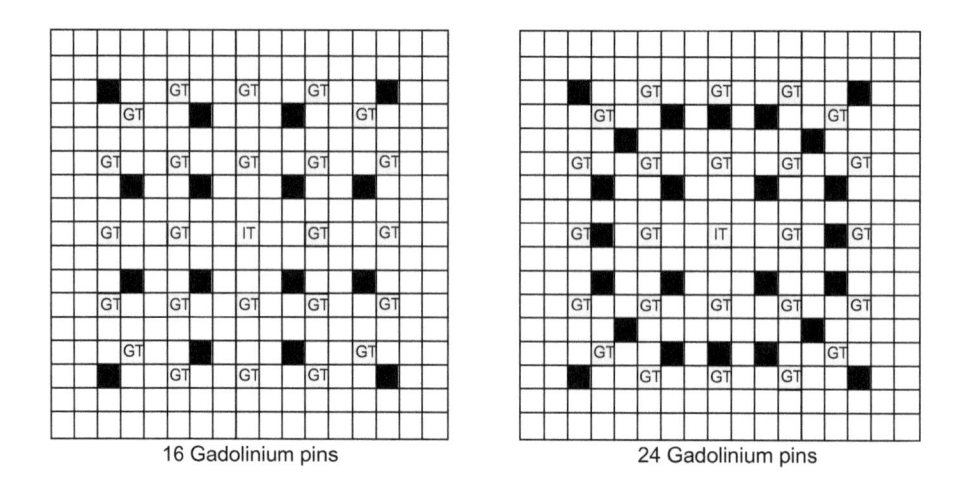

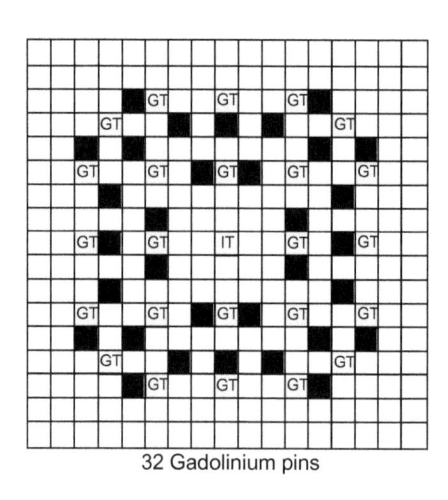

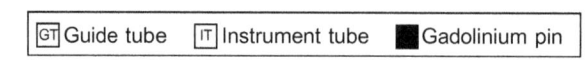

Fig. 4.3 Examples of burnable poison distribution designs.

higher reactivity at the center of the core), and as such, higher BP loadings are required.

Typically the fresh fuel is loaded in a checkerboard configuration (toward the center as explained earlier), and the highest ranking in reactivity of the previously irradiated fuels is then loaded into the core design according to k-infinity ranking. Three-dimensional reactor analysis tools (such as CASMO-SIMULATE) are then used to calculate the assembly and pin power distribution and to determine if the constraints are met for the fuel. If the peaking is not met, then the fuel can be "shuffled" to improve the power peaking. This can be done either by the use of the designer's skill and experience or by using loading pattern optimization tools. At the same time as assessing the core loading for power peaking, the energy requirements (cycle length) are also checked to ensure that there is sufficient enrichment loaded in the fresh fuel assemblies. The number and content of the BPs can be changed if sufficiently low peaking cannot be achieved, but the designer also has to be aware that once the BPs deplete in the core the power distribution will change and can lead to higher pin power peaking during the cycle than at the start.

For the designer the other key criterion during this stage of the design process is the economics of the required fuel. The main variables that affect the fuel costs for a given cycle of operation will be (a) the number of fresh fuel assemblies required and (b) their enrichment. Since the fuel is loaded in quarter core symmetry, an additional new fuel assembly required in a quarter core means the purchase of four assemblies, an increase of the order of several million dollars in fuel price. Similarly, if the core is designed such that there is too much radial or axial leakage (particularly relevant to the small cores of the iPWRs), then additional enrichment in the fuel will be required to achieve the required cycle length and fuel discharge burnup. This results in an increase in not only the enrichment costs but also the uranium ore required to be purchased.

Reducing the number of fuel and BP rod types is also a means by which the designer can improve the fuel economics. Good practice involves keeping the number of fissile enrichments to a minimum across the assemblies, rather than having multiple enrichments, either within an assembly or within a reload batch of fuel. For a reload of fuel, two or three enrichments would not be unusual. Similarly, for the BP types and enrichments of fissile material in the BP rods and the BP loading itself, minimizing the variations is key to the fuel costs. In this case, good practice may be to use only one BP type (e.g., gadolinium) and, for example, only three different weight percentages across all fuel designs, for example, four, six, and eight. Also the designer must ensure that the BP burns out fully, that is, there is no residual absorption of the BP in the fuel, which is particularly important for highly poisoned cores where the impact would be greater. For those iPWRs that require high BP loadings (e.g., for long cycles or single-batch cores), minimizing the residual absorption by using boron rather than gadolinium, or a combination of the two, would be warranted.

The compactness of the iPWR combined with fewer assemblies available (in terms of absolute number, variation in burnup, BP types, etc.,) makes the loading pattern design process more challenging and achievement of an optimized core design more difficult. For example, iPWRs have of the order of one-fifth of the fuel assemblies of a large PWR as shown in Fig. 4.4 (NuScale, 2013; http://www.nrc.gov/reading-rm/doc-collections/nuregs/staff/sr1793/initial/index.html#pub-info).

Once the cycle length has been achieved with pin power peaking, fuel rod, and assembly burnup within the allowed constraints, a brief safety analysis is often completed prior to the full set of detailed analyses and reports being completed. This overall saves time and money in the design process. The usual checks include the following:

- power peaking through the cycle,
- moderator temperature coefficient at hot full power and hot zero power,
- shutdown margin.

| NuScale/SMR-160 | Westinghouse SMR | AP1000 |
| 37 fuel assemblies | 89 fuel assemblies | 157 fuel assemblies |

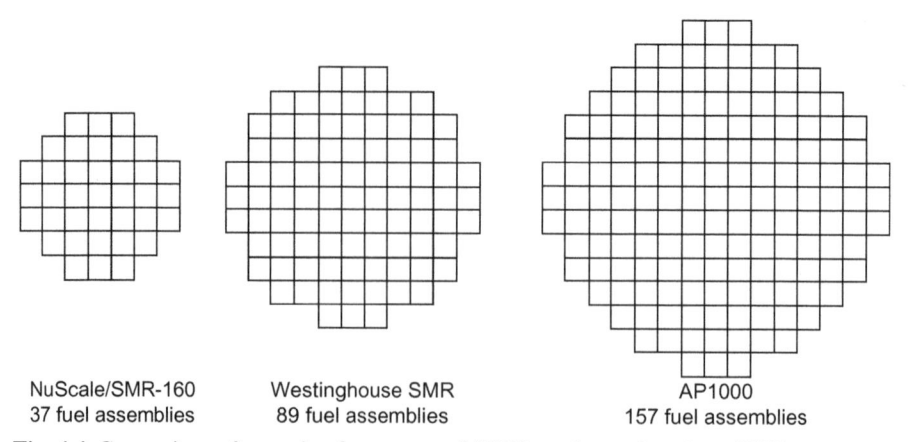

Fig. 4.4 Comparison of core size for a range of iPWRs and a modern large PWR.

However, certain iPWR designs will have their own nuances that the designers will learn are key to a successful design prior to committing the nuclear design for a full safety analysis, and so ad hoc checks will be identified.

4.3.4 Summary of the design process

Fig. 4.5 is a flow diagram that summarizes the various stages in establishing an optimized nuclear core design that have been described earlier. This is an involved and rigorous process that typically takes several weeks of man effort and iteration. The overall timescales are heavily dependent on the experience of the designers, the tools at their disposal (including consideration of setup time, run time, and ease of iteration perhaps using graphical interfaces), the experience base of previous core analysis of the same reactor design, and the nuances of the reactor design itself.

It should be noted that this is not the end of the design and analysis phase for this reactor. From here, interfaces with other elements of the design team begin (e.g., transient analysis, thermal hydraulics, mechanical design, and fuel performance), and additional iterations may be required prior to initiating the go-ahead to start manufacturing the fuel (usually approximately 12–18 months prior to refueling).

4.4 Integral pressurized water reactor (iPWR) design specifics

4.4.1 Fuel designs in the smaller cores

From a nuclear design perspective, the first iPWR design specific that needs to be considered is the compactness of the core compared with large PWRs, both in terms of the number of fuel assemblies and, in some cases, the height of the fuel assemblies. Table 4.1 provides a summary of the key nuclear design parameters for the four US iPWR designs submitted in 2011 to the US Department of Energy (DOE) first

Core and fuel technologies in integral pressurized water reactors

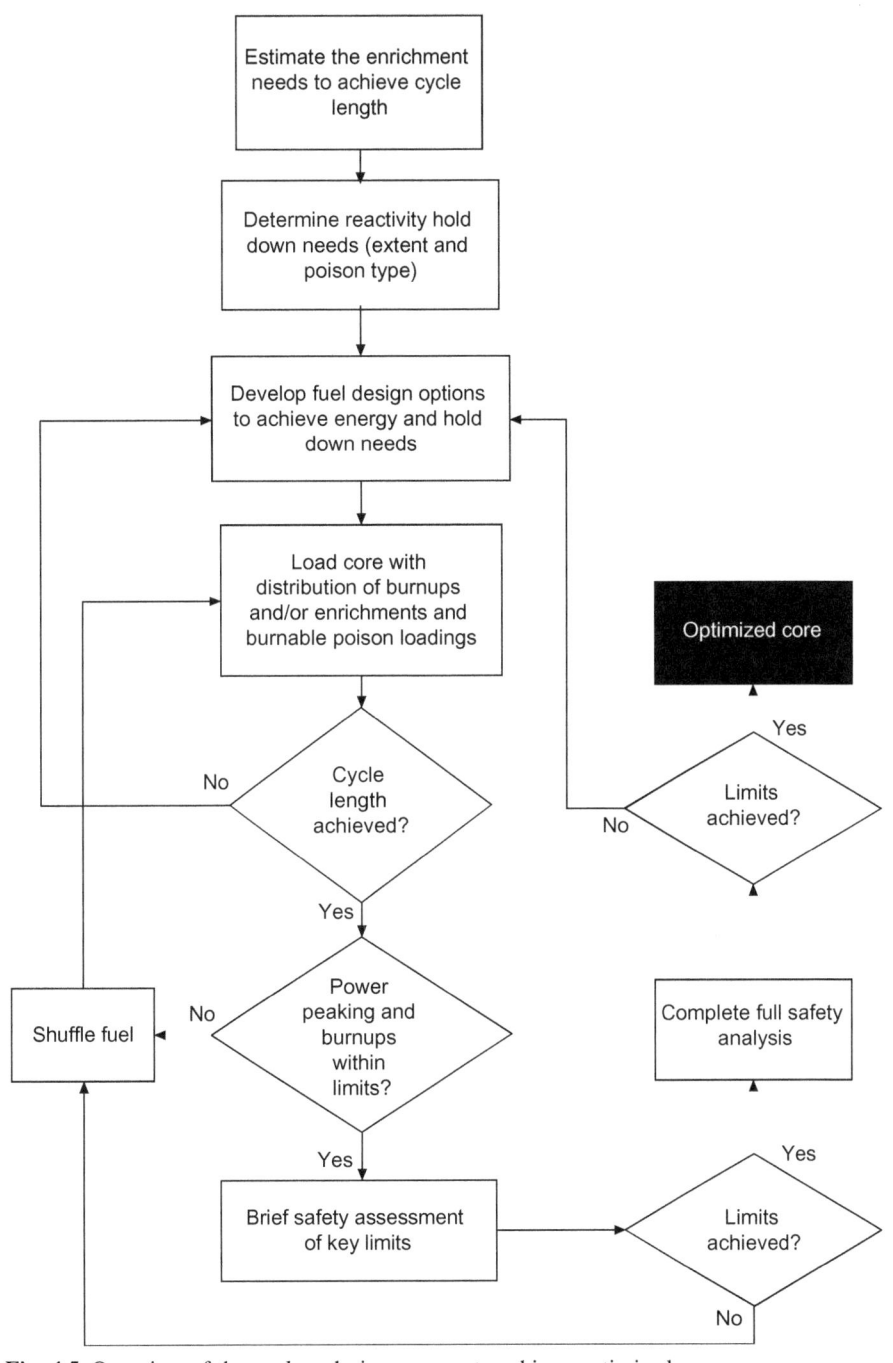

Fig. 4.5 Overview of the nuclear design process to achieve optimized core.

Table 4.1 Summary of key nuclear design parameters for a modern, large PWR (AP1000) and a range of iPWRs.

iPWR name (vendor)	AP1000 (http://www.nrc.gov/reading-rm/doc-collections/nuregs/staff/sr1793/initial/index.html#pub-info) (Westinghouse)	Westinghouse-SMR (http://www.westinghousenuclear.com/Portals/0/About/Blog/2014/Dec14NN%20Westinghouse%20SMR%20Feature.pdf; IAEA, 2018) (Westinghouse)	mPower (IAEA, 2018) (Babcock and Wilcox)	SMR-160 (Oneid, 2012; https://smrllc.com/) (Holtec)	NuScale (NuScale, 2013; IAEA, 2018) (NuScale Power)
Power					
Thermal (MWth)	3400	800	530	500	160
Electrical (MWe)	1150	225	180	160	45
Reactivity control	Control rods	Control rods	Control rods	Control rods	Control rods
	Soluble boron	Soluble boron			Soluble boron
Fuel in core					
No. of assemblies	157	89	69	37	37
Array	17 ∞ 17	17 ∞ 17	17 ∞ 17	17 ∞ 17	17 ∞ 17
Active fuel height	4.3 m/14 ft	2.4 m/8 ft	2.4 m/8 ft	3.7 m/12 ft	2.0 m/6.5 ft
Mass in core (MTHM)	85	27	20	15	9
Fissile enrichment	<5 wt% U235	<5 wt% U235	<5 wt% U235	<5 wt% U235	<5 wt% U235

Cycle length					
Month	18	24	48	48	24
Fuel demand					
MTHM per reload	36.6	Unknown	20	14.7	Unknown
MTHM per GW year	21.2		29.3	33.8	
Linear rating					
kW/m	19	14	12	14	8
kW/ft	5.8	4.2	3.6	4.2	2.4

All values are approximate.

solicitation, along with a modern large PWR, the Westinghouse AP1000 (http://www.nrc.gov/reading-rm/doc-collections/nuregs/staff/sr1793/initial/index.html#pub-info). It is important to note that although the fuel heights and overall loadings are different in each of the iPWRs, the 17 ∞ 17 array of fuel pins is consistently the same throughout (Fig. 4.6). This is primarily driven by the excellent operational performance of these fuel types in large PWRs operating today, and in particular, it builds on the extensive development work that has evolved fuel designs toward a larger number of thinner fuel pins to improve thermal margins and allow higher rod powers. These are equally important drivers in the deployment of iPWRs, hence the choice

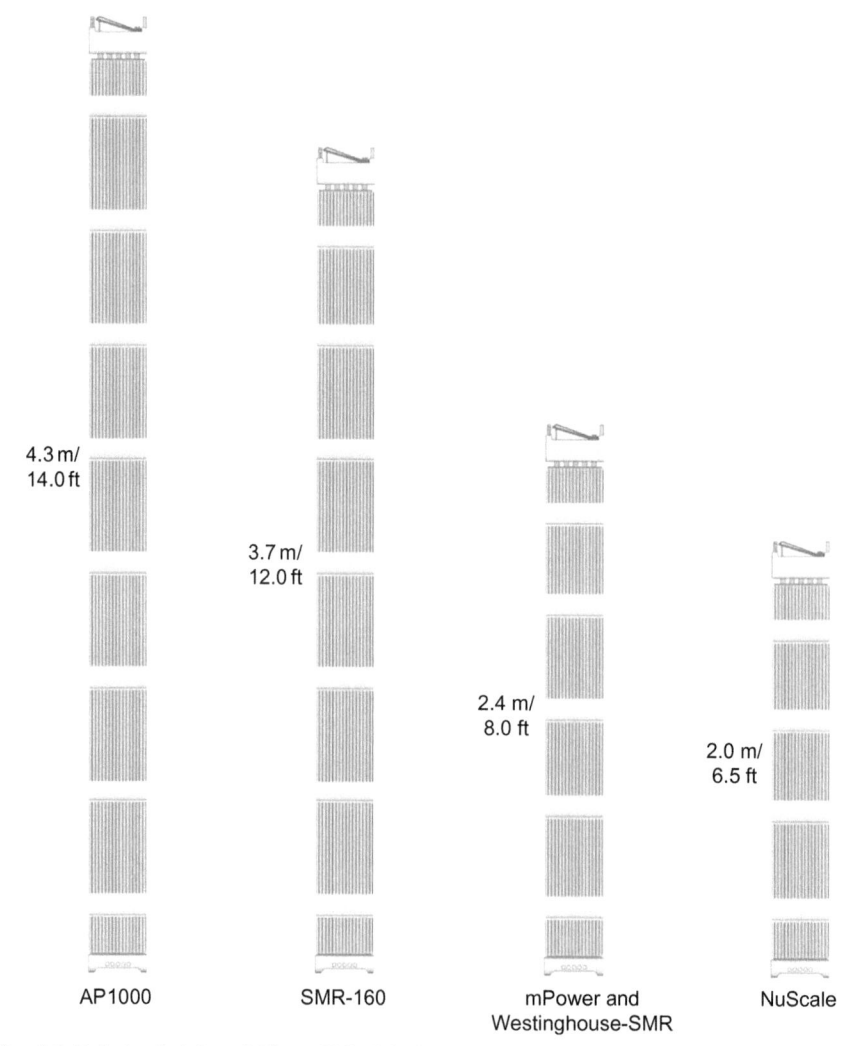

Fig. 4.6 Relative heights of 17 ∞ 17 fuel designs considered in iPWRs today (drawn to scale, but not an accurate representation of the actual fuel designs to be used).

of 17 ∞ 17 fuels. The lower linear heat ratings in the case of the NuScale and mPower iPWRs in particular will reduce fuel and cladding temperatures during operations and accidents and allow a higher relative power for the lead fuel rods during normal operations and therefore the ability to tolerate more power peaking in the core.

As shown in Fig. 4.4, a large PWR has anywhere from 157 to 193 fuel assemblies compared with iPWRs that typically have between 37 (SMR-160) (Oneid, 2012) and 89 (Westinghouse SMR) (http://www.westinghousenuclear.com/Portals/0/About/Blog/2014/Dec14NN%20Westinghouse%20SMR%20Feature.pdf). It may appear that having fewer fuel assemblies make the design of the core and the loading pattern more straightforward. But in fact, fewer assemblies mean fewer degrees of freedom to optimize the loading pattern and smooth out the reactivity and resulting power distribution across the core. The degrees of freedom are not just in terms of the fuel design and BP loadings of the fresh fuel, but once beyond the first cycle of operation, the batches of previously irradiated fuel will have a variety of burnups and hence resulting reactivities that can be distributed throughout the core to smooth the power distribution to achieve the power peaking limits.

The smaller core size also results in more neutron leakage radially (and axially, as described later in this section) and results in a more significant variation in assembly power across the core from the center where there is more fuel and hence more power to the outside of the core where lower power is generated. The radial leakage can be reduced by ensuring only the fuel with lower reactivities (generally higher burnups) are loaded on the periphery of the core or by developing and using radial reflectors in the outer core such as using stainless steel, rather than water, as used in the large PWRs. Although nonstandard in large PWRs, there have been designs (e.g., the AREVA EPR), where stainless steel reflectors have been developed and demonstrated to be effective. Additional development work on radial reflectors for iPWRs is likely to yield notable benefits, not only in the economics of the fuel but also by raising the power in the outer assemblies and hence improve the radial power variations across the core.

Neutron leakage is further exacerbated by the use of short fuel in some iPWRs. Typical large PWR fuel assemblies have between 12 and 14 ft (3.66 and 4.27 m) of fuel in 264 fuel pins, whereas several of the iPWRs have moved to a partial height version of the standard 17 ∞ 17 fuel designs, principally because full fuel heights are not required for the lower-powered reactor designs. In each case the fuel designs have been developed and optimized for the specific requirements and demands of each iPWR, for example, rating and natural convection. For example, the Westinghouse SMR and mPower iPWRs have an active height of approximately 8 ft (2.4 m), and the NuScale design is approximately 6.5 ft (2.0 m); see Table 4.1 and Fig. 4.6 for a relative comparison of fuel heights.

However, there is extensive experience in large PWRs to reduce the axial leakage by using what are known as "axial reflectors." These are simply sections of the active fuel height (typically a few inches at the top and bottom of the fuel stack) where the usual fissile enrichment is reduced or in some cases, natural (0.71 wt% U-235) or even tails (∼0.3 wt% U-235, a byproduct of the enrichment process), uranium is used. This not only improves the fuel economy by reducing the neutron leakage from the core but also reduces the costs of the fuel in terms of enrichment needs in an area of the fuel and core where there is relatively little power generated and therefore little need to enrich the fuel.

However, axial variation in the fuel stack tends to increase the cost of the fuel, not in terms of enrichment or ore costs, but simply because of the complexity of manufacturing a fuel type with more enrichments, zones, and diversity; the manufacture and in particular the quality assurance (QA) and quality control (QC) cost components are increased.

By way of illustration, based on current market prices for uranium fuels, if an increase of 0.25 wt% U-235 was required in a reload of fuel to offset either neutron leakage or inefficient use of the fuel (such as lower burnups achieved), the increase in fuel price would be of the order of a few million dollars per reload. Cost increases because of additional fuel complexity would be the same order of magnitude.

4.4.2 Use of control rods and BPs to control reactivity

Increased axial and radial complexity in fuel designs becomes even more relevant in those iPWRs that rely on control rods or heavy BP loadings rather than soluble boron in the coolant to control the excess reactivity. Unlike large PWRs where there may be a control rod located in one out of three or four assemblies, the iPWRs have a much higher density of control rods, particularly if the control rods are the means by which the excess reactivity is controlled. For example, in the mPower and SMR-160 designs, every assembly contains a control rod. In a large PWR where boron is used to control the excess reactivity, the control rods are normally fully withdrawn. But in some of the iPWRs, these control rods, and the sequence in which they are maneuvered, allow for compensation in the reduction in excess reactivity (as the fissile material depletes), for axial power changes, and for xenon variations (radially and axially)—see Table 4.1. The removal of the need for boron simplifies the chemical volume control systems and results in capital and operational cost savings, for example, waste handling of depleted soluble boron.

However, at the same time, inserting a control rod moves the power away from its location to somewhere else in the core, and excessive movements can thermally "cycle" the fuel, potentially resulting in fuel failures. There is also a need to control the rate of withdrawal due to fuel performance (such as pellet-clad mechanical interaction or thermal cycling) and for reactivity insertion limits (as explained in Section 4.2). Therefore, in terms of nuclear design, this leads to the need not only to optimize the fuel and core design but also to develop a control rod sequence for the various control rod groupings to ensure appropriate compensation in reactivity, while at the same time, also ensuring other limits, such as changes in reactivity or power ramp rates for the fuel rods, is not violated. This development of the control rod sequence includes the following:

- determining which control rods get moved together, in so-called control rod banks;
- when they get moved in or out, how far and at what rate;
- what the overlap is between one bank and the next one starting to move in/out.

Even with the use of soluble boron, similar challenges regarding control of excess reactivity are seen for those iPWRs looking to achieve long cycle lengths between refueling outages, as greater fissile loadings are required to achieve the cycle length of greater than 24 months.

Using control rod insertion throughout a cycle results in strongly varying axial power shapes; inserting control rods in an iPWR will skew the power to the bottom of the core and result in higher peaking factors. This effect is mitigated by using higher burnable poison loadings in the lower parts of the fuel or varying the fissile enrichment axially. In addition to the axial power effects, those assemblies that have the control rods inserted during the cycle tend to have their peak pin powers pushed radially to the edge of the assembly, that is, away from the inserted control rod locations. Therefore additional radial BP loadings tend to be needed to reduce the radial peaking factors. Overall, this tends to lead to a heterogeneous nuclear design in terms of fuel assemblies and areas of notable flux suppression in the fuel pins around the BPs and control rods and results in power peaking around the inserted control rods and BP locations. The disadvantage of a more heterogeneous core is a nonuniform radial and axial depletion throughout the core, which results in either inefficient use of the fuel (as burnup limits are met in some, but not all fuel batches) or difficulty in achieving power peaking limits. Furthermore, for the iPWRs that use natural circulation (e.g., NuScale and SMR-160), the nonuniformity in the power and resulting moderator density will require further thermal hydraulic analysis to ensure sufficient cooling and an accurate prediction and coupling of the neutronic and thermal hydraulic feedback. This issue, which needs to be addressed for all natural circulation designs, is exacerbated by the fuel nonuniformity.

In the case of high gadolinium loadings being used, the iPWR's nuclear designers will have to consider very carefully the fissile enrichment of the fuel pins carrying the gadolinium. As explained earlier, those pins will have a lower fissile enrichment compared with the non-BP fuel pins to assist in the licensing of the gadolinium pins, that is, lower thermal conductivity, and they should not be limiting, considering the lack of irradiation data and fuel performance validation. This will mean that to offset the lower enrichments in these BP fuel pins, the other pins will need to have their enrichment increased correspondingly. The larger the number of gadolinium rods in the core, the greater the level of additional enrichment required in the other fuel pins; thus achieving as long an operating cycle as possible, which requires average fuel enrichments as high as possible, could become in conflict with the current licensing limit of 5 wt% U-235.

In large PWRs operating today, there is extensive experience with different BP types and loadings, as well as control rod sequences and calculation of control rod worths. However, the combination and extent of some of these does result in challenges for code predictions, and the associated validation of nuclear design tools in extremely heterogeneous cores and regions of notable flux variation, in particular ensuring that BP and control rod worths, are accurately predicted. Furthermore, with numerous axial zones in the fuel designs and with significant control rod movements, advances in nodal methods may be required, including the need for variable size meshes to accommodate some of the nuclear designs.

4.4.3 Core loading

Another variation in some iPWRs that also affects the BP loadings and the complexity of the fuel design is core operation in a once through/single-batch reload, that is, all of the core is replaced each cycle of operation. mPower and SMR-160 are examples where this design option has been examined. Since the fuel is all fresh for each reload, it simplifies the nuclear design because the same fuels, loaded in the same locations with the same enrichment, BP loadings, etc., can be repeated for every cycle of operation, that is, once a truly optimized nuclear design has been produced, it is simply duplicated for every cycle and for every reactor deployed.

Fresh fuel also has a distinctive axial power profile, with a distinctive chopped cosine power distribution, peaking at the center. Therefore, if all fresh fuel is loaded into the core, in the absence of either control rods inserted or axially varying burnable poison distribution, the power peaking limits will be violated. Irradiated fuel has a much flatter axial power distribution, and in those cores with a mix of fresh and depleted fuel, the flattening influence helps to reduce axial power peaking.

All fresh fuel also means that the designer cannot rely on irradiated fuel to help in smoothing the power distribution across the core, and so, only BPs and control rods can be used. This tends to result in higher control rod and BP use in the designs rather than utilizing the lower excess reactivity that arises because some of the fuel has been irradiated. As with fresh cores in large PWRs, the use of asymmetric BPs in certain core locations may be required to achieve the power peaking limits, which also adds to the cost of the fuel.

Special attention also has to be paid to the use of a single-batch core in terms of uranium ore and enrichment utilization, by ensuring that the fuels achieve their full potential burnup. Fuel loaded on the periphery of the core will tend to have lower powers and resulting burnups compared with the leading power assemblies toward the center of the core, and since the fuel is loaded for only one cycle of operation, there is no opportunity to reload the fuel to achieve higher burnups and ensure that all of the fuel within a given batch achieve similar discharge burnups. This is an indication of how single-batch cores are not as efficient as multibatch (see Section 4.3.1).

Some of the single-batch core concepts include a cartridge-type fuel unit. This suggests (although the designs are yet to be finalized) that the fuel is loaded into the reactor as one single unit, including control rods already inserted. Although this will speed up the core loading and inspection routine and assist in the criticality safety case for fresh and spent fuel, the cartridge approach will also make it more difficult to inspect and replace any damaged fuel pins, for example, fuel inspection once at the reactor site prior to startup and rod replacement in the event of a leaker during operations.

Due to the power output required and the fuel management of the iPWRs under development today, there is a large variation in the fuel demands per reload (see Table 4.1). These limited examples indicate that the iPWRs tend to require more fuel on a per GW year-electrical basis compared with a large PWR such as AP1000. What has to also be take into account in the overall economics, however, is the outage time—for an iPWR with a three-year cycle length compared with an AP1000 with a one-and-half-year cycle length, the iPWR will have half as many outages for refueling and maintenance, which represents something of the order of 750 more days availability to produce power.

It is likely that each of the iPWR designs will have the potential to operate with a range of fuel management schemes, depending on customer and market needs, and overall economics, and further developments of the nuclear design options can continue during the development and demonstrations phases.

4.4.4 Other design considerations

Although not directly nuclear design issues, there are a few other key considerations for iPWRs that are associated to the nuclear design process. For those designs that do not use boron in the coolant, there will be a more negative MTC. This means that for a given temperature variation up the core, there will be a greater reactivity swing and a resulting power variation, with more reactivity and power at the bottom of the core. For those iPWRs that have a large temperature gradient from the bottom of the core (inlet temperature) to the top of the core (outlet temperature), this will result in large axial variations in power, and so greater control over the reactivity and axial powers will have to be taken into account in the nuclear design.

Extensive use of control rods and also BPs in some of the iPWRs will inevitably result in a harder neutron spectrum, that is, a greater proportion of the neutrons will be at higher energies because the thermal neutrons will have been captured by the absorbing materials. This has two potential consequences:

- The harder neutron spectrum will generate more plutonium in the irradiated fuel. This could have advantages because the resulting plutonium could assist in adding more fissile material to the irradiated fuel and therefore reduce the initial U-235 enrichment requirements for the fuel.
- The harder spectrum will have an impact on the overall vessel and/or reflector/barrel fluence and could have adverse effects on the lifetime of those components.

From a fuel management perspective, there are advantages of having several iPWRs on the same site and sharing facilities such as the spent fuel pool. In the cases where the fuel will be reloaded for several cycles, the nuclear designer could potentially swap the fuels between reactors and as such will have a much larger selection of fuel to choose from to attain the most economic and optimized core design. This could result in more efficient use of existing fuels (by achieving the highest burnups achievable within the limits) and the potential to purchase less fuel over the lifetime operations of several units. However, the designs with cartridge-type fuel will not be able to take advantage of the flexibility in fuel loadings.

4.5 Conclusion

Nuclear design of iPWRs is concerned primarily with the core physics characteristics, the safety and operational performance of the fuel and the core, and the reactivity control systems in the fuel and the core, with associated feedback mechanisms.

There are many similarities between the nuclear design of iPWRs and the many hundreds of operating cycles of large PWRs around the world. In particular the safety design criteria and principles are much the same, as are the design features used to achieve a viable and economic nuclear design; the fuel arrays in particular build on the experience of the commercial PWRs. Similarly the modeling tools used for the analysis are the same, carrying with them the pedigree, experience, and much of the validation required for iPWR development and deployment. This provides greater confidence in the designs being proposed, not only for the investors but also for the regulators and the operators.

However, there are several key differences in the iPWRs where challenges remain, in terms of the development of the technology, in the validation of the nuclear design concepts being proposed, and with the tools that will be used to predict accurately the iPWR's safety and operational performance.

In particular, for those designs that do not use boron in the coolant to control the excess reactivity, a greater reliance on burnable poisons and/or control rods for reactivity control is required. This produces a greater level of heterogeneity in the core, both in terms of the fuel design, but more importantly in terms of the radial and axial pin powers/peaking factors and resulting burnups. Accurate prediction of control rod and burnable poison worths and burnout rates will need to be demonstrated, along with the ability to predict accurately the fuel rod powers in those regions of the fuel and core with rapidly varying fluxes, for example, surrounding control rods and BPs.

The move to new reflector designs (primarily radially) including use of stainless steel will need to be validated for powers and fluxes in those regions.

With the use of control rods for reactivity and power control and shaping, the use of extensive BP loadings to hold down excess reactivity and control peaking, and with the resulting heterogeneous cores, many of the iPWRs share some of the characteristics of BWRs and PWRs operating today. The experience in BWR design and operations and analysis method developments to reflect accurately these characteristics could equally provide a useful insight into future requirements.

Although many of the iPWR fuel designs may be the same in terms of the 17 ∞ 17 array and the materials used for cladding, grids, etc., the need remains to qualify the fuel performance for the shorter fuel heights envisaged for several of the iPWRs. The power variations produced by control rod movements in particular will need to be analyzed to ensure no issues for either thermal cycling or pellet-clad mechanical interaction arise. Ramp rates and conditioning powers will also need to be evaluated. In addition, critical heat flux experiments are required as the linear ratings, and related correlations will be different with the shorter fuels compared with the standard height 17 ∞ 17 fuels.

In conclusion, iPWRs introduce a number of additional challenges to the nuclear designer, both in terms of the analysis process and the analysis tools used. However, the experience of large PWRs to date and to some extent the BWRs too, along with the associated validation, and design and licensing experience, and ongoing development of fuel designs and analyses today will help to address these challenges via appropriate design, validation, and testing.

References

IAEA, 2018. Advances in Small Modular Reactor Technology Developments. IAEA September 2018.

NuScale, 2013. NuScale codes and methods framework description report. (NP-TR-0812-1682-NP).

Oneid, P., 2012. SMR -160 Unconditionally Safe & Economical Green Energy Technology for the 21st Century. Holtec International, Jupitor, FL.

Key reactor system components in integral pressurized water reactors (iPWRs)

Randall J. Belles
Oak Ridge National Laboratory, Oak Ridge, TN, United States

5.1 Introduction

The component focus in this chapter is on near-term SMRs utilizing light water reactor (LWR) technology. The near-term SMRs are implementing a variety of integral pressurized water reactor (iPWR) designs. The principle iPWR reactor coolant system components are typically identical in function to similar components in larger LWRs, but the size, the number, and the location of many of these components have been significantly modified.

One simplistic way to describe an iPWR relative to the current pressurized water reactor (PWR) designs is "more water—less pipe." Because of the integral nature of the various iPWR designs, the available water volumes in an iPWR relative to the thermal power rating of the reactors is significantly increased compared with current PWR designs. Furthermore the size and length of reactor coolant system piping is significantly reduced compared with current PWR designs. This reduces the need for many of the active safety system components found in larger LWRs and generally increases the operator response time to plant upsets. A comparison of some of the major component design differences is shown in Table 5.1 (NEI, 2012).

Individual components associated with the integral reactor coolant system and external systems and components connected to the reactor coolant system are discussed individually in the following sections. The iPWR reactor coolant system components located inside the integral pressure vessel are discussed in Section 5.2. However, there are still several important reactor coolant system functions that must be fulfilled by connected components outside of the integral pressure vessel. These components are discussed in Section 5.3. A generic iPWR layout and key components are shown in Fig. 5.1 for reference to subsequent discussions. Differences from current PWR designs and commonalities with PWR designs are highlighted. Except for the option for air cooling, the secondary plant in an iPWR is very similar to the current PWR reactor fleet. Therefore, secondary plant components are fairly common and will not be discussed in detail in this chapter.

☆ This submission was written by the author acting in his own independent capacity and not on behalf of UT-Battelle, LLC, or its affiliates or successors.

Table 5.1 Comparisons of system and component design features.

LWR system/component	Current PWRs	iPWRs
Ratio of pressure vessel height to width	2–2.5	4–7
Large diameter RCS piping	Yes	No
Reactor coolant pumps	Yes	Some designs
Pressurizer	Separate vessel	Integral
Steam generators	Separate vessels	Integral
Emergency core cooling system	Active	Passive
Control rod drive mechanisms	External	Internal or external
Containment spray	Yes	No
Diesel generators	Safety related	Nonsafety related
Air-cooled secondary	No	Possible feature

5.2 Integral components

5.2.1 Pressure vessel and flange

In a current large PWR, the reactor pressure vessel holds the individual fuel assemblies, the control rods, and a significant percentage of the reactor coolant, which also acts as the moderator. Functionally the reactor pressure vessel provides one of several safety barriers to fission product release and provides support to the control rods and support to the reactor vessel internals, which support the reactor fuel and direct coolant flow within the vessel. A PWR reactor pressure vessel is characteristically a cylindrical vessel with a hemispherical bottom head and a removable, flanged and gasketed, hemispherical top head (NRC, 2006). The bottom head is welded to the cylindrical shell while, the top head is bolted to the cylindrical shell via the flanges. The vessel is nominally constructed of low-alloy carbon steel clad on the inside with a thin layer of austenitic stainless steel. The reactor pressure vessel cylinder is typically made up of a number of thick-walled ring forgings that are welded together circumferentially. The circumferential weld nearest the fuel region of the vessel is typically referred to as the beltline region. A typical PWR pressure vessel is shown in Fig. 5.2.

Current large PWR reactor pressure vessel concerns include vessel embrittlement, pressurized thermal shock, and primary water stress corrosion cracking (NRC, 2003). Neutron embrittlement occurs over time as the vessel is exposed to neutron flux. The area of greatest concern for embrittlement is the vessel beltline weld. Pressurized thermal shock (PTS) can occur when cold water is introduced while the vessel is pressurized. This creates increased thermal stress in the vessel wall. As the vessel becomes brittle over time, it can become more susceptible to cracking, especially as a result of the added stresses induced by PTS (Kang and Kupca, 2009). Primary water stress corrosion cracking is aggravated by the presence of boric acid. Primary water stress corrosion cracking has the potential to compromise reactor pressure vessel integrity in the vicinity of vessel head penetration nozzles and other low flow areas. This could lead to vessel leakage or in more extreme cases an increased potential for control rod ejection.

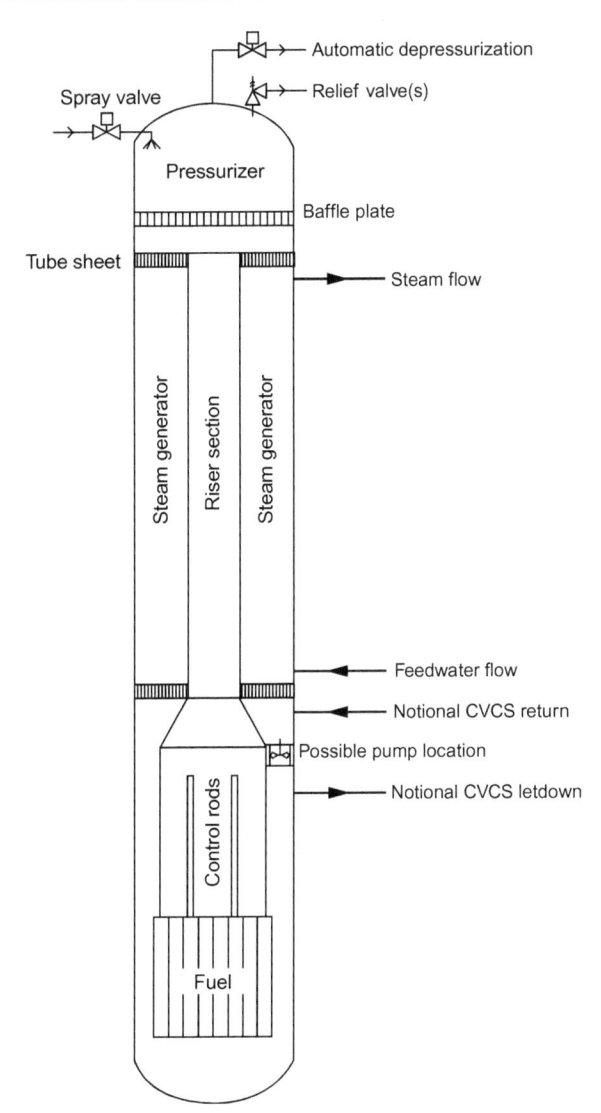

Fig. 5.1 Generic iPWR components.
Credit: R. Belles, ORNL.

Current large PWR reactor pressure vessel size will vary with the thermal power rating of the reactor. This is a function of the number of fuel assemblies and the excess reactivity necessary to achieve the desired power level and fuel cycle length. However, the ratio of cylindrical height to diameter in a large PWR reactor pressure vessel is nominally in the range of 2–2.5. The coolant design pressure of current large PWRs is normally around 2500 psi (17.2 MPa). Vessel penetrations are typical below the fuel region in a PWR vessel for instrumentation. In addition, the large inlet and outlet nozzles for supplying reactor coolant to and from the vessel are characteristically just above the fuel region in the vessel.

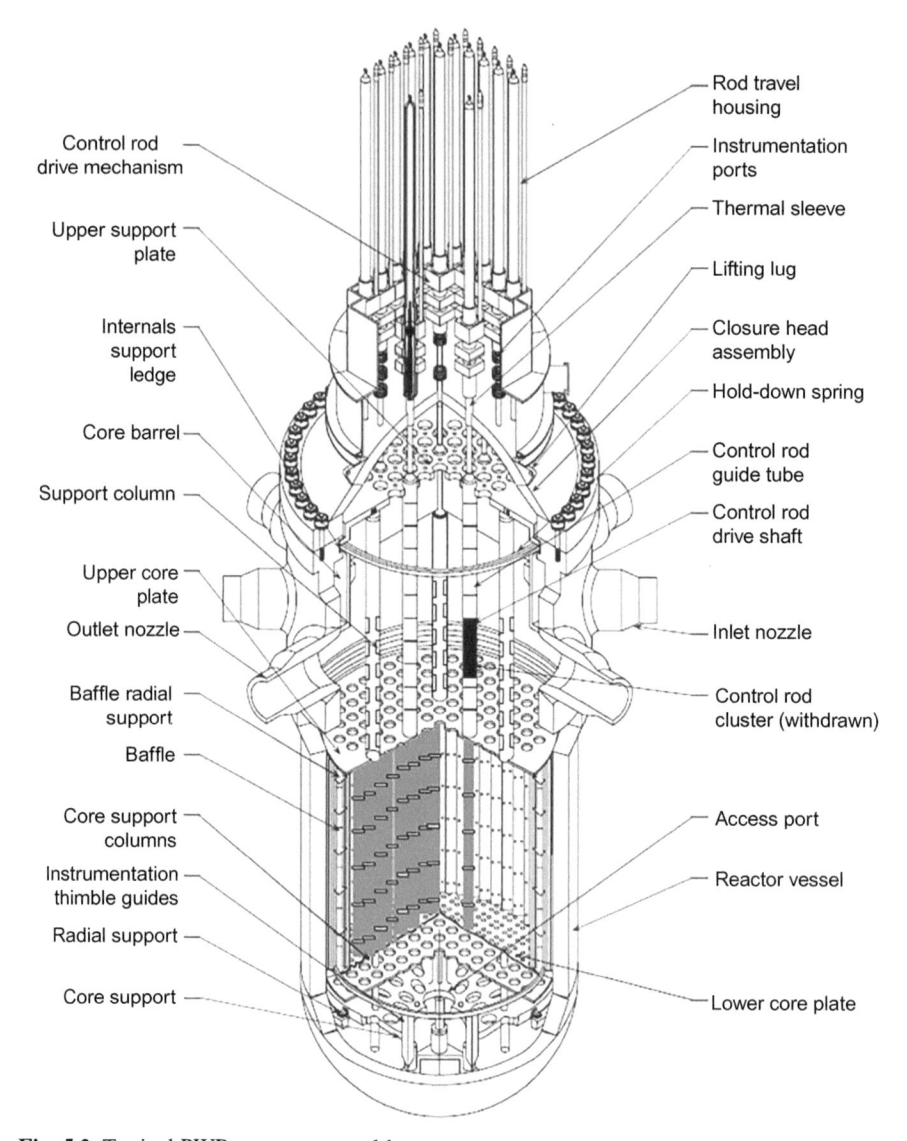

Fig. 5.2 Typical PWR pressure vessel layout.
Credit: NRC, 2006. Reactor Concepts Manual; Pressurized Water Reactor (PWR) Systems. NRC Technical Training Center, Chattanooga, TN.

As with the current large PWR designs, an iPWR reactor pressure vessel holds the individual fuel assemblies and the control rods. In addition, the iPWR reactor pressure vessel includes the pressurizer and the steam generator. Except for isolable support systems, such as the chemical and volume control system (CVCS), virtually the entire reactor coolant inventory is contained within the iPWR pressure vessel. Functionally an iPWR reactor pressure vessel performs the same roles as a current large PWR reactor pressure vessel. An iPWR pressure vessel flange is typically located above the top of the fuel and below the integral steam generator, which is significantly different than the pressure vessel head flange in a current PWR. This facilitates refueling in the tall iPWR pressure vessels and also facilitates steam generator inspection. Exceptions include the SMART iPWR, which does retain the reactor pressure vessel flange at the vessel head (Lee, 2010). The Holtec SMR-160 uses an offset steam generator, which also allows the reactor pressure vessel flange to remain at the vessel head (Oneid, 2012). Note that the Holtec design is not a true iPWR because the steam generators are flanged externally to the reactor pressure vessel and are not internal to the vessel.

The fuel in an iPWR is usually half the height of current PWR fuel. As a result the thick-walled ring forgings that make up the iPWR vessel could conceivably be stacked such that no weld exists adjacent to the fuel, thereby diminishing the vessel embrittlement concern. In addition, as can be seen in Fig. 5.1, the riser section needs to be large enough in diameter to accommodate control rod motion. Consequently the internal position of the steam generators in an iPWR design forces the reactor pressure vessel wall to be farther away from the reactor fuel than it would be in a conventional PWR. Therefore there will be added water shielding in the downcomer region of the iPWR reactor pressure vessel relative to a conventional PWR and subsequently lower fluence on the reactor pressure vessel. This would also act to diminish the vessel embrittlement concern. Integral PWR vessel materials will not differ from current PWR designs, so PTS will remain an operational concern for iPWR vessels. However, if the vessel fluence and the beltline weld embrittlement issue can be diminished in an iPWR vessel, then the overall iPWR operating window for PTS can likely be relaxed compared with a current large PWR. The International Reactor Innovative and Secure (IRIS) design, with a reactor vessel height to width ratio on the lower end of the spectrum for iPWRs, has a very large downcomer water channel that virtually eliminates the embrittlement issue and a need for an embrittlement surveillance program over the design life of the reactor vessel (Carelli et al., 2004).

Additionally, some iPWR designs are planning for internal control rod drives, and other iPWR designs are planning to exclude the use of boron chemical shim for normal operation. The presence of boric acid aggravates primary water stress corrosion; therefore designs that do not use boron for normal operations will diminish primary water stress corrosion cracking concerns for iPWR pressure vessels.

An iPWR reactor pressure vessel is smaller in volume than a conventional large PWR reactor pressure vessel. However, the ratio of cylindrical height to diameter in an iPWR pressure vessel is approximately 4–7 (NuScale, 2014; Kim, 2010; Memmott et al., 2012) to promote natural circulation for emergency operations and for normal operations in some iPWR designs. Since almost the entire reactor coolant inventory is within the reactor pressure vessel and not distributed in loop piping and other components as it is in a conventional PWR design (Fig. 5.3), natural circulation cooling is efficient and effective. The iPWR pressure vessel height to diameter ratio is much higher than the corresponding ratio for current PWRs (2.0–2.5). The increased iPWR pressure vessel height to diameter ratio is achieved by increasing the vessel

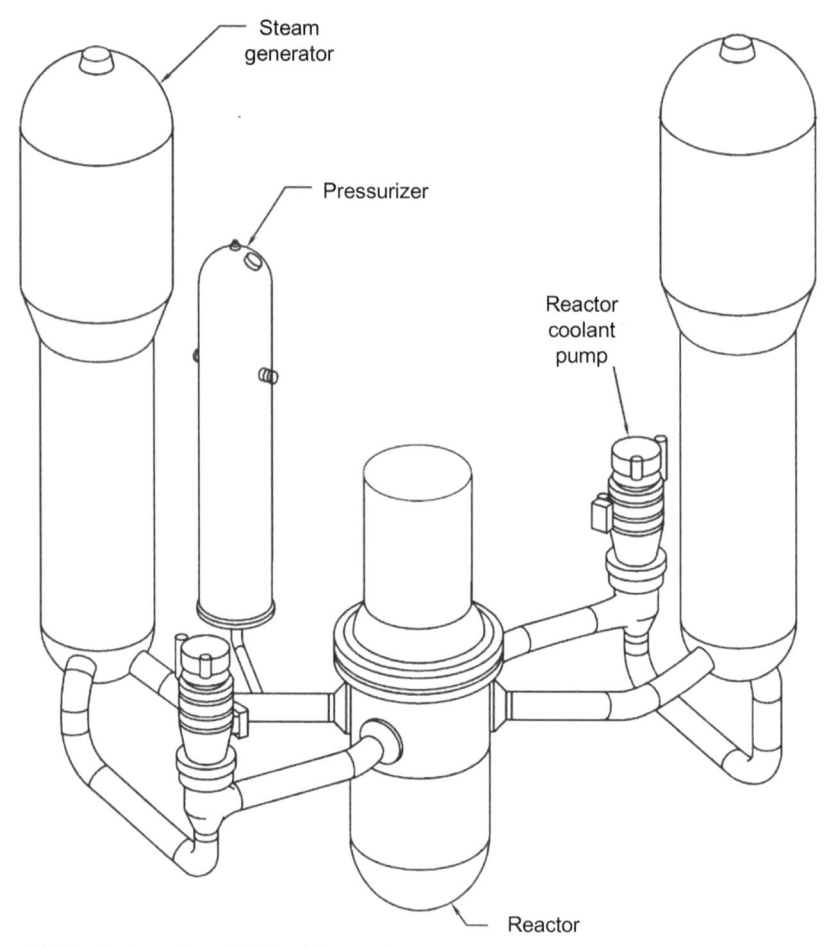

Fig. 5.3 Typical two-loop PWR piping and component configuration.
Credit: NRC, 2006. Reactor Concepts Manual; Pressurized Water Reactor (PWR) Systems. NRC Technical Training Center, Chattanooga, TN.

height and decreasing the vessel diameter, which also facilitates truck delivery of a factory built iPWR pressure vessel. In addition, because the pressure vessel includes additional components that current large PWR vessels do not, the water volume relative to core thermal power is increased considerably. The coolant design pressure of an iPWR can be equivalent to, or slightly lower than, a current large PWR design.

Vessel penetrations in an iPWR pressure vessel are significantly different than the penetrations on a PWR pressure vessel. There are no penetrations below the top of the fuel in any iPWR design—instrumentation access is always above the fuel in the iPWR designs. In addition, because there are no external coolant loops, the large PWR inlet and outlet piping penetrations, typically 27.5–29 in. (70–79 cm) (NRC, 2006), are eliminated in the iPWR designs. This effectively removes any opportunity for a large-break loss-of-coolant accident (LOCA) in the iPWR designs. The maximum vessel penetration in the iPWR designs is 10–15 times smaller than a PWR, typically 2 in. (5 cm) or less, for iPWR reactor support systems. In addition, the smaller iPWR vessel penetrations are well above the top of the reactor fuel relative to the larger PWR designs. This allows a significant amount of water to continue to be available to cool the reactor following a small-break LOCA on an iPWR design.

5.2.2 Reactor coolant system piping

In a current large PWR, the hot reactor coolant system water leaving the top of the reactor fuel is directed into the reactor coolant system hot-leg piping. The hot leg piping, nominally 29-in. piping (74 cm) (NRC, 2006), connects to a U-tube or a once-through steam generator, where heat is transferred from the primary water to the secondary water to generate steam to drive the turbine generator. Subsequently, colder primary water exits the steam generator into the intermediate leg (crossover leg) piping, nominally 31-in. piping (79 cm) (NRC, 2006), which connects to the suction side of a reactor coolant pump. The intermediate leg pipe is typically the largest diameter pipe in the reactor coolant system, and it is below the level of the top of the fuel assemblies, which can make a leak in this section of piping the most problematic of all large-break loss-of-coolant accidents. The reactor coolant pump discharges into the 27.5-in. (70 cm) (NRC, 2006) reactor coolant system cold leg, which directs primary water back into the reactor pressure vessel. Current large PWRs employ two, three, or four reactor coolant loops. A typical two-loop large PWR reactor coolant system is shown in Fig. 5.3.

Since the pressurizer and steam generator functions are integrated into the iPWR pressure vessel, all the large diameter piping associated with current large PWRs is eliminated. This eliminates any possibility of a large-break LOCA in all the iPWR designs. As a result, active emergency equipment, such as high-pressure injection pumps, associated with current large PWRs to mitigate the consequences of a large-break LOCA are also eliminated in the iPWR designs.

5.2.3 Pressurizer, heaters, spray valve, pressurizer relief tank and baffle plate

In a PWR a pressurizer functions to maintain the pressure of the primary coolant system in a range such that no boiling occurs in the primary system under normal and transient operations. In a current large PWR, the pressurizer is a separate cylindrical tank connected to the reactor coolant system piping by a surge line, nominally 10 in. (25 cm) in diameter (Datta and Jang, 2007), and a spray line, nominally 4 in. (10 cm) in diameter (NRC, 2012a). Pressure is normally controlled using heaters and a spray valve to maintain the pressure range. A balance of water and steam exist in the pressurizer space. The water level in the pressurizer provides an indication of water inventory in the reactor coolant system.

Some large PWR designs also use a power operated relief valve (PORV) connected to a pressurizer relief tank to assist in controlling pressure. The PORVs will open to reduce pressure prior to the system reaching the reactor coolant system safety valve relief set point. In large PWR designs that utilize a PORV, a stuck open PORV is a potential small-break LOCA initiating event.

A PWR pressurizer provides the surge volume for the reactor coolant system. If reactor coolant system temperature increases, less dense reactor coolant system water surges into the pressurizer, compressing the steam space and increasing the primary pressure. Steam will naturally begin to condense to lower the pressure, which may be sufficient for a small or slow transient associated with a minor reactor coolant system temperature change. More commonly, however, pressure will rise to the spray valve set point. The open spray valve then admits colder primary water from the reactor coolant system cold leg into the steam space to quickly condense steam and subsequently reduce pressure. The spray flow is driven by the discharge pressure of the reactor coolant pump. Conversely, if reactor coolant system temperature decreases, the suddenly denser reactor coolant water causes water to flow out of the pressurizer into the reactor coolant system hot leg, expanding the steam space and decreasing the primary pressure. This is controlled by use of pressurizer heaters, which heat the pressurizer water and subsequently increase pressure (NRC, 2006).

An iPWR integrates the pressurizer into the top of the reactor pressure vessel. A baffle plate or head plate with drilled openings separates the pressurizer from the reactor coolant system and acts as the surge line. Because of the integral nature of the pressurizer design, the 10-in. surge line is eliminated. The volume of an iPWR pressurizer is significantly larger than a current large PWR pressurizer relative to reactor thermal power. In the IRIS iPWR design, the pressurizer volume is about five times larger per unit of power than for a current large PWR design (Ingersoll, 2011). This larger pressurizer volume, coupled with larger reactor coolant system water inventory overall relative to reactor thermal power, provides slower pressure transients in general. This provides a number of operational benefits. First an operator will have more time to analyze changes in plant operating conditions and respond accordingly. Second the need for a fast-acting spray valve is virtually eliminated because in most cases the natural steam condensation following a slower in-surge of coolant into the large volume pressurizer will adequately maintain the reactor coolant system pressure under normal operations and expected transients. Finally the integrated location of the pressurizer in an iPWR design always provides a more direct indication to the operator of the water level above the top of the reactor fuel (IAEA, 1995).

The "normal" spray valve is eliminated in many iPWR designs because it is not essential to fine tune the system pressure as discussed in the preceding text and, in designs with no reactor coolant pumps, there is insufficient driving head in the reactor coolant system to provide adequate spray flow. Even in iPWR designs employing smaller reactor coolant pumps, the pressure differential or driving head created in the reactor coolant system is much lower than that in a current large PWR. Current large PWRs employ an "auxiliary" spray valve to back up the normal spray valve when the connected reactor coolant pump is unavailable or the normal spray valve is inoperable. The auxiliary spray valve in a current large PWR is typically driven by the charging pump discharge. The NuScale iPWR design plans to use this approach to provide for pressure reduction through pressurizer spray actuation (NuScale, 2014). Other iPWR designs will likely use this approach to include a pressurizer spray function as well. As a result the spray line is not necessarily eliminated in an iPWR design, but it is generally limited to a diameter that is less than the 4-in. size employed in current large PWRs. Pressurizer heaters in an iPWR design function the same as the heaters in current large PWR designs.

PORVs are not incorporated into the iPWR pressurizer designs. A stuck open PORV was the root cause of the Three Mile Island accident, which is eliminated in all iPWR designs. Safety relief valve discharge is directed to containment or another water storage tank inside containment. Therefore a pressurizer relief tank is not incorporated into the iPWR designs. In addition, piping to provide a nitrogen cover gas on the pressurizer relief tank and the need for a drain system and an associated pump are eliminated in the iPWR designs.

5.2.4 Pumps

Current large LWR coolant pumps function to provide forced primary coolant flow to remove the heat generated by the fission process. Existing large PWRs utilize one or two reactor coolant pumps per loop to provide the forced coolant flow through the primary system. Natural circulation flow in current large PWRs will not provide sufficient flow to remove the heat being generated during power operation. Each 6000–10,000 hp pump provides flow of approximately 100,000 gallons per minute to remove the heat generated by the reactor fuel assemblies for delivery to the steam generators. A driving head of approximately 90 psi (0.6 MPa) is generated by the reactor coolant pumps. A loss of flow from one or more reactor coolant pumps in a current large PWR results in a reactor trip (NRC, 2006).

These large pumps have seals to limit primary leakage and the seals require cooling. This provides a scenario for a small leak path and a scenario for an intersystem leakage path. Newer Generation III+ PWRs like the AP1000 design use a canned reactor coolant pump design to eliminate this possibility.

The integral nature of the iPWR pressure vessel makes the utilization of reactor coolant pumps more challenging. Several iPWR designs, including an inactive Westinghouse SMR design, plan to incorporate multiple reactor coolant pumps. Because of space limitations in the primary flow path, it is not possible to incorporate the large reactor coolant pumps used by the current PWRs. Instead, smaller reactor coolant pumps in these designs will be required to deliver the necessary flow. While specific reactor coolant pump design detail is unavailable, the resulting driving head will likely be smaller as well. Presumably a canned type pump will be incorporated. The pumps may be located in the hot leg at the top of the steam generators or in the cold leg beneath the steam generators, depending on the design. The inactive Generation mPower design called for 12 pumps (Kim, 2010), while the inactive Westinghouse design calls for 8 pumps (Memmott et al., 2012). In addition, the Korean SMART iPWR design also plans to use four canned reactor coolant pumps (Lee, 2010). The IRIS design planned to use eight spool pumps, one pump per steam generator

(Carelli et al., 2004). The remaining two active SMR designs currently under development in the United States, the NuScale iPWR design and the Holtec SMR-160 design (a PWR-based SMR), do not plan to use reactor coolant pumps (Kadak, 2017). These designs will incorporate natural circulation cooling. In addition, the Argentine CAREM iPWR design and the Japanese IMR iPWR design plan to incorporate natural circulation coolant flow for normal operation (Kadak, 2017). All the iPWR designs will be capable of removing reactor decay heat following reactor shutdown using natural circulation cooling. Therefore a dedicated AC backup is not required for the designs using reactor coolant pumps.

5.2.5 Riser

Since the iPWR steam generators are integral to the pressure vessel, an area within the vessel must be allocated for upward reactor coolant flow from the fuel assemblies to the upper tube sheet at the top of the integral steam generator(s). This area is often referred to as the riser section. It is analogous to the reactor coolant system hot leg in a current large PWR. At the top of the riser section, coolant flow will be directed down into the integral steam generator(s). In addition, the riser section provides space for the control rod drive mechanisms to operate.

5.2.6 Steam generator(s) and tube sheets

Current large PWRs utilize two to four separate large steam generators: one in each coolant loop. These steam generators are either U-tube or once-through type heat exchangers. In either case the higher pressure primary water flows inside the steam generator tubes, and the lower pressure secondary fluid is outside the tubes. In the U-tube steam generator design, dry slightly saturated steam is delivered to the turbine generator. In the once-through steam generator design, super saturated steam is delivered to the turbine generator (NRC, 2006).

Current steam generators range up to 70 ft tall and contain 3000–16,000 tubes welded to a tube sheet. The steam generator tube sheet and tubes are part of the reactor coolant system boundary. A steam generator tube rupture provides a short circuit path for primary coolant to escape containment with the secondary fluid (NRC, 2009). Steam generator tube issues include tube denting, wastage, thinning, corrosion, flow-induced vibrations, cracking and deformation of U-tube bend or of support plates, tube leakage, and fractures (Bonavigo and De Salve, 2011).

The active iPWR designs that are the most developed are planning one of two principle steam generator heat exchanger designs. The least traditional iPWR heat exchanger design is a once-through helical coil steam generator. The second iPWR heat exchanger design is a once-through straight-tube steam generator. No iPWR vendors have publicly indicated an intention to use a vertical U-tube type steam generator, which will eliminate the most significant current steam generator tube concerns regarding cracking and deformation of the U-tube bend. Helical coil steam generators provide additional heat transfer surface in a limited amount of space and the helical fluid flow generates less flow-induced vibration. Helical coil steam generators also reduce thermal stress on the feed water and steam headers generated by thermal expansion of the tubes. Likewise, there is little flow-induced tube vibration in once-through straight-tube steam generators. Therefore fewer tube supports would be required, which limits low fluid flow areas and the associated corrosion concern. However, once-through straight-tube designs do generate larger thermal stress on the feed water and steam headers due to thermal expansion of the tubes. Helical coil steam generators will likely be more costly to produce than once-through straight-tube steam generators because of the increased complexity of the design, but the cost and complexity may be worth the reduction in stress on the steam generator headers and the increase in heat transfer area compared with the once-through straight-tube steam generator design.

Variations on implementation of these designs exist. For example, the NuScale iPWR design plans to utilize two separate but intertwined helical coil steam generators with the high-pressure primary fluid outside the steam generator tubes and the lower pressure secondary fluid inside the steam generator tubes (NuScale, 2014). The SMART iPWR is designed for eight separate mini helical coil steam generators in the downcomer space around the reactor pressure vessel riser section (Lee, 2010). Likewise the CAREM iPWR is designed for 12 separate mini helical coil steam generators in the downcomer space (Mazzi, 2011). The SMART iPWR and CAREM iPWR utilize a more traditional design approach with the primary fluid inside the steam generator tubes. All three of these designs will produce superheated steam.

The inactive Generation mPower design had planned for a single once-through straight-tube steam generator that surrounded the center riser section. The primary fluid flowed inside the tubes and the secondary fluid flowed in the shell. The secondary fluids will enter and exit low in the steam generator shell; superheated steam will be produced (Kim, 2010). The inactive Westinghouse iPWR design also planned to use a single once-through straight-tube steam generator that completely surrounds the center riser section. However, in this application a separate steam drum, outside the reactor pressure vessel, will be used to remove entrapped moisture in the steam. Dry steam with minimal moisture content will be delivered to the turbine generator instead of superheated steam employed in other iPWR designs (Memmott et al., 2012).

The Holtec SMR-160 uses a steam generator and superheater that are directly flanged to the top section of the reactor pressure vessel. This has the advantage of providing more direct access to the reactor fuel during the refueling process. The Holtec design plans to exchange reactor fuel in a single cartridge replacement (Oneid, 2012).

5.2.7 Control rods and reactivity control

In all LWRs, control rods function to control the fission rate, or reactivity, by inserting or withdrawing neutron adsorbing material from the reactor fuel core. Current large PWRs typically use 17×17 fuel assemblies that include guide tubes for 24 control rod fingers, which are operated together through a spider assembly. All fuel assemblies are capable of hosting a control rod assembly, but not all fuel assemblies will be fitted with a control rod assembly in any given fuel cycle. Control rod assemblies are generally split into two groups, control groups and shutdown groups. The shutdown control rod groups are completely withdrawn from the core to provide a large source of negative reactivity to shut down the reactor in the event of an accident. The control groups are generally partially inserted into the core and are slowly withdrawn over the fuel cycle to compensate for fuel burn up and maintain operating temperature.

Current large PWRs also use a second method to control reactivity by adding soluble boric acid to the reactor coolant. Boric acid is a strong neutron absorber and is referred to as chemical shim. Current large PWRs heavily borate the reactor coolant system at the beginning of a fuel cycle and slowly dilute the boron concentration over the fuel cycle in conjunction with control rod motion to compensate for excess reactivity needed to provide a 12–24-month fuel cycle.

Many iPWR designs plan to use a half-height version of proven 17×17 array fuel assemblies. This will allow similar spider assembly control rods to be used. The iPWR cores contain fewer fuel assemblies than the current large PWRs and will typically have a higher percentage of fuel assemblies fitted with a control rod spider assembly to control excess reactivity. In addition, some iPWR designs may fit a very high percentage of the fuel assemblies with a control rod spider assembly and opt not to implement a chemical shim for normal reactor operation. However, it is anticipated that most iPWR designs will use concentrated boron as an emergency backup to shut down the reactor in the event not all the control rods are able to be inserted into the core. Therefore support systems for producing batches of boric acid will be required on most iPWR designs.

5.2.8 Control rod drive mechanisms

In all current large PWR designs, the control rod drive mechanisms (CRDMs) are external to the reactor vessel above the reactor vessel head. Prior to removing the PWR head for refueling, the control rod drive mechanisms are decoupled from the control rod spider apparatus. The control rods are then left in the respective fuel assemblies, while the head is removed.

Some iPWR designs plan to continue the practice of using control rod drive mechanisms that are external to the reactor vessel. However, because the iPWR reactor pressure vessel is much taller than current PWR pressure vessels and the iPWR pressure vessel flange is not located at the top of the pressure vessel, some design consideration will be necessary regarding how the shaft of the control rod spider is decoupled from the CRDM and protected when the upper reactor pressure vessel is removed from the lower reactor pressure vessel for refueling. Integral PWR designs planning on the use of external control rods include the SMART reactor and the NuScale reactor (Lee, 2010; NuScale, 2014).

Other iPWR designs intend to use CRDMs that are internal to the reactor pressure vessel. This will require some materials and reliability testing given the high temperature and pressure environment inside the reactor pressure vessel. In addition, electrical cabling will be necessary that penetrates through the reactor pressure vessel flange to operate the control rod drive mechanisms and provide control rod position indication. Radiation effects on internal CRDMs will be mitigated relative to large LWRs by the column of water in the taller SMR vessel. The Westinghouse SMR planned on using internal CRDMs (Memmott et al., 2012). Likewise the IRIS design had planned to use internal CRDMs (Carelli et al., 2004). The CAREM reactor uses internal hydraulic CRDMs, which will not require electrical cabling for operation, just position indication (Mazzi, 2011).

5.2.9 Automatic depressurization system valves

Older large PWRs do not directly employ an automatic depressurization system (ADS). This can make it difficult to respond to a very small loss-of-coolant accident, because coolant pressure can hang above the discharge pressure of the high-pressure injection pumps while coolant inventory is lost. Newer large Generation III+ PWRs, such as the AP1000 design, utilize automatic depressurization systems to allow primary pressure to be rapidly reduced following a loss-of-coolant accident to allow low-pressure injection systems or gravity-fed water sources to supply water to the reactor to keep the core covered.

Smaller iPWR designs incorporate an automatic depressurization system into the various protection schemes. This automatic depressurization provides the same function as it does for the newer Generation III+ passive reactor designs. In general the automatic depressurization is initiated by opening valves connected to the pressurizer steam space. The steam is typically sparged into a water-filled storage tank inside containment or collected directly within the containment for recycling back into the core to keep the fuel covered. The contents of the containment water storage tank located high in containment can be gravity fed into the reactor vessel after pressure has been reduced to atmospheric pressure.

5.2.10 Relief valves

All current large PWRs include safety relief valves to protect the reactor coolant system from exceeding the design pressure. The safety valves are connected to the separate pressurizer tank steam space via 6-in. (15 cm) pipelines (NRC, 2012b). These safety valves relieve to the pressurizer relief tank.

All iPWRs have the same need for safety relief valves to protect the reactor coolant system from exceeding the design pressure. Integral PWR safety valves are connected to the pressurizer steam space at the top of the integrated reactor pressure vessel. The connecting pipe would be much smaller than the 6-in. lines on a conventional PWR design. As with the ADS valves, the steam relieved from an iPWR is typically sparged into a water-filled storage tank inside containment or released directly into the containment environment. There is no pressurizer relief tank in an iPWR design.

5.2.11 Core basket, core barrel, core baffle

Current large PWRs utilize a core basket assembly to support the reactor pressure vessel internal assemblies. These assemblies typically include a cylindrical core barrel to separate and preheat the incoming cold leg reactor coolant from the fuel assemblies.

A thermal shield surrounds the core barrel. The core barrel directs the cold leg coolant to the bottom of the core. The fuel assemblies are supported inside the core barrel. To adapt the square fuel assemblies to the shape of the cylindrical core barrel, a core former or baffle is utilized. This arrangement will be the same for the iPWR designs.

5.2.12 Instrumentation

The ability to monitor core conditions will be of key importance for iPWR designs and will differ in implementation than current large PWR counterparts. Measurements that are currently obtained from PWR operating loops, such as reactor flow, will need to be implemented within the reactor vessel. This will be discussed further in Chapter 6.

5.3 Connected system components

5.3.1 Chemical and volume control system

An important support system for current large PWRs is the CVCS. The CVCS functions to purify the reactor coolant system, increase or decrease the coolant boron concentration, and maintain the reactor coolant inventory. A continuous flow of reactor coolant is letdown to the CVCS where it is cooled, cleaned through filters and demineralizers, reheated and pumped back into the primary system using a charging pump. The chemical shim can be adjusted by adding more borated water or adding more demineralized water back into the primary. Reactor coolant inventory can be adjusted by mismatching the letdown flow and the reinjection flow. Reactor chemistry is controlled through the CVCS volume control tank (NRC, 2006).

Because these functions are important for continued long-term operation of the reactor, all iPWR designs will require a similar support system. This will require a system external to the reactor pressure vessel with connections to the primary system in the reactor pressure vessel. These functions are not safety related, so a dedicated AC backup is not required. However, such systems will need to be isolable from the reactor coolant system. The isolation function is safety related.

Also, because a CVCS support system provides regenerative and nonregenerative heat exchangers to cool the primary fluid to protect the demineralizers, some iPWR designs may be able to take advantage of this system to also provide a means to remove decay heat after reactor shutdown. This is feasible due to the lower thermal power generated by iPWR designs. The CVCS heat exchanger in a larger PWR is not sized for this decay heat removal function.

5.3.2 Residual heat removal and auxiliary feedwater system

LWRs require the capability to cooldown following reactor shutdown to provide for system maintenance and refueling operations. Under normal operation, current large PWRs cool down in two stages. Initially the secondary system will be used to remove heat from the primary by dumping steam to the condenser via the turbine bypass system until the system temperature will no longer support sufficient boiling. At that point a forced-flow residual heat removal (RHR) system will remove reactor decay heat using a heat exchanger cooled by forced-flow safety-related component cooling water (CCW) system. In turn the CCW is cooled by the safety-related forced flow plant service water system. The RHR system is not designed to operate at the temperature and pressure required at the start of the plant cooldown, necessitating the two-step process.

In the event of a loss-of-offsite power, current large PWRs employ a diverse auxiliary feedwater (AFW) system using steam-driven, diesel-driven, or motor-driven (backed by a safety AC bus) feedwater pumps. The AFW pumps draw water from a large condensate storage tank to provide water for the continued generation of steam. Steam is dumped to the atmosphere, allowing the continued removal of decay heat from the primary. At the point where insufficient steam is generated, the RHR system and related support systems, all supported by a safety AC bus, will continue the plant cooldown to cold standby.

All iPWR designs will likely plan to use the secondary system to initially remove decay heat under normal operating conditions. Some designs may then plan to use a nonsafety related forced-flow RHR system to take over when steam is insufficient for heat removal during normal operations. However, unlike the large PWR designs, all iPWR designs will employ a passive decay heat removal system capable of removing the maximum core decay heat generation following a reactor trip. Heat will be rejected to large plant water tanks capable of removing heat for 72 h or more without refill to make up for boil off. As a result, no diesel-backed safety AC bus is required to support forced-flow RHR, CCW, or service water heat removal systems in iPWR designs. If needed, water from a tanker truck can supply additional cooling water beyond the initial 72 h; otherwise, air cooling may suffice from that point forward.

5.3.3 Emergency core cooling system and refueling water storage tank

Following a loss-of-coolant accident, current large PWRs employ an emergency core cooling system (ECCS) to mitigate core damage. This is accomplished by the injection of large amounts of cool, borated water into the reactor coolant system. The ECCS also provides highly borated water to ensure the reactor remains shutdown following the cooldown associated with a main steam line rupture. This water source is the refueling water storage tank (RWST), which is external to containment, except in the AP1000 PWR design. Subsystems taking suction from the RWST can include high-pressure injection or charging pumps, intermediate-pressure injection pumps, and the low-pressure injection or RHR system. Also, containment spray pumps take suction from the RWST to limit a containment pressure spike following a LOCA or a steam line rupture in containment. Some large PWRs also employ cold leg accumulators that inject borated water into the reactor when pressure falls below the injection set point. All these systems are safety related and are backed by the plant emergency diesel generators. Fig. 5.4 shows a typical large PWR ECCS.

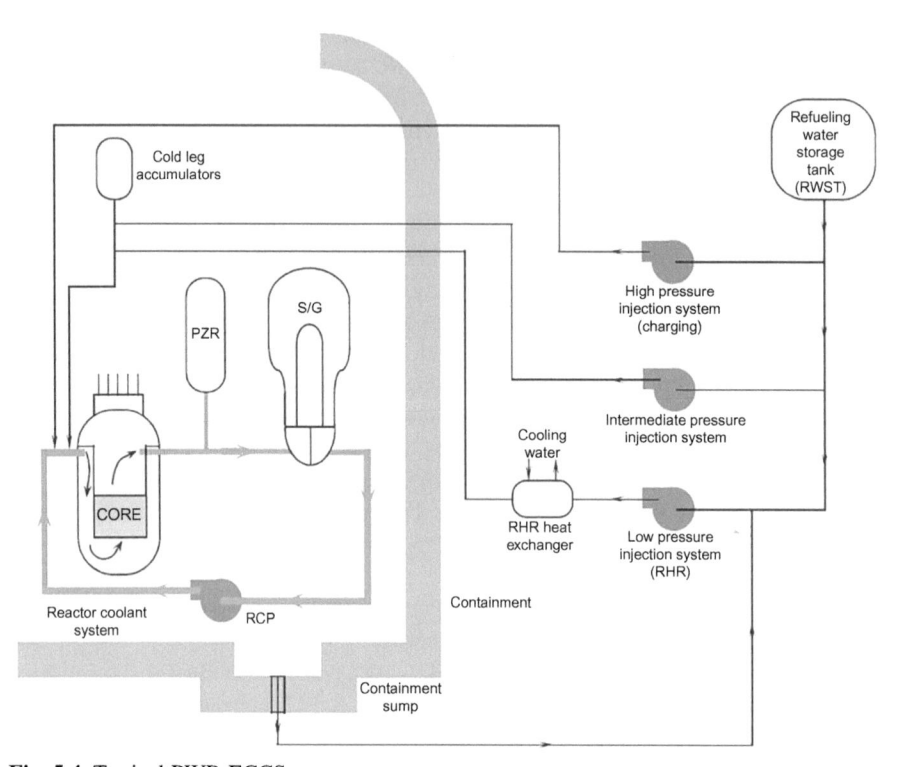

Fig. 5.4 Typical PWR ECCS components.
Credit: NRC, 2006. Reactor Concepts Manual; Pressurized Water Reactor (PWR) Systems. NRC Technical Training Center, Chattanooga, TN.

Similar to the Generation III+ AP1000 design, many iPWR designs have moved the RWST or its equivalent inside containment. The level of the RWST is well above the level of the top of the fuel. The iPWR designs have eliminated large-break LOCAs by eliminating all large-bore piping. Small-break LOCAs will likely not exceed the capacity of the CVCS charging pump. In addition, the small-bore piping connected to an iPWR reactor pressure vessel will be isolable as close as possible to the vessel, which will limit the probability of a small-break LOCA. In the event of a leak, the ADS valves will function to reduce pressure to the point where water can be gravity fed from the RWST into the reactor pressure vessel to keep the core covered. All ECCS cooling is designed to be by natural circulation. As a result the various ECCS pumps backed by emergency diesels are not required for iPWR designs.

Many iPWR designs will also use a gravity-fed boron injection tank to ensure the reactor remains subcritical following an accident. These injection tanks are located high inside the containment and not only serve as a poison source but also provide an additional source of water for emergency decay heat removal.

The iPWR containment designs are designed to be cooled passively to limit pressure spikes following an accident. Or the containment shell, such as in the NuScale design, is designed to accept significantly higher pressures than a current large PWR design (NuScale, 2014). Therefore no containment spray system is planned for the iPWR designs, except the SMART reactor (Lee, 2010).

5.3.4 External pool

The individual containment vessels of the NuScale design sit in a common open external tank referred to as the reactor pool. The NuScale containment conforms tightly to the reactor vessel without room for any additional water tanks. The NuScale reactor pool provides the emergency heat sink function provided by the internal RWST in other iPWR designs. In addition, the NuScale reactor pool provides an additional barrier to fission product release (NuScale, 2014).

The NuScale containment is maintained under a vacuum to enhance its heat transfer characteristics. Equipment to draw and maintain the vacuum will be required.

5.3.5 Control room habitability equipment

Current large PWRs require equipment to maintain control room habitability following an accident or toxic gas release. The iPWR designs have the same need for this function. While large PWRs can rely on safety-related AC-backed equipment to function in a long-term manner, the iPWR designs rely on passive systems designed to last at least 72 h. This typically requires the use of batteries and compressed air systems dedicated to this purpose.

5.3.6 Diesel generators and electrical distribution

As noted throughout this discussion, current large PWRs require a safety-related AC power bus backed by an emergency diesel generator (EDG) to provide assurance that active safety-related equipment will function when required. There are usually two EDGs per large LWR onsite. The EDGs are governed by plant technical specifications and must be tested at least once a month. Every 6 months, each EDG must be shown to come up to speed and voltage and begin the loading sequence within 10 s of receiving a start signal (NRC, 2012a). While necessary to prove the safety basis of the plant, this kind of testing can be problematical for the EDGs. In the event that one emergency diesel generator is declared inoperable, the remaining EDG must be tested within 24 h. As a result, it is not unusual for EDGs at large LWRs to be tested quite frequently, leading to higher maintenance requirements.

By removing the need for safety-related diesel generators in the various iPWR designs, the planned diesel generators become ancillary power supplies that enhance iPWR defense in depth by maintaining the availability of normal operating equipment that is not safety related. Testing of these ancillary diesels can be less frequent, and fast load testing can be relaxed such that the ancillary diesels have an opportunity to warm up for a few minutes before loading. Therefore less maintenance can be anticipated for the iPWR ancillary diesel generators.

Current large PWRs maintain EDG-backed safety-related AC buses, nominally at 480 and 4160 V AC to operate active safety equipment. In addition, current LWRs require safety-related power to operate instrumentation during and following an accident. This is provided by battery-backed 120-V DC and low voltage AC through an inverter. The iPWR designs will not require high voltage safety-related buses, because there are no active safety-related equipment components with this need. Power to reposition any iPWR safety-related valves will be required initially. Only battery-backed 120-V DC and low voltage AC through an inverter will be required to be safety-related, which will meet the needs for repositioning valves and powering instrumentation. So the electrical distribution systems of current large LWRs and iPWR designs will be similar, but the approach and amount of testing and maintenance will differ.

5.4 Future trends

The iPWR type reactors are an evolution of current PWR reactors. Future trends may include air-cooled power conversion cycles that greatly expand siting options, although with some loss of unit efficiency. Trends may also include improvements in the reliability and capability of the passive ECCS systems, leading to larger safety margins. The goal of the Holtec SMR-160 design is to replace all motors and pumps (and associated wiring, lubricating, and cooling) with gravity-driven fluid flow systems (Oneid, 2015). In addition, technology improvements may provide for the expanded use of internal reactor coolant pumps to provide primary coolant flow during normal operation augmented by natural circulation flow in shutdown situations. This could allow for better pump placement and lead to increased safety margins. Likewise, improved technology and materials, for internal components, could lead to expanded use of internal control rods, completely eliminating consideration of the rod ejection accident in all iPWR designs. Other trends could include further simplification or elimination of components. In fact the IRIS reactor design eliminated the need for an ECCS by virtue of the high-pressure spherical containment and the volume of water available in the reactor vessel (Carelli et al., 2004). This is discussed further in Chapter 3.

As piping length and size is reduced and mechanical components are eliminated, the footprint of an SMR facility overall can be reduced. Couple this with the fact that the fidelity of source term (radiation released to the environment) and radiation transport modeling continues to improve (NEI, 2012). Therefore, if SMRs with smaller facility footprints and smaller source terms can demonstrate that no accidents result in dose beyond the site boundary, then siting closer to population centers and micro-grid siting become more viable (Belles et al., 2019).

Specific uses of SMR designs beyond the standard end-product application of base load electricity production may lead to additional component considerations. Load following electricity production, water desalinization, shale oil recovery, and district heat applications is some of the possibilities that may require additional components.

Beyond this the evolution in SMRs could tend toward more advanced liquid metal-reactor, gas-cooled, and molten salt-reactor designs. Such designs are more amenable to air-cooled decay heat removal, which eliminates the need for large quantities of cooling water onsite. In addition, these advanced SMRs operate at higher temperatures, allowing them to utilize interface components to provide process heat for industrial uses. Providing process heat in lieu of generating electricity or in addition to generating electricity will introduce additional interface systems that may provide reactivity feedback to the reactor in a different manner requiring additional reactor control systems. Providing process heat services could also lead to additional accidents that may require new accident mitigation systems.

5.5 Sources of further information and advice

In the United States, the Nuclear Regulatory Commission (NRC) continues to interact with SMR vendors in preapplication discussions. The preapplication phase allows vendors to interact with the NRC prior to the formal submission of a licensing application. This allows potential licensing issues to be resolved earlier and familiarizes the NRC staff with the different iPWR designs. One outcome of the preapplication interactions is an effort by the NRC staff to take the Standard Review Plan, NUREG-0800 (NRC, 2007), and make it design specific by applying known SMR design attributes and preparing Design Specific Review Standards for these SMR designs. These Design Specific Review Standards will be public documents that will provide an excellent source of review standards for the unique application of various SMR systems and components. The NRC prepared such documents for the Generation mPower SMR design and the NuScale SMR design (NRC, 2016).

Regulatory Guide (RG) 1.232 (NRC, 2018), Guidance for Developing Principal Design Criteria for Non-Light-Water Reactors, was issued by the NRC in 2018 to provide guidance on adapting the LWR-specific general design criteria in Appendix A, "General Design Criteria for Nuclear Power Plants," of Title 10 of the *Code of Federal Regulations*, Part 50 "Domestic Licensing of Production and Utilization Facilities" (10 CFR 50) to advanced reactor designs. RG 1.232 provides guidance to non-LWR developers on the development of Principal Design Criteria for a given design and the systems and components required to meet the criteria. Although RG 1.232 does not specifically address LWR designs, advanced LWR designs, such as NuScale, could benefit from the safety interpretations and rationale provided by this document.

As industry interest in advanced non-LWRs has increased, the NRC staff has interacted with stakeholders to formulate appropriate policies for licensing and safety. The NRC has prepared draft RG-1353, which focuses on using a technology-inclusive, risk-informed, and performance-based methodology to inform the licensing basis and content of applications for nonlight water reactors. When finalized, this guidance will help advanced reactor designers select appropriate licensing basis events; classify structures, systems, and components; and assess defense in depth. As with RG 1.232, this guidance will impact the reactor systems and components required to ensure adequate protection of the public health and safety.

The NuScale SMR design submitted their design certification application to the NRC in January 2017. The safety review is underway and on schedule. The final safety evaluation is expected in late 2020. The first customer, the Utah Associated Municipal Power Systems cooperative, has identified a reactor location on Idaho National Laboratory property. Publicly available information on NRC interactions is available from the NRC website (https://www.nrc.gov/reactors/new-reactors/design-cert/nuscale.html).

The Korean design SMART reactor has received final design approval from the Korean regulatory body. Information is available on this design directly from the Korea Atomic Energy Research Institute (KAERI) at http://smart.kaeri.re.kr/pages/view/20/cooper.

References

Belles, R.J., Flanagan, G.F., Hale, R.E., Holcomb, D.E., Huning, A.J., Poore, W.P., 2019. Advanced Reactor Siting Policy Considerations. ORNL/TM-2019/1197Oak Ridge National Laboratory, Oak Ridge, TN.

Bonavigo, L., De Salve, M., 2011. Uchanin, V. (Ed.), Issues for Nuclear Power Plants Steam Generators, Steam Generator Systems: Operational Reliability and Efficiency. InTech. ISBN 978-953-307-303-3, pp. 371–392.

Carelli, M.D., Conway, L.E., Oriani, L., Petrovic, B., Lombardi, C.V., Ricotti, M.E., Barroso, A.C.O., Collado, J.M., Cinotti, L., Todreas, N.E., Grgic, D., Moraes, M.M., Boroughs, R.D., Ninokata, H., Ingersoll, D.T., Oriolo, F., 2004. The design and safety features of the IRIS reactor. Nucl. Eng. Des. 230, 151–167.

Datta, D., Jang, C., 2007. Failure probability assessment of a PWR primary system piping subcomponents under different loading conditions. In: Second International Symposium on Nuclear Power Plant Life Management, IAEA-CN-155, pp. 210–211.

IAEA, 1995. Small Reactors With Simplified Design. IAEA-TECDOC-962IAEA, Vienna, Austria.

Ingersoll, D., 2011. An overview of the safety case for small modular reactors. In: Proceedings of the ASME 2011 Small Modular Reactors Symposium. American Society of Mechanical Engineering, Washington, DC SMR2011-6586.

Kadak, A.C., 2017. A Comparison of Advanced Nuclear Technologies. Center on Global Energy Policy, Columbia University, New York, NY.

Kang, K., Kupca, L., 2009. Integrity of Reactor Pressure Vessels in Nuclear Power Plants: Assessment of Irradiation Embrittlement Effects in Reactor Pressure Vessel Steels. IAEA Nuclear Energy Series No. NP-T-3.11IAEA.

Kim, T.J., 2010. Generation mPower. ANS/DC Chapter PresentationB&W Nuclear Energy, Charlotte, NC.

Lee, W.J., 2010. The SMART Reactor, 4th Annual Asian-Pacific Nuclear Energy Forum. Korea Atomic Energy Research Institute, Daejeon.

Mazzi, R., 2011. CAREM: an innovative-integrated PWR. In: 18th International Conference on Structural Mechanics in Reactor Technology (SMiRT 18)pp. 4407–4415.

Memmott, M.J., Harkness, A.W., Van Wyk, J., 2012. Westinghouse small modular reactor nuclear steam supply system design. In: Proceedings of ICAPP '12. pp. 973–983.

NEI, 2012. Small Modular Reactor Source Terms. Nuclear Energy Institute Staff Position Paper.

NRC, 2003. Reactor Pressure Vessel Issues. NRC Office of Public Affairs, Washington, DC.

NRC, 2006. Reactor Concepts Manual; Pressurized Water Reactor (PWR) Systems. NRC Technical Training Center, Chattanooga, TN.
NRC, 2007. Standard Review Plan for the Review of Safety Analysis Reports for Nuclear Power Plants: LWR Edition. NRC, Washington, DC. https://www.nrc.gov/reading-rm/doc-collections/nuregs/staff/sr0800/.
NRC, 2009. Steam Generator Tube Issues. NRC Office of Public Affairs, Washington, DC.
NRC, 2012a. Standard Technical Specifications, Westinghouse Plants. NUREG 1431, Revision 4.0NRC Office of Nuclear Reactor Regulation, Washington, DC Section 3.8.1.
NRC, 2012b. Resolution of Generic Safety Issues: Issue 198: Hydrogen Combustion in PWR Piping (NUREG-0933, Main Report with Supplements 1–34). NRC, Washington, DC.
NRC, 2016. NuScale Design-Specific Review Standard. NRC, Washington, DC. https://www.nrc.gov/docs/ML1535/ML15355A295.html.
NRC, 2018. Guidance for Developing Principal Design Criteria for Non-Light-Water Reactors. RG 1.232NRC, Washington, DC.
NuScale, 2014. NuScale Plant Design Overview. Revision 0, RP-1114-9375NuScale Power, LLC, Corvallis, OR.
Oneid, P., 2012. SMR-160 Unconditionally Safe & Economical Green Energy Technology for the 21st Century. SMR, LLC, Jupitor, FL.
Oneid, P., 2015. Essentials of SMR-160 Small Modular Reactor. Holtec Technical Bulletin, Revision 9SMR, LLC, Jupitor, FL.

Instrumentation and control technologies for small modular reactors (SMRs)

Dara Cummins[a] and Edward (Ted) Quinn[b]
[a]Independent Contractor, Loudon, TN, United States, [b]Technology Resources, Dana Point, CA, United States

6.1 Introduction

Small modular reactors (SMRs), especially those of the integral pressurized-water reactor (iPWR) kind, are the wave of the future. Other chapters have detailed the advantages of iPWR modular designs in meeting the needs of small metropolitan centers or developing countries where the electrical grid infrastructure is not present, as well as the needs of normal power plants with the modular ability to add units. The "smallness" and the "modularity" of the new iPWR designs present several advantages to the nuclear power community.

Some of the common architecture features of an iPWR are the following:

- small capsule-like design,
- below grade installation,
- steam generators inside the reactor pressure vessel,
- small on-site crew manning for operations and maintenance (more automation),
- a multiunit control room,
- modularity and grow-ability (ability to add more modules to increase the power output for a small additional increase in infrastructure),
- more passive cooling techniques,
- fewer accident scenarios or safety cases,
- more built-in fail-safe features,
- more prognostic and diagnostic capabilities,
- low maintenance instrumentation.

With these features come challenges for the instrumentation and controls (I&C) design. iPWRs will have more automation, more redundancy, more cybersecurity, more fail-safe features, more fault-tolerant designs, more prognostic and diagnostic capability, and different measurement methods. With these design challenges, it is important to focus on I&C development early in the overall reactor design/development phase so that solutions, possibly unique solutions, can be developed and qualified early. Just as the mechanical/electrical design requires I&C compatibility, the I&C design requires mechanical/electrical modifications.

This give and take between the I&C design and the mechanical/electrical design is a necessity in these new builds and requires a high level of communication and coordination between functional divisions.

New technologies offer solutions and abilities that the old 1970s and 1980s technology did not have. The advent of digital designs and wireless communications are technological advances that should be considered in every new design. The extent to which an iPWR incorporates these technological advances will determine the amount of protective features that may need to be built into the design and the license application. In the case of microprocessor-based digital technology, a detailed defense of common cause and common mode failures must be developed. In the case of hardwired logic-based digital design, such as field programmable gate arrays (FPGAs), common cause and common mode failures are not as much of an issue, but a detailed diversity and defense in depth must still be developed. In the case of digital or wireless communications, cybersecurity threats and electromagnetic interference/radiofrequency interference (EMI/RFI) must be considered.

Traditional measurement methods may not be applicable in the new designs. New environments, submerged vessels, lack of traditional piping, and new geometries all play a part in the need for new measurement devices and methods. The traditional devices offer the safety qualification pedigree that a nuclear plant requires, but the traditional devices are designed for a traditional plant and may not be qualified or designed for the new harsher environment and/or the new smaller geometries involved. There is an emerging need for radiation-hardened, high-pressure, and temperature-qualified, submergible, smaller sensors with remote processing electronics. Although the technology is developed, the application of this technology to the new iPWR environment has not been accomplished in many cases, especially the cases of primary vessel level and flow. New qualification programs will need to be developed for the new sensor technologies.

In summary the maturation of iPWR design is bringing about a paradigm shift in instrumentation design and methodology. What used to work for large light water PWRs will not necessarily work in the iPWR environment. The new smaller dimensions, differing geometries, and harsher environments will necessitate new technologies and advanced instrumentation solutions. This chapter highlights the requirements, challenges, and potential solutions for this new I&C world.

6.1.1 Major components of an I&C system

The major components in an I&C system include the primary or sensing element, the transmitter, the processing electronics, and the actuation devices. The primary/sensing element is the device that senses the process. For example, a sensing element may be the temperature sensing device, or the pressure sensing device. The transmitter is the device that converts the physical parameter being sensed to an electronic signal that can be transferred to follow-on processing electronics. The processing electronics perform various operations on the signal, which could be a simple as a applying a gain, or as complex as a filtering algorithm. The actuation device is the device that reads the electronic signal and performs the final action. The actuation device could be a control board indicator, a computer display, or an actuation signal sent to some other mechanism such as a control valve or an electronic device. The cabling is the wiring that ties it all together. The following discussion about iPWR I&C systems is subdivided into seven categories:

- Safety System I&C
- Nuclear steam supply system (NSSS) Control Systems Instrumentation
- Balance of Plant (BOP) Instrumentation
- Diagnostics and Prognostics
- Processing Electronics
- Cabling
- Future Trends and Challenges

6.2 Safety system instrumentation and controls

Safety systems are defined as the systems necessary for the safe shutdown and cooling of the reactor during accident or transient/excursion conditions. Safety instrumentation is the instrumentation necessary to monitor the selected safety parameters and send the required actuation signals to the actuation devices necessary for safe shutdown and emergency cooling. For example, the reactor trip breakers are safety equipment designed to release the control rods into the core to stop the nuclear reaction. The instruments designed to detect a parameter excursion, process the signal, and transmit a trip signal to the reactor trip breakers are also considered safety equipment. Safety signals can trip the reactor, or they can actuate a system that will provide emergency cooling to the reactor. In the nuclear world the safety-related instrumentation is kept separate from the nonsafety instrumentation. (Safety signals can still be used as inputs for control systems but the signal is electrically isolated first so that no control system feedback can affect the safety signal and function.) Because of this separation, safety-related instrumentation has different issues and considerations from nonsafety instrumentation. The following sections discuss the techniques for safety system measurements and the unique iPWR conditions and issues involved in developing a safety I&C system for an iPWR.

6.2.1 General requirements for safety system I&C

As with traditional light water PWRs, the safety analysis and regulatory requirements drive the safety instrumentation design. The regulatory requirements stem largely from the national regulations and several other regulation guides and referenced documents. The safety analysis is the process of evaluating the plant response to anticipated occurrences, accidents, and abnormal events to ensure the safe response of the reactor. The safety analysis defines and models the expected events, steady-state outcomes, anticipated occurrences, transients, and accident conditions to ensure fuel design limits are not exceeded. It defines and models the reactor vessel pressure boundary limits and the containment pressure boundary limits to ensure they are not exceeded. It also analyzes radiation conditions to minimize the radiation dose. The design response to the safety analysis and regulatory requirements is the called the reactor protection system (RPS) and is usually defined as the set of all reactor trips (RTs), engineered safety features (ESFs), and monitoring, required to meet all safety system analyses and regulations to provide safe monitoring and safe shutdown of the reactor.

For example, the need for a reactor trip when reactor pressure drops to a certain level is derived from the safety analysis requirement to protect the core from a core damage condition known as departure from nucleate boiling (DNB). This requirement drives the need to measure reactor pressure and temperature to know the conditions associated with DNB and so protect the reactor from this potential accident condition.

The safety system instrumentation must be designed to protect the core from damage and to operate the reactor within safe limits as defined by the safety analysis and applicable regulations. For these reasons the following instrumentation requirements are generally required. Unique iPWR designs may preclude some of the instrumentation measurements itemized below or may add additional instrumentation, but for the most part the following parameters must be measured using suitable instrumentation:

- pressurizer pressure and reactor pressure
- pressurizer level
- core temperature
- reactor coolant temperature (wide range, narrow range, hot leg and cold leg)
- reactor vessel level
- steam generator level
- reactor coolant flow
- reactor water storage tank level
- feedwater flow
- main steam flow
- main steam pressure and temperature
- reactor power
- core power flux (power range, intermediate range, and source range)
- reactor coolant pump voltage and frequency
- containment pressure temperature and level

The measurement categories listed previously are used in the RPS, where RTs and ESF actuations are automatic based on sensed parameter values. Some of the sensed RPS signals are provided downstream, through isolators, to the nonsafety NSSS control system, where control actions are taken in automatic or manual to keep the reactor operating within established limits. Additional safety-related parameters that may be needed based on some iPWR designs are as follows:

- safety accumulator tank pressure,
- safety accumulator tank level,
- safety valve positions,
- feedwater flow,
- boron concentration, temperature, tank levels, and mixing amounts.

The following sections describe the traditional and new technological devices that are used and will be used to measure the parameters described in the preceding text.

6.2.2 Safety system pressure transmitters

A pressure transmitter is a device that translates physical force to an electrical signal. The most common type of force transducer uses a diaphragm, piston, bourdon tube, or bellows to sense the physical force and various strain/force-sensing devices to convert the deflection of the physical element to an electrical signal. Traditional strain-sensing devices include the following: capacitive cells, piezoresistive strain gauges, piezoelectric quartz material, and electromagnetic devices.

In the United States, companies such as Rosemount, Cameron/Barton, Foxboro, and Ultrasystems have provided transmitters that specialize in safety system pressure measurements. These transmitters may still function successfully on some iPWR designs, but many will have to be reengineered for different mounting configurations, size constraints, and environments. Many iPWR designers, when faced with a modification program, may choose to go with new technologies rather than modifying the old ones. The new technologies may offer advantages in size, redundancy, accuracy, and environmental resilience. Some of these new technologies include microelectromechanical system (MEMS) sensors, fiber optic sensors, and ultrasonic sensors.

In the optical fiber category, a company called Luna Innovations has developed and successfully tested fiber optic pressure sensors, like the one shown in Fig. 6.1, in a research reactor environment. These fiber optic pressure sensors have been shown to operate in radiation environments with flux levels much higher than those compatible with most electronic pressure sensors. With traditional technology, it is necessary to protect traditional electronic pressure transmitters from harsh radiation conditions near the core; this requires the use of long pressure sensing lines, which limit the response time to pressure transients and increase the number of wall penetrations. Luna's fiber optic pressure sensors are designed to operate in harsh environments. When these pressure sensors were combined with Etalon-based fiber optic temperature sensors providing temperature compensation, drift effects were minimized. The attractiveness of this technology for iPWRs is obvious, with the elimination of sensing lines, the minimization of penetrations, the small size of the sensor, the rapid response to pressure fluctuations, and the operability in high radiation fields. With these attributes, this technology bears merit for primary and secondary side pressure measurement (Dickerson et al., 2009).

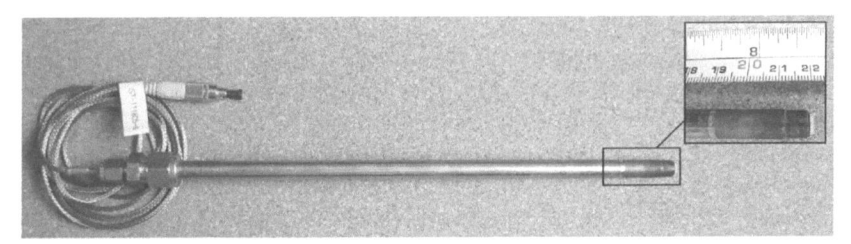

Fig. 6.1 Luna Innovations' fiber optic pressure sensor in probe housing. Courtesy of Luna Innovations Inc.

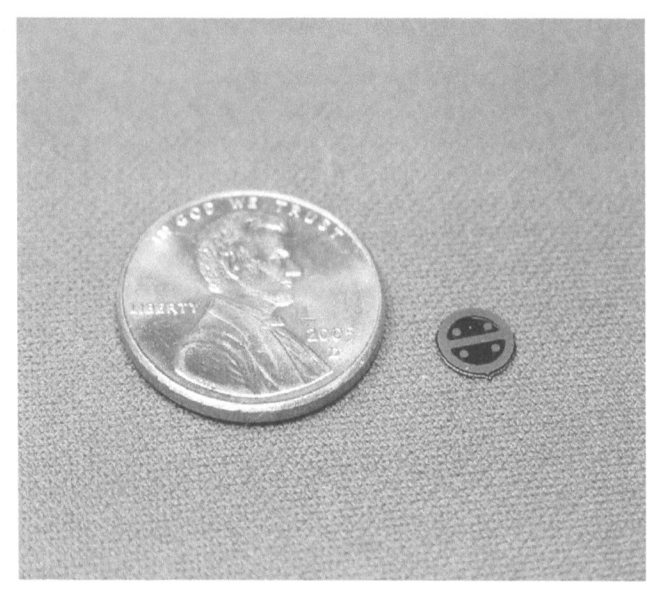

Fig. 6.2 Picture of the Sporian MEMS pressure sensor.
Courtesy of Sporian Microsystems Inc.

Another new technology for pressure sensing is the polymer-derived ceramic MEMS sensor. At the forefront of this technology, a company, Sporian Microsystems, has developed a pressure/temperature sensor made to survive high temperatures (Fig. 6.2). This technology offers a solution for pressure sensing in iPWRs due to its hardy environment survivability and its small size. The small size allows for the installation of redundant units and the measurement of pressure at many points, possibly with fewer penetrations than traditional sensors.

These new technologies have attributes like small size, heat survivability, radiation hardness, fast response, and low maintenance. These attributes are highly valued in iPWR designs for obvious reasons.

6.2.3 Safety system level transmitters

A level transmitter is a device that translates the sensed level of a vessel or tank into an electronic signal. Direct, visual, or float level detection methods are usually not possible, nor practical, for safety system measurements, so most safety system fluid level detection is accomplished with a differential pressure device. This is a device that compares the constant pressure of the reference leg to the variable pressure of the hydrostatic volume in the vessel that represents the level of the fluid. This comparison, called differential pressure (DP), is correlated and calibrated to actual fluid level in the tank or vessel.

The safety system level measurements of pressurizer level, steam generator level, safety accumulator level, and refueling water storage tank level are performed with the differential pressure method in most PWRs today. This method is desired for its ability to accommodate level measurements where the vapor/water interface between steam and liquid is not distinct.

One of the key factors affecting the use of DP methods in level detection is availability of room or space for the sensing lines. Another issue is the practical challenge of maintaining a constant reference leg pressure. In a traditional containment, there is room for the sensing lines necessary for DP-type level transmitters, and the reference leg level is maintained due to milder environmental conditions. The "room and environment" may not be available in some of the iPWR designs. For these cases, new technology may provide the much needed solutions.

Acoustic or ultrasonic signal measurement devices have been commercially developed that shoot a sound wave frequency signal from the top of the tank directly to a fluid level below. These devices use the time of the signal reflection to determine level distance. These devices could be used in tanks where there is no blocking hardware between the signal source and the fluid level and in an environment where there is a clear distinction between the liquid and gas interface. It is mechanically feasible to use this type of methodology for a refueling water storage tank, for example, where the gas to liquid level is distinct, but steam generator and pressurizer level might provide a challenge for this type of technology with a nondistinct water to vapor surface and a high degree of water vapor in the air/vapor space. The acoustic or ultrasonic devices for most safety purposes would need to be temperature compensated as the ultrasonic wave speed is affected by the temperature of the medium it penetrates (http://www.indumart.com/Level-measurement-4.pdf; http://www.industry.usa.siemens.com/automation/us/en/process-instrumentation-and-analytics/process-instrumentation/level-measurement/continuous-ultrasonic).

Another new technology that may be developed for level measurements is fiber optic sensing. A fiber optic sensor uses light (optics) to carry information about a process. In the distributed fiber sensing methodology, the optical signal transforms in an area of changing stress, strain, or temperature. These transformations are detected by the optical signal and calibrated to detect temperature differences or strain differences. Developing fiber for level measurement is in the early stages of development (Dickerson et al., 2009; EPRI, 2011, 2012; Chang et al., 2013).

Much the same as the distributed fiber measurement discussed previously, level detection can also be accomplished with strands of traditional temperature devices, such as resistance temperature devices (RTDs) or thermocouples. For this method, one solution is to distribute two strands of closely located discrete heated and unheated junction thermocouples along the length of the vessel. Level is sensed by comparing the temperature difference between the heated and unheated thermocouples, on the principle that heat transfer properties between air and water will result in varying temperature differences between the heated and unheated thermocouples, thus establishing the air (steam) to water interface. This method is only as accurate as the vertical distance between the temperature sensors.

6.2.4 Safety system temperature devices

Temperature is one of the basic measurements required in a nuclear reactor. Typical PWR temperature measurements have been accomplished using either RTDs or thermocouples. Thermocouples have been used in core temperature measurements where the temperature can be extremely hot. RTDs are typically used in the hot and cold legs of a PWR for reactor coolant temperature. Qualified versions of thermocouples and RTDs have been developed, and there are longevity data on these devices.

Another temperature measurement device, not currently used in nuclear power reactors but with some nuclear potential, is the Johnson noise thermometer (JNT). This device is based on the thermal fluctuation of a conductor. It uses the mean square thermal noise voltage generated by a resistor or capacitor to establish temperature. The application of this methodology is highly accurate for temperature measurements, but the voltage measured is in the range of microvolts and spread over a wide bandwidth. Although the use of amplifiers and digital filters enhances the signal strength, there are many practical hurdles that must be solved for an industrial use. One of the key advantages for the JNT is the elimination of typical RTD drift. A JNT device is self-calibrating, and it has the ability to maintain its accuracy indefinitely without the need for human intervention for periodic calibration. There are currently no commercial models available for purchase; however, with the ability to maintain its calibration indefinitely, the JNT may be a good new solution for iPWRs (Britton et al., 2012).

It is likely that iPWRs will continue to use thermocouple and RTD methods to measure core and coolant temperatures, but the traditional mountings and instrument size may have to be reengineered for the unique and smaller geometries involved. In these conditions, alternate temperature measurement devices might be considered.

Optical fiber technology offers another new approach for temperature sensors. This method has not been used in traditional PWR applications and represents a new option for safety system temperature measurements. Over the last decade, Luna Innovations has introduced several new fiber optic sensor technology options for making temperature measurements (Fig. 6.3). Notably, Luna is able to use a single, inexpensive, and commercially available optical fiber to make distributed measurements, characterizing thousands of temperature or strain measurements made at points along the length of the fiber in a fraction of a second. One demonstration of this technology was conducted during a short-term test in a nuclear research reactor. Temperatures were calculated from Rayleigh backscatter signals, without the use of fiber Bragg gratings, every 1 cm along a variety of commercially available single mode C-band

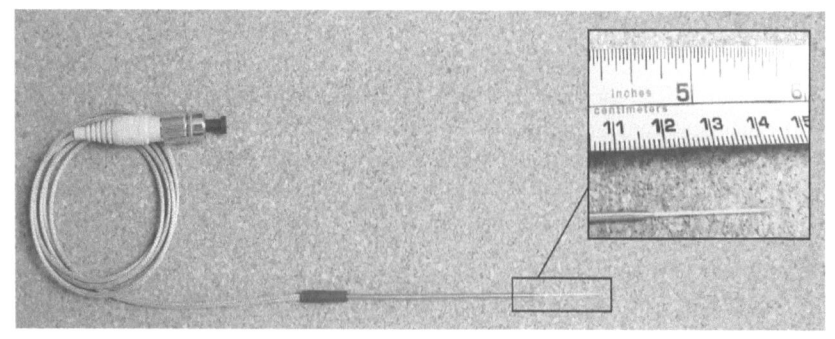

Fig. 6.3 Luna Innovations' fiber optic distributed temperature sensor in a metal capillary tubing housing.
Courtesy of Luna Innovations Inc.

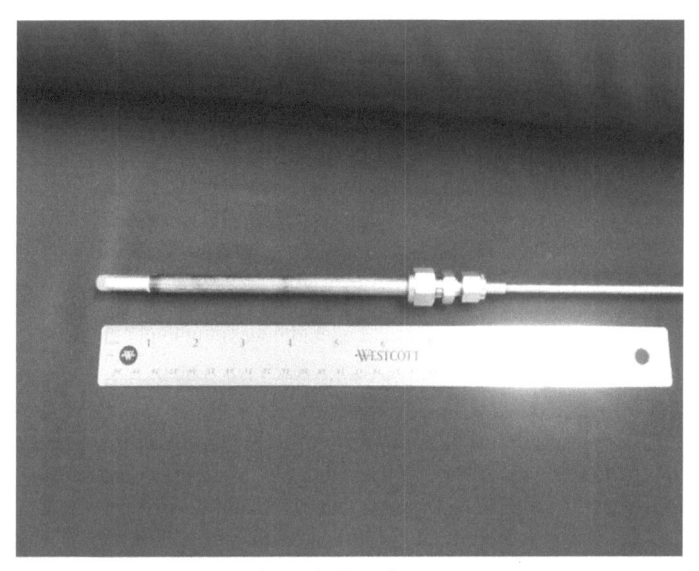

Fig. 6.4 Sporian temperature sensor in probe housing.
Courtesy of Sporian Microsystems Inc.

telecommunication optical fibers. Luna's distributed sensing technology can provide high-spatial-resolution temperature measurements in a compact format, without suffering from electromagnetic interference, and along fiber lengths up to 30 m for selected radiation environments (Sang et al., 2007).

Sporian Microsystems Inc. offers another temperature sensing option in the microelectromechanical sensor (MEMS) technology field that has a high range of temperature sensitivity (up to 1300°C) with a high degree of accuracy (Fig. 6.4). This device is composed of polymer-derived ceramic material, is extremely small, and has been tested in research reactors. Sporian is currently designing these sensors under a Department of Energy (DOE) small business contract (www.sporian.com).

6.2.5 Safety system flow transmitters

Flow measurement is required for nuclear safety systems. Reactor coolant flow and main steam flow are examples. Perhaps the most challenging of the flow measurements is reactor coolant flow. Typical PWRs use an elbow tap method to attain a DP that is then translated to flow. The elbow taps are implemented in the elbow bend of the reactor coolant pipe on the crossover piping between the steam generator and the reactor coolant pumps. The theory behind this approach is that the larger centrifugal force of the water along the outside wall of the bend is compared with the lesser force of the water traveling along the inside wall of the pipe to create a differential pressure that is related to flow by a square root relationship. This method of flow measurement must be normalized and calibrated at near full power in order to provide the necessary accuracies.

iPWR designs that have primary side piping systems may be able to use the traditional approach. Some iPWR designs may find that the geometries of the piping associated with the reactor coolant pumps (RCPs) and/or the smaller flow velocities preclude this type of measurement. In those cases, as with the nonpipe designs, new methods and approaches will need to be developed.

The NuScale design, for example, does not have reactor coolant piping or reactor coolant pumps. RCS flow is based on a natural circulation design that results in low flow velocities compared with traditional PWR designs. Both the lack of a pipe to mount the usual sensing element and the low flow velocities provide a challenge for traditional flow measurement methods. New technologies may provide the solution in these cases.

Some of the new technologies being considered for iPWRs without the traditional pipe configurations in their primary system are ultrasonic methods, transit time methods, MEM methods, and fiber optic methods. The benefits of some of these methods over the traditional DP method are that the primary sensed parameter is linear with respect to flow, as opposed to the square root relationship used in traditional DP methods. This linear relationship allows the accuracy to be consistent over the entire flow range and allows for improved accuracy at the low end of the flow range, which has not been the case in traditional DP methods. Accurate low flow measurement is going to be a requirement for several iPWR designs. The need for high accuracy at low flow conditions will drive several iPWR builders to find an alternative flow-measurement method.

Concerning primary coolant flow, some iPWR designs will have RCP pumps in their designs. It is possible and even probable that designs with RCP pumps will use RCP voltage, current, or pump speed as the primary parameter to measure flow. In a water-filled system, this approach has merit. Since "flow" is a parameter determined indirectly from other parameters, the use of pump speed/voltage/current is a viable alternative approach.

Similar to using pump speed, natural circulation designs may choose to use temperature, or delta temperature, as the method to measure reactor coolant flow. In a natural circulation design, the equivalent of the reactor coolant pump is the delta temperature across the core. In these cases the primary system delta temperature will correlate to flow directly. Although this method of correlating delta temperature to flow is not a particularly complex method, it is a new concept for both the regulator and the builder. As with all new concepts, it will take longer for the new concepts to be proven, developed, and embraced.

Flow measurement methods in applications other than the reactor coolant flow have the flexibility, in some designs, to use traditional or new technology. If the system is a safety system, the concern will be the environmental qualification, but in systems not related to safety, older more traditional methods, including DP methods using venturis or nozzles, may be possible. These methods will still have the pitfalls of lower accuracies in the low flow range, and lack of qualification for higher pressures and higher temperatures, which may force iPWR builders to use newer technology to achieve better flow measurements, even in the nonsafety systems.

6.2.6 Safety system power/flux devices

Traditional devices for measuring reactor flux and power are grouped into two categories: excore instrumentation and incore instrumentation. There are two popular radiation measuring devices: fission detectors and ion chambers. These devices have measured reactor flux and power reliably over the last 40 years.

Ion chambers, located outside the core (excore), can detect thermal neutron flux that is directly proportional to the fission rate and reactor power. Fission chamber flux detection devices, on the other hand, have a wider range than ion chambers and are more sensitive to neutrons. Both devices are typically used in traditional large PWRs. The same technology is being considered for iPWRs (Mohindra et al., 2007).

iPWRs, however, may be able to take advantage of recent size reductions in fission chambers and recent accuracy improvements. For example, micropocket fission detectors have been fabricated and tested as incore flux monitors in the 250 kW TRIGA reactor at Kansas State University. These microfission detectors have shown high performance with radiation hardness to neutrons, gamma rays, and charged radiation products (McGregor et al., 2005).

Additionally, another emerging technology in the nuclear power monitoring field is gamma thermometers. Although approved for US nuclear use as a local range power monitor since 1982, they have not been deployed widely in US PWRs. Based on the principle of temperature difference in a thermocouple type junction, which is proportional to incident gamma flux, gamma thermometers may provide an alternative option for power monitoring in the future, offering possible improvements over traditional methods in incore time response, calibration, and size (Korsaha et al., 2009).

Although iPWRs provide the opportunity for the use of new technologies for power and flux measurement, the incorporation of new technologies will require increased development time to design and qualify these new devices.

6.3 NSSS control systems instrumentation

NSSS control and indication instrumentation is not part of the safety system instrumentation; however, NSSS Instrumentation is required for day-to-day operation of the plant and is modeled in the safety analysis so that control systems can be relied upon to perform in a safe operating envelope. These control systems are also designed to handle anticipated transients in such a way as to mitigate severe consequences and return the plant to normal operating conditions. If the NSSS control systems fail to keep the reactor within its normal operating envelope, the safety system takes over and either shuts the reactor down and/or actuates safety features to protect the reactor.

6.3.1 General requirements for NSSS control system I&C

NSSS control systems are key to the daily operation of the plant. They are designed to be automatic wherever possible, with manual override options for operator control if needed. In a traditional PWR, there are over 100 NSSS control loops that provide everything from the monitoring of temperatures and automatic alarms to the full automatic control of the feedwater and main steam systems. In the iPWR world, the number of control loops will be much less than the 100 plus loops mentioned previously. This is partly due to the small nature of the design: fewer tanks, less piping, fewer valves, and less peripheral equipment. And it is partly due to the incorporation of gravity-fed or passive cooling systems. New iPWR designs have made use of the 50 plus years of existing operating experience and have improved and simplified the steam system architecture where possible, making for much simpler control systems. The following sections will discuss the unique nature of iPWR NSSS control instrumentation.

iPWR NSSS control systems will most likely use some of the same isolated safety-related instrumentation signals as found in current PWRs, but some of the instrumentation will be unique to NSSS control, like feedwater pressure and boron tank measurements, and will not require the same nuclear pedigree as safety system measurements. These measurements may likely be accomplished with the same devices that are used in traditional PWRs. The constraint in the nonsafety category instruments is primarily a size constraint. The need for smaller instruments may drive an evolution in the secondary side, nonsafety instrumentation.

On the other hand the fact that NSSS instrumentation does not have the strict Class 1E qualification that safety-related applications demand, may make the transition to advanced state-of-the-art instrumentation a natural evolution. Definition of the nonsafety I&C requirements is still evolving in most iPWR designs, so how the instrumentation is going to be developed to meet these requirements is still uncertain.

6.3.2 NSSS pressure transmitters

In many cases the NSSS pressure signals are generated by the safety system instrumentation. Pressurizer pressure and main steam pressure are two types of signals that come to the NSSS control system through isolators from the safety system. Feedwater pressure, on the other hand, is an example of a nonsafety signal that is dedicated to the NSSS function.

In the case of feedwater pressure and other NSSS dedicated sensed parameters, the iPWR designer has the flexibility to use newer designs and technologies as the evolution pedigree for safety system qualification is not required. But because some of the NSSS signals are located in accessible places with typically more space than the safety system devices, options to use the traditional sensors can be considered.

Many of the traditional suppliers of pressure sensors have modernized the signal processing (to digital in most cases) and are modifying their products with state-of-the-art processing methods. They are transitioning toward digital processing whenever possible. Also, many I&C designers are using wireless setup options whenever possible. (A discussion on wireless vs wired solutions is in Section 6.8.3.) These advances within the traditional methods give the traditional sensor manufacturers an advantage in the NSSS instrumentation market.

Some of the new technology options for NSSS pressure sensors/transmitters are the same options discussed for safety system pressure sensors, as covered in Section 6.2.2.

6.3.3 NSSS level transmitters

Level sensing and indication are a requirement for NSSS control systems. Several NSSS level measurements come from the protection system (safety system) through isolators, such as pressurizer level, steam generator level, and refueling water storage tank level. Other level indications such as boric acid tank level, volume control tank level, and accumulator tank level come directly into the NSSS processing electronics from field sensors and transmitters. It is expected that iPWR NSSS controls will have similar level signal requirements and a similar split between isolated safety signals and direct field signals as large PWRs. Those signals that come direct from the field and do not go through the safety system will have the flexibility of using newer technology for level detection, as the qualification of new devices will not be as rigorous or time consuming as a safety sensor qualification. The new technology discussion in Section 6.2.3 also applies to this section.

Another new technology that might be considered for level sensing is vibrating fork technology. This technology uses the principle of a tuning fork to detect level.

With this technology, vibrating crystals are submerged in the medium, so when the density of the medium changes, the frequency of the crystals change. Used primarily as a level switch the crystal is placed at the level of actuation in a tank, and when the medium transitions that level, the oscillation changes and instigates a switch for an actuation. The actuation in many cases is an indication or alarm, but in some cases the actuation would be a control valve. The Rosemount 2120 level switch uses this technology. Although this technology is largely for level switches, a string of vibrating forks along the length of the vessel/tank could act as a discrete level indication system, not just a switch.

As mentioned in Section 6.2.3, fiber technology offers some efficient solutions for level measurement and would likely be considered as a solution for certain tank and vessel level sensing applications.

6.3.4 NSSS temperature devices

NSSS temperature instrumentation is an open field for both new tech and older devices. The typical NSSS temperature measurements in large PWRs are the following:

- boric acid tank temperature,
- let-down heat exchanger outlet temperature,
- let-down orifice safety valve temperature,
- residual heat removal (RHR) loop return temperature,
- pressurizer liquid temperature,
- pressurizer vapor temperature,
- pressurizer surge line temperature,
- pressurizer spray line temperature,
- volume control tank temperature,
- seal water injection temperature,
- chiller related temperatures,

- charging temperature,
- pressurizer relief tank temperature,
- reactor vessel flange leak-off temperature.

Depending on design, iPWRs may require some of the same measurements. Some of these temperature measurements will originate inside the reactor vessel and will need to have a very reliable and verified temperature device pedigree. This need would tend to indicate the traditional RTD and thermocouple devices, although new technology solutions should not be abandoned, even for the harsher environments.

Some of the temperature measurements will be made away from the harsh environment that exists in or near the reactor vessel and could be an application for new technologies such as MEMs and/or fiber technology. These technologies may offer additional benefits such as easy maintainability, use-and-throw-away devices, better accuracy, more and distributed measurements, and diagnostic capabilities. The new technology is likely to be cheaper, easier to purchase, and more accurate. These economic and engineering benefits may outweigh the pedigree of the traditional RTD but will have to be evaluated on a case-by-case basis. Section 6.2.4 identifies some of the new technology options for temperature measurements.

6.3.5 NSSS flow transmitters

Flow transmitters for NSSS purposes will follow much the same process as the safety system flow devices, only without the safety-related qualification program. The devices will have to be accurate and reliable but will not have the burden of a Class 1E safety qualification.

The types of processes that will need measurement in the NSSS iPWR environment are as follows:

- feedwater flow
- steam flow
- boric acid flow
- reactor makeup water flow
- reactor coolant pump (RCP) seal flow for models with pumps
- let-down flow
- charging flow.

In the NSSS controls environment, accuracy will be most important, especially for feedwater flow. Feedwater flow is an important parameter in the calculation of the calorimetric, a calculation from which the reactor power is assessed. The more accurate the instruments, the more megawatts the plant can produce. Also the improved accuracy afforded by digital control systems will further emphasize the need for more accurate measurement devices to pair with the more accurate digital processing.

As some of these flow measurements will need to take place close to the reactor vessel, the devices may have to be more radiation, temperature, and pressure hardy than traditional NSSS sensors. This need may drive designers to newer technology, such as the ceramic-based MEMs sensor mentioned in the safety system discussion. Different variations of transit time methods and temperature-based flow methods may find their foothold here with NSSS-based applications. Further discussion on new technologies for flow measurement can be found in Section 6.2.5.

6.4 BOP instrumentation

BOP instrumentation provides additional measurements that are beyond the scope of NSSS instrumentation. BOP instrumentation deals with systems associated with the turbine generator and the electric power making part of the plant. The instrumentation parameters are the same (pressure, temperature, level, and flow), but the quality pedigree is not as rigid. In many cases the secondary side vendors supply the instrumentation and control required for their supplied equipment. For example, most instrumentation required for turbine generator control is provided by the supplier of the turbine generator system. In many traditional PWRs the vendors supply the equipment and all related instrumentation and controls.

Condenser-related measurements are another BOP instrumentation category. The condenser itself requires temperature, pressure, and level measurements. Additionally, condenser cooling systems require temperature and flow measurements. The pumps and piping that recycle the condenser water into the feedwater system need flow measurements to control flow through flow control valves.

Some heating, ventilation, and air conditioning (HVAC) system measurements may also be grouped into the BOP instrumentation category. To the extent that cooling water is recycled to provide cooling for the HVAC in certain plant areas, the HVAC-related cooling systems will need to have control systems for cooling water control valves. The HVAC systems may also need to control or throttle air flow and may require air flow or temperature sensors to facilitate this function.

It is expected that some of the same instrumentation used in traditional PWRs will be used for iPWR BOP instrumentation. In some cases, size constraints will demand a newer device or a redesigned device, but since the temperature, pressure, and radiation environments for BOP instrumentation are in general not as extreme as those for RPS and NSSS environments, the flexibility to use newer technology will allow for more instrumentation options.

6.5 Diagnostics and prognostics

The traditional instrumentation in a nuclear power plant consists of temperature, level, flow, pressure, and power. Almost all actuation and monitoring signals are derived from these five measurements. Recently, however, diagnostic measurements have become increasingly more important. With the advent of new unobtrusive ways to determine the health of plant equipment, the diagnostic measurement field is becoming the fastest growing instrumentation field in nuclear plants. The capability to catch a problem before it leads to a failure is a powerful capability in a nuclear plant. It is expected that iPWRs will take advantage of the latest in diagnostic technology, especially in the field of prognostics.

Diagnostic signals are not likely to become safety-related measurements, but they will become more prevalent in future large-scale and small-scale reactor designs. For large reactors, they are important for license extension justification; for iPWRs, they are important for containment entrapped systems, as the access to containment (at power) during an operating cycle is not possible in several iPWR designs.

Diagnostic technology has provided a significant advantage for nuclear plants over the past few years. Many decisions about current and future system health have been determined on diagnostic evidence. Currently, one of the most common of diagnostic measurements is temperature, especially with rotating equipment. Thermography "guns" have provided additional insight for electrical components that may be malfunctioning. Oil samples are another precursor to failure in rotating equipment.

Vibration measurements on rotating, oscillating, or even stationary equipment can provide valuable insight on the equipment's health. While all these measurements are valuable to a large traditional PWR, most of them require a maintenance technician to go out to the equipment and take measurements or samples. The key diagnostic measurement in an iPWR is going to be embedded or in-place sensors with automatic processing and indication, as design and operating constraints may prevent the human interface during power operations.

A manufacturing technique called shape deposition manufacturing (SDM) is a technique for embedding thin film sensors or fiber optic sensors for the continuous measurement of temperature and strain in a metal vessel or casing. Reactor vessels or other critical structures may have these sensors embedded during manufacturing for detection of increased strain or precursors to cracking (Li et al., 2000). Another technology that will be considered for some equipment is the use of fiber sensors for strain measurement (www.lunainc.com).

The field of diagnostic measurement and testing is developing quickly. The advantages for iPWRs are obvious. With the iPWR paradigm for less staff and the fact that between cycle maintenance and testing may not be possible, diagnostic and prognostic tools are necessary. Embedded sensors and the automation of these sensors are important for iPWRs.

Good resources for prognostic information can be found in IEEE publications and IAEA publications. Grant work funded by the DOE has produced several approaches that have value (Bond, 2011).

6.6 Processing electronics

Downstream of the measuring devices are the processing electronics. The processing electronics are necessary to convert the primary sensed measurement to the final measurement. The processing function can be performed by analog devices, digital hardware, and/or digital software. Most large nuclear plants use analog processing since they were designed and built in the 1970s and 1980s; however, when obsolescence and maintenance issues force replacement of the analog hardware, many plants are upgrading to digital technology. The digital systems that are most widely available are based on microprocessors that use software to implement the application needed.

Less widely used, but gaining in popularity for safety systems, are digital systems that are based on hardware logic gates that do not have executable software running in the final implementation. These devices include, but are not limited to, application specific integrated circuits (ASICs), field programmable logic devices (FPGAs), and complex programmable logic devices (CPLDs). Even though these logic devices do not have run-time executable software, they are still considered software-based programmable devices by the Nuclear Regulatory Commission (NRC), since software tools are used to program these logic devices.

In traditional PWRs the nuclear safety quality transmitters perform the translation from the sensing element to the final measured signal in the transmitter itself, which is usually located near the sensing element, sometimes on the pipe above the primary element or near sensing lines piped into the transmitter device. Further processing is done remote to the transmitter location in a mild ambient environment. Some of the remote processing involves the filtering and rescaling of the measurement.

The sensing element and the initial conversion to an electronic signal is usually analog (e.g., 4–20 mA), as most measured parameters start as an analog process. At some point in the processing, the signal can be translated to a digital signal through conversion in an analog-to-digital converter (ADC). In many cases, placing the ADC in the vicinity of the measurement is not possible due to the harsh environmental conditions. In most cases, this conversion will happen in an electronics cabinet that is located in a mild ambient environment.

Analog processing is well known and well understood in nuclear plant environments. It is a reliable technique and has been used for many years. It is appealing to both plant and regulatory agencies, as it has a demonstrated safety record. Analog devices are usually very simple in design and only implement one function. This allows the system to be fully tested and minimizes the probability of a latent design defect. Digital systems, on the other hand, can implement many functions and are far more complex. Except for very simple digital devices, 100% testability is nearly impossible so a digital common mode/cause failure (CCF) must be evaluated and addressed.

For safety-related systems the NRC (Branch Technical Position [BTP 7-19]) requires one of the following design attributes:

- A system is sufficiently simple such that every possible combination of inputs and every possible sequence of device states are tested and all outputs are verified for every case (100% tested).
- If sufficient diversity exists in the protection system, then the potential for CCF within the channels can be considered appropriately addressed without further action.

Justifying to the NRC that the system has sufficient diversity has been a challenge and introduces risk with upgrading to digital systems.

Digital technology has been adopted and used almost exclusively in all nonnuclear industries and digital systems are being selected for replacement in the nuclear industry for non-safety related applications. The ease of making a change by altering a line of code, or touching a touch screen has usurped many of the advantages of analog processing. Digital processing through programming is easier to change, easier to adjust (tuning parameters), easier to troubleshoot, and more end-user friendly than many analog processing schemes.

For nuclear safety-related systems, digital technology has been very slow to be adopted due to the added regulatory risk and cost to address the complexity and common cause failure (CCF) issues.

CCF is defined as a failure that occurs in a device or software and affects every implementation of that device or software. For example, an error in a line of code in computer software that is in all redundant channels of a protection system could cause all redundant channels to fail such that the protection system would not actuate when required. The potential for common mode/cause failures is a primary concern for safety systems. Digital systems are much more complex than analog systems and are not 100% testable in most cases and require a diverse solution. For insight into how the regulatory agencies evaluate CCFs, see US Nuclear Regulatory Commission Standard Review Plan NUREG 0800, Appendix 7.0-A, "Review Process for Digital Instrumentation and Control Systems" (US Nuclear Regulatory Commission, n. d.). This regulation applies to the current large traditional nuclear power plants and iPWRs under design and construction.

One of the ways that some nuclear plants and iPWR designers are addressing the regulator's concern for CCFs is to implement a diverse digital logic system (not executable software). One of the systems at the forefront of this digital technology is field-programmable gate array (FPGA) technology. FPGA technology uses logic blocks, like AND gates, OR gates, flip-flops, and clock signals to accomplish the signal processing. This digital logic is programmed on an integrated circuit that has an array of logic devices. While software tools are used to configure the FPGA with the desired application, once the FPGA has been configured, its function and configuration are fixed, and it can perform its function with no software interface required. The programming process is analogous to programming an EPROM.

Although the programmable software necessary to program the FPGA circuit carries with it the CCF concerns present with any programmable safety system software, implementation diversification is a way to mitigate these CCF concerns. In the FPGA case, there are three basic types of FPGA implementation technologies; static RAM, antifuse (one time programmable (OTP)), and flash EPROM. There is software diversity in the toolsets and hardware diversity in the FPGA gate structures between the FPGA technologies that can be used to address CCFs in a digital system. For example, in a typical protection system, there are four redundant channels for a process variable. If a static RAM FPGA is used in two of the redundant channels and an antifuse or flash-based FPGA is used in the other two redundant channels, then a failure in the static RAM implementation would not affect the other two redundant channels that use the other FPGA Technology.

Another concern with software-based digital systems is the control and security of the code. To be approved for nuclear safety systems, the software and configuration of the system must be documented and controlled. The code must be controlled to ensure it is not tampered with and will not allow unauthorized versions to be implemented. To change an analog system, the circuit must be physically altered. This change would involve gaining access to the facility, the cabinet, and the circuit board. Software changes and the introduction of malware, on the other hand, can be accomplished remotely, and, while access can be controlled, it is not as straightforward as the physical control of an analog system.

The FPGA technology described previously does not operate on executable software. After the circuit board is programmed with the application specific logic, it is essentially a fixed digital logic circuit. Similar to an analog system, the FPGA system can be designed such that the FPGA circuit board is required to be removed from the system to modify its logic and thus cannot be modified remotely.

Safety-related digital systems are required to be deterministic where the behavior is entirely determined by its initial state and inputs. Software-based solutions can be deterministic but require special operating systems and hardware. FPGA logic design uses state machines and when it is in one predefined state at a particular point in time, it is a deterministic solution. Additionally, since the FPGA uses logic gates, many tasks can be done in parallel, which allows them to offer higher processing power and speeds. FPGAs can provide fast response times and performance improvement over a microprocessor-based system, and they require less power to operate.

FPGA technology is a great solution for iPWRs as it offers solutions for the regulator's concerns, and it is cutting-edge technology that offers many advantages over analog solutions and software-based digital solutions.

6.7 Cabling

Cabling is the glue that connects the I&C signals together. There is cabling to the transmitter, cabling from the transmitter to the rack electronics, and cabling from the rack electronics to the end device. Recent regulatory interest has resulted in a new field of diagnostic measurements to verify cable integrity. Some of these techniques advocate proactive measures with sensors embedded in the cable itself; other methods require baseline tests for future comparisons, so cable health and monitoring must be considered up front in the design.

Periodic cable testing is likely to be a requirement for iPWRs. Most likely, the iPWRs will take a baseline set of tests and then retest at regular intervals and compare results. Some of the tests that are recommend today for I&C cable testing are as follows:

- time domain reflectometry (TDR)
- reverse TDR
- frequency domain reflectometry (FDR)
- insulation resistance (IR)
- inductance, capacitance, and impedance measurements
- visual inspections;
- insulation hardness (indenter modulus)
- partial discharge, for the higher-voltage cables.

From these tests, it is possible to determine the location of hard faults, degraded areas, moisture intrusion, insulation degradation, and poor cable connection locations. The location and the identification of degraded areas of cable are key factors in timing the replacement of cable sections in advance of failure.

In addition to cable testing, embedding sensors in the cable insulation to determine insulation aging and degradation is a new technology that may be incorporated in the manufacturing process of cables in the near future. US NRC Regulatory Guide 1.218 (US NRC, 2010) summarizes the current regulation on periodic cable testing and health/aging management of the plant's cables. This guide covers I&C cables and medium voltage cables that support large pumps and valves. Some sections in NUREG/CR-7000 provide additional guidance on cable health and monitoring. The iPWR design should incorporate cable testing features to ensure proper cable health and aging management for the future.

Devices associated with cabling must also be considered in designing robust cabling systems. It is likely that cable and wiring connectors, as well as the containment penetrations, will need new engineering design, and development as the expected environments and geometries will be very different from those traditionally encountered.

6.8 Future trends and challenges

The small scale and unique environments of iPWRs open the instrumentation field to new and innovative solutions. The temptation will be to fall back on what has been used before, and in scenarios where this works, perhaps this will be the path of least resistance. But because of design peculiarities and nonstandard physical environments, new instrumentation or redesigned instrumentation systems will likely be required. In addition to qualifying these new designs and new approaches for new measurement technologies, the iPWR design engineer will need to coordinate with the regulator to accommodate the regulator's concerns and requirements at the same time.

The sections below identify some of the known challenges and trends in the I&C world for the iPWR designer.

6.8.1 Licensing challenges in advanced SMR design

6.8.1.1 Overview

In the evolution toward iPWRs, designers are moving toward the use of more complex information and control systems with significantly more sensory instruments to assist in operational monitoring and decision-making. In some cases, local automation for smaller iPWRs is being proposed that would reduce, or even remove, the need for local operator presence. This means that I&C capabilities in nuclear facilities are migrating toward those that are more typical of conventional industrial sectors (e.g., aerospace, transportation, and conventional power generation) but are challenging the traditional regulatory requirements and guidance in the nuclear sector.

The regulatory reviews in the United States and other countries have focus areas that are more specific to iPWR reviews than the traditional larger facilities. Some of these areas are covered in the following sections.

6.8.1.2 Use of probabilistic risk (safety) assessments in licensing iPWR's

Many countries have regulatory mechanisms in place for a proponent to apply risk informed insights when addressing regulatory requirements, as documented in IAEA (2016). These risk informed insights are used to develop technical and regulatory cases (via a graded approach) that demonstrate the intent of regulatory requirements in the design. For I&C architectures, risk-informed design becomes more challenging when considering the sharing of signals under new operating models, or with much higher levels of automation in the control philosophy.

The licensing process for digital I&C systems in the United States has primarily been based on deterministic engineering criteria and is supported by reliability modeling. If an advanced digital system is to be credited in the plant probability risk assessment (PRA) for supporting any number of risk-informed decisions, then the quality of the PRA modeling must be established by showing that applicable NRC regulatory requirements are met. The challenges in moving forward with higher usage of PRA for use in risk informed applications include (1) the challenges of modeling digital I&C systems in a PRA; (2) the lack of experience of modeling digital I&C systems, including modeling self-diagnostic and online monitoring (OLM) features, as part of product reliability model; and (3) the necessary features of digital I&C systems for use in reliability and PRA modeling that credit for self-diagnostics and OLM features. Ongoing work has been documented in DOE (2019a,b) and addresses new methods of quantitative reliability modeling and assessment of automated coverage for use in regulatory acceptance.

In summary the major challenges in quantifying digital systems reliability for the diverse types of technologies and system architectures used in iPWR design are related to the qualification of analytical tools for estimating failure frequencies for I&C systems and addressing uncertainties in these estimates.

6.8.1.3 Advances in safety system end-state architecture through simplification

To achieve significant iPWR operations and maintenance (O&M) cost reductions, in addition to resolving obsolescence and reliability issues from previous legacy I&C safety systems, an advanced digital architecture is being developed for a number of iPWR designs. This architecture must be capable of enabling the following types of cost savings:

- reduced technical specification and associated surveillance tests;
- reduced maintenance and testing costs through reduced instrumentation calibrations, like the use of memory checks and self-diagnostics in lieu of testing;

- capabilities to convert a substantial portion of the field logic devices to control and protection software, negating the need for hardware devices like actuators, as used in current nuclear plants;
- high reliability through fail-over redundancy;
- minimal platforms to reduce the number and type of maintenance procedures, qualified maintenance tasks, work order models, etc.;
- reduced design basis operational events through safer reactor plant design, thereby reducing generation losses, safety challenges, and regulatory impacts.

DOE (2018) documents the development of an end-state architecture meeting the requirements listed previously and including the following characteristics:

- A plant overview architecture in the main control room that uses nonsafety video display units (VDU) to control the plant in all operational modes including safety system operation and all plant processes using prioritized alarm methods.
- A high level of I&C architecture integration, while maintaining sufficient segmentation to comply with safety criteria.
- An I&C architecture that mitigates or eliminates common cause failure (CCF) even with shared hardware resources and common software.
- Significant simplification in safety system architecture through the use of field-programmable gate array (FPGA) and other similar logic devices, as currently used in other industries such as aerospace and petrochemical. These devices lend themselves to more robust methods of verification and validation as document in Gibson et al. (2019) involving model-based analysis tools.

6.8.1.4 Protection against common cause failure in iPWR I&C design

Branch Technical Position BTP-7-19 (NRC, 2016) has been updated on a regular basis by the USNRC to provide analysis methods for the review of susceptibility to common cause failure (CCF) in digital systems by verifying the following:

- Adequate diversity has been provided in a design to meet the criteria established by NRC guidance.
- Adequate defense in depth has been provided in a design to meet the criteria established by NRC guidance.
- Displays and manual controls for plant critical safety functions initiated by operator action are diverse from digital systems used in the automatic portion of the protection systems.

The analysis in NRC (2016), addresses four echelons of defense including the control systems, the reactor trip system, the engineered safety feature system, and monitoring and indication system, which follows a best-estimate analysis method. This analysis has been performed frequently for digital upgrades to current operating plants and to advanced plants including iPWRs going thru the licensing process. IAEA (2016) addresses similar requirements for CCF analysis for the international community.

In the case of iPWR design, simplification that can be introduced as discussed in Section 6.8.1.2 can also provide defensive measures to address CCF that may allow the elimination of the diverse shutdown system, which has been required in a number of traditional operating plants to mitigate CCF. The iPWR digital architecture options today have much more robust capabilities to address CCF by hardware/software design. At the same time the use of shared systems between modules and the entry of larger numbers of digital devices into the supply chain of all systems (called embedded digital devices [EDD]), requires additional attention in the analysis phase for CCF.

6.8.1.5 Safety classification of passive nuclear power plant electrical systems

As documented in NuScale (n.d.) and the associated vendor submittals, if an iPWR was designed so that no electrical equipment was "essential" and did not meet the definition of Class 1E (i.e., the reactor plant design did not include safety-related equipment dependent on electrical power), then the design would not require Class 1E AC or DC power systems. Where no Class 1E equipment is used, the basic requirements for qualifying Class 1E electrical equipment and interfaces, which are provided in IEEE Std. 323 (IEEE 323-1974, 1974), are inapplicable. In NuScale (n.d.) the applicant provided this justification and the NRC approved.

NRC Regulatory Guide 1.97 (NRC, 2006) defines postaccident monitoring instrumentation that monitors Type A, B, and C variables and the associated Class 1E electrical system that powers these.

- Type A variables provide the primary information required to allow main control room operators to take manual actions for which no automatic control is provided.
- Type B variables provide primary information to the control room operators to assess plant safety functions.
- Type C variables provide primary information to the control room operators to indicate the potential for breach or the actual breach of fission product barriers (e.g., fuel cladding, reactor coolant primary boundary, and containment pressure boundary).

As part of its review, the NRC staff will consider whether the safety system design to provide accident monitoring instrumentation would require instrumentation to be powered by a Class 1E electrical system for type B and C variables. In the case of the iPWR described in NuScale (n.d.), instruments that monitor Type A variables were not part of the iPWR design, due to the numerous passive safety features inherent in the design.

As documented in NuScale (n.d.) the NRC found that the augmented design, qualification and QA provisions of the non-Class 1E power sources were an acceptable alternative to the guidance in Reg Guide 1.97 for Class B and C variables. This is the first case documented for a reduction in qualification of these devices and represents a major cost savings while maintaining quality due to the passive design features.

6.8.1.6 Cybersecurity for iPWRs

As documented in NRC Regulatory Guide 5.71 (NRC, 2010) and in IEC-62645 (International Electrotechnical Commission (IEC), 2019) including the Design Basis Threat (DBT), the functions requiring protection include safety-related and important-to-safety functions, security functions, and emergency preparedness functions (including offsite communications). The NRC guidance specifies that "…digital computer and communication systems and networks associated with these functions must be protected from cyber attacks that would adversely impact the integrity or confidentiality of data and software; deny access to systems, services or data; or provide an adverse impact to the operation of systems, networks and associated equipment."

As documented in Section 6.8.1.3, simplification in the safety system architecture for iPWRs also contributes in many cases to the reduction in susceptibility to cyberattacks. The use of FPGA and similar technologies provide simpler architectures that are more resistant to cybersecurity vulnerabilities since there is no programmable software involved in the operation of the digital assets.

Physical security related to the instrumentation of iPWRs will also need special attention. These security issues will need to address (1) the use and application of wireless technology inside the facility (refer to Section 6.8.4 below) and (2) the adequacy of computer security measures between modules and remote control locations. Both physical security and cybersecurity are emerging issues in many recent production, generation, and transmission designs, with iPWRs being no different.

6.8.2 Safety system instrumentation: Old versus new

Traditional PWR safety qualified pressure transmitters are devices that are able to supply an electronic signal that is proportional to an absolute pressure measurement or a differential pressure measurement. Nuclear versions of these devices were originally manufactured and qualified in the 1970s and 1980s but have had various features modernized for recent advancements in electronics and materials. The most important feature of the traditional device is the qualification of that device. A Class 1E safety qualification for any transmitter means that the device has been through a series of tests and longevity assessments that prove it is operational and survivable for nuclear conditions. Specifically, it means that the device can meet the radiological, temperature, pressure, and longevity requirements for a nuclear reactor environment. Specifics on the qualification requirements for Class 1E equipment is contained in IEEE 323-1974 (IEEE 323-1974, 1974), 1983 (IEEE 323-1983, 1983), and 2003(IEEE 323-2003, 2003) and IEEE 344-1975 (IEEE 344-1975, 1975), 1987 (IEEE 344-1987, 1987), 2004 (IEEE 344-2004, 2004), and 2013 (IEEE 344-2013, 2013).

The issue with the traditional transmitters is that they were designed for an environment unlike the one that will be experienced in an iPWR design. In a traditional PWR the containment environment, where most safety classified traditional pressure transmitter electronics are mounted, is a large, airy environment with relatively low ambient temperature and pressure, and with easy access for maintenance and repair of equipment, whereas most iPWR containments are designed as capsules that closely envelop the reactor vessel. These capsules have limited access with small enclosed environments of air, vacuum, or water that may include high ambient temperature and pressure. The location and mounting of these pressure transmitters will be quite different from the traditional approach, and the limited access for maintenance or repair will need to be factored into the design.

In some cases, there will be no traditional pipe mountings for the sensing elements, and the use of sensing lines may not be possible in all cases. These hurdles in instrumentation, going from the traditional PWR to the iPWR, will necessitate the redesign of traditional models and/or the development of new technology.

The quandary in which iPWR instrumentation designers find themselves is that the least risk-adverse approach, from a licensing and schedule perspective, is the one that uses instrumentation with a qualification pedigree and has worked in the past for traditional PWRs. However, the many changes in iPWR mechanical and physical design mean that the traditional devices may not function correctly or fit physically in this new environment and design. This dilemma forces the I&C designer to look outside the "traditional box" for answers.

New technologies, discussed in other sections, offer some solutions and some problems. Some new technologies offer smaller packaging, submersible options, more reliable manufacturing, flexible-mounting options, remote electrical processing, and fewer maintenance-intensive options. It is easily seen how these features would be welcomed by the iPWR I&C designer. The major drawback is the lack of device qualification and the lack of longevity field data that substantiates the survivability of these devices in the actual environment. As with the generation II reactors that populate many countries today, the iPWR I&C systems will undoubtedly experience the same growing pains as those experienced in the 1970s, as new designs are tried, and lessons are learned.

As mentioned previously, digital technology is expected to expand greatly in iPWRs. The same concern for common mode and common cause failure potential with software-based digital systems will continue to be an issue for iPWRs as it is with the large PWRs. The FPGA technology (Section 6.6) and technology like it that provide the reliability of digital devices without the need for executable software has the appeal to take the forefront of safety and NSSS system processing in iPWRs.

6.8.3 Instrumentation in nonsafety systems

The nonsafety instrumentation systems are wide open for technological advancements. Software controlled nonsafety digital systems are attractive to the utility owner and do not hold as much concern with the regulator as safety systems. (US Regulator action is trending to more demanding requirements for NSSS digital control systems, as common mode failures in NSSS control systems have implications on the safety system response of the plant.)

The new high-tech devices such as fiber-based sensors and ultrasonic sensors have the attractiveness for use throughout the NSSS and BOP systems. The small scale of many iPWRs provides an opportunity to use more state-of-the-art devices with better accuracy, better ease of installation, lower maintenance, better availability, etc.

The I&C designer's dilemma for instrumentation for the iPWR nonsafety systems is whether to use new technology with less of a nuclear track record or to use the traditional nuclear instrumentation with years of nuclear experience and pedigree. Some iPWR BOP systems and some iPWR NSSS systems could utilize the same instrumentation as a large traditional nuclear plant. The degree of similarity of the iPWR BOP systems to the existing large PWR BOP systems would determine the potential to use traditional instrumentation; however, the new and unique nature of many iPWR designs would create the opportunity for new instrumentation applications. It is likely that new iPWR nonsafety designs will utilize some new technology already on the market, but not currently used at the existing large PWRs.

6.8.4 Wireless versus wired solutions

Wireless systems are popular in today's high-tech environment, but they will always pose a problem for a nuclear safety system. The concern for wireless safety systems is mostly security. As anyone who works with wireless systems can verify, any system that is wireless is vulnerable. It is vulnerable to jamming, it is vulnerable to hacking, and it is a potential interferer for other signals. Even with encryption techniques and spread spectrum techniques, it is unlikely that the current technology in wireless systems will be implemented for primary signal communications in safety systems (Keebler and Berger, n.d.; Korsaha et al., 2009).

Another constraint arguing against wireless communications in safety systems is channel separation, including safety to nonsafety separation. With wireless communication, channel separation is more difficult to prove/ensure than wired communications.

Wireless networking and programming of nonsafety I&C devices is a different matter. The wireless HART system is a wireless communication system employed currently in large PWR plants, which allows transmitters that are equipped with the system to communicate information to an end user or device. This programming and control system protocol is encrypted and authenticated and has little risk of being hacked or sabotaged. Since it is not responsible for the primary safety communication (for safe plant shutdown), it is free to be used throughout NSSS and BOP control systems. This technology allows plant maintenance personnel to set up, tweak, and network transmitters remotely, and will most likely be used regularly in nonsafety iPWR systems, for ease of access reasons (Wireless HART Technology, n.d.).

BOP instrumentation has the most flexibility for wireless communications. Although the wireless solution still has to be tested for signal integrity, immunity, and noninterference with other signals, the potential to use wireless solutions for sensor-to-processing applications is greater in BOP systems than in the NSSS or safety system applications.

Wireless voice communications have been widely used in nuclear power plants for many decades. From the handheld radio to cell phones, wireless communications are a continuing trend. As many applications in the new iPWRs will be digital and digital on a smaller scale than ever before, there will be more intense testing and qualifying of wireless voice communication devices than before. The potential of a wireless voice communication device to interfere with a safety or control signal is a condition that must be fully tested and verified in any nuclear plant environment. The wireless communication of voice or ancillary data is very different from the wireless communication of safety parameter signals. While wireless voice communications are used and will continue to be used in the future, complete testing of these wireless systems will also continue to a necessity for electromagnetic compatibility (EMC) reasons.

Wireless plant computer data are already a popular technique in power plants. As long as the plant computer data are not relied upon for safety system processing, wireless plant computer data are a beneficial application. Currently, diagnostic information, such as vibration data and temperature data, is sensed in the field and transmitted through wireless hubs to a computer system. It is expected that this trend will continue in the iPWR environment (Keebler and Berger, n.d.; Korsaha et al., 2009).

In summary the I&C designer will need to weigh the benefits of wireless capabilities, with the associated reactor safety issues and wireless EMI/RFI issues, to determine the extent of wireless use in new iPWRs. To aid the I&C designer in this area, the following publication documents the concerns and issues in wireless capabilities in nuclear plants: Electric Power Research Institute (EPRI) report TR-102323-R3 published in November 2004, and EPRI report ID# 1011960 published in 2005.

6.9 Conclusion

iPWRs are a new and elegant incorporation of 50 years of operation in nuclear power plants. Lessons learned and evolutionary improvements are being and have been incorporated into the new designs. The instrumentation design is in its design phase for many iPWR developments at the current time. Many traditional I&C options are still being considered for iPWRs, but some of these traditional options will not work in many of the newer iPWR designs, either because of the geometries involved or because of the environment. It is up to the iPWR I&C designer to find the appropriate I&C solutions to the challenges of pressure, flux, level, temperature, and flow measurement and the associated processing. The opportunity to use state-of-the-art technologies that offer improved accuracy, ease of installation, ease of maintenance, and less drift should not be passed up, even if it means new qualification programs. Economics will bear out the use of new technologies, but only if a long-term view is used. The next few years should tell the story.

References

Bond, L., 2011. Prognostics and life beyond 60 y ears for nuclear power plants. In: 2011 IEEE Conference on Prognostics and Health Management.

Britton Jr., C.L., Roberts, M., Bull, N.D., Holcomb, D.E., Wood, R.T., 2012. Johnson Noise Thermometry for Advanced Small Modular Reactors. September 2012, ORNL/TM-2012/346.

Chang, Y.-T., Yen, C.-T., Wu, Y.-S., Cheng, H.-C., 2013. Using a fiber loop and fiber bragg grating as a fiber optic sensor to simultaneously measure temperature and displacement. Optomechatronics.

Dickerson, B.D., Davis, M.A., Palmer, M.E., Fielder, R.S., 2009. Temperature-compensated fiber optic pressure sensor tested under combined high-temperature and high fluence. In: Sixth American Nuclear Society International Topical Meeting on Nuclear Plant Instrumentation, Control, and Human-Machine Interface Technologies, NPIC&HMIT 2009,

Session 'Diagnostics and Predictive Maintenance', Knoxville, Tennessee, April 5–9, 2009, on CD-ROM. American Nuclear Society, LaGrange Park, IL.

DOE, June 2018. Light Water Reactor Sustainability Program Report INL/EXT-18-45683, Revision 0, Strategy for Implementation of Safety-Related Digital I&C Systems.

DOE, June 2019a. Light Water Reactor Sustainability Program Report, INL/EXT-19-54251, Technical Specification Surveillance Interval Extension of Digital Equipment in Nuclear Power Plants.

DOE, September 2019b. Light Water Reactor Sustainability Program Report, INL/EXT-19-55799, Addressing Nuclear I&C Modernization Through Application of Techniques Employed in Other Industries.

EPRI, May 26, 2011. Plant Engineering: EPRI Document on Fiber Bragg Grating Monitoring of Flow Accelerated Corrosion. see EPRI website for an abstract free to EPRI members. Product ID: 1023189.

EPRI, 2012. ref: EPRI doc on Fiber Bragg Grating monitoring of Flow Accelerated Corrosion.

Gibson, M., Elks, C., Tantawy, A., Hite, R., Gautham, S., Jayakumar, A., Deloglos, C., July 2019. Final Technical Report, M2CA-15-CA-EPRI-0703-0221, Achieving Verifiable and High Integrity Instrumentation and Control Systems Through Complexity Awareness and Constrained Designs.

IAEA, 2016. SSG-39, Design of Instrumentation and Control Systems for Nuclear Power Plants.

IEEE 323-1974, 1974. IEEE Standard for Qualifying Class 1E Equipment for Nuclear Power Generating Stations. Institute of Electrical and Electronics Engineers, Inc.

IEEE 323-1983, 1983. IEEE Standard for Qualifying Class 1E Equipment for Nuclear Power Generating Stations. Institute of Electrical and Electronics Engineers, Inc.

IEEE 323-2003, 2003. IEEE Standard for Qualifying Class 1E Equipment for Nuclear Power Generating Stations. Institute of Electrical and Electronics Engineers, Inc.

IEEE 344-1975, 1975. IEEE Recommended Practice for Seismic Qualification of Class 1E Equipment for Nuclear Power Generating Stations. Institute of Electrical and Electronics Engineers, Inc.

IEEE 344-1987, 1987. IEEE Recommended Practice for Seismic Qualification of Class 1E Equipment for Nuclear Power Generating Stations. Institute of Electrical and Electronics Engineers, Inc.

IEEE 344-2004, 2004. IEEE Recommended Practice for Seismic Qualification of Class 1E Equipment for Nuclear Power Generating Stations. Institute of Electrical and Electronics Engineers, Inc.

IEEE 344-2013, 2013. IEEE Recommended Practice for Seismic Qualification of Class 1E Equipment for Nuclear Power Generating Stations. Institute of Electrical and Electronics Engineers, Inc.

International Electrotechnical Commission (IEC), 2019. Standards IEC-62645, Nuclear Power Plants—Instrumentation and control systems—Requirements for Security Programmes for Computer-Based Systems.

Keebler, P., Berger, S., Managing the use of wireless devices in nuclear power plants. IN-Compliance Magazine. www.incompliancemag.comindex.php?ophion=com_content& view=article&id=855:managing-the-use-of-wireless-devises-in-nuclear-power-plants& catid=26:design&itemid=130.

Korsaha, K., Holcomb, D.E., Muhlheim, M.D., Mullens, J.A., Loebla, A., Bobrek, M., Howlader, M.K., Killough, S.M., Moore, M.R., Ewing, P.D., Sharpe, M., Shourbaji, A.A., Cetiner, S.M., Wilson Jr., T.L., Kisner, R.A., 2009. Instrumentation and Controls in Nuclear Power Plants: An emerging Technologies Update. NUREG/CR 6992 (October 2009).

Li, X., Golnas, A., Prinz, F.B., 2000. Shape deposition manufacturing of smart metallic structures with embedded sensors. In: Proc. SPIE 3986, Smart Structures and Materials 2000: Sensory Phenomena and Measurement Instrumentation for Smart Structures and Materials, 160, June 12, 2000. https://doi.org/10.1117/12.388103.

McGregor, D.S., Ohmes, M.F., Ortiz, R.E., Sabbir Ahmed, A.S.M., Shultis, J.K., 2005. Micropocket Fission Detectors (MPFD) for In-core Neutron Flux Monitoring. S.M.A.R.T. Laboratory, Department of Mechanical and Nuclear Engineering, Kansas State University, Manhattan, KS. Available from 18 August 2005. http://www.mne.kso.edo/~jks/papers/MPFD/pdf.

Mohindra, V., Vartolomei, M.A., McDonald, A., 2007. Fission chambers for CANDU SDS neutronic trip applications. In: Presented at the 28th annual Canadian Nuclear Society (CNS) Conference, June 2007.

NRC, 2006. Regulatory Guide 1.97, Rev 4, Criteria for Accident Monitoring Instrumentation for Nuclear Power Plants.

NRC, 2010. Regulatory Guide 5.71, Cybersecurity Programs for Nuclear Facilities.

NRC, 2016. Branch Technical Position BTP-7-19, Rev. 7, Guidance for Evaluation of Diversity and Defense-In-Depth in Digital Computer-Based Instrumentation and Control Systems.

NuScale, n.d. Safety Classification of Passive Nuclear Power Plant Electrical Systems, Safety Evaluation for Topical Report 0815-16497, Revision 1.

Sang, A., Gifford, D., Dickerson, B., Fielder, R., Froggatt, M., 2007. One centimeter spatial resolution temperature measurements in a nuclear reactor using Rayleigh scatter in optical fiber. In: Proceeding of Third European Workshop on Optical Fiber Sensors (EWOFS07), Napoli, Italy, July 4–7, (2007) Available from: http://lunainc.com/wp-content/uploads/2012/08/Temperature-Measurements-in-a-Nuclear-Reactor.pdf.

US NRC, January 2010. Regulatory Guide 1.128, NUREG/CR-7000, sections 3 and 4.6.

US Nuclear Regulatory Commission, n.d. Review Process for Digital Instrumentation and Control Systems, Standard Review Plan NUREG 0800, Appendix 7.0-A.

Wireless HART Technology, n.d., http://www.hartcomm.org/protocol/wihart/wireless_overview.html.

Human-system interfaces in small modular reactors (SMRs)

Jacques Hugo
Jacques Hugo Associates, Pretoria, South Africa

7.1 Introduction

Current and emerging nuclear power plant (NPP) design strategies include ambitious goals of reliability and safety. It is expected that these goals would be met partly by judicious implementation of new materials, new technologies, and new concepts of operations. The general consensus is that current levels of safety and performance could only be enhanced by smaller reactor units designed for a high level of passive or inherent safety in the event of malfunctions and that efficiency could only be improved by extending the energy output of the reactors to more diverse customers. To meet these challenges, several emerging designs for SMRs make provision for non-electrical applications in their concepts of operations. These new concepts require advanced technologies like integrated instrumentation and control systems to support thermal-hydraulic processes associated with different fuels, different reactor coolants, and different product streams (e.g., steam, process heat, electricity generation, and fuel reprocessing in various combinations). New technologies are also required to support new concepts of operations like modular plant operation, high levels of automation, and reduced staffing.

SMRs can reduce operating and maintenance cost through innovative concepts of operations, which include higher levels of automation, reduced staffing, online refueling of separate modules, remote monitoring of operations, and advanced human-system interface (HSI) technologies to support error-resistant operations. HSIs will also play an important role as part of resilient control systems that aim to limit the occurrence and effect of human error while also contributing to overall plant performance. However, these advantages will not be achieved without far more rigorous attention to the roles and functions of both humans and systems. Because of this, designers of the new generation of power plants need to include human factors considerations in the overall engineering process right from the start of the project and specifically in the selection and deployment of HSI devices. In many older NPPs, human factors considerations were often an afterthought in design (Three Mile Island being a classical example). Designers of new power plants however have an opportunity to eliminate many of the errors of the past by integrating human factors early in the design process and by following the excellent guidance currently available from the sources listed in the Reference section of this chapter.

Unlike the past generation of analog control and display devices found in all current NPPs, modern HSIs may not necessarily be designed exclusively for the nuclear industry. Many of the new devices are consumer appliances that are functionally suitable for deployment in the control room or outside in the plant. However, although most modern appliances are reasonably robust, there may be several instances where the nuclear plant's technical and environmental conditions would impose requirements that are not generally met by standard consumer devices. For example, there would be requirements for ruggedization of handheld devices, seismic and vibration protection of mounted devices, protection from electromagnetic interference, protection against cyberattacks, and many more. Also, consideration of the roles of humans and machines requires more critical analysis of how the implementation of new technology would affect the way functions are allocated between humans and systems than ever before.

Although new HSI technologies have the potential to significantly improve operator performance in the field and in the control room, the nuclear industry lacks well-defined criteria to ensure that new displays and controls would support human performance and also ensure operational effectiveness and safety. Without such criteria to guide the selection and deployment of new HSI technologies, designers may unwittingly create opportunities for error.

The simple rule that instrumentation and control engineers will have to learn in NPP design is that functions should not be automated just because technology makes it possible. Instead, automation decisions should be based upon a rational trade-off between the contributions that either system capabilities or human factors principles, or a combination of the two, would make to operational effectiveness and safety.

To understand these trade-offs, system engineers and human factors engineers should cooperate in understanding the context and conditions where new technologies would be applied most effectively. This includes interdependent aspects such as human performance requirements and process and system characteristics like reliability, quality, and usability. In combination, these aspects would help to determine the most appropriate selection of HSI technologies.

In anticipation of requirements that may be unique to advanced SMR designs, designers are now facing a number of tough requirements. For example, they must identify the advantages and disadvantages of technologies and accommodate them in the power plant design. This must include the context of use (e.g., the operational domain, such as control room, field operations, maintenance, or materials handling, and the operational conditions that determine the nature of the operator's task). It also requires consideration of specific human factors constraints (perceptual limitations, memory, workload, human reliability, situation awareness, and performance shaping factors), safety requirements, and the projected lifetime of products.

One of the requirements that demands considerable insight into the nature of HSI and operational requirements is to determine the optimal interaction modalities for different operational contexts. This needs to take into account spatial and physical work space characteristics and collaborative functions such as crew-system coordination, contextual adaptation, and means of communication to support shared situation awareness. Designers would also have to consider alternative perceptual and interaction modalities offered by new technologies like touch and voice interaction to simplify information access, communication, and decision-making and to reduce errors. Ultimately, they have to determine how new technology characteristics affect human performance and therefore the need for advanced capabilities to support new power plant requirements, such as reducing operational and maintenance costs by reducing the number of operators needed to manage control room tasks. This requirement leads to questions about the need for adaptive automation, computational intelligence, operator support systems, and other methods of reducing complexity, to optimize human-automation interaction. Where appropriate nuclear operating experience of advanced technologies is lacking, designers may have to resort to obtaining research information to resolve these issues.

In spite of all the requirements that will be imposed on designers to verify and validate their choice of technologies, there is already ample evidence in other industries of the benefits of the advanced technologies described in this chapter. These HSIs offer support for substantial improvement in the safety and economics of all nuclear power plants. The SMRs that are the subject of this Handbook promise to be safer and more economical plants that will reach the market in the next decade in various countries. That is just one reason why the adoption of the HSIs described here is a logical approach in current SMRs and other advanced designs.

Nevertheless, designers cannot simply assume that any new technology would contribute to safety or better human performance. Addressing issues of automation, function allocation, error reduction, and overall operator efficiency is still a major challenge. To address those challenges, three main topics are discussed in this chapter:

- The technical characteristics of HSIs for a new generation of NPPs and the human factors considerations associated with them.
- Implementation and design strategies: special considerations for the selection and deployment of advanced technologies in NPPs, whether modernized, new, conventional, or first of a kind (FOAK), including strategies for the integration of human factors and regulatory aspects into systems engineering processes.
- Future trends: how technologies are likely to develop over the next 10–15 years and how this will affect design choices for the nuclear industry.

7.2 Human-system interfaces for small modular reactors

The US Nuclear Regulatory Commission's (NRC) review guidance on HSIs, NUREG-0700, defines the HSI as "that part of the nuclear power plant through which personnel interact to perform their functions and tasks. Major HSIs include alarms, information displays, controls, and procedures" (O'Hara et al., 2002). The HSI is used to manipulate a device or system, to request and display stored data, or to actuate a single process or various preprogrammed command routines. HSIs can be organized into workstations consisting of consoles and panels, and the arrangement of workstations and supporting equipment could be organized into physical work areas such as a main control room (MCR), remote shutdown station, local control station (LCS), technical support center (TSC), and emergency operations facility (EOF). The HSI could also be characterized in terms of the environmental conditions in which the HSIs are used, including radiation, temperature, humidity, ventilation, illumination, and noise.

The NUREG-0700 definition is generally valid for HSIs currently in use, but it does not take into account the latest advances in HSI hardware and software. (At the time of writing, this guidance has been updated to Revision 3 to include review guidance on digital systems and HSIs.) The following characteristics would generally be associated with "advanced HSIs."

7.2.1 Hardware features

The physical characteristics of new HSIs include devices that support multimodal interaction, such as touch screens, gesture interaction, speech recognition and synthesis, haptic input and output (i.e., technologies that use touch and tactile feedback to enable HSI interaction), and even direct body-machine interfaces (sensors). Advanced display and interaction features already available and under development make use of handheld devices, head-mounted displays, large overview displays, three-dimensional (3D) displays (with or without glasses), and motion and position tracking. To support such extensive interaction capabilities, the whole system is typically driven by high-performance numerical and graphic processors for demanding applications such as high-resolution displays and computationally intensive applications like processing and trending of large amounts of plant data. (Several of the terms mentioned earlier may be unfamiliar to some readers; they will be explained in a later section.)

7.2.2 Software criteria

The main characteristic of new HSI software platforms is that it typically forms part of the plant's distributed control system (DCS) software. The DCS is the system that is used for overall plant automation, and the HSI forms part of the "front end" that enables the operator to interact with the plant through a hierarchy of controls and displays. This system typically allows development of the functionality and displays of the HSI with minimal low-level programming while allowing some end-user customization. It also supports full object-oriented and component-based programming, which ensures consistency of functionality, layout, and appearance of objects throughout the HSI. Systems like this also support standardized documentation and code handling formats like XML. In advanced applications, as discussed later, it would support advanced computational methods like neural/semantic networks, pattern recognition, and real-time and faster-than-real-time simulation.

7.2.3 Functional criteria

Functional features of advanced HSIs include standardized and user-configurable displays. However, the most important feature would be the organization of the whole HSI as an operator-centric or task-based system with embedded operator support, including various levels of computer-based procedures. Owing to the inherent complexity of advanced automation systems, the HSI must support intuitive navigation through a display architecture derived from proper task analysis. Advanced HSIs would also provide error-tolerant and resilient operation, support adaptive automation schemes, and provide integrated multimedia communication.

7.3 The state of HSI technology in existing nuclear power plants

Many new SMR designs are still in the conceptual or preliminary design phase, and typically, very little information for HSI design and device selection is available early in the project life cycle. Nevertheless, it is possible to generalize the characteristics of much of HSI technology that would be used in the plant. This is not so much because of similarities in new designs, but rather because of the state of the art in HSI technology. In the past, there was a certain degree of customization of instruments and controls (I&C) for specific control rooms, but this customization was more in the layout of the control room and the control boards. Most of the conventional instruments and controls (traditional "light box" alarm annunciators, panel-mounted switches, knobs, dials, and gauges) were devices that were designed to strict industry standards for reliability and robustness. However, for the foreseeable future, we can expect implementation of devices derived from consumer and commercial products, but that are fast becoming standard in many industries: high-resolution flat panel displays, touch screens, wireless handheld computers that can serve as both input and display devices, and a range of static and mobile devices designed to improve supervisory control, improve situation awareness, and enhance operator performance and reliability.

In most industries, we find that advanced automation systems have the potential to enhance the safety of workers and equipment; enhance monitoring of process variables through improved sensing; control and display capabilities; increase system reliability, resilience, and availability; and reduce the need for human operators for functions that can be achieved more efficiently through automation.

In contrast the nuclear industry has yet to reap the full benefits of advanced technologies. There are several reasons for this backlog, but at the same time, there are many reasons why a transition to advanced technology is not only inevitable but also highly desirable. Even a brief examination of the current state of the art of emerging instrumentation and control and also human interface technologies would quickly reveal the reasons for this trend.

In most existing plants, surveillance, testing, inspection, and monitoring of plant performance are all dependent on human operators and are all labor-intensive activities. This is not surprising, given the current state of technology in the majority of older plants. Traditional I&C and display technology in most plants older than 20 years consist of fixed analog devices, as mentioned earlier. The control boards and panels in the control room are typically arranged in a horseshoe configuration, and very often the controls used for control actions and the gauges where the results of such actions must be observed are widely separated on the boards and panels throughout the control room. The result is that the operator has to move around a lot to collect information from diverse sources. At the same time the operator has to keep a lot of information in his or her head while performing a procedure. Under abnormal or emergency conditions, this can produce significant workload and stress, and it is easy to see how this kind of HSI could become a potential source of human error. Indeed, there is ample evidence of the critical importance of well-designed HSIs from the accidents at Three Mile Island II, Chernobyl, and Fukushima Daiichi, all of which had inefficient analog HSIs.

Innovations in HSI technologies have the potential to alleviate or even eliminate many of the problems associated with analog I&C. Various strategies to upgrade I&C systems, including modernization of control rooms, have been emerging over the past decade (Korsah et al., 2009). These strategies range from the most common "like-for-like" replacement of systems (e.g., replacing alarm light boxes with flat panel monitors that still display alarms as conventional alarm tiles), to comprehensive human factors engineering (HFE) studies combined with systems engineering projects that consider all technical and human aspects of the new or upgraded systems.

Since most new reactor designs will employ FOAK technology (technology that has not been used in the older generation of NPPs), they have the opportunity to avoid the problems of outdated I&C and HSIs (obsolescence, unavailability, costly maintenance, and so on). However, there are still significant risks associated with FOAK designs. These risks include challenges of integration and inadequate consideration of the changing role of the operator, coupled with the possible need to define new models of human-automation collaboration, the need for integrated system validation, and many more.

Advanced technologies cannot be placed in the hands of the operator without considering how this will affect his or her task and performance. This means that designers should be intimately familiar with the characteristics of technologies, not only individual devices but also devices coupled, integrated, or interfaced with other new and older devices. An understanding of how the introduction of new technologies may affect operator behavior and performance is crucial to the success of an NPP development project in the short term and the safe and efficient operation of the plant in the long term.

7.4 Advanced HSIs and the human factors challenges

7.4.1 Purpose and objectives of advanced HSIs

As indicated in the earlier definition, the primary purpose of the HSI is to provide the operator with a means to monitor and control the plant and to restore it to a safe state when adverse conditions occur. The successful accomplishment of this objective will satisfy five important human performance goals that all contribute to the safe and efficient operation of the plant: (1) reduce complexity, (2) reduce error and improve human reliability, (3) improve usability, (4) reduce operator workload, and (5) improve situation awareness.

Achieving these objectives relies heavily upon the most effective information and communication technologies available. These technologies have the potential to improve many of the shortcomings of the old generation of analog HSIs (i.e., "hard controls and instruments," such as buttons, switches, and gauges) found in most NPPs. However, such improvements are dependent on a focus on human factors principles in human-technology interaction. Advanced automation systems are beginning to allow a more dynamic collaboration between humans and systems. We can no longer regard the complex relationship between humans and systems as "people versus technology," which was often the result of the classical function allocation approach. That outdated approach was based on attempts to implement "HABA-MABA (human are better at-machines are better at)" principles derived from Fitts' List (Hoffman et al., 2002). Rather, it now is more appropriate to focus on the total sociotechnical system as a "joint cognitive system." Woods and Hollnagel (2006) and Lintern (2007) describe a cognitive system as one that performs the cognitive work functions of knowing, understanding, planning, deciding, problem solving, analyzing, synthesizing, assessing, and judging, as they are fully integrated with perceiving and acting. In a particular work environment in the power plant, the entity that performs *perceiving* and *acting* functions would be the human agent. This implies that the control room and the entities within it could be characterized as a joint cognitive system that functions in a distributed way and involves relevant parts of the environment, the physical, mental, and cultural processes of people and the technical artifacts. The joint cognitive system viewpoint emphasizes the cognitive functions that human operators and technologies accomplish in collaboration. It allows human factors analysts and designers to analyze the system at different levels of detail, starting from the entire sociotechnical system of the NPP down to specific functions of an HSI that would have the ability to support the operator's cognitive functions.

7.4.2 Human factors challenges of HSIs

Ever since the Three Mile Island accident in 1979, human factors issues were largely associated with the design of the main control room. The possibilities of applying HFE principles to improve human performance were limited to a great extent by the constraints of discrete, analog instruments, and controls. However, due to new capabilities offered by technologies like advanced sensors and automation systems, new NPP designs are now expected to introduce fundamental changes, not only in the design of the control room but also in the role of the operator and the tools they use to monitor and control the plant. This could be regarded as a natural evolution for the industry, but it will require engineers and designers to rethink many tried and tested concepts and assumptions. For example, control center structures need to be remodeled to make provision for new types of console and panel layouts, large screen displays, new communication media, and even different crew structures. This will require a clear shift in the definition of the control room, its controls and instruments, its support structures, and also the location of the control room in the plant.

Technically, it has become possible to control the plant from a remote location, but it will be a challenge to prove the reliability of such a scheme under all operational conditions. In addition to the changes in the physical and functional architecture of control rooms, we can also expect to see changes in the allocation of operational functions to humans and systems. The mere fact that future operators will deal with computer-based "soft controls" and a multitude of high-resolution displays will already change their roles and mode of interaction with the plant (Stevens et al., 2019 have clearly demonstrated this in the design of the NuScale control room).

Where we today understand the operating crew as consisting of reactor operators, senior reactor operators, and supervisors whose roles are largely determined by operating procedures, future operating crews may be regarded rather as part of the joint human-technology system, which in turn is part of the bigger sociotechnical system of the plant. The reason for this lies in how the operator's responsibilities and interaction with the plant will change. The shift will be more than just a role change due to increasing levels of automation or an increasing supervisory role where operators' primary function will be to monitor plant status and only to intervene if actual operation deviates from set points. Rather, there are now increasing possibilities for operators to perform "predictive control" by examining past data, predicting future behavior of processes by means of extrapolation and real-time simulation, and performing corrective actions before an event is likely to occur.

A further shift in the role of the operator will be an increase in the scope of responsibility and collaboration. For example, the scope of control and monitoring functions could increase from just operations, to include maintenance, production planning, and even design and optimization.

All of these changes represent a paradigm shift for the nuclear industry, and it is almost entirely because of the advancement of automation and HSI technology. The changes have immediate implications for engineers who have to reconcile technological requirements with human abilities and limitations. There can be little doubt that automation is key to achieving cost-effective operations in future nuclear energy systems, but humans will continue to play as important a role in future systems as in today's safe nuclear power plants. We can expect a different sort of HSI from that of today's plants, but one in which the operator and crew are still able to intervene when necessary and otherwise oversee automation in many aspects of plant operation. This will require development of more "intelligent" forms of automation and adaptive interface capabilities to facilitate near-autonomous operation and efficient human-system collaboration.

The following are some of the most important considerations that need to be included in SMR power plant engineering and design strategies:

- The joint human-technology system must be defined in terms of the dynamic allocation of functions between the humans and the automation system.
- The human-technology system is not static and will require new rules and procedures for allowing a minimum number of operators to control multiple modules concurrently. Even for single module plants, it is possible that higher levels of automation will require fewer operators in the control room. However, regulators are unlikely to accept an unconventional staffing design without some kind of proof of concept. For new plants, this proof could be in the form of simulations or predictive computational models that provide reliable data on operator performance under various plant conditions (Persensky et al., 2005).

- Task support requirements—owing to the dynamic nature of the collaborative human-system relationship and the variable levels of complexity at different levels of automation, there will be variable requirements for task support. In principle the lower the level of automation, the more the operator's involvement in plant control, and thus the more support is required (usually in the form of operating procedures), especially for nonroutine tasks. HSIs that are designed to optimize human performance need to be concerned with fundamental collaborative functions such as coordination, adaptation, and communicating shared awareness within the total sociotechnical system. This goes beyond the present usage of computerized procedure systems, decision support, databases, data mining systems, and various devices to deliver this information to the user. The usability requirements for task support systems, especially those that use new HSIs, must include measures of the trust the operator places in the technology (Hugo, 2004). These considerations suggest that HSIs can be examined from many different perspectives, but when we consider the challenges of emerging power plant designs, there are two main themes that seem to influence most considerations for future implementation:
  ○ New HSI technologies offer innovative interaction modalities such as gesture control, augmented reality, remote control, and telepresence. Designers need to provide, or obtain, sufficient evidence that these new concepts are conducive to usability and will support improved human performance.
  ○ The applicability of advanced HSIs in the nuclear field is a particularly interesting question because the nuclear industry has been relatively stagnant for a long time. As a result, practices, standards, procedures, and technologies have become so entrenched that utilities, vendors, regulators, and other stakeholders have to make extraordinary efforts to justify and validate the use of new technologies. Even if those technologies had already shown proof of concept in other industries, the strict regulations and standards of the nuclear industry make implementation of any new technology an exceptional challenge. More about this in Sections 7.5 and 7.10.2.

The rest of the chapter will cover the most important aspects of HSI technology, starting with a description of the architecture or taxonomy of HSIs as they are typically deployed in "modern" power plants. The following sections will also describe the range of technologies becoming available to designers, the technical capabilities they offer to support human performance, and the use and potential of a range of nontraditional HSIs, such as virtual and augmented reality systems, haptic devices, and gesture controllers.

7.5 Differences in the treatment of HSIs in the nuclear industry

There are several reasons why the treatment of HSIs in the nuclear industry is different from other industries. For example, there are probably more regulations, guidelines, and standards for the consideration of human factors than in most other industries. SMRs in general and nonlight water reactor (non-LWR) designs in particular face even more challenges. For example, due to the emphasis on new technologies, higher levels of automation, new function allocations, and the quest for minimal staffing and lower O&M costs, designers need to cope with a large number of rules, regulations, standards, and guidelines. These include, for example, US Code of Federal Regulations like 10 CFR 50.54 (Conditions of Licenses), NUREG-0800 (Standard Review Plan), the requirements for the review of an organization's HFE program (NUREG-0711), guidelines for HSI design, such as NUREG-0700 (Human-System Interface Design Review Guidelines), ISO 11064 (1999) (Ergonomic Design of Control Centres) and IEC 60964 (Nuclear Power Plants—Control Room Design), and requirements for the integration of HFE in other engineering processes as described in IEEE 1023 (IEEE Recommended Practice for the Application of Human Factors Engineering to Systems, Equipment, and Facilities of Nuclear Power Generating Stations and Other Nuclear Facilities).

There are also many regulations dealing with occupational health and safety, building regulations, and several more, but regulations that deal with the control room and the HSI are less well known and are the source of much concern for I&C engineers and control room designers. The design and acceptance of digital systems with modern HSIs remain a significant challenge for the nuclear industry. However, the NRC has developed approaches and guidance for regulatory review and acceptance of the new technologies. HSI modernization is addressed in the NRC guidance—recent revisions to Chapter 18 of NUREG-0800, and NUREG-0711 address modernization of existing plant control rooms and HSIs in addition to new designs. Also, modern digital interfaces are covered in the recent revision to the NRC's detailed HSI review guidelines, NUREG 0700.

A strategy that many plants have adopted is to make changes first to nonsafety systems—this allows an opportunity to ensure that the necessary processes are in place and to gain experience with the processes and the newer digital technologies before making changes to the safety systems, which carry higher licensing risk. It also allows plants a low-risk way of evaluating how well the new digital I&C and HSI hardware and software work for their crews.

Because a large part of advanced reactor design would be FOAK engineering, human factors engineers need to cope with many organizational, technical, regulatory, and methodological questions that are new to the nuclear industry. All of these regulatory and best practice expectations lead to probably the biggest challenges that SMR designers face: the procurement of suitable technologies that would satisfy the stringent regulatory requirements. Most advanced HSIs would likely need qualification for nuclear application (e.g., quality, reliability, and seismic qualification). This is accompanied by the complex integration of human factors in the systems engineering process throughout the project life cycle. Developing a human factors engineering program at the planning stage ensures that it is compliant with the expectations of NUREG-0711, 10 CFR 50.54, and other requirements. This mechanism of obtaining NRC review of the program at the earliest opportunity can reduce the amount of licensing effort and risk associated with the downstream modifications. Also an early look at potential 10 CFR 50.59 (changes, tests, and experiments) issues associated with changes that will affect operator performance can help minimize licensing risk.

Because of all of these requirements, we can expect as much, if not more, regulatory oversight for SMR design projects in the form of regular, mandatory audits, quality and safety management requirements, and intensive verification and validation of, for example, HSI designs and human performance with computer-based operating procedures. Regulators will also expect to see evidence of human factors work in provisions made to protect the public, workers, and the environment. This includes, for example, attention to situation awareness, safety culture, human reliability, workload, and performance shaping factors. All of this contributes to long lead times, not only for engineering and design, but also particularly for the licensing processes.

It should also be emphasized that HFE practitioners in the nuclear industry are not excluded from scrutiny and public opinion. It is thus also important for HFE people to show how they contribute to safety to counter misconceptions about hazards (opinions often based on incomplete or outdated information or disinformation by antagonists).

7.6 How to identify and select advanced HSIs: Five dimensions

As mentioned before, designers need more formal methods to choose the most appropriate technology for a specific task. Bad decisions at the design stage are often the source of later operational problems that are very hard to fix.

One practical way to examine the viability of technology for future implementation is to consider the interaction features associated with that technology. Another way is to consider the characteristics of the work domain and the context of use. This would include the practical, economic, organizational, operational, regulatory, and contextual considerations.

In examining the key entities, principles and relationships between technologies, and the contexts where technologies are used, five key dimensions or contexts are identified that could serve as a framework for evaluation and deployment of HSI technologies for new designs. Fig. 7.1 identifies these dimensions and specific HSI topics and elements associated with each. The five dimensions or contexts shown (human factors, technology, operations, organization, and regulations) are defined in the following sections.

7.6.1 Dimension 1: The human factors context

The human factors considerations include aspects of human-system interaction, such as perceptual requirements (visual, auditory, discrimination, etc.), cognitive (workload, attention, situation awareness, etc.), and physical handling (physical ergonomics and anthropometry). Evaluation of the human factors aspects of technology is complex, because it is often difficult to observe and measure cognitive behavior. The most common methodology to determine the relationship between technology characteristics and human performance is usability testing. This helps to determine the effectiveness, efficiency, and satisfaction with which specific users can perform specific tasks with specific systems in a given context.

Human performance can be affected positively or negatively by technology, and this is determined by a number of factors: the work and environmental context within which a tool is used, the experience and skill with which a person can use the tool, the design of the tool, and how usable it is in the given task context. Optimal human

HSI dimensions			
Human factors	Cognitive ergonomics		
	Physical ergonomics		
	Performance Criteria	Workload	
		Situation awareness	
		Performance shaping factors	
	Reliability		
Technology context	Vendor info		
	Market readiness		
	Standards compliance		
	Cost of ownership		
		Usability	Effectiveness
			Efficiency
			Safety
			Learnability
			Satisfaction
	Technical performance	Operability	
		Maintainability	
		Accessibility	
		Reliability	
		Accuracy	
		Durability	
		Simplicity	
	Technology forecasts		
Operational context	Production		
	System integration		
	Environment		
	Automation level		
	Maintenance		
	Safety	Alarms	
		Spatially dedicated continuously visible displays	
		Minimum inventory instrumentations	
Organizational context	Style guides		
	Procurement policy		
	Staffing		
	Shift work		
	Standards		
Regulatory context	Code of federal regulations		
	NUREGs (nuclear regulations)		
	Regulatory guides		
	Operator licensing		
	Minimum inventory instrumentation		

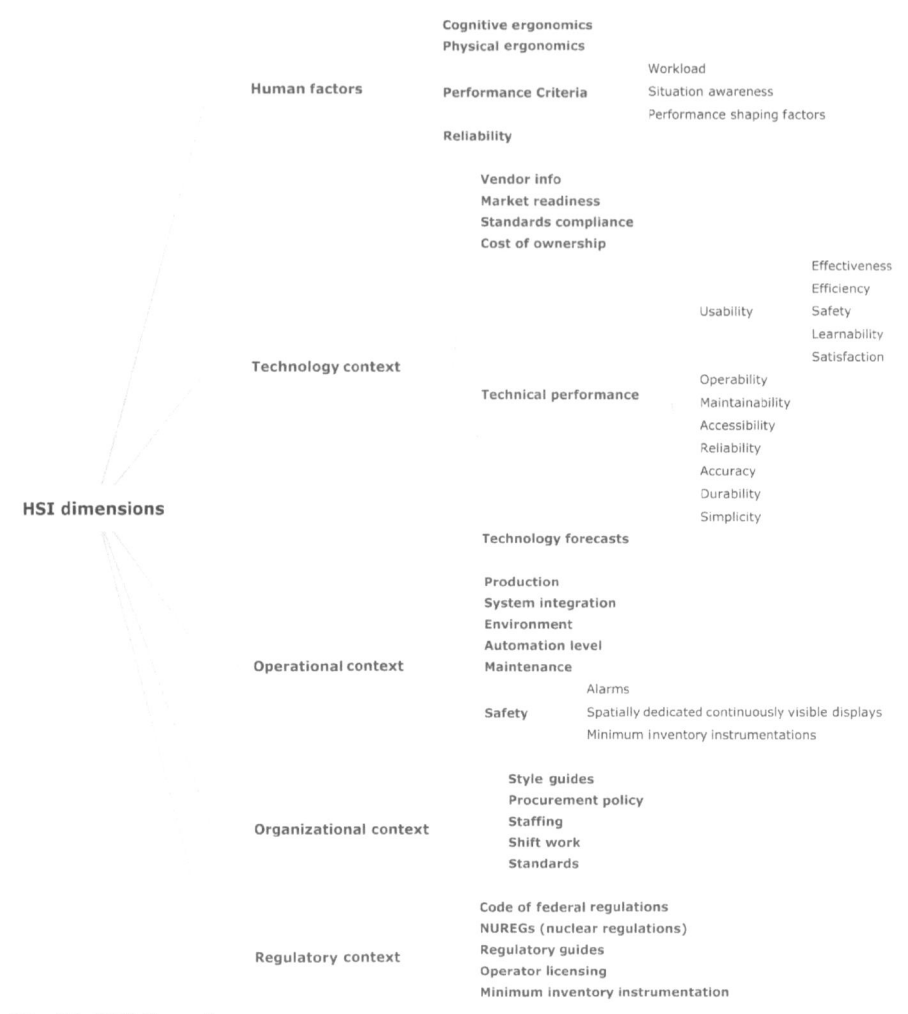

Fig. 7.1 HSI dimensions.

performance could therefore be expressed in terms of criteria such as comfort, efficiency, situation awareness, low workload, and low error probability. It also means that human performance can often be improved by providing some form of task support to help in reducing operator workload, reducing visual and cognitive complexity of the HSI, and making information more accessible.

7.6.2 Dimension 2: Technology characteristics

Technology characteristics can be grouped into two main categories—technical characteristics and context of use.

7.6.2.1 Technical characteristics

These are concerned mainly with the advantages and disadvantages of specific technologies and the human factors considerations. These would include the performance criteria listed in the succeeding text. Most advantages and disadvantages of technology can be evaluated objectively simply by comparing features and performance measurements. However, such measures are often meaningless if not performed with reference to the specific context of use, described in the succeeding text.

Other practical aspects of technology choices would include the following:

- total cost of ownership (including cost to maintain and replace)
- availability (is the product actually availability and how long will it take to procure?)
- technical performance (how does the product perform on specific measures such as accuracy, sensitivity, resolution, and reliability?)
- environmental constraints for operation (can the product withstand rough handling, dirty environments, etc.?)
- compliance [does the technology meet applicable standards, such as the Occupational Safety and Health Administration (OSHA) Act?]
- training required (how difficult is it to learn?)
- system variables (processing power, memory size, and power requirements)

In addition to these characteristics, we can identify a number of complementary categories that must also be considered when selecting technologies. These application criteria are called the engineering "ilities":

- *Usability*—As defined in ISO 9241-11 (1998), the effectiveness, efficiency, satisfaction, and safety with which specific users can perform specific tasks in defined contexts.
- *Maintainability*—This is the ease with which a product can be maintained. It is also a characteristic of design related to the amount of time taken to return a failed piece of equipment back to its normal operable state.
- *Accessibility*—This is a measure of how easy it is to obtain the device, tool, function, or information when needed.
- *Operability*—This refers to a device being ready for use or to be placed into service or the ability to keep equipment or system in a safe and reliable, functional condition.
- *Reliability*—This is the ability to consistently perform an intended or required function and mission and the probability that a given item will perform its intended function for a given period of time under a given set of conditions.
- *Durability*—How well a device is able to perform and withstand a variety of conditions by resisting stress or prolonged use.
- *Simplicity*—How easy it is to understand, explain, and operate a device or system. We can also associate simplicity with intuitiveness and usability of a device or concept.

7.6.2.2 Context of use

Context of use refers to the actual conditions under which a given HSI is used in a normal day-to-day working situation. More specifically, it refers to the specific operational scenarios where the HSI is likely to be used. This includes the roles and tasks of users—operators, technicians, engineers, and managers. The attributes of tasks that will influence technology choices include experience, knowledge, skills, and abilities.

The usefulness and applicability of technology will then be a combination of its perceived usefulness, ease of use, and applicability to the intended environment and users. It is therefore important to carry out usability tests, prototyping sessions, meetings, and user studies in the particular context of use to ensure a valid basis for technology selection.

7.6.3 Dimension 3: Operational requirements

Operational requirements and objectives for HSIs are typically derived from the need for cost-effective power production, nuclear safety, technical requirements imposed by plant processes, the level of automation, and the need to protect workers, the public, and the environment. HFE work during the early phases of plant design should focus on how systems will be operated by the users, including interfaces and their interoperability with other systems. These requirements will establish how well and under what conditions the system must perform and will thus form part of the basis for HSI design and technology selection.

7.6.4 Dimension 4: The organizational context

The organizational requirements for technology selection are often found in policies, standards, design style guides, cost-economic considerations, and vendor preferences. These requirements may also influence operational requirements. Control room staffing is one of the aspects of power plant operations that is strictly regulated by US Code of Federal Regulations, and any deviation from the regulation is subject to scrutiny and proof of concept. This has a direct influence on the way control rooms, HSIs, and automation systems are designed.

As indicated before, the underlying assumption of new designs is that higher levels of automation will enable multimodule and single-module plants to be controlled by fewer operators than are required for conventional LWR or boiling water reactor (BWR) plants. When this is verified, it may lead to staffing strategies where a single operator may be able to handle multiple modules or multiple processes under normal operating conditions, as in the case of the NuScale design (LaFerriere et al., 2019). In addition, the physical layout of the plant, the reduced need for LCSs and manual controls in the plant, and the availability of remote surveillance equipment may mean that fewer field operators will be required.

However, in the absence of sufficient operating experience and proven technical bases, the only reliable way to estimate the number of operators required under varying operational conditions would be to use advanced task analysis and modeling methods and tools such as computational human performance modeling (Hugo and Gertman, 2012). Predictive human performance information produced in this way could eventually be verified in the full scope plant simulator.

7.6.5 Dimension 5: The regulatory context

Current NRC regulations were developed based on traditional large LWR designs. Current requirements related to the human role in the plant deal primarily with avoiding human error and improving human reliability in normal and abnormal operational conditions. This includes requirements for control room staffing, criteria for evaluation of HSIs, and conducting HFE activities in the power plant. Some provision is currently made for new designs, for example, for minimum staffing of the NPP, as described in the Code of Federal Regulations, 10 CFR 50.54. However, most new reactor designs and particularly SMR designs differ substantially from traditional designs in a number of aspects, including size and number of reactors, inherent passive safety systems, fuel type, and coolant type. These differences present unique issues in terms of licensing and regulation.

Although current NRC guidance provides a general framework for conducting design-specific reviews, the review of control room and HSI designs and staffing plans and potential exemption requests is expected to be challenging for SMR designs, despite the claim that passive safety systems would eliminate safety-related operator actions (LaFerriere et al., 2019). This is because of the differences between the new reactor designs and previously licensed reactor designs and also because of a lack of research and design data to provide an adequate technical basis for decisions. Initial evaluation conducted by the NRC has identified a number of differences between SMR and other advanced reactor design and operating philosophies and the designs for larger reactors currently licensed or being evaluated for licensing. These differences include the following:

- SMRs may require different operator tasks. The task requirements will include operating multiple units in different modes of operation. A major challenge will be to identify tasks that may be omitted and those that could substantially affect operator workload. This challenge has been clearly demonstrated by early demonstrations of the NuScale prototype control room (Stevens et al., 2019).
- Very limited operational experience will be available to use as a resource, especially if these designs are FOAK. The use and observation of simulator activities will be important to verify the task analyses and staffing plans. Parallels in other industries may be useful, if they exist.
- Integration challenges exist in not only defining tasks required for operating the unit, but also for interacting with other on-site maintenance and support organizations for multiple units.
- The skill set for control room operators, especially those required to manage more than one product, may require a different distribution of qualifications (e.g., more senior reactor operators, fewer reactor operators, or even a new class of personnel).
- For some advanced SMR designs, operators will face the challenge of supervising the operation of additional units as they are placed online. As the number of modules increases, the demands on the operators will change and potentially the number of operators required for safe operation (i.e., multiple staffing plans may be needed to address the addition of more units during the construction period or subsequent operating periods).

These challenges indicate that HSIs and their selection and deployment cannot be considered in isolation of the tasks and environments associated with them. It has to be an integral part of the HFE process, which, in turn, has to be integrated with the rest of the design organization's engineering processes.

The licensing issues related to SMRs are covered elsewhere in this Handbook, but for the human factors engineer, it is essential to find an early resolution of regulatory issues regarding the use of new HSIs (see also Section 7.5 and 7.10.2.)

Early resolution will enable designers to incorporate appropriate changes during the development of their concepts of operation, designs, task analyses, and staffing plans before submitting a design review or license application. It will also support the NRC staff's review of the design and license applications.

7.7 Operational domains of HSIs

The nature of HSIs for advanced reactors can be better understood if they are characterized in terms of the "operational domains" where they are used and also the architecture of HSIs themselves. We can define "operational domain" as the physical, structural, logical, or functional characteristics that distinguish different areas in the plant where work is performed and where humans interact with technology.

There are nine distinct work domains in modern NPPs where HSIs will play an important role. Some of these are dedicated and enclosed areas; other areas inside or outside the plant have variable boundaries within which functions are performed:

- Control room—This is an enclosed area, often in close proximity to the reactor and turbine building.
- LCSs throughout the plant, typically consisting of one or more small control panels.
- Materials and waste fuel handling. Forklifts, cranes, and similar tools are typically found in these domains.
- Refueling operations, using specialized equipment to handle radioactive materials.
- Maintenance inside and outside the plant, using a range of conventional and specialized tools.
- Outage Control Center (OCC), characterized by many desktop computers, large displays, printers, planning boards, and communication equipment.
- Fuel processing installations, characterized by specialized equipment to handle hazardous materials, such as robotic manipulators (not all SMRs will have this domain).
- Technical Support Center (TSC). This center is typically somewhere on site and, like the outage control center, would have not only large status displays but also limited HSIs that provide access to some of the displays found in the control room.
- Emergency Operations Facility (EOF). This facility is located at a more remote location outside the plant perimeter and would also have access to data from the control room.

Most of these domains have a greater or lesser degree of interdependence, overlap, or redundancy, as shown in Table 7.1. The control room dominates in terms of the range and number of HSIs applied in that environment and the interfaces with other domains. The intersections between domains indicate the potential relationships or interdependence between domains. Another domain, fuel processing plants, could feature strongly in future at plants using fast breeder reactors and fuel reprocessing.

Human-system interfaces in small modular reactors (SMRs)

Table 7.1 HSI operational domains.

	Main control room	Local control stations	Remote shutdown facility	Waste and materials handling	Plant maintenance facilities	Outage control center	Technical support center	Emergency operations facility	Refueling operations facility
Fuel processing facility				1	1	1			1
Operations support center	1		1			1	1	1	
Refueling operations facility	1	1	1		1	1			
Emergency operations facility	1	2	1			1	1		
Technical support center	2	2	1			1	1		
Outage control center	2		1		3				
Plant maintenance facilities	1	2	1	1					
Waste and materials handling		1							
Remote shutdown facility	1	1							
Local control stations	3								

Frequency/importance of interface: 3 = high and 1 = low.

The interfaces between the control room and the overlapping domains consist primarily of status displays and communication devices. These interfaces enable the operating crew to maintain situation awareness of all activities throughout the plant and during all conditions.

This table illustrates the most important interfaces between the various domains and the estimated frequency and importance of these links. Where no number is shown at the intersection, there is no clear dependency or interaction.

For most operations (normal and abnormal), the MCR is the dominant work domain in terms of the number and type of functions and the range and number of HSIs applied in that environment. Some operating domains may functionally overlap with each other or with the MCR, such as the local control stations throughout the plant, materials handling, refueling operations, waste handling, and maintenance and outage management. The links and overlaps between the domains also suggest the importance and frequency of communication that may be required between them.

Two other domains interface with the control room only during upset or emergency conditions: the TSC and the EOF. The most important domains are described briefly in the succeeding text.

7.7.1 Control and monitoring centers

Plant control and monitoring functions are performed in two main areas: MCR and LCSs.

7.7.1.1 Main control room

A control room is generally understood to be the nerve center of the plant, and it often forms part of a *control center* that could also house some of the operational domains shown in Table 7.1, for example, the TSC and the OCC. The MCR in older nuclear power plants is dedicated to the control of a single unit; new plant designs and especially SMRs with their compact footprint are more likely to have a single MCR for multiple modules (see "Multimodule control rooms" in the succeeding text). Some emerging designs envisage a single control room for up to 12 modules (Stevens et al., 2019; LaFerriere et al., 2019). This kind of control room will be larger than today's control rooms for a single unit, but owing to the level of integration and automation, they may actually reduce the complexity of the overall instrumentation and control architecture by allowing common systems to share a single operator console (O'Hara et al., 2008).

It is normally assumed that a central control room is necessary as part of a strategy to rationalize plant operations, to minimize duplication of equipment, and to optimize the capability of automation systems. Central control rooms for modern plants are also considered to enhance communication between units, enable better coordination of plant-wide operations and maintenance, and provide a more effective response to upsets.

As indicated earlier, an important consideration for new designs would be the location of the control center. Traditionally the MCR is located somewhere on the "nuclear island," which normally consists of the containment, including reactor, steam generator, and primary coolant circuits. Because the nuclear island is seismically qualified and provides backup systems like electrical supply and heating, ventilation, and air conditioning (HVAC), this was typically the choice for the location of the control room. In fact, this is still one of the strictest criteria for control room habitability described in NUREG-0800 (2007), but it is also a major contributor to the cost of the control room.

As mentioned before, it is assumed that new reactors will use more passive safety designs like negative temperature coefficient of reactivity, natural circulation of coolant, or less need for active controls and fewer active protection systems like forced cooling circuits. Designers should now determine if requirements for seismic-qualified control systems and HSIs will change and if this means that the control room need not be on the nuclear island. Designers should also determine if the availability and reliability of wireless technology and fiber optics are sufficient justification for having the MCR remote from the reactors. Other important considerations would be if the need for operator response to certain events would still force location of the control room to be near the reactor. Due to the current strict NRC regulations (see NUREG-0696, 1981), proving these new concepts of operation is likely to be an importance challenge for designers.

7.7.1.2 Multimodule control rooms

Unlike large nuclear power plants that typically have one control room for each unit, compact plant designs like SMRs are more likely to operate multiple modules from a single central control room. Plants like NuScale that employ multimodule control rooms will inherit a number of characteristics from NPP control rooms as we know them today. As in the past the primary purpose of the control room and the HSIs within it will still be to enable the operator to control the plant safely and effectively. They are also used to monitor and direct complex operational activities, such as optimizing the combined output of the modules or units (Stevens et al., 2019; LaFerriere et al., 2019).

The most likely characteristics of multimodule control rooms would include the following:

- The need for a high level of automation, integration and synchronization of systems, and optimization of output. This suggests a single control room from where a minimal crew can manage the entire plant while still being able to control and monitor the operation of individual modules and systems.
- The use of advanced HSIs to simplify the display of complex system functioning and to minimize the safety critical, potentially high-consequence nature of the control task.
- The potential change in the central role of the control room operator to that of system supervisor.
- The existence of new regulatory measures to govern the control room procedures and interface technologies.

7.7.1.3 LCSs

O'Hara (2002) defines an LCS as "A place outside of the main control room where operators interact with the plant. LCSs may include multifunction workstations and panels, as well as operator interfaces, such as controls (e.g., valves, switches, and breakers) and displays (e.g., meters and VDUs)." NUREG/CR-6146 (Brown et al., 1994) also identifies multifunction and single-function LCSs:

- A *multifunction* LCS is any operator interface used for process control not located inside the control room and not consisting solely of manually operated valves or circuit breakers.
- A *single-function* LCS is defined as any operator interface, excluding multifunction control panels, that is not located in the control room. This type includes all controls (valves, switches, and breakers) and displays (meters, gauges, and monitors) operated or consulted during normal, abnormal, or emergency operations.

We can expect that many manual actions that were common with LCSs will be replaced by automated control to eliminate many of the discrete controls mentioned earlier. Instead, operators will be able to use digital controls and more advanced displays to control and monitor local processes (Brown et al., 1994).

7.7.2 Materials and waste fuel handling

In nuclear plants, as in almost all industrial plants, there is always a lot of materials handling that requires various common and also sophisticated tools. HSIs for material handling include simple overhead cranes, forklifts, and even programmable robotic systems. Advanced HSIs will play an increasingly important role in fuel and material handling systems, especially for handling hazardous materials like low- and high-level radioactive waste. Several technologies are becoming prominent in new means to improve reliability and safety in hazardous applications. These include graphic visualization for remote controls that will take traditional methods of materials handling to a new level. Devices envisaged include robots or remote-controlled machines, augmented reality, and eye- and position-tracking systems for tasks such as planning for surveillance and maintenance in plant areas where environmental conditions such as radioactivity, heat, cold, dust, or toxic materials would prevent or limit human access. In addition to visual monitoring and material handling systems, HSIs would have the ability to notify operators immediately if there is a problem in the system, including jams, misaligned sensors, worn bearings, or related issues that cause system degradation. These kinds of technologies will thus offer significant benefits for system and human performance.

7.7.3 Outage control center

Some of the biggest challenges during large plant outages are to manage the multitude of resources and maintain a high level of situation awareness to ensure the continued safety of plant personnel and to ensure that equipment is protected. Some nuclear utilities have refined the outage process to a fine art, but all struggle with the need to manage the complex communication processes during outages. Multimedia and wireless communication technologies are already proving to be an indispensable boon to outage teams, and this is likely to become a standard feature of outage management in future. In particular the need for collaboration and information sharing is satisfied by a range of information displays in the outage control center. Large interactive displays (called "smart boards") and collaborative work support systems allow real-time access to information, schematics, procedures, and all kinds of scheduling and resource information. This is augmented by a variety of handheld information and communication devices like tablets, smartphones, handheld computers, barcode readers, and cameras.

Advanced HSIs and communication devices will also help to eliminate or reduce distractions in the control room caused by maintenance personnel traffic, noise, nuisance alarms, and noncritical activities while helping the operators to maintain situation awareness. For the maintenance teams, technologies mentioned earlier will help to minimize down time and improve communication and resource management.

7.7.4 Emergency operating facility

Many of the advantages mentioned for outage control centers also apply to the EOF and the TSC. As indicated before, the EOF is normally outside the perimeter of the power plant, and it serves as the management and coordination center for the emergency staff that will operate from there in the event of an emergency at the plant. Advanced HSIs will help to manage information about important plant parameters and radiological conditions in the plant and its immediate surroundings.

7.7.5 Technical support center

The TSC is an on-site facility located close to the control room—according to NUREG-0696 (1981), the TSC must be located within a 2-min walk of the main control room. During upset and emergency conditions, it provides technical support to plant management and the reactor operating personnel located in the control room. Advanced HSIs with diagnostic features will be important here too to help TSC personnel analyze the plant conditions before and throughout the course of an accident or upset condition.

7.8 HSI technology classification

There are many different ways to classify HSI technology, depending on the context, as described before. The terms "human-system interface" and "human-system interaction" both suggest that either a technology-centric or a human-centric classification would be possible. This section provides a simple taxonomy that describes both perspectives.

7.8.1 Interaction modalities

In the past, it was easy to classify HSIs as either input (keyboards, switches, mice, etc.) or output devices (displays, gauges, or printers). With the convergence of modern HSIs, it is no longer that simple—many devices are now combining input and output on the same device (tablets, smartphones, etc.). Even distinguishing between hardware and software is becoming increasingly difficult because many devices have embedded software and the device is rather considered in terms of the functions that the user can perform with it. It is now more sensible to classify HSIs in terms of the mode of interaction or "interaction modality."

Interaction modality can be described as a means of communication between the human and the system or device. The term "communication" implies the process of exchanging information between the human and the system primarily through the visual, auditory, speech, and touch senses. All HSI technologies can be categorized according to the human sense for which the device is designed. Most devices rely on only two or three of the most common senses used to obtain information from the environment: vision, hearing, and touch. Some technologies can combine these senses into one device; more advanced devices can also enable interaction through other senses, such as speech, smell, motion, or even kinesthesia or proprioception. (Kinesthesia is the subliminal awareness of the position and movement of parts of the body by means of proprioceptory organs in the muscles and joints; Hale, 2006.) When multiple modalities are available, that is, when more than one sense can be used for some tasks or parts of tasks, the system is said to offer multimodal interaction functions. A system that is based on only one modality (e.g., a read-only device like a fixed gauge) is called unimodal.

When technology types are categorized in terms of the human sense for which they are designed, it is possible to classify interaction modality as either of the following:

- Input—perceiving information produced by the system through a device that allows a human to observe it by means of one or more senses, such as visual, auditory, or tactile.
- Output—performing an action with a specific device that would cause the system to perform a function. This output in turn becomes the input to the system in the form of discrete actuations (e.g., key presses) or continuous actions (using a mouse or similar device to select or manipulate objects on a display).

Based upon the primary senses used in interacting with a device, HSIs can now be divided into three categories: visual, auditory, and mechanical motion. Devices associated with these modalities would be either input or output devices (i.e., devices accepting user input or providing output to the user) or hybrid devices where both input and output are combined in the same device.

In the control room and any of the operational domains described earlier, a multimodal interface acts as a facilitator of human-system interaction via two or more modes of input that go beyond the traditional keyboard and mouse. Multimodal HSIs can incorporate different combinations of speech, gesture, gaze, touch, and other nonconventional modes of input. Touch and gesture have become the most commonly supported combinations of input methods, as seen in the rapid development of tablets and smartphone devices (Oviatt, 2003). These are already making an appearance in control rooms for noncontrol applications, like procedure following and calculations, but they are likely to become much more prominent in future SMRs, provided that they can be proven reliable.

These combined modalities open up a vast world of possibilities to interact with the work environment. It is already possible, for example, to interact with displays, not only with both hands but also with all fingers simultaneously or by various combinations of "hand waving."

Based on the description earlier, it is now possible to define four classes of technology: visual technologies for visual perception, audio technologies for audio perception, mechanical control devices for providing input to a system, and hybrid devices for multimodal interaction.

7.8.2 Visual interfaces

HSIs that employ vision as interaction modality are the most common devices to present information to the operator. This mode of interaction can be unidirectional only (device to observer) or interactive. New developments use cameras and sensors to detect user gaze and motion and use this to create an interactive dimension for displays. Examples are already found in entertainment-related devices like Microsoft's Xbox Kinect, Nintendo Wii, Sony PlayStation Move, and the Leap Motion gesture controllers. Other devices use gaze detection to determine where the user is looking and use these data to display contextual information or to enable users to navigate through the system by gaze only.

While standard desktop flat screen displays will continue to be the most common means to display information for everyday use, a variety of advanced visual devices are becoming attractive options for textual, graphical, and video information. Three types can be identified, ranging from large screen displays that are already available to more sophisticated technologies that will be available in consumer markets within the next few years.

7.8.2.1 Large screen displays

Large, high-density, high-resolution, and high-definition displays are already common in consumer and commercial markets, and many process and manufacturing industries are also using a variety of these displays. Typical applications include multimonitor configurations, tiled flat panels and also projection-based systems that can display images several meters wide (Ni et al., 2006). In conventional NPP control rooms, the implementation of these large displays present numerous technical difficulties, mainly because of the lack of space in an area that was originally designed for large, hardwired consoles without any digital displays. This is where designers of new plants have a distinct advantage: they can design state-of-the-art control rooms without trying to retrofit advanced HSIs into cramped spaces.

Although large displays may seem an attractive option to overcome the distributed nature of information typical of older control rooms, designers need to consider that this will not necessarily address the fundamental question of ensuring that operators benefit from increased size and resolution. Vendor hype often leads designers to assume too easily that large displays will automatically outperform small ones (Ni et al., 2006). Before equipping a control room with a multitude of large displays, human factors engineers should understand under what conditions increased size and resolution may be advantageous and how they may contribute to situation awareness. In many cases a number of standard-size displays on the operator's workstation may be more effective than a large overview display.

7.8.2.2 Wearable displays

Wearable and head-mounted displays of various types have been prototyped and tested by the military for many years. Devices range from large, heavy, full-immersion, head-mounted virtual reality displays used for specialist training to lightweight, see-through devices used for augmented reality applications. This technology is now finally becoming a commercial reality also in consumer markets. Devices like Google Glass are already used widely in manufacturing industries, but it would also offer significant opportunities to simplify common control room tasks in the nuclear industry. Obvious applications would include continually monitoring alarm annunciators while performing other tasks or having computer-based procedures available with a simple voice command.

Virtual reality has a long history in visualization and interacting with 3D environments, for example, visualizing and verifying designs long before they are built. Wearable devices like augmented reality headsets that superimpose virtual objects and information on the user's view of the real world can enable operators to perform tasks without the need for printed documentation or other support. In this way, information about the user's surrounding real world also becomes interactive and digitally manipulable. This technology is already being used in some industries to support maintenance and assembly tasks.

7.8.2.3 3D displays

A number of technologies that enable users to view objects and environments in three dimensions are becoming common in the consumer and professional media markets. They range from reasonably simple devices that require the user to use glasses to view images, to sophisticated volumetric holographic displays.

So-called 3D displays or stereo displays present normal two-dimensional (2D) images offset laterally by a small amount and displayed separately to the left and right eye. Both of these 2D offset images are then combined in the brain and create the perception of depth. Various types of devices can be used with television sets, gaming devices, and movies.

Another technique, the "polarized 3D system," uses two superimposed images that are viewed through two polarizing filters that are oriented differently. The filter passes only that light that is similarly polarized and blocks the light polarized differently, enabling each eye to see a different image. This produces a 3D effect by projecting the same scene into both eyes but depicted from slightly different perspectives.

Holographic display technology has the ability to create the illusion of 3D objects in volumetric space. Holographic devices are developing fast but will be out of reach of the average consumer for several more years. Also, all of the advanced 3D display technologies still have limitations, especially in display quality and resolution. It is unlikely that we will see real 3D or holographic displays in the nuclear industry within the next 20 years.

7.8.3 Auditory interfaces

Audio-based interaction between human and HSI is not new, but it is now developing rapidly to provide more powerful and reliable means of obtaining information or performing control actions. Alarm sounds are the most common use of audio technology and most control rooms use coded and modulated sounds to enable operators to distinguish different conditions.

Speech recognition is also not a new technology, and it has been applied in many systems with varying degrees of success. Speech recognition has never been used in control rooms, but this technology is also becoming more accurate and reliable. However, although important advances have been made, especially in the ability of such systems to recognize natural language, it remains one of the least reliable interaction methods. Nevertheless, the latest research shows significant improvements in the technology. Although still not perfect, speech recognition technology, for example, Amazon's Alexa, Apple's Siri, and Microsoft's Cortana are achieving error rates of 5.1% and 5%–10% with background noise (Globalme Language and Technology, 2019). Although this is a promising development, this kind of interface is still slow and unreliable and unlikely to be used in mission-critical applications. Research continues, and this may eventually become an option for "busy hands, busy eyes" applications. Both speaker-independent and speaker-dependent speech recognition with background noise cancelation might become more viable options for certain types of control commands in future, especially for fieldwork and maintenance where hand-free operation is often desirable.

Another auditory device that may become important for voice communication, for example, between the control room and field operators in the plant, especially in noisy areas, is bone conduction audio. This technology provides real-time information in "busy-hands busy-eyes" tasks. It maintains sound clarity in very noisy environments because the eardrum is bypassed and sound is passed directly to inner ear. It is especially important where there is a need to enhance the presentation of written or graphical information and to notify users about a particular condition without the need for a display. It is also useful where coded audio signals may convey more information than a single alarm tone or for operators with some hearing impairment.

7.8.4 Control devices and mechanical interaction

In older control rooms, discrete control input devices (i.e., devices that depend upon mechanical motion) are still the most common means for the operator to interact with the plant's systems. These devices are limited to relatively primitive devices like buttons, switches, and levers. As I&C and HSI technology evolved, it has become possible to control plant components by means of direct manipulation devices like a computer mouse, joystick, keyboard, touch screen, or trackball. As computing power continues to increase, we can expect to see even more sophisticated devices finding their way into the control room and other work areas. In the not-too distant future, we can expect to see fixed and mobile devices that allow not only direct interaction through touch and force feedback but also indirect interaction through gesture, speech, and gaze.

7.8.5 Hybrid interfaces for multimodal interaction

7.8.5.1 Gesture interaction

Gesture interaction is a way for computers to interpret purposeful human motions, thereby creating a bridge between machines and humans that allows a richer interaction experience than the primitive input methods of keyboard and mouse. Using gesture recognition, operators can literally point a finger at the computer screen to interact directly with objects, without actually touching the screen. This technology is still in its infancy, but several devices and applications are beginning to appear. Devices like the Xbox Kinect and the Leap Motion gesture controller are likely to become mainstream options very quickly. In the short term, this probably will not make conventional input devices such as mice, keyboards, and touch screens redundant, but will be added to the range of HSIs to allow operators more flexibility in interacting with plant systems. [For an extensive discussion of gesture interaction, the reader is referred to the work of Bill Buxton, a Microsoft researcher (Buxton, 2011).]

7.8.5.2 Haptic interaction

A range of advanced sensors embedded in HSIs allow operators to expand their ability to sense the state of the environment and the behavior of artifacts within the environment by means of haptic devices or "tangible interfaces." Such devices take advantage of the sense of touch to convey a range of information by applying forces such as vibration, force feedback, and sensing location and motion. This tactile stimulation can be used to assist in the detection of changing conditions or orientation of objects that the operator cannot handle manually due to hazards such as heat or radiation. In more advanced devices, it can create the illusion of virtual objects and the ability to control them in a computer simulation, to control such virtual objects, and also to enhance the remote control of machines and devices (telerobotics). Again the reader is referred to Buxton's work (Buxton, 2011, Chapters 8, 9, 13, and 14).

A common example of haptic interaction in the form of vibratory feedback is found in the Sony, Xbox, and Nintendo game controllers mentioned before. Haptic devices may also incorporate tactile sensors that measure forces exerted by the user on the interface (Jones and Sarter, 2008). It is easy to imagine how operators would be able to use a device like this to "feel" the bearing vibration of a turbine while monitoring the spin-up process!

7.8.5.3 Brain interaction

Recent state-of-the-art developments promise to offer interaction possibilities considered impossible just a few years ago. For example, direct brain-machine interaction has long been considered science fiction (think of the 1984 novel *Neuromancer* by William Gibson or the 1999 movie *The Matrix*!), but it is fast becoming a reality. Even simple game-oriented devices like Emotiv Systems' *Epoc Neuroheadset* demonstrate impressive capabilities to control devices and software (Fakhruzzaman et al., 2015). Other groundbreaking developments are a noninvasive neural interface from CTRL-Labs that reimagines the relationship between humans and machines with new, intuitive control schemes (WIRED, 2019) and an artificial intelligence-enabled cortical implant from Elon Musk's Neuralink Corporation (www.neuralink.com). Devices like these are paving the way for designers to dramatically enhance interactivity and the level of immersion in the application by, for example, enabling the system to respond to a user's facial expressions and adjusting the application's behavior dynamically in response to user emotions such as frustration or excitement and enabling users to manipulate objects in an application or even turn them on or off or change their state by simply using the power of their thoughts. This is reality and no longer science fiction; it is not too hard to imagine that these devices will find their way into certain applications in industry within 20–30 years.

7.8.5.4 Intelligent and adaptive HSIs

Although the term "intelligent" is perhaps a misnomer in HSIs, it is nevertheless an important development. This is a class of artificial intelligence-enabled technology that mimics certain aspects of human reasoning and behavior. Such systems employ statistical and probabilistic methods in conjunction with neural networks, databases, rules, and a variety of sensors to approximate human traits of reasoning, knowledge, planning, learning, communication, perception, and the ability to manipulate abstract or concrete objects (Ehlert, 2003). Software systems that are able to perform such functions could be called intelligent software agents. When this forms part of the HSI, such an agent would act in collaboration with the operator, for example, to detect certain patterns of operator responses in his or her use of the HSI, such as the need to perform a calculation. It would then either autonomously perform the function for the operator or submit the result to the operator for approval. More sophisticated agents equipped with cameras and sensors could even detect stress and workload from the operator's voice and facial expression and offer to activate specific operator support functions.

Other sensor technologies that are already common in many industries are now also slowly being deployed in the nuclear industry, For example, radio-frequency identification tags (RFID) and global satellite positioning systems (GPS) are being used to locate personnel and components in the plant.

7.9 HSI architecture and functions

The HSI in older NPPs has always been a reasonably complex system, but it was possible to describe it in fairly simple terms as consisting of control boards, panels, gauges, controls, and alarm annunciators. However, with increasing automation and availability of digital I&C systems, the HSI in newer NPPs has also become progressively more complex. The HSI in SMRs like NuScale will be a system with many functions, components, and interfaces to other systems and environments. Even a superficial review of its many components will show that not only the HSI is in fact a hierarchy of high- and low-level components, but also many of the components at the same level are linked in some way (LaFerriere et al., 2019). It is also possible to describe this structure from different viewpoints, depending upon, for example, whether it is a safety- or nonsafety-related system and whether it is used in operations or in maintenance. It is also possible to describe it as either an abstract functional or a physical structure.

Another way to make sense of the HSI architecture is to distinguish between its functions and its physical architecture. The physical architecture not only consists of the concrete computer components but also includes the operating environment (control rooms and other workspaces described in Section 7.7) and all the hardware within it. These physical components make it possible for the operating crew to perform physical interactions to perform all tasks in the work environment.

This architecture is not definitive and could be structured and described in a number of different ways. The following description is presented as a starting point for engineers and designers involved in the definition of I&C and HSI requirements. Several terms and nomenclature are commonly used to describe the physical and functional structure of the HSI. In the interest of consistency, it is recommended that designers adopt the ISO 11064-5 (Ergonomic Design of Control Rooms-Displays and Controls) terminology shown in the succeeding text. This structure can be regarded as a hierarchy of HSI display and interaction components. This excludes physical input devices, with the exception of touch screens, which are regarded as hybrid output/input devices. The hierarchy, representing all elements from abstract to concrete, is depicted in Table 7.2:

Note that the term "display" in this hierarchy is synonymous with the generic term "visual display unit (VDU)" as it is used in NUREG-0700. The term "screen" is synonymous with "display page" used in NUREG-0700 and is often used simply as "screen" in design documents when it refers to a single functional display instance of a specific system, such as a "turbine control system screen." This should not be confused with "page" as shown here, where it is simply a defined set of information on a single screen. For practical purposes, it can be regarded as synonymous with "screen."

Table 7.2 HSI elements.

1.	**Control room**—the physical space from where the plant is controlled and monitored						
	2.	**Consoles/control boards**—the hard surfaces onto which monitors are mounted. There can be multiple consoles in the control room					
		3.	**Monitor/visual display unit**—there can be multiple VDUs on the console				
			4.	**Screen**—display instance of a specific system			
				5.	**Display page**—a defined set of information that is intended to be displayed on a single screen		
					6.	**Window**—an independently controllable area on the display screen used to present objects and/or conduct a dialog with a user (a single window can occupy an entire display screen)	
						7.	**Format/layout**—the organization of display elements, for example, mimic, chart, trend curve, and faceplate
							8. **Element**—any discrete display object, for example, icons, labels, system symbols (pumps, valves, etc.), control buttons, check boxes, data values, and entry fields

Adapted from Table 2, Chapter 7, Handbook of Small Modular Nuclear Reactors, 1st edition.

Note also that the term "monitor" when used as a noun normally refers to the complete physical device, specifically its configuration, location, make, etc. This would be the hardware that is used for Element 1 and its subordinate elements shown in the diagram earlier. When used as a verb, "monitor" means "The periodic oversight of a process in order to detect the presence of signals that indicate changes in predetermined operating limits or the occurrence of a specified condition."

For detailed descriptions of each element and its function, refer to the full standard ISO 11064-5.

Further descriptions of the standardized HSI may be found on p. 17 in [ANSI/ISA-101-01-2015, 2015] (*Human Machine Interfaces for Process Automation Systems*).

7.10 Implementation and design strategies

As mentioned earlier, selecting and implementing HSIs are only one small part of the overall engineering effort of the NPP. However, the control room and the HSI are so tightly integrated with the overall architecture of the plant and, together with the operators, play such a vital role in plant efficiency and safety that it should be treated with the same rigorous engineering discipline as all other parts of the plant. Designers therefore need to attend to all areas of analysis, design, and implementation that affect human interaction with plant and systems. We know that operator performance in a control room may be equally influenced by the design of the control system software, the system architecture, the physical architecture of the control room, or the design of procedures and documentation. To address these issues, it is necessary to integrate human factors considerations into the processes used for the design of the technical system. NUREG-0711 emphasizes the importance of following a formal engineering process also for all human factors elements of the plant and to provide traceable documentation for all design decisions. This requires designers to compare the final HSIs, procedures, and training with the detailed description and specifications of the design to verify that they conform to the planned design resulting from the HFE design process and verification and validation activities (O'Hara et al., 2012).

This section describes the key aspects of an integrated approach to HFE that will ensure that the technology and design choices will actually serve the purpose for which they were selected.

7.10.1 Integration of human factors engineering in systems engineering

The implementation of advanced HSIs, as part of the complex sociotechnical system of the NPP, requires an integrated approach based on the requirements of both human and system components. One of the most reliable and effective ways to accomplish either a progressive HSI upgrade or a completely new design is the integration of HFE with the systems engineering process (SEP) (Hugo, 2012). Systems engineering (SE) is the discipline needed to deliver coherent, cost-effective systems of whatever nature, but HFE adds an important dimension by helping to integrate the whole system, that is, human *plus* equipment. The integration of human factors into the SEP considers the role of humans in the selection of HSI technology and ultimately in the operation and maintenance of the plant at every stage of the system life cycle.

This systematic approach will ensure first of all that the human performance information necessary for engineering design, technology selection, and development processes is acquired or developed even before the project starts. Secondly, it will ensure human factors evaluation of systems and operations throughout the project life cycle to identify problems and help engineers to define cost-effective solutions to achieve human and system performance enhancements. A lot of project case studies have proven that it is cheaper and more cost-effective to integrate human requirements early in the project rather than later (Hugo, 2012).

7.10.2 Regulatory requirements

As indicated in Section 7.5, the regulatory review process for SMRs will in most respects be the same as it has been for conventional NPPs, and designers of new plants will be required to comply with the requirements described in NUREG-0711 (O'Hara et al., 2012). Specifically, they will be required to demonstrate that they have followed a risk-informed process to ensure that safety considerations have been adequately addressed in the HFE process (Boring, 2010; Hugo, 2012).

7.10.3 Standards and design guidance

There are far too many standards and guidance documents to review in this chapter, but the most important ones are undoubtedly those already mentioned in Section 7.5. The NRC is reviewing and updating many of its regulations and requirements to make provision for reviewing design submittals for a new generation of SMRs. Already we are seeing an updated version of NUREG-0711 (O'Hara et al., 2012), which addresses emergent aspects of plant modifications and important human actions.

As indicated earlier, NUREG-0700 has recently been updated to Revision 3 and contains new review guidance on digital systems and HSIs (O'Hara et al., 2012).

7.10.4 Design considerations

As the nuclear power industry transitions to next-generation plant design, processes, and advanced control rooms, many new technologies become available to ensure efficient and safe operations. A key component of control room operations is the human operator, who must ultimately use the new technology. All the exciting opportunities offered by HSI technologies confront human factors engineers with some challenging requirements and decisions:

- In-depth task analysis, simulation, and prototyping are essential for successful implementation. Next-generation control rooms will offer opportunities to exploit advances in automation system technology that potentially will allow multimodule control rooms to operate with minimal direct human intervention. However, it is vitally important to establish the optimal level of automation for individual system functions and for the overall control system. Designers must reconcile an automation philosophy of "automate as much as possible" with the overriding requirement to "keep operators in the loop." A rational task allocation strategy must lead not only to an either/or strategy of allocating function to humans or machines but also to a dynamic, productive collaboration between them. There are many indications that the traditional function allocation methods are no longer adequate to determine what functions can or cannot be automated and the reasons why (Wright et al., 2000).

- The variable roles of operators (e.g., under reduced plant capacity with varying numbers of operational modules or during maintenance or emergency conditions) must be analyzed, and human factors engineers must ensure timely feedback to engineering design from function and task allocation analyses. This strategy must include consideration of the integration of multiple automated systems in the control room, how multiple failures of such automated systems should be handled, and how appropriate task support could be provided to the operator during all modes of operation. An advanced automation system not only will require an adaptive operator support system to match an adaptive automation system (Sheridan, 2002) but also will require extensive self-diagnosis of systems, error correction, intelligent alarm handling, and automated event reporting. All of this must be done without leaving the operator out of the control loop.
- Control room design, especially multimodule control rooms, requires sophisticated analysis, simulation, and design methods and tools. In particular, high-fidelity simulators are essential for validation of designs, procedures, and operator performance (LaFerriere et al., 2019). Once designers have an operational control room, they are going to face another challenge: how to keep operators vigilant and engaged in a monitoring role. We can accept that the operator will be involved in a significant amount of extra work related to a FOAK plant during the early stages of operation (e.g., tests and diagnostics). Nevertheless, there is little doubt that under normal operating conditions there will be very little for the operator to do, unless we change his or her traditional role. Vigilance and situation awareness under low workload conditions and decisions about the number and duration of shifts are serious concerns still to be resolved through detailed task analysis and studies of human reliability and human performance under various conditions.
- In addition to the changing role of operators, there will be special requirements for the development of a new generation of operating procedures, especially for variable modes of operation. Module-specific operating procedures versus plant-common procedures may require special treatment, and this needs to be resolved long before the plant is commissioned. Again a full-scope simulator will prove indispensable in the verification and validation of many aspects of the operator's task.

While the many options offered by automation and HSI technology are very exciting, the reality is that designers need to make design decisions that will ensure a licensable, commercially viable plant. Ultimately, they need to focus on what is needed to obtain utility, public, and regulatory acceptance of different levels of automation and operator performance in single and multimodule plants. Many new HSI technologies, such as augmented reality, intelligent agents, or handheld HSIs, may be of questionable value in the control room because of lack of proof of concept in the nuclear industry. However, the integrated nature of advanced automation systems will dictate an overall integrated approach to the design of the HSI. While advanced HSI technologies may be less important than building safe and commercially viable plants, designers should nevertheless recognize that such technologies offer an opportunity to significantly improve human performance and reliability in the plant and ultimately the effectiveness and efficiency of the whole sociotechnical system.

7.11 Future trends

The only clear thing about the future of HSI technologies is that there will be different development paths for various technologies to support the licensing and commercialization process of new NPPs and especially SMRs. With more and more devices to choose from, designers will often be forced to adopt a suboptimal design solution by using technologies that are not cutting-edge, but that have already passed the proof-of-concept stage.

Since a few advanced SMR designs will only see the light as operating plants by about 2026, it would be realistic to envision the evolving state of technology between now and about 10–15 years from now. From current research and trends emerging in consumer, commercial, and industrial markets, we can identify a number of the most significant technologies that might have a greater or lesser impact on future NPPs 2022–32 and beyond. These trends are shown in Table 7.3.

It is easy to see how many of these technologies overlap. As mentioned before, convergence is in fact the one overarching characteristic of current technology trends. Developers have been working on integrating more and more functions into input and output devices to be combined in the same physical device to enable different interaction modalities. Tablets and smartphones are typical examples of input and output devices that used to be two distinct technologies, like keyboards and visual display units (Cheskin Research, 2002; Silberglitt et al., 2006). Another growing trend is to develop devices that can adapt to the user's physical location and usage context (home, street, office, factory, or countryside). Devices like Google's augmented reality Glass wearable display are an example of emerging technologies that attempt to take this several steps further by interacting with the environment and continually providing users with contextual information.

This convergence is going to be particularly relevant for future SMR operating concepts where changing operator roles will require more powerful means of accessing and manipulating operational information. The discussion of the HSI architecture referred to the future "operator support system," which would be an ideal application of these hybrid and synergistic technologies.

However, in spite of the overriding convergence trend, we know from past events that technology development is not linear, especially when disruptive technologies like 3D printing and artificial intelligence emerge. It will only take breakthroughs like powerful artificial intelligence systems or quantum computing to upset all current forecasts. Nevertheless, we can expect improved reliability, resilience, adaptability, and information accessibility offered by the HSI technology convergence and functional synergy to be the strongest driving force in design decisions for future control rooms and HSIs. The ability of these technologies to deliver text, audio, and video material over the same wired, wireless, or fiber optic connections is rapidly making the conventional HSI devices of today's power plants obsolete. We can expect the

Table 7.3 Technology forecast.

Timeline and technology readiness		SMR application potential
Now	Handheld communications and computing	Tablets and smartphones are already available in many commercial and industrial applications and slowly finding its way into nuclear. Future applications include remote monitoring and control for control rooms as well as hazardous environments
	Cloud computing and data storage	Already widespread in business and domestic applications but in future would provide a complete automation system and communication service rather than the current distributed control system employing programmed logic devices throughout the plant. The key transition is moving from a fixed client server architecture to a distributed architecture with local and global intelligence
	Large touch screens	These are becoming more ubiquitous throughout the plant but especially in control rooms, technical support centers, and emergency operations centers
2022	Wearable HSIs	Wrist-mounted computers are already available in industrial applications but are migrating to head-mounted displays for a variety of field operations. In the control room, they could replace the conventional alarm annunciators and other displays that need to be monitored constantly
	Flexible and e-ink displays	Flexible displays can be put on flexible materials for use in rollable or bendable displays to provide durable information devices for industrial environments. They could be used for applications like special instruments, for documents, and even large schematics like electrical schematics, making it unnecessary for workers to carry many documents in the plant
	Augmented reality (AR)	AR combines real and virtual data to provide an interface between real plant, operators, technicians, and the digital plant. Simple applications are already available in some niche applications. In future, many maintenance and surveillance tasks will be improved by AR. In the nuclear plant, technicians can point a camera to an on-site piece of equipment, and the system will

Continued

Table 7.3 Continued

Timeline and technology readiness		SMR application potential
2024	3D displays	match it to the digital map of the plant and verify it is in its designated location and state, freeing the staff from cumbersome documents and procedures and also providing contextual information
		These devices require the user to use glasses to view images and range from flat panel displays to large virtual reality displays like computer-assisted virtual environments (CAVE). Some are already available in commercial and consumer devices and will soon be acceptable for certain nuclear energy applications. They will make it easy for plant personnel to visualize systems and environments that cannot normally be observed without exposure to various hazards
	Bio- and haptic feedback devices	A range of advanced sensors embedded in HSIs allow operators to expand their ability to sense the state of the environment and the behavior of artifacts within the environment by means of haptic devices or "tangible interfaces." Such devices can convey a range of information through vibration, touchless gesture, force feedback, and motion sensing
	Volumetric (holographic) displays	Holographic display technology can create the illusion of three-dimensional objects in space and is thus an extension of 3D displays. These displays will enable objects to be viewed from different directions
2026	Robotics, machine vision, and telepresence	Robotic systems that are capable of self-locomotion and equipped with machine vision are becoming available for various mission-critical applications, especially where humans cannot be exposed to hazardous conditions. These systems can be used for surveillance, inspection, and safe recovery actions
	Context-aware computing and intelligent agents	Some simple applications are already available, like tablets switching display orientation as the user turns it or adjusting brightness according to ambient illumination. Advanced HSIs can use sensors to detect situations and classify them as contexts and then react to changes in the environment. They can adapt

Table 7.3 Continued

Timeline and technology readiness		SMR application potential
	Intelligent agents	their functions and appearance according to the location of use, the user, changes in plant or process condition, etc. This will provide a level of operator support not available with conventional HSIs
		Intelligent agents will perform operator task support functions such as allowing operators to delegate repetitive tasks, monitoring specific alarms, processing, and summarizing complex data and make operational recommendations
2028	Integrated HSIs for remote plant monitoring and operation	Advanced automation systems will enable operators to monitor distributed power generation equipment from remote locations. Advanced HSIs and telemetric will offer a wide range of data monitoring, logging capabilities, and remote communications for cogeneration plants
	Quantum computing and cryptography	Massive computing power will enable total automation of power plants and will cause dramatic changes in the human role. This will also be the key to secure computing, making remote operation of power plants safe from cyberattacks
2032	Brain interfaces	The brain-computer interface (BCI) and neural link devices are becoming so advanced that it is set to create a whole new symbiotic relationship between man and machine. This could be an effective interface for remote-controlled robotic and telepresence systems

operator of the future to be surrounded, inside and outside the control room, by a multilevel, convergent, media-rich world where all modes of computation, information presentation, and communication are available to adapt to normal and emergency operating conditions.

Most instrumentation and control engineers assume implicitly that future plants will be highly automated. If we accept that technology advances (e.g., large-scale integration of networked intelligent sensors and control systems) will force higher levels of automation, then we can safely assume that the role and function of the operator will ultimately change. This chapter is not the place to discuss the nature of that change, but we can safely assume that operators will perform more supervisory functions and less hands-on control tasks. Human factors engineers would be wise to plan ahead to avoid technology dictating this change. They should work more closely with systems engineers than ever before to ensure that automation decisions are not based solely on the capability of advanced I&C technologies, but on a productive collaboration between humans and systems. In principle, this means that functions should be automated only if it will improve reliability, efficiency, and safety without compromising the operator's situation awareness and ability to intervene when necessary. This ability to intervene should be designed into the system in such a way that it will exploit those complex phenomena and capabilities that still make humans superior to machines: coping with uncertainty and conflicting indications, applying rules of thumb, rapid visual recognition of objects, or identifying and matching complex visual or auditory patterns and translating it into action. In contrast, operators should not be expected to perform complex mathematical calculations and to perform functions that cause undue cognitive strain, increased workload and increased error probability, or tasks that are too expensive or dangerous for human operators.

7.12 Conclusion

The monitoring, control, surveillance, inspection, test, and maintenance of NPPs are all labor-intensive activities. Because of this the nuclear industry has long been driven by the goal of continuously improving productivity while striving to maintain or enhance operational safety. SMRs currently being developed provide an ideal platform where advanced automation and HSI technologies will make achieving these goals possible.

The use of new HSI technologies for a new generation of nuclear plants is largely uncharted territory for all stakeholders. The same could be said of most new reactor designs, except that nuclear engineering can rely on far more operating experience than the deployment of new HSI technologies in this industry. Engineers, operations managers, human factors practitioners, regulators, operators, and many others will be continuously confronted by very little proof of concept for new technologies.

HSI is a broad subject that covers many topics ranging from abstract theories of human cognition and perception to the selection of specific hardware and software for specific tasks. The technologies with their increased simplicity, intuitiveness, and flexibility of use not only are launching a new era in human-computer interaction, but also are going to stimulate a revolution in the way that operators interact with nuclear power systems. This is why advanced I&C systems and HSIs will make a significant contribution to greater automation and improved human performance for SMRs. This chapter has illustrated how the application of advanced HSIs would also contribute to the smaller physical size, or footprint, of new, compact SMRs and to the operational effectiveness of the whole plant. SMRs and all new nuclear power installations will benefit from the potential of the new generation of HSIs to reduce crew workload, improve reliability, reduce the risk of human error, and improve overall system resilience and safety.

The first step in achieving the optimum human-system interface is to design the plant to exploit new technology not only for passive safety features but also for new means for the operator to obtain information and interact with the plant, thereby reducing the active contributions needed from both humans and machines during operation. This will ultimately improve the performance of both human and system. However, in spite of the promise of new, powerful technology, no design can ever be perfect, and this effectively ensures that there is a continuing role for the operator for the foreseeable future. Nevertheless, it is inevitable that automation will eventually take over the more prescriptive tasks in the power plant, which is likely to change the role of the operator to that of a situation manager—an innovator to manage the unexpected.

This chapter has barely scratched the surface of these topics, but it will provide SMR designers with the background information and basic tools to identify some of the challenges and opportunities offered by a new generation of HSIs for modern nuclear power plants. It should also help all role players understand the characteristics, functions, and place of new HSIs in the plant and how to apply these principles in their strategies for selecting the most appropriate systems.

References

ANSI/ISA-101-01-2015, 2015. Human Machine Interfaces for Process Automation Systems. International Society of Automation.

Boring, R.L., 2010. Human reliability analysis for design: using reliability methods for human factors issues. In: Seventh American Nuclear Society International Topical Meeting on Nuclear Plant Instrumentation, Control and Human–Machine Interface Technologies, NPIC&HMIT 2010, Las Vegas, Nevada, November 7–11, 2010. American Nuclear Society, LaGrange Park, IL.

Brown, W.S., Higgins, J.C., O'Hara, J.M., 1994. Local control stations: human engineering issues and insights. NUREG/CR-6146 (2011), US Nuclear Regulatory Commission, Washington, DC.

Buxton, B., 2011. Gesture-based interaction. In: Haptic Input. eBook, http://www.billbuxton.com/input14.Gesture.pdf (Accessed 10 February 2013) (Chapter 14).

Cheskin Research, 2002. Designing Digital Experiences for Youth. Market Insights Series, pp. 8–9. Fall.

Ehlert, P.A.M., 2003. Intelligent user interfaces: introduction and survey. Technical report, Delft Technical University, Delft, Netherlands.

Fakhruzzaman, M.N., Riksakomara, E., Suryotrisongko, H., 2015. EEG wave identification in human brain with Emotiv EPOC for motor imagery. Prog. Comput. Sci. 72 (2015), 269–276.

Globalme Language & Technology, 2019. A Complete Guide to Speech Recognition Technology. Available at: https://www.globalme.net/blog/the-present-future-of-speech-recognition.

Hale, K.S.D., 2006. Enhancing Situation Awareness Through Haptic Interaction in Virtual Environments Training Systems. (PhD dissertation),University of Central Florida.

Hoffman, R.R., Feltovich, P.J., Ford, K.M., Woods, D., Klein, G., Feltovich, A., 2002. A rose by any other name… would probably be given an acronym. IEEE Intell. Syst. 17 (4), 72–80.

Hugo, J., 2004. Design requirements for an integrated task support system for advanced human–system interfaces. In: Proceedings of the Fourth American Nuclear Society International Topical Meeting on Nuclear Plant Instrumentation, Controls and Human-Machine Interface Technologies. NPIC&HMIT 2010. American Nuclear Society, LaGrange Park, IL.

Hugo, J., 2012. Towards a unified HFE process for the nuclear industry. In: Proceedings of the Eighth American Nuclear Society International Topical Meeting on Nuclear Plant Instrumentation, Control, and Human–Machine Interface Technologies. NPIC&HMIT, San Diego, CA.

Hugo, J., Gertman, D., 2012. The use of computational human performance modeling as task analysis tool. In: Proceedings of the Eighth American Nuclear Society International Topical Meeting on Nuclear Plant Instrumentation, Control, and Human–Machine Interface Technologies. NPIC&HMIT, San Diego, CA.

ISO 9241-11, 1998. Ergonomic Requirements for Office Work With Visual Display Terminals (VDTs)'—Part 11: Guidance on Usability. International Standards Organization.

Jones, L.A., Sarter, N.B., 2008. Tactile displays: guidance for their design and application. Hum. Factors 50, 90–111.

Korsah, K., Holcomb, D.E., Muhlheim, M.D., et al., 2009. Instrumentation and controls in nuclear power plants: an emerging technologies update. NUREG/CR-6992, US Nuclear Regulatory Commission.

LaFerriere, K., Stevens, J., Flamand, R., 2019. Small modular reactor human-system interface and control room layout design. In: Proceedings of the Human Factors and Ergonomics Society, October 2019 Annual Meeting, Seattle, WA, pp. 516–520.

Lintern, G., 2007. What is a cognitive system? In: Proceedings of the Fourteenth International Symposium on Aviation Psychology, Dayton, OH, pp. 398–402.

Ni, T., Schmidt, G.S., Staadt, O.G., Livingston, M.A., Ball, R., May, R., 2006. A survey of large high-resolution display technologies, techniques, and applications. In: Proceedings of VR '06 Proceedings of the IEEE Conference on Virtual Reality. IEEE Computer Society, Washington, DC, pp. 223–236.

NUREG-0696, 1981. Functional Criteria for Emergency Response Facilities. US Nuclear Regulatory Commission, Washington, DC.

NUREG-0800, 2007. Standard Review Plan for the Review of Safety Analysis Reports for Nuclear Power Plants. US Nuclear Regulatory Commission, Washington, DC.

O'Hara, J., Brown, W.S., Lewis, P.M., Persensky, J.J., 2002. Human system interface design review guidelines. NUREG 0700 Rev2, US Nuclear Regulatory Commission, Washington, DC.

O'Hara, J., Higgins, J., Brown, W., 2008. Human factors considerations with respect to emerging technology in nuclear power plants. NUREG/CR-6947, US Nuclear Regulatory Commission, Washington, DC.

O'Hara, J., Higgins, J.C., Fleger, S.A., Pieringer, P.A., 2012. Human factors engineering program review model. NUREG-0711, Revision 3, US Nuclear Regulatory Commission, Washington, DC, USA.

Oviatt, S., 2003. Multimodal interfaces. In: Jacko, J., Sears, A. (Eds.), Handbook of Human-Computer Interaction. Erlbaum, Mahwah, NJ, pp. 286–304.

Persensky, J.J., Szabo, A., Plott, C., Engh, T., Barnes, V., 2005. Guidance for assessing exemption requests from the nuclear power plant licensed operator staffing requirements specified in 10 CFR 50.54(m). NUREG-1791, US Nuclear Regulatory Commission, Washington, DC.

Sheridan, T., 2002. Humans and Automation: System Design and Research Issues. Wiley Interscience, New York.

Silberglitt, R., Anton, P.S., Howell, R., Wong, A., 2006. The global technology revolution 2020: in-depth analyses: bio/nano/materials/information trends, drivers, barriers, and social implications. TR 303-NIC, 2006, RAND Corporation, Santa Monica, CA. Available at: www.rand.org/pubs/technical_reports/TR303/.

Stevens, J., LaFerriere, K., Flamand, R., 2019. Small modular reactor control room workstation demonstration. In: Proceedings of the Human Factors and Ergonomics Society October 2019 Annual Meeting, Seattle, WA, pp. 479–480.

WIRED, 2019. Brain-Machine Interface Isn't Sci-Fi Anymore. https://www.wired.com/story/brain-machine-interface-isnt-sci-fi-anymore. September 2019.

Woods, D., Hollnagel, E., 2006. Joint Cognitive Systems: Patterns in Cognitive Systems Engineering. Taylor and Francis, New York.

Wright, P., Dearden, A., Fields, B., 2000. Function allocation: a perspective from studies of work practice. Int. J. Hum. Comput. Stud. 52 (2), 335–355.

ISO 11064 (Parts 1-7). 1999, 2000, 2005, 2006, 2008, 2013. Ergonomic Design of Control Centers. International Organization for Standardization.

Safety of integral pressurized water reactors (iPWRs)

Bojan Petrovic
Georgia Institute of Technology, Atlanta, GA, United States

8.1 Introduction

Potential for enhanced safety is one of the hallmarks of SMRs designed with an integral primary circuit. (Further discussion on SMRs with integral primary circuit, in particular iPWRs, may be found in Chapter 3.) The lower power level of SMRs facilitates decay heat removal and results in a lower source term. The integral primary circuit configuration of iPWRs is more conducive to implementation of inherent safety characteristics. Combined with passive safety systems, it increases the probability of their adequate performance under different conditions, including the loss of off-site power. However, the limited experience database requires experimental efforts to address unique characteristics of iPWR SMRs (integral configuration with primary components). Integral test facilities are typically constructed to study new phenomena, to validate analysis methods, and to ultimately support design licensing. The enhanced safety characteristics of SMR, through PRA and risk-informed approach, may build a case to allow reducing the size of the emergency planning zone. The compact design improves the synergy of safety and security functions. With their unique characteristics, this class of reactors not only presents new and unique research challenges but also offers new and unique opportunities.

8.1.1 Key features of SMR/iPWRs relevant for safety

With respect to safety, large and small power reactors are expected to satisfy essentially the same technical, regulatory, and licensing requirements. However, a key question is whether there are inherent properties related to the power level that impact safety that would favor large or small reactors (IAEA, 1995, 2005a, 2009; Ingersoll, 2009; Petrovic et al., 2012). Considerations related to two important safety challenges (decay heat and source term) together with some additional aspects are discussed later:

- *Decay heat.* The most critical safety consideration of nuclear reactors is the removal of decay heat after a shutdown in off-normal conditions, in particular during a loss of off-site power event. Under the assumption of comparable power density, based on first principles, one can argue that SMRs are inherently safer in that respect. Namely, power is produced in the core and depends on the core volume, V, whereas, if everything else fails (in the absence of any other mechanism), it is ultimately dissipated through the vessel having surface area S. The larger the S/V ratio, the higher the inherent heat removal capability. Since volume increases faster than surface (third vs second power of linear size), this factor is smaller for large

reactors, and larger for SMRs, that is, SMRs inherently can reject a higher fraction of decay heat through their vessel.
- *Source term.* Assuming a similar specific power (W/gHM) and core-average fuel burnup, an SMR will have proportionally smaller source term than a large unit. However, a proportionally larger number of SMR modules will be needed to deliver the same total power. One could argue that a simultaneous failure of multiple units is unlikely, but events at Fukushima Daiichi NPP show that an unforeseen common mode failure is indeed possible, even if not likely. One should note that the common mode failure will at most make the source term for multiple SMRs as large as that of a large unit, but not worse. Moreover, due to their finer "power granularity," there is in principle more flexibility in siting (grouping or dispersing) SMRs.
- *Integral primary system.* Integral configuration is indeed what defines iPWRs. The feasibility of this design choice is facilitated by smaller core size and lower power level of SMRs, typically limited to about 300 MWe. Many (but not all) SMR designs have selected to implement integral configuration. Notably a concept of a large (1 GWe) integral PWR has been developed (Petrovic, 2014; ANE, 2017), aiming to achieve similar safety benefits as SMRs, but the implementation is more complex and requires use of advanced fuels and development of a nontraditional power conversion system.
- *Active and passive safety.* Systems providing safety functions will be discussed in more detail in the next section. Nevertheless, due to their importance, we introduce here the distinction between active and passive safety systems. Conceptually the latter are based on the laws of nature and therefore expected to be more likely to perform their function under a variety of off-normal conditions. While both small and large reactor designs may incorporate passive safety systems, practice shows that SMRs are more conducive to implementation of passive systems. A comprehensive review of proposed SMR concepts with passive safety would be cumbersome; instead a few representative examples are discussed later in this chapter. Examples of large PWRs with passive safety systems include the Westinghouse AP1000 (Schulz, 2006) and the I^{2}S-LWR concept (Petrovic, 2014; ANE, 2017).
- *Simplicity of safety systems.* A number of SMR concepts [cf. e.g., IRIS (Carelli et al., 2004; Petrovic et al., 2012)] have proved successful in systematically using their main defining characteristic—lower power level and size—to their advantage when implementing their safety approach. These designs tend to be based on simplicity enabled by the lower power level, thus promising to simultaneously achieve reliable safety and economics. They illustrate the key approach needed for a successful and competitive SMR design—it cannot be based on scaling down a larger design but needs to exploit unique SMR characteristics. For example, the integral primary circuit configuration—that is one of the enabling features and will be repeatedly mentioned throughout this chapter—is generally limited to SMRs due to the limiting reactor vessel size and cannot be directly extrapolated and applied to large power units, without additional new technologies.
- *Reactor footprint.* Smaller core size and lower power level also lead to a smaller NPP footprint, which in turn facilitates implementing seismic isolators and designing for inherent physical security.

Such properties and their impact are further discussed in the rest of this chapter.

8.1.2 Chapter overview

Expanding upon the unique aspects of SMR/iPWRs mentioned in the previous subsection, the rest of this chapter discusses their safety and related aspects:

- Section 8.2 presents approaches to safety (active and passive, inherent safety, and, safety by design) and specific solutions adopted in various SMRs to implement their safety approach.
- Section 8.3 discusses experimental efforts needed to address unique characteristics of SMRs, in particular the integral configuration and primary components within the primary loop, and provides examples of several integral test facilities.
- Section 8.4 focuses on probabilistic risk assessment (PRA). It illustrates the benefits of the safety-guided design and discusses the SMR potential to significantly enhance their safety indicators to allow reducing the size of the emergency planning zone, if possible to the plant site boundary.
- Section 8.5 briefly reviews security issues and in particular mutual synergy of safety and security functions.
- Section 8.6 presents future trends, in particular related design trends, research needs, improved analytic capabilities, licensing approaches, and testing and validation.
- "References" section provides references.

8.2 Approaches to safety: Active, passive, inherent safety and safety by design

While not limited to SMRs, it is worth summarizing the safety approaches, commenting in particular on inherent safety features and safety by design that will be shown to be preferentially linked to SMRs. "Inherent safety" and "safety by design" are sometimes used interchangeably. However, the former primarily reflects the presence of a feature that eliminates certain accidents, while the latter emphasizes that the design was consciously created to achieve such effect. There are other subtle differences in the use of safety-related terms in the United States and worldwide; IAEA-TECDOC-626 (IAEA, 1991) is a good source for the later, and therefore we will also cite it here. The following definitions may be applied to components and systems; we will introduce them as they apply to systems with understanding that they extend to components. Furthermore, we will focus on essence rather than precise definitions.

All NPPs incorporate a number of safety systems designed to provide safety functions and perform appropriate actions in off-normal and accident situations. According to the modus of their operation, they are divided into active and passive ones.

Active safety systems require external power, force, action, or signal. For example, decay heat removal may require an electric actuation signal, a motor-driven (or manually operated) valve to be opened, a pump to be operated to establish coolant flow, or some combination thereof. For active safety systems to operate, external power sources are required, and this can present vulnerabilities, even with multiple redundant and diverse external power sources (power lines, diesel generators, and batteries) as evidenced in the Fukushima Daiichi accident.

Passive safety systems, in contrast, operate based on the laws and forces of nature and are thus less susceptible to external impacts, that is, are ultimately less likely to fail. It is difficult to devise totally passive systems. In the US practice, this term is extended to include systems that in addition to their truly passive portion rely on stored energy to initiate the action (such as opening a valve, using battery or compressed air power, to establish natural circulation) that then proceeds based on the laws of nature.

Table 8.1 Example of passive and active systems and components classification.

Passive	Active
Reactor vessel	Pumps (except casing)
Reactor coolant system pressure boundary	Valves (except body)
Steam generators	Motors
Pressurizer	Diesel generators
Piping	Air compressors
Pump casings	Snubbers
Valve bodies	Control rod drive
Core shroud	Ventilation dampers
Component supports	Pressure transmitters
Pressure retaining boundaries	Pressure indicators
Heat exchangers	Water level indicators
Ventilation ducts	Switchgears
Containment	Cooling fans
Containment liner	Transistors
Electrical and mechanical penetrations	Batteries
Equipment hatches	Breakers
Seismic Category I structures	Relay
Electrical cables and connections	Switches
Cable trays	Power inverters
Electrical cabinets	Circuit boards
	Battery chargers
	Power supplies

Source: CFR, 2010. 10 CFR 54.21(a)(1)(i) in Code of Federal Regulations, Chapter 10 Energy, Parts 51 to 199, U.S. Government Printing Office (Rev. Jan. 1, 2010).

Table 8.1 illustrates the division into active and passive systems/components by providing an example of a specific NRC classification of structures and components listed in this case for the purpose of aging management review. It is adapted from 10 CFR 54.21(a)(1)(i), "Structures and Components Subject to Aging Management Review" (CFR, 2010).

A more precise distinction may be achieved (IAEA, 1991) by considering the *level of passivity*, ranging from Category A, which denotes passive systems requiring no signals, external forces, or power sources and involving no moving parts or moving fluids, to the least passive Category D, that requires or allows the following:

- Energy must only be obtained from stored energy sources such as batteries or compressed air or elevated fluids, excluding continuously generated power such as normal AC power from continuously rotating or reciprocating machinery.
- Active components are limited to controls, instrumentation, and valves, but valves used to initiate safety system operation must be single action relying on stored energy.
- Manual initiation is excluded.

An example is the emergency core cooling system, based on gravity-driven fluid circulation, initiated by fail-safe logic actuating battery-powered electric valves. For the complete definition of all categories, see the reference (IAEA, 1991).

Older so-called Generation II NPPs rely primarily on active safety systems. Many modern NPPs implement a combination of active and passive safety systems (IAEA, 2004). The Westinghouse Gen-III+ AP1000 is a passive safety plant, meaning that all safety systems are passive (Schulz, 2006).

It should be noted that passive systems may still fail, for example, a wall intended to function as a safety separation barrier may be destroyed in an earthquake; a pipe in a natural circulation loop may be crushed. This is where the *inherent (or intrinsic) safety* and the *safety-by-design principle* become important. One should differentiate between an overall inherently safe NPP and inherently safe features. Quoting IAEA-TECDOC-626 (IAEA, 1991): "*Inherent Safety* refers to the achievement of safety through the elimination or exclusion of inherent hazards through the fundamental conceptual design choices made for the nuclear plant." The reference continues on to emphasize: "Potential inherent hazards in a nuclear power plant include radioactive fission products and their associated decay heat, excess reactivity and its associated potential for power excursions, and energy releases due to high temperatures, high pressures and energetic chemical reactions. Elimination of all these hazards is required to make a nuclear power plant inherently safe. For practical power reactor sizes this appears to be impossible. Therefore the unqualified use of 'inherently safe' should be avoided for an entire nuclear power plant or its reactor." We will therefore use "inherent safety" to imply inherent safety feature(s), not the overall inherent safety.

A similar concept is expressed by the *safety-by-design* term, which emphasizes the fact that conscious design and engineering choices may lead to elimination of initiators for certain accidents or classes of accidents and thus to elimination of possibility for those classes of accidents to occur. Clearly, it is then not necessary to deal with hypothetical consequences of such accidents. That is, corresponding safety systems are not necessary, and the whole system becomes simpler, safer, and more economical, all at the same time. The safety-by-design principle is present to a certain degree in all (viable) reactor designs, more so in SMRs and in particular iPWR SMRs for the reasons that we will discuss. Arguably, it was most systematically pursued from the very beginning and implemented to a high degree in the IRIS design (Carelli et al., 2004; Petrovic et al., 2012), which may serve to illustrate many specific points. Table 8.2 summarizes implementation of safety by design (i.e., inherent safety features) in IRIS, but individual features are common with other iPWR SMRs. The intent of the safety-by-design approach in IRIS was to eliminate or reduce in severity—by design—as many of the Class IV accidents as possible.

Examining the table, it may be observed that out of the eight Class IV accidents typically considered in large loop PWRs, three have been eliminated (large break LOCA, control rod ejection, and reactor coolant pump shaft break), and four have been reduced in severity (reactor coolant pump seizure, steam generator tube rupture, steam system piping failure, and feedwater system pipe break). The first two are a direct consequence of the integral configuration, while for the remaining five accidents the positive impact is partly due to or supported by the integral configuration. Obviously the integral configuration by itself does not automatically resolve potential safety concerns, but it does facilitate addressing or eliminating many of them.

Table 8.2 Implementation of safety by design in IRIS.

IRIS design characteristic	Safety implication	Positively impacted accidents and events	Class IV design basis accidents	Safety-by-design impact on Class IV accident
Integral layout	No large primary piping	Large break LOCAs	Large break LOCA	Eliminated
Large, tall vessel	Increased water inventory Increased natural circulation	Other LOCAs Decrease in heat removal		
	Accommodates internal control rod drive mechanisms	Control rod ejection Head penetrations failure	Spectrum of control rod ejection accidents	Eliminated
Heat removal from inside the vessel	Depressurizes primary system by condensation and not by loss of mass	Other LOCAs		
	Effective heat removal by steam generator and emergency heat removal system	Other LOCAs All events requiring effective cooldown Anticipated transient without scram (ATWS)		
Reduced size, higher design pressure containment	Reduced driving force through primary opening	Other LOCAs		

Design feature	Safety-by-design implications	Accidents affected	Effect on accident
Multiple, integral, shaftless coolant pumps	No shaft Decreased importance of single pump failure	Shaft seizure/break Locked rotor	Reactor coolant pump shaft break Reactor coolant pump seizure	Eliminated Downgraded
High design-pressure steam generator system	No steam generator safety valves Primary system cannot overpressure secondary system Feedwater/steam piping designed for full reactor coolant system pressure reduces piping failure probability	Steam generator tube rupture Steam line break Feed line break	Steam generator tube rupture Steam system piping failure	Downgraded Downgraded
Once through steam generators	Limited water inventory	Feed line break Steam line break	Feedwater system pipe break	Downgraded
Integral pressurizer	Large pressurizer volume/reactor power	Overheating events, including feed line break ATWS		
Spent fuel pool underground	Security increased	Malicious external acts	Fuel handling accidents	Unaffected

Source: Petrovic, B., Ricotti, M., Monti, S., Cavlina, N., Ninokata, H., 2012. The pioneering role of IRIS in resurgence of small modular reactors (SMRs). Nucl. Technol. 178, 126–152.

Most PWR SMR designs base many of their inherent safety features on their integral design, that is, they are of the iPWR type (IAEA, 2012a). This is not by chance; it results from the synergy of two fundamental factors:

- PWRs are high-pressure systems and thus very sensitive to any leak or breach of the primary boundary. In contrast, for example, low-pressure lead or liquid salt-cooled systems would tend to self-plug leaks due to the solidification of the coolant, and this issue is not nearly as critical for them. The integral configuration eliminates external piping and multiple pressure vessels, thus eliminating or minimizing the probability of such events.
- Since the integral primary circuit configuration tends to significantly increase the pressure vessel size, it is generally not feasible for a large PWR, with an already large vessel, while it is generally feasible for an SMR with power level up to a few hundred MWe. However, an approach aiming to enable large power integral PWRs and thus facilitate extending SMR safety characteristics to GWe power level reactors has been successfully pursued and implemented into the I^2S-LWR concept (Petrovic, 2014; ANE, 2017).

An overview of representative approaches and passive and inherent safety features pursued in different iPWRs follows. We note that some also may apply or be used in large loop PWRs and some may not be fully passive, but instead have elements enabling passive safety or include specific safety approaches typically found in iPWR SMRs.

- *Integral primary circuit, integral configuration, and integral vessel layout.* As already discussed, several safety features are driven or assisted by integral configuration:
  - inherent elimination of large break LOCA;
  - inherent elimination of control rod ejection (with internal CRDMs enabled by integral configuration);
  - better response to transients (with a large primary inventory and increased pressurizer volume to power ratio, typical of integrated pressurizers);
  - immersed pumps may eliminate some primary pressure boundary penetrations and associated failure modes;
  - novel internal steam generators and primary heat exchangers eliminate or reduce in severity certain associated failure modes, such as the steam generator tube rupture (if primary is in the shell), or steam-line/feed-line break;
  - confinement of primary coolant to reactor vessel;
  - overall more compact design due to elimination of external loops.
- *Guard vessel*, reducing the impact of the primary boundary breach, but potentially impacting economics.
- *Natural circulation-based heat removal in normal (power) operation,* eliminating main coolant pumps and the possibility of their failure, typically limited to low-power systems.
- *Natural circulation-based decay heat removal in off-normal conditions,* eliminating the need for auxiliary pumps and external power, considered in many designs.
- *Soluble boron-free core.* This reduces corrosion, eliminates the need for a coolant volume control system (CVCS), and associated piping with primary pressure boundary (pressure vessel) penetrations. Of course, large PWRs may also be designed for soluble boron-free operation. However, cost considerations favor boron-free operation of (in particular smaller) SMRs and use of soluble boron in larger plant. This is due to the fact that the CVCS cost does not depend strongly on size, while the number of required control rods and their cost does. The cost breakeven point is design dependent but estimated to be in the 100- to 200-MWe range.

- *Increased operational margin,* sometimes at a cost of power downrating, which may impact economics.
- *Enhanced self-regulation.* Typically achieved through enhanced negative temperature and power feedback, frequently in conjunction with a soluble boron-free core, it provides a self-regulating and self-stabilizing effect. It is of particular interest for smaller units where load-follow operation is intended. It has to be considered, however, that a very strong negative feedback may have negative safety implications due to the reactivity insertion in cooldown scenarios.
- *Long life core.* It reduces the probability of refueling accidents and refueling outages and associated penalty and may increase proliferation resistance. However, it needs to be carefully assessed against the needed higher enrichment or reduced power density, offsetting these positive impacts. There is usually an economically optimum balance.
- *Very low-power reactors (with a small core fuel inventory)* may reduce the source term to the level that a small emergency planning zone (EPZ) is deterministically defendable, such that no off-site EPZ is needed.
- *Increased heat capacity of ultimate heat sink,* typically implemented through a combination of smaller power units and large pool, providing increased or indefinite decay heat removal capability and thus grace period after a hypothetical accident.
- *Inerted containment* to prevent hydrogen explosion (easier implemented for compact SMR designs).
- *Passive containment heat removal system* as another layer of defense.
- *Coupled pressure vessel and containment vessel response to LOCA events,* aiming to limit the loss of coolant inventory.
- Application of the *traditional defense in depth (DID) implemented through use of passive features* as much as possible.
- *Passive reactivity control systems*, including passive shutdown systems.
- *Near-zero self-regulating excess reactivity*, eliminating the possibility of prompt criticality, but usually limited to (very) low-power systems.
- *Enhanced seismic isolation.* Not specific to SMRs, but economically more feasible for compact iPWR designs (cf. e.g., Petrovic et al., 2012, Maronati et al., 2018).

It is instructive to review the implementation of the earlier features to specific iPWR designs. Eight specific designs have been selected among the many SMR designs that have been proposed over the last several decades. It would have been impossible and ineffective to include all or most of them. Instead a narrow selection aiming to be somewhat representative is presented here, together with the rationale for selection, while the reader interested in further designs is advised to examine review papers and several IAEA TECDOC's and OECD Handbooks that are being periodically prepared to capture all or most of the then-current designs under development (Ingersoll, 2009; IAEA, 1995, 2005a, 2006, 2012a; OECD, 1991, 2011; DOE, 2001).

The following rationale was used to guide our selection:

- The specific SMR design and safety concept had to be developed to a certain degree of completeness, maturity, and integration. Otherwise, it is possible to claim attractive features that however may or may not work together when the integrated design is analyzed.
- The selection focuses on power plants providing at least 40 MWe. Several interesting very low-power (typically <10 MWe) iPWR concepts have been proposed, but they are geared for a very specific purpose.

- The selection includes two prominent historical US designs that pushed the envelope and attracted broad attention by their overall safety characteristics and maturity of the design, SIR in 1990s (OECD, 1991; Matzie et al., 1992) and IRIS in 2000s (Carelli et al., 2004; IAEA, 2012a; Petrovic et al., 2012).
- The selection includes two prominent non-US design that have been under development for some time and are actively pursued, SMART (Chapter 18) (IAEA, 2005a, 2012a) and CAREM (Chapter 14) (IAEA, 2005a, 2012a) and additionally RITM-200 (Chapter 19) (IAEA, 2012a).
- The selection includes three of the four SMR designs proposed to the US DOE funding solicitations in 2012 and 2013: mPower (Halfinger and Haggerty, 2012; IAEA, 2012a; Azad, 2012), NuScale (Reyes, 2012; IAEA, 2012a; Ingersoll, 2012), and Westinghouse SMR (W-SMR) (IAEA, 2012a; Kindred, 2012). There was minimal information publicly available on the fourth design, SMR-160 or HI-SMUR [Holtec Inherently-Safe Modular Underground Reactor (HI-SMUR, n.d.)]. In the meantime, all design effort has been suspended for both the mPower and W-SMR designs.
- Of particular, interest is the NuScale design that is undergoing Design Certification with NRC and progressing toward construction of the first plant. More and updated information may be found at the NuScale technical publications webpages (NuScale, n.d.) and at the NRC NuScale Design Certification Application webpages (NRC DCA, n.d.).

Table 8.3 summarizes safety-related characteristics for the selected integral designs. The table is intended for illustrative purposes only to show the safety benefits of iPWR SMR designs as a whole and not intended for comparison between the designs, since it is nearly impossible and would be very lengthy to capture all features implemented in all specific designs.

It should be noted that while some of the listed and tabulated features have simultaneously positive impact on safety and economics (e.g., those passive safety systems that improve the safety while simplifying the design, thus making it more economical), other may challenge economics and require careful considerations and trade-off studies. Examples include power downrating that increases the capital cost and needs to be compensated by some compensating economic benefits, or which may enable certain design features not viable for current systems. While economics is discussed in a separate chapter, the reader is reminded here that the impact of safety choices on economics has always to be considered.

8.3 Testing of SMR components and systems

Most iPWR SMR designs aim to balance the use of proven LWR technology with the novel solutions necessary to exploit unique characteristics and potential advantages of SMRs. Thus it comes as no surprise that developing, analyzing, licensing, and ultimately deploying an SMR require testing and validation beyond that necessary for a "traditional" large loop PWR.

Testing of components and systems may be hierarchically divided into the following:

1. engineering development tests,
2. separate effects component tests,

Table 8.3 Summary of safety-related characteristics for selected iPWR designs.

	CAREM25	IRIS	mPOWER	NuScale	RITM-200	SIR	SMART	W-SMR
	Argentina	International	USA	USA	Russian Federation	USA, UK	Republic of Korea	USA
	CNEA	Consortium	Babcock & Wilcox	NuScale Power	OKBM Afrikantov	Combustion Engineering	KAERI	Westinghouse
Power level (MWth)	100	1000	530	160	175	1000	330	800
Power level (MWe)	27	335	180	50	50	320	100	225
Primary circuit circulation	Natural	Forced	Forced	Natural	Forced	Forced	Forced	Forced
Fully internal pumps	n/a	Yes	No	n/a	No	No	No	No
Soluble boron-free core	Yes	No	Yes	No	Yes	Yes	No	No
Internal CRDMs	Hydraulic	Electromagnetic	Electrohydraulic	No	No	No	No	Electromagnetic
Safety systems	Passive	Passive	Passive	Passive	(*)	Passive	Active and passive	Passive
DHRS	Passive	Passive	Passive	Passive	(*)	Passive	Passive	Passive
CDF	Target 1.0×10^{-7}	$\sim 1.0 \times 10^{-8}$	$\sim 10^{-8}$	$\sim 1.0 \times 10^{-8}$	(*)	(*)	1.0×10^{-6} target	(*)
LERF	Target 1.0×10^{-8}	$\sim 1.0 \times 10^{-9}$	(*)	(*)	(*)	(*)	Target 1.0×10^{-7} 1.0×10^{-8}	(*)

Notes: Single unit power level quoted. Most designs consider multimodule siting. In particular, NuScale reference design integrates 12 modules for a 600 MWe total; CDF, core damage frequency, per reactor-year; LERF, large and early release frequency, per reactor year; n/a, not applicable; (*), information not available in considered references.
Based on public information, mainly from the IAEA ARIS database (International Atomic Energy Agency, 2012a. Status of Small and Medium Sized Reactor Designs, A Supplement to the IAEA Advanced Reactors Information System (ARIS). IAEA, Vienna (September). Retrieved from: http://www.iaea.org/NuclearPower/Downloadable/Technology/files/SMR-booklet.pdf).

3. integral effects tests,
4. integral primary configuration test facilities,
5. testing of prototypes.

Specific objectives and approaches of each are further discussed later.

Engineering development tests aim to demonstrate feasibility and verify engineering capability before fabricating final components. *Separate effects component tests* examine the design, fabrication, operation, and qualification of large-scale prototypic components. They may include accelerated aging, irradiation, and seismic testing and establish performance of components. They also provide data for validation of computer codes. *Integral effects tests* examine and demonstrate integrated performance of combined systems or features, typically in a scaled setup. They may be used to show interaction between systems, including safety and nonsafety systems. They also may provide thermal-hydraulic performance parameters for models and analysis and be used to validate codes. *Integral primary configuration test facilities* are clearly of particular interest for iPWR SMRs since the integral primary configuration is the main difference to the existing, operating loop PWRs. They employ electrically heated rods to simulate the reactor core and are used to demonstrate the overall normal and off-normal performance of the whole system and validate system codes. For the same reason, several iPWR concepts are considering or have decided to build a scaled-down *prototype*. A prototype uses nuclear heat, that is, it is a critical reactor. Notably, CAREM-25 (IAEA, 2012a) is a 27-MWe prototype for a larger 100- to 200-MWe commercial version. According to IAEA (2012a), site excavation work for CAREM-25 was completed at the end of August 2012, and construction has begun. As quoted by the World Nuclear News, WNN (http://www.world-nuclear-news.org/) in December 2013, contract for the reactor vessel has been awarded, and, according to the CNEA, the unit is currently scheduled to begin cold testing in 2016 and receive its first fuel load in the second half of 2017. However, while progress was reported in 2017 and 2018, the fuel load and start seem to have been postponed to 2020 or later.

Engineering development tests for iPWR SMRs are driven by the novel components or components used in a novel way, as compared with loop PWRs. Depending on the specific iPWR design, this may include, among others, fuel and fuel assembly, internal control rod drive mechanism (electromagnetic or hydraulic), fully immersed pumps, integral steam generators (frequently with helical coils or multistage), integral pressurizer or self-pressurizing systems, and novel instrumentation to address specific needs of integral systems, for example, flow measurements and nuclear safety instrumentation (Petrovic et al., 2005).

Separate effects component tests also reflect iPWR SMR technical features. They may be related or extend engineering development tests beyond the development and may consider operation and qualification of large-scale prototypic components, including accelerated aging and irradiation testing. Some typical examples include the following:

- verifying heat transfer characteristics if novel heat exchanger types are considered for primary or decay heat removal,

- steam generators long-term inspectability and maintenance,
- main coolant pump operability and long-term performance,
- in-core instrumentation performance and long-term operability,
- fuel assembly performance (vibrations and seismic response).

Integral effects tests are intended to verify coupled performance and interaction between systems and may include the following (depending on the iPWR design):

- testing coupled performance of steam generator and emergency heat removal system (EHRS),
- reactor vessel and containment or guard vessel interaction,
- reactor coolant system and automated depressurization system (ADS) coupled performance.

Testing is necessary but expensive and complex. This may be illustrated by the following examples:

NuScale has performed a number of tests on components, such as steam generators, CRDMs, and fuel bundles (Ingersoll, NE50), and has built an integral test facility. NuScale has also commissioned a multimodule control room simulator laboratory with 12 independent module simulators, to demonstrate multimodule operation, since the basic configuration of NuScale power plant will include twelve 60 MWe modules (Ingersoll, 2012; NuScale, n.d.). The NuScale Technical Validation webpage accessed on 11/2019 listed the following testing and validation efforts and facilities (cited from the website):

- NuScale integral system test (NIST-1) facility located at Oregon State University in Corvallis, Oregon
- Critical Heat Flux testing at Stern Laboratories in Hamilton, Ontario Canada
- Helical coil steam generator testing at SIET SpA in Piacenza, Italy
- Fuels testing at AREVA's Richland Test Facility (RTF) in Richland, Washington
- Critical heat flux testing at AREVA's KATHY loop in Karlstein, Germany
- Control rod assembly (CRA) drop/shaft alignment testing at AREVA's KOPRA facility in Erlangen, Germany
- Steam generator flow-induced vibration (FIV) testing at AREVA's PETER Loop in Erlangen, Germany
- Control rod assembly guide tube (CRAGT) FIV at AREVA's MAGALY facility in Le Creusot, France

mPower invested over $100M in component testing program and building the integrated system test (IST) facility. It had conducted or planned to conduct tests on reactor coolant pumps, CRDMs, fuel, steam generator, and emergency high-pressure condenser (Azad, 2012; mPower, n.d., IST, n.d.). However, the mPower development efforts were stopped in 2017. The IST facility remained potentially usable, as stated on the facility webpage: "The current IST facility represents an investment of over $30 million. High fidelity systems, process equipment, control and data acquisition system, and process instrumentation comprise the facility enabling separate effects and integrated system testing" (IST, n.d.).

Westinghouse performed a series of tests related to W-SMR. For example, it built two full-scale fuel assemblies and tested internal CRDMs for its W-SMR (Kindred, 2012). However, as other reactor designs and projects seem to have gained higher priority, no further W-SMR tests were recently presented or announced by the company.

Of particular importance for iPWR are *phenomena related to natural circulation* (IAEA, 2005b). This is due to the fact that most of the concepts incorporate passive, natural circulation-driven, decay heat removal systems. Some of the lower-power concepts (e.g., CAREM-25, 27 MWe; NuScale, 60 MWe) also employ natural circulation for heat removal in normal full-power operation. A number of tests, related to specific passive systems, are planned aiming to better understand the phenomena, demonstrate performance of the systems, and validate codes and methodologies needed for design and licensing.

As already mentioned, *integral configuration test facilities* are indispensable for validation of system models and accurate simulation of the whole system and as the basis for licensing. Proper scaling is perhaps the most important preparation task for integral testing; it allows replicating correct physics and validating codes even though the geometry itself is not necessarily prototypical in all aspects, and the tests are electrically not nuclear heated. A hierarchical, two-tiered scaling analysis (H2TS) (Zuber, 1991) provides a framework for systematical decomposition that includes the system, subsystems, modules, constituents, geometrical configurations, physical phases (gas, liquid, and solid), fields, and phenomena. Some examples of such facilities are provided in the succeeding text.

8.3.1 IRIS SPES3 facility

The IRIS consortium designed the SPES3 integral test facility at SIET, Piacenza, Italy. Since substantial data are publicly available (Carelli et al., 2009), it will be used to illustrate some of the design choices. The facility is characterized by full height and prototypic fluid properties (pressure, temperature, and pressure drops). Volume ratio is 1:100; area ratio is the same, 1:100, to maintain the same residence times.

The facility simulates the primary circuit (integral vessel), secondary loop, and the containment system of the IRIS reactor. Comparison of additional selected parameters between the IRIS and SPES3 is shown in Table 8.4. Table 8.4 also indicates which IRIS systems are being represented. A 3D view of the SPES3 facility is shown in Fig. 8.1, while the SPES3 pressure vessel is shown in Fig. 8.2 (Carelli et al., 2009). Further details are available in Carelli et al. (2009).

8.3.2 NuScale integral system test (NIST)

NIST is a one-third scale (height) facility located at OSU in Corvallis, Oregon, USA, that replicates the entire NuScale Power Module and reactor building pool. Electrically heated simulated core enables achieving prototypic conditions, that is, operating temperature and pressure. Stability testing is performed to verify that stability of natural circulation will be maintained throughout the range of expected operating conditions. The facility implements extensive instrumentation and advanced data acquisition and control to enable the measurements necessary for safety code and reactor design validation. Tests have been performed to validate computer models including thermal efficiency, performance, and safety calculations (Reyes, 2010; Houser et al., 2013). The stainless steel facility operates at full system pressure and

Table 8.4 IRIS and SPES3 characteristic comparison.

System, component, or characteristic	IRIS	SPES3
Primary circuit integral reactor vessel (RPV) configuration with all primary components	Yes	Yes, except the pumps are external
System pressure (at pressurizer)	15.5 MPa	15.5 MPa
Core thermal power (MW)	1000	6.5
Core inlet temperature (K)	566	566
Core outlet temperature (K)	603	603
Pumps	8	1
Secondary loops	4	3
Helical steam generators (SG)	8	3
Steam generator height (m)	8.2	8.2
Steam generator tubes per SG	~700	14, 14, 28
Steam generator tube average length (m)	32	32
Emergency boration system (EBT)	2	2
Emergency heat removal system (EHRS)	4	3
Refueling water storage tank (RWST)	2	2
Dry well	1	1
Pressure suppression system (PSS)	2	2
Long-term gravity makeup system (LGMS)	2	2
Quench tank (QT)	1	1
Automatic depressurization system (ADS) trains	3	2
Containment system	Yes	Yes

Adapted from Carelli, M., Conway, L., Dzodzo, M., Maioli, A., Oriani, L., Storrick, G., Petrovic, B., Achilli, A., Cattadori, G., Congiu, C., Ferri, R., Ricotti, M., Santini, L., Bianchi, F., Meloni, P., Monti, S., Benamati, G., Berra, F., Grgic, D., Yoder, G., Alemberti, A., 2009. The SPES3 experimental facility design for the IRIS reactor simulation. Sci. Technol. Nucl. Install. 2009. https://doi.org/10.11552009/579430.

temperature and simulates reactor vessel, rod bundle, core shroud with riser, pressurizer, sump recirculation valves, helical coil steam generator, feedwater pump, containment vessel, and containment cooling pool. NIST scaling ratios are summarized in Table 8.5 (Reyes, 2010). A photo and schematic of the NIST containment and pool and pressure vessel is shown in Fig. 8.3 (Houser et al., 2013).

8.3.3 SMART integral test loop (SMART-ITL) facility

SMART-ITL located at KAERI, Republic of Korea, is a four-loop full-height test facility operating at prototypic conditions (pressure and temperature) and with the area ratio of 1:49 (NEI, 2013; Park, 2011). Its maximum power is 2 MW, or about 30% of the full power scaled down by area ratio. The facility consists of a primary system (simulating integral reactor vessel) with four steam generators, a four-loop secondary system that incorporates four trains of the passive residual heat removal system, each with a heat exchanger, an emergency cooldown tank and a makeup tank, together with valves and connecting pipes. SMART-ITL also incorporates several auxiliary systems. Further information is available in NEI (2013).

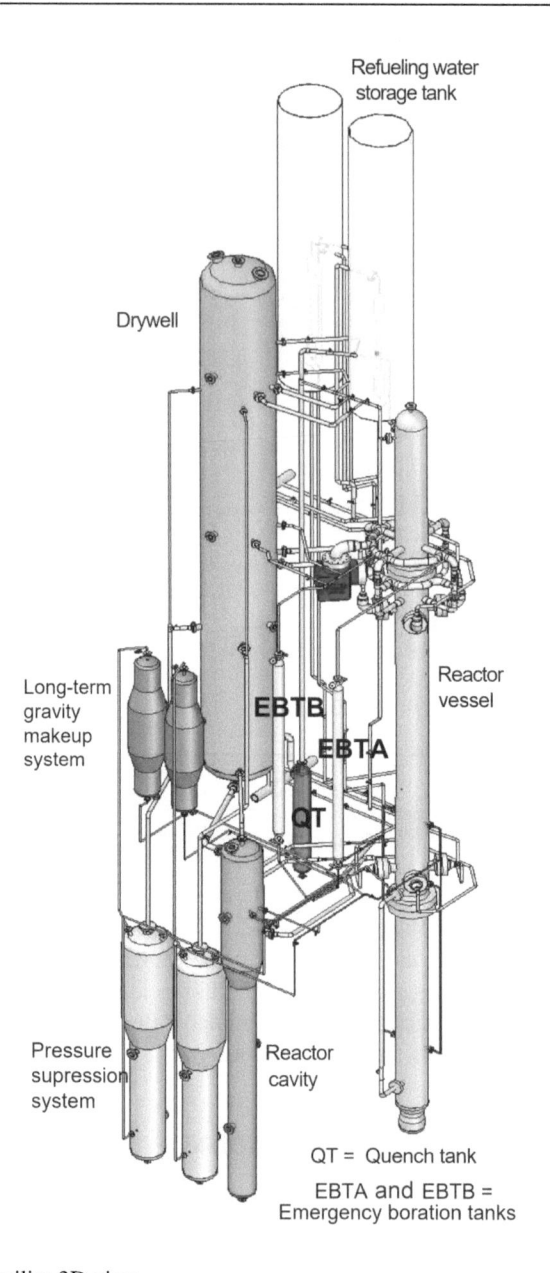

Fig. 8.1 SPES3 facility 3D view.
Adapted from Carelli, M., Conway, L., Dzodzo, M., Maioli, A., Oriani, L., Storrick, G., Petrovic, B., Achilli, A., Cattadori, G., Congiu, C., Ferri, R., Ricotti, M., Santini, L., Bianchi, F., Meloni, P., Monti, S., Benamati, G., Berra, F., Grgic, D., Yoder, G., Alemberti, A., 2009. The SPES3 experimental facility design for the IRIS reactor simulation. Sci. Technol. Nucl. Install. 2009. https://doi.org/10.11552009/579430.

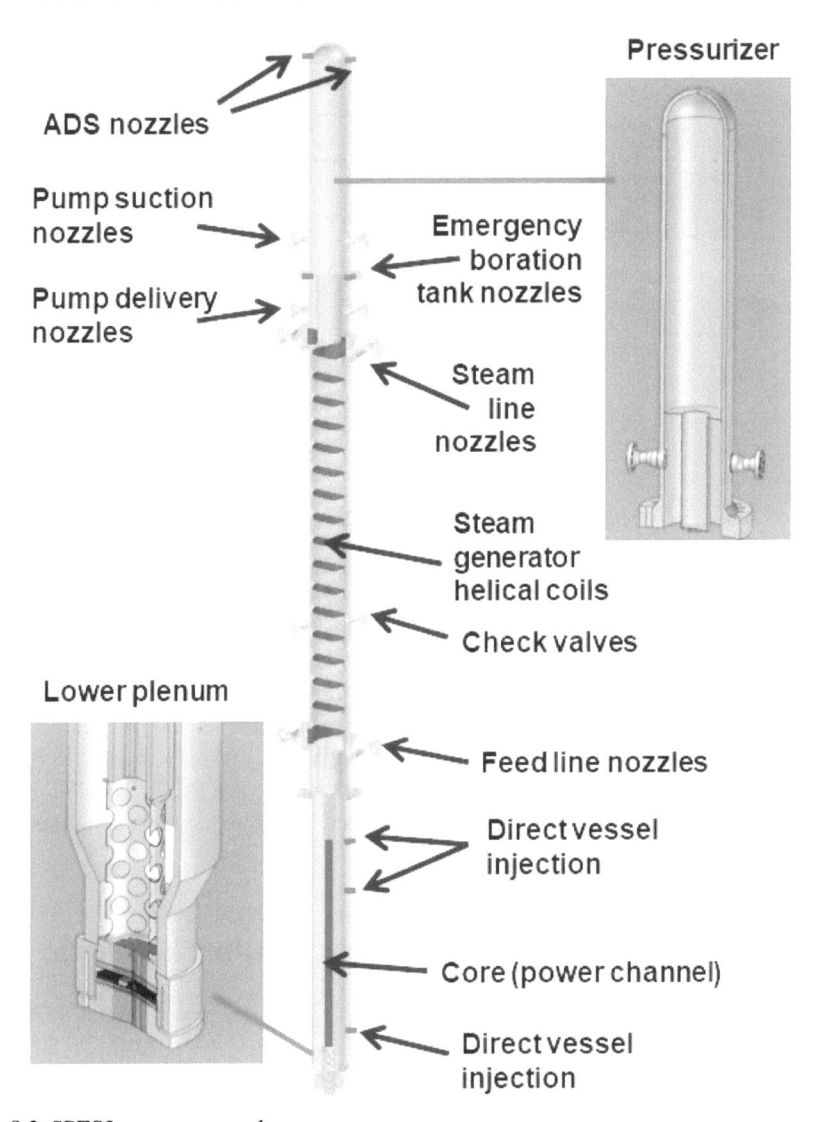

Fig. 8.2 SPES3 pressure vessel.
Adapted from Carelli, M., Conway, L., Dzodzo, M., Maioli, A., Oriani, L., Storrick, G., Petrovic, B., Achilli, A., Cattadori, G., Congiu, C., Ferri, R., Ricotti, M., Santini, L., Bianchi, F., Meloni, P., Monti, S., Benamati, G., Berra, F., Grgic, D., Yoder, G., Alemberti, A., 2009. The SPES3 experimental facility design for the IRIS reactor simulation. Sci. Technol. Nucl. Install. 2009. https://doi.org/10.11552009/579430.

Table 8.5 NIST scaling ratios.

Parameter	Ratio
Height	1:3.1
Pressure	1:1
Temperature	1:1
Cross-sectional area	1:82
Volume	1:255
Power	1:255
Velocity	1:3.1
Residence time	1:1

Source: Reyes, J.N., 2010. NuScale integral system test facility. In: Presentation to US NRC, June 2, 2010 (Retrieved from Adams, ML101520619).

8.3.4 B&W integrated system test (IST) facility

The B&W mPower integrated system test (IST) facility located in Bedford County, Virginia, USA, was constructed to include a scaled prototype of the B&W mPower reactor. All of the technical features of the B&W mPower integral reactor were included in the IST. The multiyear testing was planned to collect data to verify the reactor design and safety performance and support B&W's licensing activities with the Nuclear Regulatory Commission. The reactor prototype was installed in July 2011 and reached full operating conditions in July 2012 (mPower, n.d., IST, n.d.). However, the mPower development efforts were stopped in 2017. The IST facility remains potentially usable for separate effects and integrated system testing of other reactor concepts (IST, n.d.).

8.4 Probabilistic risk assessment (PRA)/probabilistic safety assessment (PSA)

While the topic of probabilistic risk assessment (PRA) or probabilistic safety assessment (PSA) is very broad and complex, in this section, we only aim to addresses some of the SMR-specific issues and achievements. We start with some general comments to provide a common ground and avoid misunderstandings stemming from the somewhat different licensing regulations and terminology used in different countries.

The US NRC virtual reading room NRC (n.d.) and IAEA (2012b) provide useful general information. As defined by NRC (n.d.), a design basis accident (DBA) is a "postulated accident that a nuclear facility must be designed and built to withstand without loss to the systems, structures, and components necessary to ensure public health and safety" (NRC, n.d.). Simplistically, these accidents are postulated assuming a single failure of a safety system, or a single "credible" event, and need to be deterministically "dealt with" with the remaining (diverse and redundant) safety-grade systems only, without the need and thus without taking credit for any action

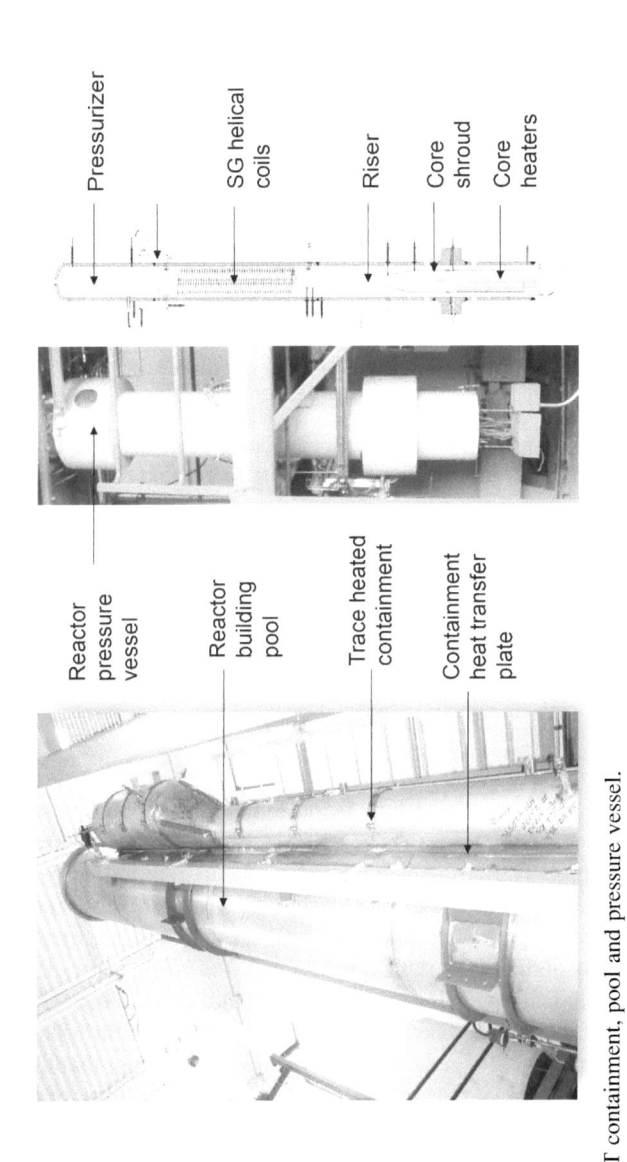

Fig. 8.3 NIST containment, pool and pressure vessel. Courtesy of NuScale.

of nonsafety-grade systems (which presumably could be less reliable). On the other hand, there is always a probability of multiple failures and events, termed beyond design basis events (BDBE). Identifying such BDB event trees and evaluating probabilities that ultimately they may lead to core damage, or even to radioactivity release, and estimating the consequences to the public, is performed by the PRA/PSA. For further discussion on Level I, Level II, and Level III PRA/PSA, the reader is directed to any PRA/PSA textbook. We note that in the PRA/PSA space, the action of nonsafety-grade systems may be taken into account (credited) with appropriate probabilities.

Some of the topics and features specific to iPWR SMRs that have impact on PRA/PSA are discussed later. Most of them are relevant to safety in general and not only to PRA/PSA but may therefore be discussed also in other sections of this chapter.

8.4.1 Defense in depth (DID)

SMRs may provide additional features, levels of safety, or barriers in the DID approach. Some SMRs have implemented one or more of the following DID-supporting features:

- Containment with higher design pressure that may provide an additional barrier or at least extend the time to radioactivity release. This is facilitated by more compact reactor and containment design.
- Passive cooling of the containment, avoiding its failure in certain loss of off-site power (LOOP) scenarios.
- Fully underground placement, potentially delaying and ameliorating a release of radioactivity.
- Significantly increased grace period may also be viewed as DID since it may allow a more systematic and better organized evacuation in case that all previous barriers fail.

8.4.2 Improved probabilistic safety indicators

Use of inherent safety features and safety by design eliminates certain accident initiators. Use of passive safety systems makes their functioning more probable in case that something does happen and removes dependence on an external power source. The combined impact reflected in most SMR designs is to significantly reduce the core damage frequency (CDF, per reactor year) and large and early release frequency (LERF, per reactor-year) probabilities. Thus CDF values of 1.0×10^{-7}–1.0×10^{-8} per reactor year are typically claimed for SMRs, with LERF probabilities usually at least one order of magnitude lower than CDF. This, if proven correct, would allow operating 10,000 reactors for 100 years (thus 10^6 reactor years) with only a small probability of an accident leading to core damage and practically negligible probability of any radiation release. In contrast the Generation II CDF values of 1.0×10^{-4}–1.0×10^{-5} imply that for 400 reactors operating over 50 years (thus 2×10^4 reactor-years), a CDF event should not be excluded, and they in fact did happen. While the health impact of all commercial nuclear accidents combined is in fact significantly smaller than the health effects that would have been caused by producing the same amount of electricity by fossil fuel, the accidents have negatively impacted public opinion, and iPWR SMRs (and advanced reactors in general) could help address that public concern.

8.4.3 PRA-guided design

In the PRA-guided design or safety-driven design, PRA is instituted from the very inception of the concept development and is used to continually and iteratively evaluate and inform the design and identify safety beneficial design changes. Obviously, any reactor development can benefit from the PRA-guided design, but there are specific technical and practical considerations that make SMRs more likely to benefit from this approach than evolutionary advanced large LWRs. Essentially, SMR designs tend to start with less well-defined details and less preconceived solutions and are therefore more open to consider even significant modifications. At the same time, they are more conducive to novel design solutions (such as the integral reactor vessel) needed to address the identified safety weak spots.

The PRA-guided design was implemented for the IRIS reactor (Kling et al., 2005). Fig. 8.4 illustrates this approach. The process is conceptually straightforward, but in practice, it may involve many redesign iterations.

By consistently applying PRA to enhance and extend its safety-by-design approach, IRIS has lowered the predicted core damage frequency (CDF) to below 10^{-7} events/reactor year and large early release frequency (LERF) to below 10^{-9} events/reactor year. PRA analyses and main design modifications are indicated in Fig. 8.5.

The CDF for the initial design, after reviewing dominant cut sets, was estimated to be $\sim 2 \times 10^{-6}$, a respectable number, but far from the IRIS target. In the next phase, marked in Fig. 8.5 as Step 1, sensitivity cases on individual significant factors

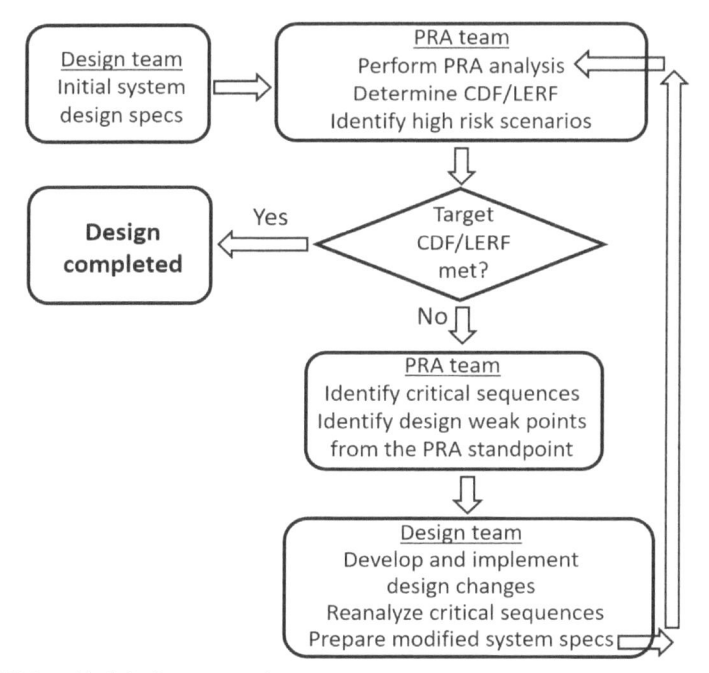

Fig. 8.4 PRA-guided design approach.

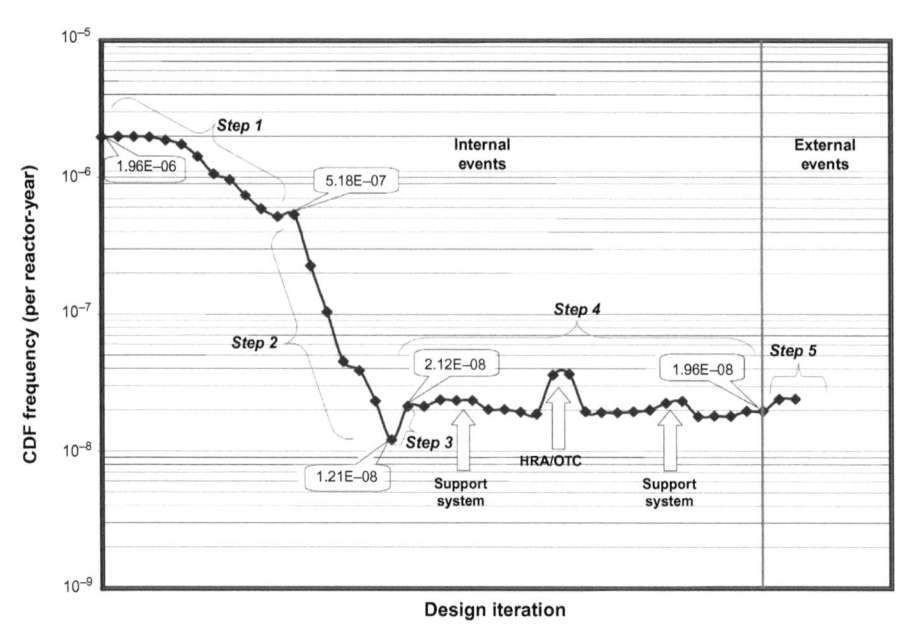

Fig. 8.5 CDF evolution in IRIS PRA guided design.
Adapted from Kling, C., Carelli, M.D., Finnicum, D., Alzbutas, R., Maioli, A., Barra, M., Ghisu, M., Leva, C., Kumagai, Y., 2005. PRA improves IRIS plant safety-by-design™. In: Proc. ICAPP'05, Seoul, Korea, May 15–19, 2005.

(test intervals, diversity, and reassessment) were performed, and the design was modified, reducing the CDF to $\sim 5 \times 10^{-7}$. That was a limit achievable by optimizing single parameters. In the next phase (Step 2), more complex design changes were evaluated to understand and improve coupled processes by simultaneous optimization of several parameters, enabling reduction of CDF to $\sim 1.2 \times 10^{-8}$. The next Step 3 accounted for higher level of design details, which now increased the CDF to $\sim 2 \times 10^{-8}$. Step 4 evaluated IRIS-specific auxiliary systems, anticipated transients without scram (ATWS), human reliability analysis, and further design details. Some weaknesses were identified that temporarily increased the estimated CDF; then the design was improved restoring the low CDF value of $\sim 2 \times 10^{-8}$. Step 5 indicates initial evaluation of external events. Every point on the graph represents an iteration including a PRA and a design modification. After implementing the PRA-suggested modifications to the reactor system layout, the preliminary PRA level 1 analysis estimated the CDF due to internal events (including ATWS) to be about 2×10^{-8}, more than one order of magnitude lower than in current advanced LWRs. Such improvement would not have been possible by mere "engineering judgment," without the benefit of systematic PRA-guided improvements and without the design characteristics of iPWR SMRs.

Clearly, PRA limited to only internal events would be misleading, and in fact, the external events frequently become the limiting factors in iPWR SMRs, but the approach may be extended to external events as well, for example, in Alzbutas et al. (2005).

8.4.4 Use of PRA/PSA to support eliminating off-site emergency planning zone (EPZ) for SMRs

The requirements and regulations related to the off-site emergency planning are different in different countries, but the overall considerations and impact are similar. The impact of an emergency planning zone (EPZ) extending beyond the nuclear power plant site boundary has both social and economic consequences. Socially, it projects the image of nuclear power being different than other industries with potential of impacting population outside the site boundary. (This is true of course for many other industries, but it is not necessarily recognized in public perception.) Economically a large EPZ has a large associated cost, resulting from the need for redundant evacuation routes, control of population density, etc. Large nuclear power plants, frequently built in multiples, may accommodate this penalty and offset it by their large power output.

SMRs, on the other hand, are ideally suited for a range of power outputs, for cogeneration (including district heating), for placement closer to the end user (thus within higher population areas), and for more diffuse siting. At the same time, their enhanced safety characteristics promote the possibility of licensing without the requirement for off-site emergency response, in other words, collapsing the EPZ to the site boundary.

Indeed, SMRs provide the technical basis for implementing such a change in EPZ:

- They have a smaller source term than large reactors, at least on the per reactor basis. While a common-cause multiple unit failure cannot be excluded, probabilistically, it is not very likely.
- iPWR SMRs tend to have smaller CDF and LERF; typical CDF values are on the order of 10^{-7}, significantly less than in the current Gen-II plants.
- Smaller source term combined with smaller CDF/LERF results in significantly reduced probability and severity of consequences.

Thus, from the purely technical viewpoint, it should be justifiable to reduce the size of the EPZ for SMRs, and, with adequate safety characteristics, this reduction would result in collapsing the EPZ to the site boundary (i.e., no off-site emergency planning is needed).

In practice the situation is more complex. The current US regulation, for example, institutes a uniform 10 mile EPZ size without regard to the plant specifics. Thus a new regulation for SMRs is needed where the EPZ size would be based on the risk-based or risk-informed approach; in the United States the latter is more likely to be acceptable. This issue has been raised several times in the past. The reader is encouraged to consult Carelli et al. (2008) for an example of a proposed technical approach of EPZ evaluation.

The recent strong interest in SMRs has led the NRC to explore the possibility of a generic resolution to emergency planning requirement issues for SMRs. The staff proposed a new approach for SMRs in SECY-11-0152 dated October 28, 2011 (NRC, 2011). The NRC webpage "Emergency Preparedness Rulemaking with Regard to SMRs and Other New Technologies" (NRC, 2015) contains updates and further information on the current status of this effort. More recently, NEI-has been leading the effort to establish a risk-based methodology for EPZ. This has been used by TVA and accepted by the NRC to conclude that a NuScale plant sited at TVA's Clinch River site could justify a site boundary EPZ.

In short, evaluating, accepting, and implementing the intrinsic capability of SMRs to limit their EPZ to the site boundary are one of the top-level targets pursued by the SMR developers and one of key technical enablers for economic competitiveness of SMRs.

8.4.5 Seismic isolators

Issues related to seismic events would warrant a separate chapter. Instead, we only point to one specific SMR feature. Due to their compact design, it is technically and economically feasible for most SMRs to place the nuclear island on seismic isolators and thus limit the impact of seismic events and significantly improve the PRA indicators. This in turn could further support the reduced EPZ.

8.4.6 Safety challenges of iPWR SMRs

In addition to their potential safety advantages, iPWR SMRs also pose specific safety-related challenges. The main ones are briefly summarized in the succeeding text:

- While in principle the reliability of passive systems and components should be higher, the actual experience-based reliability data are generally scarce. Thus a large uncertainty may need to be included in PRA/PSA.
- In particular, natural circulation plays a prominent role in the safety approach and safety systems of many iPWR SMRs, but there is much less operational experience with natural circulation in nuclear power plants and consequently some phenomena are not as well quantified as for the traditional forced-flow (pumped) systems. A good reference on the topic is IAEA (2012c).
- Smaller-power, multiunit sites will likely require a common control room to achieve economic competitiveness. All implications for safety performance and safety analyses are not well known at this time. However, NuScale has commissioned two 12-module control room simulators to model operation, train reactor operators, and demonstrate successful performance.
- Multiple SMR units at the same site will have smaller source term each than a large unit but similar total source term for a similar total power. While not likely, common mode failure cannot be completely excluded. There is no clear consensus how to treat and evaluate this situation in technical or regulatory space.
- SMRs are more conducive for cogeneration applications. Their market niche will likely include desalination, district heating, and other industrial processes where the PWR temperature is sufficient. Conceivably, there may be coupling and feedback between the nuclear and nonnuclear portion with an impact on safety performance. However, this has not been analyzed in detail, and actual experience is lacking.

8.5 Security as it relates to safety

Most SMRs (including in particular iPWR SMRs) claim some level of enhanced security compared with current large loop LWRs. This is justified by the following features that are favorable or easier to implement for SMRs:

Table 8.6 Synergy of safety and security features.

Feature	Safety impact	Security impact
Integral vessel	No external primary pipes, elimination of LB LOCA	Compact design, feasible partial, or full underground siting with enhanced physical protection
Compact containment	High design pressure; coupled vessel-containment performance in some designs limits coolant inventory loss	Compact design, feasible partial, or full underground siting with enhanced physical protection
Inherent safety features, passive safety, safety by design	Eliminates/reduces possibility of nonintentional initiation of certain accidents	Eliminates/reduces possibility of malevolent initiation of certain accidents
Compact rector building	Seismic isolation feasible	Compact design, feasible partial or full underground siting with enhanced physical protection

- Protection from external physical treats to the plant, such as airplane crash, by full or partial underground placement of a plant combined with reinforced outer structure(s). This is easier done for a small than for a large plant and typically includes placing at least the reactor and used fuel pool below grade.
- Access control and prevention of unauthorized access and intrusion by full or partial underground placement, limiting the number of access points and potential intrusion points (potentially to a single entry point). At the same time, this reduces the necessary surveillance and protection force and thus has a positive economic impact.
- More difficult access to safety-relevant equipment, again due to underground placement.
- Inherent increased resilience of passive safety systems to sabotage or intentional maloperation. Clearly, systems with less components, functioning on forces of nature, are more difficult to perturb and disable.
- Inherent safety features—accidents that cannot occur in the first place cannot be initiated, malevolently or otherwise.

Potential synergistic use of the safety and security characteristics is illustrated in Table 8.6.

Table 8.6 Synergy of safety and security features.

8.6 Future trends

Future trends for iPWRs are expected to focus on safety and economics, that is, those features that simultaneously enhance safety and economics.

In this author's view the most important design trends and features include the following:

- addressing safety from the very start, through safety-driven design (safety by design);
- using specific iPWR SMR characteristics to turn them into safety advantages;
- implementing inherent safety features;
- reliance on (exclusively) passive safety systems;
- introduction of additional features, levels of safety, and barriers to promote DID;
- partial or full below-grade placement for security reasons;
- licensing with eliminated or reduced off-site emergency planning zone (EPZ);
- extending postaccident grace period, with a larger and easier to replenish (or unlimited), ultimate heat sink;
- seismic isolation;
- advanced I&C to support safety in operation and status monitoring in off-normal conditions;
- advanced I&C for diagnostics/prognostics;
- health monitoring for reduced and optimized O&M;
- fuel with enhanced accident tolerance.

Additionally, low-power level systems will likely consider in their design:

- natural circulation for heat removal in normal operation;
- soluble boron-free operation (economically preferred for lower power levels);
- very long core life, or so-called "battery" approach;
- near-zero self-regulating excess reactivity, eliminating the possibility of prompt criticality, but usually limited to (very) low-power systems [such as ELENA, 3.3 MWth system envisioned for district heating (IAEA, 2012a)].

Related needs and challenges to enable or support desired safety features in the area of improved analytic capabilities and licensing include the following:

- *Improving our understanding of passive systems and their vulnerabilities and failure probabilities needed for PRA.* Currently, while the PRA approach is well established, specific probabilities and their uncertainties are not yet well quantified for iPWR SMRs.
- *Better understanding and addressing the potential for common mode failure and potential for negative mutual impact of multiple units at the same site.* As demonstrated by the Fukushima Daiichi accident, this is not a mere theoretical possibility. It requires developing reliable approaches to avoid such occurrences; otherwise, SMRs may not be able to fully claim (e.g., in licensing) the benefits of the smaller source term per unit.
- *Risk-informed licensing.* This will provide a framework for reducing the emergency planning zone based on adequate risk estimate rather than prescriptive values (Carelli et al., 2008).

Additionally, testing and validation will be necessary to address specific unique features:

- *Testing and experimental validation of natural circulation phenomena and integral primary configuration.* This will provide confidence in current models and support licensing.
- *Validation of advanced analytical methods.* Some novel or unique features (novel fuel, different flow regime than in current reactors, novel components, etc.) will require new simulation codes or methodologies that will need to be developed and validated based on carefully devised testing.

Some of the research needs include the following:

- *Developing and improving reliable passive decay heat removal approaches, with unlimited grace period.* This addresses THE main challenge—to ensure decay heat removal in ALL accident situations.

- *Developing effective "firewalls" between the nuclear and nonnuclear portion in cogeneration applications.* SMRs are more suitable for cogeneration applications, which require effective separation.
- *Developing novel components with improved performance.* Some examples include novel fuel forms and designs, internal CRDMs and steam generators, integral pressurizer, and fully immersed main coolant pumps.

Furthermore, considering trends in traditional PWRs and in non-LWR SMRs, the following research trends, even if currently not pursued for iPWR SMRs, could be expected to impact the next generation of iPWR SMRs:

- *Fuel with enhanced accident tolerance, such as coated particle type* (e.g., *TRISO*), *originally developed for VHTR, now as fully ceramic microencapsulated (FCM) considered for LWR.* This fuel would potentially add margin in certain safety aspects, but it has significant challenges of its own related to fabrication and cost. The author does not expect to see it applied to the first wave of iPWR SMRs.
- *Use of other-than-oxide fuel, such as nitride or silicide.* One possible driver is the desire for achieving a longer cycle, which could be justified for specific purposes, but it has its own development cost and challenges.
- *Amplified negative feedback.* Proposed for several high-temperature reactors and in some very low-power iPWRs [e.g., ELENA (IAEA, 2012a)], primarily to be used for self-regulation. However, one needs to consider cooling reactivity insertion, to make sure it is not introducing a more negative than positive safety impact for general applications.
- *Low-pressure systems.* Inspired by lead- or molten salt-cooled reactors that offer attractive safety characteristics, this is clearly not applicable for mainstream iPWRs; however, specialized iPWR applications may be tempted to consider significantly lowering the operating pressure.

References

Alzbutas, R., Augutis, J., Maioli, A., Finnicum, D., Carelli, M.D., Petrovic, B., Kling, C., Kumagai, Y., 2005. External events analysis and probabilistic risk assessment application for iris plant design™. In: Proc. 13th International Conference on Nuclear Engineering (ICONE-13), Beijing, China, May 16–20, 2005 (Paper ICONE13-50409).

ANE, 2017. Special issue on "Integral Inherently Safe Light Water Reactor Project (I^2S-LWR)" Ann. Nucl. Energy. 100 (Part 1).

Azad, A., 2012. Generation mPower. In: Presentation at the Georgia Tech Nuclear Engineering 50th Anniversary Celebration. Georgia Tech, Atlanta, GA. November 1.

Carelli, M.D., Conway, L.E., Oriani, L., Petrovic, B., Lombardi, C.V., Ricotti, M.E., Barroso, A.C.O., Collado, J.M., Cinotti, L., Todreas, N.E., Grgic, D., Moraes, M.M., Boroughs, R.D., Ninokata, H., Ingersoll, D.T., Oriolo, F., 2004. The design and safety features of the iris reactor. Nucl. Eng. Des. 230, 151–167.

Carelli, M.D., Petrovic, B., Ferroni, P., 2008. IRIS safety-by-design™ and its implication to lessen emergency planning requirements. Intl. J. Risk Assess. Manage. 8 (1/2), 123–136.

Carelli, M., Conway, L., Dzodzo, M., Maioli, A., Oriani, L., Storrick, G., Petrovic, B., Achilli, A., Cattadori, G., Congiu, C., Ferri, R., Ricotti, M., Santini, L., Bianchi, F., Meloni, P., Monti, S., Benamati, G., Berra, F., Grgic, D., Yoder, G., Alemberti, A., 2009. The SPES3 experimental facility design for the IRIS reactor simulation. Sci. Technol. Nucl. Install. 2009. https://doi.org/10.11552009/579430.

CFR, 2010. 10 CFR 54.21(a)(1)(i) in Code of Federal Regulations, Chapter 10 Energy, Parts 51 to 199, U.S. Government Printing Office (Rev. Jan. 1, 2010).

DOE, 2001. Report to Congress on Small Modular Nuclear Reactors (May 2001).

Halfinger, J.A., Haggerty, M.D., 2012. The B&W mPower scalable, practical nuclear reactor design. Nucl. Technol. 178, 164–169.

HI-SMUR, n.d. Information Available at: www.smrllc.com.

Houser, R., Young, E., Rasmussen, A., 2013. Overview of NuScale testing program. Trans. Am. Nucl. Soc. 109, 153–163.

Ingersoll, D.T., 2009. Deliberately small reactors and the second nuclear era. Prog. Nucl. Energy 51, 589–603.

Ingersoll, D., 2012. NuScale: expanding the possibilities for nuclear energy. In: Presentation at the Georgia Tech Nuclear Engineering 50th Anniversary Celebration. Georgia Tech, Atlanta, GA. November 1.

International Atomic Energy Agency, 1991. Safety Related Terms for Advanced Nuclear Plants (IAEA-TECDOC-626), Vienna.

International Atomic Energy Agency, 1995. Design and Development Status of Small and Medium Reactor Systems 1995 (IAEA-TECDOC-881). IAEA, Vienna.

International Atomic Energy Agency, 2004. Status of Advanced Light Water Reactor Designs 2004 (IAEA-TECDOC-1391). IAEA, Vienna.

International Atomic Energy Agency, 2005a. Innovative Small and Medium Sized Reactors: Design Features, Safety Approaches and R&D Trends (IAEA-TECDOC-1451). IAEA, Vienna.

International Atomic Energy Agency, 2005b. Natural Circulation in Water Cooled Nuclear Power Plants Phenomena, Models, and Methodology for System Reliability Assessments (IAEA-TECDOC-1474). IAEA, Vienna.

International Atomic Energy Agency, 2006. Status of Innovative Small and Medium Sized Reactor Designs 2005; Reactors With Conventional Refuelling Schemes (IAEA-TECDOC-1485). IAEA, Vienna.

International Atomic Energy Agency, 2009. Design Features to Achieve Defense in Depth in Small and Medium Sized Reactors, IAEA Nuclear Energy Series No. NP-T-2.2. IAEA, Vienna.

International Atomic Energy Agency, 2012a. Status of Small and Medium Sized Reactor Designs, A supplement to the IAEA Advanced Reactors Information System (ARIS). IAEA, Vienna. September. Retrieved from: http://www.iaea.org/NuclearPower/Downloadable/Technology/files/SMR-booklet.pdf.

International Atomic Energy Agency, 2012b. Safety of Nuclear Power Plants: Design, IAEA Safety Standards Series SSR-2/1 (Publication 15434). IAEA, Vienna.

International Atomic Energy Agency, 2012c. Natural Circulation Phenomena and Modelling for Advanced Water Cooled Reactor (IAEA-TECDOC-1677). IAEA, Vienna.

IST, n.d. Information on the IST Facility at: http://www.caer-ist.org/.

Kindred, T., 2012. Westinghouse SMR. In: Presentation at the Georgia Tech Nuclear Engineering 50th Anniversary Celebration. Georgia Tech, Atlanta, GA. November 1.

Kling, C., Carelli, M.D., Finnicum, D., Alzbutas, R., Maioli, A., Barra, M., Ghisu, M., Leva, C., Kumagai, Y., 2005. PRA improves IRIS plant safety-by-design™. In: Proc. ICAPP'05, Seoul, Korea, May 15–19, 2005.

Maronati, G., Petrovic, B., Ferroni, P., 2018. Assessing I2S-LWR economic competitiveness using systematic differential capital cost evaluation methodology. Ann. Nucl. Energy 2, 100 (Available online 19 June).

Matzie, R.A., Longo, J., Bradbury, R.B., Teare, K.R., Hayns, M.R., 1992. Design of the safe integral reactor. Nucl. Eng. Des. 136, 73–83.

mPower, n.d. Information Available at: http://www.babcock.com/products/Pages/IST-Facility.aspx.

NRC, 2011. Development of an Emergency Planning and Preparedness Framework for Small Modular Reactors (SECY-11-0152). NRC. October 28, 2011.

NRC, 2015. Webpage "Emergency Preparedness Rulemaking with Regard to Small Modular Reactors and Other New Technologies (RIN: 3150-AJ68; NRC Docket ID: NRC-2015-0225)." https://www.nrc.gov/reactors/new-reactors/regs-guides-comm/ep-smr-other.html.

NRC, n.d. http://www.nrc.gov/reading-rm/basic-ref/glossary/design-basis-accident.html.

NRC DCA, n.d. Webpage With Links to Documents Related to NuScale Design Certification Application: https://www.nrc.gov/reactors/new-reactors/design-cert/nuscale.html.

Nuclear Engineering International, 2013. Feature Article. December 2013, Retrieved from: http://www.world-nuclear-news.org/.

NuScale, n.d. Information Available at: https://www.nuscalepower.com. Selected Technical Publications Available at: https://www.nuscalepower.com/technology/technical-publications. Technical Validation and Testing is Listed at: https://www.nuscalepower.com/technology/technology-validation.

OECD/NEA, 1991. Small and Medium Reactors. vols. 1 and 2. OECD, Paris.

OECD/NEA, 2011. Current Status, Technical Feasibility and Economics of Nuclear Reactors. OECD, Paris.

Park, K.B., 2011. SMART, an early deployable integral reactor for multi-purpose application. In: Presented at INPRO Forum, October 10–14, 2011. IAEA, Vienna. Retrieved from: www.iaea.org/INPRO/3rd_Dialogue_Forum/08.Park.pdf.

Petrovic, B., 2014. Integral inherently safe light water reactor (I^2S-LWR) concept: extending SMR safety features to larger power output. In: Proc. Intl. Congress on Advances in Nuclear Power Plants 2014 (ICAPP'2014), Charlotte, NC, April 6–9, 2014.

Petrovic, B., Conti, D., Storrick, G.D., Oriani, L., Conway, L.E., 2005. Instrumentation Needs for Integral Primary System Reactors (IPSRs) (Final Report STD-AR-05-01). Westinghouse Electric Company.

Petrovic, B., Ricotti, M., Monti, S., Cavlina And, N., Ninokata, H., 2012. The pioneering role of IRIS in resurgence of small modular reactors (SMRs). Nucl. Technol. 178, 126–152.

Reyes, J.N., 2010. NuScale integral system test facility. In: Presentation to US NRC, June 2, 2010 (Retrieved from Adams, ML101520619).

Reyes, J.N., 2012. NuScale plant safety in response to safety events. Nucl. Technol. 178, 153–163.

Schulz, T.L., 2006. Westinghouse AP1000 advanced passive pant. Nucl. Eng. Des. 236, 1547–1557.

Zuber, N., 1991. An integrated structure and scaling methodology for severe accident technical issue resolution. In: Appendix D: A Hierarchical, Two-Tiered Scaling Analysis. U.S. Nuclear Regulatory Commission (NUREG/CR-5809).

Proliferation resistance and physical protection (PR&PP) in small modular reactors (SMRs)

Lap-Yan Cheng and Robert A. Bari
Brookhaven National Laboratory, Upton, NY, United States

9.1 Introduction

This section defines proliferation resistance (PR) and physical protection (PP) and describes why and how PR&PP matter for SMRs. A Gen IV International Forum (GIF) webinar on PR and PP (Gen IV International Forum, 2018) provides an overview of the background, development, and applications of the evaluation methodology developed by the GIF PR&PP Working Group for nuclear energy systems. It also presents the context for the need to address PR&PP issues early in the design stage, or the PR&PP-by-design paradigm.

9.1.1 Definitions of PR&PP for small modular reactors (SMRs)

The definitions (Gen IV International Forum, 2011a) of PR and PP that apply to most nuclear energy systems also apply to small modular reactors (SMRs). They are as follows:

- **Proliferation resistance** is that characteristic of an SMR that impedes the diversion or undeclared production of nuclear material or misuse of technology by the host state seeking to acquire nuclear weapons or other nuclear explosive devices.
- **Physical protection (robustness)** is that characteristic of an SMR that impedes the theft of materials suitable for nuclear explosives or radiation dispersal devices (RDDs) and the sabotage of facilities and transportation by subnational entities and other nonhost state adversaries.

Fig. 9.1 illustrates the methodological approach at its most basic. For a given system, analysts define a set of *challenges*, analyze *system response* to these challenges, and assess *outcomes*.

The challenges to the SMR are the threats posed by potential proliferant states and by subnational adversaries. The technical and institutional characteristics of the SMR

[☆] The material provided in this chapter relies heavily on the work (Gen IV International Forum, 2011a) done by the PR&PP Working Group of the Generation IV International Forum. The authors of this chapter have been, for the past decade, the current and past international cochairman of that organization. This manuscript has been authored by an employee of Brookhaven Science Associates, LLC, under contract DE-SC0012704 with the US Department of Energy.

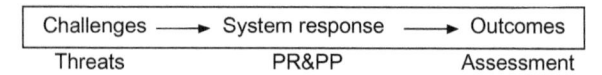

```
Challenges  ──▶  System response  ──▶  Outcomes
Threats           PR&PP                 Assessment
```

Fig. 9.1 Basic framework for the PR&PP evaluation methodology.

systems are used to evaluate the response of the system and determine its *resistance* to proliferation threats and *robustness* against sabotage and terrorism threats. The outcomes of the system response are expressed in terms of PR&PP *measures* and assessed.

The evaluation methodology assumes that an SMR has been at least conceptualized or designed, including both the intrinsic and extrinsic protective features of the system. Intrinsic features include the physical and engineering aspects of the system; extrinsic features include institutional aspects such as safeguards and external barriers. A major thrust of the PR&PP evaluation is to elucidate the interactions between the intrinsic and the extrinsic features, study their interplay, and then guide the path toward an optimized design. IAEA report STR-332 (IAEA, 2002) lists some examples of intrinsic and extrinsic proliferation resistance features and measures.

The structure for the PR&PP evaluation can be applied to the entire fuel cycle or to specific elements of the chosen fuel cycle (reactor, front end, or back end of the particular fuel cycle under consideration). The methodology is organized as a *progressive* approach to allow evaluations to become more detailed and more representative as system design progresses. PR&PP evaluations should be performed at the earliest stages of design when flow diagrams are first developed to systematically integrate proliferation resistance and physical protection robustness into the designs of SMRs along with the other possible high-level technology goals such as safety and reliability and economics. This approach provides early useful feedback to designers, program policy makers, and external stakeholders from basic process selection (e.g., recycling process and type of fuel), to detailed layout of equipment and structures, and to facility demonstration testing.

9.1.2 The importance of PR&PP for SMRs

The currently proposed SMRs (IAEA, 2018) have new design features and technologies that may require new tools and measures for safeguards and security. Some safeguards and security considerations for SMRs may be different from those for a large reactor (IAEA, 2014).

For example, there may be issues associated with the fuels such that the existing accountancy tools and measures may need to be modified or further developed for reactors using nonconventional fuel types, such as particle fuel in a high-temperature gas-cooled reactor (HTGR) and mixed fuel (uranium and plutonium) in a fast reactor. Further, issues may arise about new fuel loading schemes, as reactor cores with extremely long lifetimes may require innovative surveillance tools and measures, and long-life sealed core replacement may present novel accountancy challenges.

In evaluating the proliferation risk of SMRs, it is necessary to consider competing attributes. For example, the increased risk from using higher enriched uranium in some

SMR designs might be mitigated by the less frequent refueling that results in reducing the higher-risk period of handling fuel out of the core (Houses of Parliament, 2018). Yet, some SMR designs rely on fast spectrum core that reduces uranium contents in the fuel but requires fissile material from reprocessed spent fuel. In terms of PR risk, the reduction in uranium enrichment requirements is counterbalanced by the separation of fissile material from reprocessed spent nuclear fuel (Glaser et al., 2017).

International safeguards (IAEA, 2016) typically verify the operator's declaration of activities with nuclear material. These declarations address the receipts, shipments, storage, movement, and production of nuclear material. Inspections depend on the material type and whether the material is irradiated. The IAEA state-level approach (IAEA, 2013) will in addition consider the technical capabilities of the state including the possible existence of other nuclear activities (including commercial or academic R&D) and the location of the facilities.

Safeguard considerations will take into consideration differences in various factors, that is, the composition, physical form, and quantity of the nuclear material; the accessibility to the nuclear material, whether the reactor facility is operated continuously and how the reactor facility is refueled; the location and mobility of the reactor facility, and the existence and locations of other nuclear facilities in a given state.

For example, for nuclear materials that are normally not available to the host state, if the vendor state delivers a sealed unit that operates until the vendor replaces it with a new one, at some later time, and there is no storage capability for used fuel in the host state nor equipment to handle used fuel, this could raise issues, such as whether not only the reactor can be "sealed" by IAEA and treated as an item but also remote monitoring of the seal can readily detect any attempt to open the reactor. When comparing equivalent generating capacity, that is, many SMRs with the same total capacity with one large LWR, inspection issues would deal with whether SMRs will be colocated or separated at different sites. Additional issues would deal with their refueling schemes and whether they would be different and whether there would be separate used fuel storage for each module.

In the case of many small reactors in remote (e.g., arctic) separated locations compared with one large centrally located reactor with a large electric grid, it would be necessary to consider inspector ease of access to the remote site and the possibility of building an electrical distribution grid (Whitlock and Sprinkle, 2012). Other considerations are load-dependent versus base-load reactors, stand-alone sole source of energy supply, offshore SMRs on floating barges tied to a state/regional grid, and control versus ownership. In the context of PR&PP concerns, some aspects of SMR designs that pose improvements and challenges include fissile inventory, core life and refueling, high burnup fuel, digital instruments and control, underground placement, sealed core, higher enriched fuel, breeders, excess reactivity, fuel element size, coolant opacity, and floating reactors (Prasad et al., 2015).

The following considerations could apply to any new installation, including SMRs:

- Fuel leasing or supply arrangements that avoid on-site storage of fresh and/or used fuel.
- The isolation of the site or mobility of the reactor (sea or rail) might be a factor. Consideration should be given to access issues for both inspectorate and the adversary.

- Remote monitoring: there should be discussions between the operator/state/IAEA about small reactors that evaluate the potential of remote monitoring, including transmission of data off site.
- Will there be a different approach to physical protection and how might that affect the safeguards tools?
- Will the site or nearby sites have more or less ancillary equipment like hot cells, pin replacement capability, fuel storage, or nuclear research activities?
- Will the containment features be shared by multiple units? will there be underground containment?

The following discussion pertaining to physical security is derived in part from ideas in a white paper by the Nuclear Energy Institute (NEI, 2012) but with some change in emphasis.

Some of the same features that are being included in the design of SMRs as *safety* improvements may also improve their protection against physical threats. One feature common to some SMR designs is a compact reactor coolant boundary, contained mainly within the reactor pressure vessel (RPV). This feature may enhance the safety of light-water reactor-type SMRs, because large break loss-of-coolant accidents (LOCAs) may not need to be considered for these reactor types. This could also be potentially advantageous against deliberate acts.

Some SMRs may have several passive physical barriers and simplicity in systems required for safe shutdown. These may include such features as RPVs and containment vessels located underwater or below grade, the reactor building located partially or completely below grade, and fewer safe shutdown systems and components requiring physical protection. The below-grade installation of some SMRs may provide additional security benefits, such as minimizing aircraft impact, limiting access to vital areas, and the communication ability of adversaries. These features may provide a means of enhancing security system effectiveness against radiological sabotage. Use of the traditional multilayered defensive approach of deterrence, detection, assessment, delay, and interdiction can potentially be used effectively for physical protection of SMRs. Deterrence, detection, and delay concepts could be addressed in the early design phase of a facility to provide enough response time for on-site security force response. The ability to rely on an effective on-site response to a security threat is a potentially important factor that should be considered at the initial conceptual design phase to ensure enough intruder delays are included.

Examples of methods for extending adversary delay times, which in principle also apply to large plants that can be incorporated into SMR designs, include the following:

- locating and configuring vital components so that gaining access to these components is extremely difficult and time consuming for an intruder,
- locating and configuring critical safety systems so that there is no capability to destroy a target set from a single location,
- incorporating multiple layers of delay barriers against intruders and minimizing the number of access points to areas containing vital assets.

To the extent that SMRs are in the early stages of design or conceptual development, the earlier bullet items could be considered without the need to do potentially costly retrofits if these are considered after a plant is built.

One should also consider physical security system design options that minimize human involvement in security events (i.e., lower security risk profile), minimize impact of necessary future system modifications, and maximize adversary delay times. Examples include the following:

- designing the facility with minimum access points and multiple passive barriers based on a defense-in-depth approach to physical security;
- using redundant detection, assessment, and delay systems;
- using modular capabilities in physical security systems to minimize impact on station security staffing for system maintenance and upgrades needed to address system technology obsolescence and potential future increased design basis threats.

Again, to the extent that SMRs are in the early stages of design or conceptual development, the earlier bullet items could be considered without the need for potentially costly retrofits if these are considered after a plant is built.

A regulatory issue of importance related to physical security of SMRs is security staffing. Security staffing directly impacts operations and maintenance (O&M) costs and as such constitutes a significant financial burden over the life of the facility. SMRs are significantly smaller in size and system complexity that translates to improving security efficiency.

Key features of the physical protection programs that affect staffing expectations for nuclear facilities include the following:

- defense in depth using graded physical protection areas: increasing protection and control as vital equipment is neared (with well-defined boundaries),
- access authorization programs,
- robustness of intrusion barriers,
- alarm assessment to distinguish between false or nuisance alarms and actual intrusions and to initiate response,
- likely response to intrusions.

As discussed earlier, some general considerations for PR&PP assessments of SMRs are as follows:

- Smaller power reactors have smaller radiological inventories and thus potentially smaller releases during off-normal conditions.
- Smaller reactors have a smaller physical footprint, which can potentially lead to a smaller security force and fewer needs for surveillance and otherwise reduces the target area size.
- Some designs have potentially longer fuel cycles, which can potentially lead to fuel being inaccessible for longer periods of time.
- Smaller designs have potentially constrained ingress and egress, which makes detection and monitoring simpler.
- Ease of fuel assembly transport for some designs, which needs to be recognized and assessed.
- Higher enrichment levels for some SMRs relative to conventional light-water reactors, which need to be recognized and assessed.
- Remote locations of facilities present new challenges for inspections.

- Transportable facilities present unique technical and institutional issues relative to stationary facilities.
- Potential for a reduced security force.
- Potential for a reduced emergency planning scope.

The discussion given earlier applies generally to many SMRs currently under consideration. For the integral pressurized water reactors (iPWRs), which is the focus of this section of the Handbook, there are some distinguishing characteristics that may be pertinent for nonproliferation and security. Since the iPWRs are based on the larger, more familiar PWRs, it can be anticipated that international safeguards for the iPWRs could be developed from the basis of the current safeguards for large PWRs. Less conventional SMRs, of more novel designs, may require the development of additional specialized safeguards approaches. For iPWRs that have fuel enrichments similar to large PWRs, the material attractiveness considerations should be similar.

If the iPWRs operate with closed reactor vessels in a manner similar to large PWRs and have comparable fueling periods, this then offers a comparable barrier to fuel accessibility. If the iPWR can maintain a small physical footprint relative to other SMR designs, then this could be a security advantage.

9.2 Methods of analysis

9.2.1 The basic evaluation approach

The basic evaluation approach developed by the Generation IV International Forum's (GIF) working group on proliferation resistance and physical protection comprises definition of a set of threats or challenges, evaluation of the system's response to these challenges, and expression of outcomes in terms of measures.

A progressive approach permits broad application of the PR&PP evaluation to SMRs. SMRs assessed for PR&PP can range from systems under development to fully designed and operating systems. The scope and complexity of the assessment should be appropriate to the level of detailed design information available and the level of detail with which the threats can be specified.

The main steps to be performed in each component of the approach are illustrated in Fig. 9.2 and discussed in the following sections.

9.2.2 Definition of challenges

The initial step in the PR&PP assessment is the definition of the challenges, that is, of the threats considered within the scope of the evaluation. To be comprehensive a full suite of potential threats, referred to as the reference *threat set* (RTS), must be recognized and evaluated. If a subset of the threat space is to be the focus of a specific case study, the subset must be explicitly defined. Threats evolve over time; therefore system designs must be based on reasonable assumptions about the spectrum of threats to which facilities and materials in the system could be subjected over their full life cycles.

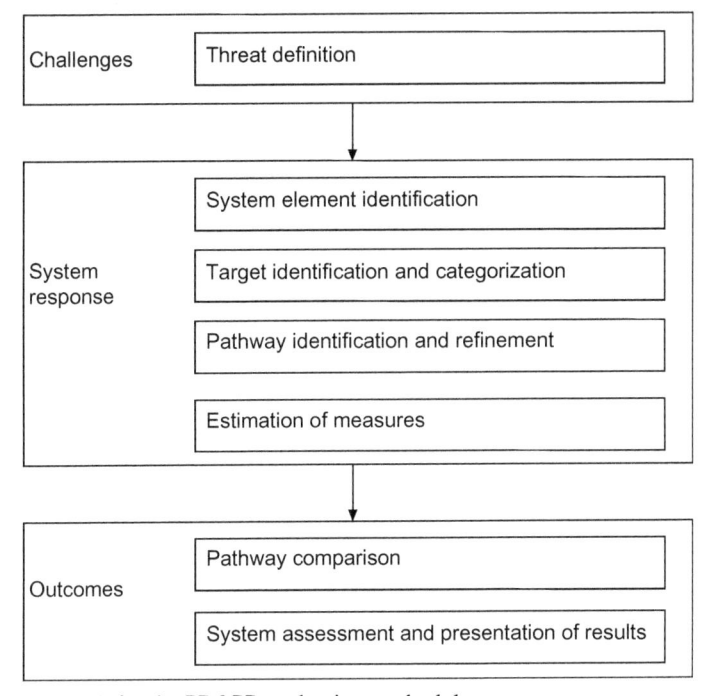

Fig. 9.2 Framework for the PR&PP evaluation methodology.

The level of detail in threat definition must be appropriate to the level of information available regarding design and deployment.

The definition of a specific PR&PP threat requires information both about the actor and the actor's strategy. Here, actor is defined by the following factors:

- type (e.g., host state and subnational)
- capabilities
- objectives
- strategies.

9.3 System response and outcomes

To evaluate the response of an SMR to proliferation, theft, and sabotage threats, analysts need to consider both technical and institutional characteristics of the SMR. The system response is evaluated using a pathway analysis method. Pathways are defined as potential sequences of events followed by actors to achieve their objectives of proliferation, theft, or sabotage.

Before analyzing pathways, it is important to define the system under consideration and identify its main elements. After identification of the system elements, it is possible to identify and categorize potential targets for each of the threats and identify pathways for those targets. The steps used to evaluate the system response are illustrated in Fig. 9.3 and discussed later.

```
┌─────────────────────────────────────────────────────────────────┐
│ System element identification                                    │
├─────────────────────────────────────────────────────────────────┤
│ Target identification and categorization                         │
├──┬──────────────────────────────────────────────────────────────┤
│  │ Pathway identification                                        │
│S │ and refinement       ┌──────────────────┬──────────────────┐ │
│y │                      │ Internal segment │ External segment │ │
│s │                      │ identification   │ identification   │ │
│t │                      ├──────────────────┴──────────────────┤ │
│e │                      │       Pathway connection            │ │
│m │                      └─────────────────────────────────────┘ │
│  ├──────────────────────────────────────────────────────────────┤
│r │ Estimation of measures                                        │
│e │                      ┌─────────────────────────────────────┐ │
│s │                      │ Estimation of measures for segments │ │
│p │                      ├─────────────────────────────────────┤ │
│  │                      │ Estimation of measures              │ │
│  │                      └─────────────────────────────────────┘ │
└──┴──────────────────────────────────────────────────────────────┘
```

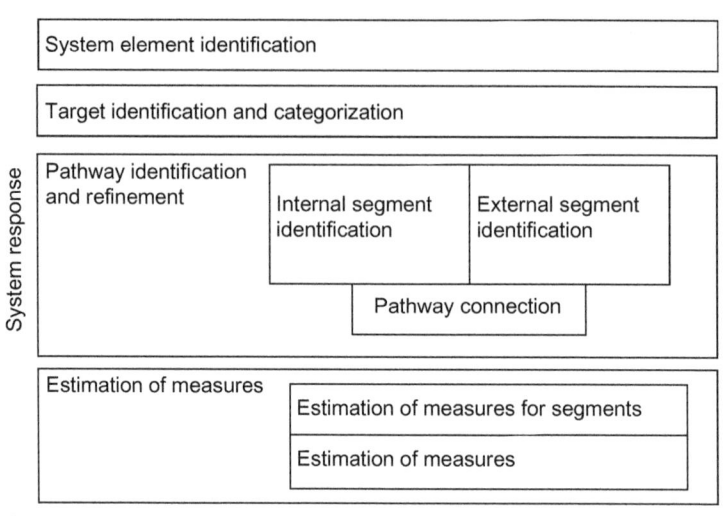

Fig. 9.3 System response steps.

9.3.1 System element identification

The boundaries of the system, which will limit the scope of the evaluation, must be clearly defined. Then the analyst must identify the system elements. The term *system element* is formally defined as a subsystem of the SMR; at the analyst's discretion a system element can comprise a facility (not just a building, but a facility in the systems engineering sense), part of a facility, a collection of facilities, or transportation within the identified SMR.

9.3.2 Target identification and categorization

Targets are the interface between the actors, and the SMR and are the basis for the definition of pathways. Clear, comprehensive target identification is an essential part of a PR or PP assessment. Targets can include nuclear or radiological material and processes, equipment, and information.

9.3.3 Pathway identification and refinement

Pathways are built around targets and are composed of segments. For coarse pathway analysis a *segment* consists of the *action* to be performed on the system. A complete PR pathway includes all actions for acquisition of material from the SMR, processing of the material into a form directly usable in weapons, and fabrication of the weapon. Each of these three general segments may comprise multiple refined subsegments. A PP pathway would involve similar general segments for theft of fissile or radiological material. For sabotage, general segments would include access to the target equipment, damaging or disabling the equipment, and the subsequent system response potentially resulting in a radioactive release.

To generate a credible set of pathways, a systematic method comprehensible to subject matter experts must be used. The analysts must provide confidence that all credible pathways are identified. However, the analysts must also avoid or dismiss pathways, after proper justification and documentation, that are obviously not credible or that do not contribute to the overall evaluation of the SMR.

Progressive refinement can proceed in two ways: Segments representing actions can be broken down into smaller subsegments, and characteristics can be added to segment descriptions to facilitate more accurate estimates of the measures.

9.3.4 Estimation of measures

The outcomes of the system's response are expressed in high-level measures for PR&PP, defined as follows:

9.3.4.1 Proliferation resistance

- *Proliferation technical difficulty*—The inherent difficulty arising from the need for technical sophistication, including material-handling capabilities, required to overcome the multiple barriers to proliferation.
- *Proliferation cost*—The economic and staffing investment required to overcome the multiple technical barriers to proliferation, including the use of existing or new facilities.
- *Proliferation time*—The minimum time required to overcome the multiple barriers to proliferation (i.e., the total time planned by the host state for the project).
- *Fissile material type*—A categorization of material based on the degree to which its characteristics affect its utility for use in nuclear explosives.
- *Detection probability*—The cumulative probability of detecting the action described by a segment or pathway.
- *Detection resource efficiency*—The staffing, equipment, and funding required to apply international safeguards to the SMR.

These measures are similar to those adopted in most assessments of PR (see discussion in Gen IV International Forum, 2011a).

9.3.4.2 Physical protection

- *Probability of adversary success*—The probability that an adversary will successfully complete a pathway and generate a consequence.
- *Consequences*—The effects resulting from the successful completion of the adversary's intended action described by a pathway, including the effects of mitigation measures.
- *Physical protection resources*—The staffing, capabilities, and costs required to provide PP, such as background screening, detection, interruption, and neutralization, and the sensitivity of these resources to changes in the threat sophistication and capability.

Measures can be estimated with qualitative and quantitative approaches, which can include documented engineering judgment and formal expert elicitation. Measures can also be estimated using probabilistic methods (such as Markov chains and event trees) and two-sided simulation methods (such as war-gaming techniques).

Gen IV International Forum (2011a) reviewed several system analysis techniques relevant for PR&PP studies.

9.3.5 Outcomes

To determine the outcomes of the system's response to a threat, analysts compare pathways and assess the system to integrate findings and interpret results.

9.3.5.1 Pathway comparison

Analysts perform a pathway analysis by considering multiple pathway segments. In general, measures are estimated for the individual segments of a pathway and must then be aggregated to yield a net measure for the pathway. Although measures for different pathways may be aggregated, it is generally more valuable to use the measures to identify the most vulnerable pathways. The objective of the system evaluation is then the identification of the most vulnerable pathways and the measures associated with them.

9.3.5.2 System assessment and presentation of results

The final steps in PR&PP evaluations are to integrate the findings of the analysis and interpret the results to arrive at an assessment of the SMR. Results include best estimates for descriptors that characterize the results, distributions reflecting the uncertainty associated with those estimates, and appropriate displays to communicate uncertainties.

9.4 Steps in the Generation IV International Forum (GIF) evaluation process

The GIF evaluation process includes nine specific steps that are organized into four main activities:

- D—Defining the work
- M—Managing the process
- P—Performing the work
- R—Reporting the work.

Each step is primarily associated with one of these activities. The nine steps are thoroughly explained in Fig. 9.4 and the accompanying text in the succeeding text. Clearly, some level of management is associated with each step. Reporting cannot all be done at the end but must be generated as the work progresses; thus the process is iterative, and sometimes the steps are concurrent. Note that the steps in the process are numbered in the order they are first performed (as shown in Fig. 9.4) but grouped for discussion under the four main activity areas described earlier.

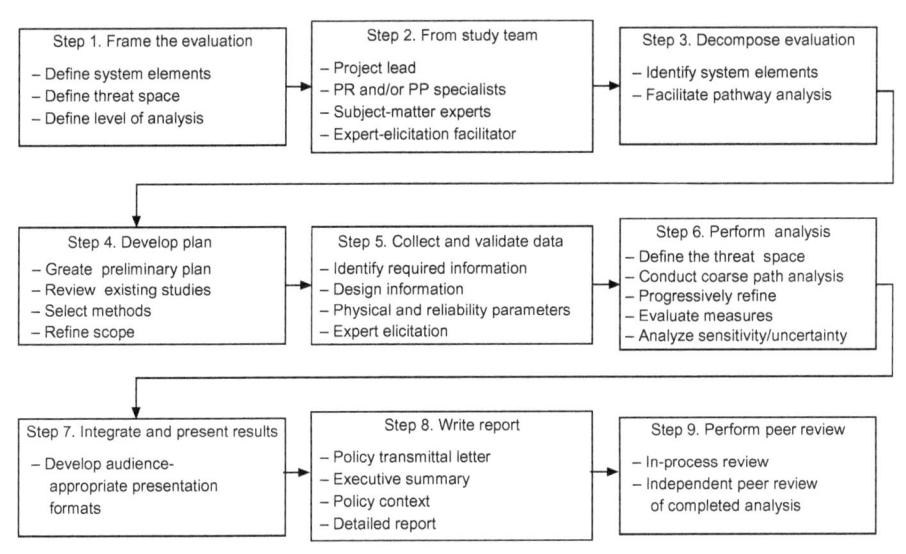

```
┌─────────────────────────────┐  ┌─────────────────────────────┐  ┌─────────────────────────────┐
│ Step 1. Frame the evaluation│  │ Step 2. From study team     │  │ Step 3. Decompose evaluation│
│ – Define system elements    │  │ – Project lead              │  │ – Identify system elements  │
│ – Define threat space       │  │ – PR and/or PP specialists  │  │ – Facilitate pathway analysis│
│ – Define level of analysis  │  │ – Subject-matter experts    │  │                             │
│                             │  │ – Expert-elicitation facilitator│                          │
└─────────────────────────────┘  └─────────────────────────────┘  └─────────────────────────────┘

┌─────────────────────────────┐  ┌─────────────────────────────┐  ┌─────────────────────────────┐
│ Step 4. Develop plan        │  │ Step 5. Collect and validate data│ Step 6. Perform analysis │
│ – Greate preliminary plan   │  │ – Identify required information│ – Define the threat space  │
│ – Review existing studies   │  │ – Design information        │  │ – Conduct coarse path analysis│
│ – Select methods            │  │ – Physical and reliability parameters│ – Progressively refine│
│ – Refine scope              │  │ – Expert elicitation        │  │ – Evaluate measures         │
│                             │  │                             │  │ – Analyze sensitivity/uncertainty│
└─────────────────────────────┘  └─────────────────────────────┘  └─────────────────────────────┘

┌─────────────────────────────┐  ┌─────────────────────────────┐  ┌─────────────────────────────┐
│ Step 7. Integrate and present results│ Step 8. Write report  │  │ Step 9. Perform peer review │
│ – Develop audience-         │  │ – Policy transmittal letter │  │ – In-process review         │
│   appropriate presentation  │  │ – Executive summary         │  │ – Independent peer review   │
│   formats                   │  │ – Policy context            │  │   of completed analysis     │
│                             │  │ – Detailed report           │  │                             │
└─────────────────────────────┘  └─────────────────────────────┘  └─────────────────────────────┘
```

Fig. 9.4 Steps in the evaluation process.

9.4.1 Main activities D and M: Defining the work and managing the process (steps 1, 2, 4, and 9)

To ensure the completeness and adequacy of results, one must structure the problem systematically, assemble an expert analysis team, and ensure competent peer review. The specific steps associated with activities D and M are described later.

9.4.1.1 Step 1: Frame the evaluation clearly and concisely (activity D)

The process of framing an evaluation requires close interaction between the analysts and the sponsors to specify the scope, particularly the system elements (facilities, processes, and materials) and the range and definition of threats. The institutional context in which safeguards and other international controls would be implemented (national and international safeguards requirements and regulatory guidance, etc.) must also be specified in sufficient detail.

The process allows for evaluation to be performed at many levels, depending on the sponsor's needs. From preconceptual design to a fully operational facility, the evaluation can and must become progressively more detailed. Time frame can also dictate the depth of analysis; quick and coarse evaluations may be needed when answers are required within weeks or months and, for some problems, potentially even sooner. Such shortcuts, however, entail a higher degree of uncertainty in the results.

9.4.1.2 Step 2: Form a study team that provides the required expertise (activity M)

The team should include experts in all required and relevant technical areas, as well as expertise in conducting the elicitation in an unbiased manner, with full description of the range of opinions. An example of an expert elicitation process for performing this work is provided in Gen IV International Forum (2011a).

9.4.1.3 Step 4: Develop a plan describing the approach and desired results (activity M)

Before undertaking this major analysis, the evaluation plan should be thoroughly developed, reviewed, and documented. In addition, the staff resources, costs, schedule, form of the results, and documentation must be clearly defined. Milestones should be developed, particularly for regular reporting to sponsors. A detailed plan for the conduct and use of peer reviews is also important to ensure quality. While developing the plan and implementing the information gathering and analysis tasks, coordination with safety evaluation, safeguards, and physical security work for the SMR could provide significant benefits.

9.4.1.4 Step 9: Commission peer reviews (activity M)

Any evaluation used to support decision-making or planned for wide distribution should include a peer review to ensure product quality. Two types of peer review have been widely used and provide different types of support:

- in-process peer review/steering committee,
- independent peer review of the completed analysis.

In-process peer review brings an expert group of practitioners and decision-makers into the process at regular intervals—perhaps once per quarter—to be briefed on the status of work and any known problem areas. Independent peer review allows objectivity through review of the finished product by independent outside experts who have not been involved in the evaluation. Both types of peer review have a potential role in proliferation resistance analysis.

9.4.2 Main activity P: Performing the work (steps 3, 5, 6, and 7)

Four steps are involved in this main activity. Steps 3 and 5 prepare for the required analysis, while the bulk of the analysis occurs under step 6, followed by integration of results for presentation in step 7. Details of the proposed process are summarized in Gen IV International Forum (2011a).

9.4.2.1 Step 3: Decompose the problem into manageable elements (main activity P)

This step decomposes the SMR into a set of system elements and threats to permit pathway analysis. Expert judgment may be used to identify system elements and threats that will be covered under qualitative, coarse pathway analysis, and those that should be subjected to progressive refinement with quantitative analysis.

9.4.2.2 Step 5: Collect and validate input data (main activity P)

The quantities and sources of input data depend on the scope of analysis. Validation of input data implies either the independent review of the data sources or examination of the consistency and basis for expert elicitation. If information and input data used in the analysis come from classified or sensitive sources, the analyst must ensure that this information is protected appropriately, including the possibility of classified or sensitive evaluation results. Most important in this step is a strong interface with facility designers. Designers should be key members of the proliferation resistance and physical protection evaluation team. Later, when the evaluation is applied to operating facilities, members of the operations team should also be included.

9.4.2.3 Step 6: Perform analysis (main activity P)

The actual evaluation is a multistage process. It addresses the system response and outcomes parts of the methodological approach. The process is summarized in Appendix A of Gen IV International Forum PR&PP (2011a).

9.4.2.4 Step 7: Integrate results for presentation (main activity P)

The presentation of results must be carried out carefully. In this process the analysts should reference and consider previous studies and should apply the best available analysis tools to generate results and prepare the output in an optimal form for presentation to designers, program policy makers, and external stakeholders.

9.4.2.5 Step 8: Write the report (main activity R)

As noted previously, reporting to the sponsors should be an ongoing process, and elements of the final report may be generated in draft form throughout the process. Ultimately the analysts must provide the results in a form that can be understood by the user, thereby enabling the user to draw appropriate conclusions. If the report contains classified or sensitive information, it may be necessary to abstract an unclassified summary. Gen IV International Forum (2011a) describes the information to be included in the report.

9.5 Lessons learned from performing proliferation resistance and physical protection (PR&PP)

9.5.1 Example sodium fast reactor (ESFR) case study

The PR&PP methodology was developed and tested with the help of an example design. This was the example sodium fast reactor (ESFR). The design is described in detail in Gen IV International Forum (2009) and consists of four 300-MWe sodium reactors and associated fuel-processing facilities. It is based on the integral fast reactor design that was in development in the 1990s. While this section of the SMR Handbook is focused on iPWRs, the lessons learned from the use of the PR&PP methodology on other reactor concepts are sufficiently generic that they can mostly apply to iPWRs as well. The basic element in common is that the analysis would be performed on a novel reactor concept that is still in the design phase.

Basic lessons learned from the Case Study Gen IV International Forum (2009) included the following:

- Each PR&PP evaluation should start with a qualitative analysis allowing scoping of the assumed threats and identification of targets, system elements, etc.
- Detailed guidance for qualitative analyses should be included in the methodology.
- Access to proper technical expertise on the system design and on safeguards and physical protection measures is essential for a PR&PP evaluation.
- The use of formal expert elicitation techniques can ensure accountability and traceability of the results and consistency in the analysis.
- Qualitative analysis offers valuable results, even at the preliminary design level.
- Greater standardization of the methodology and its use is needed.

In addition, during the evaluation process, the analyst must frequently introduce assumptions about details of the system design that are not yet available at early design stages, for example, the delay time that a door or portal might generate for a PP adversary. As the study progressed the PR&PP working group realized that when these assumptions are documented, they can provide the basis for establishing functional requirements and design bases documentation for a system at the conceptual design stage. By documenting these assumptions as design bases information, the detailed design of the SMR can be assured of producing a design that is consistent with the PR&PP performance predicted in the initial conceptual design evaluation (or, if the assumptions cannot be realized in detailed design, the original PR&PP evaluations must be modified appropriately).

9.5.2 Insights from interaction with GIF System Steering Committees (SSCs)

The interaction between the PR&PP Working Group and the GIF System Steering Committees (SSCs) has provided insights on the type of reactor system information that is necessary and useful to collect before one begins a PR&PP evaluation. These interactions resulted in the publication of a PRPPWG and GIF SSCs joint report including white papers delineating the main PR&PP features of the six GIF reactor technologies (GIF, 2011b). The report addressed also several cross-cutting issues (such as the fuel type and refueling modes, coolants, moderators, fuel cycle architectures, safeguards, safety, and economics) common to the GIF technologies. The report identified the importance of considering not only the reactor but also the fuel cycle needed by the reactor. The fuel cycle might not always require an in-depth PR&PP analysis, but the context might help in identifying potential PR&PP sensitivities in the front end or the back end of the fuel cycle as required by the GIF options under consideration. While the focus of the GIF is on six advanced reactor concepts of various designs, some alternative concepts within these designs are potentially SMRs (e.g., there are a sodium fast reactor design of low power and a lead fast reactor design of low power). Furthermore, all GIF concepts are in the design phase, which is much the same situation as with SMRs.

It is important to include information on major reactor parameters such as power, efficiencies considered, coolant, moderator (if any), power density values, fuel materials (this could be covered under fuel cycles), inlet and outlet conditions, coolant pressure, and neutron energy spectrum, for all design options under consideration.

Also useful is a high-level description of the type, or types, of fuel cycles that are unique to the reactor system and its major design options. A material flow diagram is valuable if available. Discussion should include mention of major waste streams that might contain weapon-usable material or be used to conceal diversion of weapon-usable material.

For a given SMR design, information that is particularly important to PR&PP will include potential fuel types (including high-level characteristics of fresh and spent fuel), fuel storage and transport methods, safety approach and associated vital equipment (for confinement of radioactivity and other hazards, reactivity control, decay heat removal, and exclusion of external events), and approach to physical arrangement as it affects access control and material accounting for fuel (a potential theft target) and access control to vital equipment (a potential sabotage target). Key high-level information to define or develop about the system elements is as follows:

- What material types exist or can exist within a system element?
- What operations are envisioned to occur in a system element, and whether (and how) these operations can be modified or misused?
- What kind of material movement is envisioned to occur normally in and out of a system element?
- What safeguards and security are envisioned to exist in the system element?

Potential adversary targets can be identified for the defined system elements. All system elements can be considered or only those that are judged to contain attractive adversary targets. Potential adversary targets are identified by considering material factors, facility factors, and safeguards considerations. Material factors include property attributes that can be determined from process flow sheets, such as isotopic compositions, physical forms, inventories, and flow rates. Facility factors include basic characteristics of equipment functions and facility operations, potential for facility/equipment misuse, facility/equipment accessibility, etc. Safeguards considerations include, for example, the ability of safeguards systems to detect illicit activities, facility accessibility to safeguards inspectors, availability of process information to safeguards inspectors, adequacy of containment and surveillance systems to detect diversion or misuse, and the degree of incorporation of safeguards into process design and operation.

A multilaboratory team of US subject matter experts, including several members of the PR&PP Working Group, used the PR&PP evaluation methodology as the basis for a technical evaluation of the comparative proliferation potential associated with four generic reactor types in a variety of fuel cycle implementations. These are a sodium fast reactor, a high-temperature gas reactor, a heavy water reactor, and a light-water reactor. The evaluation team undertook a systematic assessment, capturing critical assumptions and identifying inherent uncertainties in the analysis. A summary of the study was presented at the Institute of Nuclear Materials Management (INMM) 51st Annual Meeting (Zentner et al., 2010).

The relevance of the insights varies based on the various stakeholders of a PR&PP evaluation: policy makers, system designers, and the safeguards and physical protection communities.

For policy makers:

- An assessment of the proliferation potential of a particular reactor design in nuclear energy system should consider the system's overall architecture, accounting for the availability and flow of nuclear material in the front and back end of the fuel cycle.

For designers:

- Of the five PR measures, the designer will directly influence three: detection probability (DP), detection efficiency (DE), and material type (MT).
- To enhance DP and DE, designers can incorporate features in the design to facilitate easier, more efficient, and effective safeguards for inspection and monitoring. For example, minimizing the number of entry and exit points for fuel transfer between system elements will enhance material containment, protection, and accountancy (MCP&A), thus partially compensating for any lack of knowledge continuity by visual inspection during a fuel transfer.
- Material type for PR is related to the chosen composition of the nuclear material. The designer can optimize the design either to reduce the material's attractiveness (e.g., increase burnup in the uranium fuel to raise the fraction of Pu-238, thereby lowering the quality of plutonium in the spent fuel) or to make postacquisition processing of the material more complex, indirectly increasing the technical difficulty for the proliferator.

For safeguards inspectors:

- Augmenting inspections for handling and storing fresh and spent fuel would reduce proliferation potential.
- Enhanced inspection of fresh fuel would reduce the proliferation potential of covert diversion and misuse.
- Optimizing MT and material movement pathways to facilitate accountability measurements can make verification more effective and efficient.

9.6 Physical security

Consideration of physical security (PS) around SMRs is an integral part of evaluating the ability of SMRs to withstand the threats of, theft, and sabotage. In addressing PP issues, security is the equivalent of safeguards for PR issues. The following discussion is derived in part from a US Nuclear Regulatory Commission (USNRC) document, SECY-18-0076 (USNRC, 2018a) pertaining to recent developments in establishing a regulatory framework for the PS of SMRs.

Protection of plant features that provide vital safety functions, such as cooling of the reactor core, is the focus of the current physical security framework for large LWRs. The loss of plant features providing these safety functions can lead to damage to a reactor core or spent nuclear fuel, with a subsequent release of radioactive materials. SMR designs are expected to include attributes that result in smaller and slower releases of fission products following the loss of safety functions (see Section 9.1.2 for a discussion of some of these features related to SMRs in general). Accordingly, these designs may warrant different physical security requirements commensurate with the risks posed by the technology.

In the United States, Title 10 of the *Code of Federal Regulations* (10 CFR) Part 73, "Physical Protection of Plants and Materials," includes requirements for the physical security of power reactors. All applicants for a construction permit (CP) or operating license (OL) for a production or utilization facility under 10 CFR Part 50 and all combined license (COL) applications submitted under 10 CFR Part 52 must provide plans for satisfying the applicable requirements in 10 CFR Part 73. The USNRC staff noted in SECY-10-0034 (USNRC, 2010) that establishing physical security requirements and guidance for SMRs and non-LWR were key policy issues of high importance. In a subsequent report to the commission, SECY-11-0184 (USNRC, 2011), the USNRC staff concluded that the current security regulatory framework is adequate for SMRs, including related elements of the nuclear fuel cycle.

In a white paper (NEI, 2015), the NEI proposed a consequence-based physical security framework for SMRs and other new technologies (basically non-LWRs). The paper focused on the requirements related to protect against acts of radiological sabotage and discussed security by design features of many SMRs that could result in reduction in risk. The white paper suggested that a consequence-based security framework could demonstrate that the engineered safety features provide a level of protection exceeding the capabilities of the design basis threat. In that case the on-site response force would not be required to have the capability to interdict and neutralize the intruders but only to detect, assess, and communicate any security breaches to a designated off-site response force (i.e., local law enforcement).

The design basis threat (DBT) is to assess the capability of the physical protection program, including physical barriers, system design feature, and response force to detect, assess, interdict, and neutralize threats of theft and radiological sabotage. The threats are directed against target sets, defined as a minimum combination of equipment or operator actions that, if all are prevented from performing their intended safety function or prevented from being accomplished, would likely result in significant core damage or spent fuel sabotage barring extraordinary actions by plant operators. One of the technical bases for establishing reduced PS requirements for SMRs is the demonstration of incorporating security into designs to reduce reliance on human actions. For example, in the NEI White Paper: Proposed Physical Security Requirements for Advanced Reactor Technologies (NEI, 2016), three performance-based criteria have been proposed for determining the applicability of alternative security requirements for a specific design or facility:

(1) Uses a reactor technology that is not susceptible to significant core damage and spent fuel sabotage,
(2) Does not have an achievable target set,
(3) Has engineered safety and security features that allow for implementation of mitigation strategies to prevent significant core damage and spent fuel sabotage if a target set is compromised, destroyed, or rendered nonfunctional.

Once an advanced reactor design has been demonstrated to possess the performance capabilities underlying the earlier criteria, it is reasonable to expect with high confidence that through engineered safety and security measures, the response of the reactor facility to a radiological sabotage would result in no or delayed radiological consequences such that off-site resources could address the security event and the potential loss of safety functions resulting from the DBT.

Based on the current USNRC regulations, the approach for nuclear power plants to meet PS requirements is largely prescriptive planning standards and is technology specific, namely, large LWRs. In a performance-based approach to PS rulemaking, performance and results are the primary basis for regulatory decision-making, and the licensee has the flexibility to determine how to meet the established performance criteria for an effective PS program. The USNRC staff laid out in SECY-18-0076 (USNRC, 2018a) options for revising regulations and guidance related to PS for advanced reactors, including light-water SMRs and non-LWRs. The USNRC staff recommends pursuing a limited-scope rulemaking and using a performance-based and consequence-oriented approach to further assess and, if appropriate, revise a limited set of USNRC regulations and guidance to provide an alternative to current PS requirements for license applicants of advanced reactors, including SMRs. Subsequently in SRM-SECY-18-0076, "Staff Requirements—SECY-18-0076—Options and Recommendation for Physical Security for Advanced Reactors," (USNRC, 2018b), the commission directed the staff to proceed with the staff's recommended limited-scope rulemaking. The most likely focus of this limited-scope rulemaking would be to evaluate an alternative to the prescribed minimum number of armed responders currently defined in 10 CFR 73.55(k). Another potential area is the prescriptive requirements in 10 CFR 73.55 for on-site secondary alarm stations.

The performance-based approach has the flexibility of implementing an effective physical protection system by incorporating many combinations of physical protection measures while recognizing that each reactor facility and its operational circumstances may be different. The SMR designs are expected to include attributes (e.g., smaller core or passive safety systems) that result in smaller and slower releases of fission products following a loss of safety functions from malfunctions or malicious acts. The delayed radiological consequence affords more time to detect, asses, notify, and respond to an intrusion. The prolonged coping time facilitates alternative methods or approaches to include increased reliance on engineered systems that reduce reliance on operational requirements and staffing to meet the intent of regulatory requirements.

9.7 Future trends

The PR&PP methodology provides a framework to answer a wide variety of nonproliferation and security-related questions for SMRs and to optimize these systems to enhance their ability to withstand the threats of proliferation, theft, and sabotage. The PR&PP methodology provides the tools to assess SMRs with respect to the nonproliferation and security.

PR&PP analysis is intended to be performed, at least at a qualitative level, from the earliest stages of system design, at the level where initial flow diagrams and physical arrangement drawings are developed, and simultaneously with initial hazards identification and safety analysis. The methodology facilitates the early consideration of physical security and proliferation resistance because the structure of the PR&PP methodology bears strong similarity to safety analysis.

The PR&PP methodology adopts the structure of systematically identifying the nonproliferation and security challenges a system may face, evaluating the system response to these challenges and comparing outcomes. The outcomes are expressed in terms of measures, which reflect the primary information that a proliferant state or an adversary would consider in selecting strategies and pathways to achieve its objectives. By understanding those features of a facility or system that could provide more attractive pathways, the designer can introduce barriers that systematically make these pathways less attractive. When this reduction may not be possible, the analyst can highlight where special institutional measures may be required to provide appropriate levels of security.

Beyond requiring that a systematic process be used to identify threats, analyze the system response, and compare the resulting outcomes, the PR&PP methodology provides a high degree of flexibility to the analyst, subject to the requirement that the results of studies receive appropriate peer review. For this reason, it is anticipated that approaches to performing PR&PP evaluations will evolve over time, as the literature and examples of PR&PP evaluations expand. Different tools for identifying targets, evaluating system response and uncertainty, comparing pathway outcomes, and presenting results can be expected to increase in number, as will the range of questions that can be answered and insights gained from PR&PP studies.

The evolving nature of PR&PP evaluations during the design cycle of nuclear energy systems is illustrated by an initiative to update the white papers (GIF, 2011b) developed jointly by the PRPPWG and the GIF SSCs. The update process (Cheng et al., 2018) provides an opportunity to apply the concept of PR&PP by design to the current systems under consideration by GIF (some systems are nonlight-water SMRs). When done properly, PR&PP by design would achieve the technical objectives of having a system that has better PR&PP *intrinsic features* (e.g., system engineering and process features that naturally make a proliferation/nuclear material theft/sabotage effort unattractive). The system would also exhibit a better *safeguardability* [here defined as "the degree of ease with which a nuclear system can be effectively and efficiently put under international nuclear safeguards" (GIF, 2011a)]. The appendices in IAEA (2014) include information on the example of a procedure to support a facility's analysis of the safeguarding situation in support of safeguards by design (SBD) and a template listing required proliferation resistance-related design information. The SBD is generally recognized as an approach wherein international safeguards are fully integrated into the design process of a nuclear facility—from initial planning through design, construction, operation, and decommissioning. In a workshop (IAEA, 2009), safeguards and facility design experts convened to discuss basic principles to design and implement an SBD process and to identify fundamental design features deemed critical to the effective and efficient implementation of IAEA safeguards.

The trend of building future SMRs with inherent safety and security features as well as accommodations for effective and efficient safeguards highlights the importance of considering interfaces of safety, security, and safeguards at all stages of the life cycle of a nuclear plant. A challenge to integrate safety, security, and safeguards (3S) is to identify 3S interfaces to enable the optimization of synergistic effects and the minimization of potential conflicts (Khalil et al., 2012). It is feasible to realize cases in which measures intended to improve safety could have a negative effect on security or vice versa. A task force under the auspices of the IAEA (IAEA, 2010) explored a better understanding of the interfaces between safety and security at nuclear power plants and discussed the means to achieve both objectives in an optimal fashion. There are examples of technical synergies between nuclear security and safeguards in the regulatory control of new nuclear power plants (Martikka et al., 2018).

9.8 Sources of further information and advice

The reader is encouraged to see the collection of journal articles on PR&PP that is contained in a special issue of the American Nuclear Society's *Nuclear Technology* (Volume 179, Number 1, 2012). This journal issue contains numerous articles on methods and applications of PR&PP and related approaches. The US National Academy of Sciences has issued a review of methods for proliferation risk assessment and its applications to decision-making. This report was issued in 2013 and can be obtained at http://www.nap.edu/catalog.php?record_id=18335. An online monograph (Rosenthal et al., 2019) provides a comprehensive introduction to the international safeguards systems and the history of the International Atomic Energy Agency. The publications in the IAEA Nuclear Energy Series provide information in the areas of nuclear power, nuclear fuel cycle, radioactive waste management and decommissioning, and on general issues that are relevant to all of the aforementioned areas. The IAEA Nuclear Security Series of publications address nuclear security issues relating to the prevention and detection of, and response to, theft, sabotage, unauthorized access and illegal transfer or other malicious acts involving nuclear material and other radioactive substances and their associated facilities. The World Institute for Nuclear Security (WINS, https://wins.org), through its publications, workshops training, and webinars, provides an international forum in which to share and promote the implementation of best security practices among those with responsibilities for managing the security of nuclear and other radioactive material.

The IAEA has developed manuals for use by member countries on assessment approaches for nonproliferation and security aspects of innovative reactors. It has issued a report on "Options to Enhance Proliferation Resistance of Innovative Small and Medium Sized Reactors" (see IAEA, 2014). Another recently issued report by the IAEA is International Safeguards in Nuclear Facility Design and Construction, IAEA Nuclear Energy Series No. NP-T-2.8, 2013, which discusses how international safeguards concepts can be introduced during the design phase of a nuclear facility. As part in a series of facility-specific safeguards by design guidance publications, *International Safeguards in the Design of Nuclear Reactors*, IAEA Nuclear Energy Series No. NP-T-2.9, 2014, is applicable to the design and construction of nuclear power reactors, and to research reactors.

The IAEA has also developed guidance for proliferation resistance assessments. It can be obtained at "Guidance for the Application of an Assessment Methodology for Innovative Nuclear Energy Systems: INPRO Manual—Proliferation Resistance," IAEA-TECDOC-1575 Rev. 1, November 2008.

The IAEA offers 18 free online learning modules in nuclear security. Each module is based on the IAEA Nuclear Security Series and other guidance documents (https://www.iaea.org/topics/security-of-nuclear-and-other-radioactive-material/nuclear-security-e-learning). The 18 modules are divided into four topical categories: Cross-cutting Topics, Nuclear Security of Materials and Facilities, Nuclear Security of Material out of Regulatory Control, and Information and Computer Security.

Finally, three major reports on the PR&PP methodology and a bibliography can be found at the website: https://www.gen-4.org/gif/jcms/c_9365/pr-pp. The reports are (1) the report on the evaluation methodology itself (Revision 6); (2) the case study for the example sodium fast reactor, which is a four-module small reactor; and (3) the joint study performed by the PR&PP working group and the designers of each of the Generation IV designs, some of which are small reactors. The PR&PP working group also maintains and updates annually a bibliography, collecting articles, papers, and reports in open literature on topics related to the GIF PR&PP methodology.

References

Cheng, L., Cojazzi, G.G.M., Renda, G., et al., 2019. The GIF proliferation resistance and physical protection methodology applied to GEN IV system designs: an update. In: Proceedings of ESARDA Symposium 2019, 41th Annual Meeting, Stresa (VB), Italy, 14–16 May.

Gen IV International Forum, 2009. PR&PP Evaluation: ESFR Full System Case Study Report (GIF/PRPPWG/2009/002). October 2009, https://www.gen-4.org/gif/jcms/c_40415/esfr-case-study-report.

Gen IV International Forum, 2011a. PR&PP Methodology, Revision 6. September 15, 2011, https://www.gen-4.org/gif/jcms/c_40413/evaluation-methodology-for-proliferation-resistance-and-physical-protection-of-generation-iv-nuclear-energy-systems-rev-6.

Gen IV International Forum, 2011b. Proliferation Resistance and Physical Protection of the Six Generation IV Nuclear Energy Systems (GIF/PRPPWG/2011/002). July 15, 2011, https://www.gen-4.org/gif/jcms/c_40414/proliferation-resistance-and-physical-protection-of-the-six-generation-iv-nuclear-energy-systems.

Gen IV International Forum, 2018. GIF Webinar Series 20: Proliferation Resistance and Physical Protection of Gen IV Reactor. May 22, 2018, https://www.gen-4.org/gif/jcms/c_84279/webinars.

Glaser, A., Hopkins, L.B., Ramana, M.V., 2017. Resource requirements and proliferation risks associated with small modular reactors. Nucl. Technol. 184 (1), 121–129.

Houses of Parliament, 2018. Small Modular Nuclear Reactors. UK Parliament Office of Science & Technology. (POSTNOTE 580), July 2018. See https://researchbriefings.files.parliament.uk/documents/POST-PN-0580/POST-PN-0580.pdf.

IAEA, 2002. Proliferation Resistance Fundamentals for Future Nuclear Energy Systems. IAEA Department of Safeguards. (STR-332) December 2002.

IAEA, 2009. Facility Design and Plant Operation Features That Facilitate the Implementation of IAEA Safeguards. IAEA Department of Safeguards. (STR-360) February 2009.

IAEA, 2010. The Interface Between Safety and Security at Nuclear Power Plants. IAEA Report by the International Nuclear Safety Group. (INSAG-24) 2010.

IAEA, 2013. The Conceptualization and Development of Safeguards Implementation at the State Level (Report by the Director General to the Board of Governors GOV/2013/38). August 13, 2013. See, for example: https://armscontrollaw.files.wordpress.com/2012/06/state-level-safeguards-concept-report-august-2013.pdf.

IAEA, 2014. The proliferation resistance aspects of SMRs are discussed. In: Options to Enhance Proliferation Resistance of Innovative Small and Medium Sized Reactors. IAEA Nuclear Energy Series. (Report No. NP-T-1.11). An author of this chapter was a main contributor to that report.

IAEA, 2016. IAEA Safeguards—Delivering Effective Nuclear Verification for World Peace. June 2016: see https://www.iaea.org/sites/default/files/16/08/iaea_safeguards_introductory_leaflet.pdf.

IAEA, 2018. Advances in Small Modular Reactor Technology Developments. See https://aris.iaea.org/Publications/SMR-Book_2018.pdf.

Khalil, H., et al., 2012. Challenges to integration of safety and reliability with proliferation resistance and physical protection for generation IV nuclear energy systems. Nucl. Technol. 179, 112–116.

Martikka, E., et al., 2018. Technical synergies between safeguards and security. In: Maiani, L. et al., (Ed.), International Cooperation for Enhancing Nuclear Safety, Security, Safeguards and Non-Proliferation—60 Years of IAEA and EURATOM. Springer Proceedings in Physics 206, https://doi.org/10.1007/978-3-662-57366-2_11.

NEI, 2012. NEI Position Paper on Physical Security for Small Modular Reactors. July 31, 2012; see https://pbadupws.nrc.gov/docs/ML1222/ML12221A197.pdf.

NEI, 2015. White Paper on Proposed Consequence-Based Physical Security Framework for Small Modular Reactors and Other New Technologies. October 2015.

NEI, 2016. White Paper on Proposed Physical Security Requirements for Advanced Reactor Technologies. December 14, 2016.

Prasad, S., Abdulla, A., Morgan, M., Azevedo, I., 2015. Nonproliferation improvements and challenges presented by small modular reactors. Prog. Nucl. Energy 80, 102–109.

Rosenthal, M., et al., 2019. Deterring Nuclear Proliferation—The Importance of IAEA Safeguards, second ed. Brookhaven Science Associates, LLC. https://www.bnl.gov/NNS/IAEAtextbook.php.

USNRC, 2010. Potential Policy, Licensing, and Key Technical Issues for Small Modular Nuclear Reactor Designs. March 28, 2010 (ADAMS Accession No. ML093290245, SECY-10-0034).

USNRC, 2011. Security Regulatory Framework for Certifying, Approving, and Licensing Small Modular Nuclear Reactors. December 29, 2011 (ADAMS Accession No. ML112991113, SECY-11-0184).

USNRC, 2018a. Options and Recommendation for Physical Security for Advanced Reactors. August 1, 2018 (ADAMS Accession No. ML18052B032, SECY-18-0076).

USNRC, 2018b. Staff Requirements—SECY-18-0076—Options and Recommendation for Physical Security for Advanced Reactors. November 19, 2018, (ADAMS Accession No. ML18324A478, SRM-SECY-18-0076).

Whitlock, J., Sprinkle, J., 2012. Proliferation resistance considerations for remote small modular reactors. AECL Nucl. Rev. 1 (2), 10–12.

Zentner, M., Therios, I., Bari, R., Cheng, L., Yue, M., Wigeland, R., Hassberger, J., Boyer, B., Pilat, J., Rochau, G., Cleary, V., 2010. An expert elicitation based study of the proliferation resistance of a suite of nuclear power plants. In: Proceedings of the 51th Institute of Nuclear Materials Management Annual Meeting, Baltimore, MD, July 11–15.

Part Three

Implementation and applications

Economics and financing of small modular reactors (SMRs)

S. Boarin[a], M. Mancini[a], M. Ricotti[a], and G. Locatelli[b]
[a]Politecnico di Milano, Milan, Italy, [b]University of Lincoln, Lincoln, UK

10.1 Introduction

A description of the economic and industrial potential features of small modular reactors (SMRs) was given in 2010 by the US Secretary of Energy (Chu, 2010):

> '[…] Small modular reactors would be less than one-third the size of current plants. They have compact designs and could be made in factories and transported to sites by truck or rail. SMRs would be ready to 'plug and play' upon arrival. If commercially successful, SMRs would significantly expand the options for nuclear power and its applications. Their small size makes them suitable to small electric grids so they are a good option for locations that cannot accommodate large-scale plants. The modular construction process would make them more affordable by reducing capital costs and construction times. Their size would also increase flexibility for utilities since they could add units as demand changes, or use them for on-site replacement of aging fossil fuel plants. […] These SMRs are based on proven Light Water Reactor technologies and could be deployed in about 10 years'.

The goal of this chapter is to present the most relevant economic and competitive aspects related to the SMR concept.

10.1.1 Basic definitions and concepts

The nuclear sector commonly clusters nuclear power plant (NPP) life-cycle costs as capital cost, operation and maintenance, fuel and decommissioning.

- *Total capital investment cost (or capital cost)*: an all-inclusive plant capital cost, or lump-sum up-front cost. This cost is the base construction cost plus contingency, escalation, interest during construction (IDC), owner's cost (including utility's start-up cost), commissioning (non-utility start-up cost), and initial fuel core costs for a reactor (EMWG, 2007).
- *Operation and maintenance (O&M)*: costs inclusive of, but not limited to: (i) actions focused on scheduling, procedures, and work/systems control and optimization; (ii) performance of routine, preventive, predictive, scheduled and unscheduled actions aimed at preventing equipment failure or decline with the goal of increasing efficiency, reliability, and safety (Sullivan et al., 2010).

- *Fuel cost*: the sum of the costs for the fissile/fertile materials (natural uranium, low-enrichment uranium, highly enriched uranium, mixed oxide fuel, uranium-thorium, etc.) and the enrichment process of the fuel in fissile materials, plus other materials used in the fuel assemblies (zirconium, graphite, etc.), services required to produce the needed materials (mining, milling, conversion, enrichment, fabrication), fuel fabrication, shipment and handling, costs of spent-fuel disposal or reprocessing and waste (including low-level, high-level and transuranic waste) disposal.
- *Decommissioning*: costs for the administrative and technical actions taken to allow the removal of some or all of the regulatory controls from a facility. The actions will ensure the long-term protection of the public and the environment, and typically include reducing the levels of residual radionuclides in the materials and on the site of the facility, to allow the materials' safe recycling, reuse, or disposal as 'exempt waste' or as 'radioactive waste' and to allow the release of the site for unrestricted use or other use (IAEA, 2007a).

These costs contribute in different ways to the economics of a NPP. In general it is possible to compare them using one of the most important indicator for policy makers. This indicator, usually called levelized unit electricity cost (LUEC) or levelized cost of electricity (LCOE), represents a unit generation cost of electricity, accounting for all the NPP life-cycle costs and is expressed in terms of energy currency, typically [$/KW h]. For both large reactors (LRs) and SMRs the capital cost is the main component (50–75%) of the LCOE, followed by O&M and fuel, as shown in Table 10.1. From this consideration arises the opportunity to analyse in detail the nature of the capital cost item. In accordance with the glossary provided by EMWG (2007), Table 2 lists and clusters all the main accounts included in the capital cost.

If the construction time increases, almost all the cost items, apart from the equipment, are affected by such increase. In particular, the cost items affected by a time schedule increase are as follows:

- Labour cost: on the construction site of a reactor plant, thousands of people are employed.
- Rent fees for building infrastructures (e.g. special cranes).
- Escalation: the amount of all the cost items tends to increase because of a generalized inflation mechanism; the inflation rate may specifically relate to the price dynamics of the main inputs, such as structural materials, energy, etc.

Table 10.1 LCOE cost components

	OTA (1993)	DOE (2005)	MacKerron et al. (2005)	Williams & Miller (2006)	Gallanti & Parozzi (2006)	Locatelli & Mancini (2010b)
Capital costs	62%	71.9%	60–75%	48.7%	68%	59%
O&M costs	12%	11.19%	5–10%	23.25%	13%	24%
Fuel costs	26%	16.91%	8–15%	27.22%	15%	13
Decommissioning costs	0%	0%	1–5%	0.84%	4%	5%

Sources: Carelli *et al.* (2008a); Locatelli and Mancini (2010b).

Table 10.2 Example of Code of Accounts for capital costs

Account number	Account title	Account number	Account title
1	*Capitalized Pre-Construction Costs*	39	Contingency on Indirect Services
11	Land and Land Rights	**Base Construction Cost**	
12	Site Permits		
13	Plant Licensing	*4*	*Capitalized Owner's Costs*
14	Plant Permits	41	Staff Recruitment and Training
15	Plant Studies		
16	Plant Reports	42	Staff Housing
17	Other Pre-Construction Costs	43	Staff Salary-Related Costs
19	Contingency on Pre-Construction Costs	44	Other Owner's Capitalized Costs
2	*Capitalized Direct Costs*	49	Contingency on Owner's Costs
21	Structures and Improvements	*5*	*Capitalized Supplementary Costs*
22	Reactor Equipment		
23	Turbine Generator Equipment	51	Shipping and Transportation Costs
24	Electrical Equipment		
25	Heat Rejection System	52	Spare Parts
26	Miscellaneous Equipment	53	Taxes
27	Special Materials	54	Insurance
28	Simulator	55	Initial Fuel Core Load
29	Contingency on Direct Costs	58	Decommissioning Costs
Direct cost		59	Contingency on Supplementary Costs
3	*Capitalized Indirect Services Costs*	**Overnight Construction Cost**	
31	Field Indirect Costs	*6*	*Capitalized Financial Costs*
32	Construction Supervision		
33	Commissioning and Start-Up Costs	61	Escalation
		62	Fees
34	Demonstration Test Run	63	Interest During Construction
Total Field Cost			
		69	Contingency on Financial Costs
35	Design Services Offsite		
36	PM/CM Services Offsite		
37	Design Services Onsite	**Total Capital Investment Cost**	
38	PM/CM Services Onsite		

From EMWG, 2007. Economic Modeling Working Group – Cost Estimating guidelines for generation IV nuclear energy systems, Revision 4.2. GIF/EMWG/2007/004, OECD. http://www.gen-4.org/Technology/horizontal/EMWG_Guidelines.pdf. Courtesy of the Generation IV International Forum.

- Interest during construction: financial costs related to the capital remuneration increase with the investment duration.

In addition to the cost increase, each day of construction schedule delay represents a loss in terms of missed electricity generation and potential revenues.

Once construction and commissioning are completed, the NPP enters the operation mode. In this phase almost all the costs are fixed (Parsons and Du, 2009). A large part of the operation and fuel costs is independent of the electricity generated (fixed costs). Even if the plant has a low capacity factor, the labour cost, which is the main component of O&M costs, does not change, and neither does most of the maintenance cost.

As a consequence of the essentially fixed nuclear generation costs, the NPP manager's interest is generally to run the plant at its target (i.e. nominal) capacity. For this reason nuclear power is most suited for base load production.

10.1.2 Construction cost estimation

To evaluate the construction costs that, as said, represent the main component of nuclear LCOE, two approaches are usually adopted: top-down cost estimation and bottom-up cost estimation.

- *Top-down estimation*. The cost is calculated starting from a reference, known cost value, then considering the most important cost drivers that characterize the economics of that specific technology to derive scaled or proportional costs. Regarding the power plant industry, these drivers are: the plant size, the number of units to build, the site location, etc. This procedure is particularly appropriate when the plant design is still in the early phase of development, or when the plant design is characterized by a high level of complexity and number of systems as to make the cost estimation of each of them a hard task with a decrease in the reliability of the end result. An application of top-down cost estimation to SMRs is presented in Carelli *et al.* (2010).
- *Bottom-up cost estimation*. The cost analysis is carried out at 'component level' and the final cost is the sum of all of the costs related to the components manufacturing, assembling, operation, etc.

Once estimated through the above-mentioned procedures, the life-cycle costs, together with the cost of financing (equity and debt) and tax burdens, may be elaborated to perform a discounted cash flow (DCF) analysis. The DCF analysis provides the most relevant indicators of economic performance, such as the internal rate of return (IRR), the net present Value (NPV), the (LUEC) and the payback time (PBT) (see Figure 10.1). Several studies indicate that optimism in the cost estimation of large projects, such as civil and transport infrastructures, power plants, etc., is a common characteristic. This phenomenon may be observed in the case of NPPs, which are historically characterised by delay in construction and cost escalation (Locatelli and Mancini, 2012a). In order to provide a reliable cost estimation of SMRs, it is important to understand why the estimations of NPP costs, as well as large engineering projects, are usually so inaccurate and how to improve this process. Under this perspective, Flyvbjerg *et al.* (2003) show that the availability and reliability of data about

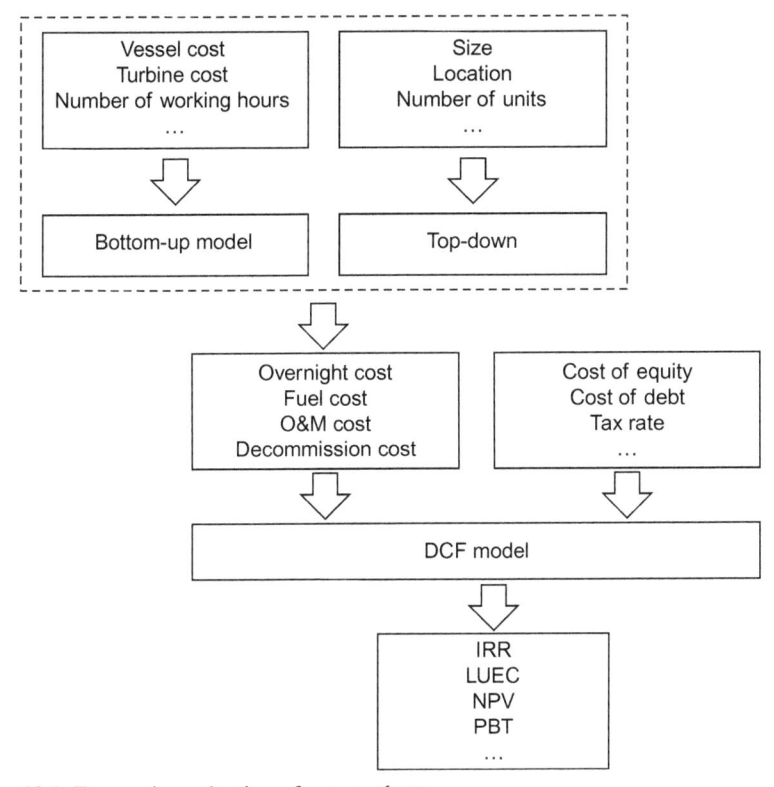

Figure 10.1 Economic evaluation of power plants.

large projects affect the estimation. The authors identify two macro-categories of causes to explain inaccuracy in the cost forecast: (i) inadequacy of the methodologies and (ii) strategic data manipulation. The latter, combined with 'optimism bias', is responsible for most of the cost escalation.

10.2 Investment and risk factors

The investment decision in an industrial activity largely depends on the capability of the project to adequately recover and remunerate the initial capital expenditure. The uncertainty of the capital cost estimation, i.e. the initial investment, as mentioned in Section 10.1.1, affects the ability to make a reliable estimation of the investment profitability. The uncertainty affects the scenario conditions, the project realization and operation; as a result, the stream of income generated by the project is also affected by uncertainty. Therefore, expected profitability has a degree of risk embedded and a series of different possible outcomes, depending on the realization of stochastic variables. Investment in liberalized electricity markets, as in most of the European and North American countries, compels investors to include uncertainty in their business plan analysis and to give risk as much relevance as profitability into their decision making. The key variables to the financial performance of the investment project are 'forecast' in order to get a reasonable estimation of the project profitability and economic soundness. All this considered, NPPs represent a long-term investment with deferred payouts. Moreover, the nuclear industry is very capital-intensive. This means that a high up-front capital investment is needed to set up the project and a long payback period is needed to recover the capital expenditure.

The longer this period, the higher is the probability that the scenario conditions may evolve in a different, unfavourable way, as compared to the forecasts. As an example, market price of electricity might be driven downwards by unexpected market dynamics; unexpected operating or design drawbacks might also undermine plant availability. A capital-intensive investment requires the full exploitation of its operating capability and an income stream as stable as possible. On a long-term horizon, a low volatility in a variable trend might translate into a widespread range of realizations of the variable value. This condition is common to every capital-intensive industry. Nevertheless, some risk factors are specific or particularly sensitive to the nuclear industry: typically, the public acceptance, the political support in the long-term energy strategy, the activity of safety and regulatory agencies.

For these reasons nuclear investment is usually perceived as the riskier investment option among the power generation technologies (Figure 10.2). Clearly, risk is not the only or the most relevant criterion in the selection of a power generation technology. Besides risk and cost, other strategic and economic issues are included in a technology investment evaluation, such as the power generation independence, the power density (as compared to the land occupation), the power supply stability (baseload), the electricity price stability, etc. The key risk factors affecting a nuclear investment project are tentatively listed and classified in Table 10.3.

Capital cost and construction lead-time have pre-eminent importance. Construction time and cost overruns are considered to be the most adverse occurrence able to undermine the nuclear power economics. Throughout the construction period, the project will be exposed to commodity price risk, vendor credit risk, engineering and construction contract performance risk, supply chain risk, sovereign risk, regulatory risk, etc. The construction phase is the most affected by the investment risk. The magnitude of a project overrun is often difficult to estimate while construction proceeds and even more difficult to rein in (Dolley, 2008). The ability to estimate construction cost in the past has proved very limited, as confirmed by US data reported in Table 10.4.

Thus, financing of nuclear power is affected by risk perception. Risk has a cost, which is transferred to the cost of capital in terms of 'risk premium', as a remuneration for possible negative outcomes (Damodaran, 2011). The 'rating' associated with an investment project represents the probability of financial default: as far as the 'rating' is low (i.e. the risk is high), the risk premium applied on the cost of capital (equity or debt) is high. Therefore nuclear projects usually bear high cost of money, compared with other energy sources. For this reason and for the long debt duration, IDCs represent a relevant part of the capital cost (Figure 10.3): any increase in the cost of capital would be a significant burden on the project economics. Besides risk premium considerations, IDCs are also heavily affected by the construction period, where

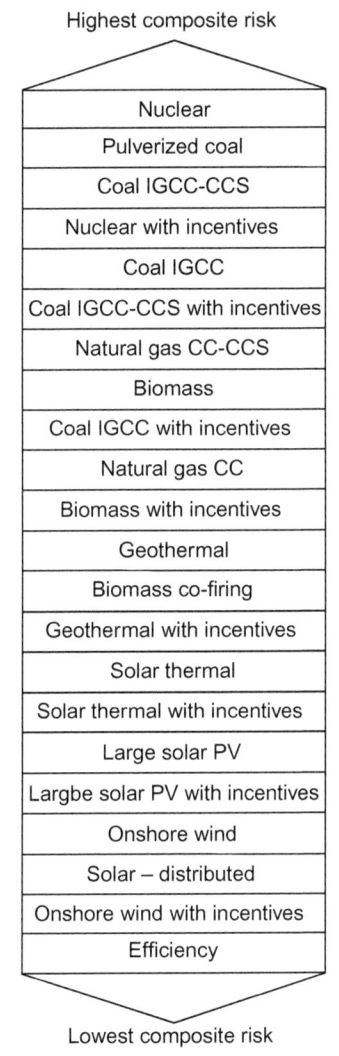

Figure 10.2 Risk ranking of generation resources for new power plants (Binz *et al.*, 2012). Key: IGCC: integrated gasification combined cycle; CC: carbon capture; CCS: carbon capture and separation.

Ranking from highest composite risk to lowest composite risk:
- Nuclear
- Pulverized coal
- Coal IGCC-CCS
- Nuclear with incentives
- Coal IGCC
- Coal IGCC-CCS with incentives
- Natural gas CC-CCS
- Biomass
- Coal IGCC with incentives
- Natural gas CC
- Biomass with incentives
- Geothermal
- Biomass co-firing
- Geothermal with incentives
- Solar thermal
- Solar thermal with incentives
- Large solar PV
- Largbe solar PV with incentives
- Onshore wind
- Solar – distributed
- Onshore wind with incentives
- Efficiency

financial exposure is the highest and the project pays compound interests on the invested capital.

Hence nuclear business risk derives from:

- capital intensive nature, with huge sunk costs and high financial exposure during very long PBTs;
- very long-term market forecast reliability;
- unexpected external unfavourable events (such as natural events, public acceptance/political support withdrawal) or intrinsic drawbacks to the project economics (such as construction time and cost overruns, operating unavailability).

Table 10.3 Main risk factors of capital-intensive and nuclear-specific industry

Risk factors, common to capital-intensive industries	Risk factors, nuclear-specific
• Complex and highly capital intensive: high up-front capital costs • Cost uncertainty • Completion risks: construction supply chain risks • Long lead times (engineering & construction, etc.) and long payback periods • Sensitive to interest rates • Plant reliability/availability/load factor • Market price of output (i.e. electricity)	• Unstable public support • Negative public acceptance • Regulatory/policy risks (revised safety measures) • Decommissioning and waste cost/liabilities

Table 10.4 Projected and actual construction costs for US nuclear power plants

Construction starts		Average overnight costs		
Year initiated	Number of plants	Utilities' projections (thousands of dollars per MW)	Actual (thousands of dollars per MW)	Overrun (percent)
1966–1967	11	612	1279	109
1968–1969	26	741	2180	194
1970–1971	12	829	2889	248
1972–1973	7	1220	3882	218
1974–1975	14	1963	4817	281
1976–1977	5	1630	4377	169
Overall average	13	938	2959	207

From Congressional Budget Office (CBO) 2008. Nuclear Power's Role in Generating Electricity. CBO, Washington, DC. Based on data from Energy Information Administration. An Analysis of Nuclear Power Plant Construction Costs. Technical Report DOE/EIA-0485 (January 1, 1986).

SMRs may represent a valuable option to mitigate several among the risk factors previously discussed. Due to their features, SMRs are able to reduce the severity of many risk factors in pre-construction, supply chain, construction and operating phases (Locatelli *et al.*, 2011a). An IAEA investigation (Barkatullah, 2011) on the topic reached the conclusion that SMRs may present mitigation factors against some major financing challenges of nuclear power (Figure 10.4). In particular, lower up-front investment of an SMR and low construction lead-time are key features able to decrease the financial risk of the investment. These are discussed in the following sections.

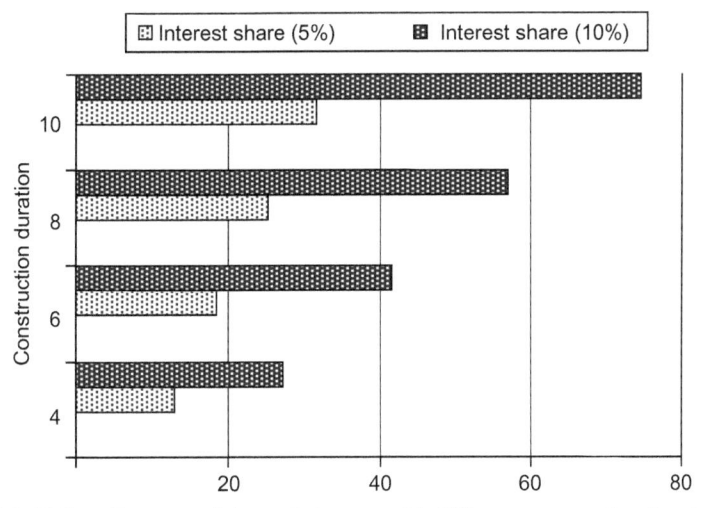

Figure 10.3 IDC as % on overnight capital costs, with different construction duration (years) and cost of money (5%, 10%) (Barkatullah, 2011).

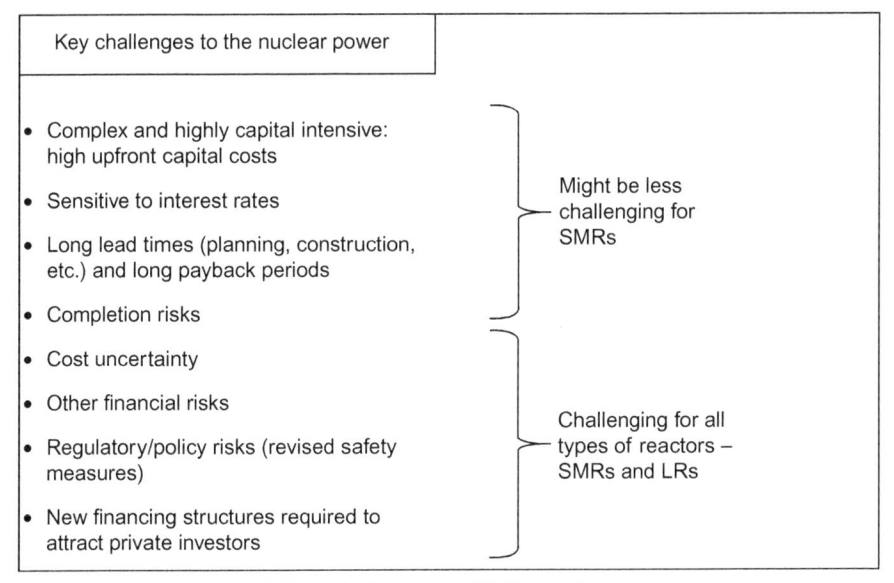

Key challenges to the nuclear power
• Complex and highly capital intensive: high upfront capital costs
• Sensitive to interest rates
• Long lead times (planning, construction, etc.) and long payback periods
• Completion risks
• Cost uncertainty
• Other financial risks
• Regulatory/policy risks (revised safety measures)
• New financing structures required to attract private investors

- Might be less challenging for SMRs (first four items)
- Challenging for all types of reactors – SMRs and LRs (last four items)

Figure 10.4 Risk factors: differential impact on SMRs and large reactors. Adapted from Barkatullah (2011).

10.2.1 Reduced up-front investment and business risk diversification

SMRs may represent a viable option to decrease the average capital at risk in the nuclear business, with respect to LR projects. Financial risk is related to the amount of invested capital. Banks usually apply credit risk control through loans portfolio diversification. The same applies to the shareholder investor (e.g. a utility). Very high capital exposure in a single project represents a stress on the balance sheet and a relevant financial and industrial risk exposure, so that a nuclear generation project could be viewed as a 'bet the farm' endeavour for a shareholder utility, due to the size of the investment and the length of time needed to commission a nuclear power facility.

A model has been proposed for relating the risk premium to the risk size (Goldberg and Rosner, 2011). The assumption is that the risk premium associated with a project is a function of the wealth of the sponsoring entity, as might be measured by, for example, NPV and debt to equity ratio. The mathematical expression for this relationship shows risk premium rising at an exponential rate as the size of the project approaches the size of the investor-firm.

If the investment size in different base load technologies is compared with the average annual revenues of a utility (Figure 10.5), it becomes evident that SMRs should be viewed more favourably by the investor community and bear lower risk premium than very large reactors. (Examples of current annual revenues for some US utilities: Exelon – $23.5 billion, Duke Energy – $19.6 billion.)

A rating methodology reported in Table 10.5 shows that business diversification in low versus high risk (i.e. nuclear) businesses is among the risk metrics considered in the evaluation of the merit of credit of a company. In this sense, SMRs allow a better industrial risk diversification, on account of a limited investment on total capital budget. In case of small-sized market or reduced capital budget availability, by including SMR in a portfolio mix it is possible to grant a business diversification, that would be pre-empted by a large plant, reducing the investment risk (Locatelli and Mancini, 2011a).

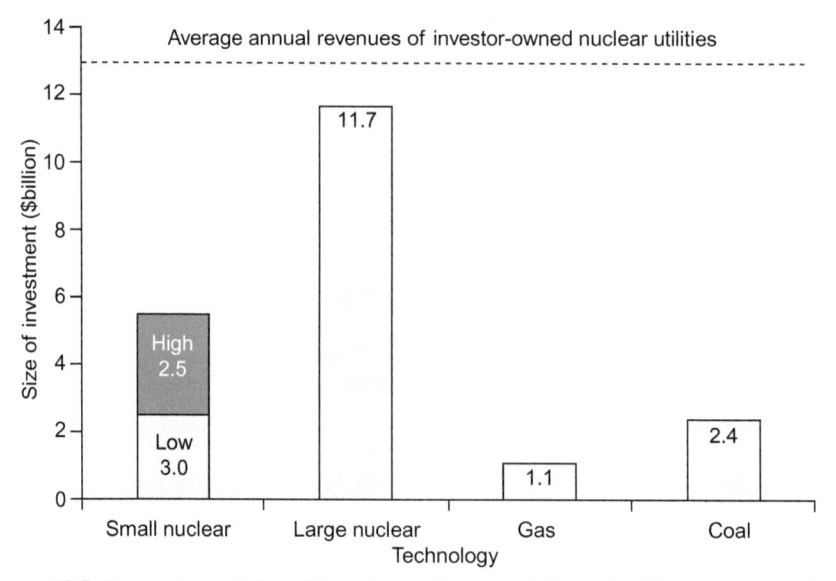

Figure 10.5 Comparison of size of investment (i.e., overnight cost) with average annual revenues of investor-owned nuclear utilities. 'Large nuclear' investment represents twin-unit GW-scale plant (Goldberg and Rosner, 2011).

Table 10.5 Moody's rating methodology for electric utilities

Broad rating factors	Broad factor weighting	Rating sub-factor	Sub-factor weighting
Rating factor weighting – regulated electric utilities			
Regulatory framework	25%		25%
Ability to recover costs and earn returns	25%		25%
Diversification	10%	Market Position	5%
		Generation and Fuel Diversity	5%
Financial strength, liquidity and key financial metrics	40%	Liquidity	10%
		CFO pre-WC/Debt	7.50%
		CFO pre-WC+Interest/Interest	7.50%
		CFO pre-WC – Dividends/Debt	7.50%
		Debt/Capitalization or Debt/Regulated Asset Value	7.50%
Rating factor weighting – unregulated electric utilities			
Market assessment, scale and competitive position	25%	Size and scale	15%
		Competitie position and market structure	10%
Cash flow predictability of busines model	25%	Fuel strategy and mix	5%
		Degree of integration and hedging strategy	5%
		Capital requirements and operational performance	5%
		Contribution from low-risk/high-risk business	10%
Financial policy	10%		
Financial strength metrics	40%	Cash flow/debt	12.5%
		Cash flow interest coverage	10%
		Retained cash/debt	12.5%
		Free cash flow/debt	5%

Source: Goldberg and Rosner (2011).

10.2.2 Control of construction lead times and costs

One of the main concerns for investors is unexpected delays during construction of a NPP and the related cost escalation. Faced by the above-mentioned risks, several investors stand 'frozen' and wait and see the market evolution, the strategies of their competitors or wait for a more mature phase of a specific reactor plant concept to exploit cost reduction and learning accumulation.

As argued by IAEA (Barkatullah, 2011), reduced plant size and complexity and design simplifications, enabled by the SMRs, should allow:

- better control on shorter construction lead-time – leaner project management (e.g. higher factory-fabrication content, modularization of reactors);
- lower supply chain risks – increased number of suppliers and reduced need of special and *ad hoc* manufacturing and installations;
- better control on construction costs – if plant complexity of gigawatt electric (GWe)-scale nuclear plants has been a driver of cost escalation (Grubler, 2010), SMRs should enable economies from standardization and accelerated learning. The ability to meet cost projection should also improve.

10.2.3 Control over market risk

Multiple SMRs represent both a 'modular' design concept and a 'modular' investment model: multiple SMRs may offer the investor a step-by-step entry in the nuclear market. As long as multiple SMRs are deployed with a staggered schedule, the investor has the option to expand, defer or even abandon a nuclear project, to adjust the investment strategy in order to catch early market opportunity or to edge a market unexpected downturn. The investment involves sequential steps with multiple 'go' or 'not-to-go' decisions that allow management to respond to changes in the market or in the regulation environment, or to adapt to technological breakthroughs. The risk edging capability of a modular investment such as multiple, staggered SMRs is enhanced compared with a monolithic LR. This flexibility against future uncertainty can be measured by the real option analysis and exploited to face the investment risks (Locatelli *et al.*, 2012).

10.3 Capital costs and economy of scale

Economic competitiveness of a power generation technology depends on the ability to provide electricity with affordable LCOE, and/or of repaying the investors by means of adequate cash inflows, granting a minimum acceptable capital remuneration compared to the level of risk and to the PBT duration. Given the relevance of the capital costs in the nuclear electricity generation cost (i.e. given the 'capital intensity' of the nuclear investment), capital cost, including overnight construction and financial costs, has a relevant impact on the key economic performance indicators.

With very few SMR projects under construction and no actual data on overnight actual costs, cost estimation of SMRs is usually performed on a top-down basis, as recommended in Section 10.1.2, starting from available information on large, advanced pressurized water reactor (PWR) units, as a starting reference cost. (The only SMR reactors under construction in 2014 were CAREM in Argentina, HTR-PM in China and the twin barge-mounted KLT-40S in Russia (Akademik Lomonosov), planned to be located near Vilyuchinsk. Construction was started in

2007 and, owing to some economic-financial problems, the plant is now expected to be completed in 2016 (http://www.world-nuclear.org/info/inf33.html)). Carelli *et al.* (2010) present a parametric methodology to compute the overnight construction cost of SMRs, based on the application of dimensionless coefficients, related to the most important differential economic features between SMRs and LRs: e.g. expected learning effect, degree of modularization, co-siting economies and simplified design. Many of these factors are dependent on the number of units built on the same site and on the plant output size.

The above-mentioned design-related economies, learning effects on costs, plant modularization, and co-siting economies account for an expected reduction of multiple plant construction cost. Based on these factors, it is estimated that the SMR economic paradigm might bring unit construction cost in line with expected costs of Generation III+(GENIII+) large PWRs, thus overcoming the loss of economy of scale. Figure 10.6 provides a qualitative sketch of SMR economic features' recovering the loss of economy of scale on unit construction cost, as far as multiple SMRs are considered an alternative investment opportunity to LR power stations, with the same overall power at site-level.

Actual information on LRs under construction (in western countries) gives evidence of relevant time-schedule and cost overruns. It must be highlighted that this comparison applies on SMR versus large NPP *expected* costs. This means that capital cost overruns, which seem to systematically affect *actual* costs of large NPP projects, are not considered. When *actual* costs of construction are considered, it is expected that, as stated in Section 10.2.2, SMRs might have better control on construction schedule and costs, and higher probability to meet capital budgeting. The main assumption is that, the simpler the design, the easier the procurement, manufacturing and assembling process and the project management.

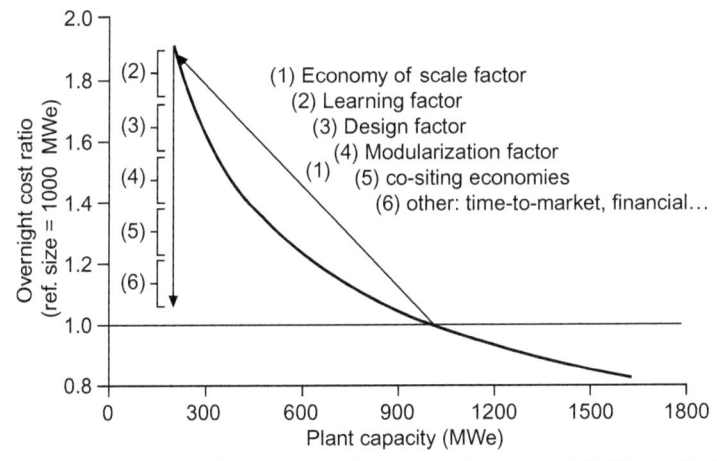

Figure 10.6 Top-down estimation of overnight construction costs of SMR: qualitative trend (Barenghi *et al.*, 2012).

In any case, a possible cost overrun on a single SMR unit would necessarily have a lower incidence on total investment than in an LR, due to the lower cost of a single SMR.

If a time/cost schedule mismatch affected the initial SMR unit(s), the simple fact of fractioning a nuclear power station into multiple smaller units makes unlikely that such a mismatch could be repeated on all of the units, on account of learning and improved practices in the supply process, construction work and project management.

Thus, as far as *actual* construction costs are considered, including possible cost overruns and financial interests escalation, SMRs might improve their cost competitiveness against LRs, as compared to the mere theoretical expectations.

Economy of scale has been the key driver of the nuclear industry over the past. The evolution of nuclear power technology is characterized by a constant trend in the output size increase. The US utilities converged to 1000–1400 MWe sized plants, French NPPs were scaled from 950 to 1550 MWe in the 1971–1999 period, up to the recent 1600 MWe European Pressurized Reactor (EPR). As a capital-intensive industry, nuclear power generation technology pursued the economy of scale law, to decrease the incidence of fixed costs over a higher output base.

In principle, SMRs are heavily penalized by the loss of economy of scale: applying a typical scale exponential law (usually with coefficient in the range 0.6–0.7), a standalone 335 MWe SMR may bear 70% cost increase on a unit base (€/kWe) over a 1340 MWe LR (Carelli *et al.*, 2007a). SMR units with smaller size would bear a greater penalty (up to 350%); this should be recovered by other means, in order to uphold cost competitiveness.

Nevertheless, the evidence of construction cost escalation of GW-scale reactors triggers considerations about the applicability of the economy of scale law on NPPs (Grubler, 2010): an increase in the plant scale apparently produces an increase in the intrinsic complexity, which challenges the project management and other activities in the plant design, construction and assembly. This translates in construction schedule delays and dramatic cost overruns. (For example Olkiluoto, Flamanville, South Texas Project, Vogtle and, more recently, Hinkley Point 3, Olkiluoto and Flamanville are under construction, while US-based projects are in the early site preparation phase; therefore the cost overrun have different nature in the two cases.) Projected cost and the lead time of the new projects under construction in Europe or under construction in US, have all been dramatically revised upwards, with a rate of increase per year of delay in the plant commissioning in excess of 20% (Table 10.6). This cost escalation coherent with a historical trend of construction cost increase over time (Figure 10.7).

A detailed analysis of the French NPP fleet (all PWRs) shows that construction costs and schedule have increased over time with the size of the plants (Figures 10.8 and 10.9). The French PWR program exhibited substantial real cost escalation, in spite of a unique institutional setting allowing centralized decision making, regulatory stability and dedicated efforts for standardized reactor designs. This evidence challenges the applicability of a learning economy on NPP construction, as far as 'traditional' NPP are considered, without introducing the concepts of design simplification and modularization, discussed in the following Sections 10.5.1 and 10.5.2.

Economics and financing of small modular reactors (SMRs) 255

Table 10.6 Cost increase and commissioning delays of NPP currently under construction

	Initial cost estimate	Revised cost estimate	Delay on commissioning
Olkiluoto 3 (Finland)	3 Bn€	8.5 Bn€	From 2009 to 2018
Flamanville (France)	3.3 Bn€	8.5 Bn€	From 2012 to 2016
Levy County (US)	5 Bn€	24 Bn€	From 2016 to 2024
South Texas Project (US)	5.4 Bn€	18.2 Bn€	Expected by 2006, then project abandoned in 2011
Hinkley Point (UK)	10 Bn£	16 Bn£	Commissioning delayed from 2017 to 2033

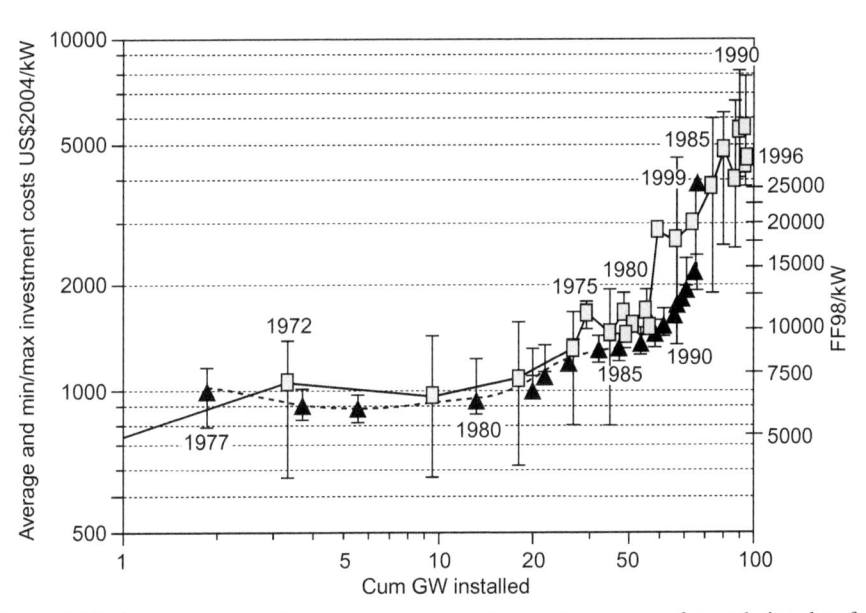

Figure 10.7 Average and min/max reactor construction costs per year of completion date for US and France versus cumulative capacity completed (Grubler, 2010).

10.4 Capital costs and multiple units

When a fleet of multiple NPP units are considered, some competitive factors intervene, to reduce the incidence of capital cost on the electricity generated. These factors are enabled and provide their best effect by the deployment of successive NPP of the same type on the same site. These factors are introduced in this chapter, despite not

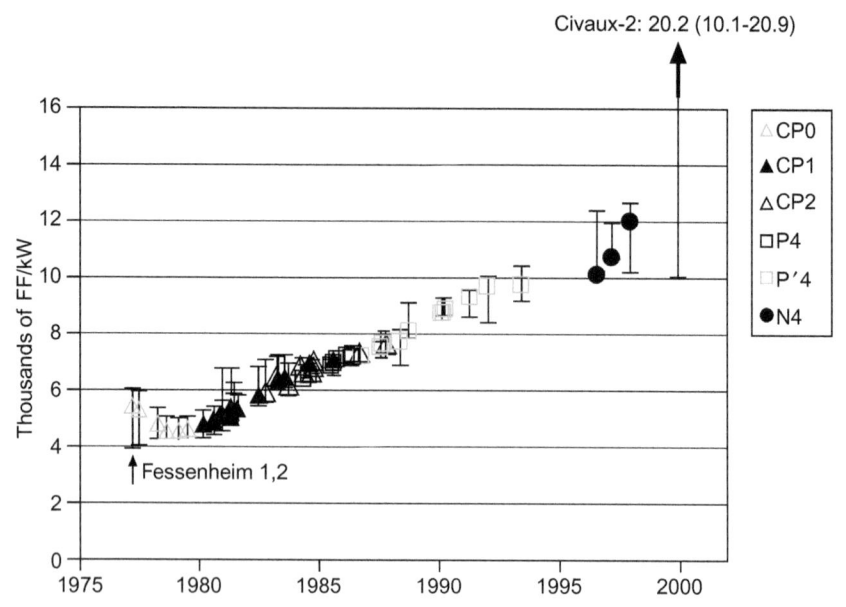

Figure 10.8 Inferred specific reactor construction costs (1000FF98/kW) per French PWR reactor sorted by reactor type and completion date (year of criticality), best guess and min/max uncertainty ranges of estimates (Grubler, 2010).

being specific to the SMR plant category, because they are expected to play a relevant role in the SMR economic competitiveness paradigm.

10.4.1 Learning

The contribution of learning (Boarin *et al.*, 2012) applies at various levels: a better work organization on the same site, where the personnel have already had experience in the construction and assembling of previous NPP modules; a learning component in factory fabrication of the equipment; a learning component in the utilization of materials and equipment by more skilled workers, etc. A scale-up of the plant output and the attempt to introduce an original French design, i.e. the N4 reactors, towards the end of the program may only partially explain such an occurrence.

Lovins (1986) presented an interesting theoretical framework, referred to as the Bupp–Derian–Komanoff–Taylor hypothesis, that suggests that with increasing application ('doing'), the complexity of the technology inevitably increases, leading to inherent cost escalation trends that limit or reverse 'learning' (cost reduction) possibilities. In other words, the technology scale-up can lead to an inevitable increase in systems complexity that translates into real-cost escalation, or 'negative learning'. Nevertheless, learning effects have been recorded in technology-advanced industries (Frischtak, 1994); learning effect description was first published by an aeronautical engineer (Wright, 1936).

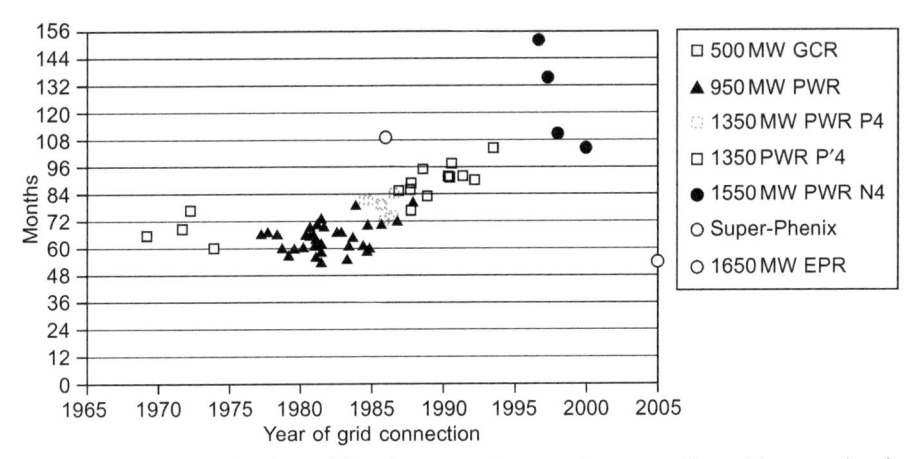

Figure 10.9 Construction time of French reactors (construction start to first grid-connection, in months). Data on 1650 MWe EPR reactor in Flamanville is a projection submitted by the French authorities to the IAEA.
Source: IAEA Power Reactor Information System (PRIS) www.iaea.org/pris. GCR = graphite gas reactors (Grubler, 2010).

The learning effect is also visible in the Korean NPP fleet deployment costs: learning accumulation has played an undeniable effect on a progressive cost decrease (Figure 10.10). KHNP, the owner of all 21 of South Korea's operating nuclear power reactors, has held a licensee relationship with Westinghouse since the late 1980s when the US-based company supplied the 945 MWe System 80 nuclear steam supply for Yonggwang 3 and 4. After that, KHNP was able to develop variants of System 80 for its own requirements under technology transfer terms in the license agreement. After introducing domestic innovations and updating technology over time, KHNP came up with the Korean Standard Nuclear Plant (KNSP), then the OPR-1000. The current APR-1400 technology represents a further evolution of that design. The construction and power generation costs of the APR-1400 are reported to be 10% lower than those of OPR-1000 units.

Korean NPP is the evidence that learning economy may apply to construction costs: in this case, learning effect was achieved through a concentrated construction (Figure 10.11), with the deployment of twin/multiple units on the same site and by avoiding substantial design modification in order to attain PWR plant standardization and control design complexity. (Wolsong NPP are PHWR-CANDU, provided by AECL, as the only exception to the PWR design.)

It may be argued that in principle, learning accumulation is expected to determine a construction cost and time-progressive decrease of successive NPP units, as it was in the Korean NPP fleet. Nevertheless, as far as western countries are considered, in the real world there is often no evidence of cost and time benefits in large NPP deployment programs. That is why simpler and smaller NPPs, with design modularity and high content of factory fabrication, have a higher chance of controlling complexity and exploiting standardization, enabling learning accumulation on both construction and assembling phases. SMRs are expected to benefit from anticipated learning effects, mostly arising from the construction and assembling of multiple units on the same site. Given the power size of a nuclear site, more SMR units should be fabricated and installed than LRs, with improved chances to learn. General learning accumulation may be recorded at the engineering procurement and construction (EPC) level residing in the human resources knowledge and approach to the project management, and to the organization and procurement issues, such as supplier selection. This learning applies independently of the site location of the new NPP and is therefore indicated as 'worldwide' learning in Figure 10.12. In addition, site-level learning accumulation is also applicable on successive NPP units built on the same site, residing in the best, refined practices and actions by local staff. The magnitude of the two effects is comparable (Boarin and Ricotti, 2011a). The learning effect is destined to

Figure 10.10 Overnight capital costs (in 2005 US dollars; exchange rate 1025 Won/US$) and construction duration (from first concrete to initial critically) of Korean NPP.
YGN = Yonggwang; UCN = Ulchin (Matzie, 2005).

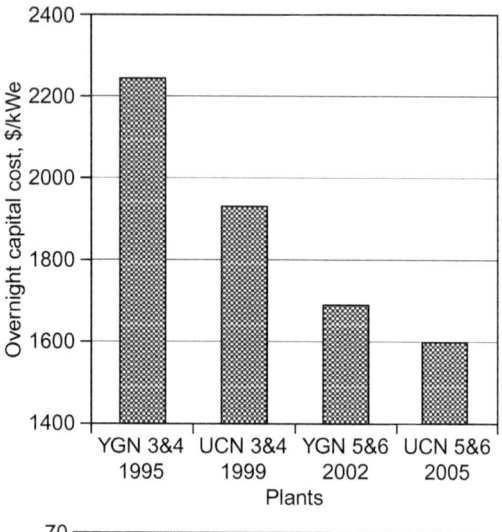

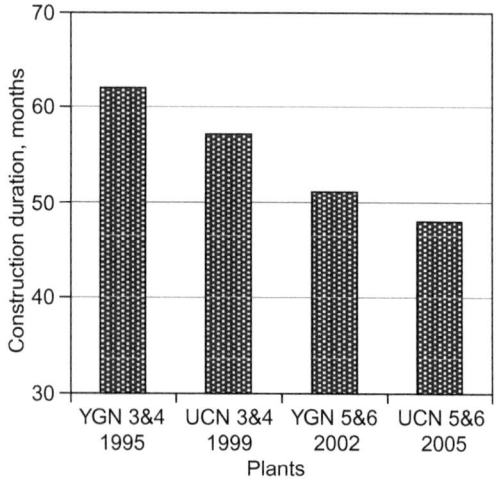

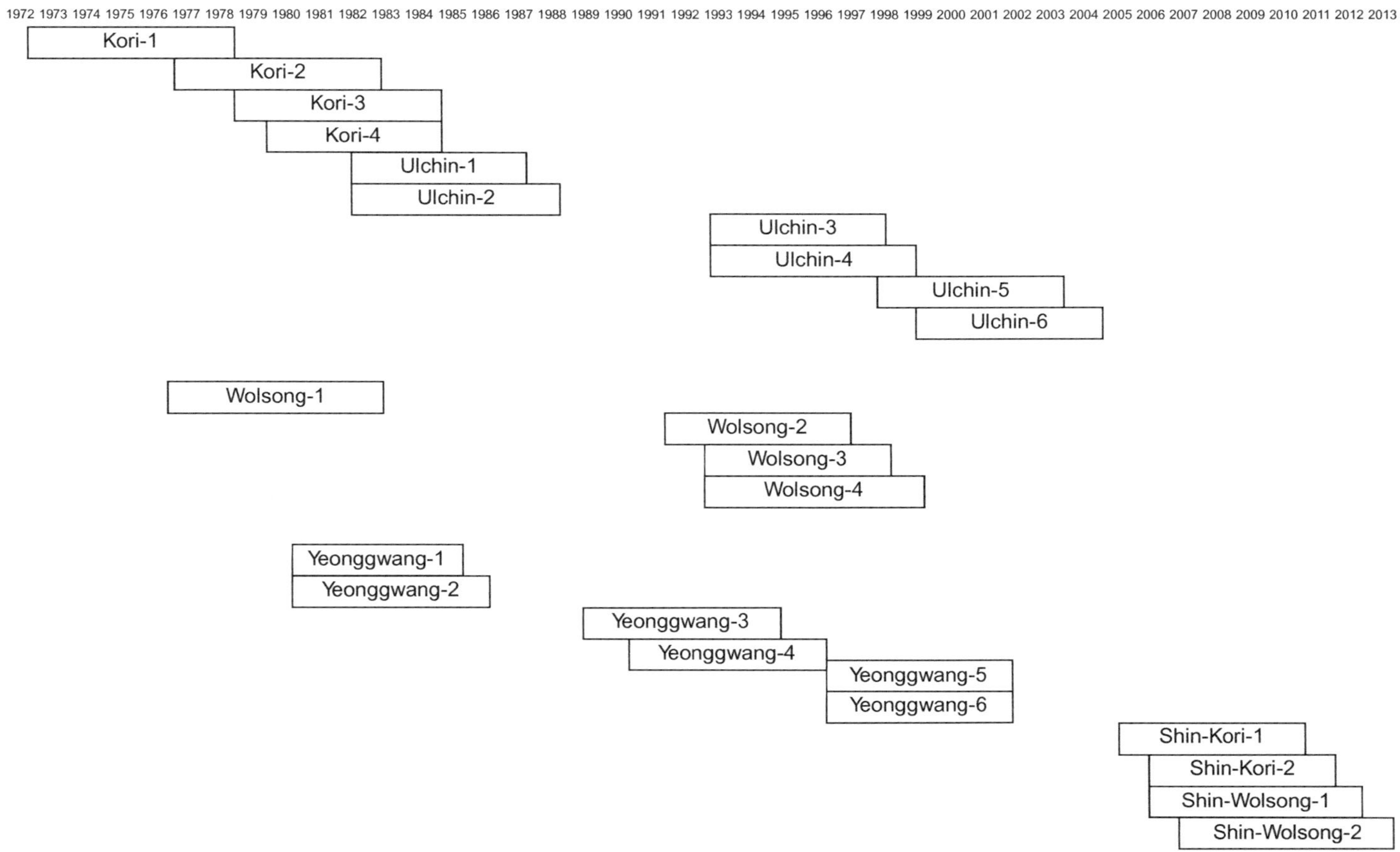

Figure 10.11 NPP deployment program in South Korea (data from http://globalenergyobservatory.org, accessed Feb. 2013).

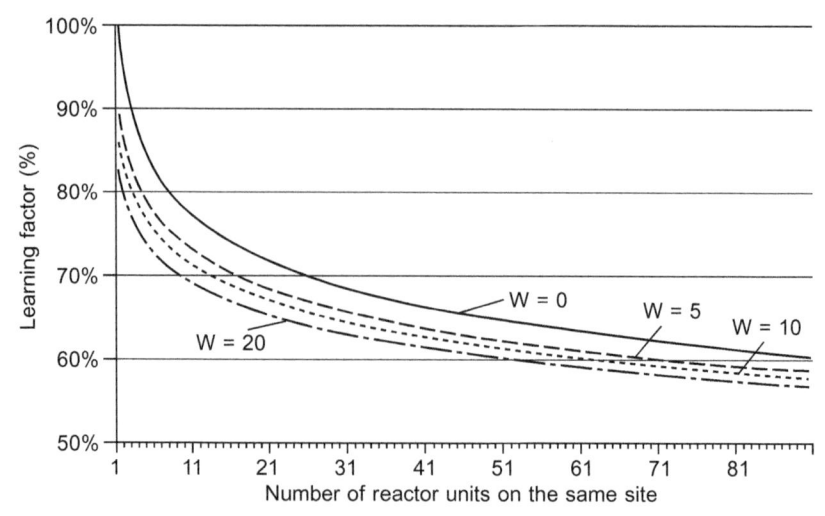

Figure 10.12 Magnitude of overall learning effect (on site + worldwide). W indicates the number of NPP of the same type already built in the world, in other site locations; along the same curve, the magnitude of the 'on-site' learning accumulation may be appreciated, while curves with different value of W indicate the magnitude of the 'worldwide' learning effect. D&D = decontamination and decommissioning (Boarin and Ricotti, 2011a).

fade out over the first five to seven units (Carelli *et al.*, 2010). For this reason, in a mature phase of the market, worldwide learning is not a differential factor for SMRs and LRs, while SMRs keep the benefit from the on-site learning accumulation, which applies in case of multiple units built on the same site.

10.4.2 Co-siting economies

Co-siting economies arise from the cost-sharing of some common structures, systems and services by multiple units built on the same site, decreasing the incidence of some fixed costs and, thus, the penalty of the loss of economy of scale (Boarin *et al.*, 2012).

10.5 Capital costs and size-specific factors

10.5.1 Modularization

The realization of NPP encompasses the phases of site preparation, construction and start-up. Traditionally, the construction phase of a NPP was performed on site, with specialized workers erecting all the civil structures, nuclear island and balance of plant (BoP) systems starting from raw material and main equipment. Every NPP construction was nearly hand-crafted, specific to the site and the plant design. Conversely, the SMR plant layout may be conceived from its design phase in a number of sub-systems or 'modules' that may be fabricated in a parallel way and then shipped and assembled on-site. The construction of the SMR plant systems on-site is reduced and the fabrication activity tends to be shifted in factory, with the following main benefits:

- controlled work conditions and improved quality standards;
- possibility to apply mini-serial production, fostering the learning accumulation and decreasing the overhead cost of the production lines;
- use of less specialized personnel on-site;
- in principle, reduction of the construction schedule due to the shift from series to parallel activities;
- as a consequence of the previous, lower financial cost escalation during construction.

On account of the smaller size of their components and systems, SMRs can achieve higher degrees of design modularization (Carelli *et al.*, 2007b, 2010). The indivisibility of some subsystems and their large scale in the LR plants compels their construction on-site, while the lower physical size of SMRs allows a greater number of systems to be factory-fabricated and then shipped to the site. Modularization requires more project management effort and transportation complexity. Communication and cooperation between suppliers and contractor has to be accurate, in order to create the schedule and ensure synchrony of the shipments. Modularization turns into real cost and time advantages as long as these additional burdens are counterbalanced by a plant layout simplification and a plant design conceived *ad hoc* to implement and ease the modularization.

10.5.2 Design factor

While modularization deals with a design and fabrication methodology, design factor is related to the specific and peculiar features and enhancements of a given design concept, in order to meet operating requirements with optimized safety, simplicity and economics. Large plants have been optimized for their particular power output. In designing a plant with smaller output, it does not necessarily make sense to just scale down a large system. Usually SMRs are not a mere re-sizing of larger units; they do not represent a way back, but, on the contrary, a further progress in the technology evolution path.

At a smaller size, different design concepts might be possible, which could lead to a more significant capital cost reduction than simple application of the scaling laws from large design would predict (Hayns and Shepherd, 1991). SMR economic rationale also lies on the enhanced passive safety features and design simplifications, often enabled by a small plant scale. The 300–400 MWe safe integral reactor (SIR) in the 1990s and the international reactor innovative and secure (IRIS) in early 2000s paved the way to the understanding of an innovative technological and economic paradigm.

Most Gen III+ reactor designs include some features that may be regarded as passive (i.e. relying on physical laws and not on human intervention for the activation), but small-scale plants can take maximum advantage of such features, due to their physically smaller size or lower power densities, and consequential lower power output. As a result, the elimination of some engineered safety systems might be possible and/or the safety downgrading of some other components. Revised, simplified and more cost-effective plant layout becomes possible, with favourable impact on costs (Carelli *et al.*, 2008a, 2008b).

Along with such design-related cost benefits, the SMR exploit the economics of small 'mass production'. SMRs are conceived to take the maximum advantage from standardization and economy of replication (Kutznetsov and Lokhov, 2011), also referred to as the 'economy of multiples' paradigm. Moreover, SMRs may encompass a broad range of reactor unit sizes. In principle, the lower the size, the higher the loss of economy of scale to be compensated, and the loss of cost effectiveness in terms of generation cost (Figure 10.13).

SMRs rely on the 'economy of multiples' but also on the 'economy of small' in the sense that design-related cost savings are necessary to recover economic competitiveness. The smaller the reactor unit size, the higher must be the design cost savings in order to have the same generation costs as LRs. Some general considerations on cost reduction by design may be drawn from several innovative SMR features, such as the integration of primary loop into the reactor vessel, with the elimination of large loss-of-coolant accident (LOCA), the wide use of passive safety systems with natural circulation of coolant in case of accident, and the elimination of some active components and safety systems. Nevertheless, the design-saving factor that is expected to decrease construction costs of SMRs is strictly dependent on the specific reactor concept. A more reliable estimate could come from a bottom-up cost analysis, referred to the specific plant layout and technical features. In the absence of this information, the economic analysis may consider the design-saving factor as a 'target' value to be achieved in order to equalize the SMR and LR projects' profitability. Thus the economic analysis might offer the manufacturer a sort of indication on a technical and economical goal for the SMR design (Boarin and Ricotti, 2011b). As a consequence, 'very

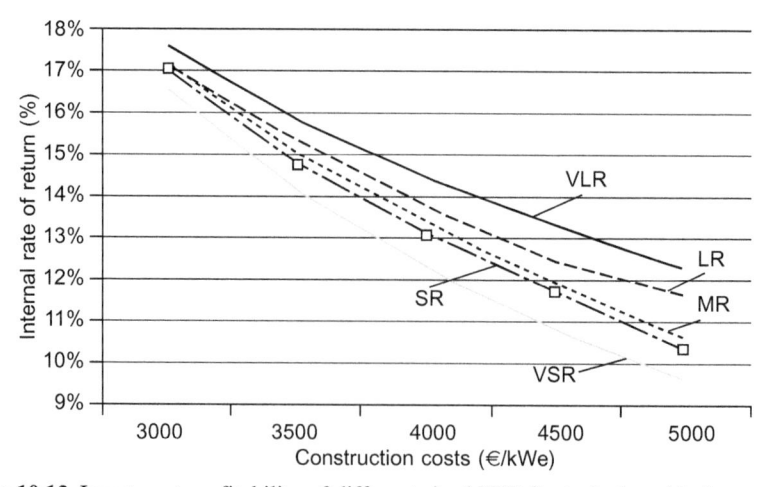

Figure 10.13 Investment profitability of different sized NPP fleets deployed in large sites: VLR = very large reactor (1500 MWe), LR = large reactor (1000 MWe), MR = medium size reactor (350 MWe), SR = small reactor (150 MWe), VSR = very small reactor (50 MWe) (Boarin and Ricotti, 2011b).

Economics and financing of small modular reactors (SMRs) 263

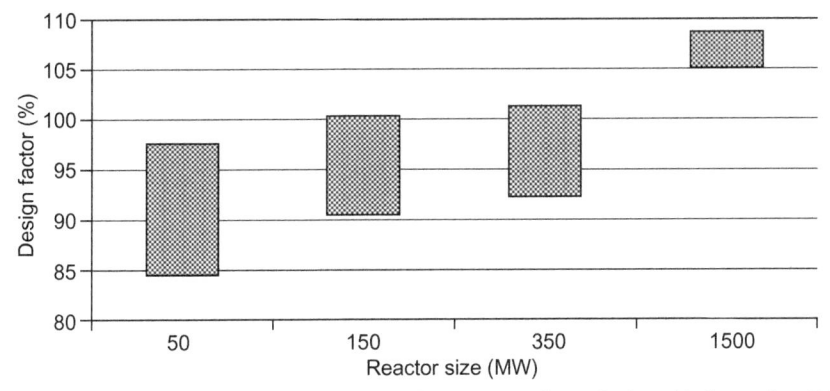

Figure 10.14 Design saving factor ranges of different SMR fleets deployed in large sites (4500 MWe), compared to a very large reactor unit (Boarin and Ricotti, 2011b).

small' reactors (VSR) must come up with additional saving factors (Figure 10.14). Rather, VSRs do not really compete in the same SMR playground since they have other unique requirements, e.g. emphasis on total capital cost, rather than on cost per KW installed, and may have unique applications, such as very small or scattered user areas.

10.6 Competitiveness of multiple small modular reactors (SMRs) versus large reactors

10.6.1 Deterministic scenarios

The economic analysis and comparison between SMRs and LRs, has given great emphasis to the capital costs that dominate nuclear generation costs, as a very capital-intensive technology. The cost comparison between LRs and multiple SMRs has been assessed based on very conservative assumptions that almost disregard savings by design at SMRs. Under this assumption and considering ideal or expected construction costs and schedule for LRs (i.e. no delays and no cost overruns), scenario analysis of alternative LR and multiple SMR projects confirms a comparable or higher economic performance of LRs, essentially due to the economy of scale on construction costs. On average, investment IRR and profitability index (PI) of LRs are 1–1.5% higher than SMRs (Boarin and Ricotti, 2009). This slight difference, applied on a relevant project investment value, translates into a significant project value increase. This holds in deterministic scenario conditions, with conservative assumptions on SMRs and ideal assumption for LRs, with no uncertainties affecting the scenario evolution (Boarin and Ricotti, 2011a).

Nonetheless, multiple SMRs have economic features that make them competitive with large NPPs under different perspectives than mere profitability. Multiple SMRs offer financial benefits that encompass intrinsic investment modularity. Investment modularity and scalability are intrinsic features of multiple SMRs that allow adaptation of the investment program to the electricity and financial market evolution. Current projected schedules of SMRs are in the range of four years for the first-of-a-kind (FOAK) and down to two years for a n-th-of-a-kind (NOAK), in some designs. This shorter construction time is due mainly to smaller size, simpler design, increased modularization, higher degree of factory fabrication and series fabrication of components.

The shorter construction schedule and the smaller output size make SMRs more readily adaptable to market conditions, both temporally and spatially. The shorter lead times and the plant capacity allow to split the investment in a closer proximity to the market evolution: if not needed, the construction of an additional SMR unit can be avoided whereas a monolithic LR investment may result in an unexpected overcapacity installed. Whereas market conditions are highly uncertain, the SMR modularity translates in adaptability; the investment flexibility in the plant deployment has an associated economic value, which is caught by real option analysis. It is demonstrated that this economic value is positive and accounts for the possibility of avoiding financial losses in market downturn and reaping early revenues in favourable market conditions. The chance to better cope with the probability of a change in the economic environment reduces the gap of competitiveness between LRs and SMRs (Locatelli *et al.*, 2012).

A short construction schedule limits the financial cost escalation during the construction period. During construction, when no revenues allow the capital repayment, financial interests are compound over a growing invested capital base, increasing exponentially. This is the reason why, assuming the same total overnight construction cost as large units, multiple SMR projects pay lower IDC than LR projects (Carelli *et al.*, 2007b; Boarin *et al.*, 2012).

Shorter PBT of each SMR unit allows to get a cash in-flow from the sale of power generated by early units. Average outstanding capital exposure may be relieved by suitable staggered deployment of successive units, and cash flow from early units may be employed to finance the construction of later units on the site. This capability to self-generate the sources of financing is not available to a single large NPP project and is a valuable option to limit up-front capital requirements: the relevant share of total capital investment cost may be provided by self-financing (Figures 10.15 and 10.16).

SMRs' investment scalability is a key value driver: by staggering the investment effort over time, the average capital-at risk and IDCs are decreased. Cash out-flow profile during the construction phase is smoother for SMRs (Figure 10.17). These features make of SMRs an affordable investment option by investors with financial constraints, despite the conservative assumption of higher total capital investment cost.

10.6.2 Introducing uncertainty in the economic analysis

On account of the investment modularization, multiple SMRs offer greater stability in their financial performance, faced with unfavourable boundary conditions: lower average invested capital accounts for lower interest capitalization and lower risk of

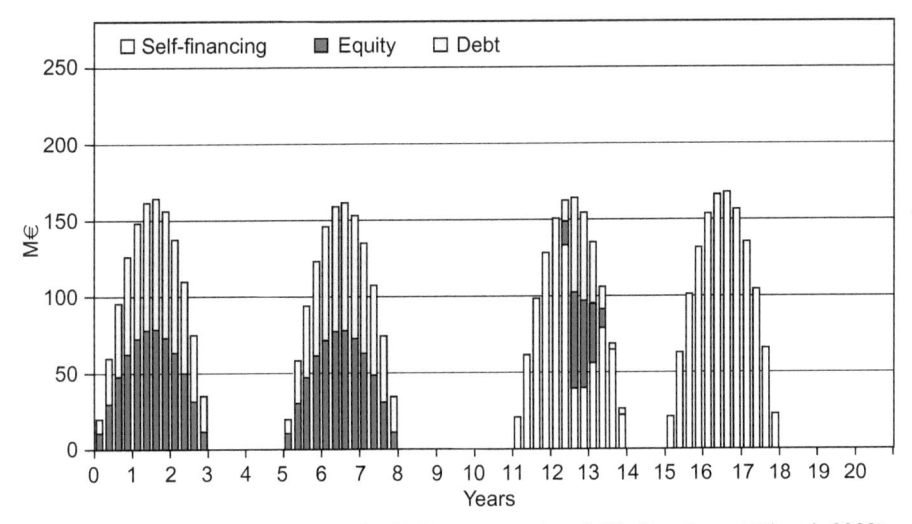

Economics and financing of small modular reactors (SMRs)   265

Figure 10.15 Sources of financing for SMRs construction (M€) (Boarin and Ricotti, 2009).

financial default. All these features are particularly valuable in the so-called 'merchant' scenarios, based on the rules of competition in liberalized electricity and capital markets, and characterized by the high cost of financing. Analysis and simulations in these conditions show that the gap in cost-effectiveness becomes narrower. With a high cost of equity and increasing cost of debt, there is a point where economic performance of SMRs overtakes that of LRs (Figure 10.18), on account of SMRs' capability to limit IDC escalation.

When deterministic and predictable scenarios are considered, assuming the construction schedule is respected, LRs normally show better economic performance based on economy of scale and lower overnight construction costs: PI and IRR are

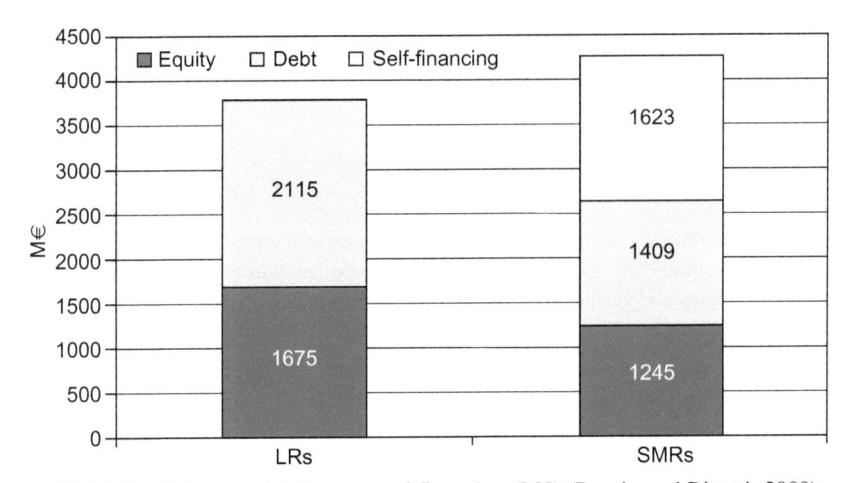

Figure 10.16 Breakdown of total sources of financing (M€) (Boarin and Ricotti, 2009).

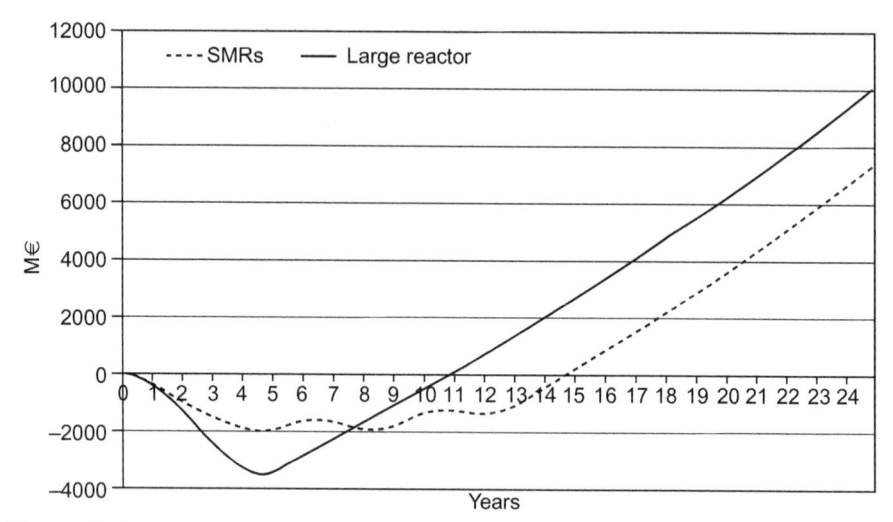

Figure 10.17 Cumulated cash flows of one LR and four equivalent SMR projects (Boarin and Ricotti, 2009).

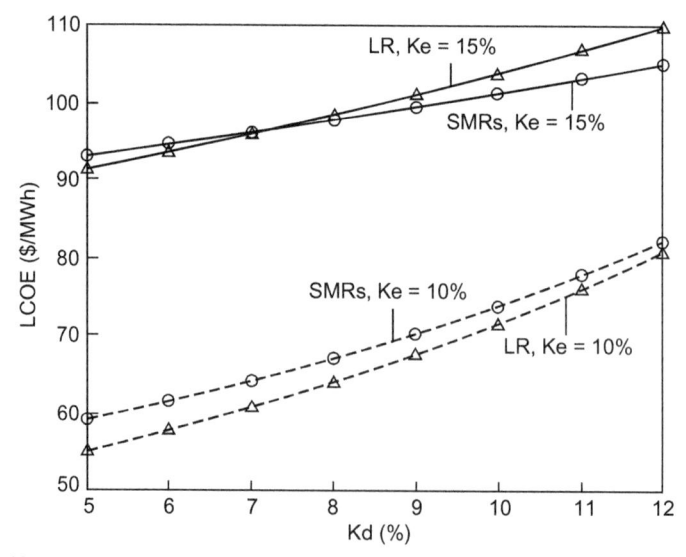

Figure 10.18 Levelized cost of electricity (LCOE) trend at increasing cost of debt Kd, at different cost of equity Ke: 'merchant case' = solid lines; 'supported case' (meaning with risk mitigation strategies and strong public support) = dashed lines (Boarin et al., 2012).

higher and, accordingly, generation cost is lower. But when scenario conditions become stochastic and uncertainties are included in the analysis, multiple SMRs may record higher mean profitability than LRs. In particular, assuming the possibility of a stochastic delay event affecting the construction schedule of both LR and multiple SMR projects, the calculated profitability distribution shows more favourable data

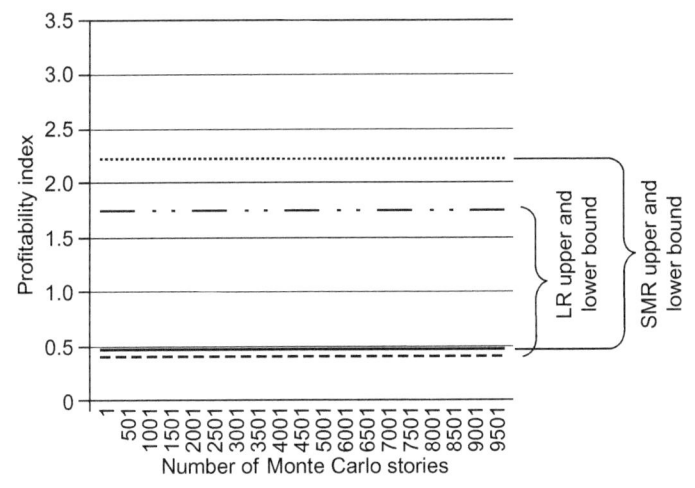

Figure 10.19 Profitability index of one LR vs. four SMRs in stochastic scenario analysis (Monte Carlo simulation, 10 000 stories).
Adapted from Boarin and Ricotti (2011a).

dispersion for SMRs toward positive values, meaning that SMRs have a greater chance of performing better in terms of profitability than LRs (Figure 10.19; Boarin and Ricotti, 2011a). A sensitivity analysis on the main economic and financial parameters shows that SMRs have a better capability to perform in changed scenario conditions (Figure 10.20).

10.6.3 SMRs and operating costs

While economic research usually concentrates on capital cost as the dominant driver of the economic competitiveness, operating costs have much lower impact on generation costs. Few estimates are available on SMR O&M costs and fuel costs. Nonetheless some trends and general considerations may be argued:

- The designers of advanced SMRs often indicate that O&M costs might be lower than those of LRs, owing to a stronger reliance of SMRs on passive safety features and to the resulting decrease in the number and complexity of safety features (Kuznetsov and Lokhov, 2011).
- Economy of scale, co-siting economies and learning influence operating costs of multiple SMRs as in the case of construction costs; comparing an LR of 1340 MWe with a fleet of four SMRs of 335 MWe each, the penalty of SMRs on O&M costs due to the loss of economy of scale is mitigated by co-siting and learning effects and the corresponding overall cost increase on LR is limited to +19%; a learning effect on O&M activities of multiple SMRs is also confirmed (Carelli et al., 2008a).
- In general SMRs offer poor neutron economy due to lower reactor core dimensions, which translates into higher fuel cost incidence on generation costs.

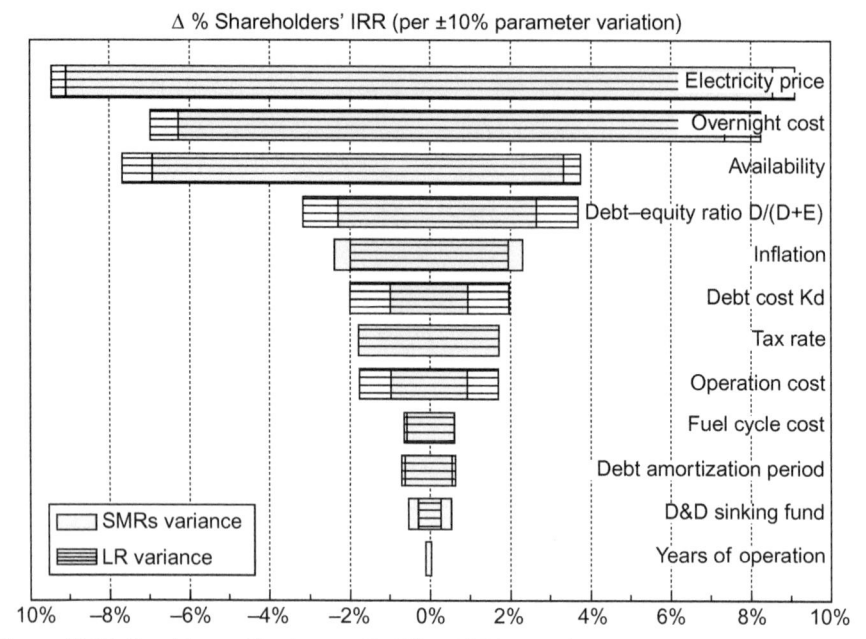

Figure 10.20 Sensitivity of project profitability (IRR) to main parameter input data variation for a merchant case (Boarin *et al.*, 2012).

- It is expected that long refueling schemes of some SMRs may increase specific fuel costs, due to a less effective fuel utilization, as compared to SMRs with conventional refueling intervals (IAEA, 2006, 2007b).
- Moreover, it is expected that for barge-mounted SMRs the sum of O&M and fuel costs is 50% higher than land-based SMRs, mainly due to a large O&M required by the barge.

Data information on decontamination and decommissioning (D&D) costs of advanced, modular SMR are not available from experience. One possible unbiased way to calculate them is to perform a statistical analysis of the data available from past decommissioning projects. Historical records show that there are several cost drivers that determine the decommissioning cost. Specifically those critical in the comparison between SMR and LR are: plant size, number of units in the site and decommissioning strategy ('immediate decommissioning' or 'deferred decommissioning'). Multiple regression analysis is a powerful tool applicable to these kinds of analysis which is able to quantify exactly the impact of each cost driver; it allows for an in-depth examination of the trend correlation between the dependent and the explanatory variables. The result of this statistical analysis is that the economy of scale also applies to the plant decommissioning activities and represents a disadvantage for SMRs (Locatelli and Mancini, 2010b); the D&D cost for a medium-sized SMR unit may be three times higher than for a large plant. On the other hand, co-siting economies should decrease D&D costs for parallel dismantling of twin units.

It is worth stressing that historical data are related only to GEN I and GEN II reactors (both large and small), not to modern GEN III+ reactors and SMRs. Regarding

SMRs, the design layout simplifications and reduced number of components should drive a cost reduction. In the same way as high content of factory fabrication should decrease construction costs by decreasing on-site assembling activities, modular and factory-assembled reactors should be dismantled in a sub-system that could be transported back to a centralized factory, where operations should be cheaper than on-site dismantling (Kuznetsov and Lokhov, 2011; IAEA, 2007b).

As a general, final comment, it can be stated that technical savings from design simplification and standardization and co-siting economies are the competing forces that play against the loss of economy of scale. The balance between these factors should be evaluated on a project-specific basis and supported by data information from actual experience.

10.6.4 Conclusion: the 'economy of multiples'

As seen, multiple SMRs on the same site may be considered as an investment option alternative to a power station based on LRs with the same overall power output. The SMR investment case bears a loss of economy of scale which may be mitigated by some specific cost benefits. These economic benefits, presented in the previous sections, are enhanced by deploying multiple units on a same site. On the construction side, learning accumulation, modularization and co-siting economies are expected to be fostered by the multiple units 'philosophy' and the 'mini-serial production' of a number of smaller and simpler plant units. In addition, design simplification is expected to further contribute to cost-reduction of SMRs, but its evaluation is strictly plant-specific and deserves further analysis and approaches.

On the investment side, the fractioning of total investment into multiple smaller batches may represent a risk mitigation factor against possible cost/time overruns and an opportunity to adapt the investment plan and the power installed rate to the market conditions. All these economic features may be summarized into an 'Economy of Multiple' concept that may counterbalance the 'Economy of Scale' philosophy, especially when uncertainty is introduced in the analysis, affecting market conditions or construction process time schedule.

Some concepts apply to the operating and decommissioning cost as well, with a loss of economy of scale to be partially recovered by the simplification of operating or dismantling procedures. SMR design simplification has a relevant impact both on construction and decommissioning costs and any economic assessment that does not take fully into account such issues tends to be very conservative against SMR.

10.7 Competitiveness of SMRs versus other generation technologies

There are specific niche markets or applications where SMRs are the only applicable option as an NPP. Given their lower capital requirements and small size, which makes them suitable for small electrical grids, SMRs can effectively address the energy needs of small newcomer countries or remote and scattered areas. Their smaller size may better fit co-generation purposes and other energy applications. In these situations, comparison with large units is not applicable. Considering their output size, a 300–400 MWe SMR plant might also be considered as an alternative generation technology to fossil-fueled, base-load small–medium plants, such as coal and combined cycle gas turbine (CCGT).

According to NEA/OECD (2011), nuclear power is generally competitive with many other technologies (coal-fired plants, gas-fired plants, renewables) in Brazil, Japan, Republic of Korea, Russian Federation and the United States. Similarly, some SMRs are expected to be competitive with several projects of coal-fired, gas-fired and renewable plants of various types, including those of small to medium-sized capacity (below 700 MWe). A Monte Carlo analysis comparing SMRs with coal and gas-fired plants (Locatelli and Mancini, 2010a) stresses the fundamental role of the carbon Tax, or the CO_2 sequestration cost, on the competitiveness of nuclear generation cost. Without a carbon tax, coal and CCGT may be more attractive then SMRs, in terms of NPV and LCOE. The carbon tax may dramatically increase the generation cost of coal-fired and CCGT plants and transfer its uncertain value to the overall uncertainty of the investment return of fossil-fueled plants, increasing SMR competitiveness (Figure 10.21).

In the open literature, several studies deal with the application of portfolio theory to power generation sector, but only few compare large and small power plants from this point of view. In Locatelli *et al.* (2011b) and Locatelli and Mancini (2011a) the investigation of the best combinations of base load power plants, for an investor on the basis of different scenarios, is carried out. As far as different Carbon tax and electricity prices are considered, the IRR and the LCOE are calculated using Montecarlo simulations for three base load technologies: nuclear, coal and CCGT. Different plant sizes are considered: for nuclear plants 335 and 1340 MWe, for coal plants 335 and 670 MWe and for CCGT plants 250 and 500 MWe.

Three markets are investigated, referred to as large grid (30 GWe), corresponding to a national-level utility, a medium grid (10 GWe), corresponding to a regional-level

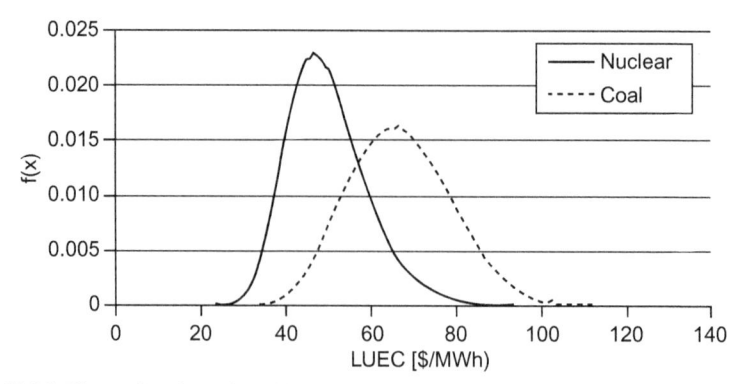

Figure 10.21 Uncertainty introduced by the carbon tax on the coal plant's LCOE (Locatelli and Mancini, 2010a).

utility, and a small grid (2 GWe), corresponding to a municipality or an island. For each market two types of portfolio are considered: (i) all possible combinations of large plants only and, (ii) small plants combinations only. In both cases the maximum site size is 1340 MWe, i.e. the size of a stand-alone large nuclear power plant, hence economy of scale and economy of multiples are taken into account.

In order to identify the best power plants portfolio mix from the investor point of view, the IRR and LUEC have been assumed as a metric of the investment performance (higher IRR; lower LUEC). The mean value of such indicators, arising from their own specific probability distributions, has been assessed against their respective standard deviation.

The results show that the nuclear power plants play a fundamental role in portfolio generation and become a convenient option when the carbon tax is included in the economic evaluation. Based on the above-mentioned criteria of IRR and LCOE, large plants may represent the best investment option where large new power capacity is required and small plants are competitive when small power installations are required. In order to achieve the highest profitability with the lowest risk, it is necessary to build several plants of different types and, in the case of small grids, this is possible only with small power plants. Although the choices of the investor will be subjected to the specific needs and the risk attitude, guidelines can be drawn to facilitate the selection process:

- large plant portfolios usually have better performance than small plant portfolios according to the LCOE indicator;
- small plant portfolios may have comparable performance with respect to large plant portfolios, according to the IRR indicator;
- in case of large markets (>10 GWe), large plant portfolios are the best alternative in most cases;
- in the case of small size markets (2 GWe), small plant portfolios are able to provide a lower investment risk than large plant portfolios for both IRR and LCOE indicators;
- the optimal mix is largely made up with nuclear power plants when a medium/high cost of CO_2 emissions or a low electricity price apply;
- in the absence of a carbon tax, the best performances are provided by coal-fired plants;
- an increase in the electricity price or a reduction of the carbon tax decreases the gap between the small and the large plant efficiency frontiers.

10.8 External factors

The analysis of other non-design or even non-technological features (external factors) may have a tremendous impact on the deployment strategy of SMRs. An 'external factor' is a factor usually not monetary and not directly considered within the investment evaluation, because it is not under the direct control of the investor. However it may strongly influence the life cycle and the attractiveness and the feasibility of the project itself. These external factors (Figure 10.22), such as security of fuel supply, public acceptance and environmental aspects, have been tentatively clustered and investigated (Mancini et al., 2009; Locatelli and Mancini, 2011b).

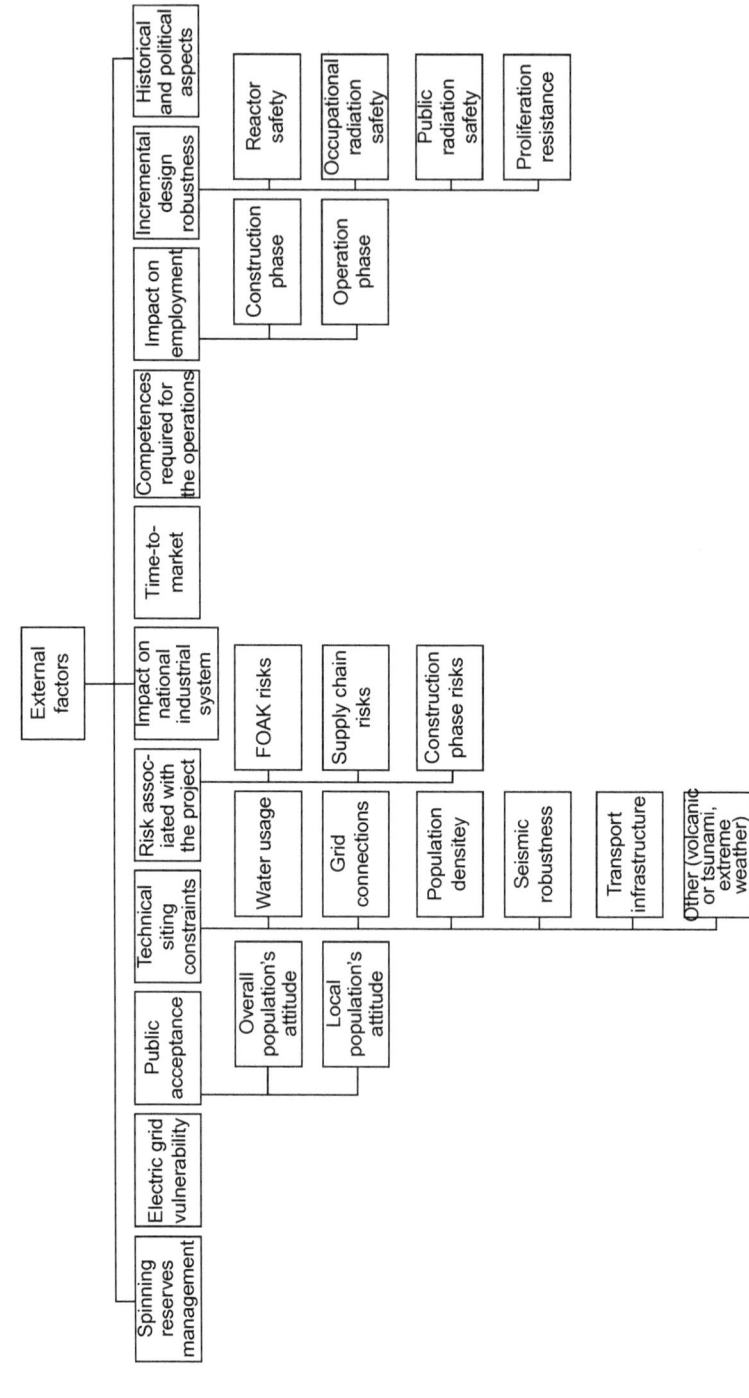

Figure 10.22 External factors to be considered in the comparison between LRs and SMRs.

The preliminary results indicate that SMRs may better fit all the factors. However it is important to point out that the 'not in my back yard' (NIMBY) syndrome limits the possibility of spreading SMRs in different sites, to exploit the advantages of a decentralized generation model like better grid stability. Therefore, a plausible scenario for many countries is the concentration of multiple SMRs in each site. Even in this configuration, the SMRs should reap many advantages through all the life cycle. During the planning and construction phases, more sites than for LRs can be exploited because more sites are suitable to SMR deployment; SMR time-to-market is shorter, and fewer risks are associated to the construction phase as well as increased benefits for local industries. In the operation phase, SMRs provide more job positions and do not require additional costs in terms of 'spinning reserves'.

The external factors could be integrated with the monetary factors to perform a holistic evaluation, through a six-step methodology (Locatelli and Mancini, 2012b):

1. Identification of relevant attributes for evaluation and selection, looking at the specific country taken into consideration.
2. Definition of measurement and evaluation process of each attribute (quantitative or qualitative, monetary or not, etc); each NPP design has to be evaluated on each attribute.
3. Definition of attribute's hierarchical structure as required by a fuzzy analytical hierarchy process (AHP) application.
4. Expert elicitation to get attribute weights; each expert has to fill in a questionnaire of pairwise comparisons between attributes or groups of them. Fuzzy AHP permits judgements through linguistic variables (Yang and Chen, 2004).
5. Pairwise comparison matrices from different decision makers are aggregated through the geometric mean method presented in Kuo *et al.* (2002). Buckley's method (Buckley, 1985) is then applied to the hierarchical structure and to get final attributes weights; these are fuzzy sets, so a decoding process is needed to obtain crisp values, the most common being the centroid method (Opricovic and Tzeng, 2004).
6. The TOPSIS method is applied for the final integration, looking at the five steps in Opricovic and Tzeng (2004).

10.9 Future trends

Most of the SMR projects are not a mere scaling of LRs, exploiting original design features. Several of them are still in the design phase and are conceived with a deployment strategy that benefits from investment modularity.

In the current developing phase of SMR concepts, due investigation of the potential, positive features is required and some complementary research efforts are needed. As far as SMR economic, financial and deployment issues are concerned, the main R&D efforts should be focused on the cost estimation (bottom-up approach) and on the estimation of the deployment flexibility value (by means of the real options analysis).

- *Bottom-up approach:* a more robust cost estimation is needed, with specific focus on design-based economies and enhancements, along with the SMR design development and more details on the structures, systems, components, layout, etc. A bottom-up approach is a suitable, alternative methodology to a similarity-based top-down approach, to estimate and

assess construction costs. O&M costs should also be included, once the SMR operating strategy and related licensing issues are addressed and known, e.g. referring to crew requirements and multiple SMR modules operated by a single control room.
- *Real options:* the application of this methodology, complementary to the DCF analysis, is able to catch the value of the investment flexibility. This approach perfectly fits the modularity features of SMRs. Modular investments give the opportunity to delay, anticipate, stop or accelerate the deployment plan, according to the time evolution of the boundary conditions for the business, given, for example, by the energy market, the regulatory framework, the political as well as the macro-economic environment. This opportunity is valuable as direct cash inflows and should be taken into account to get a correct evaluation of the SMRs investment.

Real options are the most suitable tool to evaluate the economic potential of co-generation of non-electric products. Co-generation may be used as an option to adapt the SMR's power generation to the load curve without losing economic value from the nuclear investment and without stressing the primary loop thermo-mechanics. Intrinsic modularity of multiple SMRs is particularly suitable to enhance the overall site generation flexibility, by devoting the thermal power generated by some SMR units to the co-generation processes.

10.10 Sources of further information and advice

Blyth, W., 2006. Factoring risk into investment decisions. UK Energy Research Centre, London. working paper.

Bowers, H.I., Fuller, L.C., Myers, M.L., 1983. A.2. Summary of Literature Review of Cost-Size Scaling of Steam-Electric Power Plants. Trends in Nuclear Power Plant Capital Investment Cost Estimates 1976 to 1982. Oak Ridge National Laboratory, TN, USA. ORNL/TM-8898.

Bowers, H.I., Fuller, L.C., Myers, M.L., 1987. Cost Estimating Relationships for Nuclear Power Plant Operation and Maintenance. Oak Ridge National Laboratory, TN, USA. ORNL/TM-10563.

Carayannis, E.G., 1996. Re-engineering high risk, high complexity industries through multiple level technological learning. A case study of the world nuclear power industry. Journal of Engineering and Technology Management 12 (4), 301–318.

David, P.A., Rothwell, G.S., 1996. Measuring standardization: an application to the American and French nuclear power industries. European Journal of Political Economy 12 (2), 291–308.

David, P.A., Rothwell, G.S., 1996. Standardization, diversity and learning: strategies for the coevolution of technology and industrial capacity. International Journal of Industrial Organization 14 (2), 181–201.

Feretic, D., Tomsic, Z., 2005. Probabilistic analysis of electrical energy costs comparing production costs for gas, coal and nuclear power plants. Energy Policy 33, 5–13.

Finon, D., Roques, F., 2008. In: Contractual and Financing Arrangements for New Nuclear Investment in Liberalised Markets: Which Efficient Combination? CeSSA Working Paper, European Regulation Forum on Supply Activities.

Gollier, C., Proul, D., Thais, F., Walgenwitz, G., 2005. Choice of nuclear power investments under price uncertainty: valuing modularity. Energy Economics 27, 665–685.

Ingersoll, D.T., 2009. Deliberately small reactors and the second nuclear era. Progress in Nuclear Energy 51, 589–603.

Kazachkovskii, O.D., 2001. Calculation of the economic parameters of a nuclear power plant. Atomic Energy 90 (4), 329–336.

Kennedy, D., 2007. New nuclear power generation in the UK: cost benefit analysis. Energy Policy 35, 3701–3716.

Koomey, J., Hultmanc, N.E., 2007. A reactor-level analysis of busbar costs for US nuclear plants, 1970–2005. Energy Policy 35 (11), 5630.

Krautmann, A., Solow, J.L., 1988. Economies of scale in nuclear power generation. Southern Economic Journal 55, 70–85.

Langlois, R.N., 2002. Modularity in technology and organization. Journal of Economic Behavior and Organization 49, 19–37.

Lapp, C.W., Golay, M.W., 1997. Modular design and construction techniques for nuclear power plants. Nuclear Engineering and Design 172, 327–349.

Marshall, J.M., Navarro, P., 1991. Costs of nuclear power plant construction: theory and new evidence. RAND Journal of Economics 22 (1), 148–154.

Oxera, 2005. Financing the Nuclear Option: Modeling the Cost of New Build. *Agenda: advancing economics in* business.www.oxera.com.

Phung, D.L., 1987. Theory and Evidence for Using the Economy-of-Scale Law in Power Plant Economics. Oak Ridge National Laboratory, Oak Ridge, TN, USA. ORNL/TM-10195.

Reid, L., 2003. Modeling Modularity Impacts on Nuclear Power Plant Costs. ORNL, Oak Ridge, TN.

Roques, F.A., Nuttall, W.J., Newbery, D.M., 2006. Using Probabilistic Analysis to Value Power Generation Investments under Uncertainty. University of Cambridge. CWPE 0650 and EPRG 065.

Schock, R.N., Brown, N.W., Smith, C.F., 2001. In: Nuclear Power, Small Nuclear Technology, and the Role of Technical Innovation: an Assessment. Working paper for the Baker Energy Forum. Rice University, Houston, TX.

Takizawa, S., Suzuki, A., 2004. Analysis of the decision to invest for constructing a nuclear power plant under regulation of electricity price. Decision Support Systems 37, 449–456.

Toth, F.L., Rogner, H.-H., 2006. Oil and nuclear power: past, present, and future. Energy Economics 28, 1–25.

References

Barenghi, S., Boarin, S., Ricotti, M.E., 2012. In: Investment in Different Sized SMRs: Economic Evaluation of Stochastic Scenarios by INCAS Code.Proceedings of ICAPP'12, paper 12322, Chicago.

Barkatullah, N., 2011. Possible Financing Schemes for Current and Near Term Nuclear Power Project. In: Workshop on Technology Assessment of Small and Medium-sized Reactors for Near Term Deployment, IAEA-PESS. http://www.iaea.org/NuclearPower/Downloadable/Meetings/2011/2011-12-05-12-09-WS-NPTD/Day-4/23_IAEA_Barkatullah-Economics_SMRDec2011.pdf. (accessed Aug. 2013).

Binz, R., Sedano, R., Furey, D., Mullen, D., 2012. Practicing Risk-Aware Electricity Regulation: What Every State Regulator Needs to Know. In: CERES Report, Ronald J. Binz Public Policy Consulting.

Boarin, S., Ricotti, M.E., 2009. Cost and Profitability Analysis of Modular SMRs in Different Deployment Scenarios. In: Proceedings of the 17th Int. Conf. on Nuclear Engineering – ICONE17, paper 75741, Brussels.

Boarin, S., Ricotti, M.E., 2011a. Risk Analysis of Nuclear Power Investments in Large Versus Small–Medium Sized Reactors. In: Proceedings of Nuclear Power Europe, paper 1545, Milan.

Boarin, S., Ricotti, M.E., 2011b. Multiple Nuclear Power Plants Investment Scenarios: Economy of Multiples and Economy of Scale Impact on Different Plant Sizes. In: Proceedings of ICAPP 2011, paper 11193, Nice.

Boarin, S., Locatelli, G., Mancini, M., Ricotti, M.E., 2012. Financial case studies on small–medium sized modular reactors. Nuclear Technology 178 (2), 218–232.

Buckley, J.J., 1985. Ranking alternatives using fuzzy numbers. Fuzzy Sets and Systems 15, 1–31.

Carelli, M.D., Petrovic, B., Mycoff, C.W., Trucco, P., Ricotti, M.E., Locatelli, G., 2007a. In: Economic Comparison of Different Size Nuclear Reactors.Proceedings of LAS/ANS Symposium, Cancun, Mexico.

Carelli, M.D., Petrovic, B., Mycoff, C.W., Trucco, P., Ricotti, M.E., Locatelli, G., 2007b. Smaller Sized Reactors can be Economically Attractive. In: Proceedings of ICAPP 2007, Paper 7569, Nice.

Carelli, M.D., Mycoff, C.W., Garrone, P., Locatelli, G., Mancini, M., Ricotti, M.E., Trianni, A., Trucco, P., 2008a. In: Competitiveness of Small-medium, New Generation Reactors: A Comparative Study on Capital and O&M costs. Proceedings of 16th Int. Conf. on Nuclear Engineering-ICONE 16, Paper 48931. ASME, Orlando, FL.

Carelli, M.D., Garrone, P., Locatelli, G., Mancini, M., Trucco, P., Ricotti, M.E., 2008b. Chap.6. Nuclear Renaissance: An Assessment of Small–Medium, Integrated and Modular plants. In: Garrone, P. (Ed.), Investments and Service quality in the Electricity Industry. Franco Angeli.

Carelli, M.D., Garrone, P., Locatelli, G., Mancini, M., Mycoff, M., Trucco, P., Ricotti, M.E., 2010. Economic features of integral, modular, small-to-medium size reactors. Progress in Nuclear Energy 52, 403–414.

Chu, S., 2010. The Wall Street Journal.March 23.

Damodaran, A., 2011. Chap.3 *The Basics of Risk*. In: Applied Corporate Finance, third ed. John Wiley & Sons.

DOE, 2005. Electric Power Annual 2005. DOE/EIA-0348.

Dolley, S., 2008. Options available to lessen risk of nuclear investing, analysts say. Nucleonics Week 49, 47.

EMWG, 2007. Economic Modeling Working Group – Cost Estimating guidelines for generation IV nuclear energy systems, Revision 4.2. GIF/EMWG/2007/004. OECD. http://www.gen-4.org/Technology/horizontal/EMWG_Guidelines.pdf. (accessed Aug. 2013).

Flyvbjerg, B., Bruzelius, N., Rothengatter, W., 2003. Megaprojects and Risk: An Anatomy of Ambition. Cambridge University Press.

Frischtak, C.R., 1994. Learning and technical progress in the commuter aircraft industry: an analysis of Embraer's experience. Research Policy 23 (5), 601–612.

Gallanti, M., Parozzi, F., 2006. Valutazione dei costi di produzione dell'energia elettrica da nucleare. Energia 3, 60–70.

Goldberg, S., Rosner, R., 2011. Small Modular Reactors – Key to Future Nuclear Power Generation in the U.S. Energy Policy Institute at Chicago, The Harris School of Public Policy Studies. http://epic.uchicago.edu/i/publication/SMR_Final_White_Paper_7-11.pdf. (accessed Aug. 2013).

Grubler, A., 2010. The costs of the French nuclear scale-up: A case of negative learning by doing. Energy Policy 38, 5174–5188.

Hayns, M.R., Shepherd, J., 1991. SIR reducing size can reduce cost. Nuclear Energy 30 (2), 85–93.
IAEA, 2006. Status of Innovative Small and Medium Sized Reactor Designs 2005: Reactors with Conventional Refuelling Schemes. IAEA-TECDOC-1485, Vienna.
IAEA, 2007a. Safety glossary–Terminology used in nuclear safety and radiation protection. http://www-pub.iaea.org/MTCD/publications/PDF/Pub1290_web.pdf. (accessed Aug. 2013).
IAEA, 2007b. Status of Small Reactor Designs Without On-Site Refuelling. IAEA-TECDOC-1536. http://www-pub.iaea.org/MTCD/Publications/PDF/te_1536_web.pdf. (accessed Aug. 2013).
Kuo, R.J., Chi, S.C., Kao, S.S., 2002. A decision support system for selecting convenience store location through integration of fuzzy AHP and artificial neural networks. Computers in Industry 47, 199–214.
Kuznetsov, V., Lokhov, A., 2011. Current Status, Technical Feasibility and Economics of Small Nuclear Reactors. OECD-Nuclear Energy Agency, Paris. http://www.oecd-nea.org/ndd/reports/2011/current-status-small-reactors.pdf. (accessed Aug. 2013).
Locatelli, G., Mancini, M., 2010a. Small–medium sized nuclear coal and gas power plant: A probabilistic analysis of their financial performances and influence of CO_2 cost. Energy Policy 10 (38), 6360–6374.
Locatelli, G., Mancini, M., 2010b. Competitiveness of small-medium, new generation reactors: a comparative study on decommissioning. Journal of Engineering for Gas Turbines and Power 10 (132), 102906.
Locatelli, G., Mancini, M., 2011a. Large and small baseload power plants: Drivers to define the optimal portfolios. Energy Policy 39 (12), 7762–7775.
Locatelli, G., Mancini, M., 2011b. The role of the reactor size for an investment in the nuclear sector: An evaluation of not financial parameters. Progress in Nuclear Energy 53 (2), 212–222.
Locatelli, G., Mancini, M., 2012a. Looking back to see the future: Building nuclear power plants in Europe. Construction Management and Economics 30, 623–637.
Locatelli, G., Mancini, M., 2012b. A framework for the selection of the right nuclear power plant. International Journal of Production Research 50 (17), 4753–4766.
Locatelli, G., Mancini, M., Vannucci, C., 2011a. In: How the Plant Size Impacts on Project Risks: The Nuclear Power Reactors Case Study.Proceedings of IPMA-25th World Congress, Brisbane, Queensland.
Locatelli, G., Mancini, M., Agosti, D., 2011b. In: Nuclear Power Plants in the Base Load Generation Portfolios: A Probabilistic Study of the Size Role.International Congress on Advances in Nuclear Power Plants - ICAPP 2011.
Locatelli, G., Mancini, M., Ruiz, F., Solana, D., 2012. In: Using Real Options to Evaluate the Flexibility in the Deployment of SMR.Proceedings of ICAPP'12, paper 12233, Chicago.
Lovins, A.B., 1986. The origins of the nuclear fiasco. In: Byrne, J., Rich, D. (Eds.), The Politics of Energy Research and Development. Transaction Books, pp. 7–34.
MacKerron, G., et al., 2005. Paper 4 – Economics of Nuclear Power: A Report to the Sustainable Development Commission. In: University of Sussex. http://www.sd-commission.org.uk/data/files/publications/Nuclear-paper4-Economics.pdf. (accessed Aug. 2013).
Mancini, M., Locatelli, G., Tammaro, S., 2009. Impact of the External Factors in the Nuclear Field: A Comparison Between Small Medium Reactors vs. Large Reactors. In: Proceedings of the 17th Int. Conf. on Nuclear Engineering-ICONE 17, paper 75689, Brussels.

Matzie, R.A., 2005. Building New Nuclear Plants to Cost and Schedule – An International Perspective. Royal Academy of Engineering, Energy Seminars. http://www.raeng.org.uk/events/pdf/archive/Energy_Seminars_Nuclear_Fission_Regis_Matzie.pdf. (accessed Aug. 2013).

NEA/OECD, 2011. Current Status, Technical Feasibility and Economics of Small Nuclear Reactors. http://www.oecd-nea.org.

Opricovic, S., Tzeng, G.H., 2004. Compromise solution by MCDM methods: A comparative analysis of VIKOR and TOPSIS. European Journal of Operational Research 156, 445–455.

OTA – U.S. Congress Office of Technology Assessment, 1993. Aging Nuclear Power Plants: Managing Plant Life and Decommissioning. OTA-E-575, Washington, DC. http://www.fas.org/sgp/othergov/doe/lanl/lib-www/ota/00418532.pdf. (accessed Aug. 2013).

Parsons, J.E., Du, Y., 2009. Update on the Cost of Nuclear Power, MIT-CEEPR. http://web.mit.edu/ceepr/www/publications/workingpapers/2009-004.pdf. (accessed Aug. 2013).

Sullivan, G.P., Pugh, R., Melendez, A.P., Hunt, W.D., 2010. Operations & Maintenance Best Practices – A Guide to Achieving Operational Efficiency. Pacific Northwest National Laboratory. http://www1.eere.energy.gov/femp/pdfs/OandM.pdf. (accessed Aug 2013).

Williams, K.A., Miller, K., 2006. A User's Guide for G4 – ECONS: a Generic EXCEL-based Model for Computation of the Projected Levelized Unit Electricity Cost (LUEC) from Generation IV Reactor Systems. Generation IV International Forum.

Wright, T.P., 1936. Factors Affecting the Cost of Airplanes. Journal of the Aeronautical Sciences 3 (4), 122–128.

Yang, C.C., Chen, B.S., 2004. Key quality performance evaluation using fuzzy AHP. Journal of the Chinese Institute of Industrial Engineers 21 (6), 543–550.

Licensing of small modular reactors (SMRs)

11

Richard L. Black
Consultant, McLean, VA, United States

11.1 Introduction

The licensing and deployment of any small modular reactor (SMR) are dependent on whether (a) there is a significant market demand for new nuclear power, (b) SMR technology can be developed in a timely and cost-effective manner to meet the demand, and (c) the SMR technology can be licensed effectively. As described in this handbook, enhanced safety, improved security, and flexibility in siting and application are all factors that have created a market demand. Assuming a market demand for SMRs and an ability to finance, licensing becomes the next risk factor. Effective licensing is dependent, in part, on the maturity of the SMR technology. New and unproven nuclear technologies might present a challenge to effective licensing. Accordingly, near-term licensing and deployment of SMRs are focused on the proven light water reactor (LWR) technology. However as noted in Part Two, in recent years, non-LWR SMR advanced nuclear technologies have emerged, but none of these designs are sufficiently developed for licensing reviews.

The commercial deployment of SMRs will be a global enterprise. Vendors will apply for licensing approval of their designs in the country of design origin (i.e., where the vendor is located). The approved SMR design will then be manufactured largely in the country of origin, marketed globally, and licensed for operation in the country of deployment. Various degrees of and requirements for local content for SMR systems and components might present a challenge to regulatory authorities to make sure that these safety systems and components meet the safety intent of the SMR design. The licensing/regulatory authority must be able to license and regulate SMRs in a manner that reasonably assures all safety, environmental, regulatory, and policy issues are addressed and resolved, particularly in the post-Fukushima environment. Importantly the licensing authority must be able to assess the enhanced safety characteristics of SMR designs to support approval or certification of these advanced reactor technologies and their subsequent licensing. Enhanced safety designs and the associated significantly reduced risk to the public afford the licensing authority an ability to license SMR designs based on risk and safety assessments that support the designs.

Several countries have begun regulatory reviews of SMR designs. This chapter will discuss several approaches for SMR licensing that are permitted by the US regulatory process promulgated in new licensing procedures of the US Nuclear Regulatory Commission (NRC). This discussion is provided to the extent that they might be relevant to

SMR licensing in other countries. In addition, this chapter discusses how several key SMR generic licensing issues were addressed and resolved by the NRC as an example to highlight how these safety and policy issues can be addressed by other regulatory authorities. The effective licensing of SMRs as a global enterprise will be aided by international collaborations and assistance. These collaborations, both by industry and regulatory authorities, will provide a strategy and framework to assist in the safe and effective licensing of SMRs for worldwide deployment.

11.2 US Nuclear Regulatory Commission (NRC) licensing of small modular reactors (SMRs): An example

SMRs are defined broadly to include a range of technologies, particularly a range of diverse fuels and coolants. US SMR reactor technology currently chosen for NRC licensing and near-term commercial deployment (2022–25) is based on LWR technology. Most of these SMRs will be integral pressurized water reactors (iPWRs) as discussed in Part Two of this handbook. LWRs have a well-established framework of regulatory requirements, a technical basis for these requirements, and supporting regulatory guidance that provides acceptable approaches for meeting NRC requirements. The NRC uses a standard review plan (SRP), NUREG-0800 (n.d.), to review licensing applications for these reactor designs. NUREG-0800 was revised in January 2014 to provide general review guidance for LWR SMRs. The NRC will require design-specific SRPs for the licensing of SMRs. These SRPs are under development by the NRC and SMR vendors. The SRP for the front-runner NuScale design was published in 2014 to support the submittal of the NuScale design certification application completed in January 2017. Additionally, the NRC has a well-established set of validated analytical codes and methods and a well-established infrastructure for conducting safety research needed to support its independent safety review of an LWR plant design and the technical adequacy of a licensing application. It should be emphasized that the near-term SMRs, particularly the iPWRs, will be subject to the same licensing process and the same safety requirements and standards as new large LWRs resulting in no diminution of safety.

New SMRs can be licensed under either of two existing regulatory approaches. The first approach is the traditional "two-step" process described in Title 10, Part 50 (n.d.), "Domestic Licensing of Production and Utilization Facilities," of the Code of Federal Regulations (10 CFR Part 50), which requires first a construction permit (CP) and then a separate operating license (OL). The second approach is the new "one- step" licensing process described in 10 CFR Part 52 (n.d.), "Licenses, Certifications, and Approvals for Nuclear Power Plants," which incorporates a combined construction and operating license (COL). NuScale, as first of a kind (FOAK) design, is licensing its SMR design under the Part 52, one-step licensing process. Certification of the NuScale design provides certainty that the safety aspects of the design are approved by the NRC if no changes are made to the design.

NRC's final safety evaluation report (SER) of the NuScale SMR design is expected to be issued in mid-to-late 2020. As required by 10 CFR Part 52 and discussed in more detail later, the final step to certify a design is to initiate a public rulemaking proceeding and issue a final rule after consideration of all comments that certifies the design meets all regulatory requirements. Subsequent license applicants referencing the certified design without revisions are assured that NRC will approve/certify their safety application. This approach reduces the inherent risk of safety licensing of unique and customized reactor designs and technology.

Key to the licensing of any new reactor design, including SMRs, is the approach that the regulatory authority will utilize in assessing the safety basis for the reactor and associated safety systems. Regardless of the licensing process chosen, the licensing authority will need to consider the relative merits of a deterministic versus a risk-informed performance-based approach to assess and approve the safety case for SMRs.

11.2.1 Alternatives for SMR licensing

As indicated earlier, new SMRs in the United States can be licensed by the NRC under either of two existing regulatory approaches that require a CP, an OL, or a combined CP/OL (COL). Licensing options will be briefly outlined later to the extent one or more of the options might be similar to licensing processes in other countries. Important to note in any licensing option is whether (a) the design submitted is preliminary or final, (b) the design submitted has been previously approved or certified by a competent and capable regulatory authority, and (c) the chosen site has been previously characterized and approved [in the United States through an early site permit (ESP) process] or is seeking approval. Fundamentally, any chosen licensing strategy will depend on the quality and completeness of the design and safety information, the status of the safety review of the reactor design, and the status of the review of the safety and environmental characteristics of the site.

While cultural and political considerations may vary between countries regarding nuclear regulation and the safety, environmental, and economic considerations that support licensing, fundamental safety issues and reviews are fairly consistent and generally follow the NRC model. However, the NRC has two unique licensing processes that require consideration by other regulatory authorities depending on the culture of regulation in each country. Regulatory culture connotes how each regulatory authority considers public involvement, the ability of the applicant to demonstrate and document its licensing case, and the transparency and documentation of licensing decisions. First the NRC licensing process has a requirement that reactor designs that will be certified through 10 CFR Part 52, Subpart B, must go through a rulemaking process to provide an opportunity for public involvement. No other country requires a rulemaking process for a reactor design to be certified by a cognizant regulatory authority. A design certification rule (DCR) is intended to freeze the design in a more final and formal manner. Any changes to the design must be done by amendment to the DCR that requires a formal and lengthy rulemaking process. Second, another licensing option in the United States is for a utility or other end user to "bank" a potential SMR site through the ESP process under 10 CFR Part 52. If an ESP is granted by the NRC, then the applicant can file a subsequent COL application that would reference both the ESP and the DCR. Presumably, if a license applicant relies on an ESP and a certified design, the licensing pathway should be short and efficient. An ESP has a limited time for use (within 10–20 years), but it can be transferred to another potential applicant that might want to use the site for a nuclear power plant.

- *Option 1: Preliminary design provided and site needs approval.* This option begins with the submittal of an application for a CP that contains site safety information, a complete environmental report (ER), preliminary design information, and a preliminary plan for operational programs. Once the CP is issued, an OL application is submitted that provides final design information and operational programs (with implementation schedules). Upon review and approval of the OL and completion of construction, the NRC grants authorization to load fuel.
- *Option 2: Final design provided and site needs approval.* This option starts with the submittal of an application for a CP that contains site safety information, a complete ER, final design information, and a preliminary plan for operational programs. Once the site is approved and the CP is issued, an OL application is submitted with a description of the operational programs (with implementation schedules) and confirmation that safety configurations and systems are as described in the final design application. Once construction is complete and the OL is issued, the NRC grants authorization to load fuel.
- *Option 3: Design approved/certified and site needs approval.* This option is the same as option 2 except that the COL application references a certified design, that is, a DCR. The COL further contains site safety information, a complete ER, final design information with inspections, tests, analyses, and acceptance criteria (ITAAC), and a description of the operational programs (with implementation schedules). Once the site is approved and the COL is issued and ITAAC are met, the NRC grants authorization to load fuel.
- *Option 4: design approved/certified and site approved.* This option is the same as option 3 in that a COL application is submitted that not only references the DCR but also references a previously approved site. Once the COL is issued and ITAAC are met, the NRC would grant authorization to load fuel.

The NRC and the US nuclear industry see significant advantages in the Part 52 process, and it is the preferred, if not the only, licensing process for all applications that are not FOAK designs. Advantages to this process and key considerations for international licensing are the following:

- standardized design that will remain final and stable for most applications,
- early identification and resolution of all licensing issues,
- few licensing exemptions required,
- public and transparent licensing process,
- predictable and efficient licensing process that reduces financial risk.

These licensing advantages can benefit the international application, licensing, and deployment of US-certified or other regulatory-approved SMR designs as will be discussed later in this chapter.

11.2.2 Use of deterministic or risk-informed approaches for licensing SMRs

New SMR LWR designs offer significantly enhanced safety, security, and simplicity of design as described earlier in this handbook. Both the industry and the NRC recognize that many of the existing technical requirements would be applicable to these new designs. Traditionally, LWRs were licensed using deterministic engineering judgment and analysis to prove the safety case and establish the licensing basis. However, with the significant improvements in safety design, the NRC is permitting greater emphasis on the use of probabilistic risk assessment (PRA) techniques and risk insights to establish the licensing basis for SMRs. All US SMRs seeking design certification will develop and use design-specific PRAs to support their licensing basis.

While the NRC has indicated that it will permit greater use of PRAs to support and establish a licensing basis, the use of the PRA would be commensurate with the quality and completeness of the design and attendant PRA presented with the application. Depending on the quality and completeness, the NRC might use the PRA and risk insights to complement a deterministic analysis to establish a licensing basis (including the selection of licensing basis events). Or it might rely more heavily on the PRA and use a deterministic engineering judgment and analysis to complement the PRA. In the post-Fukushima environment, the quality of a design-specific PRA coupled with the use of deterministic engineering judgment and analysis becomes crucial for the licensing of SMRs that either seek relief from traditional LWR safety requirements or are used to support revised licensing requirements. The use of PRA risk information for licensing of SMR LWR designs is particularly appropriate because the quality and completeness of the PRA is bolstered by the maturity of the LWR design and the extent and richness of the operating history information.

Industry and the NRC are developing more risk-informed approaches for the licensing of non-LWR SMRs. These new approaches will be discussed in Section 11.3.

11.2.3 SMR-specific licensing and policy issues

In the United States the NRC, the nuclear industry, and other stakeholders have been collaborating and addressing potential licensing, policy, and technical issues for SMR LWR designs since 2009. These issues result mainly from the key differences between the new SMR designs and current large LWR designs (such as size, moderator, coolant/cooling systems, fuel design, and projected operational parameters). But they also result from proposed review approaches and modifications to current policies and practices. The NRC staff addressed the key licensing and policy issues in SECY-10-0034 (March 28, 2010), and the following provides a brief description of those key issues.

11.2.3.1 Control room staffing

Most SMR designs are intended to use multiple modules at one site as a complete "reference" application of the technology. These designs also contemplate the control of multiple reactors from one control room. Current US NRC requirements for operator staffing outlined in 10 CFR 50.54(m) prescribe the number of operators required for each unit and for each control room. As an example, for three operating nuclear power units at one site, NRC regulations and guidelines assume there are at least two control rooms and a total of eight licensed operators. These requirements are based on the operation of a large LWR. They are not based on new SMR designs that are safer and simpler in operation and shutdown mode. Also the regulation does not address a situation where three or more units are controlled from a single control room.

The NRC staff reviewed preapplication submittals from SMR vendors on operator staffing. The staff considered proposed operator staffing based on control room designs and new technologies, proposed human factors and instrument and controls, and proposed research and development in this area (both by the domestic and the international community) to resolve the issue. In SECY-11-0098 (July 22, 2011), the NRC staff recommended a two-step approach to this issue: (1) in the near term allow SMRs to deviate from current staffing requirements through the exemption process and (2) conduct further assessments and propose revisions to NRC staffing requirements based on human factors engineering (HFE) analysis of staffing needs and design. The NRC concluded that adequate guidance is available to evaluate plant-specific operator staffing exemption requests based on factors such as the applicant's concept of operations, task analysis, and staffing plan. This issue will be considered in the future on a plant-specific basis, but experience will be evaluated to determine whether additional requirements and guidance are required. Some international SMR applications are considering remote operation of a reactor from a considerable distance. This application is not considered in this chapter since the type of SMR that might support remote operation is not being considered for near-term licensing and deployment.

11.2.3.2 Security requirements

Current SMR designs are not only safer and simpler but also more secure. Security and safeguards are built into the design features including primary safety systems located underground, smaller size (target), greater redundancy, and more passive features. This gives an opportunity to determine the appropriate design basis threat, develop effective emergency preparedness plans, and integrate protection measures for (a) physical security for detection, deterrence, and defense; (b) cybersecurity; and (c) material control and accounting (MC&A). SMR designs are expected to integrate all aspects of security, including security staffing and size of the protected area, into final design for NRC review.

In SECY-11-0184, "Security Regulatory Framework for Certifying, Approving, and Licensing Small Modular Nuclear Reactors," (December 29, 2011), the NRC considered security requirements for SMRs and concluded that current NRC regulations allow SMR designers and potential applicants to propose alternative methods or approaches to meet the performance-based and prescriptive security requirements based on the unique characteristics of a particular SMR design without necessitating regulatory exemptions.

The US nuclear industry wanted more certainty in licensing requirements and regulatory reviews for security based on enhanced SMR designs. Through interactions with the Nuclear Energy Institute (NEI), the industry submitted a white paper, "Proposed Physical Security Requirements for Advanced Reactors Technologies" in December 2016. This paper proposed "an approach to security that appropriately considers the enhanced safety and security incorporated into these designs and provides a more effective and efficient means to protect public health and safety." After consideration of this NEI white paper and discussions with and input from other stakeholders, the NRC staff issued SECY-18-0076 (August 1, 2018), "Options and Recommendation for Physical Security for Advanced Reactors." This SECY provided the NRC Commission with options on possible regulatory and guidance changes for security requirements for advanced reactors, including LWR and non-LWR SMRs. The NRC Commission directed the staff to initiate rulemaking based on these recommendations.

11.2.3.3 Source term for SMRs

Source term refers to the amount of radioactivity (quantities, timing, physical and chemical forms, and thermal energy) that might be released in the event of a postulated accident and is used to calculate the resulting off-site and onsite doses. Accident source terms are used for the assessment of the effectiveness of the containment and plant mitigation safety features, site suitability, and emergency planning and site exclusion zones. Accident source terms are expected to be reduced for SMRs due to their smaller fuel inventory and passive safety features that reduce the probability and consequence of postulated accidents. SMR applicants are also considering using a mechanistic source term method to determine accident dose. A mechanistic source term is developed using best-estimate phenomenological models of the transport of the fission products from the fuel through the reactor coolant system and all holdup volumes and barriers, taking into account mitigation features. For SMRs the resulting dose to the public, accounting for design-specific features and accident progression timelines, will produce lower source terms and resulting doses to workers and public.

A reduced source term would facilitate siting SMRs in locations where nuclear plants have not been traditionally sited, for example, as replacements for aging fossil units, on military installations, or collocated with industrial facilities for process heat applications. This issue is also closely related to the emergency planning requirements for SMRs.

In SECY-16-0012, "Accident Source Terms and Siting for Small Modular Reactors and Non-Light Water Reactors" (February 7, 2016), the NRC staff considered the unique safety features and reduced fission product inventory of SMRs and noted that near-term SMR applicants will likely follow the existing regulations and guidance to develop specific design basis accidents (DBA) scenarios, coupled with more realistic mechanistic methods to model the accident progression and develop the source terms for these scenarios. Evaluation of the mechanistic methods will be an important part of the staff's review, but the staff stated the approach proposed by near-term SMR design certification applicants does not currently appear to raise any policy issues.

11.2.3.4 Emergency planning

The NRC staff analyzed the potential for reduced EPZ sizes in SECY-11-0152 (October 28, 2011). It considered it acceptable for SMR vendors to establish an appropriately reduced EPZ size based on the factors noted earlier regarding smaller source terms, slower fission product releases, and reduced potential for dose consequences to populations in the vicinity of the plant. The staff noted a need to further review the "modularity" and "colocation" of SMR reactors. This review will focus on the fact that several SMR designs will be based on multiple reactors being located together in a common facility. Vendors are designing these SMRs so that the reactor modules will be independent—that is, an accident in one module will not initiate or exacerbate an event at another module. This independence permits smaller source terms and potential releases to the public. Based on these design considerations, the NRC noted it would consider a scalable EPZ size based on off-site dose to the public and emergency planning measures within the EPZ based on transient time and existing capabilities of federal, state, and local response organizations. This approach would not result in a reduction to public health and safety. The NRC also indicated that if would work with industry to consider what changes would be required to policy, requirements, or guidance related to emergency planning for enhanced designs.

In SECY-15-0077 (May 29, 2015), "Options for Emergency Preparedness for Small Modular Reactors and Other New Technologies," the NRC provided options for EP for LWR and non-LWR SMRs. The NRC Commission directed the staff to initiate a rulemaking for new EP requirements for advanced reactors and other nuclear technologies. The staff issued a draft EP regulation in the *Federal Register* on April 13, 2017 and, after consideration of stakeholder and public input, issued a proposed draft rule on August 1, 2018. This rule is still under consideration by the Commission and in the public rulemaking process.

Emergency planning size and response measures are important for international consideration and licensing. The NRC's evaluation of this issue based on off-site dose considerations of new designs and appropriate response measures for the population at risk is instructive for international licensing. International licensing considerations for EPZ size and response measures are unique for extremely remote sites. The NRC has not considered extremely remote sites since there is no pending SMR application with a remote site under consideration.

Important to the US evaluation of reduced EPZ size and measures to be taken within the EPZ is that response capabilities of off-site personnel and emergency organizations were bolstered significantly in response to the terrorists' events of September 11, 2001. Federal and state response capabilities are bolstered by better equipment, communications, coordination, and training—all of which are important capabilities to respond to a nuclear event. Additionally, in the post-Fukushima environment, nuclear power operators are considering better coordination and integration of capabilities to respond to an operating plant event. International emergency planning licensing should also consider these factors on a country or regional basis.

11.2.3.5 Multiple-module licensing

Since most US SMR vendors will apply for a "reference" plant design to be certified by the NRC with multiple modules included in the reference design, the NRC had to consider how it should issue the license(s) to the reference plant. Should the NRC issue a facility license to the multiple-modular plant, or should it issue a license for each module? The NRC's consideration of this issue was addressed in SECY-11-0079 (June 12, 2011). The NRC staff recommended that the reference plant can be reviewed and issued a design certification for all modules. NRC will permit a single design certification if the modules are a generic design that will support a single licensing review, SER, and public hearing.

However, it decided that each module should be issued a separate license with attendant-specific technical (but largely generic) specifications. This recommendation was mainly based on the phased deployment of modules and the operation and maintenance of each module over its operating life—it may be deployed, operated, maintained, and decommissioned separately from each other module. International licensing of a multiple-module plant should consider and address these same licensing issues. See discussion of Master Facility License in Section 11.5.2.

11.2.3.6 Manufacturing license

SMR modules will be manufactured in a factory-like setting, and then the modules will be transported to the site for final fabrication and installation. Manufacturing in this manner offers advantages in quality and efficiency through replication, assembly-line construction, and the maintenance of a stable and skilled workforce. NRC requirements in 10 CFR 52.167 permit the issuance of a manufacturing license with necessary and sufficient ITAACs to ensure the reactor is manufactured and operated in accordance with its license. Although no application has been submitted for a manufacturing license under Part 52, the US nuclear industry recommends that the NRC develop further licensing guidance on these requirements and how ITAAC provisions will be applied both in the plant and at the site. NRC requirements also do not address the export of SMR modules since these reactors will be licensed by the importing or host country. The importing country and customer must meet all US law and regulatory export requirements, including NRC's requirements in 10 CFR 110 (n.d.), "Export and Import of Nuclear Equipment and Material," and DOE's requirements in 10 CFR 810 (n.d.), "Assistance to Foreign Atomic Energy Activities." These requirements are in a constant process of reconsideration and revision and should be considered by both exporters of US SMR technologies and importing countries and companies.

11.2.3.7 Timeliness of SMR licensing

Timely and effective licensing of SMRs is crucial for several reasons. First, utility decision-makers or other applicants must have confidence in a stable, predictable, and timely licensing process that will recognize the need to balance the right of all stakeholders to have effective input versus the need for timely and appropriate decisions. Second, common wisdom and logic would suggest that SMRs, which are based on proven LWR technology coupled with safety and security enhancements that are demonstrated through PRA results, testing, and validation, should be channeled into a licensing review that recognizes proven results and risk insights as opposed to a review that is risk averse and demands another layer of validation. Third, timeliness of licensing decisions is critical to investment and financial decisions and results for successful deployment. Fourth, governmental energy policy makers and the public must have confidence that the licensing process will result in timely decisions that can support a government's energy policy. Delayed nuclear licensing decisions and commercial deployment may result in governmental policies that enhance other energy options to the detriment of a balanced energy policy.

The NRC has been working with industry, DOE, and other stakeholders to develop guidance for a more flexible regulatory review process for advanced reactors. See discussion in Section 11.3.

11.2.3.8 Mitigation of licensing risk

Progress between NRC and the nuclear industry on these key licensing and policy issues was an important consideration in determining whether the United States would sponsor and fund an SMR program through the US Department of Energy (DOE). The SMR program would be structured as a public/private cost-share program to help mitigate the financing and licensing risk of FOAK reactor technology.

The United States decided to fund and launch a competitive cost-share SMR program based on its assessment of three important risk factors that impact FOAK reactor deployment. First, since SMRs are a new nuclear technology, is there a market and demand for this technology? Second, do capital markets or end users have the capabilities and desire to finance this FOAK technology and follow-on commercial projects? Third, does the regulatory authority—in this case the NRC in the United States—have the resources and capabilities to license enhanced SMR designs and their unique safety and regulatory issues in a timely and effective manner?

While no risk can be totally mitigated, the United States decided to establish the DOE SMR program to provide resources and capabilities to help mitigate the financial and licensing risks of the FOAK SMRs. The US decision to establish the SMR program was based, in part, on the significant progress in addressing and resolving the key licensing issues and potential financial constraints. In the United States the capital markets were noting uncertainty about financing new nuclear power plants (NPPs) because (a) the cost of natural gas in the United States would make it difficult for NPPs in electrical markets to compete with the price of electricity generated by natural gas plants, (b) energy conservation and other economic measures were reducing demand for electricity in relation to historical demand curves, (c) the cost of renewable energy (wind and solar) was decreasing and reasonably competitive, and (d) the political environment for NPPs was unpredictable based on heightened interest in renewables and the uncertainty of the licensing of the Yucca Mountain waste disposal facility in Nevada. Financing of SMRs in global markets will also encounter the same risk factors of being competitive with other energy sources and being supported by political policy and regulatory authorities.

11.3 Non-LWR advanced reactor SMR licensing

As noted in other chapters of this handbook, non-LWR SMR designs are currently proposed and are in various stages of design, development, licensing, and commercialization. There is general agreement that these SMR designs theoretically have higher safety margins and operating resiliency than LWR SMRs. Hence, some studies suggest that these non-LWR designs should present an opportunity for a more performance-based, risk-informed licensing approach.

The optimal licensing strategy for non-LWR SMR new reactor designs will vary significantly based on the reactor technology, its design maturity, and its expected deployment for energy or industrial applications. Various proposals by regulatory agencies, industry groups, and nongovernmental organizations have suggested different pathways that could be used to license new designs and applications. Perhaps an important first step in providing an appropriate licensing strategy for these advanced designs is to develop principle design criteria (PDC) for non-LWR technology. The NRC evaluated whether the general design criteria (GDC) in current regulations apply to these advanced designs. After consideration of comments and interactions with stakeholders, the NRC issued final Regulatory Guide (RG) 1.232, "Guidance for Developing Principal Design Criteria for Non-Light-Water Reactors" (April 3, 2018). It provides guidance to reactor designers, applicants, and licensees of non-LWR designs on developing PDC for any non-LWR design subject to either Part 50 or Part 52 licensing.

In general, this developed guidance suggests that licensing can best be accomplished, based in large part on time and resource factors, using existing regulatory requirements and processes. Under NRC licensing, either under Part 50 or 52, non-LWR SMRs should proceed with their application and NRC review using phased processes such as topical reports, standard design approvals, and standard design certification as the optimal reactor licensing strategy. The combination of risk-informed and performance-based design criteria will provide the greatest flexibility for both designers and regulators in the licensing of advanced reactors. Additionally, future design criteria should be developed in an internationally harmonized way and written in a technology-neutral matter to ensure their applicability to all advanced reactor types. See also "Strategies for Advanced Reactor Licensing," a report by the Nuclear Innovation Alliance (April 2016).

To increase regulatory effectiveness through collaborative work on the technical reviews of advanced reactor and SMR technologies, the NRC and the Canadian Nuclear Safety Commission (CNSC) signed a Memorandum of Cooperation (MOC) on August 15, 2019. The MOC allows the CNSC to take NRC review results into account should an applicant propose to use a reactor design currently under review or previously reviewed by the NRC. Similarly the NRC may consider insights from CSNC prelicensing vendor design reviews and licensing review processes. Terrestrial Energy's integral molten salt reactor was selected by the NRC and the CSNC as the first advanced, non-LWR SMR technology to be subject to a joint technical review under the MOC.

The NRC staff is continuing to work with the Nuclear Energy Institute (NEI) and industry to develop more specific design guidance for non-LWR SMRs. NEI published Working Draft 18-04, "Risk-Informed Performance-Based Guidance for Non-Light Water Reactor Licensing Basis Development" (September 28, 2018). The NRC published Regulatory Guide 1.233, "Guidance for a Technology-Inclusive, Risk-Informed, and Performance-Based Methodology to Inform the Licensing Basis and Content of Applications for Licenses, Certifications, and Approvals for

Non-Light Water Reactors" (June 2020). It focuses on key areas of the design and licensing of advanced reactors, such as the selection of licensing basis events; classification of structures, systems, and components; and assessing defense in depth.

11.4 Industry codes and standards to support SMR licensing

SMRs will be licensed and deployed in the global market. However, the deployment of SMRs as a global enterprise is not supported by an international licensing/certification framework that permits a "plug and play" environment similar to electronics such as TVs, computers, and smart phones. The current nuclear licensing strategy requires any SMR design, regardless of the pedigree and robustness of licensing in country of origin, to be licensed in the country of deployment. Contrast this licensing process with that of aircraft approval. In essence an aircraft approved for airworthiness by a competent approval authority is recognized worldwide through an international convention. While the international community has recognized the need to harmonize licensing processes and practices to the extent practicable, the reactor design in the country of origin still is licensed with significant reference to and adoption of that country's approved industry codes and standards.

The nuclear energy community and standards development organizations (SDOs) [e.g., American Society of Mechanical Engineers (ASME), Institute of Electrical and Electronics Engineers (IEEE), American Nuclear Society (ANS), American Concrete Institute (ACI), ASTM International, and American Welding Society (AWS)] recognize that the development of international nuclear codes and standards through the consensus process will facilitate worldwide licensing and deployment of SMRs. SDOs are spending much time and resources to support the development of international standards for all industries. The intent for worldwide consensus standards supporting SMR designs is that licensing authorities will adopt or reference these standards in the licensing basis for the designs. Adoption or reference of international standards as acceptable methods to meet a country's licensing requirements will streamline the licensing process and facilitate the end goal of a global enterprise for SMRs.

A framework of international licensing based in part on the adoption and use of international codes and standards is a laudable goal. The NRC references approximately 520 standards in its regulations, regulatory guides, and the staff's SRP, and its staff participates in hundreds of SDO committees. The NRC regularly reviews consensus standards developed by these SDOs and, if appropriate, endorses them in its regulations, regulatory guides, and the SRP. On a 5-year cycle, approximately 425 regulatory guides, the most common source of referenced consensus standards, are reevaluated to determine whether they need updating, including the endorsement of new or revised consensus standards. More frequent revisions may occur based on technical evolutions and users' needs.

The development of new and revised nuclear industry codes and standards was stalled over the past several decades commensurate with the hiatus of new NPP deployment. If new nuclear designs and technologies are not deployed commercially,

SDOs have little business incentive to expend time and resources on new nuclear standards. However, new technologies that were applicable to the nuclear industry (digital instrumentation, wireless sensors, new materials and fabrication techniques, laser welding, etc.) were being developed and supported by new industry standards that were adopted worldwide. The development and licensing of SMRs present an opportunity to bring together representatives of the nuclear industry, SDOs, subject matter experts, academia, and national and international governmental organizations to facilitate and coordinate the timely identification, development, or revision of standards that support new nuclear designs, licensing, operation, fabrication, and deployment. In addition, there are codes and standards activities in cross-cutting areas that are relatively technology neutral in that the standards involve new materials, techniques, or methods that applicable to essentially all reactor technologies for use in new design or construction.

11.5 International strategy and framework for SMR licensing

Energy needs, and particularly clean energy needs, across the globe have created a market and demand for SMRs. The promise of SMRs, both with enhanced safety and increased flexibility in applications and financing, has created a global market and potential enterprise that must be supported by more streamlined processes for licensing and deployment. Embarking countries that desire or require nuclear energy will probably demand, solicit, and procure a nuclear technology that is proven and has been licensed by a competent regulatory authority. Embarking nations are risk averse and resource limited—they want a proven technology and previously licensed/approved design to mitigate a financial constraint and promote private or public investment. Embarking nations also have limited regulatory resources and capabilities—a certified/approved design will permit a licensing process that can leverage the proven and licensed design. The international nuclear energy community must work together to identify and develop a more effective framework for licensing of these previously approved/certified SMRs. Existing frameworks include the International Atomic Energy Agency (IAEA), International Project on Innovative Nuclear Reactors and Fuel Cycle (INPRO), the International Framework for Nuclear Energy Cooperation (IFNEC), the Cooperation in Reactor Design Evaluation and Licensing (CORDEL) Working Group, and the NRC's International Regulatory Development Program (IRDP).

11.5.1 Development of international codes and standards

A high priority should be the development of international nuclear codes and standards that can be adopted and referenced by sovereign licensing authorities. The development of codes and standards to support new non-LWR SMRs is especially important as new fuels, technologies, materials, and advanced manufacturing including 3D printing, artificial intelligence to analyze data, and supply chain improvements are expected to be deployed. As discussed earlier the standards development process engages industry, academia, interest groups, and regulators in a global innovative process. The coupling of nuclear data with new technologies and creative thinking bolstered by enhanced computer analytical capabilities should result in standards that address new SMR designs and technologies. Perhaps a key is to get the "stalled" nuclear licensing and deployment process kick started by new innovative thinking, analysis of worldwide data, and application of enhanced information into new standards that embrace risk-informed licensing. The consensus standards development process will support efficient licensing by the sovereign authority. Critical to the adoption or reference of new standards by regulatory bodies is that they participate in the development and consensus process for new nuclear codes and standards.

11.5.2 International harmonization of licensing processes and practices

IAEA member state regulatory bodies have the responsibility (per the IAEA Safety Fundamentals) to ensure that the national regulatory framework for safety is established and implemented to regulate the use of nuclear power. The regulatory framework in each country is developed using the national legal framework and considers both the IAEA safety framework and inputs from stakeholders such as industry, scientific bodies, government, and the public. As a result, differences between national frameworks can and likely will always exist. For this reason, harmonization of most requirements and guidance globally will remain a significant long-term and complex challenge that will require significant cooperative investments by member state governments. The regulatory bodies play a partial but important role in this discussion.

Nuclear regulators and the IAEA are participating in the Multinational Design Evaluation Program (MDEP). Fifteen nations and the IAEA participate in MDEP, which is facilitated by OECD/NEA. MDEP is a program where regulatory organizations jointly cooperate in sharing information about the review of specific new reactor designs. These new designs require detailed evaluation of their safety and development of ITAAC for their manufacturing and construction. Since new reactor designs, including SMRs, will be exported and a portion of their systems and components may be manufactured outside the country of origin, it is important that all regulatory authorities interact to share relevant information, experience, and expertise. Working groups have been formed under MDEP to cooperate internationally on issues that are common to the licensing of new designs such as vendor inspection coordination and digital instrumentation and control systems. MDEP also supports activities to facilitate the international convergence of codes, standards, and safety goals. However, the MDEP Technical Steering Committee decided not to expand the scope of licensing issues to include SMRs.

IAEA and member states recognized the need to develop international guidance on how to license new SMR designs more effectively. IAEA recommended in its 6th INPRO Dialogue Forum, "Licensing and Safety Issues for Small and Medium-Sized

Reactors" (July 29–August 2, 2013), that guidance on the application of a graded approach for the licensing and regulation of SMRs be developed. All sovereign nations embarking on SMR deployment must have a licensing process and capabilities that provide reasonable assurance that the operation of a nuclear power plant in the country will be safe and secure. This process must be open and transparent to all stakeholders. However, licensing capabilities to fully review the safety basis may be limited in many countries. Therefore IAEA guidance on the appropriate grading of the scope and depth of a licensing safety review, particularly with a previously approved/certified SMR, is a high priority.

The SMR Regulators' Forum was formed in 2015 as a 2-year pilot project to capture good practices and understand key challenges that are emerging in SMR regulatory discussions. The following countries are Forum members: Canada, China, Finland, France, Korea, Russia, United Kingdom, and the United States. The project would enable regulators to inform changes, if necessary, to their requirements and regulatory practices.

In its pilot project report, "Considering the Application of a Graded Approachs, Defence-in-Depth and Emergency Planning Zone Size for Small Modular Reactors" (January 2018), the Forum considered existing member SMR regulatory processes and practices and addressed the following three major issues for both light water and nonlight water SMR reactors:

- **Graded approach**: Current SMR designs seek to relax regulatory requirements based on enhanced design considerations and improved safety and security analyses. Therefore there is a need for regulators to consider and apply a graded approach to licensing. This approach would permit both the license applicant and the regulator to limit the scope and depth of analysis and review based on the SMR design and safety considerations. It is recognized that existing regulatory requirements and tools can successfully be applied to grading practices. However, there is no consistent application of grading on specific issues. Thus one key conclusion of this report is that significant benefit could be gained if the IAEA was to issue guidance on graded approach.
- **Defense in depth (DiD)**: A number of SMR designers and license applicants are proposing alternate ways to address DiD in their designs. The Forum looked at these approaches and attempted to develop in the pilot project report common positions around certain regulatory practices to ensure that the fundamental principles of DiD are maintained. It was agreed that the DiD concept, although developed for large nuclear plants, is valid for SMRs. However, SMR design specifics including facility size, modular design, and novel technologies raise questions on the application of DiD principles. The Forum found it not possible to express detailed requirements at this stage because the spectrum of SMRs is very large and the lack of information about SMR designs and technical details.
- **Emergency planning zones (EPZs)**: On the basis of the alleged characteristics of SMRs, smaller EPZs are being proposed by some SMR vendors. The Forum in its report examined existing practices and strategies for understanding how flexible (i.e., risk informed) EPZs are established to have a common position on this issue. EPZ size for SMRs can be scaled based on the technology, novel features, and specific design characteristics. The existing IAEA safety requirements and methodology, in general, for determining the EPZ size are effective in establishing emergency planning zones and distances, specifically, IAEA Safety Standard Series No. GSR Part 7 and associated lower level publications.

In January 2007 the World Nuclear Association established the Cooperation in Reactor Design Evaluation and Licensing (CORDEL) Working Group to coordinate discussions on reactor safety standards between the nuclear industry, nuclear regulators, and other interested institutions interested in nuclear power. The Small Modular Reactor Ad hoc Group (SMRAG) was created by CORDEL in September 2013 to establish a path toward harmonizing global SMR licensing and deployment. The SMRAG issued a report, Facilitating International Licensing of Small Modular Reactors (August 2015). The report notes that while some progress in standardization of licensing and harmonization of regulatory requirements has been achieved in international programs, an internationally standardized approach for licensing of SMRs that are globally deployed, ideally in fleets, needs further thought and development.

Clearly a fundamental constraint to harmonization is that each country's licensing process is prescribed by requirements established by the national regulatory authority derived from national laws. Amending the laws or changing nuclear regulations and requirements is a difficult government undertaking. Regardless the uniqueness of SMR designs, passive safety systems and modular fabrication present a great opportunity to explore the simplification and standardization of some aspects of licensing. The SMRAG report highlights some of these aspects, briefly discussed later.

11.5.2.1 The international transfer of a reactor module certification

The modular fabrication of reactor modules for SMRs in a factory setting consistent with an approved design certification enables a more efficient licensing process in the country of deployment. The deploying host country should be able to eliminate or simplify the need to conduct a safety review and approval of each module. Based on the fabrication of identical modules, the deploying country should be able to review the application to verify there are no changes to the design. Subsequent design changes by the SMR vendor after certification can be handled through a formal change management process by both the vendor and the operator. The report notes that this simplified licensing approach could be easily adopted to "newcomer" countries but would require a case-by-case study in existing nuclear countries to ensure consistency with laws and regulations. As noted previously in this chapter, regardless of simplification of safety reviews of certified designs, other environmental, site, security, and emergency preparedness requirements would still be required.

11.5.2.2 Master Facility License

A Master Facility License is defined by the report as a "legal document issued by the regulatory body granting authorization to perform specified activities related to a facility or activity." The license would apply to one or more SMRs at the facility. Modifications to the authorized SMR might be reviewed as part of the design certification process or change management program. Safety issues that are applicable to the facility or site would be reviewed under the Master Facility License. The report concludes that a Master Facility License would have several beneficial features that would minimize or reduce repetition in the licensing process.

The SMRAG report also discusses other licensing issues and challenges such as (a) ensuring that SMRs adopt more recent design extension conditions that result from design and safety lessons learned from Fukushima and other incidents; (b) SMR fabrication facilities adopt and modify, if needed, industry codes and standards that apply to nuclear suppliers and fabrications for current nuclear power plants; and (c) process and responsibility for design change management of modifications for certified designs— for example, responsibility of vendor or operator, particularly if a manufacturing license is issued. The report's conclusion is that the safety, security, economics, fabrication, and deployment of SMRs worldwide present a great opportunity to rethink the harmonization of international nuclear licensing approaches. This Handbook endorses that conclusion.

A recent Massachusetts Institute of Technology (MIT) study, "The Future of Nuclear Energy in a Carbon-Constrained World" (2018), also highlighted the need for international harmonization and standardization and recommended the following:

> *Regulatory requirements for advanced reactors should be coordinated and aligned internationally to enable international deployment of commercial reactor designs, and to standardize and ensure a high level of safety worldwide. National differences in safety regulations due to accepted cultural practices make it difficult to develop a universally accepted regulatory licensing regime. But certain basic standards for nuclear safety should be maintained internationally due to the far-reaching environmental and social/political effects of nuclear plant operation. Initial international agreement on specific topics (e.g., station blackout resiliency) and joint licensing evaluations could advance discussions about undertaking reciprocal reactor design evaluations between nations or standardizing international safety requirements.*

11.5.3 International certification of SMRs

At the 6th INPRO Dialogue Forum, the IAEA recommended consideration of international certification of SMRs. This long-term recommendation must recognize and integrate with the preservation of sovereign authority for licensing. The aviation industry is recognized as a credible model for international certification of airplanes. It may be a difficult task for the international nuclear community to develop a similar certification process simply because NPPs are sited at a fixed location within the boundaries of a sovereign nation with governmental responsibility to license and regulate to protect its citizens and neighbors. Difficult as this task may be, the standardized design of SMRs that will be manufactured and deployed internationally presents an opportunity to assess areas of licensing that may benefit from international certification. Possible licensing areas for international certification include the following:

- SMR operators;
- SMR manufacturing facilities, equipment, and processes;
- SMR standardized operation and maintenance.

The IAEA must take a lead role in assessing international certification projects. Member states need to participate in and approve any certification process. The IAEA must develop the capabilities and processes to review, approve, and inspect international certification requirements. The global deployment of standardized reactor designs that offer significantly enhanced safety and security features provides a great opportunity to integrate traditional country-specific licensing processes into a more international licensing framework.

11.5.4 International cooperation to assess worldwide operating data

Worldwide informational data support many industries and decisions. As an example the US National Weather Service recently significantly increased the accuracy and long-range prediction capability of its weather forecasts by accessing the data from over 2000 global surface weather stations. Before access to these data, it was only able to predict the weather in 3-day forecasts. Based on global data from surface stations and satellites (much of it gathered by Google), it now predicts the weather in 10–16 day forecasts with significantly increased accuracy. Sharing of worldwide weather data enhanced both US and European computer models based on improved algorithms resulting from the data to more accurately predict weather conditions.

Similarly the World Association of Nuclear Operators (WANO) analyzes the performance data from 120 members who operate 430 nuclear reactors worldwide. Since its inception in 1989, WANO has collected, screened, and analyzed millions of hours of operating data and performance and provided its members with lessons learned and basic and superficial performance trends. SMR designers and regulators are relying on more performance-based and risk-informed insights for grading of licensing standards. A detailed assessment of worldwide operating data should provide a better foundation to determine what structures, systems, and components (SSCs) are important to safety.

While this data could be important for risk-based PRAs, it also could better inform licensing decisions. The assessment of this worldwide data must go beyond WANO's performance assessments—it has to review the data and other data from other sources such as manufacturers and suppliers to develop algorithms and models that predict failure of SSCs under a range of operating conditions. For example, if an analysis of worldwide data shows that a certain safety-related component only fails in certain conditions of heat and pressure and those conditions do not exist in some SMR designs, then a significantly reduced risk profile can be assigned to that component by the vendor's PRA. This reduced risk profile can be reviewed by the regulatory authority to better inform both deterministic and risk-based licensing decisions to determine safety SSCs.

The nuclear industry has a tremendous amount of engineering, scientific, operating, and maintenance data. However, the nuclear industry lags other industries in its ability to analyze that data in an innovative way to determine how design development and licensing can rely on that data to improve design and licensing decisions.

The nuclear industry also lags other industries in improving international collaborations to establish standards that apply worldwide data to enhance connectivity for global enterprises and "fleets" of products. The scope, depth, and significance of worldwide nuclear data should encourage the industry and regulators to engage "new" thinking on how to optimally apply the data for informed designs and licensing decisions.

11.6 Conclusion

SMRs hold the promise for successful commercial deployment in many and diverse global markets. SMRs offer enhanced safety, security, and flexibility for all applications. Most energy and environmental governmental policies support this clean energy alternative. However, this energy alternative and commercial promise must be advanced by enhanced and informed licensing that recognizes the advantages of the SMR safety design, fabrication quality, reduced public risk, and deployment flexibility.

This chapter provides a strategy and framework, based in part on US NRC SMR licensing processes and decisions, to support effective and timely licensing. It offers recommendations on how this new reactor technology can be licensed in a collaborative international framework but still recognizes the regulatory responsibilities of each sovereign regulatory authority. Successful licensing pathways must be collaborative, based on the uniformity of SMR designs and manufacturing, and yet reflect the safety and siting considerations that are unique for each application. Enhanced SMR characteristics present a new nuclear licensing paradigm that can shift traditional sovereign regulatory authority responsibilities to an international strategy and framework for certification of approved and licensed SMR designs and attendant fabrication, operation, and maintenance processes. This licensing strategy will rely on more performance-based, risk-informed information that will be largely developed in an international collaborate framework.

References

10 CFR 110, n.d. Export and Import of Nuclear Equipment and Material.
10 CFR 810, n.d. Assistance to Foreign Atomic Energy Activities.
10 CFR Part 52, n.d. Licenses, Certifications, and Approvals for Nuclear Power Plants.
IAEA 6th INPRO Dialogue Forum, 2013. Licensing and Safety Issues for Small and Medium-Sized Reactors. 29 July–2 August.
Massachusetts Institute of Technology (MIT), 2018. The Future of Nuclear Power in a Carbon-Constrained World.
NRC NUREG-0800, n.d. Standard Review Plan for the Review of Safety Analysis Reports for Nuclear Power Plants.
NRC RG 1.232, April 3, 2018. Guidance for Developing Principal Design Criteria for Non-Light Water Reactors.
NRC SECY-10-0034, March 28, 2010. Potential Policy, Licensing and Key Technical Issues for Small Modular Reactor Designs.

NRC SECY-11-0079, June 12, 2011. License Structure for Multi-Module Facilities Related to Small Modular Nuclear Power Reactors.

NRC SECY-11-0098, July 22, 2011. Operator Staffing for Small or Multi-Module Nuclear Power Plant Facilities.

NRC SECY-11-0184, 2011. Security Regulatory Framework for Certifying, Approving and Licensing Small Modular Reactors. December 29.

NRC SECY-11-0152, October 28, 2011. Development of an Emergency Planning and Preparedness Framework for Small Modular Reactors.

NRC SECY-15-0077, 2015. Options for Emergency Preparedness for Small Modular Reactors and Other New Technologies. May 29.

NRC SECY-16-0012, 2016. Accident Source Terms and Siting for Small Modular Reactors and Non-Light Water Reactors. February 7.

NRC SECY-18-0076, 2018. Options and Recommendations for Physical Security for Advanced Reactors. August 1.

Nuclear Energy Institute (NEI), 2018. Working Draft 18-04, Risk-Informed Performance-Based Guidance for Non-Light Water Reactor Licensing Basis Development. September 28.

Title 10, Part 50, n.d. Domestic Licensing of Production and Utilization Facilities, US Code of Federal Regulations (CFR).

Construction methods for small modular reactors (SMRs) 12

N. Town and S. Lawler
Rolls-Royce plc, Derby, UK

12.1 Introduction

For an assessment to be made of the supply chain and manufacturing options available for small reactors, the fundamental requirements for cost-effective deployment need to be explored and understood.

12.1.1 Economic development

From the dawn of nuclear power there has been a concerted effort to increase the economic performance of each installation; that is maximising the revenue created from each plant verses capital investment incurred. Broadly, the improvement to the economic performance can be classified into three areas:

- design;
- manufacturing;
- operation.

The resulting change in economics from any of these can be observed in changes to the overnight cost of capital (expressed in $/kW), that is the capital cost of the plant expressed in $ per kW of electrical power output.

The ambition to improve the economic situation has resulted in a drive to increase the power output from individual generating units, and this is not specific to nuclear installations. However it can also be observed that for each generating option there are economic ceilings where increased capacity no longer delivers a lower cost of operation. For coal this point came in the early 1970s, as shown in Figure 12.1. Equally the relationship between power and cost is defined with a non-linear characterisation; for example the additional cost to increase the plant size from 600 to 1200 MWe is not double the original investment, as shown in Figure 12.2.

So what are the limits to going bigger and specifically does this present an opportunity for the small reactor?

12.1.2 Limitations with existing technologies

There are two key limitations to increased sizing of large reactors:

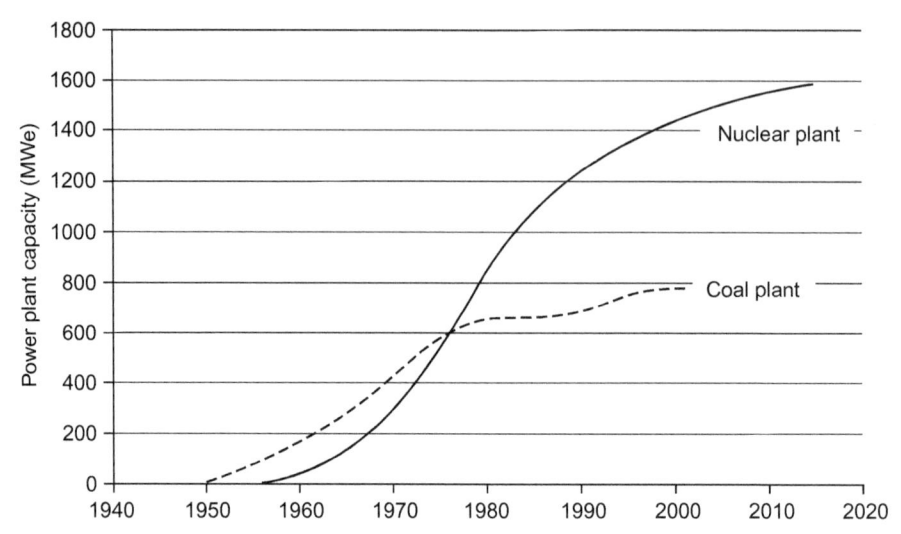

Figure 12.1 Graph showing the increase in capacity from individual units through time.

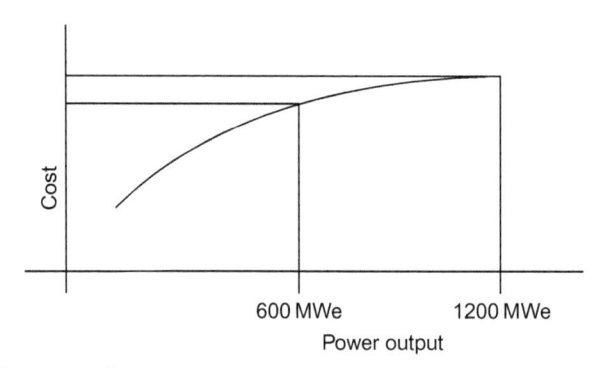

Figure 12.2 Illustrative plant cost vs. power output.

- Transmission and distributions networks limit the options for deploying large amounts of additional capacity from a single location.
- Manufacturing processes limit the size of key vessel and components.

Existing grid capacity has grown from fossil-fired generation sources and has defined the grid structures. Increasing the size of single nuclear generating capacity past this point will incur transmission infrastructure upgrade costs or will potentially entail running the distribution network with increased line losses.

Current manufacturing capacity and capabilities are challenged to deliver ever larger vessels. The global manufacturing supply chain is already at the edge of the envelope to produce components of the physical dimensions required by plants today (e.g. AP1000, CPR1000, EPR and VVER).

With these two constraints the nuclear power generation portfolio has previously been unable to challenge the economies of scale that the increasingly larger installations provide. It is against this backdrop that the small reactor can play to a different set of strengths, flipping the economics of scale and delivering on economics of volume. With the opportunity to deliver physically smaller components at volume the supply chain paradigm changes. For example, if the supply chain can be restructured with elements that closer resembled other engineered commodities from the aerospace, automotive, or even white goods sectors it could deliver nuclear power at a significantly lower cost. This is the key attribute that has to be understood before supply chain and manufacturing options can be established.

12.1.3 Understanding the opportunity

The small reactor has two attributes that can meet this cost challenge. Firstly, the smaller physical size of the vessels permits larger assemblies to be transported to site in a complete and tested condition. The implication is that a significantly higher portion of the plant can be built and tested in a factory prior to delivery.

The second facet is that small reactors are deployed in larger volumes to meet the same capacity requirement. This larger volume justifies investment in manufacturing techniques that larger, low volume nuclear plants have been unable to justify previously. Building plants at these projected volumes present a challenge significantly different to anything the nuclear industry has encountered previously.

12.1.4 Challenges for industry: step or incremental change?

So how should industry rise to this challenge: evolution or revolution? Evolution will drive existing processes harder. Revolutionary change will introduce new techniques. New techniques often imply greater risk. A way through this could be to look at techniques that have been validated in other industrial sectors. The small reactor as a volume product is a revolutionary step for the nuclear sector. There is the opportunity to pick up and emulate the lean manufacturing techniques that have driven efficiency in other manufacturing sectors. The rigor and discipline that accompany lean manufacture will cascade down into sub-tier vendors. The key challenge to the success of the revolution is the culture of the supply base. Can existing supplier bases change their organisational structures and cultures to deliver nuclear components in this manner or will the demand be satisfied by new incumbents who already have the appropriate operating culture? Having acknowledged the organisational cultural change that is required, this chapter will focus on the revolutionary challenge from an engineering perspective.

The two halves of this revolutionary challenge will be explored independently, technologies and techniques around component fabrication, and then a platform for integrating these in a production environment.

12.2 Options for manufacturing

Manufacturing processes are characterised and applied depending on the volume of product they are delivering. Each manufacturing process aims to maximise efficiencies, and minimise production costs. With high volume manufacture the assembly line has been established as offering the best solution. Popular examples can be found in automotive applications. A lot of the early production line thinking emerged in this industrial sector, and is credited to Henry Ford and the application of the production line methodology for the Ford Model T. The production line demands a pace, repeating the same sequence of activities at the same location on each manufactured unit. The small reactor has opportunities for factory build, with vessels and components that are physically smaller than conventional nuclear plant. Standardisation of design similarly aligns with the capabilities of a production line approach – 'Any customer can have a car painted any colour that he wants so long as it is black' (Ford and Crowther, 1922). Equally the repetition of sequenced activities fits with the assurance that nuclear manufacturing requires.

However, the potential application of the conventional production line approach is compromised for the small reactor for two fundamental reasons. The first is volume and profile of sales and the ramp-up of those sales over time. The second is level of activity required to assemble each unit factored against the volume of units flowing down the production line.

12.2.1 Volume and profile of sales build-up

The production line is based around a continuous flow of product through a factory and has been successfully demonstrated for food products, telecoms products, cars and aircraft. However, for these products the market volume of sales is assured before the factory capital investment is made. With a product that is the physical size of a small reactor vessel the investment in workstations and tooling is significant. Equally for a small reactor, manufactured in a factory at a specific volume of units per year in a repetitive manner, it can be conceived that there is an economic case to support a level of investment in jigs and tooling. This appetite for facility investment is tempered by the realism of the market demand. With an emerging customer base there is no basis to justify the capital investment on facilities at the outset – compromising the appeal of a pure production line implementation. There are three potential solutions to this;

1. Investment based on rational industrial business risk – 'Step of faith'.
2. State involvement – state ownership of the same business risk.
3. Alternative implementation methods.

The step of faith is a little lighthearted; however, it makes the point that all business investment contains an element of risk. The risk with lower levels of customer orders prior to establishing a small reactor facility could restrict the net level of investment. This restriction could in turn limit the ability of the factory to deliver product at mature volumes in an effective and cost-efficient manner in the longer term.

The second option offered is state involvement. The options for state involvement will also vary from region to region across the globe. The US Department of Energy (DOE) has recognised this with their four-phase approach to the acceleration of small reactor deployment. Their third phase recognises this 'step of faith' transition from early technology adopters to full-scale production. This state stimulus can come in many forms ranging from a public–private partnership to a commitment to power purchase.

The third option is an alternative manufacturing flowline assembly. Flowline is an interim option between a full-scale production line and individual batch manufacturing, and incorporates the best features of both approaches. With batch production manufacture the unit is produced in a series of stages, each stage building on the preceding stage. Running repeat operations in batches does recover some of the operating efficiency. The point with batch production is that the item remains stationary throughout and the workstation is reconfigured around it for each operation. The time to set up and re-configure between each of these operations leads to delays. For assured processes differences from set-up to set-up cannot be tolerated. Batch manufacture gives greater opportunity for batch to batch tolerance variation, which is a bad thing. However, it also creates a window for product customisation. In our circumstance this could be the difference between a generator protection system that operates at 60 Hz instead of 50 Hz.

12.2.2 The flowline

The 'flowline principle' is a process that could be used for the manufacture of small reactors. The flowline approach fits in between batch and assembly line manufacturing as shown in Figure 3. The history of nuclear manufacturing has evolved from a base of bespoke component manufacturing. This manufacturing base cannot support the volumes projected for small reactors. A realistic alternative is the development of a hybrid manufacturing method that sits between batch production and production line. This hybrid is termed flowline and has been successfully applied in a number of high-technology, high-integrity, intermediate-volume businesses.

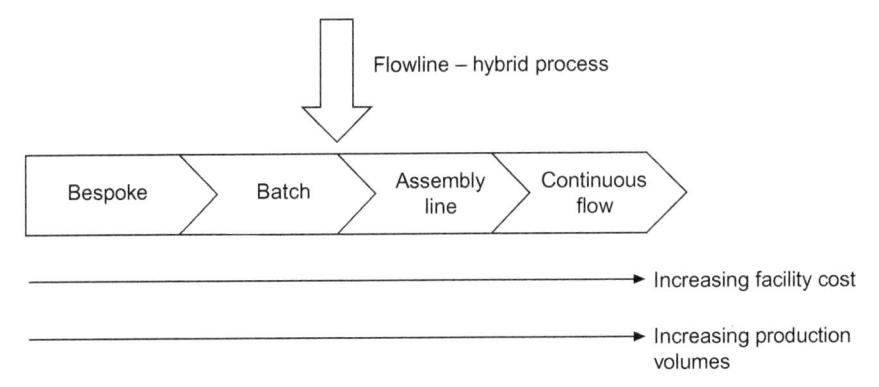

Figure 12.3 Application volume for flowline manufacture.

The flowline is an approach where the unit in work is moved in a number of discrete steps. Each step is called a station and each station has a number of predefined steps to be performed at each point. The tooling for each activity is held at each station, eliminating the need for changes in set-up. The work piece is moved from station to station when the work is complete. The work performed at each station is planned and timed so that each station completes at the same time and then the line can increment along. For the small reactor this technique is ideal. The flowline can be scaled up and down to meet the market growth. Figure 12.4 shows how the activities can be grouped at the different stations to optimise the step time across all four stations.

This growth scaling can be achieved by changing the number of steps that are performed in each workstation. If the number of stations is one then the flowline is operating as a bespoke manufacturing cell, at the other extreme if the number of stations were infinite then the flowline would be operating like a production line.

This attribute of scalable production capability is important in delivering the mature nth-of-a-kind (NOAK) costs. For a small reactor the first-of-a-kind (FOAK) to NOAK cost transition is important, where customer commitment will be based against perceived costings. It is important that there is an achievable path to reduce the costs between the initial and mature production units. The flowline can be seen to deliver these cost reduction opportunities in a scalable manner, as shown in Figure 12.5.

To achieve an increase in production volumes the number of stations on the flowline can be increased and the number of activities performed at each station

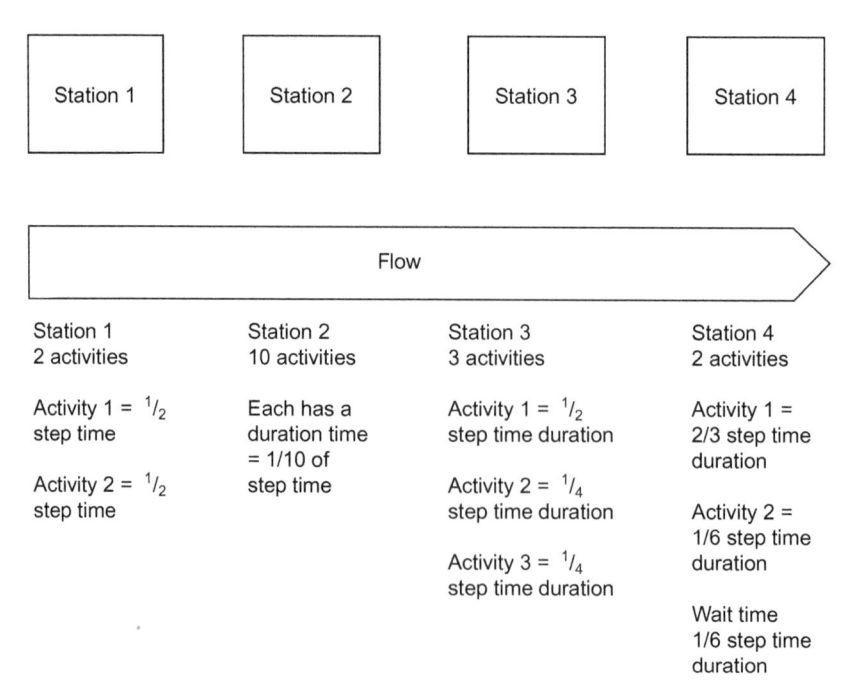

Station 1	Station 2	Station 3	Station 4
2 activities	10 activities	3 activities	2 activities
Activity 1 = $1/2$ step time	Each has a duration time = 1/10 of step time	Activity 1 = $1/2$ step time duration	Activity 1 = 2/3 step time duration
Activity 2 = $1/2$ step time		Activity 2 = $1/4$ step time duration	Activity 2 = 1/6 step time duration
		Activity 3 = $1/4$ step time duration	Wait time 1/6 step time duration

Figure 12.4 Illustration of flowline activity allocation.

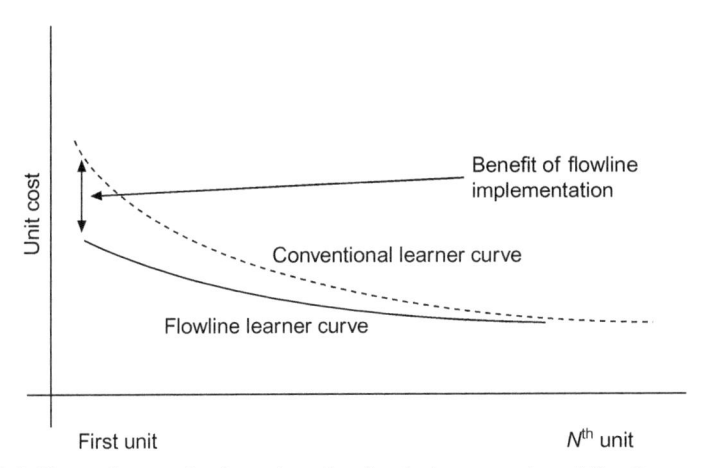

Figure 12.5 Illustrative graph of cost benefit of early incorporation of flowline concept.

can be reduced. This is a unique attribute of the flowline that enables it to be scaled up as the market demand grows. For example, say that the initial units are built on an eight-station flowline with 40 operations at each station. If the number of stations was increased to ten the number of operations at each station can be reduced to 32. The underlying assumption in this hypothetical example is that each station already has the maximum number of activities at each station.

The deployment of the flowline concept has been shown to be a well-established process, proven to be capable of replication in different geographical locations and industrial segments. It is suited for complex assemblies with an intermediate production volume. The flowline concept adopts a single-track flow system to the assembly of a product, using individual workstations or process steps to progress the assembly procedure. The step flow moves down the line are defined by the term Takt (cycle) times. Each activity within each station is mapped in terms of its Takt time. To ensure the greatest level of repeatability and predictability from the final product at each step minimum inspection is preferred. With the incorporation of 'design for assembly' processes a lot of inspection can be eliminated. This is part of the product simplification and standardisation activities. There is a bias toward automated online validation and testing of components or assemblies during the assembly process.

Certain key features are essential to the success of any flowline system. Single product family, no rework, stable supply of parts/assemblies/kits and a high reliability of workstation equipment, tools, etc. With a single product family there is a consistent approach to the work that allows for standardised work instructions to be created and scheduled for specific stations. This is turn requires a stable parts supply. Units are assembled in a uniform and predictable manner. Kits of parts can be 'kitted' away from the flowline, even at an alternative location. These kits will contain all the parts for a specific assembly operation. For example for the fitting of a sensor, the sensor, the gland and seals will all be supplied as a kit along with the correct fasteners. This enables the assembly team to verify that they have all the materials to perform the installation of that component prior to commencing each step. The kitting and delivery of components onto this flowline is therefore also stable and becomes an assured process. With this stability comes to opportunity to pre-kit and collate parts in the build sequence prior to incorporation in the final module build. The final aspect is tooling stability. With known and repeatable assembly operations being performed consistently at a specific work station there is a reduction in the opportunity for errors from misplaced or misused tooling and fixtures.

Design of any flowline is heavily reliant upon the product design and its suitability to conform to flowline criteria. The application and use of process and design failure mode and effect analysis (FMEA) tools during the conceptual design of the product is fundamental to delivering a robust flowline system. As an example these manufacturing ideas have been shown to work well, incorporating simple things like baulking features in the design to ensure that parts are orientated correctly to each other. As an alternative example assembly operations can be grouped such that all fastenings of a specific size are fitted at one assembly station. This reduces the potential assembly errors. The adoption of poka-yoke (fail-safeing) of parts, the utilisation of common tools will also contribute to the success of any flowline system.

Consideration of the facility layout is of paramount importance, while employing the use of lean tools to design a fully efficient flowline capability. This capability will take into account the on-time delivery of sub-modules from suppliers and vendors into the assembly facility.

The storage of inventory in such a facility is not a feature of the efficient flowline philosophy. Therefore delivery of parts, sub-modules and kits direct to the assembly facility workstations is considered essential to support the pure implementation of this philosophy. However, industrial experience shows that an initial 'buffer' stock of critical parts is required as a new flowline is established. One option is to pre-kit parts for assembly away from the flowline and sequence them for assembly. The level of buffer inventory held may be reduced and optimized as the process finds its natural rhythm. The rhythm of the flowline is referred to as its drumbeat. The drumbeat does not change. If the volume through the line needs to increase, one of the options is to further sub-divide the operations performed at each station.

Delivery of parts from the pre-kitting can be arranged to use reusable trays or standardised trolleys. These trays and trolleys act as 'shadow boards', allowing the flowline personnel easily and quickly to validate that all the parts for their operations are present at the outset of their build sequence for their flowline workstation. Similarly they can confirm visually that all the parts for that stage of the build sequence have been incorporated. These techniques are not specific to a nuclear build and have been validated in other industrial sectors. With the historically lower build rates from large nuclear plants there has been little need or incentive to incorporate these techniques. For the first time in the nuclear sector the small reactor presents an opportunity to adopt and implement these lean manufacturing techniques. There are significant consequences for a production capability built on this philosophy. The rigor of supply of sub-assemblies and modules flows down to the sub-tier vendors. The supply of parts onto the flow line is important. This discipline cascades down the supply chain. In this respect the small reactor may not be a revival for existing nuclear suppliers, but may be demanding that the industry searches for new suppliers with alternative manufacturing capacity.

With a higher level of process automation incorporated in the assembly activity, attention is also needed to understand the opportunities afforded for automation in the manufacturing control. These technologies have not been deployed in a nuclear context previously with the smaller volumes of manufacture. With the larger volumes of product the investment to support the deployment of these techniques can be considered. The use of RFID (radio-frequency identification) tags provides a semi-automated, real-time monitoring capability of the location of parts during the assembly process. This technique is already used within the aerospace industry where the incorporation of parts are tracked, and can readily be applied in the small reactor product. There are opportunities to include additional build control techniques from other industries too; For example the use of a high-resolution local global positioning system (GPS) network within the factory can be combined with handheld scanners and some direct part marking will allow components to be scanned as they are incorporated onto the assembly. The build control database can record the incorporation of a specific part and serial numbered component. This concept can be extended to include the tooling. For example a torque wrench could contain a GPS device that automatically records the torque setting applied to a specific bolt at a specific location. Techniques such as these are not specific to nuclear applications, it is the repeatable build sequence of a volume of units that makes off-the-shelf techniques like this attractive to small reactor assembly.

In other manufacturing sectors around the globe there are already facilities employing the flowline approach, history can identify equivalent systems applied to large assemblies, not dissimilar in complexity and physical size to the small reactor product: For example, during the period of 1941–1945, the USA employed an optimised assembly process to produce 2751 Liberty ships at 18 different shipyards – each ship was of common design (pre-fabricated sections) and helped the allies win the war. The first few vessels took around 230 days to build but, based on experience and the application of continuous improvement philosophy, the average eventually dropped to 42 days.

The introduction of the flowline system has realised output efficiency improvements, including:

- reduced inventory holding;
- clear responsibilities and accountabilities are defined for each activity station;
- the provision of control and sequencing of activities;
- a defined level of inherent quality assurance.

Other simple features include:

- materials handling is positioned to avoid obstructing the walking path of the operator;
- flow racks are filled and delivered away from the flowline to avoid interrupting operator work cycles as parts are replenished;
- parts are correctly orientated for easy operation, allowing operators to use both hands simultaneously;

- parts containers are sized for ease of use by the operators, not the material handler or the supplying process.

In the design of the flowline each of the stations are timed to a common duration. The process steps and the staffing levels are then optimised around each of these steps. Experience from industry shows that 20 operations or 10 operators in each station is the sustainable limit.

12.2.3 Role of standardisation

Implicit in the flowline concept is the standardisation of product design. In a volume manufacture situation, a small reactor vendor will not be able to cope with customisation demands without destroying the economic model for the sector. The design will therefore be fed through a sealed manufacturing route. In itself this is no different from many other high-quality, assured manufacturing operations but it raises questions such as the regulator's approach to manufacture of components that could be fitted into any one of a number of reactors going through the flowline at that point. Although this issue could be seen as implicit in the NRC's Part 52 licensing regime, it is not obvious that implications such as this have actually been considered in the detail required.

12.2.4 Component sizing

The current designs of large, new-build reactors have components that require transport to site individually, for subsequent assembly on-site once they arrive. The main containment vessel of a small reactor on the other hand will fit within the specifications of a single large reactor component. An AP1000 steam generator is 22.5 m $\sim$ (74 foot) long, 5.6 m $\sim$ (18 foot) diameter and weighs 700 t $\sim$ (1.5×10^6 lb). The same component in an EPR has similar dimensions but weighs 550 t $\sim$ (1.2×10^6 lb). Compare this to a small reactor where a complete integral unit can fit within the above envelope, and weigh less than 500 t $\sim$ (1.1×10^6 lb) and it becomes apparent that from a manufacturing perspective the components can be manipulated on the flowline with relative ease. The entire small reactor unit can now be transported as a single unit direct for installation to the site, moving more tasks to the factory to complete assembly in a controlled environment.

Integral reactors often have more numerous, smaller components per plant which lends itself to a flowline approach. For example, a typical large, new-build main coolant pump will weigh 124 t (2.7×10^5 lb), which represents a significant part to lift and negotiate in a manufacturing environment. In a small reactor, however, each pump may weigh only 500 kg (1100 lb). This is of a comparable magnitude to the aero-engine flowline model, so it seems appropriate to adopt this technique to the small reactor product.

12.3 Component fabrication

The mature production volumes for small reactor plants predict tens of units being produced per year. With a standardised product approach there is a business rationale to invest in tooling and techniques that bring about both unit cost reduction and also lead time improvement. For example with ten units in manufacture each year using conventionally deployed nuclear sector manufacturing techniques, the value of work in progress would be significant, requiring capability to be duplicated to deliver the required volume. As a simplistic example, in a facility that produced ten components per year, any activity that takes more than 36 days to complete would require investment in a parallel capacity to achieve the volume. A specific example could be the weld cladding and staged inspection of a vessel.

The deployment of any new manufacturing technique, even one that has been validated in other comparable industrial sectors, will still need to satisfy the regulatory agencies prior to any plant incorporation. This point on securing regulatory acceptance has specific importance for a small reactor.

12.3.1 Additive manufacture

Additive manufacturing processes differ significantly from traditional formative and subtractive processes. Formative processes, e.g. injection moulding or casting, require the initial manufacture of long-lead, high-cost tooling or moulds into which material is injected or poured. Subtractive processes, e.g. milling or turning, typically require the procurement of a long-lead, high-cost forging from which material is removed to achieve the final component definition.

Additive manufacturing processes create components by the selective addition of material layer by layer to form the component geometry. Material is only added at each layer where specified by the computer-aided design (CAD) definition. This fundamental difference can offer a number of business and technical advantages over traditional process.

Additive manufacturing processes are often sub-divided according to the energy source used and/or the raw material delivery method. There are two energy sources widely used in industry; lasers and electron beam. There are also two material-delivery methods: 'powder bed' and 'blown powder'.

The term additive layer manufacturing (ALM) is typically used and should not be confused with rapid prototyping which infers a rapid manufacture that may or may not be correct and an end use for the component.

Laser sintering is an inaccurate and unofficial definition, and should not be used as a description for this technology. Direct metal laser sintering is a registered trademark of Electro-Optical Systems (EOS) GmbH and should not be used, as it would denote a specific vendor, not a technology or process group.

3D-printing is an increasingly common term within the industry. However, it is predominantly used when referring to a much wider group of technologies including polymer-based systems.

12.3.1.1 Benefits of ALM

The ALM process can provide a number of advantages, both business and technical, over traditional manufacturing methods. The justification for considering ALM arises from the volume of the market. For a small reactor with product lifespans of 40 to 60 years it is clear that no single supply chain can support the surge of production volumes at product launch and then sustainment for replacement parts towards the tail end of the product life with the same supply chain. This again is a new challenge for the nuclear industry that has evolved around a culture of bespoke engineering, where support for a plant through life from the same bespoke supply base is acceptable. For a small reactor delivered as a volume product the sustaining engineering challenge to the manufacturing supply base is different. Sustaining the product through life can no longer use the same original equipment manufacturer (OEM) supply chain economically. Automotive manufacturers do not supply replacement door panels for 10-year-old vehicles using the same line that delivers for their current year models. The small reactor supply chain faces the same set of challenges. Techniques such as ALM that have lower volume applications come into play offering significant economic advantages over the initial OEM supply chain. This is part of a strategy to maintain the competitive support of the plant through its operating life.

More specifically the attributes of ALM can be considered as two groups, OEM benefits and through-life benefits. ALM–OEM benefits include the following:

- Unit and/or through-life cost saving from reduced material quantity and/or machining costs offered by the ALM process.
- Unit and/or through-life cost savings from a reduced part count, e.g. manufacturing an assembly of multiple parts as a single component.
- The availability of welding test pieces, non-destructive examination (NDE) test pieces and other assembly and manufacturing aids significantly ahead of lead units can greatly reduce development programmes.

ALM – through-life benefits include the following:

- ALM can provide an alternative strategic sourcing route to the traditional forging, casting or fabrication route. It is highly suited to the low-volume, high-quality requirements of supplying nuclear-grade components. It can therefore be used to mitigate the risk of an existing manufacturing route that is threatened and/or may not be viable in the future for through-life sustainment.
- Where the existing manufacturing route is causing significant difficulties in supporting build, due to length of the manufacturing timescales.

By manufacturing components in layers, geometries that could not previously be manufactured can now be produced. This enables a number of potential benefits:

- Optimisation of design definition for improved performance.
- Assemblies can be consolidated into a single component, thus simplifying manufacturing processes, reducing through-life costs, removing welds and fabrication, reducing inventory.
- Multi-functional parts can be produced by integrating cooling channels, electrical controls and instrumentation and/or cable management channels within structures.
- A structural integrity improvement may be realised, e.g. elimination of welds.

- High stiffness to weight ratio parts can be manufactured with internal lattice structures driven by finite element analysis (FEA) optimisation algorithms. These structures have thicker lattices where loading is high and thinner lattices where loading is low.
- Graded structures can be produced, where the material type is varied within a single structure, e.g. a tube can be manufactured that is Type 316 L stainless steel at one end and Inconel 625 at the other, without the need for a transition weld.

12.3.2 Electron beam melting (EBM)

Additive manufacture by electron beam manufactures parts by melting metal powder layer-by-layer using an electron beam in a high vacuum environment. Parts produced by electron beam melting (EBM) are usually smaller than those produced by the laser process, but the EBM process does yield components that are fully dense, void-free and extremely strong.

As with ALM, CAD geometry, a power source and metal powder is required. However unlike ALM, EBM-fabricated components do not require heat treatment because of their high densification and operating temperature of typically 700–1000 °C during the fabrication phase. EBM is used in the aerospace sector where titanium and titanium-alloy components are produced, although systems do exist which can produce components up to 450 mm × 100 mm × 100 mm in a small variety of high-value metals.

Nuclear component materials in nickel-based alloys could be manufactured by the EBM process, but a cost–benefit analysis should be made to determine the viability when compared to existing fabrication techniques.

12.3.3 Shaped metal deposition (SMD)

Shaped metal deposition (SMD) is a wire-based additive manufacturing process based on a three-dimensional tungsten-inert gas (TIG) welding system. SMD can produce components in nuclear-grade materials, without the need for special tooling, in a relatively short lead-time when compared to forging or casting routes. The SMD process was developed and patented by Rolls-Royce and was subsequently licensed for use and further development by the University of Sheffield.

The SMD system employs cold wire TIG deposition, using a tungsten cathode to weld the selected material in an inert argon atmosphere to prevent the substrate, electrode and part from reacting with atmospheric gasses. This is a very stable process, chosen for the maturity of process control. The TIG welding head is attached to a six-axis robot and produces weld on to a rotating turntable base. Weld material is built up in an additive way and the construction is monitored throughout the entire process to ensure that the weld parameters are maintained and a robust final component is produced.

Because SMD is a weld-based technique, accuracy in fabrication of the component depends on the thermal stresses induced during the welding process. Control of the thermal transfer during weld deposition can allow reduction in the final component residual stress and hence the distortion. Simulation modelling tools can be employed to predict the distortion levels and this can be fed back in to the SMD process during the deposition phase.

The wall thickness of the component is controlled by the current, travel speed and wire feed rate and also to some extent by the wire thickness. The travel speed is the product of the rotation rate on the turntable, and the rate of movement of the robotic head – the faster the travel speed, the thinner the wall thickness, but the limiting factor is the ability to maintain a robust arc at the weld head.

Cylindrical components are easily produced on the turntable but with the multi-axis robotic head, there is opportunity to produce more complex forms. One such form is shown in Figure 12.6 where two cylindrical components were formed as a prototype exercise.

The complete component is essentially a 100% weld and unlike the previous additive manufacturing systems which use metal powder to produce the component, SMD components do require post-processing heat treatment and final machining to attain the design-intent form. The SMD process is also well positioned to provide a route for adding material to an existing component such as a large vessel with a boss or nozzle. As a deployable SMD system, this solution could allow for the fabrication of more simple cylindrical forms through the traditional forging route and adding on the external features through an SMD-type technique.

12.3.4 Cladding

Cladding of nuclear components, in particular large vessels, is applied to prevent the corrosion of the vessel substrate by the aggressive nature of the operating environment. The vessel substrate material is typically a low-alloy steel such as SA508 Grade 3 Class 1, which is a commercial-grade pressure vessel steel used across the nuclear

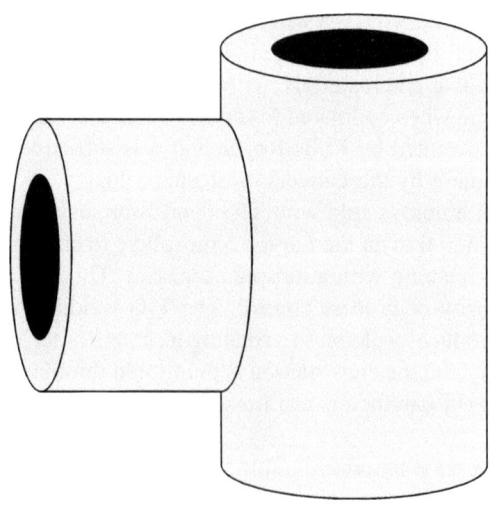

Figure 12.6 Picture of SMD trial piece.

sector. This substrate material can be covered or 'clad' with a less corrosive, more inert material such as a nickel-based alloy or a grade of stainless steel.

The current cladding technique is weld-based using wire or strip, fused to the substrate via the welding process. This is an expensive, time-consuming and materially wasteful process – to attain the necessary inert chemistry, several weld passes are made to build up the clad thickness. After each pass, a run of NDE is required to ensure that the clad is not only fully bonded to the substrate, or its previous clad layer, but also that there are no excessive defects such as cracks or high levels of porosity. The final clad layer is then machined back to attain the chemistry content and to provide a good-quality surface finish with a geometrically sound profile. In a typical cladding process, up to 60% of the laid-down clad could be machined away, making the current process wasteful.

The ideal cladding technique would lay down material onto the substrate in a single pass and would not require post-clad machining. The cost and time-savings associated with such a technique would be significant. One solution is to adopt an additive manufacturing process, known as diode laser powder deposition (DLPD). Cladding by DLPD is a deposition welding process in which a layer of metal powder (the clad) is deposited on the parent (the substrate) material. The two materials are fused by the energy provided by the laser and a metallurgical bond occurs. Figures 12.7 and 12.8 show the deposition of clad during proving trials.

This cladding technique offers high precision, a high level of automation, a robust and repeatable process and a clad–substrate interface with low dilution, thus attaining the necessary chemistry with a thin layer of clad deposit. This cladding technique is significantly quicker (see Table 12.1) than the existing weld-based technique and combined with a single-pass potential and the possibility of including an in-line

Figure 12.7 DLPD cladding in the vertical (1G) orientation.
(courtesy of IWS Fraunhofer DLPD)

Figure 12.8 Cladding in the horizontal (2G) orientation.
(courtesy of IWS Fraunhofer)

Table 12.1 Comparison of traditional and DLPD cladding techniques

	TIG clad	MIG clad	DLPD clad
Coverage rate (hours per m^2)	20	5	2

TIG, tungsten-inert gas; MIG, metal-inert gas; DLPD, diode laser powder deposition.

NDE/process feedback system as a part of the cladding process, cladding of large vessels could be reduced from weeks to hours.

12.3.5 Hot isostatic pressing (HIP)

Hot isostatic pressing (Hipping) is the consolidation and densification under high temperature and pressure of metal powder, housed within a canister that represents the final geometry desired. HIP components offer the potential for nett-shape (NS) or near-nett-shape (NNS) component fabrication with high densification of the volume, resulting in a small metallurgical grain size and thus providing superior mechanical properties to cast or even forged components.

HIPped components have been, and still are used in the automotive, aerospace and medical industries, and some experience of HIPped components has been gained in nuclear applications

The HIP process is well-established and consists of five key stages:

1. Procurement of metal powder to desired specification. This is a fundamental quality phase in the overall fabrication route. The powder must be of the right quality, with metal particle size

distribution and morphology well defined. Failure to define and attain the correct powder will result in a poor-quality HIPped product.
2. Canister modelling and fabrication, including application of deformation modelling to optimise NNS potential. This is the most labour-intensive phase of the fabrication route. Canisters are typically low-alloy steel in sheet-metal form, that are formed and welded manually. Robustness of the canister and its welds is critical – failure of the canister during the HIP cycle will result in a defective HIP cycle – so each canister undergoes extensive inspection. The adoption of the HIP process over the unit volumes presented by small reactors depends on the effective automation of canister manufacturing.
3. Loading of canister in to the HIP furnace. The canister is loaded with powder, vibrated to maximise powder-fill, evacuated then sealed prior to loading to the HIP furnace. A typical loaded HIP canister is shown in Figure 12.9.
4. HIP cycle stage – typically in excess of 96 MN/m^2 (14 000 psi) and 1200 °C. Cycle time, pressure and temperature are dependent upon powder material and canister geometry, but typically this phase in the process is a small number of hours.
5. Removal of canister post-HIP cycle, via acid pickle or machining. A typical finished HIP component is shown in Figure 12.10.

HIP furnaces are available worldwide, each with their own 'working envelope' limiting the physical size of component which can be HIPped. This has an implication for the nuclear sector where some candidate components such as large valve bodies or pressure vessels exceed even the largest HIP vessel commercially available. The largest HIP vessel in existence today is in Japan and has a working envelope of some 2 m in diameter and 4.2 m in length. Nevertheless design schemes do exist for HIP furnace

Figure 12.9 An example of a HIP canister.
(courtesy of Rolls-Royce)

Figure 12.10 An example of a post-HIP cycle component. (courtesy of Rolls-Royce)

working diameters in excess of 3.5 m which would suit a number of large nuclear-grade components.

The potential for HIPping of components to save on unit cost and lead-time is great, especially when compared with forged components. The most time-consuming phase of the HIPping cycle is the canister development, but a regular drum-beat of components offers the opportunity to drive-down cost through automation of the canister fabrication process. A further distinct advantage of the HIP process is the inherent repeatability and robustness in the material properties: large forged components are notoriously difficult to produce and will always contain defects and variations in grain structure from component to component. HIPped components will always have the same fine grain structure throughout the bulk of the component and possess isotropic mechanical properties.

HIPped components are yet to be realised in a civil nuclear arena, but experience of HIPped product in an industrial environment is being gained through the oil and gas sector where HIPped large-bore pipework has been introduced. In the marine sector there are further opportunities where HIPping has yet to be realised, and the UK's naval programme is already using some HIPped components and has gained good credibility in this part of the nuclear sector.

The small reactor programme offers a huge opportunity to introduce new HIPped components to a new reactor design, which could align well with a regulatory code case approval scheme. Retrospective introduction of HIPped components is a further opportunity but this would involve the like-for-like replacement of components and the design would not change. HIPping for small reactors offers the chance to influence the design of the components to suit the HIPping process.

12.4 Advanced joining techniques

Joining of components throughout the nuclear plant is typically performed by fusion welding, more specifically a TIG process, however there is also some local use of Electron Beam welding but this is aimed at thin-section (less than 20 mm) materials.

TIG welding of thick-section materials, such as pressure vessels, is an expensive, time-consuming practice involving extensive pre-work such as fixturing, tooling, pre-heating of components, etc. TIG welding of thick sections is also performed over a number of runs, typically up to 100 runs for sections of 140 mm or above, and multiple runs of NDE throughout the welding process. Welding, inspection and completion of a large vessel therefore takes many weeks, even months, so accounts for a vast proportion of the fabrication cost and component lead-time element.

Development of the electron beam welding process to thick-section welding offers a number of significant potential benefits: single-pass welding of thick sections (40–140 mm), no requirement for pre-heating of the components and no requirement for a weld filler material. These all offer significant process and cost savings over traditional welding systems, but the key game-changing benefit offered by electron beam welding of thick sections is the removal of inter-stage NDE. Table 12.2 illustrates how the removal of inter-stage NDE from the welding of a thick-section component can dramatically reduce the overall time in process – by nearly 75%.

In short, electron beam welding of a complete large pressure vessel, which could consist of up to five thick-section circumferential welds, could be reduced conservatively by a factor of four. However there is one significant drawback: electron beam welding requires a vacuum to eliminate beam divergence which results in poor beam focus and hence loss of welding capability, especially when considering thick sections. Current technology requires the use of a vacuum chamber which houses the

Table 12.2 Comparison of TIG and electron beam (EB) welding procedure based on a 140 mm thick circumferential weld

	Stage [hours]	
	TIG	EB
Initial set up, pre-heat time	175	120
Interstage NDE #1	40	0
Pre-heat, weld and clean	135	0
Interstage NDE #2	40	0
Pre-heat, weld and clean	135	0
Interstage NDE #3	40	0
Pre-heat, weld and clean	135	0
Final NDE	40	40
Post-weld heat treatment	15	15
Post heat-treatment NDE	40	40
Total [hours]	795	215

components to be welded as well as the electron beam system – vacuum chambers are extremely expensive to construct and are typically built to accommodate the size of components to be joined, so even small reactor nuclear vessels would require large vacuum chambers.

A viable alternative would be to bring the vacuum environment to the components to be welded. One could envisage the development of a 'jacket' which would wrap and seal around the components to be welded and so would provide a local vacuum for the electron beam. The electron beam gun could be attached to the sealing jacket, and the components rotated around the gun to perform the welding process. The key to success for this technique is around the design of the electron beam gun and the sealing jacket or ring. To further benefit from this proposal, the sealing system should employ design-for-assembly philosophies and ensure that consumables such as seals and sealing surfaces, are easily replaceable. It would be further prudent to design the whole system such that a number of electron beam gun manufacturers could be used on the system thus negating a single-source route and the risk associated with such a strategy.

12.4.1 Coatings systems

As well as fabrication of nuclear components and the drive to eliminate waste and increase manufacturing efficiency by the introduction of advanced manufacturing techniques, there is a need to improve component life within the nuclear plant. Coating systems are deployed within nuclear plant to improve wear-resistance of moving parts, such as valves, pump internals, and other moving mechanisms.

In the automotive and aerospace sectors, diamond-like-coatings (DLC) are deployed extensively, offering excellent wear resistance and durability in service. For the nuclear environment, DLCs have been seen to break down in service and de-bond from the parent substrate, thus losing its effectiveness and discharging debris into the primary circuit. Optimised DLC systems are being developed by industry, aimed specifically at the pressurized-water reactor environment.

Other coating systems are cobalt-based and are therefore not ideal for a nuclear environment: the neutron flux activates the cobalt leading to circuit contamination, which requires as low as reasonably practicable (ALARP) management during routine maintenance periods. Development of a cobalt-free coating system is under way at Rolls-Royce. This iron-based system will be in powder form and readily available for the HIP process if desired.

12.5 Supply chain implications

The impact of the flowline on the supply chain is also significant, the whole system being based on a robust material-requirements-planning (MRP) foundation. For larger suppliers the 'Kanban' system can be operated, delivering sub-assemblies or part-kits to the workstations directly on the flowline. Each part will have a unique identifier permanently marked via physical (laser) etching or nameplate, incorporating both a visually readable and electronic-scanning data-matrix identifier. During the assembly process the dot-matrix data identifier is read using a handheld matrix-reading device which interprets the matrix into defined attributes such as item part number and serial number, build record. The completed unit assembly will tally with a full inventory of parts, each logged via the data-matrix system. This provides a complete build history of the small reactor and could also be used to automatically trigger a payment milestone to the supplier when the part is incorporated onto the customer unit.

12.5.1 Deployment

The deployment options for a small reactor build on one of the key attributes of their design. The small modular reactor (SMR) factory build aspect creates a number of significant advantages during the deployment phase. Delivery of assured modules to a customer site has been claimed as a banner headline, but it is worth reviewing the aspects behind these individual claims to understand the contribution of their worth to the overall deployment:

- time to build;
- risk – schedule risk and cost over-run;
- validation – off-site pre-delivery testing;
- international aspirations;
- Critical path through site activities with modular build.

12.5.1.1 Modularity: addressing schedule and cost risk

For the large part nuclear construction project cost increases and schedule delays have their root in site uncertainty when compared to original plans and cost estimates. Modularity has been offered as the solution to these over-runs.

The Construction Industry Institute defines modularity as 'work that represents substantial offsite construction and assembly of components and areas of the finished product.' There are incremental levels of modularity that need to be considered; Graduations of modular construction are as follows:

- *Skid* – modular skid assemblies of equipment that are delivered to site fully instrumented and validated, requiring little installation and commissioning activity. Within this definition it would also be anticipated that the skids are capable of being delivered to site using conventional transportation networks.
- *Super skid* – this represents the next increment with larger portions of plant intended to be installed in a building.
- *Prefabricated* – this is the final step in the definition of modular and would extend to a fully prefabricated module comprising plant equipment in an architecturally complete housing.

Has a modular build approach been deployed already with large nuclear plants? There are elements of modular construction encompassed within large nuclear plants. The Vogtle plant built by Southern Co. in Georgia has utilised a level of off-site modular manufacture. This facility has been established to build modules for the plant in a manner that has the same overall goal. However, the modules for a large nuclear plant are essentially one-off modules, built off-site in a piece-by-piece approach. Without the repeat volume flowing through such a facility it can be argued that module build for a large nuclear plant is an off-site variation to the on-site stick build approach and falls short on delivering some of the goals for the module-build approach advocated for small reactors.

The incorporation of advanced modular construction centres around completely shop-fabricated modules. This method of manufacture is more demanding in terms of upfront planning and management. With greater emphasis on design integration, procurement, manufacturing and construction activities there is less natural opportunity for schedule variation. This modular method of site deployment marries beautifully with the flowline approach bringing about greater schedule adherence.

This modular deployment model is not without consequence. The preliminary and final design activities must remain sealed to allow time for the procurement and delivery activities to be completed. Design change at this point erodes the benefits of the advanced modular approach. The mindset of deployable small reactor manufacturing is driving the product line closer and closer towards commodity-based supply chain design.

Having taken a view on the supply and delivery end of the deployment model, there are equivalent considerations that need to be taken into account with the site implementation.

12.5.2 International perspective

There are some options, however, for wider levels of international cooperation using a globalised modular approach to the build and development of small reactors. A single product that has access to a global market will need to offer routes for international offset and localisation of content and workscope. Depending on the approach to modularity throughout the plant, i.e. both the nuclear island components and the balance of plant, it could be possible to support a localisation scope objective where the mechanical and process interface can be defined at a module boundary. In this way the modularisation within a whole power plant can open up opportunities for localisation. Equally the fixed interfaces across these module boundaries will limit options for change. History has shown that the level of control across modular interface boundaries will need to be managed under robust change control to ensure that interface compatibility problems do not arise and cause delays on site. The small reactor that is moving towards a commodity product is going to need to balance the business case requirements of localisation against the value of standardisation.

12.5.3 Power plant critical path

The level of site preparation can be greater with the commoditised module delivery. This can have the effect of moving the nuclear island off the critical path for site construction activities. This potential impact of this is typified by looking at the key dimensions of the typical steam turbine generators, as in Table 12.3.

It can be seen that the larger turbine generators quickly become significant units and sit outside of conventional road transportation envelopes. There are regional variations on transportation sizes. Packages around 5 m wide and tall can usually be

Table 12.3 Comparative dimensions for turbine generators

	50 MWe	175 MWe	250 MWe
Length	12 m (~39′)	19 m (~62′)	20 m (~65′)
Width	4 m (~13′)	8 m (~26′)	11 m (~36′)
Height	5 m (~16′)	6 m (~19′)	10 m (~32′)

accommodated as oversize loads. Moving beyond 5 m moves away from a turbine generator set that can be road transported as a single module. A turbine generator that is not modular can put the rotating machine on the critical path for the site build programme. This is substantiated with information from combined-cycle gas turbine construction programmes. From the time the last generator casings are delivered to site it is a further year until the turbine is running on its turning gear, and a further three months before the set is ready to be synchronised for the first time. It is easily conceivable in a small reactor power plant with larger turbine generators, which are not capable of road transportation as a single unit, that the turbine can sit close to the on-site programme critical path.

12.5.4 Deployment model: in service

The opportunity to deploy a small reactor to compete against in any energy portfolio is challenged by its relatively high $/kW overnight capital cost. Some benefit is derived from a factory build approach of plant modules, but this alone does not achieve the desired cost target. Similarly a position on volume production with an overhead amortised over a larger number of modules will make further inroads into the $/kW. The final element in reduction of the competitive position for a small reactor lies in the in-service deployment.

In March 2012 Jack Bailey – VP Nuclear Generation Development for TVA – offered a view on the FOAK plant in-service model. He commented that initial goals for centralised regional services could not be realised for early deployed small reactor plants. The authors would challenge that view, asserting that it is this operating model that needs to be incorporated from the outset. In the same way as a production line needs to be designed at the outset for a level of product delivery, the in-service support package needs to be designed from the outset. For example, if the small reactor is to secure a level of plant sales comparable to current closed cycle gas turbine plants then the training burden alone needs a level of innovative thinking to support the volume of operating staff that need to be qualified on the plant. Early statements about centralised engineering support need to be developed to maintain support for the plants coming on-line. Part of the market appeal for a small reactor plant is the lower capital outlay. The reduced capital outlay opens up the small reactor to a new population of nuclear owners and operators. These 'new to nuclear' owners will not have mature engineering upkeep organisations to support their small reactor plants. As a consequence the third aspect to achieve a competitive cost base for a small reactor is the deployment of a fleet model that provides a centralised support service over a number of sites.

12.6 Conclusion

It can be seen that the supply and deployment of the small reactor offers an opportunity for the nuclear industry to adopt a revolutionary supply chain model, and maybe a new group of suppliers. There are opportunities to build an alternative supply chain established around manufacturing techniques that have been validated in other sectors.

With greater volumes of units flowing through the factories the incorporation of advanced manufacturing techniques can be substantiated. These techniques in turn offer efficiencies in manufacture that are reflected in lower manufacturing costs. The factory assembly of modules can be supported in a manner that offers incremental capacity increases reflecting market certainty.

The incorporation of contemporary automation for factory build techniques delivers a greater level of automatic part tracking from standardised product assembly. This in turn opens up options for fleet management across multiple sites managed to a common template. The deployment model for a small reactor is a revolutionary step for the nuclear industry.

Reference

Ford, H., Crowther, S., 1922. My Life and Work. Kessinger Publishing.

Hybrid energy systems using small modular nuclear reactors (SMRs)

Shannon M. Bragg-Sitton
Idaho National Laboratory, Idaho Falls, ID, United States

13.1 Introduction

Large-scale nuclear reactors are traditionally operated for a singular purpose: steady-state production of dispatchable baseload electricity that is distributed broadly on the electric grid. While this implementation is key to a sustainable, reliable energy grid, small modular nuclear reactors (SMRs) offer new opportunities for increased use of clean nuclear energy for both electric and thermal applications in more locations while still accommodating the desire to support renewable production sources. There are several questions that a potential investor or utility must ask before making the decision to build a power plant, followed by the decision on where to site that plant:

- What is the specific energy requirement that needs to be met? Is it primarily thermal or electrical or a mix of both? Could a single integrated energy system meet both needs?
- Does the energy demand vary over short or long time frames (e.g., hourly, daily, or seasonal)?
- What resources are available in the region (e.g., water, land, and carbon feedstock)? Are these resources necessary to the operation of a proposed energy system? Can they be used to enhance the operation of an integrated energy system?
- Is renewable generation an attractive option in the region? Are renewables presently in use in the planned deployment location (e.g., wind, solar, hydro, geothermal, and biomass), and what is the potential for contributions from renewable energy sources to increase of the lifetime of the power plant being considered?
- Will the proposed power plant be integrated on a small or large electricity grid, or will it be operated independently to support a specific industrial need?

The answers to these and related questions provide the framework for a multipurpose implementation of nuclear energy systems that could integrate multiple resources to produce multiple output products. Cogeneration systems (single input systems, providing both thermal and electrical output) and tightly coupled, multiinput and multioutput nuclear hybrid energy systems (NHESs) can be designed to flexibly operate based on thermal and/or electrical energy demands while accommodating multiple input streams. Those input streams derive from several independent reactor units or a combination of resources such as nuclear reactors, wind turbines, solar photovoltaic (PV) panels, biofuels, and fossil fuels.

Many cities, municipalities, countries, and utilities have established or are considering establishing low- or zero-carbon emission and clean energy goals (ThirdWay, 2019). Given the strong public (and, hence, political) desire to support noncarbon-emitting sources, electricity produced by renewable sources is often treated as "must-take" on the grid. This scenario can mean that the baseload provider is required to ramp down production when there is ample production from wind or solar generation, or sell electricity at a loss. The variability of renewable plants, such as wind and solar, occurs on a relatively short timescale. This intermittency places significant demands on the dispatchable, baseload plants that also supply the grid because they are then required to vary relatively large fractions of their load over a short time. A 2012 study conducted by the Organization for Economic Co-operation and Development (OECD) Nuclear Energy Agency (NEA) reported that the substantial amount of renewables introduced on the grid in Germany led repeatedly to prices below the marginal costs of nuclear, including several instances of negative prices (OECD NEA, 2012). Extremely low and even negative price conditions are occurring at higher frequency in the United States as well, attributed to both low-cost natural gas and increasing penetration of variable renewable generators whose production outstrips demand during certain hours of the day (Mundahl, 2018). Overall, negative prices hurt consumers and are a sign of market inefficiencies. The periodic glut of electricity that drives wholesale prices down also means that conventional electricity producers (e.g., coal, hydro, and nuclear) will sell less electricity overall, making it difficult to recoup capital costs and, in some cases, revenues may even fall short of fixed annual costs for operations and maintenance. The result is that plants may be retired early—leaving the market before their license expires (OECD NEA, 2019). This early retirement trend can be detrimental to achieving the bold clean energy goals that have been established. Despite recent growth in the deployment of renewable generators around the world, the International Energy Agency (IEA) reported in May 2019 that the fraction of clean energy contributions to electricity has not changed over the last 20 years (International Energy Agency, 2019). This stagnation is attributed to the availability of low-cost natural gas in many locations, which is being paired with wind and solar energy at the expense of traditionally baseload generators, including nuclear generation. In the cases described renewable generators are connected directly to the grid, resulting in loosely coupled generation sources. This chapter considers an alternate scenario in which renewable generation would be tightly coupled with the nuclear generation source—behind the grid—to meet the grid demand as an integrated energy system while simultaneously producing other commodities with the available thermal energy.

Increasing the penetration of clean, affordable, reliable, secure, and resilient energy sources on electrical grids around the world can be accomplished by progressively establishing tightly coupled systems of distributed, dispatchable power generation assets that include a high penetration of variable renewable resources, and energy storage (thermal, electrical, and/or chemical) should be considered. Optimization and integration of these more complex and interactive power systems will require new technology with new approaches to deliver the optimized energy services across local, regional, and national boundaries. Recent advances in control systems, energy management systems, advanced informatics, and forecasting enable innovations in integrated plant design.

13.1.1 Definition of a "hybrid" energy system

Fig. 13.1 provides an overview of a generic tightly coupled hybrid energy system (HES) as a basis for discussion. Note that a simplified version of the illustrated HES could incorporate just a single input with multiple outputs (cogeneration) or could simply integrate two input sources for a single output stream (e.g., electricity). The latter single output scenario would require load-following operation of at least one of the production sources. In this scenario the thermal energy produced by the baseload source is not optimally employed, nor is the full benefit of the financial investment realized. The latter scenario is not considered a desirable configuration, but cogeneration may offer a desirable early implementation of a tightly coupled energy system without the added complexity of multiple energy sources, one of which is highly variable.

Industrial thermal energy demands are often met using dedicated thermal generation plants, most often fueled by coal or natural gas combustion. The current electricity grid employs diverse energy resources to meet demands for electricity. However, resources supporting the electricity grid are only very loosely coupled in most of the

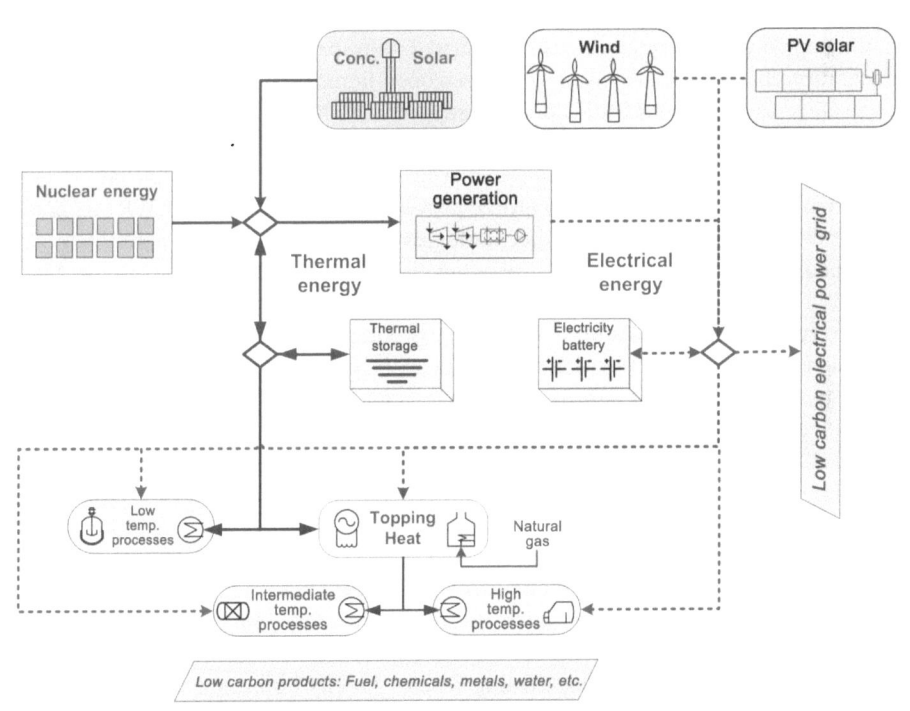

Fig. 13.1 Generalized tightly coupled nuclear hybrid energy system; the illustrated configuration would produce electricity and consumer products via process heat application. Some electricity produced may also be directed to the thermal energy application, and the coupled process may require additional feed sources (e.g., natural gas as a carbon feedstock). Topping heat may or may not be necessary for intermediate and high-temperature processes as a function of the outlet temperature of the selected nuclear reactor technology.

present implementations; each production source connects to the grid individually, and their relative input is managed on the large-scale grid as a whole. In a tightly coupled hybrid configuration, these subsystems would be coupled *behind* the electrical grid both to meet external grid demands and to meet thermal and electrical demands internal to the integrated energy system. This definition of a hybrid system, sometimes referred to as an integrated energy park, is assumed for the remainder of this chapter.

13.1.2 Key features of SMRs

SMRs are uniquely suited to tightly coupled, integrated energy applications. SMRs are distinguished by their relatively small power production (10–100 s of megawatts electric) and design for inherent, passive safety. Very small modular reactors, also referred to as microreactors, that produce less than 1 MWe up to 10 s of MWe may also be suitable for very small grids (i.e., microgrids) or small thermal energy demands. Plants could incorporate multiple SMR units at these production levels such that they can more easily be sized to meet the specific end-user demand for the output streams (e.g., electricity and thermal input to a process application) or to maximize plant thermal efficiency. The smaller per-unit size offers increased flexibility for investors (lower initial capital outlay), reduces costs associated with load-balancing, eases siting and integration challenges, and ensures increased operational flexibility. The inherent, passive safety designed into SMR concepts supports the NHES goals of system safety, resiliency, and environmental stewardship by minimizing the potential for negative consequences (e.g., radiological release) of a design basis or beyond design basis event.

Potential hybrid systems could utilize proven light water reactor (LWR) technology or proposed advanced reactor technologies that would operate at higher temperature and, hence, provide higher temperature heat for nonelectrical applications. Most currently operating LWRs produce on the order of gigawatts (GW) of electricity. Retrofit or repurposing of existing LWRs to incorporate a nonelectric output stream is considered among the potential applications of NHES technology. For example, LWR power plants in the US Midwest have been evaluated for technical and economic feasibility of the on-site production of hydrogen (Frick et al., 2019; Boardman et al., 2019). Results of these studies have indicated sufficient economic benefits that they are proceeding to demonstration. Repurposing of existing LWR plants could offer opportunity for extending the life of operating nuclear power plants that are currently experiencing the effects of competition from low-cost natural gas [potentially resulting in plant shutdown before license expiration (Szilard et al., 2017)] and increased grid penetration by renewable energy production sources that have been supported by government incentives and are benefitting from decreasing costs. However, repurposing an existing reactor facility could introduce challenges and hurdles in the license amendment process that could deter investment given the limited remaining plant life.

Many reactor designs currently considered for NHES fall into the category of SMRs (<300 MW electric power), as these plant sizes couple well with many of the process applications considered and are an excellent fit for small-scale regional grids or isolated industrial applications that demand both thermal and electrical resources. Most currently operating electrical generation plants are less than 500 MWe in capacity, so an integrated energy system on this scale could be envisioned to replace aging plants, particularly aging coal plants that have significant CO_2 emissions. Significantly smaller systems (~50 MWe) could be a good fit for NHES coupled with wind generation, as this is the approximate capacity of most individual wind farms. Because individual wind farms are affected by regional storm systems, the collective change in generation at the grid level can be on the order of GW in several tens of minutes. This magnitude of fluctuation can be challenging to accommodate, even for natural gas-fired peaking plants.

Many of the SMR plant concepts would ultimately incorporate multiple units. For such an implementation, additional capacity can be added incrementally, with individual units built in phases as necessary to meet growth in market demand. These units could be operated independently or in concert as a group depending on the overarching control strategy. Modular buildout improves the financial investment profile of the overall project, where the plant owner could choose to first construct the basic plant units (e.g., nuclear generation, energy conversion system, and electrical power generation) to establish a revenue stream, while the remainder of the plant is completed, building in the necessary interconnection points and control system structure to allow later addition of additional generation sources (e.g., renewable energy system or additional nuclear units) and thermal energy applications.

Small-scale, modular reactors incorporate significantly smaller components than large-scale plants, such that they can be factory built. Large system components for traditional, large-scale baseload nuclear plants are often built on-site and are reliant on foreign suppliers. SMR component factories could utilize a domestic supply chain and could be sited very near to the intended plant site, or components could be easily transported to the intended plant location. One might even envision a future hybrid system implementation powering a domestic SMR component factory.

Modular construction also allows alternate operational scenarios and integrated system control strategies than would be possible for a hybrid system that incorporates a single large-scale nuclear plant. In a multiunit plant in which each of those units provides a modest amount of thermal energy input, some of the input units could be dedicated to a particular output application. Other units could then be designated as "swing plants" that switch output between applications as necessary based on customer demand, economic factors, required maintenance, or refueling activities.

The siting of an SMR plant having one or more nuclear units is significantly more flexible than traditional large-scale plants. The possibility to locate SMRs in densely populated regions (due to reduced emergency planning zone) introduces the opportunity to site the plant closer to the final customer. The estimated US land availability (suitability) for small-scale versus traditional large-scale nuclear plants is discussed in Section 13.3.2. In a hybrid implementation, siting flexibility translates to siting of the industrial heat-use application near those population centers as well. By producing nonelectricity products (heat, chemicals, etc.) near the point of use, the economical attractiveness of the planned facility is increased, and the market size is enlarged (particularly as aging coal plants require replacement).

Smart grids could enable the implementation of smaller input sources, such as SMRs, by balancing the load dynamics at a local scale rather than at the large scale required by traditional, large-scale nuclear plants. In this case SMR plants could be located based on other (nongrid) subsystem requirements. Siting could be in the vicinity of the process feedstock resource (e.g., coal, natural gas, and biomass), near the end user (e.g., local community or commercial industry), or near the coupled renewable input source. Such siting would reduce transport distances for both electricity and thermal energy, thereby minimizing transmission losses. Hence, SMRs offer operational flexibility by introducing a broad range of production opportunities and simplified coupling to renewable sources and to more process applications than large-scale nuclear implementations.

Multiple deployment opportunities can be envisioned for NHESs, particularly those utilizing small modular reactors. Early implementations might provide electricity and thermal energy for independent industrial implementations without an intention to connect to the main electrical grid, allowing the system to be optimized based only on internal energy demands that are likely more predictable than external demand from the grid. Alternately an early hybrid energy park could provide electricity and heat to small, remote communities that currently rely on diesel generators that rely on regular delivery of diesel fuel trucked into the region. Later implementations might integrate the hybrid system directly to the large-scale grid while internally managing the thermal and electrical energy resources to meet the grid demand and maximize economic return. Considerations for potential hybrid system architectures are discussed throughout this chapter. The specific needs (desired commodities) of a potential customer and resources located at the intended site will aid in the design of an optimal energy system.

13.2 Principles of HESs

The current electrical grid is diversified—many input sources contribute to meet the ever-varying demands for electricity. However, in this loosely coupled, large-scale energy system, grid priority for renewables can drive baseload, dispatchable power systems to follow the remainder of the grid demand by reducing power. Although new nuclear power plants can likely operate at a power level as low as 25% of their rated capacity (older plants may be limited to something closer to 50% of rated capacity), this is not a desirable operating mode from the perspective of optimally using the invested capital and fuel loading in these plants. Baseload power plants were not designed for an operational mode that incorporates significant power cycling. Cycling of either baseload nuclear or coal plants to accommodate the variable demand that results from the introduction of highly intermittent resources on the grid can result in additional wear and tear on the plant systems, potentially increasing operations and maintenance costs and potentially shortening the life of the plant. The impact of LWR plant cycling is being assessed by the Electric Power Research Institute (EPRI) Flexible Power Operations Technical Advisory Group based on the experience of multiple plants operated by EPRI member utilities (EPRI, 2017). Although no significant impact has yet been observed, decreased plant revenue certainly impacts the economic viability of the plant.

A truly "hybrid" system would be tightly coupled, requiring individual subsystems to be operated in an integrated fashion to respond appropriately to grid-level transients while optimizing the energy use and minimizing cycling within the integrated system. This generalized architecture description then begs the question of what subsystems make sense in an integrated system and the optimal capacity for each subsystem. Using established performance criteria the vast possibilities for hybridization must be narrowed to solutions that are technically feasible and can efficiently and reliably meet the need for a selected region or industrial application.

13.2.1 Potential nuclear architectures

Incorporating a fission-based power source in a multioutput system (e.g., electricity and process heat) can offer significant advantages over carbon-based production sources, such as coal or natural gas, including reduced atmospheric waste streams (e.g., carbon or other gaseous emissions) and reduced impact on environmental resources (e.g., land usage or modification, permanent withdrawals of fresh water, and thermal emissions). In an integrated multioutput system, thermal energy from the nuclear subsystem can be diverted to industrial applications in times of low electricity demand from the grid or during times of high renewable energy input to electricity generation. High-temperature, high-quality heat from advanced reactor concepts might be used for high-temperature industrial processes, such as hydrogen production or synthetic fuels production. Low-temperature heat from either advanced or LWR systems could be applied to district heating, desalination processes, or low-temperature biomass pretreatment and ethanol production. Subcritical steam can also be superheated through process heat recuperation, chemical heat pumps, or topping heat prior to being directed toward high-temperature heat applications, allowing even LWRs to compete in these markets. Research to advance proposed heat augmentation techniques is ongoing (Sabharwall et al., 2013; Satmon et al., 2017).

For the current discussion a dynamic NHES describes an integrated energy complex composed of one or more nuclear reactors coupled to renewable power generation sources (e.g., wind, solar, and geothermal) and possibly linked to the production of one or more chemicals, fuels, or commodity manufacturing plants. Various exchanges between thermal, electrical, mechanical, and chemical energy could make it possible to produce, store, and deliver the highest value products to the market at the right time. Hydrogen, for example, can be generated by intermittent thermal/electrical output of a power plant instead of reducing boiler output during periods of reduced power demand on the grid. This hydrogen could supply either a captive or merchant market, depending on geographic location and market factors. Alternatively, hydrogen could be stored and dispensed to an emerging fleet of hydrogen fuel-cell vehicles, to generate power for a microgrid using a set of solid oxide fuel cells, or to firm up the variable power output from the wind or solar power plant. Other process-oriented heat applications for synthetic fuels and chemical production have been developed and evaluated by the Idaho National Laboratory (INL) in support of the Industrial Alliance for the next-generation nuclear plant (NGNP) (Nelson et al., 2011a,b) and more recently in support of the Department of Energy Office of Nuclear Energy (DOE-NE) Integrated Energy Systems (IES) program within Crosscutting Technologies Development and the Light Water Reactor Sustainability (LWRS) program (Frick et al., 2019; Boardman et al., 2019; Epiney et al., 2019).

A multiinput and multioutput system is the ultimate goal for a nuclear hybrid energy park architecture. These systems would be composed of two or more energy conversion subsystems that are traditionally separate but would be physically coupled in a hybrid system to produce outputs by dynamically integrating energy and material flows among energy production and delivery systems. These couplings would occur behind the electrical transmission bus such that all subsystems share interconnections and the system would be operated via a unified control system resulting in a single, highly dynamic, and responsive system that interacts with the electrical grid.

13.2.2 System efficiency through "load-dynamic" operation

A NHES would operate *dynamically*. In this sense the system could be considered "load following," in that the electricity produced would match the grid demand; however, the system is perhaps better described as "load dynamic." Rather than modify the system production to match grid demand by varying the reactor power level (which would describe a load-following *reactor*), the system would be designed and controlled to maintain reactor power level while shifting output from the grid to the alternate output stream—such that the reactor itself is "load dynamic." Hence a nominally baseload nuclear power plant would be used flexibly within the system to produce electricity to mitigate either variable renewable input to the integrated system or variable electricity demand from the grid. Increased efficiency can thus be achieved by matching electrical output to demand while utilizing excess generation capacity for the production of other commodities when it is available or storing it for later use.

The envisioned system operation offers a new approach to system optimization from a financial perspective, resulting from the opportunity to produce revenue from a variety of product streams while avoiding inefficiencies of underutilized capital. This approach recognizes and takes advantage of the benefits of operating a nuclear power plant at its nameplate capacity in a steady-state fashion to reduce wear on nuclear system components. It allows the nuclear plant to operate at a high capacity factor to recoup the relatively high cost of capital compared with fossil fuel plants while taking advantage of the relatively low recurring cost of uranium fuel sources.

13.3 Evaluating the merit of proposed hybrid system architectures

Evaluation of a proposed HES architecture should include assessment of performance or attractiveness in several broad areas, including (1) technical feasibility, (2) overall system economics, (3) environmental impacts, (4) production reliability/on-demand availability, (5) system resiliency, (6) system security, and (7) overall public or political attractiveness. To understand and evaluate the potential merits of a tightly coupled hybrid system, it is necessary to define characteristics that exemplify "good" system architecture.

13.3.1 Technical feasibility

Technical feasibility is, in the most basic sense, defined by system operability, measured in terms of online operating capacity factor and thermal efficiency (as a percent of the energy resources converted to a more useable energy currency or product). Power plants typically aim to achieve at least an 85% capacity factor, in consideration of grid demand cycles and seasonal plant outages for equipment maintenance. The capacity factor for the current fleet of LWRs in the United States has ranged from approximately 86% to over 93% (over the entire US nuclear fleet) from 2006 to 2019 (NEI, 2020). The chemical industry strives to establish the highest capacity factor possible, sometimes reaching 98%–99% of nameplate capacity for annual operation of the plant. Subsidized mandates (viz., production tax credits) to build wind power have decreased the capacity factor of many baseload power plants, including some nuclear plants, to around 50%.

Some key technical parameters that affect the feasibility of a given combination of subsystems in a hybridized plant include (from the perspective of the reactor) reactor outlet temperature (ROT), reactor inlet temperature (RIT), heat flux, heat capacity, and peak operating temperature (and, in some cases, minimum operating temperature) of either the thermal hydraulic system materials or the heat transport fluid. These characteristics dictate materials of construction, determine the optimum power generation cycle, and dictate what process applications may best couple with reactor-produced thermal energy.

When the operating schedules of two or more subsystems are codependent, then plant design and operating schedules must account for startup and synchronization of all subsystems. Energy storage subsystems that may be coupled on both the thermal and electrical branches of the system may be necessary buffered to ensure smooth shutdown of one or the other facility in the event of an off-normal interruption of operations or for planned maintenance. The feasibility of cycling unit operations also needs to be considered, taking into account thermal energy production ramp rates, lag in energy delivery systems, the effects of mechanical and thermal stresses, and electrical hysteresis effects on battery storage units.

Technical feasibility is also impacted by the ability to modularize unit operations. As discussed in Section 13.1.2, SMRs are ideally suited (and designed for) modularization. Multiple modules in a single plant add complexity to the physical integration, system operations, and control. However, modular implementations also offer the opportunity to increase capacity more easily with increasing demand, to attain higher integrated system capacity factors while allowing individual subsystems to be taken offline as needed for maintenance or refueling, and provide additional operational flexibility based on changing market trends.

13.3.2 Overall system economics

Economic assessment of a given NHES architecture should involve a number of inputs, including market size, profitability, total capital investment, operations and maintenance costs, and manufacturing costs (which include fuel and other fixed or variable costs). The overall system economics vary when additional revenue-producing subsystems are integrated with the SMR. Economic analysis becomes particularly complex when considering the time variability in the cost of electricity, which is also dependent on the production source (taking into account any grid priority and feed-in tariffs for renewables), variability in electricity demand, potential future carbon taxes, and revenue from nonelectricity products (e.g., methanol).

Private industry requires sound economic analysis and reasonable assurance of value—both present and future—before adopting and implementing innovative energy generation and dissemination strategies. Achieving a low uncertainty estimation on the economic return of hybridized SMRs requires additional research and development activities for some of the proposed subsystems (e.g., advanced, nonwater-cooled reactor designs) but more specifically for the functional integration of those subsystems, system monitoring technology, and control system architecture. Even in the case of a hybrid system architecture that employs off-the-shelf subsystems with demonstrated performance history, the integration of those subsystems is a departure from the known operational space that introduces economic uncertainty and risk. Some of the considered subsystems operate on very different timescales and with very different characteristics, requiring demonstration of system interaction under transient conditions to verify that the physical interfaces (hardware coupling), control system hierarchy, and control implementation are functional across the range of operating conditions.

13.3.3 Environmental impacts

A clear objective of NHES is to better utilize our natural resources, reducing withdrawals from the environment and impact on the environment due to waste stream emissions. One aspect of this environmental stewardship is the reduction of greenhouse gas (GHG) emissions. This is commonly measured in terms of the equivalent amount of CO_2 emitted when all GHGs are summed and adjusted based on their energy adsorption capacity.

Total water withdrawal for an ecosystem is becoming increasingly important, especially in consideration of the impact of erratic weather patterns and atmospheric temperature excursions that affect the available surface water at a given location. In arid climates where surface waters are practically nonexistent or where ground water withdrawals have already drained subterranean aquifers, water conservation is an essential requirement. Higher temperature advanced reactors (nonwater cooled) operate at higher efficiencies than traditional LWR systems, resulting in less waste heat that must be rejected to the environment. This not only reduces the thermal output to the environment but also reduces the water loss in the condenser cooling loop.

Land withdrawals are defined as permanent alteration to land that results from installation of the facility, such as modification of mountain tops to install wind capacity, disposal of coal ash waste that results from coal combustion (typically dumped into landfills or ash ponds near coal-fired power plants), or clearing land based on the required plant footprint and associated emergency planning zone. These modifications can impact an ecosystem just as much as waste emissions, including long-term life-cycle impacts on resident species. The adverse effects of rejecting waste heat to the environment are an additional environmental consideration.

Environmental impacts could be due to resource or feedstock extraction or production; impact of emissions, including both gaseous emissions and rejection of high-temperature water; or impact of energy transmission requirements. Electricity production located at some distance from the end user may require the installation of a new long-distance high-voltage power lines in some regions, but rail transportation of an energy *product* (e.g., hydrogen, synthetic fuels) might have fewer regulatory constraints and lower environmental impact, although such transport may require additional steps (e.g., H_2 compression).

Environmental stewardship requires that the buildout of modern energy infrastructure consider the long-term outlook for scarce resources, such as water, land, and carbon feedstock. Unfortunately the long-term horizon is often overlooked or underemphasized. Evaluation and design of tightly coupled, small-scale HESs focus on selecting system configurations that are environmentally responsible while meeting the demand for a variety of energy commodities.

13.3.4 Production reliability

Reliability impacts the economic performance of a plant by increasing the capacity factor and avoiding large costs that operators face for sudden loss of energy supply (both electricity and heat). Furthermore, customers have come to expect on-demand availability of high-quality energy products. Therefore energy systems should be judged on their ability to respond to market demands, on a minute, hourly, daily, weekly, and seasonal basis. A hybrid system must be capable of making temporal adjustments on timescales that ensure power quality standards on the grid (i.e., power factor, voltage, frequency, and phase); other hybrid system products can be selected based on the ability of the production subsystem to handle variation in production rates and schedules. One of the key advantages that a hybrid system could offer is the additional possibility to redirect energy production to more lucrative products given market situation changes in a short time while continuing to meet grid demand. This flexibility provides a clear benefit in terms of economic performance and capability to cope with fluctuations in demand.

13.3.5 System resiliency and sustainability

The long-term resiliency and sustainability of a hybrid system are important but difficult attributes to quantify. NHES that can adapt and compete in an evolving market should account for potential future risks and opportunities. This type of predictive analysis, while based on less certain assumptions, is key to making sound investment decisions that will influence future technology implementations.

Classes of risks and opportunities that should be considered include the following:

(a) Changes in the regulatory framework, with particular attention to environmental protection.
(b) Volatility and long-term availability of raw supplies (e.g., fuel, process application feedstock, and chemical compounds).
(c) Reliance of a community or industry on energy generation or chemical production means exposure to supply chain risks.
(d) Scarcity in the raw material supply or high fluctuation in feedstock prices that subsequently give rise to volatility in the market and ultimately translates to higher overall prices for the final products.

Generally, these situations represent very low probability events, such as scarcity of oil or gas supply, but the consequences of such a shortage could be so dramatic that the overall risk is not negligible.

Portfolio diversification may be important to the long-term sustainability of an energy solution. A hybrid system may offer opportunity for diversification of the current supplier set for the commodities produced by the system. The need for diversification applies to the broad scope of energy commodities, such as providing an economically viable, domestic option for transportation fuels, or more narrowly to diversifying the energy source for electricity production. Diversification is achieved only if the compared energy sources are not correlated. Successful diversification increases the long-term sustainability of the candidate system and may be crucial to the stabilization of long-term energy prices.

In many cases advanced SMRs represent an alternative route to deliver the same products as are currently available (e.g., methanol) by means of different starting materials, contributing to stabilization of the final product by positive diversification. Low sensitivity of system efficiency to the ultimate heat distribution (end use) offers system versatility, acting as risk mitigation for unforeseen changes in the definition of the system boundary condition (e.g., variation of the demand volume and heat-to-energy ratio or heat usage).

Hybrid systems naturally diversify customers. For example, if a much cheaper source of electricity is identified by a customer, reducing the electricity demand from the NHES plant, the NHES owner/operator may be capable of adapting production to refocus on heat utilization processes to respond to the demand changes. This operating mode introduces stability in the economic performance and capability to exploit emerging or growing markets.

13.3.6 System security

System security refers to vulnerability to physical and cyberattacks, including vulnerability of both the installed facility and the supply chain (equipment feedstock). An integrated energy system based on domestically secure natural resources would be preferred over one having external dependencies due to the associated risk factors.

A secure, reliable control system must also be designed and demonstrated for complex NHES. A resilient control system will ensure that the integrated energy facility is safe from cyberattack. Robust control also ensures safe, predictable response of the integrated system to off-normal conditions, such as physical attack or natural phenomena including earthquake or flood.

13.3.7 Overall public or political acceptance

As with any large project, attention must be given to society values, such as an appetite for renewable energy, concerns for nuclear materials and waste management, proliferation concerns, and general safety issues. Public acceptance is influenced by perceived positive (or negative) impact of a new facility, industry, or system on the local community. This impact could be quantified in terms of the number of jobs created and sustained by the proposed project, or the annual gross domestic product increase attributed to the project. These and other community economic considerations have an important effect on public acceptance and support for a proposed project, related political decisions, zoning, and facility siting. SMRs offer the possibility to be located closer to population centers and the final customer. The small plant size, coupled with inherent safety systems, reduces the likelihood that a severe accident would occur, and the smaller inventory of radioactive material relative to large-scale LWRs reduces the potential for release in the event of an accident. This reduced risk is reflected in a reduced emergency planning zone, allowing these plants to be located closer to the energy user. The US NRC has approved an early site permit application by the Tennessee Valley Authority, which includes a reduced emergency planning zone based on a risk assessment for the NuScale SMR design (US NRC, 2020). By producing nonelectricity products (heat, chemicals, etc.) near the point of use, the economic attractiveness of the planned facility is increased, and the market size is enlarged (particularly as aging coal plants require replacement).

NHES presents a unique acceptance challenge in regulatory space. Colocation of nuclear and chemical subsystems that are subject to different licensing and regulatory bodies is new territory for regulators. Siting and operation of a nuclear plant require adherence to a cadre of regulations that define the necessary emergency planning zone, the minimum number of operators per reactor, and various other details. Chemical plants are similarly regulated with regard to emergency planning zone, emissions, etc. If implemented in their current form, the independent nuclear and chemical plant regulations would prohibit siting the plants within a single energy park facility. Joint permitting is essential to a tightly coupled hybrid facility, but it could also be a significant factor in gaining public and political acceptance. These challenges may entail development of a new licensing process with the US Nuclear Regulatory Commission and the appropriate chemical industry regulatory bodies to enact appropriate risk assessments that have not previously been considered. An interim solution could be independent siting of each subsystem with transmission lines crossing the site boundaries. However, this configuration would not offer significant improvement over the current hybrid grid configuration, as it would not allow for tight coupling and integrated control of the various subsystems.

13.4 The when, why, and how of SMR hybridization

Potential hybrid systems could utilize proven LWR technology or proposed advanced reactor technologies that would operate at higher temperature and, hence, provide higher-temperature heat for nonelectrical applications. Small modular reactors offer unique opportunities for hybrid system applications due to their small size, modular implementation, operational flexibility, and investment flexibility. Potential SMR hybrid implementations are explored below with a focus on system siting and nuclear-renewable integration. Detailed discussion of potential nonelectrical applications that could be integrated in such a system is provided in Section 13.5.

13.4.1 Emerging electricity markets

The emergence of "smart" electricity grids and small modular reactors may have significant impact on future electricity markets. Significant progress in data acquisition systems and information processing have led to smart grids capable of smoothing load/demand curves and integrating energy storage systems, allowing a baseload system to operate at a stable level for longer periods of time in a predictive fashion (based on anticipated grid demand determined from historical trends and current state estimators) rather than varying production based on fluctuations in renewable energy availability. These advances are also key to reliable operation of hybrid system architecture.

Smart grids could enable the implementation of smaller input sources, such as SMRs, by balancing the load dynamics at a local scale rather than at the large scale required by traditional, large-scale nuclear plants. In this case SMRs could be sited based on other subsystem requirements. Siting could be in the vicinity of the process feedstock (e.g., coal, natural gas, and biomass), near the end user (e.g., local community or commercial industry), or near the coupled renewable input source. Such siting would reduce transport distances for both electricity and thermal energy, thereby minimizing transmission losses. Hence, SMRs offer operational flexibility, not to mention the already discussed investment flexibility, by introducing a broad range of production opportunities and simplified coupling to renewable sources and coupling to many more process applications than large-scale nuclear implementations.

The 2012 OECD NEA report on nuclear and renewables (OECD NEA, 2012) makes several recommendations regarding full representation (transparency) of power generation costs at the system level and the need for low-carbon energy technologies in current and future energy markets; these recommendations are again upheld in the 2019 system cost study by NEA (OECD NEA, 2019). Specifically, this report points to the need for low-carbon technologies to supplement variable renewables as renewables continue to grow to a larger grid share. While the report goes on to note that nuclear energy can provide flexible, low-carbon backup capacity as the renewable penetration on the grid increases, the introduction of SMRs in a hybridized system takes this recommendation a step further. Direct integration of low-carbon, baseload, dispatchable energy (e.g., nuclear) with low-carbon renewables allows policy goals for reduced GHG emissions to be met while retaining high capacity factors for the baseload source via load-dynamic operation (operating at steady-state power to produce multiple products). The NEA report further recommends establishing flexibility at the system level for future low-carbon systems. Accomplishing this flexibility will require increased load-following capability in dispatchable energy sources, such as nuclear, expanded energy storage systems, and increased responsiveness to demand changes. Again the proposed HES configurations would meet this need at a local level via load-dynamic operation (vs load-following) while reliably meeting the grid demand.

13.4.2 Overview of SMR concepts considered for hybrid application

Multiple SMR concepts are currently being studied and are in various stages of development. These concepts range from advanced LWR concepts, specifically integral PWRs (iPWRs), to advanced concepts such as the high-temperature gas-cooled reactor (HTGR), fluoride salt-cooled high-temperature reactor (FHR), and liquid metal-cooled reactor (LMR) concepts, all of which were identified (in a conceptual sense) as advanced reactor concepts warranting further study by the Generation IV International Forum (GIF) at a meeting held in 2000 (OECD Nuclear Energy Agency, 2010). Microreactors (ranging in size from 100 s of kWe to tens of MWe) cooled by liquid metal heat pipes are also becoming prevalent among microreactor technology developers. The fundamental neutronic and thermal hydraulic differences that exist between a water-cooled, thermal spectrum reactor, cooled using either natural convection or forced convection, thermal spectrum reactors using alternate coolants (HTGR and FHR), and fast spectrum reactor concepts (LMR) could result in significant differences in technically feasible HES architectures and their associated system operating parameters. The operating temperature for each of these concepts is the primary driver in the available options for coupling to different process applications and establishes the potential need for temperature-boosting technologies. Key features for each of these reactor types are summarized in Table 13.1.

In defining operating strategies for an integrated hybrid system, which by definition should be designed to accommodate variability, it should be noted that the preferred approach is to operate the reactor or reactors at the nameplate power level over long periods of time (i.e., between standard maintenance and refueling cycles) to minimize stress on the system components, to maximize core life, and to maximize revenue. Cycling from approximately 65% to 100% is achievable for much of the core life of LWRs (EPRI, 2017), but cycling is not the standard approach to system operation. To maintain robust reactor performance at a steady power level, the anticipated cycling of power demand and the subsequent need to switch between production of electricity and direct use of thermal energy should be buffered from the reactor via the interface design. Buffering might be accomplished by use of energy storage systems and possibly be aided by selection of a reactor design that would have a

Table 13.1 Summary of reactor classes generally considered for SMR designs in hybrid implementations.

General reactor concept	Coolant	Reactor outlet temperature (°C)	Key features (SMR concept)
Light water reactor (LWR)	Water	~300°C	Integral pressurized water reactors (most designs) Multimodule plants Natural circulation emergency cooling (some concepts also include natural circulation for primary cooling as well)
Liquid metal reactor (LMR)	Sodium Lead Lead-bismuth	~500–550°C	Near atmospheric pressure Fast neutron spectrum High heat capacity coolant Faster response time than other SMR concepts
Fluoride salt cooled high-temperature reactor (FHR)	Fluoride salt: FLiBe primary, KF-ZrF$_4$ secondary	FOAK ~700°C NOAK ~850–1000°C	Near atmospheric pressure Solid fuel (e.g., TRISO) Increased operating temperature requires material advances (OECD Nuclear Energy Agency, 2010)
High-temperature gas-cooled reactor (HTGR)	Helium	FOAK ~750°C NOAK ~900–950°C	Graphite moderator Pebble bed or prismatic core (US NRC, 2020) High-pressure system (~6.4–9 MPa) Pressurized coolant Increased operating temperature requires material advances

FOAK, first-of-a-kind implementation (first reactor build); *NOAK*, *n*th-of-a-kind implementation.

protracted response time to changes on the secondary side of the system (e.g., one that is least coupled to the balance of plant). Significant analysis must be performed to assess the overall operational performance characteristics of reactor concepts that could fit within the NHES model, including system response times (e.g., as it is affected by reactivity feedback coefficients, thermal capacity of the coolant, and requirement for an intermediate heat exchanger), reactor coupling with the balance of plant, and possible load-leveling schemes.

13.4.3 System siting and resource integration

Increasing penetration of renewable sources requires rapid adaptation of the current energy infrastructure. Renewable energy comprises a heterogeneous set of options, but for illustrative purposes, this discussion focuses on wind and solar energy. These resources are considered mature and are being deployed at a significant rate in the United States and Europe. The National Renewable Energy Laboratory (NREL) maintains an interactive prospector for wind energy opportunities in the United States (NREL, 2020). A solar prospector tool was similarly developed by NREL for solar energy opportunities, but this tool was retired in 2016; working closely with NREL, the SolarAnywhere Data website was instead developed by Clean Power Research (Dise, 2016). INL similarly developed a virtual prospector tool for water resources, supporting potential future hydropower projects (Hall et al., 2006).

The NREL wind prospector tool (NREL, 2020) can be used to identify regions having significant wind generation potential in the United States, with the greatest potential in the Midwest and coastal regions. Similarly the SolarAnywhere tool (Dise, 2016) can be used to identify regions in the United States having the greatest solar energy potential, with the Southwest and Western regions of the United States having the highest solar intensity. Incorporation of large amounts of wind and solar energy into a utility system has required load-following or load-leveling capability to utilize these variable resources when they are available. Based on the output of the various projects, the average wind capacity factor in 2018 exceeded 40% among projects build in recent years, reaching a fleet-wide average of 35%. These capacity factors have been increasing from as low as 23.5% for projects installed from 1998 to 2001 to 41.9% for projects installed from 2014 to 2017 (Wiser and Bolinger, 2018). Capacity factors for utility-scale solar PV systems range widely, from 12.1% to 34.8% (median 25.2%), reflecting differences in insolation in each deployment region (Bolinger et al., 2019). While there is some cause and effect relationship between solar availability and demand (at least for hot, sunny days where the peak solar insolation coincides with the daily and summer peaks), there is no similar cause and effect relationship between wind availability and electricity demand.

Oak Ridge National Laboratory (ORNL) has developed a tool for evaluating siting options for new electrical generation using geographic information system (GIS) data sources and spatial modeling capabilities. Referred to as Oak Ridge Siting Analysis for power Generation Expansion (OR-SAGE), the objective of this tool was to use industry-accepted parameters for screening potential reactor sites and to apply GIS data sources and spatial modeling capabilities to evaluate suitability of those sites (Mays et al., 2012). The tool considers various input data, such as proximity of electrical transmission lines, population density, regional seismicity, water sources, protected lands, ground slope, hazards (landslide and flood plain), and proximity of external hazardous operations. The tool then establishes criteria for each parameter to assess the feasibility of siting a large or small reactor in a specific location. This method allows for broad regional siting studies to narrow down to regions of interest, candidate areas, candidate sites, and finally to a preferred site. Fig. 13.2 shows results from OR-SAGE identifying potential sites for large ($\sim$1600 MWe) and small

Fig. 13.2 Possible siting options for small and large reactors based on OR-SAGE analysis (Mays and Using, 2011). Analysis results indicate approximately 24% of the US land area would be suitable for siting of a small reactor, whereas 13% would be suitable for a large reactor. Courtesy G. Mays, ORNL.

(~350 MWe) reactors (new capacity). When overlaid with renewable resource availability, possible sites for a hybrid nuclear-renewable plant become evident.

13.4.4 Nuclear-renewable integration

Current electrical grids are not inherently well suited to accommodate high penetration of variable renewables. Three significant challenges exist:

(1) The geographical location of significant renewable energy resources are often removed from the major population centers, and hydropower has limited geographical options for additional buildout.
(2) High variability characterizes wind and solar generation resources, and hydropower is highly seasonal.
(3) The dispatchability of wind and solar resources is limited (meaning it cannot be throttled up or down on demand), requiring coupling to energy storage or other dispatchable resources.

For illustrative purposes, projected wind turbine generation rates for 1 week are shown in Fig. 13.3 for a wind farm in Wyoming where a 40% annual capacity factor has been practically realized. There are periods of high or low output that last for days, but significant changes in the generation rate can occur over a very short time (an hour or two). There are also periods of intermediate levels of generation.

There are several strategies for operating a hybridized power generation system to smooth the variability of wind generation. In the first and conceptually simplest, the primary heat source in the hybrid system (e.g., an iPWR or advanced SMR) and the

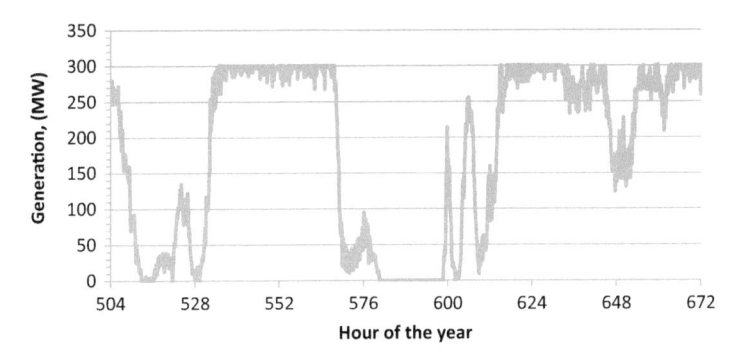

Fig. 13.3 Actual output of a wind farm with 300 MWe total installed capacity over a 7-day period at the end of January.

wind farm are operated together to produce a combined electrical output that is constant, essentially mimicking baseload generation. For the example wind farm shown in Fig. 13.3, this would correspond to making 300 MWe at all times, or as close to this objective as possible. This type of operation would allow the integrated hybrid system to replace an existing baseload 300 MWe coal-fired power plant that was shut down either by obsolescence or to avoid the cost of future CO_2 capture requirements or emissions penalties.

A difficulty with this approach is that because of wind's low availability, on the order of 30%–5% for regions considered attractive, the thermal generation source in the hybrid system must generate power for the balance of the operation, or 55%–70% of the total energy generated. This, in turn, means that the primary heat source of the hybrid system delivers heat to the coupled process application only when the wind blows, 30%–5% of the expected annual total, and during some of that time only reduced heat delivery is available. This level of operation would not be economical for the coupled process plant.

One alternative operating basis is to always supply a specified minimum amount of heat to the process plant, so its operating rate might vary between, for example, 70% and 100% of its nominal capacity as needed to offset wind fluctuations. This leads to exceptionally large process plants coupled to relatively small wind farms, both outside of desirable ranges. Alternatively an auxiliary fossil-fired steam generator (a 3rd input system) could provide heat to the process plant when nuclear heat is not available. However, it would have to run 55%–7% of the time, largely countering the CO_2 emission advantages of a nuclear-based hybrid system.

Another strategy is to address only the problematic high-frequency components of wind variability. Power grids are well adapted for handling the diurnal variations in demand, so wind variability on that timescale can be handled similarly using the same equipment as long as the wind capacity is no more than approximately the amount of diurnal cycling, typically about 30% of the daily peak. In this operating mode, by switching heat output between the power system and the process plant, the HES would generate electricity such that the total of the wind-derived and the nuclear-derived electricity does not vary faster than a specified rate. The ability of an SMR hybrid system to compensate for rapid changes in wind generation depends on how fast steam can be switched between the electrical generation and process plants, not on how fast the reactor itself can respond to a transient. In the case of a rapid drop in wind generation, the electric output from the nuclear plant would rise rapidly to moderate the drop in wind generation.

The following section addresses possible applications that could be coupled to a HES to maximize the use of the thermal energy generated while still meeting the electricity demand. While some industrial processes would utilize only thermal energy, others require both thermal and electrical energy input from the integrated system.

13.5 Coupling reactor thermal output to nonelectric applications

A primary motive for nuclear-renewable HESs is the efficient alternative use of the generated heat when it is not needed for electric power production due to low net demand conditions. The proposed load-dynamic behavior of a system that incorporates variable renewable power generation introduces system complexities as a result of timing (when the heat is available), timescales (required response rate), and the large amount of excess heat that must be diverted to industrial applications. Industrial processes can potentially be designed to absorb the heat at timescales aligned with heat availability. Operating temperatures for each of the "classes" of proposed SMRs result in different options for coupling to the selected process applications (McKellar, 2010a). Table 13.2 provides a brief overview of possible process applications and the corresponding operating temperatures.

13.5.1 General considerations

A number of integration issues must be considered when coupling process heat applications to nuclear reactors. Key considerations include the following:

- reactor outlet temperature
- reactor inlet temperature
- fluid composition
- pressure of primary coolant, heat transfer loop, and process heat application
- primary coolant heat capacity
- tritium migration or other radionuclide contamination

The ROT defines the temperature and heat that can be provided to the process heat application. Low-temperature reactors can still provide process heat to most applications but will require temperature amplification via heat recuperation or high-temperature topping heat from fossil fuels and/or electric resistance heating to couple with higher-temperature process heat applications. GHG emissions can still be reduced relative to traditional generation sources in these cases. Process heat application studies under the NGNP program found that a ROT of 825–850°C would provide sufficient heat to handle most applications. However, coupling a high-temperature

Table 13.2 Mapping of process applications to potential energy sources based on approximate process temperature requirements.

Selected processes	Process temperature (°C)	Candidate reactor class for heat input
District heating Seawater desalination Feedstock drying, minerals concentration	80–200	LWR or waste heat from higher temperature reactor
Petroleum refining	250–550	LWR, LMR, FHR
Oil shale and oil sands processing	300–600	LMR, FHR
Cogeneration of electricity and steam	350–800	LMR, FHR, HTGR, LWR with heat augmentation
Steam reforming of natural gas or coal (methanol production); synthetic gas production	500–900	LMR, FHR, HTGR, LWR with heat augmentation
Hydrogen production via HTSE Hydrogen production via steam methane reforming Calcination Coal gasification	800–1000	FHR, HTGR, LMR, or LWR with heat augmentation

Note that some of the higher-temperature processes could be accomplished with lower-temperature input heat sources via temperature amplification steps.

reactor to a low-temperature process heat application can reduce the overall process efficiency due to inefficient use of the high-temperature heat. If the heat required by the process is a sufficiently low temperature, then utilization of waste heat from a high-temperature reactor could be a viable design option. Optimal coupling of subsystems based on temperature outlet/inlet requirements may result in the most efficient use of the reactor thermal output, but regional needs may sway the selection of a system configuration based on economic factors.

Each generalized reactor type has a specified temperature difference (rise) and pressure difference (drop) across the core. For instance a helium-cooled HTGR has a nominal temperature rise of 350–400°C, while the core temperature rise for a fluoride salt-cooled high-temperature reactor is on the order of 100–150°C. The temperature of the heat transfer fluid returned from the process heat application must be at or below the RIT. If it is at a higher temperature, heat must either be rejected to the environment or used in an additional process (i.e., power production using a bottoming cycle). A smaller core temperature rise corresponds to a higher-average temperature process heat than designs having a larger core temperature rise, generally leading to a reduction in the number of reactors needed to provide the high-temperature heat.

The coolant and secondary heat transfer loop fluid composition can affect which process heat applications may be integrated. The melting temperature of molten salts and liquid metals (e.g., sodium) must be considered when applied to process heat applications. Some applications would return the working fluid at temperatures below the coolant solidification temperatures, which could plug heat exchangers and piping without careful design of the heat transfer interfaces and potentially boosting the temperature of the return fluid. Coolants such as water, steam, carbon dioxide, and helium do not have such issues, although condensation must be considered.

Process heat applications require heat not only at specific temperatures but also specific pressures. Low-pressure coolants, such as liquid metals and molten salts, provide ideal pressure conditions at the core, but large pressure differences may occur at the process heat exchangers where high temperature and pressure differences must be considered in the heat exchanger design. High-pressure coolants, such as helium and pressurized water, may have the same effect if used for low-pressure process heat applications.

Process heat applications generally require a secondary heat transfer loop to isolate the primary core coolant loop from the process heat application. A primary purpose of this secondary loop is to reduce the migration of tritium from the reactor core to the process application. Additional heat transfer loops may be added to further reduce potential tritium migration, but each loop reduces the process heat temperature based on the temperature difference required to transfer heat across the additional heat exchangers. Isolation via heat transfer loops also reduces any potential for radioactive contamination of the end product under off-normal conditions (e.g., in the event of a steam generator tube rupture).

Some process heat applications use the heat rejected from the power conversion cycle of the reactor. Low-temperature applications, such as thermally driven seawater desalination and district heating, may be able to use the heat rejected from the condenser of a Rankine cycle. The temperatures needed for these applications are higher than the usual rejection heat temperature such that some heat augmentation may be required. Heat may alternately be extracted from the low-pressure turbines at a higher temperature; this reduces the electric power generated, but the overall system thermal efficiency is increased due to the additional heat utilization.

13.5.2 Overview of process heat applications

A number of process heat applications have been considered for hybrid systems integration. Many of these applications have been analyzed under the NGNP and advanced reactor concepts programs (Nelson et al., 2011a,b; McKellar, 2010b; Wood, 2010a,b,c, 2012; Gandrik, 2011, 2012; Robertson, 2011a,b) for cogeneration applications with the HTGR concept. Other applications are currently being addressed via the DOE-NE IES and LWRS programs (Frick et al., 2019; Boardman et al., 2019; Epiney et al., 2019). Key applications of interest to NGNP included the following:

- hydrogen production
    - high-temperature steam electrolysis (HTSE)
    - steam methane reforming
- coal-to-gasoline production via methanol production
- natural gas-to-gasoline production via methanol production
- coal-to-diesel production via Fischer-Tropsch synthesis

- natural gas-to-diesel production via Fischer-Tropsch synthesis
- ammonia production
  - hydrogen and nitrogen from steam methane reforming
  - hydrogen and nitrogen from HTSE
  - hydrogen from HTSE and nitrogen from air separation unit
- steam-assisted gravity drainage applied to oil sands
- oil shale processing (ex situ or in situ)
- olefins via methanol production
- seawater desalination
  - reverse osmosis
  - multistage flash distillation
  - multieffect distillation

Small nuclear reactors are already used for district heating applications in some remote locations. Other possible applications that could be coupled in a hybrid configuration that were not evaluated under NGNP could include metal refining and hydrogen production via thermal chemical water splitting. The DOE-NE IES and LWRS programs have evaluated the use of LWRs to support HTSE and water desalination via reverse osmosis.

13.5.2.1 Hydrogen production

Hydrogen can be produced through several processes, including reforming technologies and water-splitting technologies. Production technologies briefly described here include low-temperature electrolysis (LTE), HTSE, and nuclear-integrated steam methane reforming. Hydrogen is a key element for making fuels and other industrial chemicals; recent studies suggest a potential 6- to 10-fold increase in the hydrogen market by 2050 (Ruth, 2019).

The primary hydrogen industry is currently making hydrogen from natural gas via steam reforming. Water and methane provide feedstock for the process in which some of the methane is used to make steam and the remainder is combined with the steam to create hydrogen and carbon dioxide. In the traditional case, natural gas would be used to drive the steam reforming to produce hydrogen. Reforming technologies generally produce hydrogen at the lowest cost due to the current low cost of fossil fuels, such as natural gas; hence, reforming provides a target price point for alternative carbon-free hydrogen production technologies. For the nuclear-integrated steam methane reforming case, process heat from the reactor would be added to the steam reformer (McKellar, 2010b; Wood, 2010a).

Hydrogen can also be produced using water-splitting technologies or electrolysis (HydroGEN, 2019). LTE is driven solely by electricity. This process is accomplished requires only electricity, either placing electrodes in an electrolytic solution or using membranes to separate the hydrogen from the oxygen. While LTE is commercially available and, hence, is a viable option for integrated systems, HTSE provides opportunity for increased efficiency in hydrogen production through coupling with a nuclear reactor. HTSE uses the high-temperature heat and electricity to split water into hydrogen and oxygen, where the additional heat reduces the amount of electrical work needed (Nelson et al., 2011b; Laurencin and Mougin, 2015; O'Brien et al., 2010). Solid oxide electrolysis cells are used to electrochemically separate the hydrogen and oxygen from steam at temperatures around 800°C. Although the temperature of the steam is high, LWRs can be used by employing heat augmentation techniques, such as resistive heating or chemical heat pumps.

Hydrogen is an important component for many industrial processes. The DOE Hydrogen at Scale (H2@Scale) program, managed by the Office of Energy Efficiency and Renewable Energy, is supporting early stage R&D on hydrogen production, transport, and storage to support the growing hydrogen market (HydroGEN, 2019). The conventional method of steam methane reforming is the least expensive hydrogen production method but produces carbon dioxide and requires constant operation. Hydrogen from electrolysis processes can be turned on and off readily and, for HESs, offers a means to integrate intermittent renewable resources. HTSE has higher production efficiency than LTE but requires some technology maturation to be ready for full commercial adoption. While thermal energy input is required for HTSE, heat recuperation within the electrolysis process greatly reduces the amount needed. Only 10%–15% of the heat from a reactor is needed to maintain HTSE temperatures. The remainder of the heat is converted to electricity to run the electrolysis cells. As long as the cells are maintained at the desired temperature, the electrolysis process can be turned on and off as needed; also, this must still be demonstrated to determine any limitations to cycling of the electrolysis cells. HTSE also has no carbon dioxide emissions.

13.5.2.2 Natural gas or coal to gasoline via methanol production

Synthetic gasoline (syngas) may be produced from natural gas or coal via the conventional methanol-gasoline process. These processes produce methanol as an intermediate product, synthetic gasoline and liquefied petroleum gas as end products, and significant GHG emissions. The conventional coal and natural gas cases differ in the approach to generate syngas; however, from the syngas to gasoline via methanol production, the processes are the same.

The nuclear heat integration cases for methanol to gasoline are slightly different in their integration points for the coal and natural gas feedstocks. For the coal-to-gasoline process, hydrogen from nuclear-driven HTSE is used in the gasification process. Power from the reactor is used for compression and sulfur removal. For the natural gas-to-gasoline process, nuclear heat is used for the reforming process, and electric power is used for compression. Nuclear integration provides significant reduction in carbon dioxide emissions relative to the conventional (nonnuclear) case (Nelson et al., 2011a; Gandrik, 2012; Wood, 2010b).

13.5.2.3 Coal and natural gas-to-diesel production via Fischer-Tropsch

Synthetic diesel may be produced from coal or natural gas through the Fischer-Tropsch process. The Fischer-Tropsch reaction converts syngas, hydrogen, and carbon dioxide to a liquid fuel by using iron- or cobalt-based catalysts. The processes for coal-to-diesel and natural gas-to-diesel differ both in syngas production and diesel production.

For the nuclear integrated cases, nuclear-generated heat and electricity are applied to the HTSE process to produce hydrogen for the coal case. Nuclear heat is applied to the reforming, CO_2 and sulfur removal, hydrotreating, and product upgrading processes for the natural gas case. Nuclear integration provides significant reduction in coal consumption and CO_2 emissions and modest decreases in natural gas consumption (Nelson et al., 2011a; Gandrik, 2012).

13.5.2.4 Ammonia production

Conventional ammonia production uses natural gas, steam, and air to produce hydrogen and nitrogen using a two-step steam methane reforming process. The ammonia is synthesized using the Bosch process and by adding carbon dioxide from the reforming process to produce ammonia derivatives such as urea, nitric acid, and ammonium nitrate.

Three types of nuclear-integrated ammonia production are envisioned. The first case adds nuclear heat in the sulfur removal process and the primary reformer. The nuclear subsystem also provides power for compression, refrigeration, fans, and expansion. The second process bypasses the reformer and uses nuclear-generated hydrogen via HTSE. The hydrogen is used directly in the ammonia production process and is also combusted with air to produce the nitrogen needed for the process. Power is also applied to compression, fans, and expanders as in the previous case. The final case again uses HTSE to produce hydrogen, but the nitrogen is provided by an air separation unit, which requires power from the reactor. Ammonia production with nuclear integration would provide modest to large decreases in natural gas consumption (depending on the specific case) and very significant decreases in CO_2 emissions (up to 99% reduction relative to traditional processes) (Nelson et al., 2011a; Wood, 2010c).

13.5.2.5 Water desalination

Conventional water desalination uses low-temperature heat to purify seawater. Three types of desalination are considered: multistage flash distillation, multieffect distillation, and reverse osmosis. The multistage flash distillation is a process where incoming seawater is pumped to a higher pressure and heated to near boiling. Through a series of stages, the seawater pressure is decreased to generate vapor that is condensed by the incoming seawater. The multieffect distillation process uses a steam heat source and a series of evaporators at successively lower pressures to create purified water. The reverse osmosis process uses membranes to separate the pure water from the brine. In this case only electricity is required to pump water through the membranes. The conventional water purification approach could use steam from the bottoming cycle of a natural gas combined power cycle as the steam source. For the nuclear-integrated cases, the steam would come from the low-pressure turbine. As stated previously the temperatures needed for water desalination are higher than the conventional heat rejection temperature of the power cycle. Therefore power generation would need to be decreased to compensate, but the steam extracted from this point in the cycle would be used for the desalination process. These combined trends would result in an overall increase in the efficiency of the combined system (Dise, 2016).

13.5.2.6 Steam-assisted gravity drainage

The recovery of bitumen from oil sands deposits using steam is called the steam-assisted gravity drainage (SAGD) process. Traditionally, steam is generated in a boiler fired by natural gas. In the nuclear-integrated case, the traditional natural gas combustion would be replaced by reactor-produced high-temperature steam. The steam is injected into the oil sands deposit where it heats the bitumen and allows it to be brought to the surface. Bitumen is blended with naphtha to produce dilbit, which is sent to a refinery for upgrading and conversion to transportation fuels and other petroleum products. In addition to dilbit, the conventional process generates significant GHG emissions (Nelson et al., 2011a; Gandrik, 2011).

13.5.2.7 Oil shale

The high-temperature conversion of kerogen, the organic portion of oil shale, into natural gas and shale oil in the subsurface is called the in situ retort process. Traditionally, steam is generated in a boiler fired by natural gas. A closed-loop injection-and-return piping system recirculates the steam from the boiler to the subsurface and back. In the subsurface the heat from the circulating steam is transferred by conduction through the pipe wall into the oil shale. As the shale heats the kerogen is converted to natural gas and shale oil, which are transported to the surface by the pressure generated during the conversion through production wells. A portion of the produced natural gas is used to generate steam in the gas-fired boiler. The CO_2 that is generated in the process is released to the atmosphere. A nuclear-integrated case would utilize reactor-derived heat in place of the natural gas combustor to dramatically decrease carbon dioxide emissions and dramatically increase natural gas production (Nelson et al., 2011b; Robertson, 2011a).

The high-temperature conversion of kerogen, the organic portion of oil shale, into natural gas and shale oil above ground is called the ex situ retort process. Oil shale ore is mined from locations near the retort kiln. All mining equipment and machinery is powered by electricity purchased from the grid. Mined ore is fed into the rotating, horizontally oriented kiln for retorting. The shale enters the processor and is heated from ambient temperature to about 500°C. During the retort process the kerogen decomposes into hydrocarbon gases, shale oil, and a carbonaceous residue called char. The combustion of the char, which occurs at approximately 750°C, provides the heat necessary for preheating and retorting the shale oil and releases CO_2. At that temperature the carbonate material in the shale substrate decomposes and releases additional CO_2. The CO_2 generated by the process is released to the atmosphere. Spent shale, which has been retorted and combusted, cools as it transfers its heat to the incoming ore and is further cooled by water spray as it exits the processor. When the condensable raw shale oil product leaves the processor kiln, it must be stabilized for transport via pipeline to a refinery. The oil is upgraded by hydrotreating, which lowers its pour point to acceptable pipeline limits and reduces the nitrogen and sulfur concentrations. The nuclear-integrated configuration would provide heat to the kiln and electricity throughout the process, significantly decreasing carbon dioxide emissions (Robertson, 2011b).

13.5.2.8 Olefins via methanol production

Olefins are made from a light olefin (ethylene and propylene). They are the backbone of the petrochemical market and are consumed in the production of plastics. The conventional process uses coal to make syngas that is converted into methanol. The methanol is used in the olefin synthesis process to create ethylene and propylene with a variety of natural gas components. The nuclear-integrated process would add heat and power to create hydrogen using HTSE, significantly decreasing carbon dioxide emissions (Wood, 2012).

13.5.3 Hybrid configuration selection and optimization

Preliminary, generalized analyses of proposed hybrid system architectures were conducted by researchers at INL in the early 2010s to provide initial insight into the value of system hybridization (see additional resources listed at the end of this chapter). More detailed analyses of tightly coupled HES require detailed subsystem models to accurately predict dynamic system performance and to identify potential operating procedures to best manage and mitigate variable energy generation sources and to alternate between output products (e.g., electricity and a chemical process). Detailed, dynamic subsystem models have supported more rigorous analyses for the use of current fleet LWRs for hydrogen production (Frick et al., 2019; Boardman et al., 2019) and water desalination via reverse osmosis (Epiney et al., 2019). The modeling, analysis, and optimization framework developed via the DOE-NE IES program is currently being enhanced to support a "plug and play" approach to evaluate a larger range of hybrid system options. This framework will support representation of various reactor types and sizes, multiple renewable energy generators, various power conversion systems and heat transfer fluids, energy storage options, and additional process applications (Bragg-Sitton et al., 2020a). As the tools for dynamic hybrid system analysis become more sophisticated and conceptual subsystems and integrated system designs are better defined, more accurate dynamic analyses include both technical and economic evaluation to support optimization of system design configuration, assess specific operating parameters, design integrated system control architecture, and establish instrumentation requirements for state estimation necessary for control.

13.6 Future trends

Expanded use of nuclear energy for industrial applications raises several technical challenges that can be addressed through conceptual design; system modeling and optimization; component, subsystem, and integrated system testing (nonnuclear and nuclear); and eventual deployment of the advanced energy system. Early work should be focused on identifying key challenges and then addressing those challenges via design solutions. A 2020 Roadmap for Integrated Energy Systems identifies many of these challenges across the various subsystems that could contribute to a deployed HES (Bragg-Sitton et al., 2020b). Advanced modeling and simulation tools can provide significant insight to system feasibility and anticipated performance, beginning with simplified steady-state analysis and moving to more complex dynamic analysis with more detailed subsystem models. The more detailed, dynamic system model can then be used as a virtual test bed for control system design prior to hardware implementation. However, many of the significant challenges associated with hybrid systems derive from complex integrated system control and hardware interfaces, including fast-switching smart valves necessary to divert thermal energy between output streams and connections to energy storage systems. These challenges can be addressed through a number of component and subsystem tests for model validation, followed by integrated system testing.

13.6.1 Steady-state and dynamic system modeling and simulation

Hybrid systems analyses are conducted in two stages: simplified steady-state analysis to determine technical feasibility of a candidate system architecture, followed by detailed dynamic analysis of promising integrated system configurations for system optimization, detailed performance analysis, and control system design.

Top-level system analyses are performed using steady-state modeling and analysis tools, such as Aspen Plus and HYSYS. These analyses help establish system and subsystem boundary conditions, including operational temperatures, pressures, and flow rates necessary to maintain overall energy balance, providing the essential information to determine if a proposed configuration is technically feasible. Introduction of preliminary economic analysis tools then allows a system designer to determine the potential economic value (return on investment) associated with the proposed integrated system configuration when deployed in a specific geographic location and energy market.

Dynamic system modeling integrates more detailed component and subsystem models, using validated component models and modeling tools where possible. INL has developed many of these models using Modelica, which are then integrated and assessed via custom built system optimization tools in the RAVEN environment (Bragg-Sitton et al., 2020a,b). For advanced subsystems, such as advanced SMR concepts employing nonwater coolants or novel heat exchanger designs, validated component models may not be available. In these instances the model developer should use reasonable assumptions for subsystem design parameters and performance behavior using validated modeling tools. Separate effects tests and subsystem testing may be used to later validate those assumptions.

Dynamic simulation of hybrid system performance is where some of the key questions regarding the reliability, resiliency, and efficiency of hybrid system operation can begin to be answered. From the perspective of the reactor subsystem, these questions might include the following:

- To what extent can the power of a single reactor module be varied to accommodate load variations? How can this modulation of reactor power level or power output be best accomplished?

- What benefits, if any, can be realized by incorporating multiple small reactor modules in the integrated system?
- What factors limit the rate and magnitude of increasing or decreasing reactor power, including the frequency or magnitude of the proposed fluctuations in system load, frequency, and voltage?
- If rapid response time is required by the balance-of-plant, what mechanisms can be used to buffer the reactor(s) from rapid transients, and how can these be implemented with high reliability?
- If steam off-take is required to support coupled processes, where is the most appropriate interface point considering operational and safety impacts?
- What steps are necessary to develop and demonstrate system-level control that will minimize individual reactor cycling?
- How will the overall system safety analysis be performed to reflect the relationship between reactor safety characteristics and the overall system safety?
- How can electricity demand changes be managed? That is, a decrease in electricity demand would trigger a switch from electricity generation to thermal power storage or direct use as thermal energy in the integrated industrial process. What type of interface (valving, controls, and operational strategies) might be required to divert reactor thermal power to the appropriate subsystem, and what is the characteristic response time associated with those components?

A well-defined dynamic simulation will provide an excellent platform for control system development. The hybrid system control must first establish control hierarchy, including priorities for electricity production (i.e., first meeting the electricity demand before considering other output streams) or allocation of the thermal energy based on the current market price for the output commodity. Second the control system requires specific state estimators (temperature, pressure, etc.) to provide input to the control algorithms, recognizing that optimal placement of instrumentation necessary to provide these data is critical to reliable control system performance. A validated simulation of the integrated system provides an excellent virtual test bed for sensitivity studies associated with the control system design and optimization of the control system architecture.

Results of the dynamic simulation will identify areas of significant uncertainty or significant sensitivity that will impact the prioritization of subsystem and integrated system testing. While modeling and simulation can provide a vast amount of understanding on the potential performance of an integrated system, there is great value in translating simulations to hardware demonstrations (particularly nonnuclear, electrically heated demonstrations) prior to building an integrated nuclear hybrid system prototype or demonstration system.

13.6.2 Component, subsystem, and integrated system testing

A tightly coupled integrated system will benefit substantially from demonstration via integrated test facility. Such a facility would allow for both physical and virtual representation of subsystem components, requiring either physical or data linkage between the subsystems. An integral test facility could support component testing, partially integrated system testing, and/or fully integrated system testing including interfacing with the electrical grid.

Significant scaled, nonnuclear, integrated system testing has been conducted for space nuclear power and propulsion systems (Bragg-Sitton et al., 2009, 2010; Bragg-Sitton, 2004), which in some cases are essentially scaled versions of the proposed terrestrial HESs. Low capacity fission power systems are currently being developed for space application via both computational modeling and experimental efforts. The National Aeronautics and Space Administration (NASA) and the Department of Energy Office of Nuclear Energy (DOE-NE) have adopted a hardware-based approach to system development intended for early identification of challenges to the system design, fabrication, assembly, and operation and to assist in design optimization. To minimize cost and development time, a nonnuclear test approach is used to demonstrate integrated system operation during the system design and development stage, offering opportunity for system reconfiguration without the radiological hazards associated with a fueled system.

For a typical NHES, initial hardware demonstration could similarly integrate a physical reactor simulator that uses electric heaters to mimic the heat generated by nuclear fuel pins. Operation of the reactor simulator would rely on a virtual component derived from system modeling to computationally simulate the neutronic response that would be observed in a fueled system using measured system temperatures as state estimators for the heater control logic. Once all the relevant feedback mechanisms are understood for a particular reactor design, appropriate instrumentation and measurement points can be selected for the nonnuclear test hardware such that the virtual reactivity feedback may be applied appropriately (Bragg-Sitton et al., 2010). Other system components could be represented by physical hardware or simulated computationally depending on the stage of facility development, the availability of a specified component or subsystem, or the relative experience with that subsystem that impacts the availability of validated computational models. Integrated system demonstration with hardware in the loop allows the researcher to evaluate integration challenges and to characterize system response time and response characteristics. "Virtual" reactor kinetics applied to the control system architecture can generate significant understanding at a system level but in a manner that allows for system reconfiguration, control system design and demonstration, operator training, test to failure, etc. without the added complexity and security inherent to a nuclear system.

Successful nonnuclear demonstration of a fully integrated NHES will provide a strong foundation for future build of a nuclear prototype. Several test facilities have been built and operated by potential SMR vendors for specific, focused purposes, including testing to characterize the performance of individual components and materials, to evaluate subsystem thermal hydraulics, and to identify and evaluate integral system effects. In many cases, these facilities are designed around a specific reactor concept and are intended for testing of traditional LWR designs, but others may be more generally applicable. Potentially relevant nonnuclear test facilities could include the various nonnuclear test facilities constructed to refine the design of the *Central Argentina de Elementos Modulares* (CAREM) reactor in Argentina, which began construction of a 25-MWe prototype nuclear unit in February 2014 (Nuclear Engineering International, 2018); Korean Atomic Energy Research Institute (KAERI) has built several facilities to verify and refine the design of the system-integrated modular advanced reactor (SMART) plant (IAEA, 2013; Chung et al., 2007); Babcock and Wilcox (B&W) constructed the integrated system test (IST) facility to verify the mPower reactor design and safety performance in support of NRC licensing activities (BWXT Center for Advanced Engineering and Research, 2015); and NuScale Power has designed and built a one-third scale, electrically heated prototype facility [the NuScale Integral System Test (NIST) facility] to demonstrate concept viability and stability (NuScale Power, 2020).

Study of the design of these test facilities may offer insight into design of an SMR HES test bed. Some of the existing test facilities might be leveraged to benchmark component and subsystem models that could be incorporated in hybrid system simulation. Additional facilities may be necessary to verify the performance or refine the design of specific interface components for hybrid systems, such as fast-switch valves that will be necessary to achieve the desired load-dynamic behavior in a tightly integrated system and buffer components, such as thermal energy and electrical energy storage technologies. Integral system testing with these components installed can then be conducted to evaluate the subsystem interactions across valves and energy storage devices to verify transient system behavior given both anticipated operational occurrences and transients associated with accident conditions.

In 2019 two private/public hydrogen production demonstration projects at currently operating nuclear plant sites were awarded through DOE Financial Assistance Funding Opportunities Announcements. A project led by FirstEnergy Services, Xcel Energy, and Arizona Public Service (APS) was awarded to demonstrate low-temperature proton-exchange membrane hydrogen production technology at the Davis-Besse Nuclear Power Station. This project will also include detailed technical and economic assessments for Xcel Energy and APS. A second hydrogen production demonstration project will be installed at an Exelon nuclear plant in the US Midwest. Once demonstrated for current fleet plants, the relative scale of the technology gap for next-generation SMRs and other advanced reactors may be significantly reduced. However, each configuration may entail different integration hardware, instrumentation, and control approaches as a function of the selected subsystem technologies. Hence, additional technology gaps may be present for each IES instance that will require some level of testing, safety analysis, and regulatory approval.

Acknowledgments

Significant contribution has been made to the topics covered in this chapter by fellow researchers at Idaho National Laboratory. Opinions expressed in this chapter are that of the author and are not attributable to the US Department of Energy or Idaho National Laboratory.

References

Boardman, R., Rabiti, C., Hancock, S., et al., 2019. Evaluation of non-electric market options for a light-water reactor in the Midwest. INL/EXT-19-55090.

Bolinger, M., Seel, J., Robson, D., 2019. Utility-Scale Solar: Empirical Trends in Project Technology, Cost, Performance, and PPA Pricing in the United States—2019 Edition. Lawrence Berkeley National Laboratory, Berkeley, CA, December.

Bragg-Sitton, S.M., 2004. Analysis of Space Reactor System Components: Investigation Through Simulation and Non-Nuclear Testing (Doctoral dissertation) University of Michigan, Ann Arbor, MI.

Bragg-Sitton, S.M., Hervol, D.S., Godfroy, T.J., 2009. Testing of an integrated reactor core simulator and power conversion system with simulated reactivity feedback. In: Proceedings of the Nuclear and Emerging Technologies for Space (NETS-2009), American Nuclear Society, June 14-19, 2009, Atlanta, GA, paper 208198.

Bragg-Sitton, S.M., Godfroy, T.J., Webster, K.L., 2010. Improving the fidelity of electrically heated nuclear systems testing using simulated neutronic feedback. Nucl. Eng. Des. 240 (10), 2745–2754.

Bragg-Sitton, S., Boardman, R., Rabiti, C., O'Brien, J., 2020a. Reimagining future energy systems: maximizing energy utilization via integrated nuclear-renewable energy systems. Int. J. Energy Res. (February). https://doi.org/10.1002/er.5207.

Bragg-Sitton, S., Rabiti, C., Boardman, R., et al., 2020b. Integrated energy systems: 2020 roadmap. INL/EXT-20-57708, March.

BWXT Center for Advanced Engineering and Research, 2015. Integrated Systems Test (IST) Facility. http://www.caer-ist.org/about-ist.html. (Accessed March 2020).

Chung, M.K., et al., 2007. Verification tests performed for development of an integral type reactor. In: Proc. Int. Conf. on Non-Electric Applications of Nuclear Power: Seawater Desalination, Hydrogen Production and Other Industrial Applications, Oarai, Japan.

Dise, S., 2016. The Future of the NREL Solar Prospector and Public Solar Data. Clean Power Research. October 3. https://www.cleanpower.com/2016/solar-prospector-sunsetting/. (Accessed March 2020).

Epiney, A., Richards, J., Hansen, J., Talbot, P., Burli, P., Rabiti, C., Bragg-Sitton, S.M., 2019. Case study: integrated nuclear-driven water desalination—providing regional potable water in Arizona. Idaho National Laboratory, INL/EXT-19-55736, September.

EPRI, 2017. Transitioning nuclear power plants to flexible power operations: experience report summary and update of approach to transition nuclear power plants to flexible power operations. EPRI report 3002010414.

Frick, K., Talbot, P., Wendt, D., et al., 2019. Evaluation of hydrogen production feasibility for a light water reactor in the Midwest. Idaho National Laboratory, INL/EXT-19-55395, September.

Gandrik, A.M., 2011. HTGR-integrated oil sands recovery via steam-assisted gravity drainage. Idaho National Laboratory Technical Evaluation TEV-704, Rev. 2, September 30.

Gandrik, A.M., 2012. HTGR-integrated coal and gas to liquids production analysis. Idaho National Laboratory Technical Evaluation TEV-672, Rev. 3, April 19.

Hall, D., Reeves, K., Brizzee, J., Lee, R., Carroll, G., Sommers, G., 2006. Feasibility assessment of the water energy resources of the United States for new low power and small hydro classes of hydroelectric plants. DOE-ID-11263, January.

HydroGEN, 2019. Advanced Water Splitting Materials. Available at: https://www.h2awsm.org/. (Accessed July 2019).

IAEA, 2013. Status Report 77, System-Integrated Modular Advanced Reactor. Available at: http://www.iaea.org/NuclearPower/Downloadable/aris/2013/29.SMART.pdf.

International Energy Agency, 2019. Nuclear Power in a Clean Energy System. May. https://www.iea.org/reports/nuclear-power-in-a-clean-energy-system.

Laurencin, J., Mougin, J., 2015. February (Chapter 6). High-temperature steam electrolysis. In: Godula-Jopek, A. (Ed.), Hydrogen Production: Electrolysis. Wiley-VCH Verlag GmbH & Co. KGaA, https://doi.org/10.1002/9783527676507.ch6.

Mays, G.T., 2011. Using GIS screening/modeling provides basis for evaluating array of siting related options and issues. In: Presentation, 2011 Utility Working Conference, Hollywood, FL, August 17.

Mays, G.T., et al., 2012. Application of spatial data modeling and geographical information systems (GIS) for identification of potential siting options for various electrical generation sources. Report Prepared for the Electric Power Research Institute (EPRI), ORNL/TM-2011/157/R1, May.

McKellar, M.G., 2010a. An analysis of the effect of reactor outlet temperature of a high temperature reactor on electric power generation, hydrogen production, and process heat. Idaho National Laboratory Technical Evaluation TEV-981, September 14.

McKellar, M.G., 2010b. Nuclear-integrated hydrogen production analysis. Idaho National Laboratory Technical Evaluation TEV-693, Rev. 1, May 15.

Mundahl, E., 2018. California Renewables and the Mystery of Negative Power Prices. InsideSources. posted to Energy, August 9. Available at: https://www.insidesources.com/california-renewables-and-the-mystery-of-negative-power-prices/.

NEI, 2020. U.S. Nuclear Generating Statistics. https://www.nei.org/resources/statistics/us-nuclear-generating-statistics. March (updated).

Nelson, L., Gandrik, A., McKellar, M., Patterson, M., Robertson, E., Wood, R., Maio, V., 2011a. Integration of high temperature gas-cooled reactors into industrial process applications. Idaho National Laboratory External Report # INL/EXT-09-16942 Rev. 3, September.

Nelson, L., Gandrik, A., McKellar, M., Patterson, M., Robertson, E., Wood, R., 2011b. Integration of high temperature gas-cooled reactors into selected industrial process applications. Idaho National Laboratory External Report # INL/EXT-11-23008, August.

NREL, 2020. National Renewable Energy Laboratory Wind Prospector. http://maps.nrel.gov/wind_prospector. (Accessed March 2020).

Nuclear Engineering International, 2018. Progress for Argentina's CAREM SMR. 9 May. Available at: https://www.neimagazine.com/news/newsprogress-for-argentinas-carem-smr-6144828.

NuScale Power, 2020. Technology Validation. https://www.nuscalepower.com/technology/technology-validation. (Accessed March 2020).

O'Brien, J.E., Stoots, C.M., Herring, J.S., 2010. (Chapter 20). High temperature electrolysis of steam. In: Hino, R., Yan, X. (Eds.), Nuclear Hydrogen Production Handbook. CRC Press, New York. ISBN 978-1-4398-1083-5.

OECD NEA, 2012. Nuclear Energy and Renewables: System Effects in Low-Carbon Electricity Systems. Available at: https://www.oecd-nea.org/ndd/pubs/2012/7056-system-effects.pdf.

OECD NEA, 2019. The Costs of Decarbonization: System Costs With High Shares of Nuclear and Renewables. Available at: https://www.oecd-nea.org/ndd/pubs/2019/7299-system-costs.pdf.

OECD Nuclear Energy Agency, 2010. Generation IV International Forum. www.gen-4.org.

Robertson, E.P., 2011a. Integration of HTGRs with an in situ oil shale operation. Idaho National Laboratory Technical Evaluation TEV-1029, Rev.1, May 16.

Robertson, E.P., 2011b. Integration of HTGRs to an ex situ oil shale retort operation. Idaho National Laboratory Technical Evaluation TEV-1091, Rev. 1, July 19.

Ruth, M., 2019. H2@Scale: hydrogen integrating energy systems, national renewable energy laboratory. NREL/PR-6A20-75422, November.

Sabharwall, P., Wendt, D., Utgikar, V., 2013. Application of chemical heat pumps for temperature amplification in nuclear hybrid energy systems for synthetic fuel production. INL/EXT-13-30463, October.

Satmon, J., Sarah, B.L., Samson, S.J., Zhu, M., 2017. Chemical heat pump (CHP) simulation, energy and exergy analysis. Int. J. New Technol. Res. 3 (1), 53–56.

Szilard, R., Sharpe, P., Borders, T., 2017. Economic and market challenges facing the U.S. nuclear commercial fleet—cost and revenue study. INL/EXT-17-42944, September. https://gain.inl.gov/SiteAssets/Teresa/Market%20Challenges%20for%20Nuclear%20Fleet-ESSAI%20Study%20Sept2017%20(1).pdf.

ThirdWay, 2019. Clean Energy Targets Are Trending. posted December 11. https://www.thirdway.org/graphic/clean-energy-targets-are-trending?utm_source=Third+Way+Subscribers&utm_campaign=19464f27a2-EMAIL_CAMPAIGN_2019_06_10_06_29_COPY_01&utm_medium=email&utm_term=0_8952c391fb-19464f27a2-195838221.

US NRC, 2020. Application Documents for the NuScale Design. https://www.nrc.gov/reactors/new-reactors/design-cert/nuscale/documents.html. (Accessed 21 February 2020).

Wiser, R., Bolinger, M., 2018. Wind technologies market report. Lawrence Berkeley National Laboratory DOE/GO-102019-5191, August 2019.

Wood, R.A., 2010a. HTGR-integrated hydrogen production via steam methane reforming (SMR) process analysis. Idaho National Laboratory Technical Evaluation TEV-953, September 15.

Wood, R.A., 2010b. Nuclear-integrated methanol-to gasoline production analysis. Idaho National Laboratory Technical Evaluation TEV-667, Rev. 2, May 15.

Wood, R.A., 2010c. Nuclear-integrated ammonia production analysis. Idaho National Laboratory Technical Evaluation TEV-666, Rev. 2, May 5.

Wood, R., 2012. Nuclear-integrated methanol-to-olefins production analysis. Idaho National Laboratory Technical Evaluation TEV-1567, July 25.

Part Four

International R&D and deployment

Small modular reactors (SMRs): The case of Argentina

14

Dario F. Delmastro
National Atomic Energy Commission and Universidad Nacional de Cuyo, San Carlos de Bariloche, Río Negro, Argentina

14.1 Introduction

About 55 years ago, Argentina decided to use a natural uranium fuel cycle for electricity production using nuclear power plants (NPPs). This decision was aimed to allow control of the whole fuel cycle. Small and medium reactors based on heavy water reactors (HWRs) were installed, and many facilities were constructed for uranium conversion, special alloy and Zircaloy tube production, fuel element production, and heavy water production.

Since the 1950s Argentina has developed and constructed research reactors. This activity was very successful, and research reactors were exported to Peru, Algeria, Egypt, and Australia.

In the early 1980s Argentina developed uranium enrichment capabilities using the gas diffusion method. This achievement facilitates the use of enriched uranium in local NPPs, and the Central Argentina de Elementos Modulares (CAREM) project started. This project consists of the development, design, and construction of small NPPs based on integrated pressurized water reactors (iPWRs). First a prototype of an electrical output of about 27 MW (CAREM 25) will be constructed to validate the innovation of the CAREM concept and then develop a commercial version. After several years of development, the CAREM project had reached such a level that the Argentine government decided to construct a CAREM prototype. Several activities have been performed, allowing the start of the CAREM prototype construction in February 2014. Currently the CAREM project is the main activity in small modular nuclear reactor research in Argentina. These activities are briefly described in this chapter.

14.2 Small modular reactor (SMR) research and development in Argentina

14.2.1 Development of research reactors

In the 1950s Argentina decided to start the design and construction of research reactors. The main purpose was the production of radioisotopes, and this also was very important for the development of engineering and construction capabilities.

Handbook of Small Modular Nuclear Reactors. https://doi.org/10.1016/B978-0-12-823916-2.00014-X
Copyright © 2021 Elsevier Ltd. All rights reserved.

The RA-1, an Argonaut-type critical facility, was entirely constructed in the country following a basic design provided by Argonne National Lab and became critical in January 1958. Later, it was redesigned locally as a 40-kW research reactor. Fuel for the RA-1 research reactor was designed and produced in the country.

In the 1960s the Atomic Energy Commission of Argentina (CNEA) designed RA-3, a 5-MW open tank reactor for radioisotope production. The reactor became critical in 1967, and its construction had a very important local participation. RA-3 was also a milestone in achieving expertise in the field of research reactors. In the late 1970s CNEA exported to Peru a 10-MW research and radioisotope production reactor. Currently, CNEA is building RA-10 a 30 MW multipurpose research reactor at Ezeiza, Argentina. RA-10 is very important for increasing the local radioisotope production.

INVAP was created in 1976 and in the following years became involved in many technological projects: uranium enrichment, research reactors, etc. A research and training reactor (RA-6) was designed and built by INVAP in Bariloche, Argentina. The company has participated in the construction of the RP-10 reactor in Peru. INVAP has also built the research reactors NUR in Algiers, Algeria; ETRR-2 reactor in Cairo, Egypt; and OPAL in Lucas Heights, Australia. Currently, INVAP is designing or participating in the construction of several research reactors like Pallas (the Netherlands), RMB (Brazil), LPRR (KSA), and RA-10 (Argentina).

14.2.2 Development of heavy water reactors

In 1965 the government requested the CNEA to prepare a feasibility report on the construction of an NPP in the region were most of the population and industry were concentrated. The CNEA's report showed the feasibility of installing a nuclear power plant, and the Atucha site was selected.

In 1966 the CNEA was authorized to start the bidding procedure, and 17 offers were received. By the end of 1967, the decision was taken. The construction of Atucha-1 started in 1968, and the plant was connected to the grid in 1974. The main supplier was Siemens AG, from Germany, and the design was a pressure vessel design, moderated and cooled by heavy water, with natural uranium fuel, based on Siemens-KWU PWRs and the German pressurized heavy water reactor MZFR. Local industry participation in the project was about 40%. The reactor has operated well, with an overall load factor of about 70%. The thermal reactor power of Atucha-1 is 1179 MW, the generator output is 357 MWe, and the net plant power totals 335 MWe.

In 1972 it was decided to construct a second NPP in Embalse, province of Córdoba. The call for bids followed the same trend of the Atucha-1 and a 50%, minimum local industry participation was requested. In March 1973 the offer of a 600 MWe natural uranium-HWR with pressure tubes, made jointly by Canadian and Italian firms, was selected. A very wide technology transfer agreement was negotiated between Argentina and Canada and signed in March 1974. Different reasons caused both delays in the construction and only partial fulfillment of the technology transfer agreement. CNEA became in 1979 the main contractor for the assembly works, and Argentine firms were responsible for the assembly of many systems. The plant entered into commercial service in 1984. The reactor has operated well, with an overall load factor of about 84%.

The reactor is of a CANDU600 type. The moderator and coolant are heavy water. The fuel channels are horizontal pressure tubes. The moderator is at low pressure and separated from the coolant. The thermal reactor power was 2.109 MW, generator output was 648 MWe, and net plant power totals 600 MWe.

The plant completed its first operating cycle on December 31, 2015, and it was shut down for a life extension and upgrade works. In 2019 it returned to service with a generator output of 683 MWe.

In 1979 a call for offers for the third NPP, to be built next to the Atucha-1, was issued. Argentina was requesting a NPP, a heavy water production plant, and the constitution of a joint company between CNEA and the supplier, who would become architect-engineer for the construction of the present and future nuclear plants in the country. The German and Swiss' offer of a 700 MWe version of the Atucha-1 pressure vessel reactor and a heavy water production plant was selected. In 1981 the architect-engineering company ENACE was formed as a joint venture with Germany, and construction of the Atucha-2 started.

Argentine suffered economic crisis, the project was delayed, and construction was formally stopped in 1994. The Argentine nuclear sector was restructured. A state-owned shareholder company, Nucleoeléctrica Argentina SA (NA-SA), was created for the operation of the two nuclear stations and the construction and subsequent operation of the third one. In August 2006 the government decided to restart the construction of Atucha-2 under NASA management. On the June 27, 2014 the plant connected for the first time to the grid. Participation of the local industry was very important.

The Atucha-2 was developed from the Atucha-1 design by Siemens-KWU. Many of the components have a conceptual design identical to those of Atucha-1, while the plant layout and other features are derived from the design of the pre-Konvoi and Konvoi 1300 plants. The thermal reactor power is 2.160 MW, the generator output is 745 MWe, and the net plant power totals 692 MWe.

In the 1980s the Empresa Nuclear Argentina de Centrales Eléctricas (ENACE) designed the ARGOS PHWR 380. Its flow diagram and main technical characteristics were practically the same as those of Atucha NPPs. It had 60 vertical hydraulically actuated absorber rods. The spent fuel pool was located within the reactor building, and different fuel options were considered. Its safety design was mainly based on probabilistic safety criteria. Also in the 1980s the Dirección de Centrales Nucleares of CNEA designed the TPA 700/300. These were both based on the CANDU 600 and aimed at allowing important local participation. The fuel of Atucha-1 was originally natural uranium, but it was substituted by 0.85% enriched uranium to reduce fuel costs. The exit burnup was increased from 6 to 11 MW d/kg U.

CNEA has developed under the Combustible Avanzado para Reactores Argentinos (CARA) project an advanced fuel element concept for HWR, specially designed to fit Argentinean fuel-cycle requirements. Atucha-1 and Embalse have quite different designs for the fuel elements. Both NPPs use on-load refueling, but they differ in the number and length of the refueled elements. In Embalse a CANDU reactor type, with a total of 12 fuel elements, 6-m-long channel, and 2 fuel elements is refueled at a time. In Atucha-1 a pressure vessel design, moderated and cooled by heavy water, each vertical channel has one single fuel bundle of about 5 m in length for its active portion, hung by its upper part. Atucha-1 is fueled on power by a fueling machine that sits above the reactor.

The main objectives of CARA are as follows:

- the fuel element that could be used in both reactor types,
- NPP performance enhancement,
- thermohydraulic and thermomechanical extra margins,
- negative void coefficients,
- cost and spent fuel reduction.

This advanced fuel element for HWR has collapsible cladding, 100 cm length and 52 fuel pins. In 2012 a Canadian patent was issued (Florido et al., 2012).

14.2.3 Development of iPWRs

The Argentine CAREM project consists of the development, design, and construction of an advanced, simple, and small NPP. CAREM design criteria, or similar ones, have been adopted by other plant designers, originating a new generation of reactor designs, of which CAREM was, chronologically, one of the first. The first step of this project is the construction of the prototype of about 27 MWe (CAREM 25). This project allows Argentina to sustain activities in the NPP design and construction area, assuring the availability of updated technology in the midterm. The design basis is supported by the cumulative experience acquired in research reactors design, construction and operation, pressurized heavy water reactors (PHWR) NPP operation, maintenance and improvement, and the finalization of the Atucha-2 and the development of advanced design solutions. CAREM has been recognized as an international near-term deployment reactor by the Generation IV International Forum (US DOE Nuclear Energy Research Advisory Committee and the Generation IV International Forum, 2002). The project is owned by CNEA, and INVAP was its main contractor until the early 2000s. Currently the project is managed by CNEA.

14.3 Integrated pressurized water reactor: CAREM

14.3.1 CAREM 25 design

CAREM 25 is an indirect cycle reactor with some distinctive and characteristic features that greatly simplify the reactor and also contribute to a higher level of safety (Boado Magan et al., 2011):

- integrated primary cooling system,
- primary cooling by natural circulation,
- self-pressurization,
- safety systems relying on passive features.

The CAREM 25 reactor pressure vessel (RPV) contains the core, the steam generators (SGs), the whole primary coolant, and the absorber rods drive mechanisms (Fig. 14.1). The RPV diameter is about 3.2 m, and the overall length is about 11 m (Delmastro et al., 2002).

The core of the prototype has 61 hexagonal cross-section fuel assemblies (FA) having about 1.4 m active length. Each fuel assembly contains 108 fuel rods, 18 guide thimbles, and an instrumentation thimble. Its components are typical of the PWR fuel assemblies. The fuel is enriched UO_2. Core reactivity is controlled by the use of Gd_2O_3 as burnable poison in specific fuel rods and movable absorbing elements (AEs) belonging to the adjust and control system. Chemical compounds are not used in the water for reactivity control during normal operation. The fuel cycle can be tailored to customer requirements, with a reference design of 390 full-power days and 50% of core replacement.

Each AE consists of a cluster of rods linked by a structural element, so the whole cluster moves as a single unit. Absorber rods fit into the guide thimbles. The absorbent

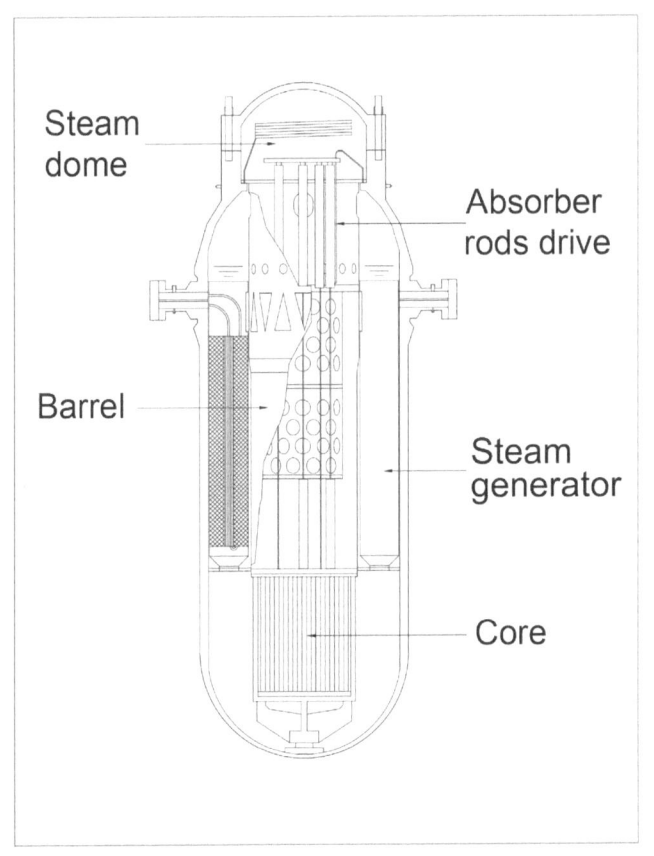

Fig. 14.1 Diagram of the CAREM 25 reactor pressure vessel and internals. Courtesy National Atomic Energy Commission (CNEA) of Argentina.

material is the commonly used Ag-In-Cd alloy. The AEs are used for reactivity control during normal operation (adjust and control system) and to produce a sudden interruption of the nuclear chain reaction when required (fast shutdown system).

Twelve identical "mini-helical" vertical SG, of the "once-through," type are placed equally distant from each other along the inner surface of the RPV. They are used to transfer heat from the primary to the secondary circuit, producing dry steam at 47 bar, with a minimum 30°C of superheating. The location of the SG above the core produces natural circulation in the primary circuit. The secondary system circulates upward within the tubes, while the primary goes in a countercurrent flow. An external shell surrounding the outer coil layer and adequate seal form the flow separation system. It guarantees that the entire stream of the primary system flows through the SG. To achieve a rather uniform pressure loss and superheating on the secondary side, the length of all tubes is equalized by changing the number of tubes per coil layer. Thus the outer coil layers will hold a larger number of tubes than the inner ones. For safety reasons, SG are designed to withstand the primary pressure without pressure in the secondary side, and the whole live steam system is designed to withstand primary pressure up to isolation valves (including the steam outlet/water inlet headers) for the case of an SG tube brake.

The natural circulation of the coolant produces different flow rates in the primary system according to the power generated (and removed). Under different power transients a self-correcting response in the flow rate is obtained.

Due to the self-pressurizing of the RPV (steam dome), the system keeps the pressure very close to the saturation pressure. At all the operating conditions, this has proved to be sufficient to guarantee a remarkable stability of the RPV pressure response. The control system is capable of keeping the reactor pressure practically at the operating set point through different transients, even in case of power ramps. The negative reactivity feedback coefficients and the large water inventory of the primary circuit combined with the self-pressurization features make this behavior possible with minimum control rod motion. It concludes that the reactor has excellent behavioral responses under operational transients.

Nuclear safety has been incorporated in CAREM 25 from the start of the design. The defense-in-depth concept has specially been considered. Many intrinsic characteristics contribute to the avoidance or mitigation of eventual accidents.

Safety systems of CAREM 25 are based on passive features and must guarantee that there is no need of active actions to mitigate the accidents during a long period. They are duplicated to fulfill the redundancy criteria. The shutdown system should be diversified to fulfill regulatory requirements.

The first shutdown system (FSS) is designed to shut down the core when an abnormality or a deviation from normal situations occurs and to maintain the core subcritical during all shutdown states. This function is achieved by dropping a total of 25 neutron-absorbing elements into the core by the action of gravity. Each neutron-absorbing element is a cluster composed of a maximum of 18 individual rods, which are together in a single unit. Each unit fits well into guide thimbles of each FA.

Internal hydraulic control rod drives (CRDs) avoid the use of mechanical shafts passing through RPV, or the extension of the primary pressure boundary, and thus eliminates any possibilities of big loss-of-coolant accidents (LOCA) since the whole device is located inside the RPV. Their design is an important development in the CAREM concept. Nine out of 25 CRD are the fast shutdown system. During normal operation, they are kept in the upper position, where the piston partially closes the outlet orifice and reduces the water flow to a leakage. The CRD of the adjust and control system is a hinged device, controlled in steps fixed in position by pulses over a base flow and designed to guarantee that each pulse will produce only one step.

Both types of device perform the SCRAM function by the same principle: "rod drops by gravity when flow is interrupted," so malfunction of any powered part of the hydraulic circuit (i.e., valve or pump failures) will cause the immediate shutdown of the reactor. CRDs of the fast shutdown system are designed using a large gap between piston and cylinder to obtain a minimum dropping time, thus taking few seconds to insert absorbing rods completely inside the core. For the adjust and control system, CRD manufacturing and assembly allowances are stricter, and clearances are narrower, but there is no stringent requirement on dropping time.

The second shutdown system is a gravity-driven injection device of borated water at high pressure. It actuates automatically when the reactor protection system detects the failure of the FSS. The system consists of an assembly of two tanks located in the upper part of the containment. The assembly is connected to the reactor vessel by two piping lines: one from the steam dome to the upper part of one tank and the other from a position below the reactor water level to the lower part of the other tank. When the system is triggered, the valves open automatically, and the borated water drains into the primary system by gravity. The discharge of the assembly produces the complete shutdown of the reactor.

The residual heat removal system has been designed to reduce the pressure on the primary system and to remove the decay heat in case of loss of heat sink. It is a simple and reliable system that uses condensing steam from the primary system in emergency condensers. The emergency condensers are heat exchangers consisting of an arrangement of parallel horizontal U-tubes between two common headers. The top header is connected to the reactor vessel steam dome, while the lower header is connected to the reactor vessel at a position below the reactor water level. The condensers are located in pools filled with cold water inside the containment building. The inlet valves in the steam line are always open, while the outlet valves are normally closed; therefore the tube bundles are filled with condensate. When the system is triggered, the outlet valves open automatically. The water drains from the tubes, and steam from the primary system enters the tube bundles and is condensed on the cold surface of the tubes. The condensate is returned to the reactor vessel, forming a natural circulation circuit. In this way, heat is removed from the reactor coolant. During the condensation process the heat is transferred to the water of the pools by a boiling process. This evaporated water is then condensed in the suppression pool of the containment.

The emergency injection system prevents core exposure in case of a LOCA. In the event of such accident, the primary system is depressurized with the help of the emergency condensers to less than 20 bar, with the water level over the top of the core. At 20 bar a low-pressure water injection system comes into operation. The system consists of two tanks with borated water connected to the RPV. The tanks are pressurized;

thus when during a LOCA, the pressure in the reactor vessel reaches 20 bar, the rupture disks break, and the flooding of the RPV starts.

Safety relief valves protect the integrity of the reactor pressure vessel against overpressure, in case of strong imbalances between the core power and the power removed from the RPV. Each valve is capable of producing 100% of the necessary relief. The blow-down pipes from the safety valves are routed to the suppression pool.

The primary system, the reactor coolant pressure boundary, safety systems, and high-pressure components of the reactor auxiliary systems are enclosed in the primary containment—a cylindrical concrete structure with an embedded steel liner. The primary containment is of pressure suppression type with two major compartments: a dry well and wet well. The lower part of wet well volume is filled with water that works as the condensation pool, and the upper part is a gas compression chamber. The building surrounding the containment has been designed in several levels.

It supports structures with the same seismic classification, allowing the integration of the RPV, the safety and reactor auxiliary systems, the spent fuels pool, and other related systems in one block.

CAREM 25 NPP has a standard steam cycle of simple design.

Technical and economic advantages are obtained with the CAREM 25 design compared with the traditional design:

- Due to the absence of large-diameter piping associated to the primary system, no large LOCA has to be handled by the safety systems. The elimination of large LOCA considerably reduces the needs in emergency core cooling system (ECCS) components, AC supply systems, etc.
- Eliminating primary pumps and pressurizer results in added safety (loss of flow accident elimination) and advantages for maintenance and availability.
- Innovative hydraulic drive control rods avoid rod ejection accidents.
- A large coolant inventory in the primary results in large thermal inertia and long response time in case of transients or accidents.
- Passive safety systems with a long grace period are incorporated.
- Shielding requirements are reduced by the elimination of gamma sources of dispersed primary piping and parts.
- The large water volume between the core and the wall leads to a very low fast neutron dose over the RPV wall.
- The ergonomic design and layout make the maintenance easier. Maintenance activities like SG tube inspection do not compete with refueling activities because it will be carried out through 12 radial plena with individual removable blind flanges.

14.3.2 CAREM developments

The CAREM project involves technological and engineering solutions, as well as several innovative design features that must be properly demonstrated during the design phase. Also specific codes used for modeling systems related with safety issues to obtain design parameters (primary cooling system, reactor core, fuel design, etc.) must be verified and validated against worldwide benchmarks and/or experimental data to build confidence in their results (Mazzi et al., 2002).

Within the CAREM project the effort has been focused mainly on the nuclear island (inside containment and safety systems) where several innovative design solutions require developments of the first stage (to assure they comply with functional requirements). This comprises mainly the reactor core cooling system (RCCS), the reactor core and fuel assembly, and the FSS. An extensive experimental plan has been prepared, including the design and construction of several experimental facilities to fulfill the project's requirements.

A high-pressure natural circulation rig, CAPCN, was constructed to perform dynamic tests of RCCS. Its purpose was mainly to study the thermohydraulic dynamic response of the CAREM primary loop, including all the coupled phenomena that may be described by one-dimensional models. This includes the assessment of the calculation codes on models of the rig and the extension of validated models to the analysis of the CAREM reactor.

The CAPCN rig resembles CAREM in the primary loop (self-pressurized natural circulation) and the SG (helical once-through), while the secondary loop is designed only to produce adequate boundary conditions. Operational parameters are reproduced for intensive quantities (pressure, temperature, void fraction, heat flux, etc.) and scaled for extensive quantities (flow, heating power, cross sections, etc.). Height was kept approximately on a 1:1 scale. The heating power may be regulated up to 300 kW. The secondary loop pressure and cold leg temperatures are controlled through valves. The pump regulates the flow. The condenser is of air-cooled type with airflow control. The control of the actuators (heaters, valves, pumps, etc.), data acquisition, and operating follow-up were carried out from a control room, through a PC-based, multinode software (flexible enough to define any feedback loop).

Many experiments were performed to investigate the thermohydraulic response of the system in conditions similar to CAREM operational states. The influence of different parameters like vapor dome volume, hydraulic resistance, and dome nitrogen pressure was studied. Perturbations in the thermal power, heat removal, and pressure relief were applied. The dynamic responses at low pressure and temperature and with control feedback loops were also studied. It was observed that around the operating point self-pressurized natural circulation was very stable, even with important deviations on the relevant parameters. A representative group of transients were selected, to check computer models.

The SGs will be subject to a series of tests, including standard mechanical evaluations and hydraulic tests. The preparation of tests in a low-pressure rig for pressure losses, flow-induced vibrations, and general assembling behavior is ongoing.

The thermohydraulic design of CAREM reactor core was carried out using a 3-D two fluid model code. To take into account the strong coupling of the thermohydraulic and neutronic behaviors of the core, this code was linked with neutronic codes. This coupled model allows a 3-D evaluation of power and thermohydraulic parameters at any stage of the burnup cycle.

The mass flow rate in the core of the CAREM reactor is rather low compared with typical light-water reactors, and therefore correlations or experimental data available should be assessed in the range of interest. To perform this assessment, experiments were conducted at the thermohydraulic laboratories of the Institute of Physics and Power Engineering (IPPE, Obninsk, Russian Federation).

The main goal of the experimental program was to generate a substantial database to assess the prediction methodology for critical heat flux (CHF) applicable to the CAREM core, covering a wide range of thermohydraulic parameters around the point of normal operation. Most tests were performed using a low-pressure Freon rig, and results were extrapolated to water conditions through scaling models. Finally a reduced set of tests were performed in water at high pressure and temperature, to validate the method for scaling. Different test sections were assembled to simulate different regions in the fuel element and radial uniform and nonuniform power generations. A bundle with 35% of the full length was tested to obtain CHF data under average subcooled conditions. More than 250 experimental points under different conditions were obtained in the Freon loop and more than 25 points in the water loop.

An experimental test program to assess the influence of spacers on CHF, in a Freon loop, is under preparation.

The fuel assemblies and absorbing clusters were or will be subject to a series of tests, including standard mechanical evaluations and hydraulic tests. The latter comprise the following:

- tests in a low-pressure rig evaluating pressure losses, flow-induced vibrations, and general assembling behavior (performed);
- endurance tests in a high-pressure loop pointed to wear-out and fretting issues.

A fuel rod irradiation test was performed in the Halden boiling water reactor. The purpose of this test was the characterization of relevant performance aspects of the fuel, such as temperature behavior, dimensional stability, and fission gas release.

Neutronic modeling validations were made against VVER reactor geometry using experimental data from a ZR-6 Research Reactor, Hungary, and a series of benchmark data for typical PWR reactors. The RA-8 critical facility has been designed and constructed as an experimental facility to measure neutronic parameters of CAREM. Further experimental data were obtained from RA-8.

One of the most innovative systems behind the CAREM concept is the hydraulic (in-vessel) control rod drive (HCRD) mechanism. The design embraces mechanical and thermohydraulic innovative solutions, so a complete experimental program is necessary to achieve the high reliability performance jointly with low maintenance. This development plan includes the construction of several experimental facilities. Preliminary tests were first performed to prove the feasibility of the theoretical approach, to have an idea of some of the most sensitive controlling parameters and to determine spot points to be focused on during design. Tests were undertaken on a rough device with promising experimental results, and good agreement with first modeling data was obtained.

First prototype tests helped to determine preliminary operating parameters on a full-scale mechanism as a first approach toward detail engineering. These parameters include range of flow and ways to produce hydraulic pulses. Manufacturing hints that simplified and reduce costs of the first design were also found. Tests were carried out in a specially built rig, and as part of this experimental development, it was decided to separate the regulating and fast-drop requirements in different devices.

Test on a low-pressure loop were carried out with the CRD at atmospheric pressure and with feed-water temperature regulation up to low subcooling. The feedwater pipeline simulated alternative configurations of the piping layout with a second injection line (dummy) to test possible interference of pulses. The ad hoc test loop (CEM) was designed to allow automatic control of flow, pressure, and temperature, and its instrumentation produces information of operating parameters including pulse shape and timing. The tests included the characterization of the mechanism and the driving water circuit at different operating conditions and the study of abnormal situations as increase in drag forces, pump failure, loss of control on water flow or temperature, saturated water injection, suspended particle influence, and pressure "noise" in feeding line. The tests, carried out at turbulent regime, which are the closest conditions to operation obtained in this loop, showed good reliability and repetitiveness as well as sensitivity margins for the relevant variables within control capabilities of a standard system.

Finally a high-pressure loop (CAPEM) was constructed to reach the actual operating conditions ($P = 12.25$ MPa, $T \approx 326°C$). The main objectives are to verify the behavior of the mechanisms, to tune up the final controlling parameter values, and to perform endurance tests. After this stage the system under abnormal conditions, such as the behavior during RPV depressurization and simulated breakage of feeding pipes, will be tested.

Since the HCRD design adopted has no movable parts outside the RPV, it was necessary to design a special probe to measure the rod position able to withstand primary environmental conditions. The proposed design consists in a coil wired around the HCRD cylinder with an external associated circuit that measure electric reluctance variations induced by the movement of the piston-shaft (made of magnetic steel) inside the cylinder. A cold test performed showed that the system is capable of sensing one-step movement of the regulating CRD, with an acceptable accuracy. In-furnace high-temperature tests were conducted to evaluate the behavior of the system against temperature changes similar to those occurring during operational transients.

To fulfill the Argentinian regulation, CAREM 25 has two independent shutdown systems. Two independent reactor protection systems exist to actuate these shutdown systems and the safety systems. The qualification of these reactor protection systems includes the development, construction, and testing of prototypes.

14.3.3 Post-Fukushima actions

The CAREM prototype design was reviewed considering Fukushima experience. Topics like seismic requirements, loss of heat sink, and black-out were considered. The design basis earthquake was reviewed. A risk-based criterion, based in Argentine Regulations AR 3.10.1 and AR 3.1.3, was used. CAREM-25 considers in its design base the loss of heat sink and station black-out during the grace period. Provisions were considered to allow, after the grace period, core decay heat removal and containment cooling using the fire extinguishing system or an autonomous system.

14.4 Deployment of SMRs in Argentina

On December 17, 2009, the National Law 26566/2009 declared of national interest, inter alia, the design, construction, and start-up of CAREM prototype, establishing a special regime. CNEA was entrusted to complete these tasks. The Preliminary Safety Analysis Report and the Quality Manual were presented to the Argentinean Regulatory Authority in 2009.

In 2010 the ARN issued a new procedure for the licensing of prototype NPPs. Information was provided to and analyzed by the regulatory body according to this new procedure. The Universidad Tecnológica Nacional—Facultad Regional Avellaneda has performed the Environmental Impact Study of CAREM reactor prototype, and it was presented to the authorities of Province of Buenos Aires in December 2012.

Site activities such as soil studies and environmental analyses have been performed. The construction of a high-pressure and high-temperature rig for testing the innovative HCRD mechanism has been finished. This rig can also be adapted for testing the structural behavior of the FA. In 2013 the ARN and the province of Buenos Aires issued the necessary permits allowing the start of prototype construction. These permits allowed CNEA to start construction works of the CAREM 25 project, except for the Nuclear Module. The full start of construction was conditioned to accomplish a regulatory requirement to update CAREM 25 Design Report considering some changes in the layout or in civil structures of plant. The construction of CAREM prototype started in February 2014.

In December 2014, ARN issued the Authorization for the Construction of Nuclear Module of CAREM 25 (ARN, 2016).

Contracts and agreements were taken with different Argentinean stakeholders to perform detail engineering and RPV (Fig. 14.2), Fuel Assemblies, Nuclear Building, and Balance of Plant construction (Fig. 14.3). The construction of the CAREM prototype is ongoing.

14.5 Future trends

The CAREM project is the main R&D project in Argentina, and in the future, it will lead the activities related to SMRs in the country. The first step of this project is the construction of the prototype of about 27 MWe (CAREM 25). The following steps consider the prototype operation and the commercial modules' development.

For the commercial modules' development, the economic aspects are very important. CAREM power was initially fixed at 15 MWe, but the prototype power was increased to achieve a better economic performance. In general the economy of size can be used to improve the competitiveness of a given reactor configuration. But there are technical and economic constraints that limit the maximum economical size of a given concept. In integrated primary system reactors the size of the reactor pressure vessel is a relevant one. The CAREM concept economy was analyzed using a very advanced tool, and two alternatives were evaluated for the primary system flow rate (natural circulation and assisted circulation). Below 150 MWe the natural convection

Fig. 14.2 CAREM 25 reactor pressure vessel construction.
Credit: Courtesy National Atomic Energy Commission (CNEA) of Argentina.

Fig. 14.3 CAREM 25 site. Nuclear Building and Balance of Plant construction.
Credit: Courtesy National Atomic Energy Commission (CNEA) of Argentina.

option is preferred. But over that power level, the size and cost of the RPV are outside the acceptable range so the forced convection option is preferable. The maximum power achievable using pumps is about 300 MWe.

The development of a natural circulation commercial module of about 120 MWe is foreseen as the next step. This module will better benefit from the experience developed with CAREM prototype design, engineering, licensing, construction, and operation while incorporating the economic improvements arising from the economy of size. Modules optimal size studies are ongoing. Improvements, which are based on the prototype design experience, are included.

The multimodular nuclear power plant approach is also been considered for the commercial version of CAREM. A multimodular nuclear power plant composed of four modules of 120 MWe each, CAREM-480, was proposed (Delmastro et al., 2017). In this plant, several systems are shared to reduce costs. This configuration also allows the stagger construction of the modules. This enables the modules already in operation to coexist with others in the construction phase. In this way, negative cash flows are offset by positive ones, thus reducing the financial risk.

The construction of the first CAREM commercial unit at the province of Formosa was considered. Sitting studies were performed.

A further step considers the development of a forced convection module of larger power. Different options could be considered to assist flow circulation in CAREM concept. CAREM configuration has many similarities with BWRs, and the use of jet pumps was preliminary considered (Delmastro et al., 2006). In integrated primary system reactors, an annular space is available in the downcomer below the steam generators, and the jet pumps could be located there. Another option is the use of internal centrifugal canned pumps located in the lower plenum of the reactor pressure vessel. But as primary system natural circulation precludes the loss of flow accident, the use of forced circulation must take into account the potential new requirements over the thermal margins evaluation and the safety systems.

14.6 Sources of further information and advice

The International Atomic Energy Agency every year produces many technical documents related to nuclear power energy and particularly nuclear reactors and SMRs. Many of them include information related to SMR research and development and deployment in Argentina (IAEA, 1989, 2002, 2004, 2006, 2018).

References

ARN, 2016. Argentinean National Report for the Convention of Nuclear Safety, Seventh Report. Ciudad Autónoma de Buenos Aires, Argentina. August 2016.
Boado Magan, H., Delmastro, D.F., Markiewicz, M., Lopasso, E., Diez, F., Giménez, M., Rauschert, A., Halpert, S., Chocrón, M., Dezzutti, J.C., Pirani, H., Balbi, C., Fittipaldi, A., Schlamp, M., Murmis, G.M., Lis, H., 2011. CAREM project status. Sci. Technol. Nucl. Install. 2011, 140373.

Delmastro, D., Mazzi, R., Santecchia, A., Ishida, V., Gómez, S., Gómez de Soler, S., Ramilo, L., 2002. CAREM: an advanced integrated PWR. In: IAEA, Small and Medium Sized Reactors: Status and Prospects, IAEA-CSP-14/P, pp. 224–231.

Delmastro, D., Patruno, L., Masson, V., 2006. An assisted flow circulation option for integral pressure water reactors. In: Proceedings of ICONE 14, Miami.

Delmastro, D., Gil, P.C., Di Pace, M., Chocrón, M., Conti, C., Irigaray, M., 2017. CAREM escala Comercial. Rev. Comisión Nac. Energía At. 1666-103667/68, 5–9. (in Spanish). Presented at XLIII Reunión Anual AATN, Ciudad Autónoma de Buenos Aires, Argentina, November 2016.

Florido, P.C., Cirimello, R.O., Bergallo, J.E., Marino, A.C., Brasnarof, D.O., Delmastro, D.F., Gonzalez, J.H., Juanico, L.E., (CNEA), 2012. Modular fuel element adaptable to different nuclear power plants with cooling channels. Canadian patent 2307402. March 20 2012.

IAEA, 1989. Status of Advanced Technology and Design for Water Cooled Reactors: Heavy Water Reactors. IAEA-TECDOC-510, IAEA, Vienna, Austria.

IAEA, 2002. Heavy Water Reactors: Status and Projected Development. IAEA-TRS-407, IAEA, Vienna, Austria.

IAEA, 2004. Status of Advanced Light Water Reactor Designs 2004. IAEA-TECDOC-1391, IAEA, Vienna, Austria.

IAEA, 2006. Status of Innovative Small and Medium Sized Reactor Designs 2005: Reactors With Conventional Refuelling Schemes. IAEA-TECDOC-1485, IAEA, Vienna, Austria.

IAEA, 2018. Advances in Small Modular Reactor Technology Developments 2018 Edition, Austria.

Mazzi, R., Santecchia, A., Ishida, V., Delmastro, D., Gómez, S., Gómez de Soler, S., Ramilo, L., 2002. CAREM project development. In: IAEA, Small and Medium Sized Reactors: Status and Prospects, pp. 232–243. IAEA-CSP-14/P.

US DOE Nuclear Energy Research Advisory Committee and the Generation IV International Forum, 2002. A Technology Roadmap for Generation IV Nuclear Energy Systems. GIF-002-00.

Small modular reactors (SMRs): The case of Canada

Metin Yetisir
Advanced Reactor Technologies, Canadian Nuclear Laboratories, Chalk River, ON, Canada

15.1 Introduction

Canada has a long tradition of nuclear power generation using pressurized heavy-water reactors (PHWR) developed by Atomic Energy of Canada Limited (AECL), a Canadian crown corporation. The Canadian PHWR design is the Canada Deuterium Uranium (CANDU) reactor that uses fuel channels, natural uranium fuel, pressurized heavy-water coolant, and heavy-water moderator in a low-pressure calandria vessel. The majority of operating domestic CANDU units are located in Ontario with a single operating unit in the province of New Brunswick. In addition to this, prototypical work in other technology concepts in Canada led to the development and operation of various domestically developed research and demonstration reactors including Whiteshell WR-1, the NRU reactor, and a fleet of SLOWPOKE designs.

With the emergence of SMRs in the early 2000s, Canada has seen an interest from SMR vendors considering potential deployments of first-of-a-kind demonstration units in Canada. Many of these reactor technologies are nonwater-cooled advanced reactors. The main reasons for considering Canada for first-of-a-kind deployment are the potential applications for off-grid and on-grid small nuclear and, notably, the readiness of the Canadian Nuclear Safety Commission (CNSC) to regulate advanced reactors (Lee, 2018). The flexible and risk-informed regulatory framework adopted by the CNSC allows it to evaluate any type or size of reactor design on the basis of broader safety principles rather than specific technological requirements.

In addition to the regulatory framework suitable for licensing advanced reactors, there are many key elements in place to make SMRs a reality in Canada:

- provincial and federal government support,
- a well-established nuclear industry,
- an extensive nuclear supply chain,
- extensive nuclear R&D infrastructure,
- an active nuclear workforce,
- many potential on- and off-grid applications for small reactors,
- available and supportive sites for prototype deployment.

As a result, some 20 SMR vendors currently operate in Canada at various levels of engagement. As of December 2019, 12 of these vendors were in direct discussions with the CNSC.

Generating electricity is a provincial responsibility in Canada. The provinces of Ontario and New Brunswick generate roughly 50% and 30% of their electricity from nuclear power, respectively. Although all operating nuclear reactors in Canada are of the CANDU type, major power utilities in Ontario (Ontario Power Generation and Bruce Power) and New Brunswick (New Brunswick Power) are interested in new nuclear builds and are engaged with SMR technology developers. SaskPower in the province of Saskatchewan and Qulliq Energy Corporation in the Canadian territory of Nunavut have expressed an interest in SMRs as an option for their energy needs.

15.2 Canada's SMR strategy

15.2.1 SMR Roadmap

With increasing activities in Canada by SMR vendors and interest by utilities and nuclear industry, natural resources of Canada (NRCan) launched an SMR Roadmap initiative in February 2018. After a 10-month engagement process with the industry and potential end users, NRCan released the findings of four expert working groups on technology, economics and finance, indigenous and public engagement, regulatory readiness, and waste management in a report "A Call to Action: A Canadian Roadmap for Small Modular Reactors" (SMR Roadmap, 2018) in November 2018.

In partnership with interested provinces, territories, and power utilities, Natural Resources Canada has convened a roadmap to engage stakeholders on the future of small modular reactors in Canada. Through a series of expert working groups and workshops held across Canada, the roadmap has gathered feedback on the direction for the possible development and deployment of SMRs in Canada (SMR Roadmap, 2018).

The roadmap identified three major areas of applications for SMRs:

(1) on-grid electricity to replace fossil fuel plants ($\sim$150–300 MWe),
(2) heavy industry: non-GHG-emitting heat and electricity ($\sim$10 MW to $>$150 MWe),
(3) remote communities: district heating, desalination, and electricity ($<$10 MWe).

The size and characteristics of SMRs suitable for these applications are quite different. Hence, more than one type of SMR will likely be deployed to meet the needs of these varied applications in Canada. For example, larger SMRs, benefiting from economy of scale for economic competitiveness, are suitable for on-grid electricity production. Very small SMRs, sometimes referred to as microreactors, are suitable for off-grid applications as a replacement for diesel generators at remote locations. Their small footprint and transportability are important factors. Industrial applications, such as process steam for heavy industry, hydrogen production, mining, and oil extraction, require high-temperature industrial heat at $>$500°C and favor nonwater-cooled advanced SMRs that operate at temperatures greater than 500°C.

The SMR Roadmap identified key players and the roles that they can play: Federal Government, Interested Provincial and Territorial Governments, Regulator (CNSC), Atomic Energy of Canada Limited (AECL), Canadian Nuclear Laboratories (CNL), Nuclear Waste Management Organization (NWMO), Canadian Nuclear Association (CNA), CANDU Owners Group (COG), Organization of Canadian Nuclear Industries

(OCNI), and the Canadian Nuclear Supply Chain, Utilities and Owner/Operators, SMR Vendors and Technology Developers, Universities and Colleges, Research Institutions, Laboratories, End-User Industries, and Civil Society.

The roadmap provided specific recommendations in the following four thematic areas (pillars) and suggested the next steps that should be taken:

- Pillar 1: Demonstration and deployment
- Pillar 2: Policy, legislation, and regulation
- Pillar 3: Capacity, engagement, and public confidence
- Pillar 4: International partnerships and markets

The Canadian SMR Roadmap report incorporates learnings from the engagement workshops, outlines the benefits of SMRs to Canada, identifies key players for SMR deployments in Canada, and provides specific recommendations to the key players. The full Canadian SMR Roadmap report, session reports, and technical working group reports are available on the NRCan SMR Roadmap website (SMR Roadmap, 2018).

15.2.2 Case study: Province of Ontario

The province of Ontario generates more than 50% of its electricity from nuclear energy. It is committed to nuclear considering SMRs for on-grid, off-grid, and industrial applications.

A feasibility study on the potential deployment of very small modular reactors (vSMRs) at off-grid mines in Ontario was conducted for the Ontario Ministry of Energy (HATCH, 2016). The study considered nine water-cooled and advanced vSMRs with power levels ranging from 3 to 32.5 MWe and found that all vSMRs considered in the study are competitive against diesel power generation technology for remote mining applications.

The two largest power utilities in Ontario, Ontario Power Generation and Bruce Power, are actively engaged with SMR vendors and considering both off-grid applications (Nalasco, 2019) and grid-connected electricity generation (NuSCale Website, n.d.).

15.2.3 Case study: Province of New Brunswick

Although Canada has not made a next-generation SMR nuclear technology selection yet, the province of New Brunswick made a decision to focus on advanced fast neutron spectrum reactors (Sollows et al., 2019) after assessing its energy needs beyond 2030. As part of this initiative, the province of New Brunswick, through New Brunswick Energy Solutions Corporation, committed $10 million in June 2018 toward the establishment of an advanced SMR research cluster in New Brunswick. Two fast-spectrum reactor vendors, Moltex Energy (a fast-spectrum MSR developer) and ARC Nuclear (a sodium fast reactor), invested in this initiative with matching funds that will be used to support R&D of these advanced technologies. The provincial power generator NB Power and the University of New Brunswick are the other key participants.

New Brunswick is considering the site of the Point Lepreau Nuclear Generating Station for prototype SMR installation.

15.3 SMR markets and potential applications in Canada

There are a number of potential applications in Canada where SMRs can be used not only for generating electricity but also for generating steam for district heating, desalination, and as industrial process heat. Other applications may include hydrogen production and steam-assisted gravity drainage process to extract oil and power for resource extraction. The economic impact of potential off-grid applications of SMRs in Canada is assessed to be significant (Wojtaszek, 2019).

The Canadian SMR Roadmap (SMR Roadmap, 2018) identified three main areas for SMR applications. The following sections examine these three areas.

15.3.1 On-grid applications for electricity

Canada has a history of producing low greenhouse gas-emitting electricity because of its extensive water resources. Close to 60% of Canada's electricity is produced by hydroelectric dams as seen in Fig. 15.1.

To reduce GHG emissions and meet Canada's commitment to international community, Canadian government established new regulations to phase out traditional

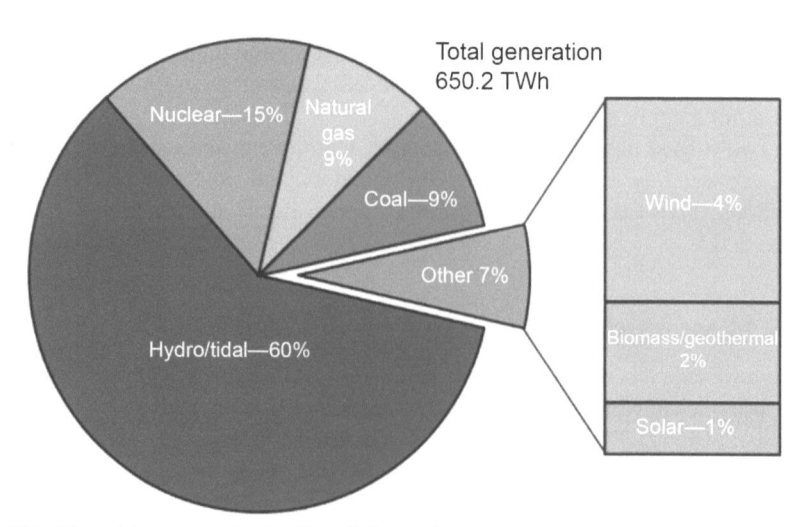

Fig. 15.1 Electricity generation in Canada by fuel type.
2017 Data from Canada Energy Regulator Website, n.d. Provincial and Territorial Energy Profiles—Canada. https://www.cer-rec.gc.ca/nrg/ntgrtd/mrkt/nrgsstmprfls/cda-eng.html (Accessed 13 September 2019).

coal-fired electricity by 2030 (Environment and Climate Change Canada Website, n.d.). With this goal, Canada is targeting to have 90% of its electricity from nonemitting sources by 2030 and cutting carbon pollution from the electricity sector by 12.8 million tons (Environment and Climate Change Canada Website, n.d.).

As part of GHG emission reduction strategies, the provinces of Ontario, New Brunswick, and Saskatchewan have expressed interest in SMRs for electricity production and actively engaged with SMR vendors. Ontario and New Brunswick already produce a significant percentage of their on-grid electricity through nuclear power generators (using CANDU reactors) and will have to address the electricity gap that will be created with the retirement of CANDU stations (Sollows et al., 2019; Financial Accountable Office of Ontario Website, n.d.). Saskatchewan indicated that it is looking at SMRs as an option to replacing its coal-fired electric generating stations (Global News Website, n.d.). Ontario is also considering off-grid SMRs for mining applications (HATCH, 2016; Nalasco, 2019).

Transportation sector is the largest GHG-emitting sector in Canada (Environment and Climate Change Canada, 2019) with 24% of the total emissions; see Fig. 15.2. Electrification of transportation will significantly help in achieving Canada's long-term GHG emission reduction targets. However, this will require significant increase in electricity demand that has to come from clean energy sources. SMRs can be part of the solution to meet this demand.

Waste and others
42 Mt CO_2 eq (6%)

Agriculture
72 Mt CO_2 eq (10%)

Oil and gas
195 Mt CO_2 eq (27%)

Heavy industy
73 Mt CO_2 eq (10%)

Total 716 Mt CO_2 equivalent

Electricity
74 Mt CO_2 eq (10%)

Buildings
85 Mt CO_2 eq (12%)

Transportation
174 Mt CO_2 eq (24%)

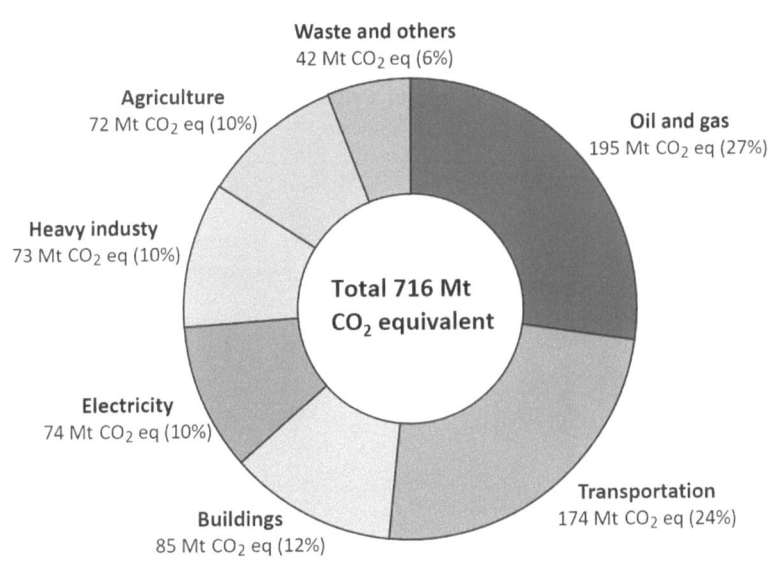

Fig. 15.2 GHG emissions by economic sector in Canada.
2017 Data from Environment and Climate Change Canada, 2019. 2019 National Inventory Report, 1990–2017: Greenhouse Gas Sources and Sinks in Canada. http://publications.gc.ca/collections/collection_2019/eccc/En81-4-1-2017-eng.pdf (Accessed 12 September 2019).

15.3.2 Heavy industry

15.3.2.1 Mining

Canada has extensive land mass with rich natural resources, mostly located at off-grid northern locations. In 2017 Canada's mining industry employed directly and indirectly 634,000 workers at 1189 mines across Canada and contributed $97 billion to Canada's GDP (Marshall, 2018). Overall, Canada's mining sector is expected to grow at an average of 1.9% per year until 2035 (The Conference Board of Canada, 2014), resulting in a total increase of 50%, equivalent to 650 new mines. The Canadian mining industry acknowledges the climate change, focused on supporting the transition to a lower-carbon future and considers nuclear power a clean energy source (Marshall, 2018).

Out of some 1200 operating or proposed mines in Canada (Marshall, 2018), 32 off-grid mines rely on diesel power generators; the remainder is connected to electricity grids. The power demands for off-grid mines in Canada range from 4 to 125 MWe. As shown in Fig. 15.3 the power requirements of 91% of these mines are between 5 and 50 MWe. More information is presented in Wojtaszek (2019). A consideration for SMRs at off-grid mining locations is the transportability of them from one location, after mining completion, to another.

15.3.2.2 Oil sands extraction

Although most oil sands projects are located near a provincial electricity grid, these projects require a significant amount of thermal power for bitumen extraction and upgrading processes. Currently, this thermal power is mostly generated via the burning of natural gas and is significant contributor to greenhouse gas emissions as shown in Fig. 15.2. Surface mining and in situ extraction of oil from underground are

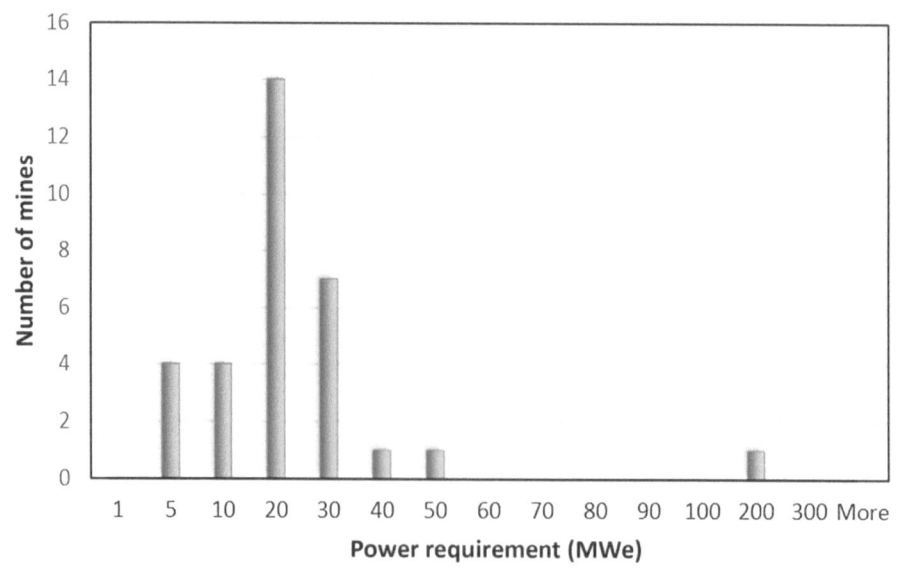

Fig. 15.3 A histogram of power requirements of off-grid mines in Canada (Wojtaszek, 2019).

resource intensive and may require high-temperature steam equivalent to more than 1000-MWt power. The analysis presented in Wojtaszek (2019) shows that more than 50 SMRs with 300 MWe (~900 MWt) power rating would be needed for bitumen extraction from oil sands if SMRs are used exclusively for that purpose. Advanced reactors with steam temperature greater than 500°C are suitable for this application.

15.3.3 Remote communities

A potential application for SMRs is electricity and steam production for district heating in off-grid remote communities. In these remote locations, power is generated mostly by diesel engines, which are sometimes operated by diesel delivered by air. Replacing diesel with clean energy sources and enhancing the quality of life at remote communities are two priority items in Canada.

In some of the Northern Territories in Canada, for example, in Nunavut or in Northwest Territories, the actual cost of generating electricity in 2015 was greater than CAD $1/kWh, more than an order of magnitude higher than the CAD ~0.10/kWh in the rest of Canada that is on the grid (Standing Senate Committee on Energy, The Environment and Natural Resources, 2015). This is shown in Fig. 15.4. This is because the electricity is generated by diesel engines often operating with air-lifted diesel fuel.

As of 2011, there were 292 remote communities in Canada, of which 176 of them are powered by diesel generators (Government of Canada, 2011). The number of remote communities has decreased from 380 to 292 over 25 years primarily as a result of grid extension and abandonment of communities due to relocation to larger villages or cities. A series of papers by University of Waterloo (Karanasios and Parker, 2017a,

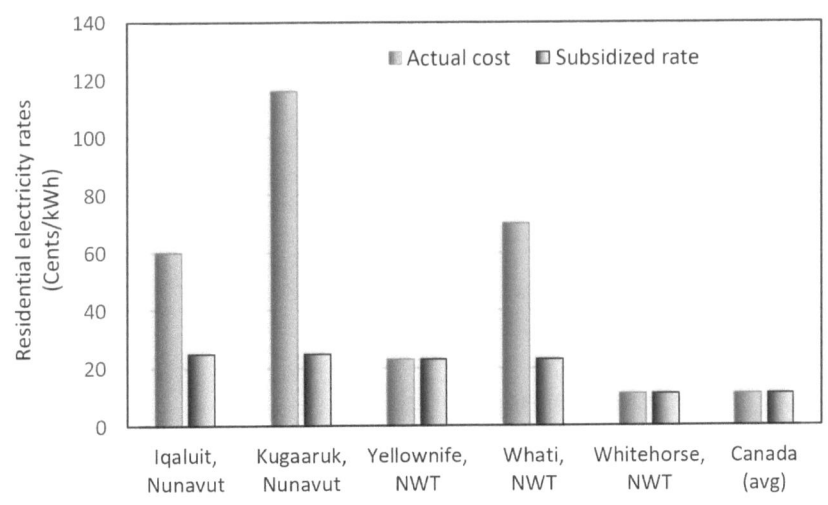

Fig. 15.4 Residential electricity rates at selected locations in Canadian Territories and rest of Canada.
Data from Standing Senate Committee on Energy, The Environment and Natural Resources, 2015. Powering Canada's Territories. http://www.parl.gc.ca/Content/SEN/Committee/412/enev/rep/rep14jun15-e.pdf (Accessed 23 August 2019).

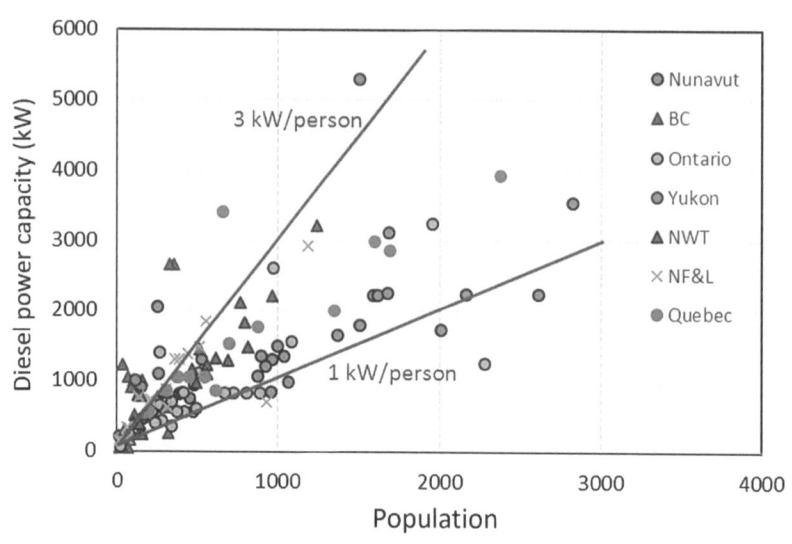

Fig. 15.5 Populations and diesel power capacity in off-grid communities in Yukon (Karanasios and Parker, 2017a), NWT (Karanasios and Parker, 2017b), Nunavut (Karanasios and Parker, 2017c), BC (Karanasios and Parker, 2017d), Ontario (Karanasios and Parker, 2017e), Quebec (Karanasios and Parker, 2017f), and NF&L (Karanasios and Parker, 2017g).

b,c,d,e,f,g) provide data on population and power needs of remote communities. The data extracted from these papers (see Fig. 15.5) indicate the following:

- Canada's remote communities range in population from less than 100 to 3000 residents.
- Electricity generated by diesel engines ranges from less than 1 to 5 MWe, with the majority of communities using less than 2 MWe.
- Typical electricity usage per person ranges from 1 to 3 kW/person. This number is affected by the utilization of the diesel generators that operate at 15%–40% of full power. In some small communities, there is significant excess power capacity, possibly due to emigration.
- The installation of SMRs may lead to the installation of electric heating systems. If fossil-fired heating systems are replaced by electric heating, electricity demand of that community may increase by a factor of almost 10 if all heating is provided electrically. However, a more likely scenario is that the community has a district heating system in place that could draw thermal power from the SMRs. Such a system would require less thermal power than the equivalent electric power for heating (Wojtaszek, 2019).

15.3.4 Other potential applications

15.3.4.1 Floating power stations and icebreakers

Canada's north has a strategic value as a source of natural resources and a valuable trade route connecting the Atlantic Ocean and the Pacific Ocean. Mines in the Arctic Archipelago and gas and oil discoveries showed the potential of this region as a source of great wealth.

The warming of Earth and, as a result, the melting of Arctic ice are making the Northwest Passage an increasingly attractive trade route. Canada considers this route its own and named it the "Canadian Northwest Passage" in 2009. However, this is not recognized by other countries including the United States and various European countries. Canada's "use it or lose it" policy requires increased presence in these seaways with icebreakers, submarines, increased Canadian population (remote Arctic communities), and military bases that can be powered by nuclear reactors of SMR size.

15.3.4.2 Military bases

The Canadian Armed Forces, including those under the North American Aerospace Defense Command (NORAD), operate from a number of locations in the North, such as Inuvik, Yellowknife, Rankin Inlet, Iqaluit, and Goose Bay. They share a number of facilities with other federal government organizations, such as a cold weather training facility at Resolute Bay, a signals intelligence facility at Canadian Forces Station Alert—the northern most permanently inhabited facility in the world—and a high Arctic weather station at Eureka. In addition, work is ongoing to complete the Nanisivik Naval Facility, which will support operations of the new Arctic Offshore Patrol Ships, and other government maritime vessels in an effort to maintain a Canadian presence in Arctic waters during the navigable season (Ministry of National Defence, 2017).

The Canadian Forces cover a large area with more than 800 buildings at over 60 sites[a] (Ministry of National Defence, 2017). A significant number of these sites are part of the North Warning System (11 long-range radar stations and 39 short-range radars) (Wikipedia, n.d.) with seasonal peak staffing levels less than 100 (Nunatsiaq News, n.d.). Using the data in Fig. 15.5, it can be concluded that the power needs of these stations are quite a bit less than 1 MWe. These potential sites could be a good fit for a microreactor.

15.3.4.3 Summary of potential Canadian applications for SMRs

The power needs of typical on- and off-grid Canadian SMR applications are summarized in Table 15.1.

Table 15.1 Typical power requirements for Canadian applications.

Military bases (population <300)	<1 MWe and steam
Remote communities (population 100–3000)	1–5 MWe and steam
Arctic sovereignty (icebreakers)	30–70 MWe
Mining	10–50 MWe
Oil extraction	>250 MWe and steam
On-grid electricity to replace fossil fuel plants	150 to >300 MWe

---

[a] Including the Joint Task Force North, headquartered in Yellowknife with detachments in Whitehorse and Iqaluit.

15.4 Canadian regulatory framework

CNSC's regulatory framework (CNSC Website, n.d.-a) broadly uses the risk-informed approach contained in the IAEA safety and security framework, and this allows for the CNSC to evaluate any type or size of reactor design on the basis of broader safety principles rather than specific technological requirements. This approach is suitable for advanced reactors that have significantly different design and safety characteristics than the well-understood water-cooled reactors and allows technology developers to propose alternate approaches that satisfy these broader safety principles (de Vos, 2019). As well, CNSC's prelicensing vendor design review (VDR) process (CNSC Website, n.d.-b) enables vendors to engage with the regulator at an early stage of the reactor development. The three-stage VDR brings increased credibility to the vendor's design while it progresses through the various stages of review with increasing depth. This has been a useful tool for SMR vendors to secure funds from investors for further development.

Prelicensing VDR (CNSC Website, n.d.-b) is an optional service offered by the CNSC to provide an assessment of a nuclear power plant design based on a vendor's reactor technology. Although this is not a required part of the licensing process for a new nuclear power plant, it provides a staged path to licensing. The CNSC's VDR includes three phases:

Phase 1: Prelicensing assessment of compliance with regulatory requirements: This review is conducted to assess if the vendor design and design processes demonstrate that Canadian regulatory requirements and expectations for design and safety analysis are being implemented. CNSC also evaluates how a systematic and quality-assured research and development program is being established to inform decision-making commensurate with novelty and safety significance. It is a useful phase for vendors to learn CNSC requirements and demonstrate that they are on track to meet them.

Phase 2: Prelicensing assessment for any potential fundamental barriers to licensing: This phase goes into further details and focuses on identifying if any potential fundamental barriers to licensing exist or are emerging with respect to the reactor design. Phase 2 serves to give the CNSC a significant level of assurance that the vendor has taken into account CNSC design requirements as the design becomes more established at the system level. Consideration is also given to the extent to which generic or outstanding safety issues have been resolved. In addition, CNSC staff conduct an audit of the design process, to verify that it has been implemented correctly and in accordance with the vendor's policies and procedures. The vendor is also expected to provide follow-up information to demonstrate how it is resolving any issues identified during Phase 1.

Phase 3: Follow-up: This phase allows the vendor to follow up on certain aspects of Phase 2 findings by seeking more information from the CNSC about a Phase 2 topic and/or asking the CNSC to review additional activities taken by the vendor toward the reactor's design readiness.

CNSC Website (n.d.-b) provides more information on CNCS's VDR process. As listed in Tables 15.2 and 15.3, there are 12 SMR vendors at various phases of (or pending) review at CNC's VDR process.

Table 15.2 Status of CNSC's prelicensing vendor design reviews in force (December 2019; CNSC Website, n.d.-b).

Vendor	Name of design and cooling type	Power (MWe)	Applied for	Review start date	Status
Terrestrial Energy Inc.	IMSR (thermal MSR)	200	Phase 1 Phase 2	April 2016 December 2018	Complete Assessment in progress
Ultra Safe Nuclear Corporation	MMR-5 and MMR-10 (HTGR)	5–10	Phase 1 Phase 2	December 2016 Pending	Complete Project start pending
LeadCold Nuclear Inc.	SEALER (lead-cooled)	3	Phase 1	January 2017	On hold at vendor's request
Advanced Reactor Concepts Ltd.	ARC-100 (sodium-cooled)	100	Phase 1	September 2017	Complete
Moltex Energy	Moltex Energy Stable Salt Reactor (fast MSR)	300	Series Phase 1 and 2	December 2017	Phase 1 assessment in progress
SMR, LLC. (A Holtec International Company)	SMR-160 (iPWR)	160	Phase 1	July 2018	Assessment in progress
NuScale Power, LLC	NuScale (iPWR)	60	Phase 2	Pending end 2019	Project start pending
U-Battery Canada Ltd.	U-Battery (HTGR)	4	Phase 1	Pending end 2019	Project start pending

Table 15.3 Status of CNSC's prelicensing vendor design reviews under development (December 2019; CNSC Website, n.d.-b).

Vendor	Name of design and cooling type	Power (MWe)	Application received	Applied for
X-Energy, LLC	Xe-100 (HTGR)	75 MWe	October 2019	To be determined
GE-Hitachi Nuclear Energy	BWRX-300 (BWR)	300	March 2019	Phase 2
Westinghouse Electric Company, LLC	eVinci (heat pipe reactor)	Various outputs up to 25 MWe	February 2018	Phase 2
StarCore Nuclear	StarCore Module (HTGR)	10	October 2016	Series Phase 1 and 2

15.5 Support for development and deployment

15.5.1 Supply chain readiness

Canada has a full spectrum of supply chain supporting the Canadian nuclear industry. The manufacturers and service providers are organized under the Organization of Canadian Nuclear Industries (OCNI). Canada's supply chain has remained and continues to remain active through refurbishment and life extension activities of CANDU reactors. OCNI was an active participant in the development of the Canadian SMR Roadmap and a supporter of SMR activities in Canada.

Most SMR technologies proposed for deployment in Canada will require new capabilities and necessitate the supply chain to adapt to new technologies. OCNI has actively challenged its members to identify and respond to the SMR Roadmap recommendations for the supply chain. Canadian engineering companies are forming partnerships with SMRs vendors: For example, LeadCold has partnered with Promation Nuclear on the development, design, and safety analysis of the SEALER reactor; Holtec partnered with SNC-Lavalin to collaborate in the development of Holtec's SMR-160 reactor; and Urenco partnered with Kinectics to develop and license the U-Battery reactor in Canada.

15.5.2 CNL's SMR demonstration siting initiative

CNL was formed in 2011 as a result of the reorganization of AECL. CNL is now a private-sector organization that operates the research laboratories at Chalk River and Whiteshell sites. After the reorganization, AECL remained a crown corporation and retained a small group of employees to oversee the operations of CNL. CNL's main mandates are to provide R&D support to the nuclear industry and to provide technical support to the regulator and the government on various nuclear technologies, including small modular reactors (SMRs).

CNL has identified a strategic goal of demonstrating the commercial viability of the SMR by 2026. For this purpose, CNL issued a request for expressions of interest in June 2017 (CNL Press Release, 2017) to gather input on siting a SMR demonstration plant at an AECL-owned and CNL-managed site in Canada. This was followed by an invitation for SMR demonstration projects in April 2018 (CNL Press Release, 2018) with two invitation uptakes planned annually. CNL's invitation process consists of four stages. In Stages 1 and 2, CNL evaluates with increased rigor the technical and business merits of the demonstration project including the proposed designs, the financial viability of the projects, and the necessary national security and integrity requirements. Stage 3 includes preliminary, nonexclusive discussions regarding land arrangements, project risk management, and contractual terms. The fourth and final stage includes construction, testing and commissioning, operation, and ultimately decommissioning of the SMR unit.

There has been significant interest by SMR vendors in CNL's invitation for SMR demonstration. As of December 2019, three proponents (Starcore, Terrestrial Energy, and U-Battery) successfully completed the "prequalification" stage (Stage 1) of

CNL's invitation and were invited to submit to Stage 2 (due diligence), and a fourth proponent (Global First Power) has entered the "contract negotiations" stage (Stage 3). The application from Global First Power (GFP), with support from OPG and Ultra Safe Nuclear Corporation (USNC), supports a proposal to deploy a micro modular reactor (MMR) plant at Chalk River in Ontario.

In 2019 March the GFP/USNC/OPG consortium submitted an application to the CNSC (World Nuclear News, 2019) for a license to prepare site, which is the first step in the CNSC licensing process. This marks an important milestone in Canada as being the first SMR license application. Shortly after this announcement an environmental assessment study was initiated for the MMR reactor (WNN Article, 2019).

15.5.3 R&D support

Until recently (early 2010s), nuclear R&D in Canada focused heavily on the existing fleet of operating CANDU reactors. CNL is the primary institution for nuclear R&D in Canada and performs nuclear R&D with funding from the government of Canada. Canadian government's divestiture of the CANDU technology coincided with the developments in SMRs and advanced reactors in Canada, and, as a result, CNL initiated R&D programs on SMRs and various advanced reactors. Currently, R&D programs at CNL address not only technology-agnostic research areas relevant to SMRs, such as passive safety, integrity of below-grade containment structures, cybersecurity, fuel manufacturing and characterization, and computational modeling tools, but also specific technology areas related to iPWRs, MSRs, and HTGRs.

In June 2019, CNL launched a new program, Canadian Nuclear Research Initiative (CNRI), to support jointly funded research projects with SMR proponents in Canada (CNL Website, n.d.). The goal of the program is to accelerate the deployment of SMRs in Canada. Although the project definition and scoping process are still ongoing, strong interest by MSR and VHTR vendors indicated that there will likely be new industry—CNL R&D collaborations on these reactor technologies.

A number of Canadian universities are doing research on nuclear power technologies including SMRs and advanced reactors. For example, University of New Brunswick (UNB) is developing facilities and expertise in fast reactors, Royal Military College (RMC) is developing an organic coolant SMR concept based on Canadian-developed technologies, and the University of Ontario Institute of Technology (UOIT) is doing research on hydrogen production using nuclear power. A consortium of universities, led by McMaster University, initiated a SMR-specific training program, small modular advanced reactor training (SMART), to meet the evolving needs of the Canadian nuclear power sector. This program includes McMaster University, University of Waterloo, UOIT, RMC, UNB, and Queen's University.

As identified by the SMR roadmap, international collaboration is recognized to be an important aspect of the successful deployment of SMRs in Canada. Hence, Canada is considering to join the Generation IV International Forum (GIF)—MSR and VHTR systems for international R&D collaborations and currently participating as an observer in these two GIF systems.

15.6 Future trends

Global movement in recognizing climate change and establishing national targets to reduce greenhouse gas (GHG) emissions is changing the energy markets. Under the Paris Agreement, Canada is committed to significantly reduce GHG emissions.

15.6.1 Greenhouse gas emissions in Canada and Canada's targets for 2030 and 2050

Paris Agreement was ratified by Canada on October 5, 2016 and entered into force on November 2016 (Government of Canada, 2019).

Under the Paris Agreement, Canada has set a target to reduce GHG emissions to 30% below 2005 levels by 2030 (from 732 Mt CO_2 equivalent in 2005 to 513 Mt CO_2 equivalent in 2030) and 90% below 2005 levels (to 73 Mt CO_2 equivalent) by 2050 (Government of Canada, n.d.). As part of its efforts to achieve these goals, the Government of Canada introduced in December 2018 regulations to phase out coal by 2030 (Government of Canada Website, n.d.). For the most part the coal power generation is being replaced by natural gas and renewables.

Fig. 15.6 shows the historical GHG emissions in Canada and projections to 2030 (Environment and Climate Change Canada Report, 2019). As seen in Fig. 15.2, transportation and oil and gas are the major contributors to GHG emissions in Canada followed by heat for buildings, electricity generation, agriculture, and heavy industry. Based on the current measures taken, GHG emissions are expected to reduce to 701 Mt CO_2 equivalent by 2030 (Office of the Parliamentary Budget Officer, 2019):

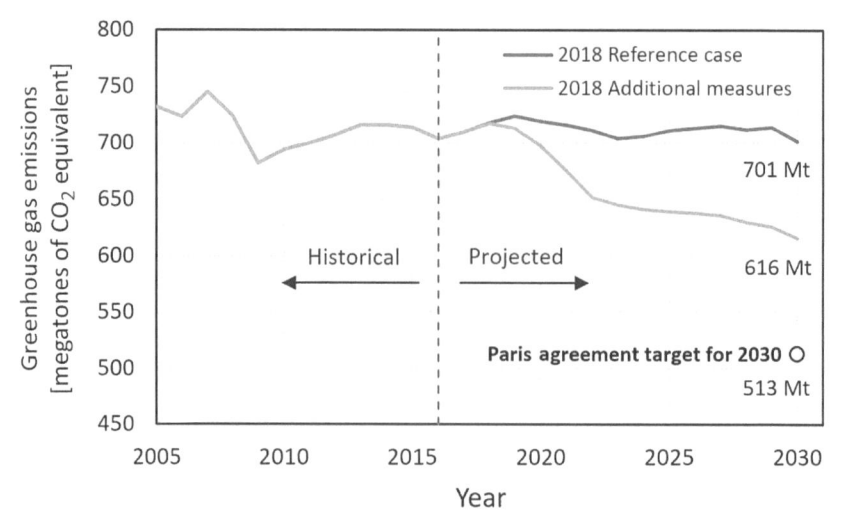

Fig. 15.6 Historical greenhouse gas emissions and projections in Canada, 2005–30 (Environment and Climate Change Canada Report, 2019).

- Electricity is expected to provide the largest reduction of 53 Mt, primarily through the accelerated phaseout of coal-fired electricity generating stations.
- Transportation is expected to reduce emissions by 32 Mt through increased fuel efficiency and an increase in the share of zero-emission vehicles (electric or hydrogen powered).
- Improved building efficiency is expected to result in a 16-Mt reduction in GHG by 2020.
- Emissions in the oil and gas sector are expected to rise by 12 Mt due to the continued growth in the oil sands subsector.
- Emissions in heavy industry are also projected to increase (by 11 Mt) as growth outpaces energy efficiency gains.

An additional 103-Mt reduction is required to achieve the 2030 Paris target. Some of this reduction is expected to come from land use, land use change, and forestry and the Western Climate Initiative (Office of the Parliamentary Budget Officer, 2019). However, additional reductions are required. Currently, policies and measures are being developed for additional GHG emission reductions. Under one of these scenarios, GHG emissions are expected to reduce to 616-Mt CO_2 equivalent by 2030. This is shown in Fig. 15.6 under the "2018 Additional Measures" scenario. Policies and measures for additional measures are under development but not yet fully implemented.

Even with additional measures, current projections fall short of Paris Agreement target, and it is not clear how Canada can meet its GHG targets without significant electricity and process heat generation by new nuclear power plants.

15.6.2 Future trends in the power generation industry

To achieve GHG emission targets, Canada will need to fundamentally transform energy production, distribution, and consumption. Some of these changes are already happening or started to happen. A snapshot of these changes is listed in the succeeding text:

- Increased penetration of renewables, that is, solar and wind, into the electricity grid: Significant reduction in the cost of solar photovoltaics and wind turbines in recent years made these technologies competitive with traditional electricity sources. A challenge with solar and wind is their intermittent nature and weather dependence. The intermittent nature of this energy source implies either a need for energy storage or another source of energy for baseload. Canada is somewhat immune to this issue because of it has significant portion of its electricity produced by hydroelectric dams. Hydroelectric dams can be used to control electricity production akin to storing energy. However, the larger the portion of the electricity produced by intermittent power sources, the more there will be a need for additional energy storage systems. SMRs in this ecosystem would also provide a way to balance the intermittent sources by providing baseload energy and load variability.
- Distributed generation of power through renewables is changing the electricity grid in Canada. Distributed technologies, in which power is generated at or near its point of use, are playing an increasingly important role in electric grid and especially important for Canada that has a large land area. "A significant increase in customer-side participation in the energy supplied and demand-side management is contributing to a paradigm shift in the power industry towards a more decentralized energy supply and bidirectional power flows. This is driving the modernization of the grid project known as Smart Grid" (Natural Resources Canada website, n.d.). SMRs would be a natural fit in this microgrid concept.

- Transportation sector has started to transition to electric vehicles. Because of the large land area and long distances between major cities in Canada, transportation sector is the largest GHG emitter in Canada. This can be seen in Fig. 15.2. Hence, total decarbonization of transportation sector (a realistic date is 2050) will have a significant impact to Canada's energy sector. An obvious, but not the only, choice for decarbonization is the use of electric vehicles, provided that the electricity used to charge these vehicles are produced from clean energy sources, such as solar, wind, and nuclear. A complementary technology to electric batteries is the hydrogen fuel cell vehicles using hydrogen as fuel. As with electric vehicles the source of energy to produce hydrogen should be from clean energy sources. With energy density significantly higher than current electric batteries, hydrogen can allow vehicles to travel farther between refueling, as well as charge in less time. However, today's primary method of hydrogen production, steam methane reforming, can create as much as 10 kg of GHG for each kg of hydrogen produced. Hydrogen can be produced by electrolysis using electricity, but the high cost of hydrogen produced this way may limit its acceptance by general population. Producing hydrogen using high-temperature steam generated by advanced nuclear plants (IAEA Nuclear Energy Series, n.d.) can reduce the cost and make hydrogen fuel cell technologies more attractive to mass market. Canada is developing a hydrogen production process using the copper-chlorine thermochemical cycle (Suppiah, n.d.) that requires steam at temperatures of 500°C and higher. These steam temperatures are compatible with operating temperatures of nonwater-cooled advanced SMRs.
- As seen in Fig. 15.2, after electricity and transportation, the two largest contributors to GHG emissions are heavy industry and the oil and gas sector. Processes such as cement production, potash production, ammonia production, and bitumen upgrading all require hydrogen as an input product to their process. Electricity, hydrogen, and high-temperature steam produced by nuclear are possible replacements for fossil fuels presently used by these sectors. Canadian mining sector is already transitioning to nonfossil energy sources and opened its first all-electric mine in Northern Ontario in September 2019 (CTV News, 2019).
- Canada, together with the United States and Japan, is leading the nuclear innovation: Clean Energy (NICE) Future initiative (NICE Future Website, n.d.) to promote a dialogue on the role that nuclear energy can play in the clean energy systems. The NICE Future initiative envisions that nuclear energy can be integrated with other clean energy technologies to create a emission-free economy that includes integrated nuclear-renewables for flexible electricity generation, hydrogen production, industrial process heat, district heating, and desalination for drinking water. SMRs and advanced Generation IV reactors can play a significant role in meeting these goals.

15.7 Conclusion

Since the early 2010s, Canada has seen a surge of SMR technology developers expressing interest to deploy SMRs in Canada. A major motivation for SMRs is the recognition of global climate change and the need to reduce GHG emissions by replacing fossil fuels with clean energy sources. As shown in Fig. 15.6, current policies and measures in Canada are not projected to meet the Paris targets for 2030 of 513 Mt of CO_2 equivalent even with the "2018 additional measures" scenario, which includes carbon tax. If Canada is to achieve the targets outlined in the Paris Agreement, a faster penetration of clean energy technologies (wind, solar, and nuclear) into

Canadian economy is needed. SMRs are expected to play a significant role in providing both heat and electricity while also responding to the increased demand and flexibility in the electricity grid. Most nonelectrical applications will require advanced SMRs that can provide steam at temperatures greater than 500°C. Hence the future nuclear energy sector in Canada will likely include a mix of advanced nuclear reactors (i.e., HTGR and MSR) in addition to the well-known water-cooled nuclear reactors.

Acknowledgments

This book chapter is prepared with contributions by G. Strati, M. Moore, D. F. Wang, D. Wojtaszek (all from CNL), and P. Chan (Royal Military Collage). Their contributions are greatly appreciated.

References

CNL Press Release, June 1, 2017. CNL Seeks Input on Small Modular Reactor Technology. http://www.cnl.ca/en/home/news-and-publications/news-releases/2017/SMR.aspx (Accessed 29 August 2019).

CNL Press Release, April 17, 2018. CNL Announces Invitation to site Canada's First Small Modular Reactor. http://www.cnl.ca/en/home/news-and-publications/news-releases/2018/cnl-announces-invitation-to-site-canada-first-smr.aspx (Accessed 29 August 2019).

CNL Website, n.d. https://www.cnl.ca/en/home/commercial/cnl-s-canadian-nuclear-research-initiative-.aspx (Accessed 29 August 2019).

CNSC Website, n.d.-a. Regulatory Framework Overview. http://nuclearsafety.gc.ca/eng/acts-and-regulations/regulatory-framework/index.cfm (Accessed 12 September 2019).

CNSC Website, n.d.-b. https://www.nuclearsafety.gc.ca/eng/reactors/power-plants/pre-licensing-vendor-design-review/index.cfm (Accessed 28 December 2019).

CTV News, September 24, 2019. Official Opening of Canada's First All-electric Mine: Northern Ontario. https://northernontario.ctvnews.ca/official-opening-of-canada-s-first-all-electric-mine-northern-ontario-1.4607600 (Accessed 5 December 2019).

de Vos, M., 2019. Regulatory context on…alternative approaches. In: Invited Presentation, 39th Annual Conference of the Canadian Nuclear Society and 43rd Annual CNS/CNA Student Conference, Ottawa, ON, Canada, June 23–26.

Environment and Climate Change Canada, 2019. 2019 National Inventory Report, 1990–2017: Greenhouse Gas Sources and Sinks in Canada. http://publications.gc.ca/collections/collection_2019/eccc/En81-4-1-2017-eng.pdf (Accessed 12 September 2019).

Environment and Climate Change Canada Report, January 2019. Progress Towards Canada's Greenhouse Gas Emissions Reduction Target. https://www.canada.ca/content/dam/eccc/documents/pdf/cesindicators/progress-towards-canada-greenhouse-gas-reduction-target/2019/progress-towards-ghg-emissions-target-en.pdf (Accessed 15 September 2019).

Environment and Climate Change Canada Website, n.d. Canada's Coal Power Phase-out Reaches Another Milestone. https://www.canada.ca/en/environment-climate-change/news/2018/12/canadas-coal-power-phase-out-reaches-another-milestone.html (Accessed 13 September 2019).

Financial Accountable Office of Ontario Website, n.d. Nuclear Refurbishment Report. https://www.fao-on.org/web/default/files/publications/Nuc%20Refurb%20nov%202017/Nuclear-Refurb-EN.pdf (Accessed 13 September 2019).

Global News Website, n.d. Saskatchewan Giving Early Consideration for Small Nuclear Reactors. https://globalnews.ca/news/5286649/saskatchewan-small-nuclear-reactors/ (Accessed 13 September 2019).

Government of Canada, 2011. Status of Remote/Off-Grid Communities in Canada. https://www.nrcan.gc.ca/sites/www.nrcan.gc.ca/files/canmetenergy/files/pubs/2013-118_en.pdf (Accessed 7 September 2019).

Government of Canada, 2019. Environmental and natural resources website. In: The Paris Agreement. https://www.canada.ca/en/environment-climate-change/services/climate-change/paris-agreement.html (Accessed 14 September 2019).

Government of Canada, n.d. Canada's Mid-Century Long-Term Low-Greenhouse Gas Development Strategy. https://unfccc.int/files/focus/long-term_strategies/application/pdf/canadas_mid-century_long-term_strategy.pdf (Accessed 14 September 2019).

Government of Canada Website, n.d. Canada's Coal Power Phase-Out Reaches Another Milestone. https://www.canada.ca/en/environment-climate-change/news/2018/12/canadas-coal-power-phase-out-reaches-another-milestone.html (Accessed 14 September 2019).

HATCH, 2016. Ontario Ministry of Energy—SMR Deployment Feasibility Study, Feasibility of the Potential Deployment of Small Modular Reactors (SRMs) in Ontario. June 2.

IAEA Nuclear Energy Series, n.d. Hydrogen Production Using Nuclear Energy. https://www-pub.iaea.org/MTCD/Publications/PDF/Pub1577_web.pdf (Accessed 15 September 2019).

Karanasios, K., Parker, P., 2017a. Recent developments in renewable energy in remote aboriginal communities, Yukon, Canada. J. Pap. Can. Econ. Dev. 17, 29–40.

Karanasios, K., Parker, P., 2017b. Recent developments in renewable energy in remote aboriginal communities, NWT, Canada. J. Pap. Can. Econ. Dev. 17, 41–53.

Karanasios, K., Parker, P., 2017c. Recent developments in renewable energy in remote aboriginal communities, Nunavut, Canada. J. Pap. Can. Econ. Dev. 17, 54–64.

Karanasios, K., Parker, P., 2017d. Recent developments in renewable energy in remote aboriginal communities, British Columbia, Canada. J. Pap. Can. Econ. Dev. 17, 65–81.

Karanasios, K., Parker, P., 2017e. Recent developments in renewable energy in remote aboriginal communities, Ontario, Canada. J. Pap. Can. Econ. Dev. 17, 82–97.

Karanasios, K., Parker, P., 2017f. Recent developments in renewable energy in remote aboriginal communities, Quebec, Canada. J. Pap. Can. Econ. Dev. 17, 98–108.

Karanasios, K., Parker, P., 2017g. Recent developments in renewable energy in remote aboriginal communities, New Foodland and Labrador, Canada. J. Pap. Can. Econ. Dev. 17, 109–118.

Lee, K.W., 2018. The Canadian Nuclear Safety Commission's Readiness to regulate small modular reactors. In: 1st International Conference on Generation IV and Small Reactors, Ottawa, ON, Canada, November 6–8.

Marshall, B., 2018. Facts & Figures of the Canadian Mining Industry—2018. The Mining Association of Canada, Ottawa, ON.

Ministry of National Defence, 2017. Strong, Secure, Engaged: Canada's Defence Policy. http://publications.gc.ca/collections/collection_2017/mdn-dnd/D2-386-2017-eng.pdf (Accessed 7 September 2019).

Nalasco, A., 2019. Ontario's perspective on small modular reactor (SMR) development. In: Ministry of Energy, Government of Ontario, Invited Presentation, 39th Annual Conference of the Canadian Nuclear Society and 43rd Annual CNS/CNA Student Conference, Ottawa, ON, Canada, June 23–26.

Natural Resources Canada website, n.d. Smart Grid. https://www.nrcan.gc.ca/energy/electricity-infrastructure/smart-grid/4565 (Accessed 15 September 2019).

NICE Future Website, n.d. http://www.cleanenergyministerial.org/initiative-clean-energy-ministerial/nuclear-innovation-clean-energy-future-nice-future (Accessed 5 January 2019).

Nunatsiaq News, n.d. NWS: It's Working People Who Make the Radar Run. http://nunatsiaq.com/stories/article/20918_nws_its_working_people_who_make_the_radar_run/ (Accessed 5 September 2019).

NuSCale Website, n.d. https://www.nuscalepower.com/projects/current-projects/canada (Accessed 15 December 2019).

Office of the Parliamentary Budget Officer, June 2019. Closing the Gap: Carbon Pricing for the Paris Target. https://www.pbo-dpb.gc.ca/web/default/files/Documents/Reports/2019/Paris_Target/Paris_Target_EN.pdf (Accessed 14 September 2019).

SMR Roadmap, 2018. A Call to Action: A Canadian Roadmap for Small Modular Reactors. November, https://smrroadmap.ca/ (Accessed 18 August 2019).

Sollows, D., McPherson, B., Thompson, P.D., 2019. Overview of the New Brunswick SMR initiative and resulting impacts for the province of New Brunswick. In: Proceedings of the 39th Annual Conference of the Canadian Nuclear Society and 43rd Annual CNS/CNA Student Conference, Ottawa, ON, Canada, June 23–26.

Standing Senate Committee on Energy, The Environment and Natural Resources, 2015. Powering Canada's Territories. http://www.parl.gc.ca/Content/SEN/Committee/412/enev/rep/rep14jun15-e.pdf (Accessed 23 August 2019).

Suppiah, S., n.d. Hydrogen Production and Utilization—A Canadian Perspective. https://snrworkshop.inl.gov/Shared%20Documents/Suppiah.pdf (Accessed 15 December 2019).

The Conference Board of Canada, 2014. Long-Term Outlook 2014: Forecast Trends by Industry (Publication 6199), Ottawa, ON.

Wikipedia, n.d. North Warning System. https://en.wikipedia.org/wiki/North_Warning_System (Accessed 15 September 2019).

WNN Article, July 17, 2019. Canada Launches Environmental Assessment of Micro Modular Reactor. https://www.world-nuclear-news.org/Articles/Canada-launches-environmental-assessment-of-Micro (Accessed 15 December 2019).

Wojtaszek, D.T., 2019. Potential off-grid markets for SMRs in Canada. CNL Nucl. Rev. J. 8 (2), 87–96.

World Nuclear News, April 2, 2019. First Canadian SMR License Application Submitted. http://world-nuclear-news.org/Articles/First-Canadian-SMR-licence-application-submitted (Accessed 13 September 2019).

Small modular reactors (SMRs): The case of China

16

Danrong Song
Nuclear Power Institute of China, Chengdu, People's Republic of China

16.1 Introduction

China's economy has been rapidly growing since the 1980s. The increased energy demand and environmental problems caused by firing fossil fuels are key challenges for China's sustainable development. Until April 2019, there are 45 nuclear power units in operation in the mainland China, total capacities 4.589 E07 kWe, 4th rank in the world after that of the United States, France, and Japan. There are 12 nuclear power units under construction in the mainland China, total capacities 1.278 E07 kWe. In the end of 2018, the total electricity generation reached 1.9 E09 kWe. The composition of primary energy production was 60.2% fossil, 18.5% hydropower, 9.7% wind power, 9.2% solar power, and 2.4% nuclear power (http://www.chinapower.com.cn/informationzxbg/20180615/113552.html). According to the "State Medium-Long Term (2005–2020) Development Programme of Nuclear Power" issued in October 2007, the total capacity of operating nuclear power plants in 2020 will be 40 GW(e) plus 18 GW(e) under construction. Considerably, increasing application of nuclear energy will greatly improve China's primary energy mix and also its air quality.

To meet the goal of the Chinese government, several Chinese national industries have developed kinds of reactor, such as CAP1400, Hualong 1, and ACP100. Among them, HTR-200 and ACP100 are small modular reactors (SMRs) that have been developed by Tsinghua University and China National Nuclear Corporation separately.

There are several reasons that have driven Chinese national industries to develop SMRs. They are suitable for small electricity grids, district heating, process heating supply, and seawater desalination. According to specific conditions, different counties have varying goals in China. In the north of China, the demand of energy for city heat consumption is several hundred million tons of coal per year, of which 10% is used for heating in the winter season. Due to air pollution in winter, SMR district heating is one of the options. In east China, energy consumption industries, such as building materials, metallurgy, and chemical engineering, have set up their own thermal power plants. The emissions constitute over 70% of the total emission in China. For most of the industries located in east China coastal areas, the serious lack of fresh water resources has become the bottleneck of economic progress. SMRs for process heating supply and seawater desalination would be a good option. In outlying areas of China, such as mountain area and islands, SMRs will be the best choice for electricity generation.

16.2 SMRs in the People's Republic (PR) of China: HTR-200

16.2.1 Introduction of HTR-200

Since the 1970s the high-temperature gas-cooled reactor (HTR) technology has been developed in China. A 10-MW(th) test reactor (HTR-10) with spherical fuel elements was constructed in 2000 and is now in operation. A number of safety-related experiments have been conducted on the HTR-10. R&D on direct cycle helium turbine technology is being carried out. Coupling a helium turbine system to the existing 10-MW (th) test reactor is foreseen. The construction of an industrial scale demonstration plant of modular HTGR (HTR-200) is one of the so-called national major science and technology special projects. The operation date of the 200 MW(e) HTR-200 will be in the year 2020.

16.2.2 Technical aspects

The HTR-200 design has the following remarkable technical features (IAEA-TECDOC-1485, 2005):

- Spherical fuel elements with tristructural isotropic type (TRISO)-coated particles are used, which have a proven capability of fission product retention under 1600°C in accidents.
- Ceramic materials, i.e., graphite and carbon bricks that are resistant to high temperatures, surround the reactor core.
- The decay heat in fuel elements is assumed to be dissipated by means of heat conduction and radiation to the outside of the reactor pressure vessel and then taken away to the ultimate heat sink by water cooling panels on the surface of the primary concrete cell. Therefore no coolant flow through the reactor core would be necessary for the decay heat removal in loss of coolant flow or loss-of-pressure accidents. The maximum temperature of fuel in accidents will be limited to 1600°C.
- Spherical fuel elements are charged and discharged continuously in a so-called "multipass" mode, which means the fuel elements pass through a reactor core several times before reaching the discharge burnup.
- Two independent reactor shutdown systems are foreseen. Both systems are assumed to be located in the graphite blocks of the side reflector. When called upon, neutron absorber elements will fall into the designated channels located in the side reflectors, driven by gravity.
- The reactor core and steam generator are housed in two steel pressure vessels, which are connected by a connecting vessel. Inside the connecting vessel a hot gas duct is mounted. All pressure-retaining components, which constitute the primary pressure boundary, are in touch with the cold helium of the reactor inlet temperature.
- Under an accident with complete loss of pressure, the primary helium inventory will be released into the atmosphere due to the radioactive material in the primary helium loop being very low and then will be no fuel failure during a loss-of-coolant accident (LOCA).
- Several HTR-PM (pebble bed module) modules could be built at one site to satisfy the power capacity demand of a utility. Some auxiliary systems and facilities could be shared among the modules.

16.2.3 Main design parameters

The main technical parameters of HTR-200 can be found in Table 16.1 and Figs. 16.1 and 16.2.

Table 16.1 The main technical parameters of HTR-200.

Parameter	Unit	Value
Capacity of the power plant	mWe	211
Number of reactors	–	2
Thermal output of each reactor	MW	250
Operating pressure of helium gas in primary loop	MPa	7.0
Core coolant inlet temperature	°C	250
Core coolant outlet temperature	°C	750
Diameter/height of active zone	m	3/11
Enrichment of new fuel pellets	%	8.9
Mass of heavy metals in each fuel pellet	g	7.0
Average power density of reactor core	MW/m^3	3.22
Maximum power density of reactor core	MW/m^3	6.57
Total number of fuel cartridges in reactor core	Each reactor	10,000
Number of fuel cycles in core	–	15
Average number of fuel cartridges put in reactor each day	–	400

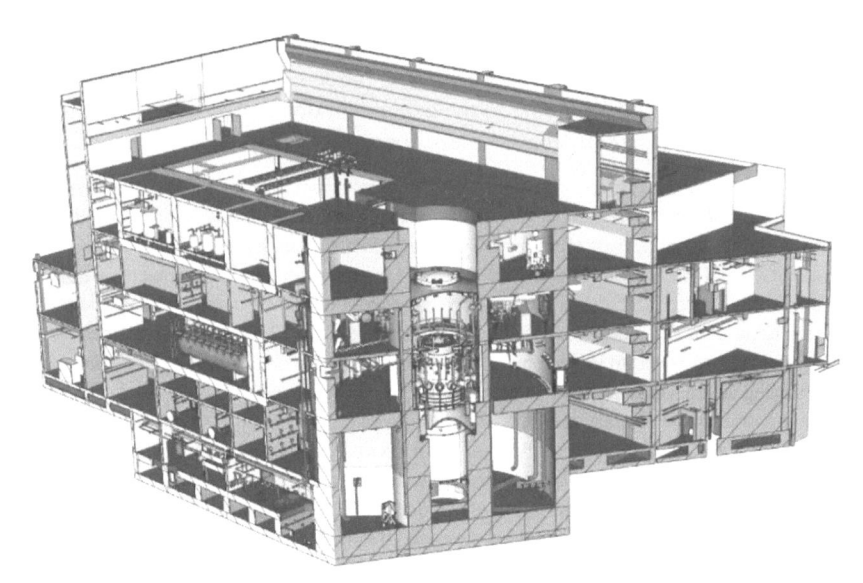

Fig. 16.1 HTR-200: two nuclear steam supply systems (NSSS) modules-general layout of plant (IAEA-TECDOC-1485, 2005).
Used with permission from International Atomic Energy Agency, 2006. Status of Innovative Small and Medium Sized Reactor Designs 2005. IAEA-TECDOC-CD-1485. IAEA, Vienna.

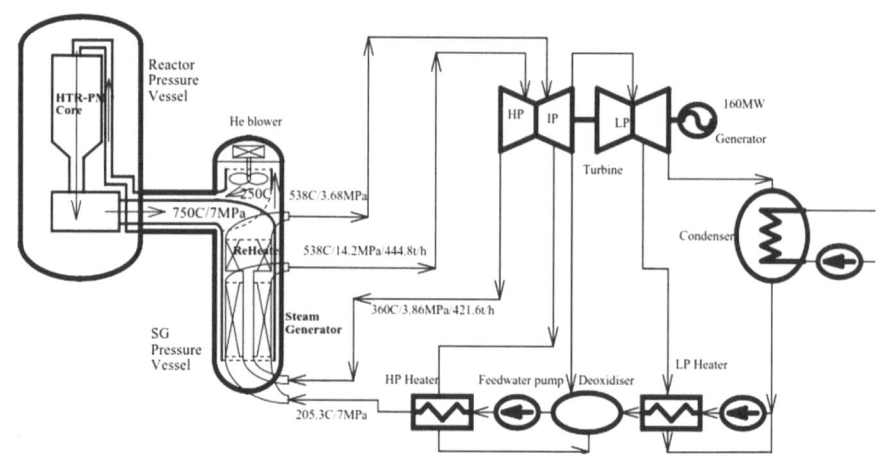

Fig. 16.2 HTR-200: plant process flow (IAEA-TECDOC-1485, 2005). *LP*, low pressure; *IP*, intermediate pressure; *HP*, high pressure.
Used with permission from International Atomic Energy Agency, 2006. Status of Innovative Small and Medium Sized Reactor Designs 2005. IAEA-TECDOC-CD-1485. IAEA, Vienna.

16.2.4 Engineered safety feature plan

The safety design philosophy of HTR-200 is to realize the required high level of safety and, at the same time, to simplify the design of the systems required only for safety purposes, to the greatest extent possible. Emergency measures outside the plant boundary should be made technically not necessary or reduced to a minimum level. The HTR-200 safety design is to a large degree based on the inherent and/or passive safety features, while still adhering to the defense-in-depth principles. The following three features characterize the basic safety concept of HTR-200:

- Radioactive materials are confined through the implementation of multiple barriers with a strong emphasis on fuel elements, especially in accidents. Fuel elements with coated particles serve as the first barrier. Every fuel kernel of about 0.5 mm diameter is coated with three layers of pyrocarbon and one silicon carbon (SiC) layer. A large number of coated particles are dispersed in the graphite matrix of 5 cm diameter to form the fuel-containing part of a fuel element, which in turn is protected by a 0.5-cm thick fuel-free graphite layer. The fuel elements used for HTR-200 were demonstrated to be capable of confining fission products within the coated particles under temperatures of ~1600°C that are not expected for any plausible accident scenario. The second barrier is the primary pressure boundary, which consists of a pressure vessel that hosts units of the primary components. The third barrier is a reactor building and some additional auxiliary buildings, which house the primary helium-containing components.
- The decay heat is automatically removed under accident conditions. In the case of an accident, the primary helium circulator is stopped. Because of the low-power density and the large heat capacity of the graphite structures, the decay heat in fuel elements will dissipate to the outside of a reactor pressure vessel by means of heat conduction and radiation within the core internal structures, without leading to unacceptable fuel temperatures.
- The overall negative temperature feedback is guaranteed under all conditions. The reactor nuclear design assures that the temperature reactivity coefficients of fuel and moderator are

always negative under all operating and accident conditions. Together with the protection action of stopping the primary helium blower, it will lead to an automatic reactor shutdown in an accident.

16.2.5 Testing and verification

A number of test facilities have been set up, for example, helium test loops and fuel handling test rigs. Experiments have been performed to test helium sealing, helium purification technologies, and fuel handling components. Fuel burnup has been researched. Progress has also been made in research of metallic alloy and insulation materials.

From April 2003 to September 2006, four experiments were completed to confirm and verify inherent safety features of modular HTRs: loss of off-site power without intervention, main helium blower shutdown without intervention, loss of main heat sink without intervention, and especially withdrawal of all rods without intervention. The residual heat of the reactor was carried out entirely passively and the reactor maintained its safety condition. All these experiments were authorized, guided, and supervised by the national nuclear safety authority.

16.3 SMRs in PR of China: ACP100

16.3.1 Introduction of ACP100

Since 2010, one type of small and medium-sized water-cooled pressurized reactor called ACP100 has been developed by China's national nuclear corporation. ACP100 is an innovative reactor based on existing PWR technology, adapting a "passive" safety system and "integrated" reactor design technology. After 8 years development, its overall design was complete, including conceptual design and basic design and preliminary safety analysis report. A number of R&D on safety-related experiments have been carried out in the past years. IAEA gave the review comments on ACP100 Generic Reactor Safety Review (GRSR) report on April 2016, the 1st SMR completion of GRSR in the world. The construction of the ACP100 will be started around the end of 2019.

16.3.2 Technical aspects

The ACP100 design has the following remarkable technical features:

- Primary system and equipment integrated layout. The maximum size of the connection pipe is 5–8 cm, whereas in a large NPP, it is 80–90 cm.
- Large primary coolant inventory.
- Small radioactivity storage quantity. The total radioactivity of an SMR is one-tenth that of large NPPs, and meanwhile a multilayer barrier is added to keep the accident source term at a low level.
- The vessel and equipment layout is beneficial for natural circulation.
- The decay heat removal is more effective: two to four times the efficiency of large NPP heat removal from the vessel surface.

- Smaller decay thermal power: one-fifth to one-tenth times of decay thermal power compared with that of a large PWR after shutdown is easier to achieve safety by way of the "passive" system.
- Reactor and spent fuel pool are underground for better protection against exterior accident and for the reduction of radioactive material release.
- No operator intervention is needed for 72 h after an accident.

- Passive severe accident prevention and mitigation action, such as for containment hydrogen eliminator and cavity flooding, is built in to ensure the integrity of pressure containment.
- The modular design technique makes it easy to control the product quality and shorten the site construction period.

16.3.3 Main design parameters

The main technical parameters of ACP100 can be found in Table 16.2.

16.3.4 General layout of the plant

According to the proposed site, the layout will be two 125 MWe integral PWR nuclear units (Fig. 16.3). To enhance capability to avoid exterior accident, the containment is protected by a thick wall outside (Fig. 16.4).

Table 16.2 The main technical parameters of ACP100.

Parameter	Unit	Value
Thermal power	MWt	385
Electrical power	~MWe	125
Design life	Years	60
Refueling period	Years	2
Coolant inlet temperature	°C	282
Coolant outlet temperature	°C	323
Coolant average temperature	°C	303
Best estimate flow	m^3/h	9500
Operation pressure	MPa	15
Fuel assembly type	–	17 × 17 square assembly
Fuel active section height	mm	2150
Fuel assembly number	–	57
Fuel enrichment	%	4.2
Drive mechanism type	–	Magnetism lifting
Control rod number	–	20
Reactivity control method	–	Control rod, solid burnable poison and boron
Steam generator type	–	Once-through steam generator
Steam generator number	–	16
Main steam temperature	°C	>290
Main steam pressure	MPa	4.5
Main steam output	t/h	590
Main feed water temperature	°C	140

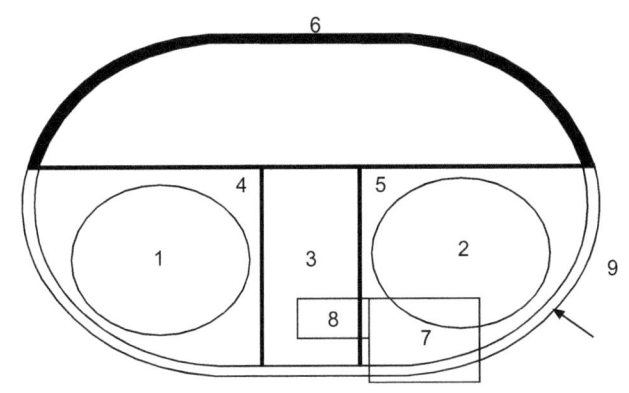

Fig. 16.3 ACP100: two modules, general layout of plant (1, reactor building 1; 2, reactor building 2; 3, refuel building; 4, connecting building 1; 5, connecting building 2; 6, auxiliary building; 7, passage building; 8, equipment handling area; and 9, antiplane containment).

16.3.5 Nuclear steam supply system

An integrated layout is used instead of a loop type (Fig. 16.5):

- The main pump and reactor pressure vessel are connected by a short pipe.
- The steam generator sets in the reactor pressure vessel.
- The nuclear steam supply system, apart from the pressurizer, is integrated in the reactor module.
- The control rod drive mechanism (CRDM), pressure vessel, reactor internals, once-through steam generator (OTSG), and canned motor pump are all mature technology.

Reactor coolant system:

- System function and composition—Four main pumps, 16 OTSG, one pressurizer.
- System description—Operation pressure 15.0 MPa, core exit temperature 325°C.

Main steam system:

- System function and composition—Main steam system, bypass system, and moisture separator reheat system.
- System description—Operation pressure 4.5 MPa, temperature 290°C, flow 590 t/h.

CRDM:

- Location: Outside the vessel.
- Step: 15.875 mm/min.
- Max travel speed: 72 mm/min.
- Electromechanical delay time: ≤ 150 ms.
- Temperature: 200°C.
- Pressure housing design life: 60 years.
- Min. cumulative step number of service: 6.0×10^6 steps.

Fuel assembly:

- 17×17 square pitch arrangement.

Fig. 16.4 ACP100: a diagram of safety systems.

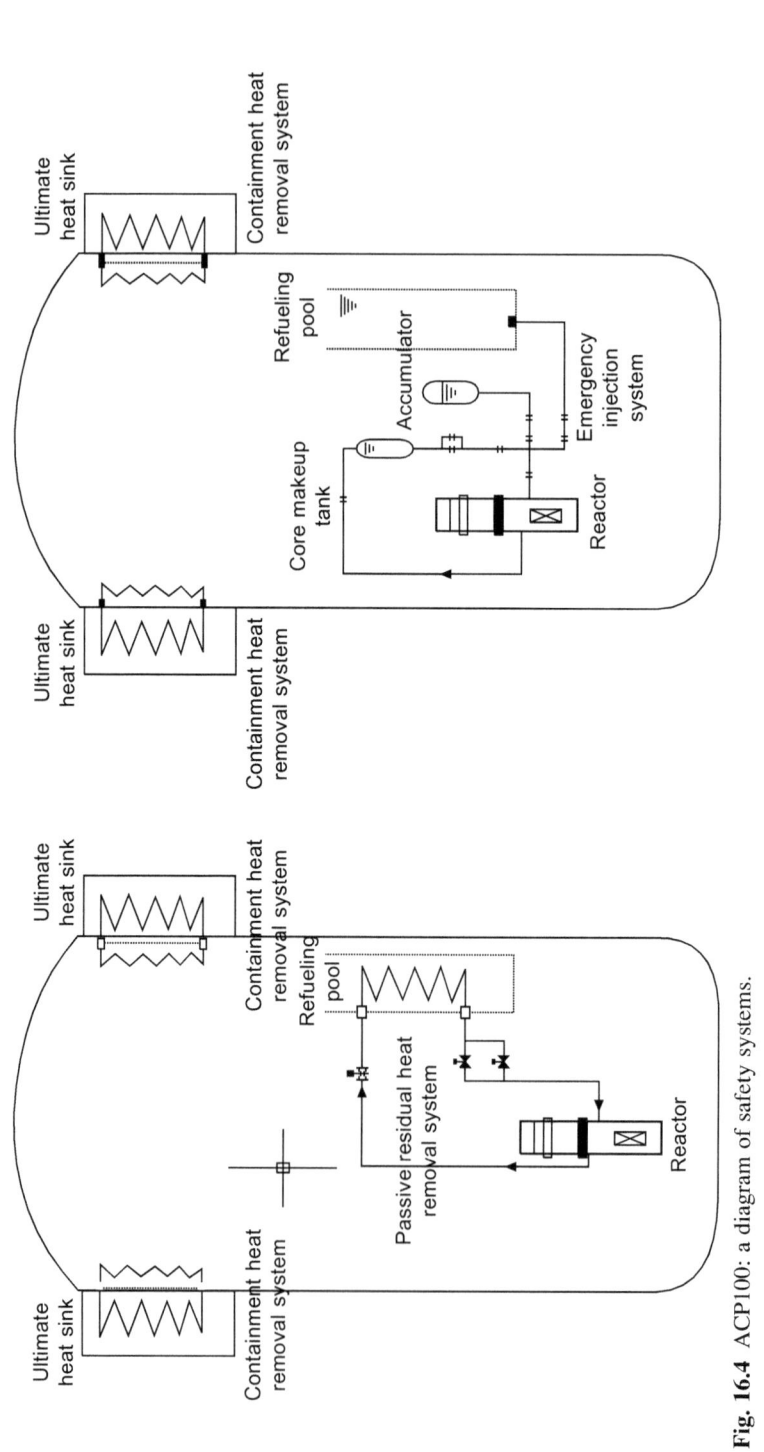

Fig. 16.5 ACP100: reactor with other main equipment connected (Zhong, 2013).

- Fuel rod: 264.
- Guide tube: 24.
- Instrumentation tube: 1.
- Total height: ~2500 mm.

16.3.6 Engineered safety feature plan

The safety design philosophy of ACP100 is to realize a high level of safety and, at the same time, to simplify the design of the systems by means of a passive engineering safety system. No emergency diesel generator is needed. The emergency measures outside the plant boundary should be made technically unnecessary or reduced to a minimum. The passive system of ACP100 include long-term residual heat removal system to cope with station blackout accident, passive core cooling system to cope with LOCA, cavity flooding during a severe accident, and a passive containment heat removal system. The passive containment heat removal system means that the heat is taken away by the gas and steam convection between containment and ultimate heat sink by natural circulation. This ensures the containment integrity under accident conditions:

- The absence of any large-diameter piping associated with the primary system removes the possibility of large-break LOCAs. The elimination of large-break LOCAs substantially reduces the necessity for emergency core cooling system components, alternate current (AC) supply systems, etc.
- The large coolant inventory in the primary circuit results in large thermal inertia and a long response time in the case of transients or accidents.
- Inherent safety features include an integrated primary coolant system, eliminating large break LOCAs; long characteristic times in the event of a transient or severe accident, due to large coolant inventory and the use of passive safety systems; and negative reactivity effects and coefficients.
- Passive safety systems are duplicated to fulfill redundancy criteria. According to Chinese nuclear safety regulations, the shutdown system is diversified. The systems include a residual heat removal system, an emergency injection system, and safety relief valves that protect the reactor pressure vessel against overpressurization in the case of strong differences between core power and the power removed from the reactor pressure vessel. All safety systems mentioned in this paragraph are passive systems.

16.3.7 Role of passive safety design features

16.3.7.1 Level 1: Prevention of abnormal operation and failure

Contributions of ACP100 inherent and passive safety features at this level are as follows: owing to the absence of large-diameter piping in the primary system, large-break LOCAs are eliminated; canned pump eliminates boron injection for pump sealing system.

16.3.7.2 Level 2: Control of abnormal operation and detection of failure

The ACP100 passive safety feature for this level is as follows: a large coolant inventory in the primary circuit results in a larger thermal inertia and in longer response times in the case of transients or accidents.

16.3.7.3 Level 3: Control of accidents within the design basis

ACP100's safety systems are based on passive features obviating the need for actions related to accident management over a long period. Long-term residual heat removal systems cope with station blackout accidents, passive core cooling systems cope with loss of coolant accidents and cavity flooding during serve accidents, and passive containment of heat removal system.

16.3.7.4 Level 4: Control of severe plant conditions, including prevention of accident progression and mitigation of consequences of severe accidents

Contributions of inherent and passive features of ACP100 at this defense in-depth level are as follows: when core uncover is assumed, only for analytic purposes, low heat-up rates of fuel elements in the exposed part of the core are predicted, if the geometry is still intact. The characteristic time of core melting is long, eventually preventing temperature excursion due to a metal-water reaction, which in turn limits the hydrogen generation rate. The hydrogen concentration in the containment is reduced by catalytic recombiners. There is sufficient floor space for cooling of molten debris, and extra layers of concrete are used to avoid direct exposure of the containment basement to debris.

16.3.7.5 Level 5: Mitigation of radiological consequences of significant release of radioactive materials

The following passive features of ACP100 make a contribution to this defense in depth level: relatively small fuel inventory, compared with larger NPPs; slower progression of accidents; and increased retention of fission products (facilitated by such features as reduced power density, increased thermal inertia, etc.); the containment is located inside the airplane protection concrete and underground building, which reduces the release of fission products due to local deposition. The ACP100 concept provides for extended accident prevention and mitigation by relying on the principles of simplicity, reliability, redundancy, and passivity.

16.3.8 Post-Fukushima actions

After the "Fukushima" accident the design and operation of NPPs should strictly comply with the requirements of the safety code. In addition, the combination of various internal and external accidents should be considered. The following accidents have been analyzed since the "Fukushima" accident:

- The loss of off-site power supply and the emergency diesel power supply. ACP100 has an integrated primary coolant system and removes residual heating to a large capacity containment pool through a heat exchanger, depending on primary coolant natural circulation. The reactor will not lose cooling in case of loss of power and can sustain the residual heat removal for 72 h.
- The combination of a LOCA and loss of all the power. After the LOCA, ACP100 will achieve core cooling, and heat will be completely removed from the containment due to the passive facility in this process; no accident aggravation will occur after loss of power.
- Safety and accident risk of the spent fuel storage pool. The spent fuel pool of the ACP100 SMR lies underground, and a standby makeup pool is set outside the plant. Fuel will not be uncovered in the event of the extreme condition of the structure of the spent fuel pool breaking under seismic conditions and the cooling lost for the loss of power.
- Core melt. ACP100 is the third-generation PWR and has inherent safety characteristics, no big LOCAs, etc., and adopts complete passive safety features that obviously reduce the

accident probability and consequences. In addition, there are the severe accident prevention and mitigation actions, such as core injection system (CIS), hydrogen recombiner, and the corresponding accident management guide.

16.3.9 Testing and verification

Six test research subjects have been done for the development of ACP100 (see Table 16.3):

- control rod drive line cold and hot test
- control rod drive line antiearthquake test
- internal vibration test research
- fuel assembly critical heat flux test research
- passive emergency core cooling system integration test
- core makeup tank (CMT) and passive residual heat removal system test research

16.4 Deployment of SMRs in PR of China

16.4.1 HTR-200

With the HTR-10 operation and experiments serving as a supporting technological test bed, the HTR-200 commercial demonstration plant is being promoted. China Huaneng Corporation, China Nuclear Engineering & Construction Corporation, and Tsinghua University have established Huaneng Nuclear Power Development Co., Ltd. After siting evaluation, conceptual design, basic design, safety analyses, and Fukushima accident safety inspection, the Chinese government officially approved the start date on December 9, 2012. From year 2016 to 2019, major equipment, including reactor pressure vessel, steam generator, ceramic internals, and head of reactor pressure vessel, have been installed. The first demonstration HTR-200 plant will be operation in the year 2020. The first demonstration HTR-200 is located in Shidao Bay, Shandong province. After the demonstration HTR-200, six-module unit (HTR-PM600), commercial 600 MWe unit can be deployed, as supplement to PWRs, such as replacing coal-fired power plant and cogeneration of steam and electricity. The feasibility study have been performed in the following sites: Sanmen, Zhejiang

Table 16.3 Verification testing and schedule of ACP100.

Code number	Name	Period
1	Control rod drive line cold and hot testing	2011–13
2	Passive emergency core cooling system integration testing	2011–13
3	Internals vibration testing	2012–13
4	Fuel assembly critical heat flux testing	2011–13
5	CMT and passive residual heat removal system testing	2011–13
6	Control rod drive line shock testing	2012–13

Province; Ruijin, Jiangxi Province; Xiapu, Fujian Province; Wan'an, Fujian Province; and Bai'an, Guangdong Province (Yujie, 2019).

16.4.2 ACP100

16.4.2.1 Licensing

The following activities have been accomplished related to licensing:

- The contract of SMR combined research with National Nuclear and Radiation Safety Center was signed in 2011.
- The following work has been achieved: National Nuclear & Radiation Safety Center gave the comments on the SMR research report of design preparation phase, a technical exchange of SMR containment design after Fukushima nuclear accident occurred, passive integration test research technical exchange occurred, and the test program was approved.
- The Q1 questions and question reply of concept design stage was completed, and the concept design was approved.
- Signed several specific research programs and standard design safety analysis combined research with National Nuclear & Radiation Safety Center in 2013.
- IAEA gave the review comments on ACP100 Generic Reactor Safety Review (GRSR) report on April 2016.
- In March, Hainan Nuclear Power Co., Ltd. officially submitted the PSAR of Hainan Changjiang SMR demonstration project to National Nuclear Safety Authority.
- In August, National Nuclear Safety Authority started to review the PSAR of Hainan Changjiang SMR demonstration project.

16.4.2.2 Site selection

The demonstration of the ACP100 nuclear power plant, with one 385-MWth reactor, will be located in Changjiang County, Hainan Province, on the south island area of China.

In July 2019, China National Nuclear Co. announced to start site excavation.

In year 2019, long-term equipment, such as reactor vessel, once-through steam generator and turbine generator signed the contract and started to manufacture.

16.5 Future trends

China, as a major player in nuclear industry, has already developed several kinds of SMR, such as HTR-200, ACP100, and some other mental cooling reactors. In the near future, China will continue its SMR development. In the light-water reactor (LWR) family, the SMR in China is focused on modular and integrated reactors, factory manufacture and installation, railway or truck transportation, emergency zone decreasing, and external accidents resistance by underground reactor building layout. In non-LWR family the SMR, such as gas-cooled reactors and sodium-cooled fast reactors, is focused on nonproliferation, nuclear fuel utilization, emergency zone cancelation, and enhancing economics. We hope two or three kinds of SMR will deploy in China after 2020.

There are several other announced SMR activities in China. China National Nuclear Co. has developed floating nuclear power plant ACP100S that was upgrade by ACP100. China General Nuclear Power Corporation has developed floating nuclear power plant ACPR50S.

Acknowledgments

I would like to thank Li Yunyi, Zhao Qiang, Qin Zhong, Zhong Fajie, and Xu Bin for useful discussions and information. I also thank the reviewers for their suggestions.

References

IAEA-TECDOC-1485, 2005. Status of Innovative Small and Medium Sized Reactor Designs. IAEA, Vienna. pp. 511, 514, 527, ISBN 92-0-101006-0, ISSN 1011-4289.
Yujie, D., 2019. Progress of HTR-PM demonstration power plant project. In: IAEA 17th INPRO Dialogue Forum, Ulsan, Republic of Korea, July 2–5.
Zhong, F., 2013. Safety features and licensing of ACP100 design. In: IAEA 6th INPRO Dialogue Forum on Global Nuclear Energy Sustainability, Vienna, Austria, July.

Small modular reactors (SMRs): The case of Japan

Tsutomo Okubo
Japan Atomic Energy Agency (retired), Oarai-Machi, Japan

17.1 Introduction

In this chapter the status of small modular reactor (SMR) R&D and deployment in Japan is described. In Japan the terminology of "SMR" normally means small- and medium-sized reactor, following the same definition as in the International Atomic Energy Agency (IAEA). Based on the IAEA's classification, the "small reactor" has a power output less than 300 MWe and the "medium reactor" between 300 and 700 MWe (IAEA, 2006). However, in this chapter, reactor concepts under the power output up to around 500 MWe are included, reflecting the SMR R&D status in Japan.

There was much SMR R&D activity in Japan in the past, especially after the accidents at Three Mile Island Unit 2 (TMI-2) in 1979 in the United States and Chernobyl Unit 4 occurred in 1986 in Ukraine. After the accidents the passive safety characteristics in the reactor safety features that were considered to be favorable were introduced into reactor safety design and then gathered a lot of attention as there was much interest in finding how they could be utilized as effectively as possible. In introducing passive safety features into the reactor design, the smaller size reactor, typically less than about 500 MWe, was in general easier than the larger size, such as more than 1000 MWe.

However, the small-sized reactors commonly have an economic defect, as it has been realized that the "scale merit"—the advantages of scale that occur with larger plants—is important in reactor design for economic reasons. Therefore, in SMR R&D, it has been a very important requirement to overcome the "scale demerit" of the smaller-sized reactors; this point will be discussed for each SMR concept. Apart from the passive safety features, there are some original and special needs or purposes, especially for small reactors, such as the variety of energy products and flexibility in design, siting, and fuel cycle options.

The SMR R&D status in Japan is described in Sections 17.2 and 17.3, and the deployment of SMRs in Japan will be discussed in Section 17.4. The future trends in Japan and the sources of further information will be given in Sections 17.5 and 17.6, respectively.

17.2 Small modular nuclear reactor (SMR) R&D in Japan

17.2.1 SMR R&D in the 1980s and 1990s

In the present section the Japanese status of SMR R&D is summarized for some representative conventional and advanced reactor concepts. In Japan, there was an important period for SMR R&D in the 1980s–90s with a lot of R&D activity. Therefore SMR R&D activities in this period are briefly overviewed in the following.

In Japan, although the first commercial reactor for electricity generation was a gas-cooled one with a small output of 166 MWe (operated in 1966–98), which was imported from England, the next commercial reactors were all light-water reactors (LWRs). The power output of the first commercial LWR in Japan was as low as 357 MWe (operation started in 1970). However, the power output was gradually increased to 600-MWe class, then 900-MWe class, and finally over 1100 MWe, based on the "scale merit" of the larger reactor. This was the same trend as that in other countries at that time, and it was because there was basically little need in Japan for small reactors for electricity generation.

After the accidents of TMI-2 in 1979 and Chernobyl Unit 4 in 1986, however, the passive safety characteristics in the reactor safety features were considered to be favorable and were used to promote the introduction of new reactors. To introduce the passive safety features, such as the natural circulation core cooling and the gravity feed emergency core cooling system (ECCS) effectively into the reactor design, taking into account balance with the economic aspects, the smaller size of the reactor, typically less than about 500 MWe, was considered more suitable than the larger size, such as more than 1000 MWe. This situation did not appear only in Japan, but was a worldwide trend after TMI-2 and Chernobyl Unit 4 accidents. The most typical and well-known reactor concept in this period was that named "Process Inherent Ultimate Safety (PIUS)" (622 MWe/3 modules) (Hannerz et al., 1986), a Swedish integral type LWR or pressurized water reactor (PWR) design concept by ABB-ATOM. This introduced passive measures, ones not requiring operator actions or external energy supplies, to provide safe operation for the reactor shutdown and decay heat removal after a transient or accident situation. This was the integral type reactor concept without primary coolant piping preventing the large-break loss-of-coolant accident (LB LOCA). That is a kind of integral PWR (iPWR) concept. Based on and extending this concept, three PWR type concepts were developed in Japan. The inherently safe and economical reactor (ISER) (210 MWe) concept was developed by the University of Tokyo and others (Oda et al., 1986), introducing a steel reactor vessel instead of the prestressed concrete reactor vessel (PCRV). The Mitsubishi intrinsic safe integrated reactor (MISIR) (300 MWe) concept was proposed (Kudo et al., 1987), and the system-integrated PWR (SPWR) (350 MWe) concept was developed by Japan Atomic Energy Research Institute (JAERI) (Sako, 1988). The Marine Reactor X (MRX) (30 MWe) was developed by JAERI for marine propulsion or local energy supply as an integral type PWR with passive safety features and a submerged reactor vessel (Kusunoki et al., 2000).

Extending the conventional loop-type PWR concept, the MS-600 (600 MWe) was proposed by Mitsubishi Heavy Industries (MHI) (Makihara et al., 1991), based on the AP (Advanced PWR)-600 (600 MWe) developed by Westinghouse, which introduced the passive ECCS and passive containment cooling system (PCCS). Also, extending the conventional loop-type boiling water reactor (BWR) concept, HSBWR- 600 (600 MWe) by Hitachi (Kataoka et al., 1988) and TOSBWR-900P (310 MWe) by Toshiba (Nagasaka et al., 1990) were proposed. They were based on the simplified BWR (SBWR) (600 MWe) concept developed by General Electric (GE), which introduced the natural circulation core cooling and the gravity-driven ECCS.

Although many SMR concepts were developed in Japan up to the 1990s as briefly described earlier, none of them could be realized or developed further. This was mainly because they could not overcome the "scale demerit" in the economic aspect and hence were not attractive to the users, that is, the electric companies. On the other hand the large LWRs of over 1000 MWe were continuously improved, including the safety features based on their operational experiences, especially under the national projects of the "improvement standardization programs for LWRs." In this way, up to around 1985, the development of advanced BWR (ABWR) and advanced PWR (APWR), which are the "third-generation" reactors, were already finished, and they were ready to be introduced after the Chernobyl Unit 4 accident. In reality, after the construction initiation in 1991, the first ABWR plant in the world with the power output of 1356 MWe was in operation as the Kashiwazaki-Kariwa Unit 6 in 1996.

17.2.2 SMR R&D after 2000

After the Chernobyl Unit 4 accident in 1986, there was little new reactor installation. In 2000 the Generation IV reactor development was proposed by the US Department of Energy (DOE). This stimulated R&D activity in Japan, including that on SMRs. At this time the economic aspects were considered to be the most important point in the development, and hence, R&D was aimed at overcoming the scale demerit to achieve the same level of the construction cost (per kilowatt-electric) and the electricity generation cost (per kilowatt-hour) as for the existing large LWRs. To accomplish high economic performance, the design simplification of the system is considered to be the most effective approach. Optimizing the simplification, three SMR design concepts were proposed and developed in Japan based on the LWR concept. They are integrated modular water reactor (IMR), compact containment boiling water reactor (CCR) and modular simplified and medium small reactor [double MS (DMS)] (Okubo, 2011). In addition to these, other SMR concepts based on the high-temperature gas-cooled reactor (HTGR) were also proposed, utilizing their excellent characteristic of the higher thermal efficiency in HTGR than in LWRs. One of them is gas turbine high-temperature reactor 300 (GTHTR300). Furthermore an SMR concept based on the sodium-cooled fast reactor (SFR) concept was also proposed, especially for local use in isolated area conditions, such as in Alaska. This is named as the 4S (supersafe, small, and simple). In the following sections, these five SMR concepts are presented briefly.

17.3 SMR technologies in Japan

17.3.1 IMR

The IMR is an iPWR concept and was proposed by MHI (Hibi et al., 2005) in cooperation with the Japan Atomic Power Company (JAPC) (Okazaki et al., 2011). The targets for the SMR design are summarized in Table 17.1. The schematic view of the reactor concept is shown in Fig. 17.1. Its major specifications are summarized in Table 17.2. Although it is based on the PWR concept, boiling of the primary coolant, that is, water, is allowed around the top area of the core (average void fraction at the core outlet is about 20%), and the core is cooled by the natural circulation of the coolant. Boiling of the core coolant is favorable to increase the driving force for the

Table 17.1 Targets for the SMR design.

Power output: 300–600 MWe
Construction cost: The same as large reactors or less
Capacity factor: More than 90%
Safety: At least as safe as existing plant
Construction period: Less than 24 months
License suitability: Suitable for the present license

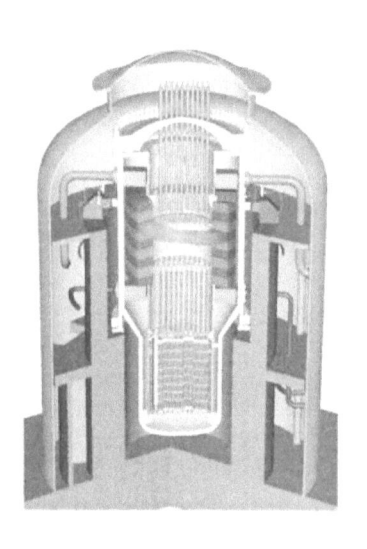

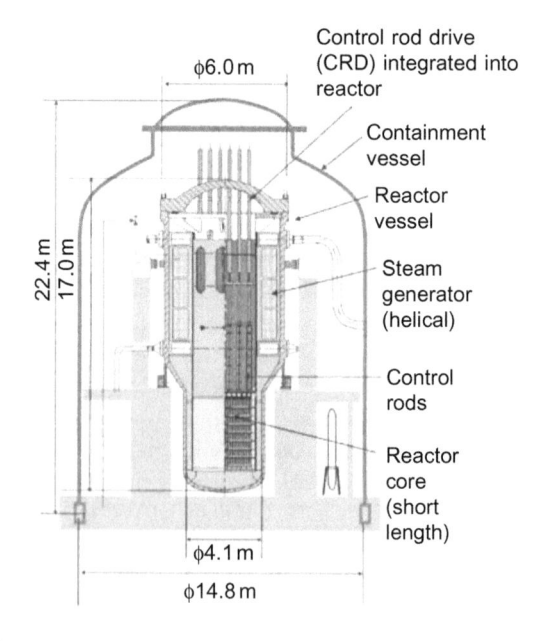

Fig. 17.1 Schematic view of an IMR.

Table 17.2 Major specifications of an IMR.

Reactor power	350 MWe
Core thermal output	1000 MWt
System pressure	15.5 MPa
Primary system inlet/outlet temperature	573/618 K (300/345°C)
Primary coolant flow rate	3.0 t/s
Steam temperature	562 K (289°C)
Steam pressure	5 MPa
Core equivalent diameter	3 m
Core height	3.7 m
Refueling interval	26 months
Core average burnup	45 GW d/t
Capacity factor	90% or more
Construction period	Less than 24 months

natural circulation core cooling. In this concept the primary cooling system and the steam generators (helical once-through type) are enclosed in the reactor pressure vessel, and hence, any large break LOCA is eliminated. In this reactor concept, several major components are eliminated, such as reactor coolant pumps, pressurizer, and primary piping.

The power output is 350 MWe. The average discharge burnup is 45 GW d/t, and the refueling interval is 26 months. Considering this longer lifetime of fuel rods and the increased coolant temperature up to 345°C, a Zr-Nb alloy is applied to the cladding tube. The system pressure is 15.5 MPa and is the same pressure as in the normal PWR. For the safety system design, the IMR is significantly simplified. That is, it does not require the ECCS and only requires the steam generator cooling system for decay heat removal. The steam generator cooling system uses pumps, driven by diesel and gas turbine, as the active equipment.

As the IMR concept is different from the PWR in the flow characteristics, two experiments were conducted to check the flow characteristics inside the reactor vessel of the IMR. One was conducted under the actual conditions of the temperature, the pressure, and the length of axial direction of IMR reactor system. Based on the results, it is confirmed that the natural circulation core cooling is available under the operation and design conditions of IMR. The other was conducted using a simulator of the main structure to confirm the three-dimensional flow characteristics in the reactor vessel. By the test, flow characteristics of the main structure inside reactor vessel could be understood, and the methods for evaluating two-phase flow behaviors were developed.

For further developments of the IMR, optimization studies will be performed. In addition, possibilities to develop the concepts with various output ranges will be considered, utilizing the technology gained in the previous development. For example, a plant with 100-MWe output has been developed based on the IMR.

17.3.2 CCR

The CCR is a BWR and was proposed by the Toshiba Corporation (Heki et al., 2005) in cooperation with the JAPC (Okazaki et al., 2011). The targets for the design are the same as in the IMR and are summarized in Table 17.1. A schematic view of the reactor is shown in Fig. 17.2. Its major specifications are summarized in Table 17.3. The core is shortened and cooled by the natural circulation of the coolant without the recirculation pumps. One of the characteristic features for the system simplification

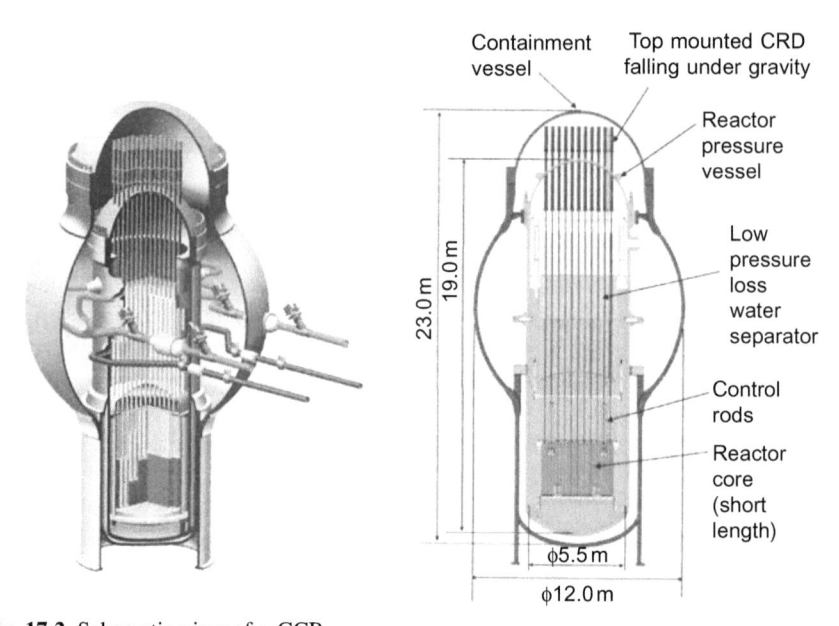

Fig. 17.2 Schematic view of a CCR.

Table 17.3 Major specifications of a CCR.

Reactor power	423 MWe
Core thermal output	1268 MWt
System pressure	7 MPa
Primary system inlet/outlet temperature	488/560 K (215/287°C)
Primary coolant flow rate	3.3 t/s
Steam temperature	560 K (287°C)
Steam pressure	7 MPa
Core equivalent diameter	3.5 m
Core height	2.2 m
Refueling interval	24 months
Core average burnup	45 GW d/t
Capacity factor	90% or more
Construction period	Less than 24 months

for this concept is the high pressure (about 4 MPa)-resistant compact containment vessel without the suppression pool. The control rod-drive mechanism is mounted at the top of the reactor vessel, and, hence, the penetration at the bottom of the reactor vessel can be eliminated.

The power output is 423 MWe. The average discharge burnup is 45 GW d/t, and the refueling interval is 24 months. The effective length of the core is shortened to 2.2 m to increase the driving force for the natural circulation core cooling. The system pressure is 7 MPa and is the same pressure as in the normal BWR. For the safety system design, the CCR concept is significantly simplified. That is, it does not require the ECCS and only requires the isolation condensers (IC) cooling system for decay heat removal. The IC system is a passive component system and is designed to remove decay heat for 3 days without any operation by the operators. As the control rod-driven mechanism of the CCR is different from that of the BWR and is also different from that in a PWR considering the operation in the steam environment, the abrasion-resistant tests for the material in the steam environment were conducted to confirm the applicability.

From the economic point of view, the estimation results on the construction cost using the same evaluation method as for the large reactor are given in Fig. 17.3 for the IMR and CCR concepts (Okazaki et al., 2011). In the figure the unit construction cost per kilowatt-electric is compared with the IMR, CCR, and ABWR cases. The values are normalized by that of the ABWR case with 1356-MWe output. It can be recognized that both the IMR and CCR concepts are at almost the same economic

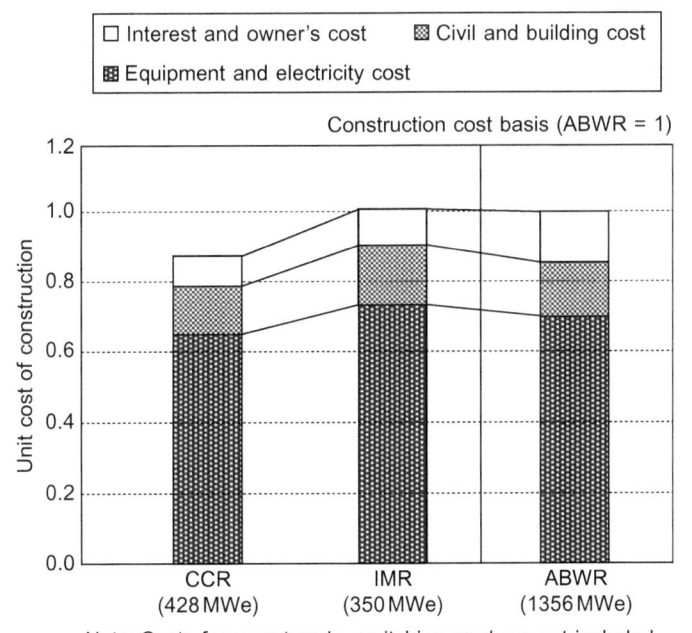

Fig. 17.3 Construction cost evaluation of an IMR and a CCR.

level as the typical large reactor of ABWR. The main reasons for this economic achievement are the simplification of the system by eliminating the systems and components and, hence, the resultant downsizing of the containment vessel. In addition, the downsizing of buildings due to the simplification of the systems and components enables the construction and civil engineering costs to be reduced. However, it should be noted that this evaluation is performed assuming that the new plant shares the port and the switching yard with the existing plant, and the additional costs for those facilities are required if the plant is established in a new site.

For further developments of CCR, optimization studies will be performed. In addition, possibilities of developing the concepts with various output range are considered, utilizing the technologies gained in the previous development. For example, a concept of a plant with 100-MWe output is developed based on the CCR concept.

17.3.3 DMS

The DMS is a typical simplified BWR and was proposed by Hitachi, Ltd. (Kawabata et al., 2008) in cooperation with the JAPC. The targets for the design are the same as in the IMR and are summarized in Table 17.1. A schematic view of the reactor is shown in Fig. 17.4, and its major specifications are summarized in Table 17.4.

The core is shortened and cooled by the natural circulation of the coolant without the recirculation pumps. One of the characteristic features for the system simplification for this concept is elimination of the steam-water separator and dryer, by

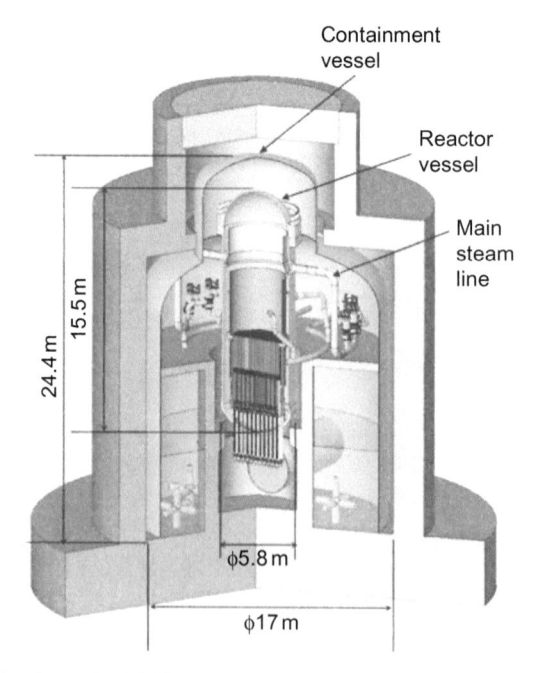

Fig. 17.4 Schematic view of a DMS.

Table 17.4 Major specifications of a DMS.

Reactor power	428 MWe
Core thermal output	1200 MWt
System pressure	7.2 MPa
Primary system inlet/outlet temperature	488/560 K (215/287°C)
Primary coolant flow rate	3.2 t/s
Steam temperature	560 K (287°C)
Steam pressure	7.2 MPa
Core equivalent diameter	4 m
Core height	2.0 m
Refueling interval	24 months
Core average burnup	45 GW d/t
Capacity factor	90% or more
Construction period	Less than 24 months

introducing a gravity separation mechanism for the steam-water separation process. Therefore the reactor pressure vessel is simplified and compact.

The power output is 428 MWe. The average discharge burnup is 45 GW d/t, and the refueling interval is 24 months. The effective length of the core is shortened to 2 m to increase the driving force for the natural circulation core cooling. The system pressure is 7.2 MPa and is the same pressure as in the normal BWR. For the safety system design, the DMS employs a simplified ECCS with passive accumulator system and hence, eliminated the high-pressure core flooder (HPCF) system and decreased about 60% of the reactor core isolation cooling (RCIC) system capacity.

17.3.4 GTHTR300

The GTHTR300 is a high-temperature helium gas-cooled modular reactor with a high thermal efficiency gas turbine cycle system and was proposed by the Japan Atomic Energy Agency (JAEA) (Kunitomi et al., 2002, 2004). The targets for the design are summarized in Table 17.5 as the user requirements. A schematic view of the reactor concept is shown in Fig. 17.5, and its major specifications are summarized in Table 17.6. The GTHTR300 is characterized by the significantly simplified system design, the high thermal efficiency, and the improved economy from utilizing the proven technologies obtained with the JAEA's high-temperature engineering test reactor (HTTR), as well as the good safety features.

The power output is 280 MWe. The average discharge burnup is 120 GW d/t, and the refueling interval is 2 years. The fuel is the block-type one with the average U enrichment of 14%. The core is the pin-in-block and annular type, and its effective height is 8 m. The core is cooled by the helium gas, and the core inlet/outlet temperatures are 860/1123 K under the system pressure of 6.8 MPa. For the safety system design, the GTHTR300 concept employs the target of "severe accident free" (Katanishi et al., 2003). This can be attained by its special characteristics as follows: There is no possibility for core melt, and the core can stand high temperatures, because the fuel is ceramic and the graphite

Table 17.5 User requirements for a GTHTR300 reactor.

Safety goal	Radioactive nuclides release shall be prevented by complete passive system
	Meet the site evaluation requirement for a current LWR
	Site evacuation shall not be necessary
Site condition	Replacement of current LWR site or new site
Seismic condition	The same as that for a next generation LWR
Fuel cycle	Once through, fuel burnup more than 100 GW d/t
Nuclear proliferation	High amount of weapon grade Pu shall not be produced
Radiation protection	0.5 man-Sv/reactor year
Radioactive waste disposal	Liquid: 1/10 of the latest LWR
	Solid: Reuse of fuel blocks, if possible
Power level	100–4000 MWe/site, Modular type 100–300 MWe/unit
Life time	60 years
Availability	More than 90%
Inspection interval	2 years
Inspection period	Less than 30 days
Economy	Capital cost: 40–50 billion yen/unit (160–200 yen/We) Electric cost: 4 yen/kW h

Source: Kunitomi, K., et al., 2002. Design study on gas turbine high temperature reactor (GTHTR300). Trans. At. Energy Soc. Jpn. 1(4), 352–360 (in Japanese).

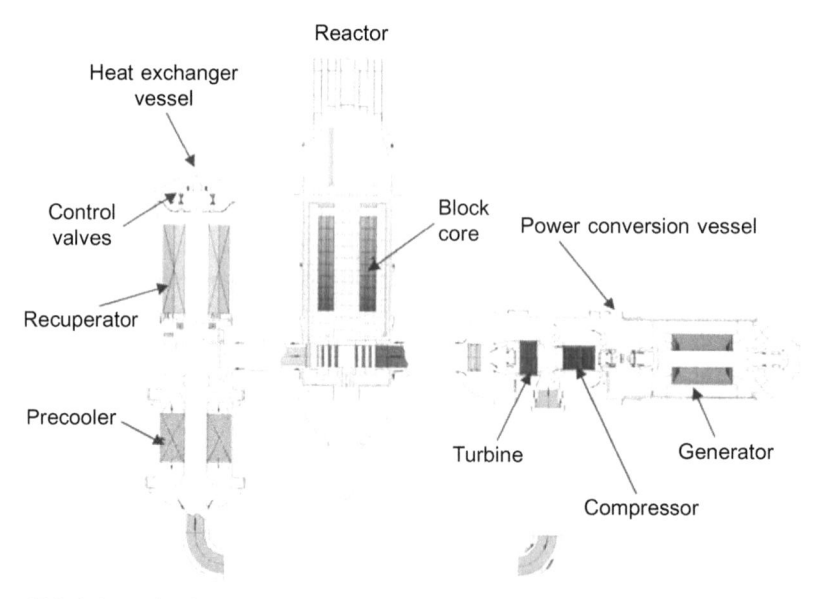

Fig. 17.5 Schematic view of GTHTR300 reactor concept.
From Kunitomi, K., et al., 2002. Design study on gas turbine high temperature reactor (GTHTR300). Trans. At. Energy Soc. Jpn. 1(4), 352–360 (in Japanese).

Table 17.6 Major specifications of a GTHTR300 reactor.

Reactor power	280 MWe
Core thermal output	600 MWt
Reactor inlet/outlet temperature	860/1123 K (587/850°C)
Coolant pressure	7 MPa
Coolant mass flow rate	438 kg/s
Core height	8 m
Inner/outer diameter of core	3.6/5.5 m
Fuel enrichment	14 wt%
Average power density	5.4 MW/m^3
Burnup period/batch	730 days
No. of batch/cycle	2
Reactor pressure vessel inner diameter	7.6 m
Gas turbine vessel inner diameter	5.7 m
Heat exchanger vessel inner diameter	5.8 m

materials and the core structures are graphite. Even if abnormal events occur, the transients are expected to be slow due to the low core power density and large core heat capacity as well as the negative reactivity feedback. In addition, the coolant, that is, the helium gas, is very stable and does not result in any mechanical energy production or release. The maximum fuel temperature can be lower than 1873 K (1600°C) even in an accident, and hence the integrity of the fuel, the coated fuel particle, is assured. Therefore containment is not necessary by adopting the confinement, which is the reactor building with a low leak rate less than about 1%/d.

The basic design was finished by the end of the 2003 fiscal year and was reviewed by electric companies, vendors, and universities. According to the preliminary cost evaluation results, the construction cost was about 200,000 yen/kWe, and the electricity generation cost was about 4.5 yen/kW h. The design of the cogeneration system of the GTHTR300C was also performed for hydrogen production.

After that, based on this concept, the basic design of a small reactor concept of HTR50S of 50-MWt output is underway (Ohashi et al., 2011) in collaboration with Kazakhstan. This concept utilizes a steam generator system under a reactor outlet temperature of 1023 K (750°C) and is intended to be realized in the 2020s. Also the MHR-50 (Mitsubishi small-sized high-temperature gas-cooled modular reactor-50) concept has been developed by MHI (Shimizu et al., 2011) in cooperation with the JAEA. This is a high-temperature helium gas-cooled modular reactor concept of 50-MWe output with a steam generator and a steam turbine cycle system, utilizing technologies verified by the JAEA's HTTR as much as possible.

The HTTR is a high-temperature helium gas-cooled test reactor with an output of 30 MWt, constructed to establish and advance the technical bases of the high-temperature gas-cooled reactor and irradiation tests in the high-temperature core (Saito et al., 1994). The first criticality was achieved in November 1998, and full power was reached with the core outlet coolant temperature of 1123 K (850°C) in December 2001. The maximum coolant temperature of 1223 K (950°C) was achieved in April 2004. The core inlet coolant temperature is 668 K (395°C). The major specifications of the HTTR are summarized in Table 17.7.

At present the reactor has been utilized for safety demonstration tests (Nakagawa et al., 2004). These tests are performed to demonstrate its inherent safety characteristics under international collaboration. Tests such as for control rod withdrawal, loss of core flow, and loss of core cooling are included. Also, to establish the technical bases for the nuclear heat utilization, related R&D for the iodine-sulfur (IS) process to produce hydrogen from the water has been performed.

17.3.5 4S

The 4S concept is a small modular SFR with the good safety features and a very long operation period without refueling. It has been proposed by the Toshiba Corporation (Tsuboi et al., 2009, 2012) in cooperation with Central Research Institute of Power Industry (CRIEPI). The features of the 4S are summarized in Table 17.8. A schematic view of the reactor concept is shown in Fig. 17.6, and its major specifications are summarized in Table 17.9. The 4S concept is characterized by the significantly simplified

Table 17.7 Major specifications of an HTTR.

Core thermal output	30 MWt
Reactor inlet/outlet coolant temperature	668/1123 or 1223 K (395/850 or 950°C)
Coolant pressure	4 MPa
Coolant mass flow rate	12.4 or 10.2 kg/s (for 850 or 950°C)
Core equivalent diameter	2.3 m
Core height	2.9 m
Fuel enrichment	3–10 (6 in average) wt%
Average power density	2.5 MW/m^3
Average/maximum linear power density	11.5/21.3 kW/m
Maximum fuel temperature	1463 or 1593 K (1190 or 1320°C for 850 or 950°C)
Average burnup	22 GW d/t

Table 17.8 Features of a 4S reactor.

Sodium-cooled fast reactor with power output of 10 or 50 MWe 30 years continuous operation without refueling
Safety design utilizing natural phenomena (automatic reactor shutdown and decay heat removal without human operation)
Reduction of maintenance introducing passive components
Enhanced safety and security by locating reactor building under ground level

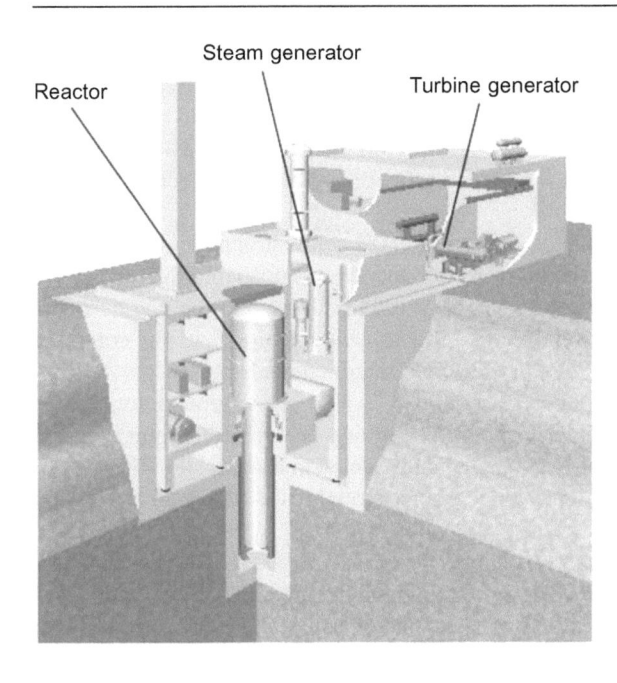

Fig. 17.6 Schematic view of 4S reactor concept.
From Tsuboi, Y., et al., 2009. Development of the 4S and related technologies (1)—plant system overview and current status. In: Proceedings of ICAPP'09, Tokyo, Japan, May 10–14, 2009 (Paper 9214).

Table 17.9 Major specifications of a 4S reactor (10 MWe design).

Electric output	10 MWe
Core thermal output	30 MWt
Number of loops	1
Primary sodium inlet/outlet temperature	628/783 K (355/510°C)
Primary sodium flow rate	547 t/h
Intermediate sodium inlet/outlet temperature	583/758 K (310/485°C)
Intermediate sodium flow rate	482 t/h
Turbine throttle conditions	
Flow rate	44.2 t/h
Pressure	10 MPa
Temperature	723 K (450°C)
Core equivalent diameter	0.95 m
Core average burnup	34 GW d/t
Fuel slug length	2.5 m
Fuel pin gas plenum length	2.7 m

system design with nonrefueling, low maintenance requirements, and high safety features; it is suitable for supplying energy to remote communities, mining sites, and so on.

The 4S is a pool-type sodium-cooled uranium metallic fuel fast reactor, setting the core, the primary electromagnetic pumps, and the intermediate heat exchanger (IHX) inside the reactor vessel. The power output is 10 or 50 MWe. The average discharge burnup is about 35 GW d/t, and the refueling interval is about 30 years. The core is narrow relative to its height (3.5 m) and is controlled by the movable six-segmented cylindrical reflector surrounding the core. The core can be shut down by fall down of the reflector to below the core region by the gravity. The reflector is able to shut the core down even if, for some reason, one of the six reflector segments is stuck.

As shown in Fig. 17.6, the reactor vessel is located below ground level, providing substantial protection against an aircraft impact and hence enhancing the security of the design. The containment system consists of the top dome and the guard vessel, which surrounds the reactor vessel and the reflector drive equipment. The core heat is transferred from the primary loop to the single intermediate heat transport system and is then exchanged in the steam generator, also located below ground, to produce the steam, which drives the conventional steam turbine generator equipment in the water-steam loop. The 4S power plant can be constructed in a small site. The overall area covering the belowground and aboveground structures is about 50 m long by 30 m wide.

The 4S has some passive safety features, such as the redundant residual heat removal system using only the natural circulation and the metallic core with the negative reactivity coefficients. It has two independent and redundant residual heat removal systems. They consist of the intermediate reactor auxiliary cooling system (IRACS), which removes decay heat using the air cooler installed in the intermediate heat transport system, and the reactor vessel auxiliary cooling system (RVACS), which removes heat transferred radially from the sodium coolant to the annulus between the guard vessel and a cylindrical steel heat collector located inside the cylindrical underground concrete wall.

17.4 Deployment of SMRs in Japan

Except for the small nuclear power stations constructed around 1970, there has been no deployment of SMRs in Japan. After the severe earthquake on March 11, 2011, followed by the severe accident at the Fukushima Dai-ichi Nuclear Power Station, deployment of nuclear power plants has basically been frozen in Japan, including deployment of SMRs in Japan. Since the policy for nuclear energy utilization has been under discussion in Japan after the Fukushima Dai-ichi accident and has not yet been determined, it is difficult to present a perspective for the SMR deployment in Japan at this time (February 2014).

However, the activities for deployment of some Japanese SMR concepts in other countries are continuing. One of them is the 4S (described in Section 17.3.5) deployment activity in Alaska, USA, which is under preliminary review by the US Nuclear Regulatory Commission. The other is the HTR50S (mentioned in Section 17.3.4) deployment activity in Kazakhstan, to be realized in the 2020s.

17.5 Future trends

As mention in the previous section, the Japanese policy for the nuclear energy utilization has been under discussion since the Fukushima Dai-ichi accident. It is difficult to present the likely future trend in Japan. After the Fukushima Dai-ichi accident, safety requirements have become more severe. Some SMR design concepts may need to be changed or checked against the new requirements.

Since the deployment of nuclear power plants is not expected to be easy in Japan, the SMR deployment abroad would be one likely future trend in Japan. In such a situation, there are two possibilities for the SMR deployment. One is the use based on the experienced LWR-based reactor. The other would be the use of new and non-LWR based reactor. It can be said that the Japanese SMR have the potential to answer both possibilities.

17.6 Sources of further information and advice

A valuable source of further information is the new revision of the IAEA-TECDOC on the SMR status in the world. The revision of the TECDOC has been issued around every 5 years. The websites of the Japanese vendors of Mitsubishi, Toshiba, and Hitachi would be other information sources, using keywords of the SMR and each reactor, such as IMR, CCR, DMS, and 4S. The JAEA website is also valuable for the gas-cooled concepts of GTHTR300 and HTR50S. Also the papers submitted to the international conferences, such as ICONE, ICAPP, and ANS topical meeting on the SMR, would be valuable.

References

Hannerz, K., et al. (1986): PIUS LWR Design Progress: IAEA Technical Committee Meeting on Advances in Light Water Reactor Technology, Washington, DC, November 17–19, 1986.

Heki, H., et al., 2005. Development status of compact containment BWR plant. In: Proceedings of ICAPP'05, Seoul, Korea, May 15–19, 2005, (Paper 5174).

Hibi, K., et al., 2005. Improvement of reactor design on integrated modular water reactor (IMR) development. In: Proceedings of ICAPP'05, Seoul, Korea, May 15–19, 2005, (Paper 5215).

IAEA, 2006. Status of Innovative Small and Medium Sized Reactor Designs 2005 (IAEA-TECDOC-1485).

Katanishi, S., et al., 2003. Safety design philosophy of gas turbine high temperature reactor (GTHTR300). Trans. At. Energy Soc. Jpn. 2 (1), 55–67, (in Japanese).

Kataoka, Y., et al., 1988. Conceptual design and thermal-hydraulics of natural circulation boiling water reactors. Nucl. Technol. 82 (2), 147–156.

Kawabata, Y., et al., 2008. The plant feature and performance of DMS (double MS: modular simplified & medium small reactor). In: Proceedings of ICONE16, Orlando, USA, May 11–14, 2008, (ICONE16-48949).

Kudo, F., et al., 1987. Preliminary study of inherently safe integrated small PWR. In: Abstracts of 1987 Annual Meeting of Atomic Energy Society of Japan, E48 (in Japanese).

Kunitomi, K., et al., 2002. Design study on gas turbine high temperature reactor (GTHTR300). Trans. At. Energy Soc. Jpn. 1 (4), 352–360, (in Japanese).

Kunitomi, K., et al., 2004. Japan's future HTR—the GTHTR300. Nucl. Eng. Des. 233, 309–327.

Kusunoki, T., et al., 2000. Design of advanced integral-type marine reactor MRX. Nucl. Eng. Des. 201, 155–175.

Makihara, Y., et al., 1991. On Mitsubishi small and medium sized reactor MS-600. In: Abstracts of 1991 Annual Meeting of Atomic Energy Society of Japan, SO_3, (in Japanese).

Nagasaka, H., et al., 1990. Study of a natural-circulation boiling water reactor with passive safety. Nucl. Technol. 92 (2), 260–268.

Nakagawa, S., et al., 2004. Safety demonstration tests using high temperature engineering test reactor. Nucl. Eng. Des. 233, 301–308.

Oda, J., et al., 1986. A conceptual design of intrinsically safe and economical reactor. In: IAEA Technical Committee Meeting on Advances in Light Water Reactor Technology, Washington, DC, November 17–19, 1986.

Ohashi, H., et al., 2011. Conceptual design of small-sized HTGR system for steam supply and electricity generation (HTR50S). In: Proceedings of ASME 2011 Small Modular Reactor Symposium, Washington, DC, September 28–30, 2011, (SMR2011-6558).

Okazaki, T., et al., 2011. A study for small-medium LWR development of JAPC. In: Proceedings of ICONE19, Osaka, Japan, October 24–25, 2011, (ICONE19-43646).

Okubo, T., 2011. Status of SMR development in Japan. In: 1st ANS SMR 2011 Conference, Washington, DC, November 1–4, 2011.

Saito, S., et al., 1994. Design of High Temperature Engineering Test Reactor (HTTR) (JAERI-Report 1332).

Sako, K., 1988. Conceptual design of SPWR. In: Proceedings on ANS International Topical Meeting on Safety of Next Generation Power Reactors, Seattle, USA, May 1–5, 1988.

Shimizu, K., et al., 2011. Small-sized high temperature reactor (MHR-50) for electricity generation: plant concept and characteristics. Prog. Nucl. Energy 53, 846–854.

Tsuboi, Y., et al., 2009. Development of the 4S and related technologies (1)—plant system overview and current status. In: Proceedings of ICAPP'09, Tokyo, Japan, May 10–14, 2009, (Paper 9214).

Tsuboi, Y., et al., 2012. Design of the 4S reactor. Nucl. Technol. 178, 201–217.

Small modular reactors (SMRs): The case of the Republic of Korea

18

Suhn Choi
Korea Atomic Energy Research Institute, Daejeon, Republic of Korea

18.1 Introduction

Various advanced types of small modular reactors (SMRs) have been under development worldwide, and a few of them are ready for construction. One beneficial advantage of an SMR is its easy receptivity of advanced design concepts and technology. Drastic safety enhancement can be achieved by adopting inherent safety features and passive safety systems. Economic improvement is pursued through system simplification, modularization, and reduction of the construction time.

For the last two decades in the Republic of Korea, the Korea Atomic Energy Research Institute (KAERI) has been working on SMR technology development for integral pressurized water reactor (iPWR), sodium-cooled fast reactor (SFR), and very high temperature gas-cooled reactor (HTGR) designs. Besides KAERI's research, several developments for the various types of SMRs have also been made by the other organizations such as KEPCO E&C, Seoul National University (SNU), KAIST, and UNIST, whose reactors are different in size, power, core, and even purpose.

System-integrated Modular Advanced ReacTor (SMART) is a promising advanced small nuclear power reactor. It is an integral type reactor with a sensible mixture of proven technologies and advanced design features. SMART aims at achieving enhanced safety and improved economics; the enhancement of safety and reliability is realized by incorporating inherent safety improving features and reliable passive safety systems. The improvement in the economics is achieved through a system simplification, component modularization, reduction of construction time, and high plant availability. The standard design approval assures the safety of the SMART system.

SMART, a small-sized integral-type pressurized-water reactor (PWR) with a rated thermal power of 364 MWt, is one of those advanced SMRs. Design characteristics contributing to the safety enhancement are basically inherent safety features such as the integral configuration of the reactor coolant system and improved natural circulation capability. By introducing a passive residual heat removal system (PRHRS) and an advanced loss-of-coolant accident (LOCA) mitigation system, significant safety enhancement is achieved.

After an extreme natural disaster led to severe damage at the Fukushima Daiichi nuclear power plant, mitigation measures and facilities to cope with severe accidents became one of the key safety issues. It is clear that maintaining continuous and proper core cooling capability after shutdown of a reactor is very important. In a series of realistic simulations, it has been shown that the SMART PRHRS working on the secondary side effectively removes decay heat and maintains the reactor in a stable condition for over 20 days without power sources and operator's mitigation action. This grace time can be indefinitely extended if the PRHRS condensate tanks accessible from outside of the containment are periodically replenished.

A small-sized reactor is generally considered to be economically less competitive than a large-sized one. However, there are many possible mechanisms for its economic improvement. In SMART, elimination of large pipes and valves due to system simplification, relative ease of a modularization, component standardization, on-shop fabrication, and direct site installation of the components are additional advantages that can contribute to the reduction of construction cost. This makes SMART more competitive in economics with large plants.

Application of SMART as an energy source for multipurpose-electricity generation, seawater desalination or district heating promises a new era of nuclear energy utilization. SMART is expected to be one of the first new nuclear power plants (NPPs) in the range of 120 MWe, which is a very useful energy for various industrial applications.

Since 1997 Korea Atomic Energy Research Institute (KAERI) has been developing sodium-cooled fast reactor (SFR) design technologies under a national nuclear R&D program. The goals of the SFR design technology development project were to secure strategic key technologies and to develop the conceptual design of the SFR, which are necessary for a reduction of a high level waste volume and toxicity. To this end, conceptual core design of KALIMER-1200 and the transuranics (TRU) burner of KALIMER-600 were developed, and a conceptual design of KALIMER-150, the national project of Prototype Gen IV sodium cooled fast reactor (PGSFR), is being performed as the main outcomes of the studies for the last two decades. KAERI is in charge of the design and the validation of the nuclear steam supply system (NSSS) and fuel development, while domestic participants are responsible for balance of the plant. Foreign partner, Argonne National Laboratory (ANL), is also supporting this development through international cooperation programs.

Two step design phase was done at the end of 2017 by issuing the specific design safety analysis dockets with design documents and safety analyses results sufficient for assessing safety of PGSFR. Verification and validation activities are being performed in parallel with the design progress to verify the safety of the PGSFR. A large-scale sodium thermohydraulic test program called STELLA is being progressed. STELLA-1 and STELLA-2 test facilities have been built to perform the sodium component tests of the decay heat removal system (DHRS) and to demonstrate the plant safety of the PGSFR, respectively. The other R&D activities are also being performed, including verification and validation of computational codes and development of the metal fuel fabrication technology.

PGSFR was determined to be 150 MWe suitable for technology demonstration and thus can be classified as a small modular reactor (SMR) and can be developed as a new nonlight water SMR in the near future.

Rapid climate changes and heavy reliance on imported fossil fuel motivated the Korean government to set up a long-term vision for transition to the hydrogen economy in 2005. One of the big challenges is how to produce a large amount of hydrogen in a clean, safe, and economic way. Among the various hydrogen-production methods, massive, safe, and economic production of hydrogen by water-splitting using a HTGR is a successful route to the hydrogen economy. Especially in Korea where the use of land is limited, nuclear hydrogen is deemed a practical solution due to its high energy density. Another advantage of nuclear hydrogen is that it is a sustainable and technology-led energy unaffected by the unrest associated with fossil fuel. Current hydrogen demand is mainly from oil refinery and chemical industries. Hydrogen is mostly produced by steam reforming, using the fossil fuel heat that emits large amount of greenhouse gases. Today in Korea, more than 1 Mtons/year of hydrogen is produced and consumed at oil refinery industries. In 2040 it is projected by the hydrogen roadmap that the 25% of total hydrogen demand will be supplied by nuclear hydrogen, which is around 3 Mtons/year even without considering the hydrogen iron ore reduction.

By virtue of its high temperature, HTGR can be used in high-temperature process heat applications including hydrogen production and high-efficiency electricity generation. The most effective application of HTGR is the huge hydrogen production in support of the hydrogen economy. To attain high temperatures an HTGR uses thermochemically stable helium as coolant and graphite as moderator and refractory coated-particle fuel, tristructural isotropic particle fuel (TRISO).

The design of the HTGR endeavors to make the maximum use of inherent safety and boasts the highest level of passive safety features possible. Safe shutdown of the reactor is ensured in an accident condition even without any operator's emergency actions, by the low-power density of the core, the inherent negative fuel temperature coefficient, and the large graphite heat capacity. The core afterheat is removed by natural phenomena such as conduction, radiation, and convection to the self-actuating reactor cavity cooling system. Such inherent safety features provide advantages in public acceptance.

KEPCO Engineering and Construction Company focuses, among a wide spectrum of SMR applications, on the option for a marine floating nuclear power plant and has set up the associated business plan and R&D roadmap. In 2019 it proposed a conceptual design of its own SMR, named BANDI-60S. Rather than competing with large land-based nuclear or fossil power plants optimized for centralized electrical grids, BANDI-60S would find its own way in rather nonconventional areas or niche applications, for instance, distributed power and heat supply to remote communities, sea water desalination for regions suffering clean water shortage, nuclear-renewable hybrid energy systems, etc.

Regional Energy Reactor 10 MWt (REX-10) is a small integral-type PWR proposed by the Regional Energy Research Institute for the Next Generation (RERI), which was led by Seoul National University (SNU). The aim of the REX-10 development project was to achieve a more stable, efficient, area-independent nuclear system for energy production. Design goals of REX-10 included not only implementing high levels of inherent safety into the reactor design to enhance the public acceptance of the innovative system but also attaining nonproliferation during all the processes of construction and operation.

KAIST proposed the micromodular reactor (KAIST-MMR) concept, which adopts a new coolant, supercritical CO_2, for the nuclear reactor cooling, and converting heat to work to achieve a full modularization of a small modular reactor (SMR). By adopting this approach the KAIST research team expects to successfully develop a nuclear power plant (or module) concept that can be fully modularized and suggest a revolutionary path to the future nuclear technology.

Unfortunately the present Korean government, President Moon administration, promised and is pursuing the denuclear energy policy in opposition to the past Korean nuclear policies. As a result, long-term research program on nuclear power plants, especially the development of new NPP including SMR programs, is greatly undermined to lose its thrust momentum.

18.2 Korean integral pressurized-water reactor: System-integrated Modular Advanced ReacTor

18.2.1 Chronicles of the SMART R&D program

KAERI started developing the System-integrated Modular Advanced ReacTor (SMART) in 1997, aiming to export it to countries with small electric grids and water supply issues. For the last 20 years, SMART-specific design methodologies and computer code systems have been developed together through a series of validation experiments and equipment verifications. From 2009 to 2012 the project of SMART technology validation and Standard Design Approval (SDA) was carried out. As a result a full set of licensing documents, including a Standard Design Safety Analysis Report (SSAR), were submitted to the Nuclear Safety and Security Commission (NSSC) at the end of December 2010. After one and half years of intensive licensing review, the SDA for SMART was officially issued on July 4, 2012 by the NSSC, in compliance with Article 12 of the Nuclear Safety Act. This may be the first license for an integral reactor in the world. A safety enhancement project on SMART was launched in 2012 and completed in February 2016. The development of SMART has taken over 300 million USD and 1700 R&D man years. Fig. 18.1 shows the SMART development program.

By successfully obtaining SDA, Korea and the Saudi Arabia signed a Memorandum of Understanding (MOU) on a SMART Partnership and Human Capability Build-up in March 2015 to construct at least two SMART in Saudi Arabia and jointly promote the reactor on the global market. Under the MOU, the two countries have conducted a 3-year preproject engineering (PPE) for construction of SMART in Saudi Arabia. The MOU also calls for the two countries to cooperate on the commercialization and promotion of the SMART to third countries. Finally the preproject engineering for SMART between Korea and Saudi Arabia has progressed for the First-of-a-Kind (FOAK) plant construction in Saudi Arabia.

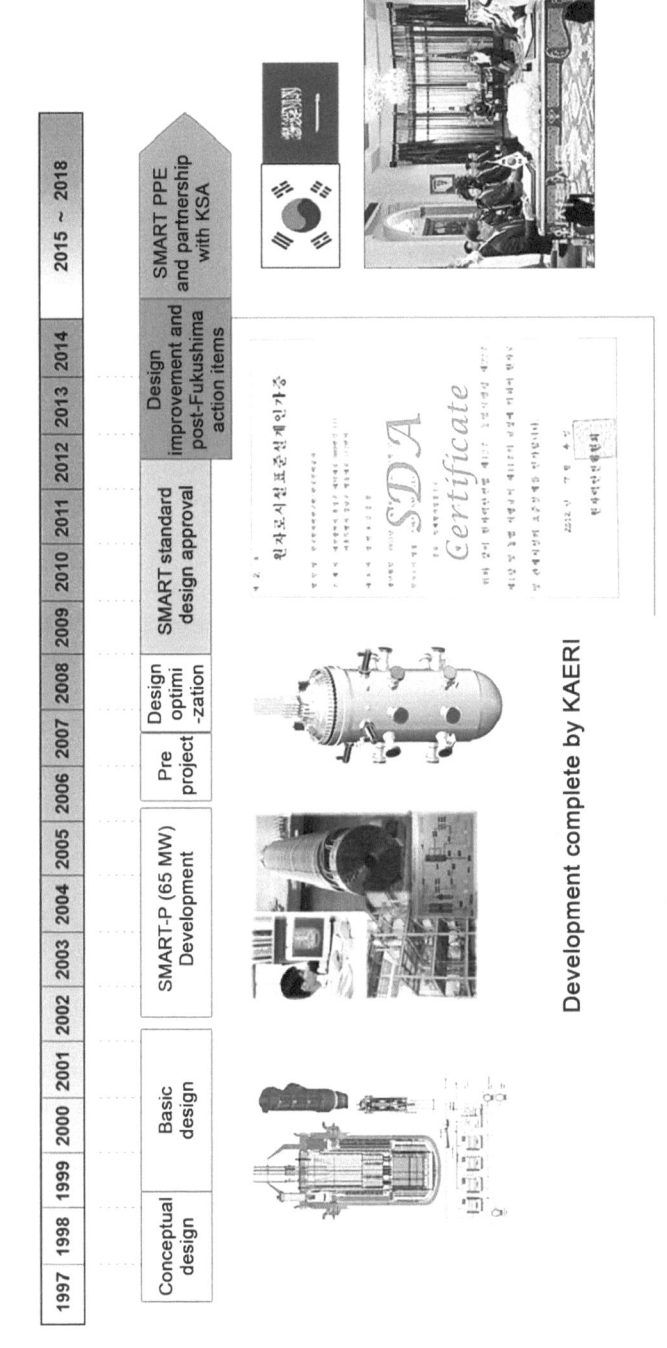

Fig. 18.1 SMART development program.

18.2.2 Design characteristics of the SMART

SMART is an integral pressurized water reactor (PWR) with a rated net electrical power of 120 MW(e) from 364 MW(th), which has been uprated from SDA design during the last several years. SMART becomes more advanced design to enhance the safety, reliability, and economics by adopting fully passive safety features. The advanced design features and technologies implemented in SMART were verified and validated during the SDA review. To enhance safety and reliability, the design configuration incorporates inherent safety features and passive safety systems. The design aims to achieve an improvement in the economics through system simplification, component modularization, reduction of construction time, and high plant availability. The advanced design features of SMART include integrated steam generators (SG), in-vessel pressurizer, and horizontally mounted canned motor pumps.

SMART is a multipurpose application reactor for electricity production and seawater desalination and is suitable for small or isolated grids. SMART has a unit output enough to meet the demands of electricity and fresh water for a city of 100,000 population.

SMART has an integral reactor coolant system (RCS) configuration that eliminates the use of large pipes for the connection of the major components. The major primary components including the core with 57 fuel assemblies, four reactor coolant pumps (RCPs), eight steam generators (SGs), and a pressurizer (PZR) are all installed in a single reactor pressure vessel (RPV). Twenty-five control rod drive mechanisms (CRDMs) are attached to nozzles on the reactor closure head. The CRDM is an electromechanical device, which provides controlled linear motion to the control rod assembly (CRA) through an extension shaft assembly (ESA). The SMART core is composed of 57 fuel assemblies, the design and performance of which are based on the proven 17×17 array with ceramic fuel rods of the commercial PWRs. Slightly enriched uranium oxide ($UO_2 < 5.0$ w/o) is used as the fuel ingredient providing a sufficient amount of reactivity required for 30–36 months cycle operation.

Due to the integral design the possibility of a large break loss of coolant accident (LBLOCA) is inherently eliminated. The SMART nuclear steam supply system (NSSS) consists of the RCS forming pressure boundary, the passive residual heat removal system (PRHRS), the passive safety injection system (PSIS), the automatic depressurization system (ADS), the chemical and volume control system (CVCS), and so on. The configuration of SMART is presented in Fig. 18.2. And detailed design parameters of SMART plant are shown in Table 18.1.

The secondary system receives superheated steam from the NSSS and uses most of the steam for electricity generation, seawater desalination, and the preheaters. The main steam pressure is controlled so as to be constant during a power operation. A load change is achieved by changing the feedwater flow rate. A seawater desalination system such as the multiple effect distillation (MED), multiple stage flash (MSF), and reverse osmosis (RO) may be used in conjunction with the secondary system by using proper interfacing methods.

ADS : automatic depressurization system
CCWS : component colling water system
CMT : core makeup tank
CPRSS : containment pressure and radioactivity suppression system
CVCS : chemical and volume control system
ECT : emergency cooldown tank
ERVC : ex-reactor vessel cooling
HEX : heat exchanger
IRWST : in-containment refueling water storage tank
PMT : PRHRS makeup tank
PRHRS : passive residual heat removal system
PSIS : passive safety injection system
PSV : pressurizer safety valve
PZR : pressurizer
RCP : reactor coolant pump
RCS : reactor coolant system
RRT : radioactive material removal tank
SC : shutdown cooling function
SG : steam generator
SIT: safety injection tank

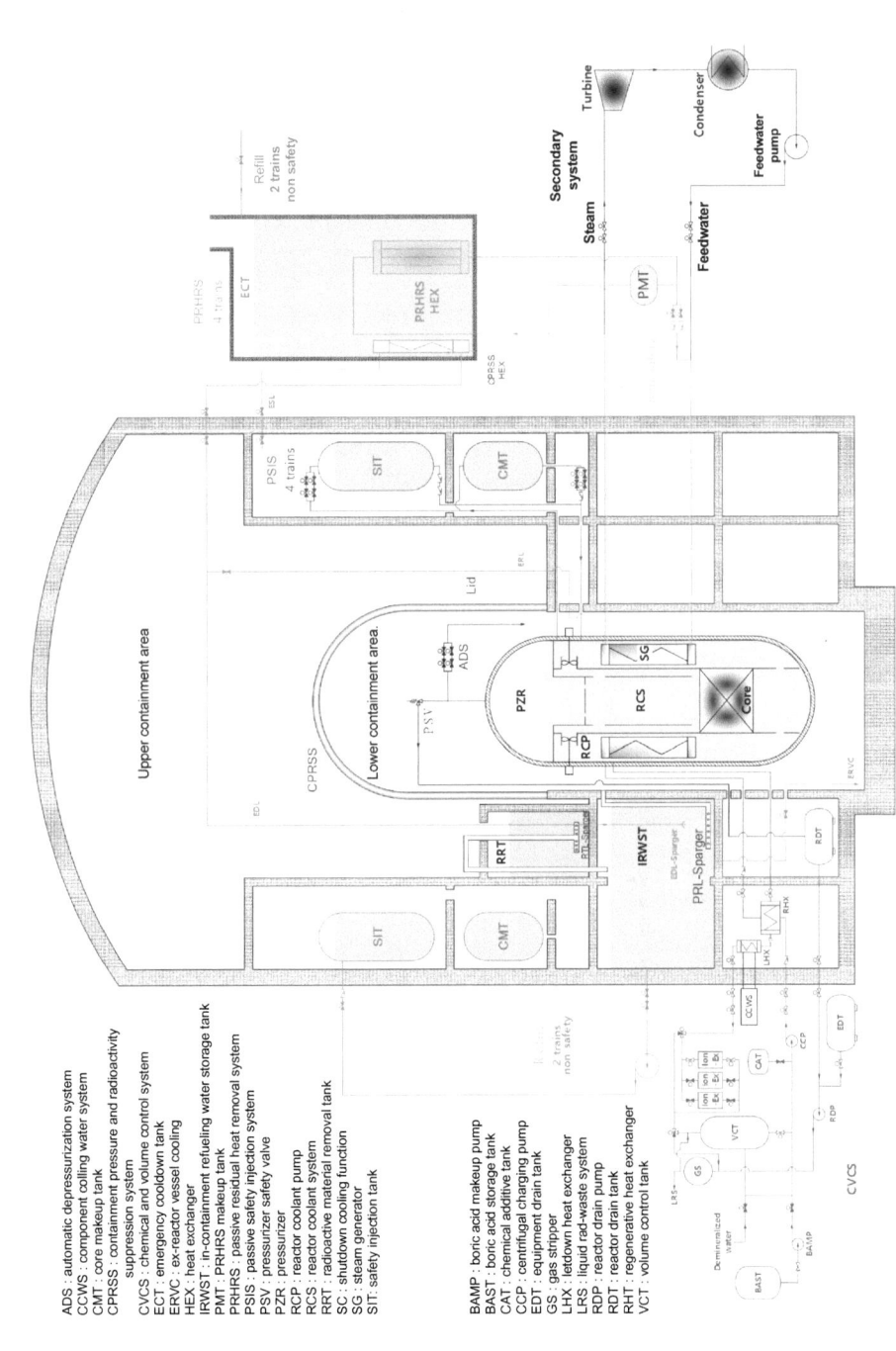

BAMP : boric acid makeup pump
BAST : boric acid storage tank
CAT : chemical additive tank
CCP : centrifugal charging pump
EDT : equipment drain tank
GS : gas stripper
LHX : letdown heat exchanger
LRS : liquid rad-waste system
RDP : reactor drain pump
RDT : reactor drain tank
RHT : regenerative heat exchanger
VCT : volume control tank

Fig. 18.2 Configuration of SMART.

Table 18.1 iPWR SMART technical data.

Description	Value
General plant data	
Reactor thermal output	364 MWth
Power plant output	Gross 120 MWe
Mode of operation	Load follow
Plant design life	60 years
Seismic design, safe shutdown earthquake in g scale (SSE[g])	0.3
Primary coolant material	Light water
Moderator material	Light water
Nonelectric applications	Desalination, district heat
Safety goals	
Core damage frequency/reactor year	<1E−6
Large early release frequency/reactor year	<1E−7
Reactor core	
Active core height	2.00 m
Equivalent core diameter	1.832 m
Fuel material	UO_2
Cladding material	Zircaloy-4
Number of fuel assemblies	57
Fuel cycle length	36 months
Primary coolant system	
Primary coolant flow rate	2090 kg/s
Reactor operating pressure	15 MPa
Core coolant inlet temperature	295.7°C
Power conversion system	
Working medium	Steam
Working medium flow rate at nominal conditions	160.8 kg/s
Working medium supply flow rate at nominal conditions	13.4 kg/s
Working medium supply temperature	200°C

Steam generator or heat exchanger	
Number	8
Tube outside diameter	17 mm
Tube material	Inconel 690
Reactor coolant pump (primary circulation system)	
Circulation type	Forced
Number of pumps	4
Pressurizer volume	61 m^3
Residual heat removal systems	Passive
Safety injection systems	Passive
Primary containment	
Overall form (spherical/cylindrical)	Cylindrical
Dimensions—diameter	44 m
Design pressure	0.42 MPa
Core catcher	None
In-vessel retention-ex-reactor vessel cooling (IVR-ERVC)	Applied
Protection against aircraft crash	Yes

18.2.2.1 Reactor coolant system

Reactor vessel assembly

The reactor vessel assembly of SMART contains its major primary systems such as fuel and core, eight SGs, a pressurizer (PZR), four RCPs, and 25 CRDMs in a single pressurized reactor vessel (PRV) as shown in Fig. 18.3. The integrated arrangement of these components enables the removal of the large size pipe connections between major reactor coolant systems and thus fundamentally eliminates the possibility of LB LOCAs. This feature, in turn, becomes a contributing factor for the safety enhancement of SMART. A large volume of the primary coolant provides large thermal inertia and a long response time and thus enhances the resistance to system transients and accidents.

Also the large free volume in the top part of the RPV located above the reactor water level is used as a PZR region. A large volume of the PZR can accommodate a wide range of pressure transients during a reactor power operation.

The primary system pressure is maintained constant due to the large PZR steam volume and a heater control without spray. The core exit temperature is programmed to maintain the primary system pressure constant during a load change. In this way the reactor always operates at its own operating pressure range matched with the system condition. Eight SGs are located at the circumferential periphery with an equal spacing inside the RPV and relatively high above the core to provide a driving force for a natural circulation of the coolant.

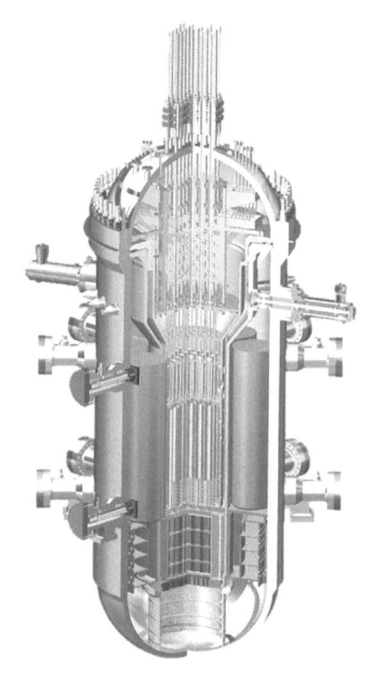

Fig. 18.3 SMART reactor vessel assembly.

Fuel assembly and core

The SMART core is composed of 57 fuel assemblies, the design and performance of which are based on the proven 17×17 array with UO_2 ceramic fuel rods in the commercial PWRs. Slightly enriched uranium oxide of less than 5 wt% is used as the fuel ingredient, providing a sufficient amount of reactivity required for a 36-month cycle operation. The fuel assembly is designed to accommodate power ramps during the load following maneuvering normally experienced in a commercial PWR.

The SMART core design providing an inherent safety is characterized by the following:

- longer cycle operation with a two-batch reload scheme,
- low core power density,
- adequate thermal margin of more than 15%,
- inherently free from Xenon oscillation instability,
- minimum rod motion for the load follows with coolant temperature control.

Steam generator cassette

SMART has eight identical SGs, which are located on the annulus formed by the RPV and the core support barrel (CSB). Each SG is of a once-through design with a number of helically coiled tubes. The primary reactor coolant flows downward in the shell side of the SG tubes, while the secondary feedwater flows upward through the inside of the tube. The secondary feedwater is evaporated in the tube and exits the SG cassette nozzle header at a 30°C superheated steam condition of 5.2 MPa. In the case of an abnormal shutdown of the reactor, the SG is used as the heat exchanger for the PRHRS, which permits an independent operation of the PRHRS from the hydraulic condition of the primary system.

Reactor coolant pump

SMART has four RCPs horizontally installed on the upper shell of the RPV. Each RCP is an integral unit consisting of a canned asynchronous three-phase motor.

Being a canned motor pump that does not require pump seals, its characteristic basically eliminates a SB LOCA associated with a pump seal failure that is one of the design basis events for conventional reactors.

The RCS transfers heat generated from the core to the secondary system through the SG and plays the role of a barrier that prevents the release of reactor coolant and radioactive materials to the reactor containment. The RCS and its supporting systems are designed with sufficient core cooling margin for protecting the reactor core from damage during all normal operation and anticipated operational occurrence (AOO). As shown in Fig. 18.3, the reactor coolant flows upward through the core, inside core support barrel (CSB) assembly including upper guide structure (UGS) barrel assembly, RCP, SG shell side, flow mixing header assembly (FMHA), lower plenum, and into the core. The forced circulation flow of the reactor coolant is formed by four RCPs installed at the upper side of the reactor pressure vessel. By this coolant circulation the core heat is delivered to the secondary system via the SGs.

18.2.2.2 Engineered safety features

Nuclear safety

Nuclear safety is the most important criterion for SMART design, assured by appropriate design, construction, and operation of a nuclear power plant. Design, construction (especially quality management) and operation regulations, requirements, and guidelines have been established by KAERI and the international nuclear industry, for example, IAEA, WENRA, US NRC, IEEE, to provide a sound basis for nuclear safety. The foundation of nuclear safety is the defense-in-depth principle, which is intensively applied in the design of nuclear power plants.

The SMART design safety philosophy is developed on the basis of the requirements of modern safety regulations accepted in Korea and IAEA safety standards. NPP safety is achieved by applying the defense-in-depth (DID) principle. This protection principle implies a system of physical barriers on the way by which the ionizing radiation and radioactive substances can be released into the environment. This system is used together with a complex of engineering and managerial measures for protecting these barriers and maintaining their effectiveness and measures for protecting the personnel, population, and the environment.

In the middle of the SMART SDA licensing process, the Fukushima accident occurred. Right after the Fukushima accident, the Korean government and regulatory body performed detailed safety audits on all the operating NPPs and new plants under construction (APR 1400), or under development (SMART) in Korea. As a result the Regulatory body requested 50 action items to enhance safety and KAERI submitted a conformity report to the regulatory body, which describes how SMART system maintains the safety against the kind of severe accident that occurred at Fukushima.

In the wake of the extreme natural disaster that led to severe damage at the Fukushima Daiichi nuclear power plant, mitigation measures and facilities to cope with severe accident become one of the key safety issues. It is clear that maintaining continuous and proper core cooling capability after shutdown of a reactor is very important. In a series of realistic simulations, it has been shown that the SMART PRHRS working on the secondary side of SG effectively removes decay heat and maintains the reactor in a stable condition for over 20 days without power sources and operator's mitigation action.

SMART safety design principles

The safety approach for design and operation of SMART is based on the defense-in-depth philosophy. Multiple barriers, such as fuel pellet, cladding, RCS pressure boundary, and finally containment, prevent radioactive releases to the environment. Multiple and diverse systems are designed to remove heat for the protection of those barriers.

The SMART defense-in-depth approach adopts proven engineering and regulatory practices that have evolved over the last 50 years of reactor design and commercial nuclear power plant licensing. The SMART defense-in-depth approach also provides for special treatment requirements in design, manufacturing, construction, testing, operations, and maintenance to compensate for uncertainties in design and analysis.

This approach provides a high degree of confidence that equipment will perform as expected. This is reflected in the SMART design, which includes conservative design features, inherent safety features, and passive engineered safety systems combined with beyond-design accident mitigation systems. The SMART design minimizes the potential for the occurrence of transients and accidents and ensures that required safety functions are satisfactorily performed should a transient or accident occur. The SMART design prevents the release of radioactive material and mitigates the consequences of postulated accidents.

The principles of multiple, independent, and concentric barriers to radiation release are applied. These principles include the use of design margin, redundancy, and diversity in the design of the structures, systems, and components that support the required safety functions and maintain the integrity and effectiveness of these barriers.

- First level: Minimization of abnormal operation and failures

    This is achieved by system simplification, minimizing components, and fully automated digital control and man-machine interface system designs. SMART instrumentation and control and man-machine interface systems are designed as a unified system of both operators and I&C system. One-man operation under normal plant operation mode is a part of the SMART control room design objectives.

- Second level: Control of abnormal operation and detection of failures

    Fully automated control and man-machine systems minimize abnormal operations. Simplified systems and minimized components make it easy to detect failures. For example, the SG inlet temperature control logic is programmed to minimize the coolant volume change for load-following operation, and the canned-motor type RCP does not require a seal injection system. These features make minimal changes in CVCS charging/letdown flow rate, and consequently the pressurizer level can be utilized to detect an unidentified coolant leakage to be increased. The SMART core monitoring system (SCOMS) and the SMART core protection system (SCOPS) contribute to the immediate response to abnormal operation and detection of system failures.

- Third level: Control of accidents within the design basis

    This is satisfied by the SMART engineered safety features-component control system (ESF-CCS), which are highly reliable and designed to function actively or passively on demand, in addition to the reactor protection systems. The passive safety injection system (PSIS) guarantees coolant inventory control during an accident, while the passive residual heat removal system (PRHRS), the automatic depressurization system (ADS), acts to lower the RCS pressure so that gravity head injection at low system pressure from safety injection tank (SIT) to be available at the early stage in LOCA, and the containment pressure and radioactivity suppression system (CPRSS) ensure safety of the plant in design basis accidents. As a final barrier for radiation confinement, a steel-lined containment behaves as an additional layer of defense in depth.

- Fourth level: Control of severe plant conditions, including prevention of accident progression and mitigation of the consequences of severe accidents

    This level ensures that radioactive releases are kept as low as practicable in the very unlikely event that the first three levels have failed. The most important objective of this level is the protection of the containment function. This level is satisfied by in-vessel retention (IVR) through External Reactor Vessel Cooling (ERVC) concepts and hydrogen control using passive autocatalytic recombines (PAR).

The high reliability and integrity are confirmed in the SMART design by taking into account single failure criteria, fail-safe, defense-in-depth, and diversity in design against common cause failure. SMART design can minimize or eliminate susceptibility to common cause failures by implementing the principles of diversity and independence among systems and components. Diversity is achieved by having two distinct-independent systems available for shutting down the reactor. Redundant and independent trains and components are installed in the safety systems such as the PRHRS, PSIS, ADS, and CPRSS. The single failure criteria are applied in the SMART safety analyses to show the capability of achieving emergency core reactivity control, emergency core and containment heat removal, and containment isolation, integrity, and atmosphere clean-up. Diversity and redundancy are also implemented in the instrumentation and control system and structure. The principle of fail-safe design is incorporated, for example, in dropping of the CRDM after loss of electrical power, and valves are designed to be fail open or fail close in the secondary, auxiliary, and safety systems. The safety goal of the SMART design requires that the core damage frequency (CDF) for internal events is less than 1.0×10^{-6}/RY and containment failure frequency by internal events is less than 1.0×10^{-7}/RY.

Description of SMART safety systems

The safety systems of SMART comprise a sensible mix of proven technologies and advanced design features and are designed to function automatically on demand. They consist of the PRHRS, PSIS, ADS, CPRSS, reactor overpressure protection system, and severe accident mitigation system. The reactor can be shut down under any circumstances by inserting control rods, or by boron injection. The passive safety systems of SMART maintain the plant in a safe shutdown condition following design basis accidents, both LOCA or non-LOCA, without safety grade AC power or operator actions for at least 72 h. The PRHRS prevents overheating and overpressurizing of the primary system in case of emergency situations where normal heat removal using the secondary system is unavailable. The reactor containment is resistant to any kind of seismic activity and can withstand possible airplane-crash incident.

The PRHRS removes the RCS heat by natural circulation in emergency situations where normal steam extraction or feedwater supply is unavailable. The PRHRS cools the RCS to the safe shutdown condition within 36 h after the accident initiation and maintains the safe shutdown condition for at least another 36 h. Therefore the safety function is performed for at least 72 h without any corrective action by operator or the aid of external AC power.

The safety function of the PRHRS is maintained continuously for a long-term period when an emergency cooldown tank (ECT) is refilled periodically by a PRHRS ECT refilling system. The PRHRS consists of four independent trains, and each train is composed of one ECT, a PRHRS heat exchanger and a PRHRS makeup tank. In the design of the PRHRS, the possibility of loss of one train by a single failure is eliminated.

The PSIS prevents core uncovery in case of a small break loss-of-coolant accident (SBLOCA) by injecting water into the RCS and also removes heat from the core. The PSIS consists of four mechanically independent trains, and each train is composed of one core makeup tank (CMT) and one safety injection tank (SIT) with related valves, instrumentation equipment, one safety injection line (SIL), and a pressure balance line (PBL). The CMTs are filled with boric acid water to provide makeup and boration functions to the RCS during LOCA and non-LOCA.

The ADS acts to lower the RCS pressure so that gravity head injection from the SITs can be available at an early stage in a LOCA. The ADS can be manually operated in the case of a total loss of secondary heat removal (TLOSHR) accident. In addition, the ADS can depressurize the RCS below the operation pressure of a refilling system within at least 72 h after LOCA, even without the aid of operator or AC power. The ADS consists of two trains, and each train is operated automatically when the water level of the CMTs reaches the designed setpoint level.

18.2.2.3 Instrumentation and controls system and control rooms

The SMART Man Machine Interface Systems (MMIS) consists of instrumentation and control (I&C) systems and control rooms. The SMART MMIS, to which digital information processing technologies and a communication network scheme are applied, has a multilayer structure. The I&C systems of SMART are composed of protection systems, control systems, monitoring systems, and instrumentation systems, as shown in Fig. 18.4. Control rooms for SMART include a main control room (MCR) and a remote shutdown room (RSR).

SMART MMIS adopts levels of defense to achieve the design of defense in depth to the arrangement of I&C systems for the purpose of operating the reactor or shutting it down and cooling it. The levels of SMART MMIS are the control system, the reactor protection system, the engineered safety feature-component control system (ESF-CCS), and the monitoring and indication system. The levels are concentrically arranged in that when the control system fails, the reactor protection system shuts down reactivity; when both the control system and the reactor protection system fail, the ESF-CCS continues to support the physical barriers to radiological release by cooling the fuel, thus allowing time for other measure to be taken by reactor operators to reduce reactivity.

18.2.3 SMART technology verification

SMART technology is a sensible mixture of conventional technologies and innovative concepts to improve the level of safety, reliability, and economics. All technology implemented into SMART should be proven or qualified by experiment, testing, or analysis. The equipment should be designed according to the applicable approved standards if possible. A SMART design verification program, including comprehensive experiments and the development of the analysis model, was planned and performed to confirm the advanced design features of SMART that are yet to be proven through the design and operation of existing PWRs.

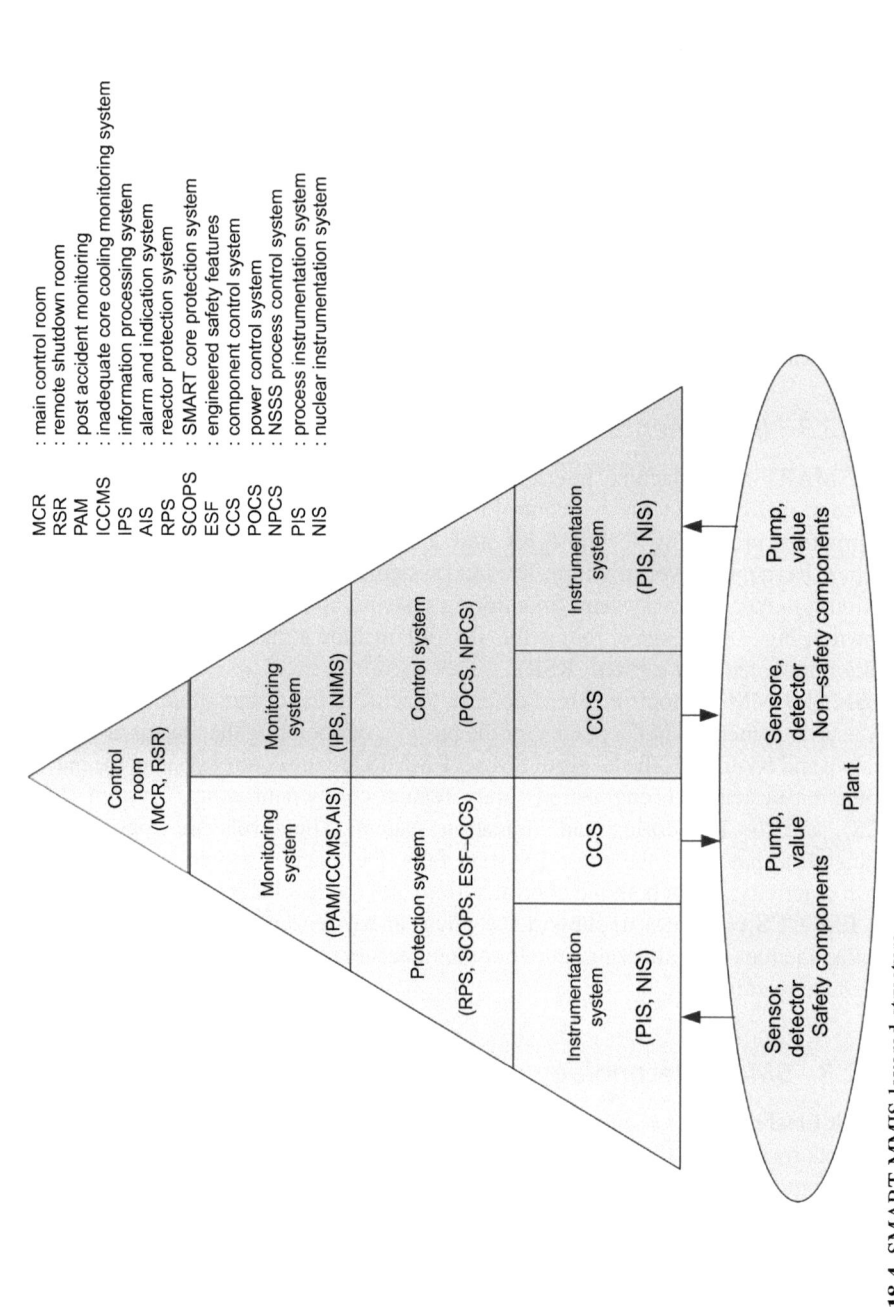

MCR	: main control room
RSR	: remote shutdown room
PAM	: post accident monitoring
ICCMS	: inadequate core cooling monitoring system
IPS	: information processing system
AIS	: alarm and indication system
RPS	: reactor protection system
SCOPS	: SMART core protection system
ESF	: engineered safety features
CCS	: component control system
POCS	: power control system
NPCS	: NSSS process control system
PIS	: process instrumentation system
NIS	: nuclear instrumentation system

Fig. 18.4 SMART MMIS layered structure.

This program includes basic thermohydraulic experiments, separate effect tests on the major components, and integral effect tests of the safety system. Basic fundamental thermohydraulic experiments were carried out during the concept development to assure the key technology of the advanced safety systems. After the SMART concept development, essential technologies are required for the development of SMART, such as a performance of a helically coiled tube steam generator. To develop these technologies, separate effect tests of SMART's major components were performed to obtain a fundamental database. Based on that, computer analysis models were developed.

18.2.3.1 Thermohydraulic test

A series of fundamental tests and experiments have been carried out throughout the SMART development phases to examine the physical phenomena related with the specific SMART design concepts. The main purpose of these experiments was twofold: to understand the thermohydraulic behavior of the specific design concepts and to obtain fundamental data to be used, in turn, for further feedback to the optimization of design. Among the experiments conducted, specific SMART design-related experiments are as follows:

- boiling heat transfer characteristics in the helically coiled steam generator tube,
- experiment for natural circulation in the integral arrangement of the reactor system,
- two-phase critical flow tests with noncondensable gases to investigate the thermohydraulic phenomena of critical flow in the presence of noncondensable gases,
- critical heat flux measurement for 5×5 test bundles simulating SMART fuel assembly,
- out-of-pile mechanical/hydraulic tests for full-scale SMART fuel assemblies,
- water chemistry and corrosion tests at a loop facility to examine the corrosion behavior and characteristics of fuel cladding, internal structural materials, and steam generator tube materials at reactor operating conditions;
- experiments on wet thermal insulation to determine the insulating effects for the internal pressurizer (PZR) design and to derive a heat transfer coefficient for the design,
- experiments on phenomena and characteristics of heat transfer through the condensing mechanism of the heat exchanger inside PRHRS tanks.

Brief introductions to the typical tests and experiments implemented to verify the SMART design characteristic are given in the succeeding text:

Critical heat flux tests

Critical heat flux (CHF) tests using SMART fuel bundles have been conducted to provide database for the development of a CHF correlation, which is essential for the evaluation of the thermal design criteria and safety analysis of SMART. A series of experiments was carried out to investigate the thermal mixing and CHF in 5×5 test bundles with uniform and nonuniform axial power shapes. A high-pressure water test loop included pressure housing with 5×5 SMART fuel simulators, high-head pump, pressurizer with noncondensable gas, heat exchangers, preheater, and mixers. The CHF experiments varied the pressure from 2.5 to 17 MPa and the mass flux from 200 to 2500 kg/m^2 s which covers the normal operating conditions and the anticipated operational occurrences (AOO) for SMART. Several hundreds of CHF data items obtained from the water loop CHF experiments were employed for the development and validation of a CHF correlation, which was applied to the thermal design and safety analysis of SMART.

Two-phase critical flow test with a non-condensable gas

The early SMART concept adopted an in-vessel pressurizer type with an inherent self-pressure regulating capability designed to operate via the thermopneumatic balance between the water, steam and nitrogen gas, which are the three fluids that fill the pressurizer. In the event of a rupture of a pipe line connected to the pressurizer at a high system pressure, a mixture of water, steam, and nitrogen is discharged through the break at critical flow conditions. The computer codes for the safety analysis of SMART need to use a verified and validated model for this critical flow. To investigate the thermal-hydraulic phenomena of the critical flow affected by the noncondensable gas entrained in the two-phase break flow, a separate effects test facility was designed and installed at the KAERI site. The test facility can be operated at a temperature of 323°C and pressure of 12 MPa, with a maximum break size of 20 mm in diameter. A nitrogen gas flow-rate of up to 0.5 kg/s can be injected and mixed with a two-phase mixture in the test section to simulate the transient behavior expected during a LOCA. The test data from the facility were used for the development and verification of the critical break flow model for SMART.

Integral effect test

The SMART-P is a pilot plant of the integral-type reactor SMART, which has new innovative design features, aimed at achieving a highly enhanced safety and improved economics. An experimental verification by an integral simulation for a transient and accident (VISTA) facility has been constructed to simulate the various transient and events of an integral reactor. The VISTA facility (Fig. 18.5) has been used to understand the thermal-hydraulic behavior including several operational transients and design basis accidents. During the past 5 years, several integral effect tests have been carried out and reported, including performance tests, reactor coolant pump (RCP) transients, power transients and heat-up or cool-down procedures, and safety-related design basis accidents. It contributes to verifying the system design of the reference plant.

18.2.3.2 Major components performance test

A performance test of the major components such as the RCP, steam generator (SG), and control rod drive mechanism (CRDM) was carried out. In the SMART standard design approval program, additional performance tests for the RCP and CRDM are scheduled to be performed to verify the final design models.

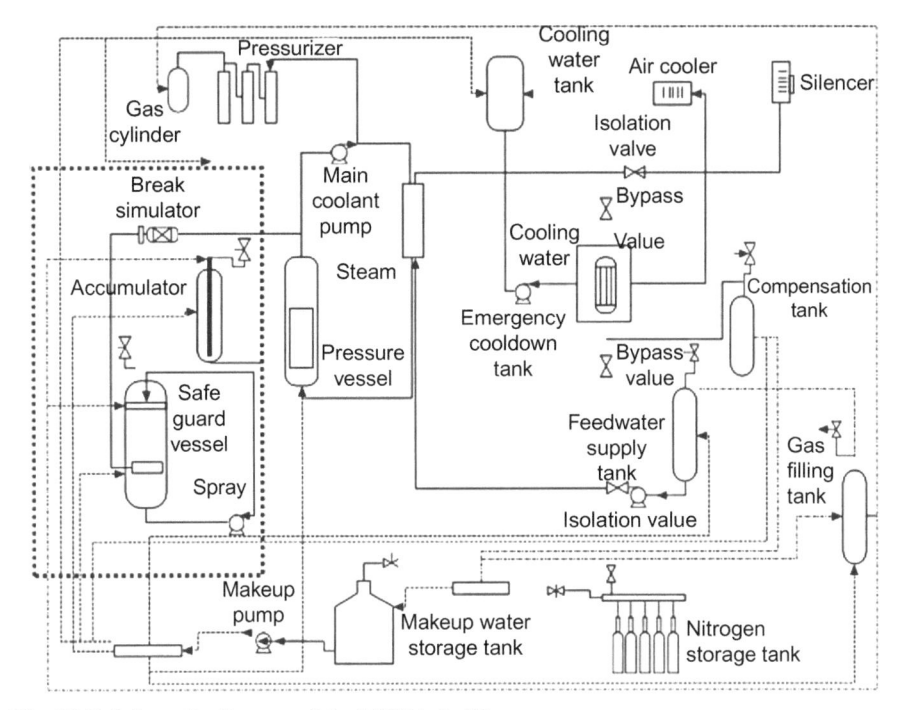

Fig. 18.5 Schematic diagram of the VISTA facility.

18.3 Development of other small modular nuclear reactor (SMR) programs in the Republic of Korea

The present Korean government, President Moon administration, promised and is pursuing the denuclear energy policy in opposition to the past Korean nuclear policies. As a result, long-term research program on nuclear power plants, especially the development of new NPP, is greatly undermined.

Over the past three decades, Korea's nuclear industry has experienced great change and growth. Through the joint design with Westinghouse Electric Company in early 1980s, the technical self-reliance of large-capacity commercial pressurized water reactor has been achieved, and based on this, it succeeded in developing SMART as a small integral nuclear reactor. It also exported APR-1400 nuclear power plants to the UAE in 2009. During this period, Korea's nuclear ecosystem has grown, working on safety research, developing design tools that are fundamental to NPP, and developing various types of reactors (various fuels and core and coolant). In particular the SMART has given a big moment for the development of the other SMRs in Korea.

Since the beginning of 2000, SMR development has been carried out in various organizations. Especially after Fukushima Daiichi accident, SMR development got focused due to its strong merits—inherent safety, economy from modularity, small grid compatibility, and versatile application. Many SMR developments have been attempted from conceptual design to preliminary design, from micro to small reactor. In addition to KAERI's SMART, there are light water reactors such as BANDI-60S of KEPCO E&C and REX-10 of Seoul National University, mini and micro in power, respectively. BANDI-60S is developed for marine use. Other than water-cooled SMR, there are also a VHTR demonstration, a high-temperature gas-cooled reactor of KAERI, and a KAIST-MMR of KAIST. In addition, a SFR demonstration of KAERI and a SMR concept of UNIST using liquid metal as a coolant are underway. These reactors will be introduced in more detail in the following sections.

18.3.1 BANDI-60S (KEPCO E&C)

18.3.1.1 Overview

Over the last 40 years, KEPCO E&C has been offering a full scope of engineering services in the conventional nuclear power business in Korea and overseas. As for the small modular reactor (SMR), however, it was only in the early 2010s that it realized the potential of the SMR in the future and began to follow up the global R&D trends and look into the key technologies.

Among a wide spectrum of SMR applications, KEPCO E&C focuses on the option for a marine floating nuclear power plant and has set up the associated business plan and R&D roadmap. In 2019 it proposed a conceptual design of its own SMR, named BANDI-60S. Rather than competing with large land-based nuclear or fossil power plants optimized for centralized electrical grids, BANDI-60S would find its own way in rather nonconventional areas or niche applications, for instance, distributed power and heat supply to remote communities, sea water desalination for regions suffering clean water shortage, nuclear-renewable hybrid energy systems, etc.

BANDI-60S is a so-called "block-type" reactor system as shown in Fig. 18.6. Main components of the primary system are directly "nozzle-to-nozzle" connected to each other without big pipes, which are to be designed and manufactured to ASME BPVC Sec. III, Class 1 NB-3300, and thereby exempted from accidental breaks. It enables us to eliminate the scenarios of large break LOCAs from the design basis as in the integral SMRs. The advantages of the block type relative to the integral one are such as the block type would be simpler and easier in surveillance and maintenance of main components, more proven in design and manufacturing, and more scalable in size.

Key design goals or attributes of BANDI-60S include the following:

- extended refueling cycle of 4–5 years,
- soluble boron-free design and operation,
- control element drive mechanism installed inside the reactor pressure vessel (in-vessel CEDM),
- in-core instrumentation lines routed from the reactor vessel upper head (top-mounted ICI),
- passive safety features for emergency core cooling and residual heat removal,
- reactor coolant pumps with leak-tight canned motors,
- a pressurizer integrated into the reactor vessel upper head region, etc.

Technologies and design details to achieve them would require a lot of time and efforts. Among them, KEPCO E&C determined the soluble boron-free design,

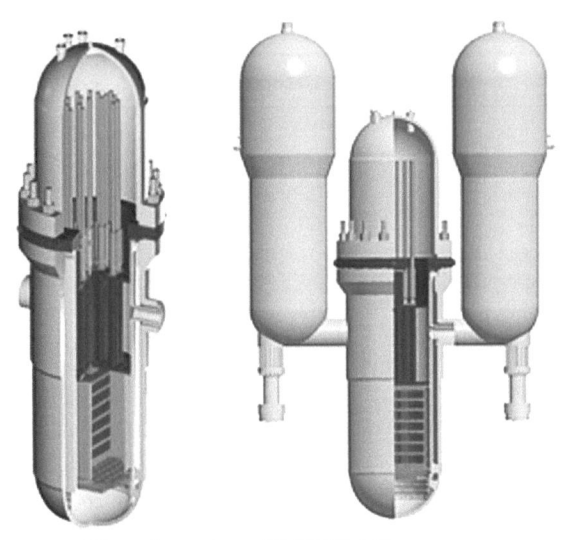

Fig. 18.6 Reactor and RCS configuration of BANDI-60S.

in-vessel CEDM, and top-mounted ICI as more crucial ones. Aligned with Korean nuclear industries, KEPCO E&C proposed an R&D project to the Korean government, which was accepted and began in 2013. With eager and close collaborative works of all participants for 5 years, the project was successfully completed in 2018. Experience and knowledge obtained then are now being fully utilized for BANDI-60S and would also be applicable to other advanced SMRs.

18.3.1.2 Future plan

Basic design concept of BANDI-60S is rooted in KEPCO E&C's technologies and experience proven through its year-long services in conventional nuclear power business. It would help minimize technical and licensing uncertainties and thus reduce the time to deployment. Critical nuclear safety functions of reactor shutdown, decay heat removal, and radiation protection will be further enhanced by adopting inherent and passive safety design features.

Along with the basic design option, new technologies will be sought and examined for their feasibility, some of which are a compact plate-shell steam generator, advanced nuclear fuels, more reliable and simpler I&C systems, automatic load following, autonomous operation, etc.

For more diverse and wider applications of BANDI-60S, there will be collaborations with the marine-shipbuilding industry. To be mounted on the marine floating nuclear power plant, its systems and components will be designed to comply with the requirements and regulations of International Maritime Organization (IMO). Applications to the nuclear-renewable hybrid energy systems will also be investigated in the long run.

Table 18.2 Design features of BANDI-60S.

Parameters	Design values and description
General	
Reactor type	PWR with a block-type system arrangement
Power thermal/electrical (MW)	200/60
Design life (year)	60
Fuel and reactor core	
Fuel type/material/enrichment (%)	17×17 square/UO_2/less than 5.0
No. of fuel assembly	52
Linear power density (kW/m)	6.65 (average)
Fuel burnup (GWd/ton)	35 (average)
Reactivity control	Control rods with in-vessel CEDM) and a secondary shutdown system (soluble boron free)
Refueling cycle (year)	4–5
Reactor coolant system	
Coolant circulation mode	Forced
Operating pressure (MPa)	15.0
Core inlet/exit temp. (°C)	290/325
RPV height/diameter (m)	11.2/2.8
Pressurizer	RV integrated/steam/heaters and spray
Reactor coolant pumps	2 RCPs with leak-tight canned motor
Steam generators	Recirculation type with U-tube or plate-shell
Steam condition/pressure (MPa)	Saturated/6.0
Safety Systems	
Safety injection	Passive
Decay heat removal	Passive

18.3.1.3 Technical data

Key design attributes of BANDI-60S are introduced in Table. 18.2.

Block-type arrangement of reactor coolant system

The RCS consists of a reactor pressure vessel, two steam generators, two reactor coolant pumps, and a pressurizer as shown in Fig. 18.6 where the reactor pressure vessel and steam generators are closely coupled by "nozzle-to-nozzle" connections. As for the steam generator (SG), the basic design is a U-tube recirculation type with moisture separators and steam dryer integrated. Its reliable performance has been proven enough through long and plentiful operations of commercial PWRs. In parallel with the basic option, a new type of SG is now being developed based on the plate-shell heat exchanger technology, which would reduce the size by more than a half. Pressurizer is integrated into the reactor vessel upper head region, providing water and steam volume much larger than those of conventional PWRs. The pressure is controlled by heaters and sprays. Reactor coolant pumps are equipped with leak-tight canned motors and attached to the steam generator lower head.

Soluble boron-free design and operation

Soluble boron-free design eliminates troublesome concerns caused by boron treatments in the conventional PWR plants. Chemical and volume control system would be significantly simplified, leading to the reduction of a plant footprint. It would also alleviate the corrosive operating conditions and reduce the liquid radioactive waste. Magnitude of the negative moderator temperature coefficient would grow further and thereby enhance the nuclear safety.

Due to the soluble boron-free design and extended refueling cycle, however, much more control rods and burnable absorbers are needed to suppress and control the excess reactivity of nuclear fuel over time. Among the whole 52 fuel assemblies in the core, a preliminary core design has assigned 26 locations for the regulating and shutdown control rods, 14 for the secondary shutdown system and 12 for the in-core instrumentations as shown in Fig. 18.7. The refueling cycle length extends to over 4 years, which would go longer if the strong load following capability that BANDI-60S would have is aggressively utilized in actual operations in the future.

A basic corrosion experiment and new water chemistry control strategy for the soluble boron-free operation was done in the aforementioned government R&D project in 2013–18. A secondary shutdown or diverse reactivity control system will be developed and adopted to BANDI-60S to comply with GDC 26 and 27 of 10 CFR 50.

In-vessel control element drive mechanism

For small reactors the control element (or rod) ejection accident would be one of the most critical safety concerns since the reactivity worth of a single control element is stronger and thus has a severer impact on the core than for large reactors. To eliminate this safety issue from the design basis, BANDI-60S will have the CEDM within the reactor pressure vessel.

However, the operating temperature, pressure, radiation, and water-submerging conditions inside the reactor vessel are highly challenging so that the functional and material integrity of the in-vessel CEDM must be demonstrated. As such the associated element technologies were developed and an integrated full-scale demonstration test was done in the aforementioned government R&D project in 2013–18.

Passive safety systems

The safety systems of BANDI-60S will rely on the forces of nature such as gravity or buoyancy. The main safety systems are a passive safety injection system (PSIS), a passive residual heat removal system (PRHRS), and a passive containment cooling system (PCCS) as depicted in Fig. 18.8.

Core makeup tank (CMT) of the PSIS makes up the RCS inventory by gravity for any loss of coolant events. If either the pressurizer level decreases, the RCS depressurizes or the containment pressure increases beyond specified set points, the isolation valves (normally closed) in the pressure balance line between the CMT and the SG inlet are automatically opened to equalize the CMT pressure with the RCS. If the RCS pressure further decreases by itself or by automatic depressurization valves (ADVs) to the containment pressure, the cooling water in the emergency core cooling

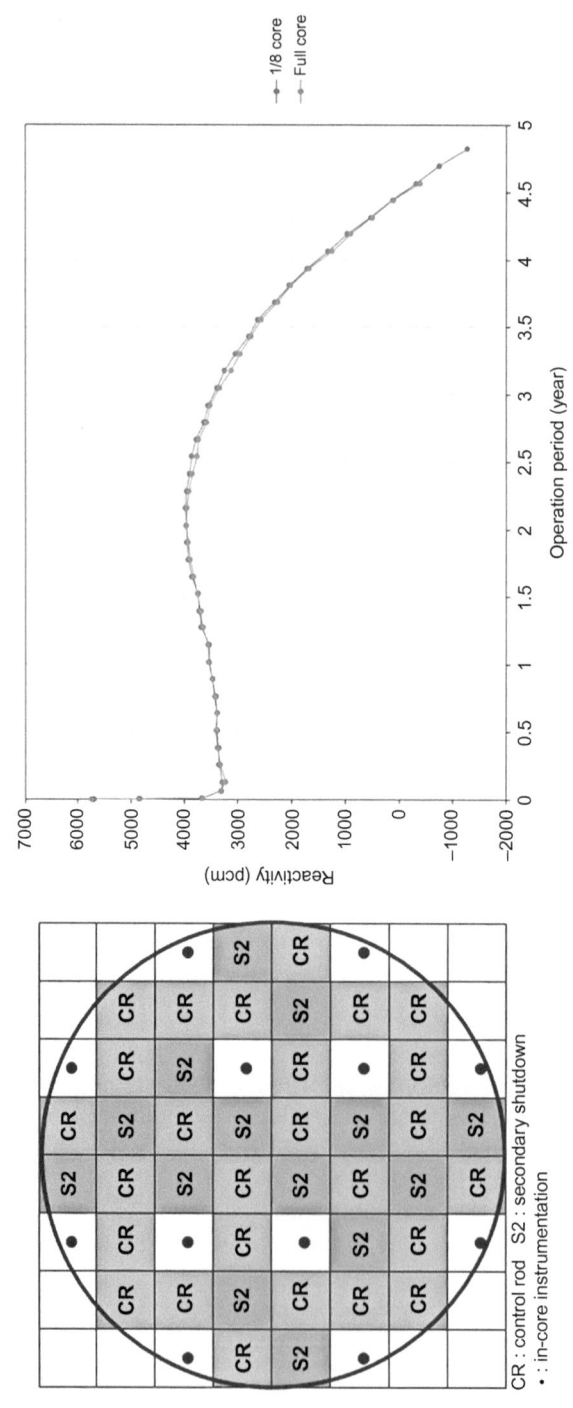

Fig. 18.7 Core arrangement and reactivity change over time of BANDI-60S.

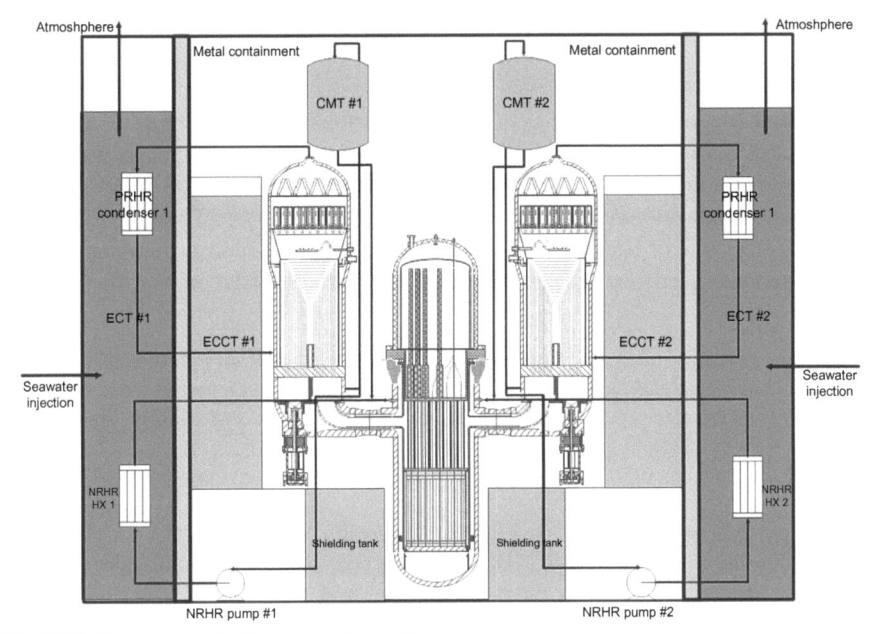

Fig. 18.8 Arrangement of main passive safety systems.

tanks (ECCTs) within the containment is injected into the core by gravity. Water spilled from the break is collected in the containment bottom, and the reactor core is eventually flooded and maintained in a long-term cooling condition by convective natural circulation.

The PRHRS removes the core decay heat and RCS sensible heat in abnormal conditions when the RCS normal cooling is not available through SGs and condenser. The PRHR heat exchangers are submerged in the emergency cool-down tanks (ECTs), which wrap around the outer surface of the containment metal wall. The elevations of reactor core and ECTs are arranged below the sea level. Steam from the SGs condenses in the PRHR heat exchangers and returns to the SGs by gravity. The RCS is cooled down to a safe shutdown state and maintained for an extended time without refilling the ECTs. If the ECT water gets too hot or the level too low during an accident, seawater is injected into the ECTs by gravity or density difference to secure the ultimate heat sink. Heat from the containment is removed to the ECTs through the metal wall by conduction.

18.3.2 REX-10 (SNU)

18.3.2.1 Overview

Regional Energy Reactor 10 MWt (REX-10) is a small integral-type PWR proposed by the Regional Energy Research Institute for the Next Generation (RERI), which was led by Seoul National University (SNU). The aim of the REX-10 development project was to achieve a more stable, efficient, area-independent nuclear system for energy production (Lee et al., 2012). Design goals of REX-10 included not only implementing high levels of inherent safety into the reactor design to enhance the public acceptance of the innovative system but also attaining nonproliferation during all the processes of construction and operation. In particular, great emphasis was placed on the assurance of passive safety features for REX-10 at the design stage.

The REX-10 development project was funded by Ministry of Trade, Industry, and Energy of Republic of Korea from 2005 to 2010. The RERI consortium consisted of nine universities, and one research institute to conduct R&Ds in five divisions as follows:

- first div.: design of nuclear systems—design of Th-fueled core, design of safety systems, experiments on natural circulation in REX-10 (Jang et al., 2011), performance test of the steam-gas pressurizer (Kim et al., 2009), development of a system code for REX-10 (Lee and Park, 2013a), etc.
- second div.: design of radiation shields/development of an environmental radioactivity monitoring system.
- third div.: unmanned remote control system (Bae et al., 2008).
- fourth div.: development of the SCADA system for operation of the regional energy system.
- fifth div.: protection system for the regional energy system.

18.3.2.2 Future plans

After the R&D project supported from Korean government completed in 2010, the research activities on the regional energy reactor has been sustained based on individual research frameworks to continue developing key technologies for REX-10. They include the following:

Thermal-hydraulic analysis of the natural circulation system in REX-10 using TASS/SMR and SPACE codes: TASS/SMR and SPACE are the thermohydraulic analysis code for an integral type PWR developed by KAERI and one for a conventional PWR developed by Korean nuclear industry, respectively (Lee and Park, 2013b).

Implementation of the dynamic motion model into the thermal-hydraulic analysis codes, TAPINS and MARS considering maritime application of the integral type reactor (Lee and Park, 2013a; Beom et al., 2019).

Fundamental thermal-hydraulic experiments under motion conditions, especially rolling conditions including natural circulation, critical heat flux, etc. (Hwang et al., 2012).

18.3.2.3 Technical data

The schematic diagram of REX-10 is depicted in Fig. 18.9. REX-10 is designed to generate the rated output of 10 MWt at a low pressure (2.0 MPa) compared with existing commercial reactors. It is an integral-type small PWR, which contains the primary components inside the reactor vessel. This layout eliminates the possibility of system depressurization by LBLOCA by virtue of the absence of large-diameter pipelines. In addition, as the coolant circulates by gravity-driven free convection without a RCP, all safety issues associated with the failure of the RCP can be eliminated.

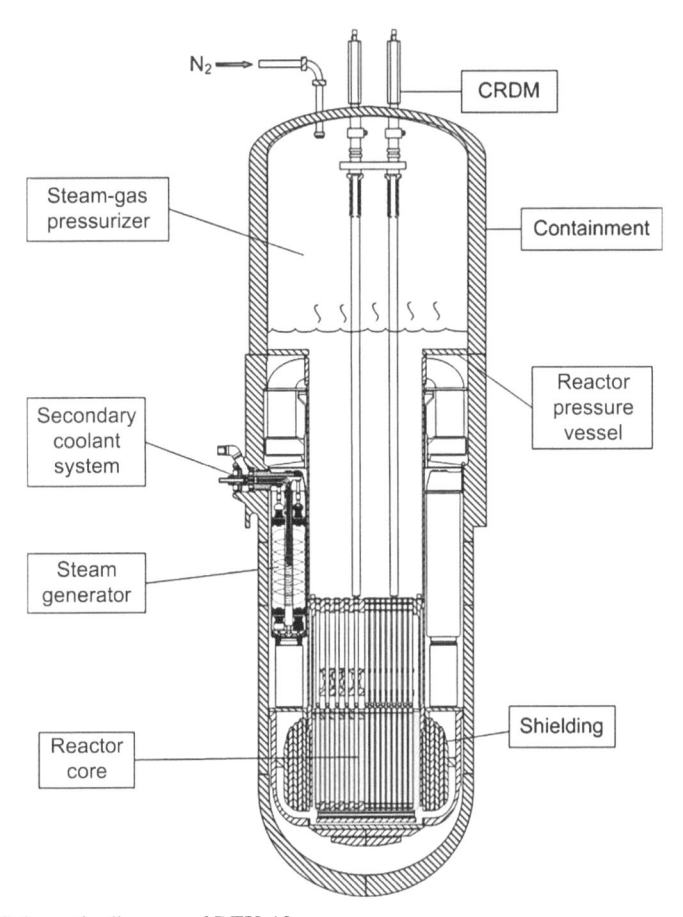

Fig. 18.9 Schematic diagram of REX-10.

The major design parameters of REX-10 are listed in Table 18.3. A cylindrical reactor vessel with a height of 5.715-m houses the core, CRDM, pressurizer, and steam generator. Placed at the bottom of the reactor pressure vessel are the 9 × 9 heterogeneous Th/UO$_2$ fuel assemblies that are used to achieve an ultralong fuel cycle of up to 20 years on the basis of the Seed-Blanket Unit (SBU) design. This thorium-based fuel has a major benefit in aspect of the intrinsic proliferation resistance since ^{232}U, formed along with the bred ^{233}U, and its daughter isotopes emit intensive alpha and gamma radiation, hindering the access to nuclear fuel. A total of 37 assemblies with an active height of 0.8 m are composed of fuel bundles with the fuel material enriched to 20 w/o. In REX-10 the intrinsic feedback capability is enhanced by refusal from soluble boron control, while excess reactivity is dealt with by burnable poison and control rods. The long riser region above the core provides sufficient head for free convection of fluid. The natural circulation capability of REX-10 was demonstrated in the scaled-down test facility named RTF, which was a full-height full-pressure facility with reduced power by 1/50.

Table 18.3 Main design parameters of REX-10.

Parameters	Design value
General information	
Reactor type	Integral PWR
Reactor power (MW)	10
Service years (year)	20
Reactor coolant system	
Cooling mode	Natural circulation
Operating pressure (MPa)	2.0
Core inlet/outlet temp. (°C)	165.0/200.0
Fuel and reactor core	
Fuel type	9 × 9 Square FA
Fuel material	Hetero Th/UO_2
No. of fuel assembly	37
Steam generator	
Type	Helical coil HX
Feedwater temperature (°C)	120.0
Steam temperature (°C)	142.0 (sat. steam)
Reactor vessel and pressurizer	
Vessel outer diameter (m)	2.272
Pressurizer type	Steam-gas PRZ
Noncondensable gas	Nitrogen

The gaseous mixture volume in the upper part of the reactor pressure vessel above the coolant level is referred to as the built-in steam-gas pressurizer. It is the most distinguishing component of REX-10 and represents the progress in incorporating passive features into the integral reactor system. In normal operation the saturation vapor corresponding to the temperature of the hot fluid in the RCS is maintained in the gas region by establishing the dynamic equilibrium with liquid region and mixed with the noncondensable gas like nitrogen. The once-through steam generator of REX-10 consists of helical tubes wrapped around the entire annulus between the core barrel and the reactor pressure vessel. The primary coolant flows downward across the tube bundles to evaporate the coolant on the tubeside. Flowing in the opposite direction, the secondary feedwater enters into the helical coil in a subcooled state; by the time it leaves the coil, it turns into saturated steam.

The containment of REX-10 is filled with water and buried underground. The large amount of water in the containment can serve not only as a barrier against the release of radioactive materials from the reactor system but also as a heat sink under accident conditions. Engineered safety systems of REX-10 are illustrated in Fig. 18.10. As a representative safety system prepared for REX-10, the PRHRS removes the decay heat in the event of reactor shutdown. Automatically put into action by the trip signal, the PRHRS condenses vapor from the steam generator by means of natural circulation through a heat exchanger, which is located higher than the S/G and submerged in the water of containment building.

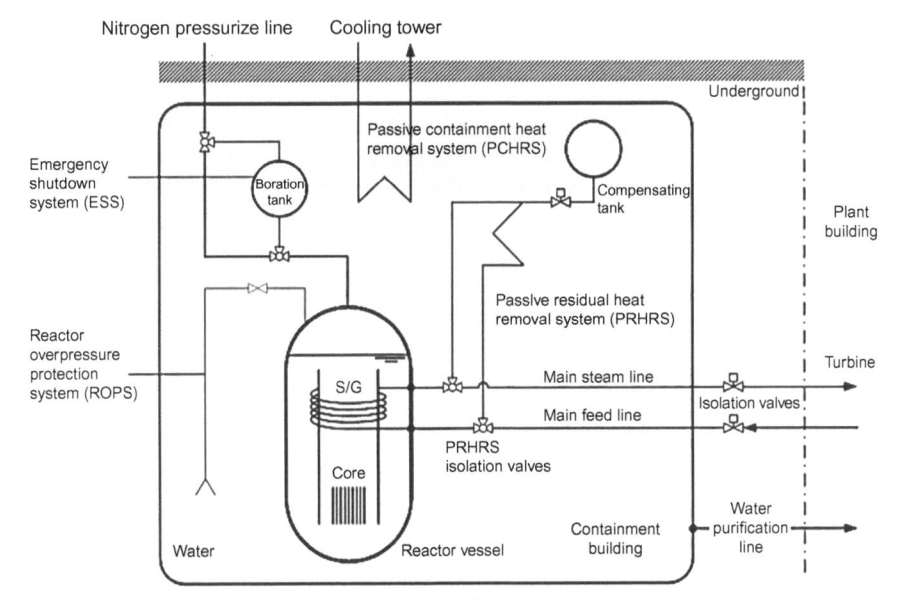

Fig. 18.10 Engineered safety systems of REX-10.

18.3.3 PGSFR (KAERI)

18.3.3.1 Overview

The spent nuclear fuel (SNF) problem has been a common concern among countries, which have utilized nuclear energy for a long time or have a plan to extend the utilization of nuclear energy. A sodium-cooled fast reactor (SFR) has been widely recognized as a technical alternative to effectively manage the SNF, owing to its transmutation capability of long lived radiotoxic nuclides included in the SNF.

The Korea Atomic Energy Research Institute (KAERI) has been developing SFR design technologies since 1997 under a national nuclear R&D program. The goals of the SFR design technology development project were to secure strategic key technologies and to develop the conceptual design of a SFR, which are necessary for a reduction of a high level waste volume and toxicity. To this end, conceptual designs of KALIMER-150, conceptual core design of KALIMER-1200, and the transuranics (TRU) burner of KALIMER-600 have been completed as the main outcomes of those studies.

The long-term development plan for the future nuclear energy systems was authorized by the Korean Atomic Energy Commission in 2008 and updated by Korea Atomic Energy Promotion Council in 2011; it includes a construction of a prototype SFR by 2028 for demonstration of TRU transmutation technologies. The national project to develop the Prototype Gen IV sodium-cooled fast reactor (PGSFR) was initiated to achieve the national mission stated in the preceding text in 2012. For this the Sodium Cooled Fast Reactor Development Agency dedicated to the PGSFR development was established in the middle of 2012. KAERI is in charge of the design and the validation of the nuclear steam supply system (NSSS) and fuel development, and domestic participants were responsible for balance of the plant system design. Argonne National Laboratory (ANL) supported KAERI with their experiences in SFR development through international cooperation programs. The electric power of PGSFR was determined to be 150 MWe suitable for technology demonstration and can be classified as a small modular reactor (SMR) in consideration of the capacity and can be developed as a new nonlight water SMR in the near future.

The first design phase of the PGSFR was done at the end of 2015 by issuing a preliminary safety information document (PSID) with design documents and safety evaluation. The second phase of the development was done at the end of 2017 by issuing the specific design safety analysis report (SDSAR) with design documents and safety analyses results sufficient for assessing safety of PGSFR. Ten topical reports for major design codes and methodologies were also published at the end of 2017 and submitted to regulatory body in 2018. All of the basic design concepts of structures, systems and components (SSC) were determined and incorporated into the specific design safety analysis report, which includes design requirements and system descriptions and the results of safety analysis for postulated accident scenarios.

To support and demonstrate the safety performance of the PGSFR design, verification and validation activities are being performed in parallel with the design progress. A large-scale sodium thermohydraulic test program called STELLA is being progressed. As the first step of the program (STELLA-1), the sodium component tests of the decay heat removal system (DHRS) has been completed, and the obtained data were used for validating computer codes for thermal sizing and system transient analysis of the DHRS components. As the second step an integral effect test loop (STELLA-2) has been started to demonstrate the plant safety and to support the PGSFR design certification. The integral test loop is to demonstrate the safety performance of the DHRS and reactor vault cooling system (RVCS) in conjunction with the primary heat transport system (PHTS). The construction of STELLA-2 facility is scheduled to be completed by the end of 2019 and the integral effect test will be completed by the end of 2020.

Various R&D activities are being performed, including verification and validation of computational codes and development of the metal fuel fabrication technology. The reactor mock-up physics experiment in the BFS facility was completed in 2015 in collaboration with Institute of Physics and Power Engineering (IPPE) in Russia. The irradiation test of advanced cladding material (FC92) and test fuel have been performed at the BOR-60 experimental fast reactor.

18.3.3.2 Future plan

In 2017, it was decided to suspend the design intensification of PGSFR in consideration of the national energy environment and select a new policy direction after 2020. The new SFR development program will be decided by reassessing future schedules and discussing rational directions based on the research outcomes so far obtained. Accordingly, Korean SFR development is focus on further improvements of strategic key technologies, the construction and validation of the integrated sodium thermohydraulic test facility (STELLA-2), and the development of the licensing environment through the review of topical reports.

18.3.3.3 Technical data

The overall design features of PGSFR can be summarized as metal fueled, pool-type sodium-cooled fast reactor with active and passive decay heat removal and shutdown systems. The major design parameters are listed in Table 18.4.

Total 217 fuel rods are arranged into the fuel assembly with hexagonal configuration. Total 112 fuel assemblies are loaded into the core in the hexagonal configuration. The active core height is 90 cm. The cycle length of uranium equilibrium core is about 290 effective full power days (EFPDs). There is neither an axial nor radial blanket to prevent additional TRU production in the blanket region. The active core is directly faced into the steel reflector.

The primary heat transport system (PHTS) is a pool type. All the structures and components of PHTS such as four intermediate heat exchangers (IHXs) and two mechanical pumps are submerged into a large sodium pool confined by double vessels, reactor vessel and containment vessel. The intermediate heat transport system (IHTS) consists of two loops with two once-through-type steam generators. The annular linear induction (ALIP) pump is used in the IHTS, and IHTS is connected to the sodium-water reaction pressure relief system (SWRPRS) to prevent overpressure of IHTS loop when sodium-water reaction occurs in the steam generator.

The decay heat is rejected to the atmosphere by decay heat removal system (DHRS). DHRS consists of four trains: two active decay heat removal systems (ADHRS) and two passive decay heat removal systems (PDHRS). The sodium to air heat exchangers are a finned tube type for the ADHRS and a helical coil type

Table 18.4 Major design parameters of PGSFR.

Parameter	Value
Core power (MWt/MWe)	392.2/150
Coolant temperatures (inlet/outlet) (°C)	390/545
Total core flow rate (kg/s)	1984.2
Fuel type (initial and transition)	U-10%Zr
Cycle length (EFPD)	290
Fuel cladding material	FC92(FMS)
Number of batch (inner/outer)	4/5
Active core height (cm)	90
Pitch to diameter ratio (P/D)	1.14
Total heavy metal inventory (ton)	7.55
Discharge burnup (avg./peak) (MW d/kg)	65.9/105.2
Average/peak linear power (W/cm)	159.1/324.6
Sodium void reactivity at EOEC ($)	−1.53

for the PDHRS. The active functions of ADHRS are provided by the blowers. The active circuit has also passive function by the natural circulation of sodium and air, and its passive heat removal capacity of the active circuit is more than 50% of designed heat removal capacity when the blowers are not operable. The PGSFR has independent and diversified safety shutdown systems, which consists of six primary control rods and three secondary shutdown rods.

The major safety design features of the PGSFR come from the use of the metal fuel, the pool-type PHTS and the sodium coolant. The PGSFR is a pool-type and is operated at atmospheric pressure; therefore there are a large margin to coolant boiling and no possibility of large pressure release accident by a pipe break. The sodium coolant and metal fuel have excellent heat transfer characteristics, which enable to reject the decay heat easily to atmosphere by natural circulation. Sodium coolant has a good compatibility with the metal structures and components. The metal fuel has a high heat conductivity, which enables to operate low fuel temperature at normal condition. The safety design approaches of the PGSFR are concentrated on the prevention of the severe accident by enhancing the inherent safety of the metal fuel and the reliability of the decay heat removal systems.

18.3.4 VHTR (KAERI)

18.3.4.1 Overview

A VHTR is an inherent safe reactor that provides heat up to 950°C. The high-temperature heat can be used for hydrogen production in an economical way when it combined with thermochemical process or high-temperature electrolysis. Nuclear hydrogen production using VHTR is a sustainable, eco-friendly and technology-led energy that can support hydrogen economy in Korea.

To prepare upcoming hydrogen economy, the Korean government launched the nuclear hydrogen key technologies development project in 2006 as a national program and approved by the Korea Atomic Energy Commission in 2008. The nuclear hydrogen program consists of two major projects: the nuclear hydrogen key technologies development project and the nuclear hydrogen development and demonstration (NHDD) project. Fig. 18.11 illustrates the nuclear hydrogen program in Korea.

The key technologies development project focused on the development and validation of key and challenging technologies for the realization of the nuclear hydrogen system. The key technologies has been focused on the development of computational tools, high-temperature experimental technology, high-temperature material database, TRISO fuel fabrication, and hydrogen production process in collaboration with Gen-IV international forum study. To launch the NHDD project, a system concept study has been carried out between 2012 and 2014 and a system concept for a demonstration plant was prepared as shown in Fig. 18.12 with design parameters. However, the NHDD project has not been launched as of 2019 because not only some technologies such as materials, design standards, and regulations are not prepared for the core outlet temperature of 950°C for economical hydrogen production, but also the hydrogen economy is not matured.

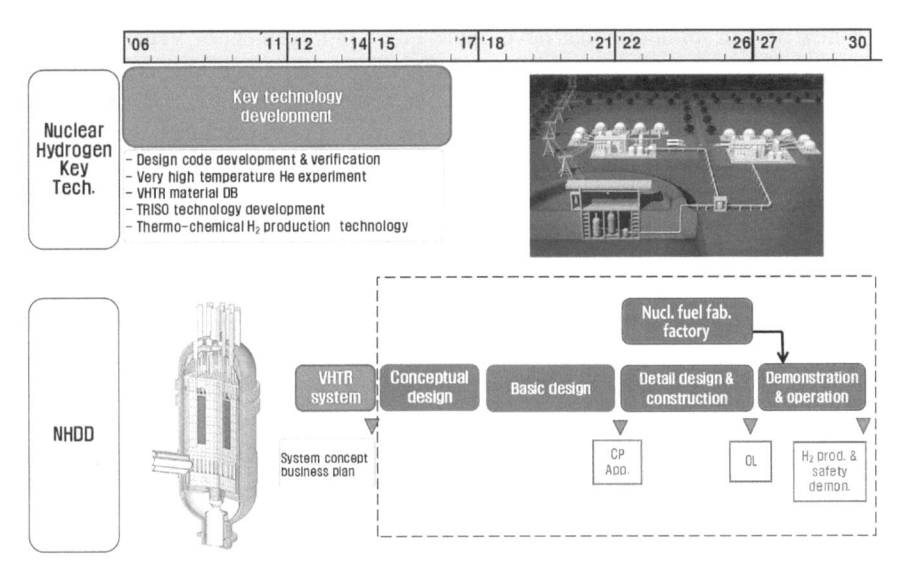

Fig. 18.11 Nuclear hydrogen program in Korea.

Thermal power	350 MW with block-type core
Reactor inlet/outlet temperature	490/950°C at 70 bars
TRISO fuel enrichment	Below 15.5 w/o
Refueling period	18 month
RPV material	SA 508/533 with vessel cooling
Hydrogen production	35,000 ton/year with S-I process

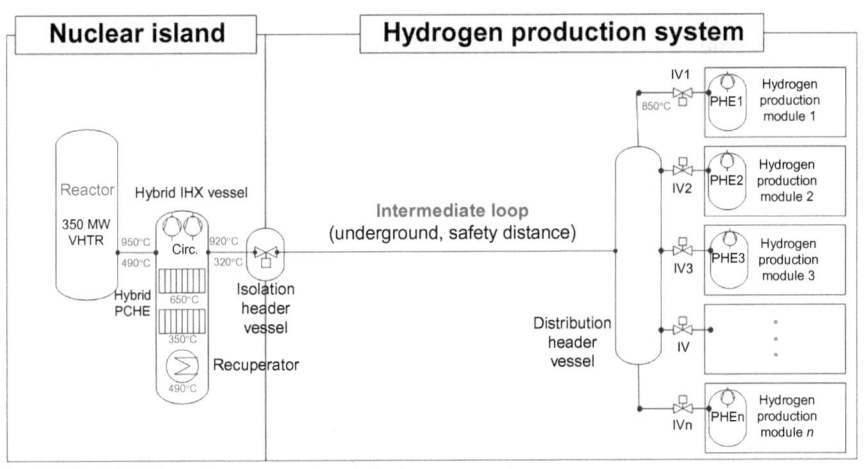

Fig. 18.12 NHDD system concept and design parameters.

A new project for continuous development and improvement of VHTR key technologies was initiated for a 3-year period in 2017. The main concerns of technology improvement are design and analysis codes, experimental tests for codes validation data, high-temperature intermediate heat exchanger, high-temperature materials database, TRISO coated particle fuel, and heat utilization.

18.3.4.2 Future plan

In 2019 the Korean government released a hydrogen economy plan with a roadmap to invigorate the hydrogen economy. The plan expects that the cumulative number of hydrogen vehicles and charging stations in 2040 are expected to 6.2 million and 1.2 thousands, respectively. In addition, the fuel cell batteries capacity will increase to 17.1 GW.

To support and realize the hydrogen economy plan, a roadmap for hydrogen technology development is under discussion, and VHTR is reviewed as one of technologies for hydrogen. Until the construction of a demonstration plant for massive hydrogen is determined, the R&D for the improvement of key technologies will continue to focus on the realization of core outlet temperature of 950°C for the time being.

18.3.5 MMR (KAIST)

18.3.5.1 Overview

A new coolant, supercritical CO_2, was proposed for the nuclear reactor cooling and converting heat to work to achieve a full modularization of a small modular reactor (SMR). By adopting this approach the KAIST research team expects to successfully develop a nuclear power plant (or module) concept that can be fully modularized and suggest a revolutionary path to the future nuclear technology. The proposed concept is named as the KAIST micromodular reactor (KAIST-MMR). Furthermore the supercritical CO_2 power conversion system that is being actively developed around the world as the future power conversion system was combined to the nuclear reactor directly to reduce the total footprint of the nuclear power module and enables to integrate the whole nuclear reactor core and the power conversion system into a single pressure vessel. This solution can greatly enhance the economics and transportability of the nuclear power plant. In addition a uranium carbide fuel is proposed for this nuclear reactor instead of uranium dioxide fuel used in the conventional water-cooled nuclear power plant so that KAIST-MMR can be operated for 20 years full power without refueling. A passive decay heat removal system was also added to the concept to cool the decay heat naturally even there is no electrical input from outside. The proposed nuclear power module, namely KAIST-MMR, is developed for an electricity source for an isolated region, distributed power generation and civilian maritime transport power source. The concept development started from 2012 until 2014 with KAIST internal funding, further progress was made with the support from Korean government support through national research foundation in the period of 2014–16.

The concept was well received appeared in multiple journal articles (Kim et al., 2015; Baik et al., 2014; Oh et al., 2018) and international magazines, and it was awarded as one of the best 10 R&D achievements of KAIST in 2014.

18.3.5.2 Future plan

Initially KAIST-MMR development focused more on the land base application. However, since the interest in civilian maritime nuclear propulsion is growing around the world to reduce the greenhouse gas emission, the application of KAIST-MMR to the ship propulsion is also under investigation. To now, KAIST-MMR showed an excellent load following performance for these aforementioned applications. In the near future the concept design phase will be finalized, which will be followed by basic design and detail design phases with the possible formation of a consortium with the organizations in interest.

18.3.5.3 Technical data

KAIST MMR is a 12 MWe gas-cooled fast reactor, using supercritical CO_2 (S-CO_2) as a coolant and working fluid of power cycle. Since the MMR is produced in a modular way and transported to the installation site easily, construction period is short and can achieve good economy. A conceptual design of S-CO_2 cooled reactor core without 20 years refueling is developed based on the uranium carbide fuel technology. A passive safety system for decay heat removal is designed and incorporated to prevent severe accidents. The target of the system is to be able to transport a single core and power conversion system module via ground transportation. To meet this target single module's total weight is minimized in the order of 100 tons (Fig. 18.13 and Table 18.5).

The core has a long life without refueling while having a compact core design. The core consists of 18 fuel assemblies, 12 drum type control elements, secondary shutdown control rod, and reflector (Fig. 18.14 and Table 18.6). The shape of fuel assembly is hexagonal type. Each channel is wire wrapped to enhance convective heat transfer. Uranium carbide is selected as fuel pellets, and oxide-dispersed steel (ODS) is chosen as a cladding material. The design life time is 20 years full power operation without refueling. Excessive reactivity is regulated with an innovative replaceable fixed absorber (RFA).

The power cycle layout (Fig. 18.15) is a simple recuperated cycle for simplicity and efficiency. The recuperator was chosen to be a Printed Circuit Heat Exchanger (PCHE) type. Friction factor and heat transfer coefficient for an S-CO_2 PCHE is developed based on experimental and CFD analyses. Ultimate heat sink was chosen to be ambient air. The heat is transported via a CO_2 cooling loop connecting precooler in the power cycle and a CO_2-Air heat exchanger cooled by air fans. Electric-driven air fans and circulation compressor are used to make forced flow in the connection loop. KAIST-MMR turbomachinery is consisted of one centrifugal compressor and one radial inflow turbine supported by magnetic bearing to have hermetic

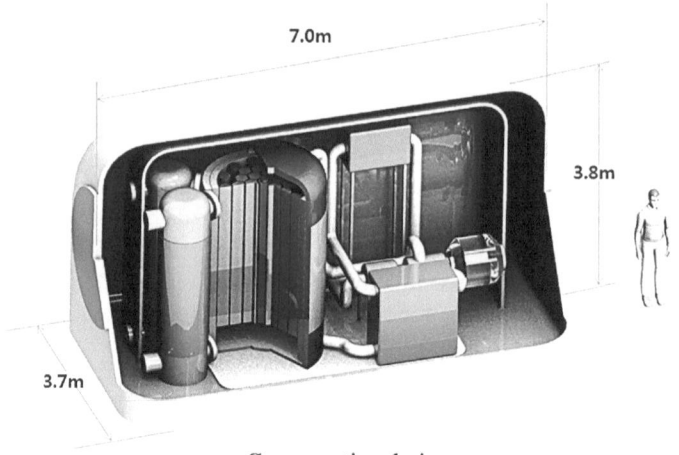

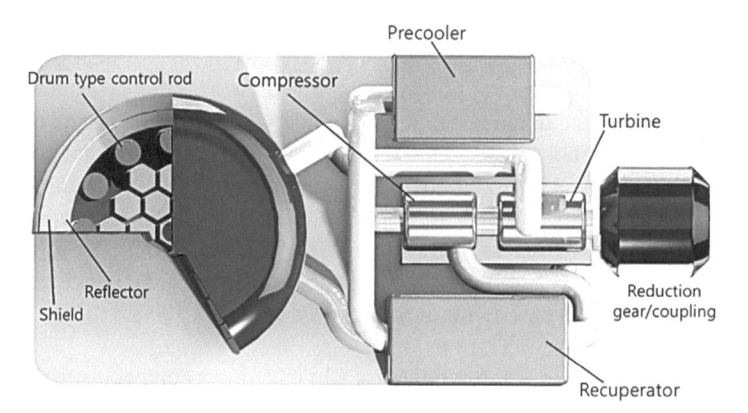

Fig. 18.13 KAIST-MMR internal layout.

Table 18.5 Overview of KAIST-MMR.

Thermal efficiency	Net electricity	Mass flow rate	Mass of total module
34.1%	11.8 MW	180.0 kg/s	155 tons

type turbomachinery. The inventory control system is integrated in the system by using the inventory tank, which works as the first containment pressure vessel as well. The whole system is in the outer containment pressure vessel again to eliminate release of radioactive material via containment bypass. In other words a double containment structure is adopted for safety and modularity of the whole system.

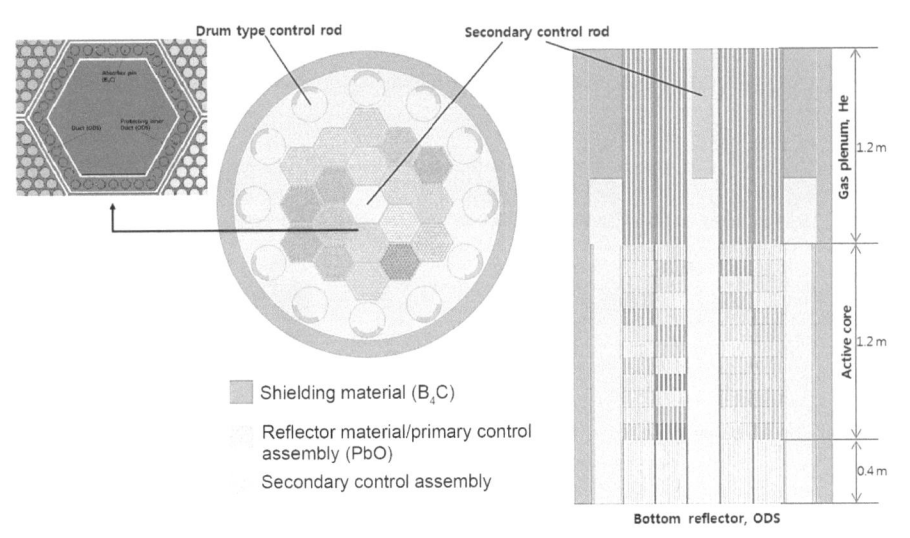

Fig. 18.14 KAIST-MMR core layout.

Table 18.6 Overview of KAIST-MMR core design parameters.

Design parameter	Value
Rector power/lifetime	36.2 MWth/20 years
Number of FAs	18
Whole core equivalent radius/height	82 cm/280 cm
Coolant pressure/speed	20 MPa/6.92 m/s
Coolant inlet/outlet temp.	655.35 K/823.15 K
Control drum material	98% TD 98 w/o enriched B4C
Control rod material	98% TD 98 w/o enriched B4C

A passive decay heat removal (PDHR) system (Fig. 18.16) is designed and incorporated in the concept. PDHR passively rejects decay heat of MMR core to ambient air through additional S-CO_2 flow path. The flow is driven by natural circulation. In normal operation the additional flow path is inactive, but it becomes activate by automatic operation of PDHR valve under accident conditions. Since PDHR is driven by natural circulation, assessing pressure drop of whole loop is the most important factor for natural circulation analysis. Thus a wire-wrapped channel pressure drop experiment was also conducted to reduce uncertainty in the pressure drop of a fuel assembly.

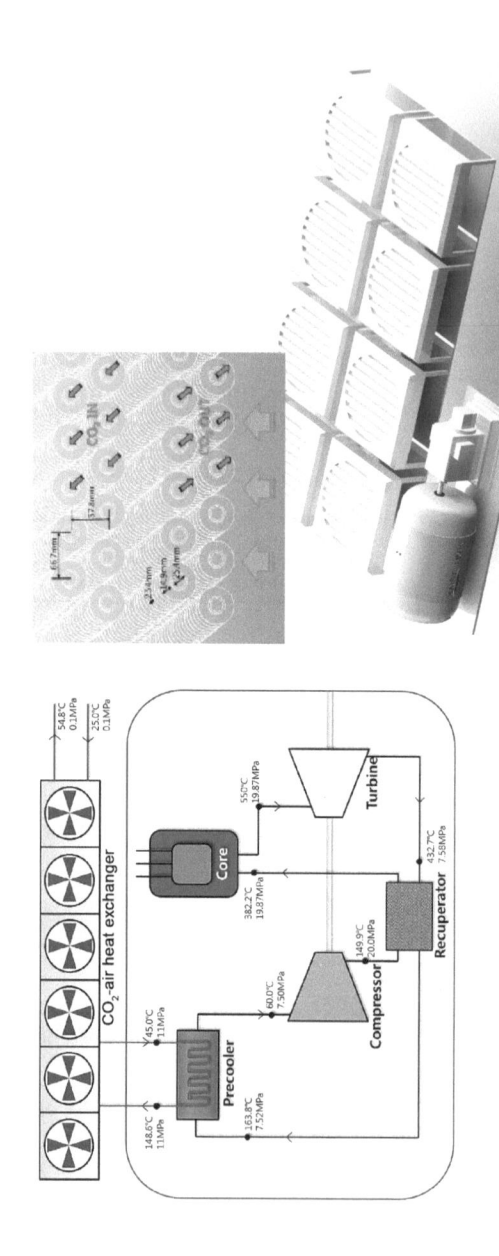

Fig. 18.15 Air-cooling layout of KAIST-MMR.

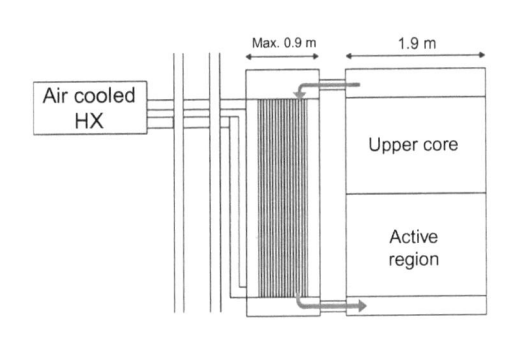

Fig. 18.16 PDHR of KAIST-MMR.

18.3.6 MINERVA (UNIST)

18.3.6.1 Overview

Ulsan Institute of Science and Technology (UNIST) has started to develop the concept of marine reactor that can be used for 40 years without supplementing fuel in the Polar in 2019. It has begun the development of a reactor that can be used as a ship's power in the marine environment. This development program will be implemented with the support of the Ministry of science and ICT over the next 4 years.

The reactor is named as Micro Nuclear Energy Research and Verification Arena (MINERVA) as a liquid metal reactor cooled by lead-bismuth. The core aims to generate output of about 20 MWe without replacing the fuel for up to 40 years. The reactor can be used as a propulsion power source for vessels such as research vessels with a lifespan of 40 years and will be able to supply necessary power to islands with a population of 200,000 by floating power generation.

The developed reactor is a small and safe SMR, so it should also be safe for corrosion and shock loads, unlike land construction.

Acknowledgment

I would like to show gratitude to Dr. B.J. Lee (BANDI: KEPCO E&C), Prof. H.K. Cho (REX-10: SNU), Prof. J.I. Lee (MMR: KAIST), Dr. M.H. Kim (VHTR: KAERI), Dr. T.W. Kim (SFR: KAERI), and J.W. Kim (SMART: KAERI) for sharing their R&D information and preparing the corresponding manuscript.

References

Bae, I.H., Na, M.G., Lee, Y.J., Park, G.-C., 2008. Calculation of the power peaking factor in a nuclear reactor using support vector regression models. Ann. Nucl. Energy 35 (12), 2200–2205.

Baik, S.J., Bae, S.J., Kim, S.G., Lee, J.I., 2014. Preliminary design of KAIST micro modular reactor with dry air cooling. In: Proceedings of the KNS 2014 Spring Meeting.

Beom, H.-K., Kim, G.-W., Park, G.-C., Cho, H.K., 2019. Verification and improvement of dynamic motion model in MARS for marine reactor thermal-hydraulic analysis under ocean condition. Nucl. Eng. Technol. 51 (5), 1231–1240.
Hwang, J.-S., Lee, Y.-G., Park, G.-C., 2012. Characteristics of critical heat flux under rolling condition for flow boiling in vertical tube. Nucl. Eng. Des. 252, 153–162.
Jang, B.-I., Kim, M.H., Jeun, G., 2011. Transient analysis of natural circulation nuclear reactor REX-10. J. Nucl. Sci. Technol. 48 (7), 1046–1056.
Kim, J.-W., Lee, Y.-G., Ahn, H.-K., Park, G.-C., 2009. Condensation heat transfer characteristics in the presence of noncondensable gas on natural convection at high pressure. Nucl. Eng. Des. 239, 688–698.
Kim, S.G., Oh, B., Baik, S.J., Yu, H., Kim, Y., Lee, J.I., 2015. Conceptual system design of a supercritical CO_2 cooled micro modular reactor. In: Proceedings of the KNS 2015 Spring Meeting.
Lee, Y.-G., Park, G.-C., 2013a. TAPINS: TAPINS: a thermal-hydraulics system code for transient analysis of a fully-passive integral PWR. Nucl. Eng. Technol. 45 (4), 439–458.
Lee, Y.-G., Park, G.-C., 2013b. Assessment of TAPINS and TASS/SMR codes and application to overpower transients in REX-10. Nucl. Eng. Des. 263, 296–307.
Lee, Y.-G., Kim, J.-W., Park, G.-C., 2012. Development of a thermal–hydraulic system code, TAPINS, for 10 MW regional energy reactor. Nucl. Eng. Des. 249, 364–378.
Oh, B.S., Kim, S.G., Cho, S.K., Lee, J.I., 2018. Start up modeling of S-CO_2 cooled KAIST micro modular reactor using extended turbomachinery performance map. In: 2nd European sCO_2 Conference.

Further reading

Caree, F., Yvon, P., Lee, W.J., Dong, Y., Tachibana, Y., Petti, D., 2009. VHTR—ongoing international projects. In: GIF Symposium—Paris (France), 9–10 September.
Carelli, M.D., 2002. Iris: an integrated international approach to design and deploy a new generation reactor, status and prospects for small and medium size reactors. In: Proc. Int. Seminar, Cairo, May 2001, IAEA-CSP-14/P, Vienna.
Chang, M.H., Hwang, Y.D., 2003. Coupling of MED-TVC with SMART for nuclear desalination. Nucl. Desalination 1 (1), 69–80.
Chang, M.H., et al., 1999. SMART—an advanced small integral PWR for nuclear desalination and power generation, Global 1999. In: Proc. Int. Conf., Jackson Hole, USA, Aug. 29–Sept. 3. ANS/ENS.
Chung, M.K., et al., 2007. Verification tests performed for development of an integral type reactor. In: Proc. Int. Conf. on Non-Electric Applications of Nuclear Power: Seawater Desalination, Hydrogen Production and other Industrial Applications, Oarai, Japan.
International Atomic Energy Agency, 1996. Design and Development Status of Small and Medium Reactor Systems 1995. IAEA-TECDOC-881IAEA, Vienna.
International Atomic Energy Agency, 2000. Desalination Economic Evaluation Program (DEEP). Computer User Manual Series No. 14IAEA, Vienna.
International Atomic Energy Agency, 2004. Status of Advanced Light Water Reactor Designs. IAEA-TECDOC-1391IAEA, Vienna.
International Atomic Energy Agency, 2006. Status of Innovative Small and Medium Sized Reactor Design 2005. IAEA-TECDOC-1485IAEA, Vienna.

International Atomic Energy Agency, 2010. Small Reactors without On-site Refueling. Neutronic Characteristics, Emergency Planning and Development Scenarios, Final Report of an IAEA Coordinated Research Project. IAEA TECDOC 1652.

Ishida, T., et al., 2000. Advanced marine reactor MRX and its application for electricity and heat co-generation. In: Small Power and Heat Generation Systems on the Basis of Propulsion and Innovative Reactor Technologies. IAEA, Vienna IAEA-TECDOC-1172.

Kim, T.W., et al., 1997. Development of PSA Workstation KIRAP. Korea Atomic Energy Research Institute, Taejon KAERI/TR-847/1997.

Kim, S.H., et al., 2002. Design development and verification of a system integrated modular PWR. In: Proc. 4th Int. Conf. on Nuclear Option in Countries With Small and Medium Electricity Grids, Croatia, June.

Kim, H.C., et al., 2003a. Safety analysis of SMART, Global environment and advanced NPPs. In: Proc. Int. Conf. Kyoto, Sept. 15–19, 2003, GENES4/ANP2003, Kyoto, Japan.

Kim, S.H., et al., 2003b. Design verification program of SMART, global environment and advanced NPPs. In: Proc. Int. Conf. Kyoto, Sept. 15–19, 2003, GENES4/ANP2003, Kyoto, Japan.

Kim, Y.-I., Lee, Y.B., Lee, C.B., Chang, J.-W., Choi, C.-W., 2013. Design concept of advanced sodium-cooled fast reactor and Related R&D in Korea. Sci. Technol. Nucl. Install. 2013, 290362.

Zee, S.K., et al., 2007. Design Report for SMART Reactor System Development, KAERI/TR-2846/2007. KAERI, Taejon.

Small modular reactors (SMRs): The case of Russia

Vladimir Kuznetsov
Consultant, Austria

19.1 Introduction

The category of small modular reactors (SMRs) includes those with an equivalent electric output less than ~300 MW(e), having a high degree of factory fabrication allowing for transportation of factory-assembled reactor modules or even the whole plant by barge, rail, or truck, and with an option to build power stations of flexible capacity through a multimodule approach. Such designs are being developed and deployed in the Russian Federation with the view of providing a secure energy supply to small regional energy systems located in remote and hard to access areas of the country where the climate is characterized by extremes and the transportation routes providing connections to the rest of the country are unreliable and available only over a short season during the year. Moreover the overall Russian strategy is to have reactors of different capacities from large and very large (1200–1500 MW(e)) to small and very small (10–150 MW(e)), including possibly medium sized (300–600 MW(e)), to cater for a variety of centralized and regional energy needs stemming from the geographical, population, industrial, and climatic conditions of the country (Velikhov, 2008).

Some 60%–70% of Russian territory is affected by permafrost that complicates large-scale construction and makes it very expensive to develop and maintain reliable transportation routes. These are the large territories in the North and East of the country, which are characterized by sparse population concentrated mostly around mining and raw material reprocessing enterprises and military bases. The temperatures in winter may be extremely low, and heat for residential needs is, therefore, in demand as well as electricity. Connections to the more populated areas of the country are seasonal (not available in winter) and unreliable (may be not available in some years due to the damage caused by the permafrost melting). In such an environment, basic requirements to an energy source are the ability to operate over long periods without the need of fuel delivery and the ability to operate in a cogeneration mode, producing heat and electricity.

The territories in the Russian North are rich in oil, gas, alumina, nickel, diamonds, and other valuable natural resources, the development of which is crucial for the still largely resource export-oriented Russian economy. One of the factors associated with mining is the life span of a mine, which is often limited to just a few decades (IAEA, 2007). In view of this, relocatable energy sources may have an advantage. In its North and East, Russia has a long coastal line, with the seaside being covered with ice over a long winter. Small and sparse settlements located along or nearby this coastal line (typically, they are settlements around local small enterprises or military bases) also need energy and heat and may benefit from autonomous reliable energy sources matching their small or not present electricity grids. Here, transportable barge-mounted nuclear cogeneration plants with the onboard fresh fuel reserves and spent fuel storage or with support from the dedicated ships designed to transport fuel and enable reloading operations are being considered as candidates (IAEA, 2009).

Finally the Russian Federation is considering a project to develop gas and oil production from the shelf of the Barents Sea, located in the country's North. Nuclear reactors may be employed for such a production, including those located ashore or on a barge or on the sea bottom to power the underwater mining plant, and for propulsion purposes on submarines delivering gas to the onshore terminal (Velikhov, 2008).

Based on data from NEA-OECD (2016), the cost of electricity generation in some areas in Russia's North and East could be an order of magnitude higher compared with the minimum one in central and Eastern Regions of the country covered with large capacity interconnected grids. High costs mean lack of the centralized grids, difficult conditions for fuel delivery, limited demand, or other specific siting conditions making it impractical or impossible to build an economy of scale large power plant. Such areas are, therefore, potential markets for SMRs.

Design development activities for civil SMRs started in Russia in the 1980s and borrowed extensively from the experience in design and operation of the marine propulsion reactors for the navy and the nuclear icebreaker fleet. This positive experience includes several different designs of pressurized water reactors (PWRs) and one reactor cooled by lead-bismuth eutectics; it amounts to not less than 6500 reactor years overall with over 260 reactor years for the reactors of nuclear icebreakers alone (Sozonyuk, 2011). For comparison, experience in the operation of all conventional land-based nuclear power plants worldwide constitutes 18,096 reactor years (PRIS, 2019).

The Russian SMR designs based on PWR technology incorporate both proven features of the previous-generation marine propulsion reactors and the state-of-the-art features of contemporary VVER type reactors. The SMR design employing lead-bismuth eutectics technology borrows both from submarine reactors of the 1980s and from the experience of the Russian sodium cooled fast reactors.

Comparing the development of SMRs in Russia with that of other countries, the following can be noted:

- The term "small modular reactors," introduced in the US programs on small reactor development in the mid-2000s, was until recently not in common use in Russia, although many of the Russian designs share common design approaches with the SMRs being currently developed elsewhere. In this, Russian activities for such SMRs were started earlier, in mid-1980s.
- Unlike some SMR designs, underground location of the reactor modules is not commonly considered in Russia. However, at least one Russian SMR design concept is being considered for seabed location.
- Russian designers commonly accept SMRs to be more expensive sources of electricity compared with the state-of-the-art large nuclear power plants (NPPs) and do not believe they may directly compete with larger plants, say, through shorter construction periods or multi-module plant configurations. Instead, they target particular niche markets where electricity

generation costs are high, where cogeneration, long refueling interval, or plant relocatability are assets, where transportation routes are seasonal, where the demand is limited and siting conditions are specific (i.e., no water in winter due to deep freezing of rivers or other water reservoirs). According to NEA-OECD (2016) and NEA-OECD (2011), similar conditions are also being observed in several other countries. The aforementioned unique standpoint is because Russian policy is to have large-, small-, and, possibly, medium-sized reactors on a complementary, not a competitive, basis. Complementarity is pursued in view of different niches for reactors of different capacity available domestically and, potentially, worldwide.
- As in the US case, Russian SMRs are being designed and licensed to operate first in their country of origin. Should the experience of their operation be positive, they could later be offered on world markets with some features tailored for the needs of such markets, for example, cogeneration option with heat production changed to cogeneration option with seawater desalination.

The objective of this chapter is to present the design and safety features, including defense-in-depth, probabilistic safety goals and, specifically, design features for protection against external event impacts, the design and operating characteristics, and the anticipated applications for a variety of SMRs being deployed or developed in the Russian Federation.

In line with this objective, Sections 19.2–19.4 present SMR designs being developed and already having been or being deployed, stemming from the Russian design organizations/companies OKBM Afrikantov (OKBM Afrikantov, 2019), NIKIET (NIKIET, 2019a), and AKME Engineering (AKME Engineering, 2019), respectively. Also highlighted are the ongoing R&D for some SMR projects.

Another objective of this chapter is to present the deployment status and prospects for the Russian SMRs, which is accomplished in Section 19.5. In this section the design, licensing, and deployment status of SMRs is highlighted, as well as the prospects for SMR deployment in the indigenous and foreign markets.

Section 19.6 presents the future trends for the barge-mounted, land-based, seabed-based, and nuclear icebreaker SMRs developed in the Russian Federation. A conclusion is drawn in Section 19.7, and, finally, Section 19.8 provides a short description of sources of further information and advice, including a brief commentary on key publications and databases of International Organizations. The chapter is concluded by a comprehensive list of references.

Finally, as it was already mentioned, Russian Federation has a potential market space not only for large and small but also for medium-sized reactors. Design concepts of such reactors are also being developed, and, although they are beyond the scope of this book, Section 19.5 lists such design concepts and points to the best available sources of further information.

19.2 OKBM Afrikantov small modular reactor (SMR) projects being deployed and developed in Russia

Small modular reactors that are being developed or had been developed by the Russian "Afrikantov Experimental Design Bureau for Mechanical Engineering" (OKBM Afrikantov) include the following:

- KLT-40S reactor for the twin-unit barge-mounted (floating) NPP with a cogeneration option,
- ABV-6E reactor for the single-unit barge-mounted or land-based NPP with a cogeneration option,
- RITM-200 reactor for new-generation multipurpose nuclear icebreakers and twin- or four-unit land-based plants and its design evolution RITM-200M for new-generation twin-unit barge-mounted plants.

All of the aforementioned designs are indirect cycle PWRs. ABV-6E and RITM-200 (as well as RITM-200M) have integral design of the primary circuit with in-vessel steam generators.

Of the designs listed, KLT-40S has been deployed within the twin-unit barge-mounted plant "Akademik Lomonosov," which started commercial operation in December 2019. The RITM-200 reactors for new-generation multipurpose nuclear icebreakers are at the final deployment stage with six reactors having been manufactured and four of them been installed on the two icebreakers with one of them having undergone sea tests with operating reactors as of the end of December 2019. Further details on design development, licensing, and deployment are provided in Section 19.5.

The design and operating characteristics of the SMRs from OKBM Afrikantov are summarized in Table 19.1 and the core and fuel design characteristics in Table 19.2. For ABV-6E the data for barge-mounted and land-based plant options are presented separately, when different. For RITM-200 the full set of data is presented for a land-based plant version, while the data for icebreaker option is presented separately, when different. Tables 19.1 and 19.2 also present some data for the newer RITM-200M design modification intended for a new-generation barge-mounted plant; these data are given in *italics*.

The most mature of the SMR designs by OKBM Afrikantov is the KLT-40S reactor of 38.5 MW(e) per module. It has a submarine prototype experience of several thousands of reactor years and an icebreaker prototype experience of nearly 300 reactor years. The KLT-40S design appears to be similar to a conventional PWR reactor with separate vessels accommodating the reactor core and internals, the main circulation pumps, the pressurizer, and the steam generators (see Fig. 19.1). Control rod drives are external and located above the reactor vessel lid. However, in the KLT-40S case, all separate vessels are compact and connected by short pipelines with vessel penetrations in the hot legs. The piping is double wall and incorporates so-called narrowing devices near the penetrations. All modules are connected with brackets fixed by bolts and nuts to restrict possible leaks in the case of a pipe rupture (not shown in Fig. 19.1). All primary coolant systems are located within the primary pressure boundary. The aforementioned features effectively reduce the scope and magnitude of possible loss-of-coolant accidents (LOCAs). The refueling is performed in batches every 30–36 months.

The ABV-6E design shown in Fig. 19.2 is also backed by some prototype operation experience, the details of which have not been disclosed. It incorporates an integral primary circuit design with in-vessel steam generators. However, the pressurizer and the control rod drives are external. The core design and dimensions are similar

Table 19.1 Design and operating characteristics of the SMRs from OKBM Afrikantov.

Characteristic	KLT-40S	ABV-6E	RITM-200 (*RITM-200M*)
Electric/thermal power, MW	2 × 38.5 (nonelectrical applications disabled)/2 × 150	(6–9)/38	2 × 50/2 × 175
Nonelectrical products	– Heat for district heating: 2 × 25 GCal/h, or – Potable water: 20,000–100,000 m³/day	– Heat for district heating: 12 GCal/h, or – Potable water: 20,000 m³/day	– 30 MW(e) of shaft power per module; – 248 t/h of steam at 295°C, 3.82 MPa
Plant configuration	Twin-unit for a barge-mounted NPP	– Single unit for a barge-mounted NPP; – Single-unit land-based plant as option	– Twin-unit nuclear icebreaker reactor; – Twin- or four-unit land-based plant – *Twin-unit barge-mounted plant*
Construction period, months/mode of operation	48/Load following (10%–100%)	48/Load following (20%–100%)	<48/Flexible load following (30-100-30)%
Thermodynamic cycle type/efficiency	Indirect Rankine steam condensing cycle/23.3%	Indirect Rankine steam condensing cycle/21%	Indirect Rankine steam condensing cycle/28.6%
Primary circulation	Forced	Natural	Forced
Primary pressure, MPa	12.7	16.2	15.7
Core inlet/outlet temperatures, °C	280/316	248/327	277/313
Mode of reactivity control in operation	– Mechanical control rods with external drives; – No liquid boron	– Mechanical control rods with external drives – Liquid boron	– Mechanical control rods with external drives – No liquid boron
Reactor vessel diameter × height, mm	2176 × 4148	2135 × 4479	3300 × 8500
Secondary pressure, MPa	3.82	3.83	3.82
SG secondary side inlet/outlet temperatures, °C	170/290	106/295	Not specified/295
Turbine type	Two condensing—extraction steam turbines, one per each reactor	Condensing—extraction steam turbine	Icebreaker: not specified, electric propulsion Land-based or *barge-mounted plant*: similar to KLT-40S

Continued

Table 19.1 Continued

Characteristic	KLT-40S	ABV-6E	RITM-200 (*RITM-200M*)
I&C system	Based on the state of the art for PWR and marine reactors	Based on the state of the art for PWR and marine reactors	Based on the state of the art for PWR and marine reactors
Containment type and dimensions, m	– Primary rectangular steel containment $12 \times 7.92 \times 12$; – Secondary containment: rectangular steel system of compartments, 15,000 m^3	– Primary cylindrical metal and water shielding tank – Secondary containment: rectangular steel system of compartments	– Hermetically sealed primary rectangular steel containment $6 \times 6 \times 15.5$ ($6.8 \times 6.8 \times 16.0$) For icebreaker and *barge-mounted plant*: – Secondary containment: rectangular steel system of compartments For land-based plant: – Solid building of reinforced concrete – Collapsible building structure of thin reinforced concrete
Plant surface area, m^2	Coast: 8000 Water area: 15,000	For barge-mounted plant: – Coast: 6000 – Water area: 10,000 For land-based plant: 20,000	Limited to the icebreaker or *barge* For land-based plant: – 10,300 for twin-unit plant – 16,500 for four-module plant

Based on data from ARIS, 2019. Technical Data, KLT-40S. Available from: https://aris.iaea.org/PDF/KLT-40S.pdf (Accessed 30 December 2019); ARIS, 2018. Advances in Small Modular Reactor Technology Developments, a Supplement to the IAEA Advanced Reactors Information System (ARIS), Vienna. Available from: https://aris.iaea.org/Publications/SMR-Book_2018.pdf (Accessed 30 December 2019), and reproduced with permission by IAEA, 2007. Status of Small Reactor Designs Without On-Site Refuelling, IAEA-TECDOC-1536, Vienna. Available from: https://www.iaea.org/publications/7772/status-of-small-reactor-designs-without-on-site-refuelling (Accessed 30 December 2019).

Table 19.2 Core and fuel design characteristics of the SMRs from OKBM Afrikantov.

Characteristic	KLT-40S	ABV-6E	RITM-200 (RITM-200M)
Electric/thermal power, MW	2×38.5 (nonelectrical applications disabled)/2×150	(6–9)/38	2×50/2×175
Core diameter × height, mm	1155×1200	1155×1200	1550×1650
Average core power density, MW/m^3	119.3	30	~45
Average fuel element linear heat rate, W/cm	140	65	Several times less than in the KLT-40S
Fuel material	UO_2 in inert matrix	UO_2 in inert matrix	UO_2 in inert matrix
Fuel element type	Smooth-rod, cylindrical	Smooth-rod, cylindrical	Smooth-rod, cylindrical
Cladding material	Zirconium alloy	Chromium-nickel alloy	Chromium-nickel alloy
Fuel element outer diameter, mm	6.8	6.9	6.8
Lattice geometry	Triangular	Triangular	Triangular
Number of fuel elements in fuel assembly	69, 72, 75	69, 72, 75	No information
Number of fuel assemblies in the core	121	121	199 (241)
Burnable absorber	Gadolinium	Gadolinium	Gadolinium
Enrichment of the reload fuel, ^{235}U weight %	18.6	18.7	<20
Interval between refuelings, months	30–36	120–144	60 (120)
Average fuel burnup, MWday/kg	45.4	Not specified	45–47
Mode of refueling	Refueling in batches on the barge	Whole core refueling at a specialized plant or by a dedicated ship	Whole core refueling at a specialized plant or by a dedicated ship

Based on data from ARIS, 2019. Technical Data, KLT-40S. Available from: https://aris.iaea.org/PDF/KLT-40S.pdf (Accessed 30 December 2019); ARIS, 2018. Advances in Small Modular Reactor Technology Developments, a Supplement to the IAEA Advanced Reactors Information System (ARIS), Vienna. Available from: https://aris.iaea.org/Publications/SMR-Book_2018.pdf (Accessed 30 December 2019) and reproduced with permission by IAEA, 2007. Status of Small Reactor Designs Without On-Site Refuelling, IAEA-TECDOC-1536, Vienna. Available from: https://www.iaea.org/publications/7772/status-of-small-reactor-designs-without-on-site-refuelling (Accessed 30 December 2019).

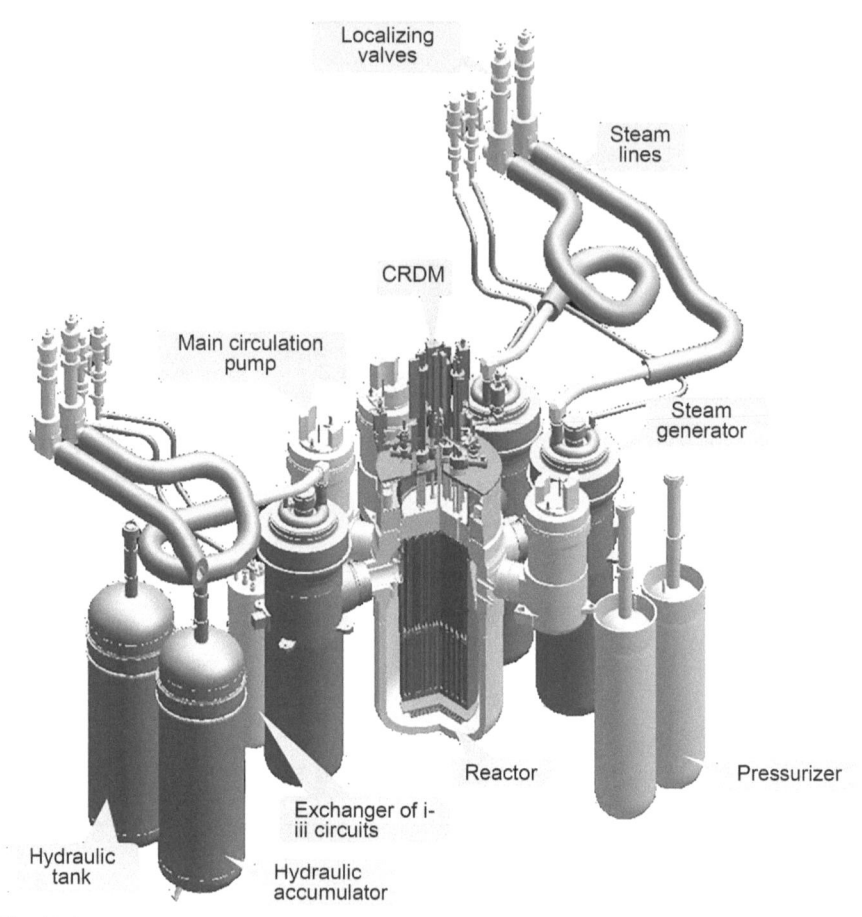

Fig. 19.1 Layout of the KLT-40S reactor.
Reproduced with permission by the IAEA from IAEA, 2009. Design Features to Achieve Defence in Depth in Small and Medium Sized Reactors, IAEA Nuclear Energy Series No. NP-T-2.2, Vienna. Available from: https://www.iaea.org/publications/8094/design-features-to-achieve-defence-in-depth-in-small-and-medium-sized-reactors-smrs (Accessed 30 December 2019).

to those of the KLT-40S. The reactor is refueled at the factory or by a dedicated refueling ship each 10–12 years (whole core refueling). To support achieving a long refueling interval, chromium-nickel alloy is used as cladding material instead of the traditional zirconium alloy. The ABV reactor employs natural circulation of the primary coolant in normal operation mode, although in some versions of this design pumps were adopted for the primary circulation (ARIS, 2018).

The RITM-200—the newest of all SMR designs presented in this section—has a per-module electrical output of 50 MW(e). It incorporates the experience in design and operation of many of the Russian marine propulsion reactors. Like ABV-6E,

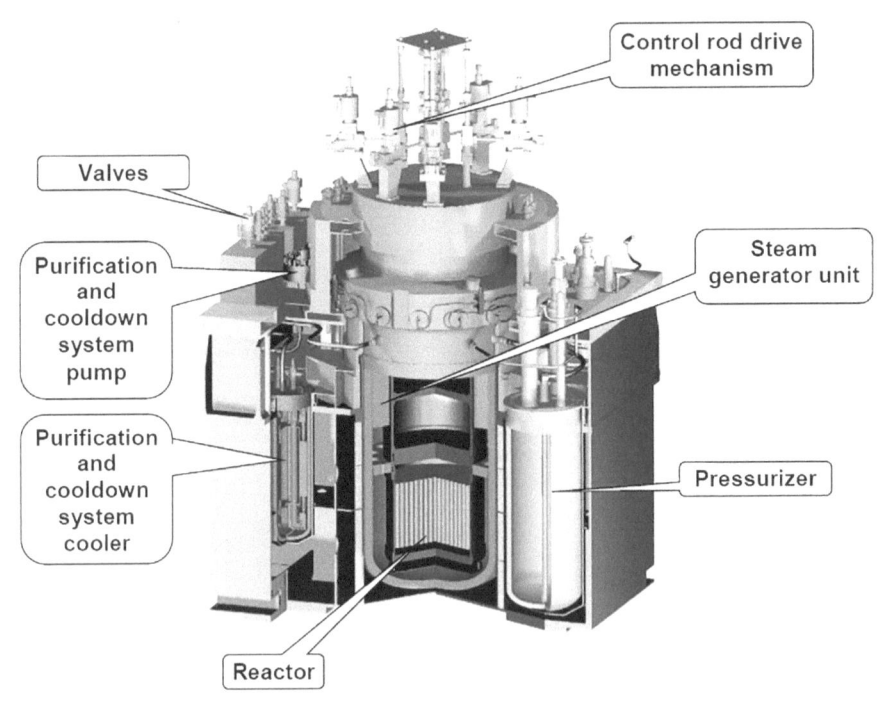

Fig. 19.2 Layout of the ABV reactor.
Reproduced with permission by the IAEA from IAEA, 2007. Status of Small Reactor Designs Without On-Site Refuelling, IAEA-TECDOC-1536, Vienna. Available from: https://www.iaea.org/publications/7772/status-of-small-reactor-designs-without-on-site-refuelling (Accessed 30 December 2019).

RITM-200 has an integral primary circuit design with in-vessel steam generators (see Fig. 19.3). The pressurizer and the control rod drives are again external. The core design and dimensions are modified against those of the KLT-40S (see Table 19.2). The reactor employs forced circulation of the primary coolant provided by the main circulation pumps, as shown in Fig. 19.3. Chromium-nickel alloy is used as cladding material instead of the traditional zirconium alloy, which inter alia increases reactor maneuverability and load-following capability (Veshnyakov, 2011).

The RITM-200M, a further design evolution of the RITM-200 reactor series, employs a larger dimeter core and a longer refueling interval of 10 years (see italic font in Tables 19.1 and 19.2).

The RITM-200 is considered for both land-based and barge-mounted plants, and detailed designs of such plants are currently under development (ARIS, 2018). For a land-based plant, two- and four-unit versions are considered. The barge-mounted plant with two RITM-200M reactors will have neither fresh and spent fuel storages nor refueling equipment. It will be refueled at a special plant or by a dedicated self-propelled ship having fresh and spent fuel storages and the refueling equipment on board.

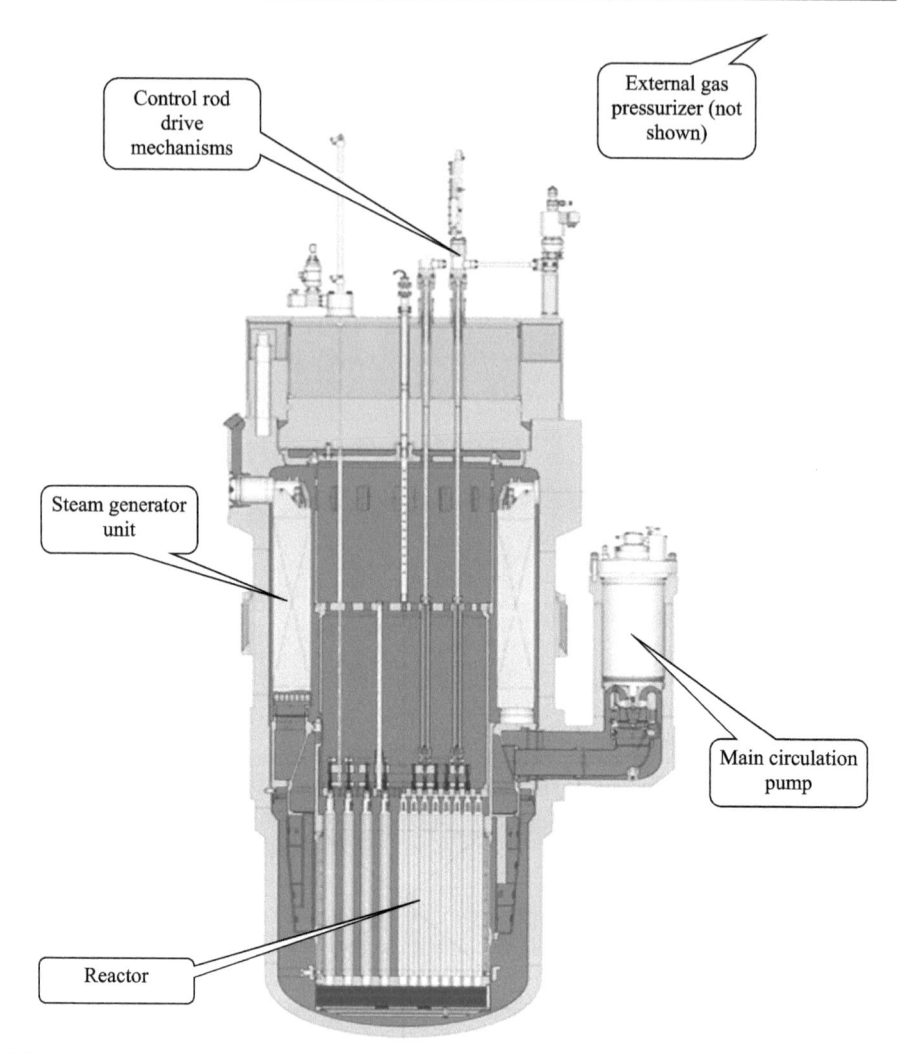

Fig. 19.3 Layout of the RITM-200 reactor.
Credit: Elsevier, 2015. In: Carelli, M.D., Ingersoll, D.T. (Eds.), Handbook of Small Modular Reactors. Woodhead Publishing Series in Energy: Number 64. Elsevier, Cambridge.

The SMR designs from OKBM Afrikantov have the following common design features:

- All reactors being designed for single-, twin-, or four-unit plant configurations have individual turbine generators in the power circuits of each of the reactors.
- Flexible cogeneration options (production of heat for residential heating or superheated process steam or seawater desalination) are being provided for in the designs as options.

Construction period is typically 4 years (perhaps, somewhat shorter for the newest RITM-200).

- All designs provide for load-following capability.
- The fuel is UO_2 in inert matrix within the cylindrical fuel elements, and the fuel enrichment is always below 20%, even for the icebreaker reactor RITM-200.
- All designs, including those with the integral primary circuit, have external control rod drives, external pumps, and external pressurizers that use gas as a working medium to ensure a slow response of the primary pressure to rapid changes of the core temperature.
- Mechanical control rods are used for reactivity control in all designs, and, additionally, liquid boron system is used in the ABV-6E reactor only.
- The average fuel burnup is within the range of 45–47 MWday/kg, that is, below the values achieved in the state-of-the-art NPPs with large reactors.
- All designs employ double containments; for RITM-200 land-based plant, a triple containment system is being designed; see Table 19.1.

The major differences between the designs are as follows:

- The KLT-40S employs an older compact modular design of the primary circuit, while the ABV-6E and the RITM-200 employ integral designs of the primary circuit with in-vessel steam generators.
- The KLT-40S is refueled in a conventional batch mode each 2.5–3 years, while the ABV-6E and the RITM-200 provide for factory refueling with a longer interval between refueling of 5–12 years.
- Chromium-nickel alloy is used as cladding material in the ABV-6E and the RITM-200 to ensure operation with a long interval between refueling and enhance load-following capability.
- Liquid boron system is used for reactivity control in the ABV-6E design, while no such system is being provided for in the KLT-40S and RITM-200 designs.

Comparison of the SMR designs from OKBM Afrikantov indicates the designers prefer to follow an evolutionary approach within which new components are introduced with caution and that well-proven components are used in new designs whenever applicable. The major new direction is related to the integral design of the primary circuit within which only steam generators are being placed inside the reactor vessel. The integral design makes it possible to increase electric output (compare KLT-40S and RITM-200 in Table 19.1) and reduces the dimensions and the mass of a nuclear island [for a single-module nuclear island, the corresponding reduction in mass is from 1770 (KLT-40S) to 1100 t (RITM-200)]. The core design of the RITM-200 also makes it possible to increase by a factor of 20 the number of cycles of power change allowed during the core's lifetime (Veshnyakov, 2011).

The overall approach to safety design is similar to that the designers attempt to eliminate or derate as many accidents as possible "by design" and then deal with the remaining ones by plausible combinations of the redundant and diverse active and passive systems (ARIS, 2018; IAEA, 2009). A short summary of the safety design features incorporated in the SMRs from OKBM Afrikantov is provided later, in line with the defense-in-depth strategy defined in the Safety Standard No. SSR-2/1 (Rev. 1) Safety of Nuclear Power Plants: Design, Specific Safety Requirements (IAEA, 2016a). The focus is on design features specific for the addressed SMRs.

According to IAEA (2016a), cited with permission by the IAEA, "the purpose of the first level of defence is to prevent deviations from normal operation and the failure of items important to safety." For this level the OKBM designs provide "sound and conservative designs, siting, construction, operation and maintenance" and, specifically (ARIS, 2018, 2019; IAEA, 2007, 2009; Veshnyakov, 2011):

- negative reactivity coefficients over the whole operation cycle;
- compact modular designs with no long pipelines (KLT-40S) or integral primary circuit designs with minimized reactor pressure vessel penetrations (ABV-6E, RITM-200, RITM-200M), to eliminate or derate certain groups of LOCA;
- eliminated liquid boron system (KLT-40S, RITM-200, RITM-200M).

"The purpose of the second level of defence is to detect and control deviations from normal operational states in order to prevent anticipated operational occurrences at the plant from escalating to accident conditions" (IAEA, 2016a). Specific systems and features are provided in the OKBM Afrikantov designs to support meeting the objective of Level 2, including the following (ARIS, 2018, 2019; IAEA, 2007; Veshnyakov, 2011):

- state-of-the-art I&C systems based on PWR and marine propulsion reactor experience;
- redundant and diverse reactor shutdown systems (e.g., control rods driven by gravity, force of springs, and mechanically);
- relatively large coolant inventory and high heat capacity of the primary circuit or nuclear installation as a whole, resulting in larger thermal inertia of the system;
- external pressurizers operated on gas medium to ensure slow primary pressure increase in transients.

"For the third level of defence, it is assumed that, although very unlikely, the escalation of certain anticipated operational occurrences or postulated initiating events might not be controlled at a preceding level and that an accident could develop. In the design of the plant, such accidents are postulated to occur. This leads to a requirement that inherent and/or engineered safety features, safety systems and procedures be capable of preventing damage to the reactor core or preventing radioactive releases requiring off-site protecting actions and returning the plant to a safe state" (IAEA, 2016a).

To meet the objective of Level 3, the OKBM designs include redundant and diverse active and passive reactor shutdown, emergency core cooling, and residual heat removal systems (ARIS, 2018, 2019; IAEA, 2007, 2009; Veshnyakov, 2011). In this, they are more on the side of the state-of-the-art Russian VVER-type reactors, as passive decay heat removal systems are seldom found in the designs of the previous-generation marine propulsion reactors.

"The purpose of the fourth level of defence is to mitigate the consequences of accidents that result from failure of the third level of defence in depth. This is achieved by preventing the progression of such accidents and mitigating the consequences of a severe accident" (IAEA, 2016a). To meet this objective, all of the OKBM Afrikantov SMR designs provide for in-vessel retention of corium without a core catcher. The following features support achieving this quality:

- relatively low core power density (all designs omit KLT-40S) (see Table 19.1),
- active and passive systems of reactor vessel cooling.

In addition to this, all designs incorporate as a minimum double containment (see Table 19.1) and active and passive containment cooling systems.

Finally, "the purpose of the fifth and final level of defence is to mitigate the radiological consequences of radioactive releases that could potentially result from accident conditions" (IAEA, 2016a). To meet this objective, all OKBM Afrikantov SMR designs provide for on-site and off-site emergency measures and off-site emergency planning zones. However, the requirements to off-site emergency planning are reduced compared with those in place for conventional NPPs with large reactors. As an example, cited with permission by the IAEA from IAEA (2009), the on-site and off-site emergency planning measures for a barge-mounted plant with the two KLT-40S reactors are as follows:

- Staff presence in the compartments adjacent to the containment and in other compartments with high radiation level to be excluded.
- To limit radiation dose to the population living within a 1 km radius from the floating NPP, depending on the actual radiation situation, some protection measures such as iodine prophylaxis or sheltering will be implemented.
- Temporary limitation will be established on the consumption of some agricultural products grown within a radius of 0.5 km from the "floating" plant, when contaminated with radioactive release.
- Evacuation of the population is not required at any distance from the "floating" NPP.

The SMR designs from OKBM Afrikantov incorporate protection against external event impacts. The plants are designed for the design earthquake frequency 10^{-2}/year and the maximum design earthquake frequency 10^{-4}/year, with maximum design earthquake 8 on the MSK scale (ARIS, 2018; IAEA, 2009). In this the equipment, machinery, systems important for safety, and their mounting are designed for 3g peak ground acceleration (PGA). They remain operable under inclination and heaving, typical of a floating NPP operation conditions (IAEA, 2009).

In addition to the aforementioned, the designs of the barge-mounted plants take into account the external events specific of a barge-mounted NPP, such as the following (cited with permission by the IAEA from IAEA, 2009):

- sinking of the floating NPP;
- grounding, including that on a rocky ground;
- collisions with other ships;
- fall of a military aircraft onto the plant from a high altitude;
- blockage of water intakes by debris from another ship;
- helicopter crash landing on a floating NPP.

All barge-mounted NPPs are being designed or have been designed to produce energy only when rigidly moored to a shore (or a shore-fixed structure) (IAEA, 2009). Plant mooring devices provide for plant retention at a tsunami wave height of up to 4 m. Barge-mounted plants are being designed to be moored in a bay protected by a dam (Sozonyuk, 2011) (see Fig. 19.4).

The probabilistic safety analysis performed for the OKBM SMR designs indicates the core damage frequency (CDF) is equal to or below 10^{-6}/year, while the large early release frequency (LERF) is equal to or below 10^{-7}/year (IAEA, 2007; ARIS, 2019).

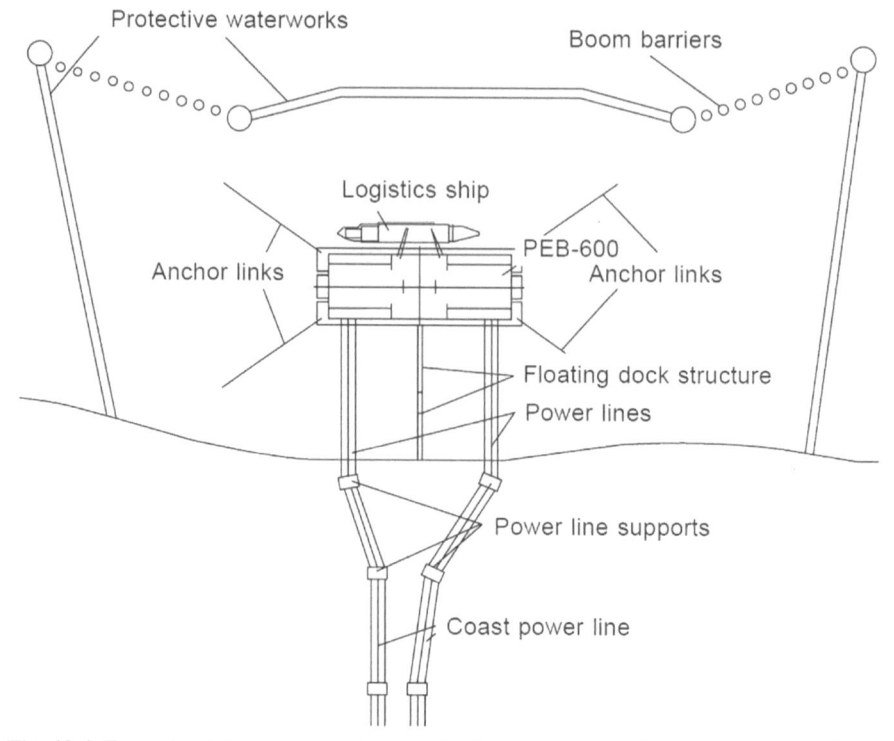

Fig. 19.4 Example of the shore structures and a dam to accommodate a barge-mounted NPP PEB-600.
Reproduced with permission by the IAEA from IAEA, 2006. Status of Innovative Small and Medium Sized Reactor Designs 2005: Reactors With Conventional Refuelling Schemes, IAEA-TECDOC-1485, Vienna. Available from: http://www-pub.iaea.org/MTCD/publications/PDF/te_1485_web.pdf (Accessed 30 December 2019).

These numbers are similar to the values typical of the state-of-the-art large NPPs with large reactors (ARIS, 2019).

The development and deployment status of SMRs from OKBM Afrikantov in Russia is presented in more detail in Section 19.5.

19.3 SMRs being developed by Joint Stock Company (JSC) NIKIET in Russia

The JSC NIKIET (Research and Development Institute of Power Engineering—RDIPE) named after N.A. Dollezhal (NIKIET, 2019a)—the enterprise belonging to the State Atomic Energy Corporation Rosatom—is a developer and manufacturer of equipment for nuclear power plants. NIKIET is the author of SMR concepts of various types and purposes (NIKIET, 2019b): NIKA (integral primary circuit PWR with a capacity of 100 MW(e)), KARAT (a family of boiling water reactors with a capacity range from 50 to 100 MW(e)), UNITERM (PWR with natural circulation in primary circuit in all operation modes, 6.6 MW(e)), SHELF (integral primary circuit PWR of 6.6 MW(e) with one of its variants intended for seabed location), VITYAZ (integral primary circuit PWR of ~1 MW(e) for transportable NPP), and ATGOR (high-temperature gas-cooled reactor with a capacity of about 1 MW(e)), also for transportable NPP. At present time, NIKIET is focused on the development of factory fueled SMRs for autonomous (remotely controlled) small nuclear power plants with a unit capacity up to 10 MW(e)—SHELF and ATGOR.

This section will present the designs of the three reactors being developed by NIKIET—UNITHERM (historically the first) and SHELF and ATGOR (for which design development is in focus currently). Short descriptions of other NIKIET SMRs, which are all at a conceptual design stage, could be found in ARIS (2018) and in NIKIET (2019b).

Historically the first of NIKIET SMR designs, named UNITHERM, has been in development since 1994 (ARIS, 2018). It is a somewhat unusual indirect cycle PWR with an intermediate heat transport system. The reactor is intended for semiautonomous operation in the course of a long refueling interval within a land-based or a barge-mounted plant. The second, more modern design named SHELF is an indirect cycle PWR intended for autonomous operation during of about 6 years within a land-based or an unmanned seabed-based NPP. Finally, ATGOR is an innovative transportable NPP with gas-cooled high-temperature reactor intended for electric and thermal energy supply to local consumers, including mitigation of the emergency situation consequences in remote regions.

The design and operating characteristics of the UNITHERM and SHELF are given in Table 19.3. Table 19.4 presents the core and fuel design characteristics for these reactors.

The UNITHERM plant is being developed to provide electricity and, when needed, other energy products to small settlements and enterprises (e.g., mining) located in remote areas with severe climatic conditions. The SHELF seabed-based plant is being developed to act as an energy source for the offshore oil and gas mining enterprises.

Fig. 19.5 shows a somewhat unusual three-circuit scheme of the UNITHERM reactor. The UNITHERM is a cogeneration plant producing heat for district heating or steam for industrial applications. In such a reactor an intermediate heat transport circuit is needed to prevent radioactivity from getting into the distribution network for district heating or industrial applications. The primary circuit operates in a natural circulation mode.

The heat generated in the reactor core is first transferred through a heat exchanger to the independent intermediate heat transport circuit located in an isolated volume within the reactor pressure vessel. Both the primary and the intermediate heat transport systems operate on natural circulation of the coolant. From the intermediate circuit, heat is then transferred to a power circuit through a steam generator located in a module outside of the reactor pressure vessel.

The intermediate circuit of UNITHERM consists of eight parallel sections (each hosting 10 subsections), each of which is a thermal siphon housed in an individual vessel with the individual steam generator section (ARIS, 2018). If one of the steam

Table 19.3 Design and operating characteristics of the SMRs from NIKIET.

Characteristic	UNITERM	SHELF
Electric/thermal power, MW	6.6/30 (in turbine condensing mode)	6.6/28.4
Nonelectrical products	Heat, up to 28 GCal/h	500 m³/h desalinated water 12 GCal/h heat
Plant configuration	Single-unit or multimodule plant, land-based or barge-mounted	– Single-unit land-based plant – Seabed-based or barge-mounted plant as an option
Construction period, months/mode of operation	Not specified/load follow autonomous (unmanned, 1 year)	<48/Load follow (20%–100%) autonomous (unmanned 11.5 months)
Thermodynamic cycle type/efficiency	Indirect Rankine cycle/22%	Indirect Rankine cycle/23.2%
Primary pressure, MPa	16.5	14.7
Core inlet/outlet temperatures, °C	249/330	270/310
Primary circulation	Natural	Forced
Mode of reactivity control in operation	Mechanical control rods with external drives and soluble boron	Mechanical control rods
Reactor diameter × height, mm	2900 × 9800	1538 × 2950
Reactor capsule height × diameter, m	Non applicable	14 × 8 (12 × 8 for land-based plant)
Secondary pressure, MPa	Intermediate circuit 3.9 Secondary (power) circuit 1.35	2.4
SG secondary side outlet temperatures, °C	Secondary (power) circuit 258	260

Turbine type	Standard turbine equipment with low steam parameters	Condensing cogeneration turbine, five stages
I&C system	Advanced systems ensuring autonomous plant operation	Advanced systems ensuring autonomous plant operation
Containment type and dimensions, m	– Primary containment made of steel – Secondary containment made of reinforced concrete	– Primary containment made of steel Ø 3.85 × 5 – Secondary containment is plant shell Ø 8 × 14 (inner) or Ø 8 × 12 (inner)
Transportable module weight, t	Not specified	<400
Plant surface area, m²	10,000	8000

Based on data from NIKIET, 2019b. https://www.nikiet.ru/index.php/2018-05-15-08-28-04/innovatsionnye-proekty/asmm (Accessed 30 December 2019); ARIS, 2018. Advances in Small Modular Reactor Technology Developments, a Supplement to the IAEA Advanced Reactors Information System (ARIS), Vienna. Available from: https://aris.iaea.org/Publications/SMR-Book_2018.pdf (Accessed 30 December 2019) and reproduced with permission by IAEA, 2007. Status of Small Reactor Designs Without On-Site Refuelling, IAEA-TECDOC-1536, Vienna. Available from: https://www.iaea.org/publications/7772/status-of-small-reactor-designs-without-on-site-refuelling (Accessed 30 December 2019).

Table 19.4 Core and fuel design characteristics of the SMRs from NIKIET.

Characteristic	UNITHERM	SHELF
Electric/thermal power, MW	6.6/30 (in turbine condensing mode)	6.6/28.4
Core diameter × height, mm	1130 × 1100	1050 × 800
Average core power density, MW/m^3	18.1	44
Average fuel element linear heat rate, W/cm	Not specified	61
Fuel material	UO_2 in zirconium matrix with silumin coating	UO_2 in zirconium matrix with silumin coating
Fuel element type	Cylindrical rod with four spacing ribs on the outer surface, self-spaced	Cylindrical rod with twisted ribs on the outer surface, self-spaced
Cladding material	Zirconium alloy 110	Zirconium alloy 110
Fuel element outer diameter, mm	Not specified	6.95
Lattice geometry	Triangular	Triangular
Number of fuel elements in fuel assembly	Not specified	Not specified
Number of fuel assemblies in the core	265	163
Burnable absorber	Boron, gadolinium	Boron, gadolinium
Enrichment of the reload fuel, ^{235}U weight %	19.75	19.7
Interval between refuelings, months	180 (15 years)	55.2 (4.6 years)
Average fuel burnup, MWday/kg	115	160
Mode of refueling	Refueling at a factory	Refueling at a factory

Based on data from NIKIET, 2019b. https://www.nikiet.ru/index.php/2018-05-15-08-28-04/innovatsionnye-proekty/asmm (Accessed 30 December 2019); ARIS, 2018. Advances in Small Modular Reactor Technology Developments, a Supplement to the IAEA Advanced Reactors Information System (ARIS), Vienna. Available from: https://aris.iaea.org/Publications/SMR-Book_2018.pdf (Accessed 30 December 2019) and reproduced with permission by IAEA, 2007. Status of Small Reactor Designs Without On-Site Refuelling, IAEA-TECDOC-1536, Vienna. Available from: https://www.iaea.org/publications/7772/status-of-small-reactor-designs-without-on-site-refuelling (Accessed 30 December 2019).

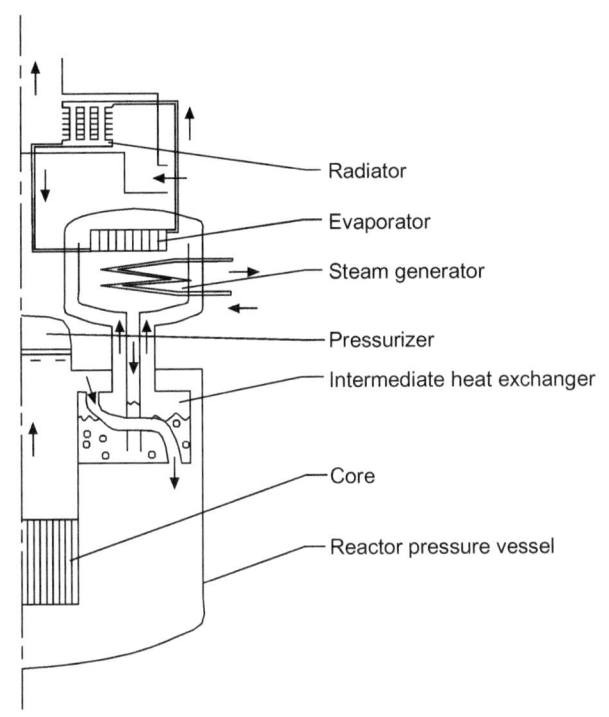

Fig. 19.5 Simplified flow diagram of the UNITHERM, reproduced with permission by the IAEA from reference (IAEA, 2007).
Credit: IAEA, 2007. Status of Small Reactor Designs Without On-Site Refuelling, IAEA-TECDOC-1536, Vienna. Available from: https://www.iaea.org/publications/7772/status-of-small-reactor-designs-without-on-site-refuelling (Accessed 30 December 2019).

generators' surfaces in one of the sections is ruptured, the corresponding section is being cut off using the lock valves on the consumer circuit. The plant needs not be stopped (it could continue operation with other sections), and the repair of the lost section could be accomplished during the scheduled repair period. When needed, the plant could also produce steam for industrial applications. For this purpose, steam is being extracted from the turbine, and the heat extraction circuit is operated on forced circulation of the medium (steam) (IAEA, 2007).

The UNITHERM reactor is being designed for autonomous operation. To ensure safety in such operation, yet another circuit shown in the upper part of Fig. 19.5 is added. This circuit, actually a purely passive safety system, consists of a continuously operated heat exchanger—evaporator and the radiator connected to evaporator, cooled by atmospheric air at ambient conditions (IAEA, 2007). This circuit is capable of bringing the reactor to a hot standby condition with no operation of the control rods and acts as a decay heat removal system in accidental conditions, for example, in the event of a loss of the normal heat removal path to the network for district heating (IAEA, 2007).

UNITHERM is being designed for operation in the severe climatic conditions of the Russian North and East, where the ambient temperatures undergo seasonal changes from −55 to +35°C. To make the reactor continuously operable under such conditions, some options can be offered as to use ammonia, ethylene glycol, or alcohol instead of water in the intermediate circuit (IAEA, 2007).

Another design of an SMR being developed by NIKIET is named SHELF (ARIS, 2018). Its concept was first formulated only in 2012, but its design development progressed quite notably ever since (ARIS, 2018). The SHELF is an encapsulated 6.6 MW (e) PWR with integral design of the primary circuit. The SHELF is being developed in two versions; one is where a single capsule contains all nuclear island components and a turbo-generator plant, and another one is where one smaller capsule hosts the nuclear island and another larger one hosts the turbo-generator plant (ARIS, 2018). The single-capsule version is intended for seabed-based location, while the two-capsule version is intended for land-based location and, with some modifications, for barge-mounted location.

General view of a land-based encapsulated SHELF nuclear island is given in Fig. 19.6. The reactor is a PWR of integral primary circuit design with in-vessel steam generators, pressurizer, control rod drives, and sealed canned pumps (graphics not available) (ARIS, 2018). The plant has two turbine generators, each connected to one of the two in-vessel steam generators. The reactor itself is immersed in a water pool located in the bottom part of the primary steel containment. This pool contains metal structures that, together with water, not only act as a radiation shielding but also take on the functions of a heat sink.

For the seabed-based variant of the SHELF NPP, the shell is designed to withstand a water depth of 300 m, although the targeted depth of a seabed site is 50–100 m. Autonomous operation is foreseen with reactor control being executed via cable from the water or land-based control center (e.g., oil platform). The electricity is supplied to the on-water or underwater user (e.g., a gas mining facility) using another cable. In a

Fig. 19.6 General view of the SHELF land-based reactor module.
Reproduced with permission by the NIKIET from NIKIET, 2019b. https://www.nikiet.ru/index.php/2018-05-15-08-28-04/innovatsionnye-proekty/asmm (Accessed 30 December 2019).

land-based plant version, the balance of plant is located in a separate capsule, and air cooling of turbine condensers is applied, which broadens siting possibilities for the plant (ARIS, 2018).

The SHELF plants are being designed for unmanned (automated and remotely controlled) operation during the whole 6-year cycle of operation between refuelings. Refuelings are assumed to be performed at specialized plants (ARIS, 2018). For the refueling the seabed-based SHELF capsule is being raised to the surface and brought to a refueling base (or a dedicated ship).

Both UNITHERM and SHELF incorporate experience from the design of small marine propulsion reactors. Specifically, they use so-called self-spaced cylindrical fuel elements with the external twisted ribs and employ no spacer grids. The fuel is borrowed from the Russian experience with marine propulsion reactors; Zr-coated dispersed UO_2 particles in a Zr matrix coated by silumin (Si-Al). Such fuel, often referred to as "cold" fuel (owing to its exceptional heat conductivity characteristics), is also capable of very high burnups exceeding those of conventional large-capacity reactors (ARIS, 2018); see Table 19.4.

Being designed for autonomous operation, the UNITHERM and SHELF reactors rely strongly on the inherent and passive safety features. The inherent features include relatively low core power density and large thermal inertia of the primary circuit owing to the relatively large inventory of the primary water coolant (IAEA, 2007; Elsevier, 2015). The reactors employ compact modular (UNITHERM) or integral primary circuit designs minimizing the list and scope of possible LOCA.

The state-of-the-art instrumentation and control (I&C) systems are being employed. The reactors have no liquid boron reactivity control systems. All safety systems in both designs are passive, including the mechanical control rods driven by gravity and the redundant and diverse passive decay heat removal systems. Specifically, UNITHERM incorporates an independent passive decay heat removal system based on an evaporator cooled by external air, as shown in the upper part of Fig. 19.5. SHELF has the bottom part of the reactor vessel immersed in a pool of water located in the bottom part of the primary steel containment. A reliable ultimate heat sink for the seabed-based SHELF plant version is provided by abundant seawater at ambient temperature.

Both reactors provide for a high level of natural circulation of the coolant. In UNITHERM the circulation in the primary and intermediate circuit is natural in the normal operation mode (IAEA, 2007), while in SHELF-forced circulation is used in normal operation mode (ARIS, 2018). However, natural circulation in SHELF is sufficient to remove the decay heat and could also remove heat from the reactor operated at 65% of the rated power (ARIS, 2012).

Both designs incorporate measures to prevent the core becoming uncovered in accidents, such as compact primary containments and reactor vessel penetrations located well above the core. Double containment is provided in both designs, as well as passive systems of the primary containment cooling.

Both design concepts incorporate provisions for protection against possible impacts of external events. The UNITHERM reactor is being designed for the seismic loads corresponding to 8–9 on the MSK 64 scale (ARIS, 2012). For land-based

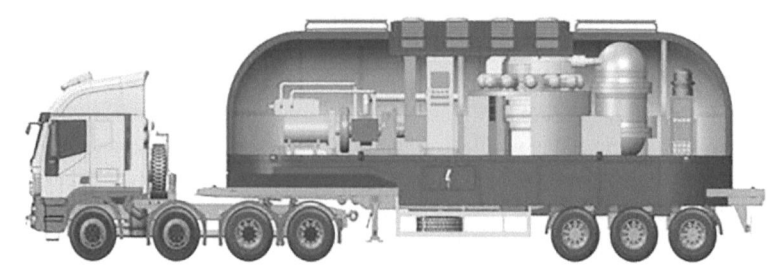

Fig. 19.7 General view of the ATGOR transportable NPP (plant version transported by truck). Reproduced with permission by the NIKIET from NIKIET, 2019b. https://www.nikiet.ru/index. php/2018-05-15-08-28-04/innovatsionnye-proekty/asmm (Accessed 30 December 2019).

SHELF the safe shutdown earthquake (SSE) corresponds to eight on MSK-64 scale (ARIS, 2018). For seabed-based SHELF the issue of protection against external even impacts is more complex, specifically because of the absence of internationally acknowledged requirements on protection against external event impacts for seabed-based NPPs. Seismic impacts may roughly be the same; however, tsunamis are not effective at depths around 100 m. Aircraft impact may be mitigated using steel nets around the plant. Resistance to torpedo attacks needs to be clarified.

Both SHELF and UNITHERM have the evaluated core damage frequency not exceeding 10^{-6}/year (ARIS, 2012, 2018). The design status of UNITHERM and SHELF is highlighted in Section 19.5.

The ATGOR transportable NPP is based on the direct cycle gas-cooled small reactor. Layout of the ATGOR NPP transported by track is shown in Fig. 19.7.

The mobile ATGOR unit is installed on automobile semitrailer and capable to produce 3.5 MW(th) and 0.4–1.2 MW(e). It can operate in a dispatch regime with the step of 200 kW(th). The ATGOR reactor lifetime is about 60 years, and the period of operation between refueling is 10 years.

The system of energy conversion in ATGOR is based on parallel gas turbine units with external combustion chambers. ATGOR can serve up to five gas turbine units with standard turbines of the Russian design and has two independent sources of heat (the main is nuclear reactor, and the second is the start-up and reserve combustion chamber operated on fossil fuel). The reactor, therefore, has a start-up capability at low temperatures down to −50°C. District heating can be provided with the use of the conventional standard steam boilers.

The deployment time for ATGOR on the site is less than 2 h. The plant can be transported by any means of transport, including rail, truck, or barge. Power unit is assembled at the site by block-modular method. All components of the unit provide for maximum level of factory manufacturing. This is estimated to yield eightfold savings in labor and capital costs.

As of mid-2019, fuel design for the ATGOR transportable plant has been developed on a conceptual level; no further details are available. It is also noted that ATGOR is deemed to be a plant of next generation, while for the first generation of transportable NPPs, it is more logical to use PWR type reactors based on a more mature technology.

19.4 SMR projects developed by JSC AKME Engineering in Russia

The JSC AKME Engineering (AKME Engineering, 2019), a joint venture of the State Atomic Energy Corporation Rosatom and the private JSC Irkutskenergo, has developed the design of the SVBR-100 reactor. The SVBR-100 is a small modular reactor of 100 MW(e) per module cooled by lead–bismuth eutectics. It is based on the experience of the propulsion reactors of the Russian alpha-class submarines that successfully operated in the time period between 1975 and 1995 and gained operational experience of 80 reactor years. The Russian Federation is the only country in the world with positive experience of reactors cooled by lead-bismuth eutectics. However, this experience so far relates to noncivilian application reactors.

The Russian program for marine propulsion reactors has resolved the two major issues relevant for lead-bismuth eutectics coolant, namely, those of corrosion-/erosion-free operation of structural materials in the coolant flow at temperatures below ~500°C with exclusion of lead oxide deposits in the primary circuit and those of safe operation with respect to generated ^{210}Po, based on data from p. 126 of *Current status, technical feasibility and economics of small nuclear reactors* (NEA-OECD, 2011) and on data from IAEA (2007). Moreover a reported accident with spillage of the primary coolant has demonstrated that extrairradiation doses to personnel could be effectively avoided during spillages and during the subsequent repair works at a factory (IAEA, 2012).

Unlike its marine propulsion prototypes in which beryllium moderators were used in the core, the SVBR-100 is a reactor with fast neutron spectrum. Generically, this means that a substantial amount of R&D would be required for qualification and licensing of this new reactor.

The design and operating characteristics of the SVBR-100 are summarized in Table 19.5. Table 19.6 gives the core and fuel design characteristics.

The SVBR-100 is being designed as a modular reactor for single-module or multimodule nuclear power plants with optional cogeneration capacity. Being a fast reactor, it could be flexible in fuel cycle options, that is, easily adjustable to operation with different types of fuel—uranium, uranium-plutonium, uranium–transuraniums, and uranium-/thorium-based oxides or nitrides—in a once-through or closed nuclear fuel cycle, effectively matching the fuel cycle options of the day (IAEA, 2007; ARIS, 2018). When operated with mixed oxide (MOX) or with mixed nitride fuel in a closed nuclear fuel cycle, SVBR-100 could ensure preservation or slight increase of its fissile inventory through an infinite number of recycles. In this case the reprocessing becomes essentially reduced to just removal of fission products and addition of some fertile material, for example, depleted uranium (Toshinsky et al., 2019).

The SVBR-100 could be used for baseload electricity generation within single or multimodule plants of different capacity. A load-following option is, in principle, possible and could be considered for future design modifications. Several conceptual studies of 4- and 16-module SVBR-100–based plants have been developed in the Russian Federation (IAEA, 2007). Some of these designs consider a partly underground location of the reactor modules that, together with other features of

Table 19.5 Design and operating characteristics of the SVBR-100 from JSC "AKME Engineering."

Characteristic	SVBR-100
Electric/thermal power, MW	100/280
Nonelectrical products	Heat or desalinated water or process steam, as an option
Plant configuration	Single-module (prototype plant), flexible multimodule plant configurations (in the future)
Construction period, months/ mode of operation	42/Baseload, load following possible
Thermodynamic cycle type/efficiency	Indirect Rankine cycle on saturated steam/36%
Primary circulation	Forced
Primary pressure, MPa	Near atmospheric + weight of the heavy lead–bismuth coolant
Core inlet/outlet temperatures, °C	340/482
Mode of reactivity control in operation	Mechanical control rods
Reactor vessel diameter × height, mm	4530 × 7550
Turbine type	Available standard equipment
Seismic design	0.5 g
I&C system	Similar to Na-cooled reactors, special coolant chemistry control
Containment type and dimensions, m	Depends on plant configuration, reinforced concrete for multimodule plants
Plant surface area, m^2	150,000 (single-module plant)

Based on data from ARIS, 2012. Status of Small and medium Sized Reactor Designs, a Supplement to the IAEA Advanced Reactors Information System (ARIS), Vienna. Available from: https://aris.iaea.org/Publications/smr-status-sep-2012.pdf (Accessed 30 December 2019); ARIS, 2018. Advances in Small Modular Reactor Technology Developments, a Supplement to the IAEA Advanced Reactors Information System (ARIS), Vienna. Available from: https://aris.iaea.org/Publications/SMR-Book_2018.pdf (Accessed 30 December 2019) and AKME Engineering, 2018. SVBR-100 (JSC AKME Engineering, Russian Federation), Booklet. AKME Engineering, Moscow.

the SVBR-100, makes these reactors the closest match to the philosophy of SMRs being pursued in the United States.

Fig. 19.8 presents the schematics of the equipment layout in the SVBR-100 primary circuit. Different from sodium-cooled fast reactors, the SVBR-100 plant has no intermediate heat transport system. This is because lead-bismuth eutectics do not react exothermically with water or air. The reactor is pool type with primary circulation provided by the pumps with externally located drives (see Fig. 19.8). There is a very small excess pressure in the primary circuit (inert gas with an excess pressure of less than 10 kPa fills in the space above the coolant-free level). The circulation scheme is optimized to prevent steam bubbles from passing through the reactor core in accidents.

Two reactor vessels are provided, the main vessel and the guard vessel. This practically excludes LOCA-type accidents. The secondary circuit operates on saturated steam so that steam separators are being included in the design.

Table 19.6 Core and fuel design characteristics of the SVBR-100 from JSC "AKME Engineering."

Characteristic	SVBR-100
Electric/thermal power, MW	100/280
Core diameter × height, mm	1645 × 900
Average core power density, MW/m^3	146
Fuel material	UO_2 ((U-Pu)O_2, UN, (U-Pu)N in future designs)
Fuel element type	Cylindrical
Cladding material	Stainless steel EP-823
Lattice geometry	Triangular
Burnable absorber	No, fast reactor
Enrichment of the reload fuel, ^{235}U weight %	<19.6%
Average fuel burnup, MWday/kg	60
Interval between refuelings, years	7–8
Mode of refueling	Whole core refueling on the site

Based on data from ARIS, 2012. Status of Small and medium Sized Reactor Designs, a Supplement to the IAEA Advanced Reactors Information System (ARIS), Vienna. Available from: https://aris.iaea.org/Publications/smr-status-sep-2012.pdf (Accessed 30 December 2019); ARIS, 2018, Advances in Small Modular Reactor Technology Developments, a Supplement to the IAEA Advanced Reactors Information System (ARIS), Vienna. Available from: https://aris.iaea.org/Publications/SMR-Book_2018.pdf (Accessed 30 December 2019) and AKME Engineering, 2018. SVBR-100 (JSC AKME Engineering, Russian Federation), Booklet. AKME Engineering, Moscow.

The relatively high melting temperature of the lead–bismuth eutectics (125°C) requires heating the reactor internals and the coolant before initial supply of the latter to the reactor vessel. It also may require heating of the shutdown reactor if the spent fuel composition in the reactor does not ensure enough of decay heat to keep the primary coolant liquid. Systems to accomplish such operations have been developed and applied in the Russian marine propulsion reactor program. Moreover a procedure for safe freezing/unfreezing of the reactor coolant based on a particular time-temperature curve has been developed and tested on a land-based facility. No further details of these technologies are available. The overall layout of the reactor module appears very compact, as shown in Fig. 19.9.

The inherent and passive safety features of SVBR-100 include the following (ARIS, 2018, 2019; IAEA, 2007, 2012):

- low-pressure primary coolant system, contributing to the prevention of LOCA (Level 1 of the defense in depth);
- very high-boiling temperature of Pb-Bi eutectics (1670°C at atmospheric pressure), double reactor vessel, relatively high temperature of Pb-Bi eutectics freezing (125°C at atmospheric pressure), and location of the reactor module in a water tank, practically excluding LOCA and limiting potential radioactivity release in accidents with core melt (Level 1 and Level 4 of the defense in depth);
- chemical inertness of Pb-Bi in air and water, preventing fires and explosions (Level 1 of the defense in depth);
- negative optimum reactivity feedbacks, including very small reactivity margin for fuel burnup achieved with MOX or mixed nitride types of fuel at the expense of high conversion

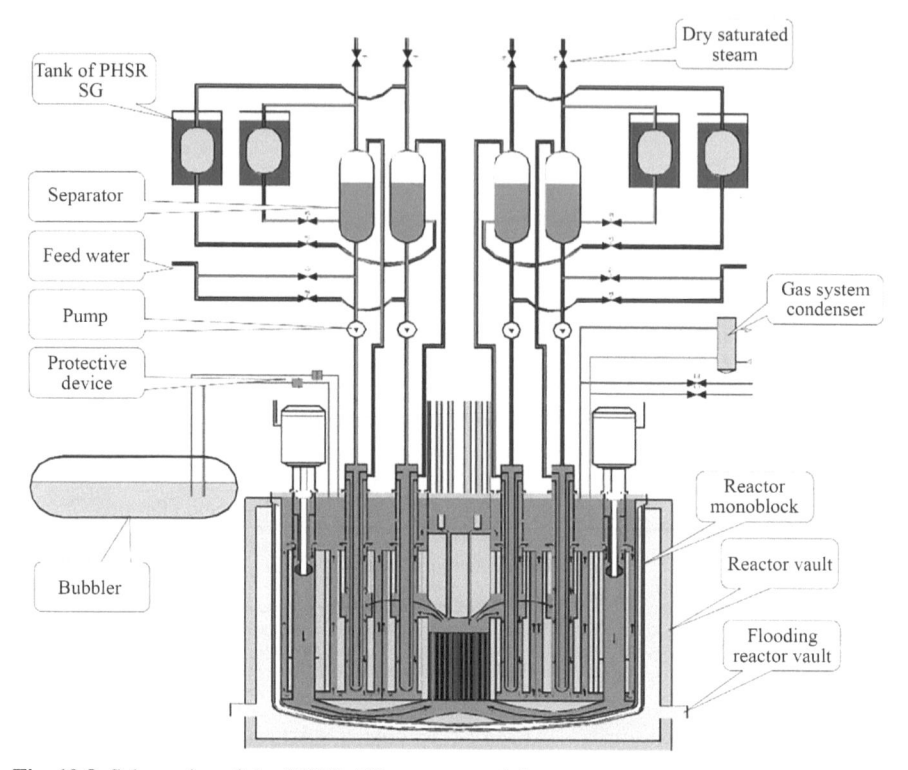

Fig. 19.8 Schematics of the SVBR-100 reactor module.
Reproduced with permission by the AKME Engineering from AKME Engineering, 2018.
SVBR-100 (JSC AKME Engineering, Russian Federation), Booklet. AKME Engineering, Moscow.

or a very small breeding ratio (1.05) in the reactor core, preventing reactivity-induced accidents (Level 1 of the defense in depth);
- a primary coolant flow path organized to prevent the possibility of steam bubbles from getting into the reactor core, to avoid prompt criticality events owing to positive void worth (Level 1 of the defense in depth);
- a pool-type design of the reactor with high heat capacity of the primary circuit, ensuring high thermal inertia in transients (Level 2 of the defense in depth);
- a high level of natural circulation of the primary coolant sufficient to remove the decay heat from the core (Level 3 of the defense in depth).

In addition to the aforementioned, the SVBR-100 incorporates shutdown systems based on mechanical control rods inserted by gravity and by the force of springs additional passive shutdown system with fusible locks without servo drive mechanisms and four diverse passive decay heat removal systems. The reactor monoblock pit is flooded with water in the design extension condition. A steam generator leak localizing system is also included in the design (gas system steam condenser and bubbler with bursting membrane) to prevent the ingress of pressurized steam from the

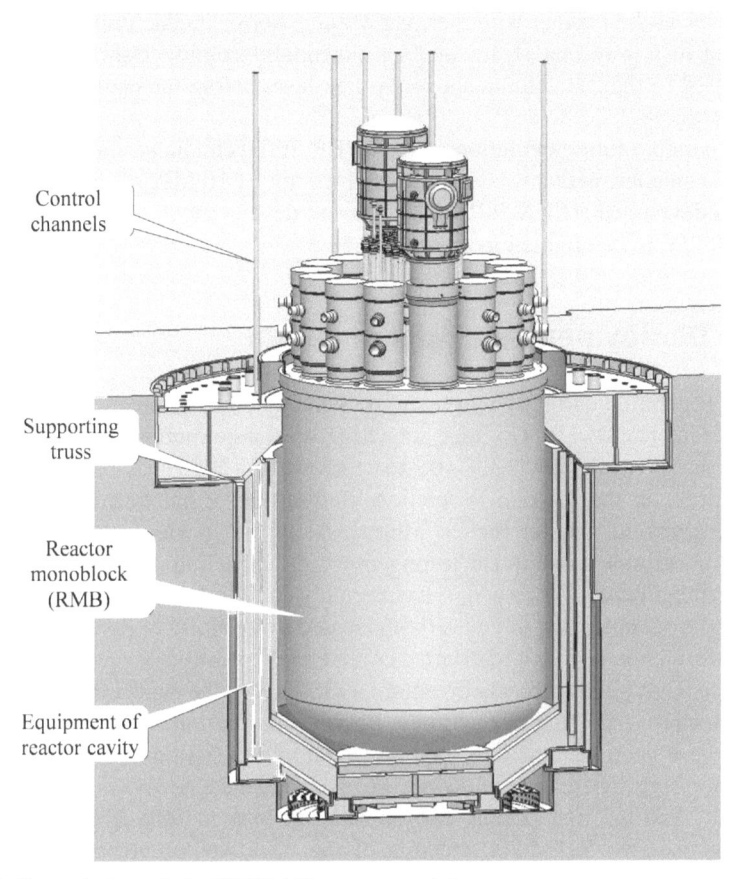

Fig. 19.9 General view of the SVBR-100 reactor module.
Reproduced with permission by the AKME Engineering from AKME Engineering, 2018. SVBR-100 (JSC AKME Engineering, Russian Federation), Booklet. AKME Engineering, Moscow.

secondary into the primary circuit owing to a steam generator tube rupture (Level 3 of the defense in depth). Reinforced concrete containment is provided to prevent hypothetical radioactive releases beyond the plant boundary. The seismic design of the SVBR-100 incorporates features to ensure safe reactor shutdown at 0.5 g PGA.

The predicted core damage frequency of 10^{-8}/year (ARIS, 2012) is primarily because Pb-Bi coolant is chemically inert with water and air, allows for primary circuit operation at near-atmospheric pressure, and has a proven capability to self-cure cracks (freezing temperature 125°C). Moreover, Pb-Bi eutectics has a very high boiling temperature (1670°C), requires no intermediate heat transport system in plant design, and ensures a high level of natural circulation in loss-of-flow accidents. Altogether the aforementioned features make it possible to develop a simple and robust small reactor design. Additionally, the integral layout of the primary circuit includes a free level of the coolant with inert gas volume above it. The primary flow path is organized in a way that steam bubbles potentially coming from steam generator are released to the gas volume above the free level before the coolant is directed to the reactor core.

As a possible future evolution of the SVBR-100 technology, the conceptual proposal of a smaller, perhaps, barge-mounted reactor of 10 MW(e), named SVBR-10, has been developed (IAEA, 2010). The current development and deployment status of SVBR-100 is highlighted in Section 19.5.

19.5 Deployment of SMRs in Russia

Russian Federation currently appears to be a champion in deployment of small modular reactors worldwide. On June 24, 2019 the barge-mounted plant "Akademik Lomonosov" with the two KLT-40S reactors (ARIS, 2018) has received an operational license in the Russian Federation. Before that, it has been assembled Saint Petersburg, towed without fuel to Murmansk, where it was fueled at the factory and has undergone a complete testing program, including full power tests. After the operating license was issued, it has been towed with fuel in the cores from Murmansk to its operating site in Pevek (both located in the North of Russia) where after a short confirmation testing it has started commercial operation on December 19, 2019.

For the RITM-200 icebreaker reactor (ARIS, 2018), the standard design of which has been approved by the State Atomic Energy Corporation "Rosatom" in 2012, six reactors have been manufactured and delivered by rail (without fuel in the cores) to Saint Petersburg. There, four of them were installed on the newly constructed two icebreakers, "Arktika" and "Ural," two on each. In May to June 2019, fuel has been loaded in the two RITM-200 reactors of the "Arktika" icebreaker. Fuel loading was followed by comprehensive testing program after which the Arktika icebreaker with operating reactors has started sea tests in the Gulf of Finland on December 12, 2019. The start-up of commercial operation for the "Arktika" icebreaker is scheduled for 2020.

In parallel with the deployment of reactors on the icebreakers, a detailed design for the optimized barge-mounted NPP with the two RITM-200M reactors (ARIS, 2018) has been completed in development in the first half of 2019. The reactor design is essentially the same as RITM-200, offering an operation interval between refueling of 5–10 years. The optimized barge-mounted NPP design with the two RITM-200M reactors employs a larger reactor core and will have no storages of fresh and spent fuel onboard. The refueling is expected to be performed at the refueling base, located either on land or on a dedicated self-propelled ship designed to transport fresh and spent nuclear fuel and incorporating refueling equipment to enable barge-mounted plant refueling at its operation site. Should the latter be implemented, this would set aside most of the legal and institutional issues associated with export deployments of barge-mounted NPPs, related to transportation of factory-fueled reactors through territorial waters of countries that are third parties to a transaction (IAEA, 2013).

For the RITM reactor family design, activities are also in progress to develop a land-based NPP with two or, optionally, four RITM-200 reactor unit and several new RITM family reactors for the new smaller and larger icebreakers.

Regarding the ABV reactor for barge-mounted and land-based nuclear power plants (ARIS, 2018), its previous design version with a shorter core lifetime was licensed many years ago, and work to develop and qualify a new core design with a longer continuous operation between refuelings has been completed. However, no decision has so far been made regarding the construction, and no activities related to licensing of the new design are being observed.

Regarding UNITHERM reactor the design development remains at a conceptual stage, although a preliminary conceptual design of the complete plant has been developed in 2015. Further development of the project is said to pose no major technological difficulties and, therefore, could be completed within 5 years, once an order for the plant is placed and adequate financing is secured. Negotiations have been in progress for a long time with potential customers in the Russian Yakutia region, but as of mid-2019, no progress toward project implementation was observed. According to the developer, active work on the project UNITHERM is suspended due to the relatively high specific indicators of the cost of the developed unit.

For the SHELF reactor (ARIS, 2018), basic designs of the plants for seabed-based and land-based locations have been completed; detailed design development is in progress. As of mid-2019, possible sites for SHELF deployment were under negotiation, and a decision on going on with licensing of one or more projects was awaited. Licensing of the plant design is scheduled for 2020–22.

For the SVBR-100 lead-bismuth eutectic-cooled reactor, detailed design has been in large part developed, and a decision has been made in 2006 on the construction of a single-module prototype of this reactor at the site in Dimitrovgrad near the "State Scientific Centre—NIIAR" (Russia) (ARIS, 2018). Public hearings were accomplished in 2014 giving a green light to the construction, and first concrete was poured on the site sometime thereafter. The license for SVBR-100 placement was issued in 2015 (ARIS, 2018). However, after that, the project has shown little to no progress, because the budget for realization of this project calculated according to the newly emplaced national economic assessment rules exceeded the original evaluation by a factor of 2.4 (RIA Novosti, 2017). Originally the SVBR-100 project was indicated as a priority in the Russian Federal Program "Nuclear power technologies of new-generation for the period 2010–2015 and for the future up to 2020," emplaced by the RF Government Order No. 50 (2010). However, the project was in 2017 excluded from this program, and the program itself was stopped in implementation ahead of time, in 2018 (RF Government order No. 1451 from 30 November 2018). Since then, different options for further project financing toward completion are being examined, but as of mid-2019, no tangible progress toward further project implementation was observed. The developers currently consider 2021 and 2025 as target dates for construction start and operation license acquisition for the single-module prototype SVBR-100.

19.6 Future trends

In Russia SMRs are not considered as possible competitors to large reactors. The overall idea is to have reactors of different capacities within a broad power range to cater to the needs of a variety of potential customers in the diverse energy demand, siting, transportation, and climatic conditions throughout the country; see Section 19.1.

The trend that clearly manifested itself since the first edition of this book was published in 2015 (Elsevier, 2015) is that of design unification. It might be boosted by relatively difficult situation in the Russian economy and the specifics of Russian nuclear industry that is strongly controlled and essentially financed by the State. With limited funding for new developments, just a couple of new SMR plant designs that not only incorporate reasonable innovations but also benefit from using many design features previously proven in operation of marine propulsion reactors are being soundly developed in several versions each to cater to a variety of siting options (land based, barge mounted, or even seabed based), requested power levels (through single or twin, or four-unit design configurations, but in some cases just through upscaling or downscaling of the design) and energy products (electricity, district heat or industrial steam, etc.). Each of such SMR plant lines is based on essentially the same reactor design, while each SMR plant version targets benefitting from being produced in small series.

The most notable example here is provided by the RITM reactor family based on the RITM-200 design of 50 MW(e) (ARIS, 2018), which is a PWR with integral primary circuit. With RITM-200 reactors, several (up to 7) new-generation icebreakers could be built (six reactors have already been manufactured and four of them installed on two icebreakers that are set afloat in Saint Petersburg). In addition to this, RITM-200B design is being developed for a small single-reactor icebreaker for local offshore operations, and design development for the upscaled RITM-400 design for the largest "Leader" nuclear icebreaker considered to be built in the Russian Far East is being finalized. At the same time, design development for the optimized barge-mounted plant with two RITM-200M reactors has been finalized, and design development for land-based power plants with RITM-200 reactors is in progress (ARIS, 2018).

Another example is that of the encapsulated SHELF reactor, which is also a PWR with integral design of the primary circuit, but with a very small power output of only 6.6 MW (ARIS, 2018). Here, three versions of the plant design different in the number of capsules are being developed for land-based, seabed-based, and barge-mounted applications, while possible siting options are being examined. In all of the aforementioned cases, the plant designs could be modified to enable cogeneration options or nonelectrical applications.

The first-of-a-kind plant "Akademik Lomonosov" with the two KLT-40S reactors, which is deemed to demonstrate the validity of technical solutions implemented in a barge-mounted NPP design concept and pave a way to future generations of such plants, is likely to be the last plant where the somewhat dated KLT-40S reactors are implemented. All future barge-mounted or land-based SMR plants in Russia are expected to accommodate newer, more efficient reactor designs, such as RITM- 200 (Sozonyuk, 2011).

All Russian SMRs are being designed to be licensed and deployed first in their country of origin—the Russian Federation—where they could cater to a variety of energy needs, specifically, in remote areas where generation costs are currently much higher than on the mainland. In the case of success, after several years of operation at rated capacity factors, some of them could be considered for deployment in other countries. Specifically, barge-mounted plants could be offered to a number of other countries for the purposes of seawater desalination (e.g., performed by a desalination plant located on a separate barge) (Sozonyuk, 2011).

19.7 Conclusion

Owing to multiyear national experience in design, deployment, and operation of marine propulsion reactors, including the reactors of nuclear icebreakers, and courtesy of a sustained support from the Government, the Russian Federation is currently a world leader in the deployment of innovative SMRs. The project that has been commercially deployed is a barge-mounted nuclear power plant "Akademik Lomonosov" with the two KLT-40S reactors. The projects at advanced deployment stage are the two nuclear icebreaker ships, "Arktika" and "Ural," each with the two RITM-200 reactors. The new RITM-200 reactor, which is a pressurized water reactor with integral design of the primary circuit, points to a current approach in SMR development pursued in Russia. Reactors of essentially the same design are currently being considered for use in the next-generation barge-mounted and land-based nuclear power plants, as well as for a variety of new icebreaker ships of different design and draught. Design development for a variety of new plants with RITM-200 reactors shows good progress. For example, detailed design has already been developed for the new optimized barge-mounted plant and the new land-based plant, both with the two RITM-200M reactors.

Another project, for which notable progress in design development is being observed, is that of the SHELF, a very small pressurized water reactor with integral design of the primary circuit. Nuclear power plant with this reactor is being developed in three versions: for sea bottom-based, land-based, and barge-mounted applications. Basic design has been developed, and siting options are being examined for a land-based plant version currently. The land-based SHELF plant version would rely on air cooling of the turbine condensers.

Different from SMRs with pressurized water reactors, the design development finalization for SVBR-100, a fast neutron spectrum SMR cooled by lead-bismuth eutectics, has come to a temporary standstill, and its deployment has been postponed to a period after 2020. The project has currently no funding, and new options to move on with its realization are being examined.

In the Russian Federation, small reactors are not viewed as direct competitors to large reactors. For small reactors, targeted are particular niche markets where generation costs are high, where cogeneration, long refueling interval, or plant relocatability are assets, where transportation routes are seasonal, where the demand is limited and siting conditions are specific (i.e., no water in winter due to deep freezing of rivers or other water reservoirs). The layout of the country offers a variety of niche opportunities for such reactors in areas where generation costs are much higher than in the rest of the country. Export deployment of the Russian SMR-based NPPs will be considered after they have been deployed domestically and demonstrated the effectiveness of their technologies.

19.8 Sources of further information

The first recommended source of further information on SMRs presented in this section is the electronic advanced reactor information system (ARIS) maintained by the International Atomic Energy Agency (IAEA) (ARIS, 2019). This database contains reasonably detailed descriptions of SMRs prepared first hand by the designers (vendors) according to a common template developed by IAEA, all in English. The adopted description structure includes applications and special features, summary table of major design and operating characteristics, fuel cycle options, economics and maintainability, safety design and safety systems, proliferation resistance and physical protection, description of the nuclear island, descriptions of the turbine plant, control room, I&C systems, and plant layout. Each description is complemented by a comprehensive list of references to national and international publications providing further details of a particular design. Of the SMRs addressed in this chapter, ARIS contains the design description of the KLT-40S reactor.

ARIS is supplemented by periodically updated brochure currently titled "Advances in Small Modular Reactor Technology Developments" (ARIS, 2018), which contains similarly structured but much shorter (4-page) descriptions of design and operating characteristics and safety features for nearly all SMR designs being developed or deployed worldwide, including all of the SMRs presented in this chapter. This brochure also contains periodically (once in 2 years) updated information on design status and targeted timeframes for licensing and deployment of SMRs.

The IAEA Nuclear Energy Series No. NP-T-2.2 issued in 2009, entitled design features to achieve defense in depth in small- and medium-sized reactors, presents detailed descriptions of the design and safety features of a number of SMRs, including siting requirements and specific external events addressed in the design (IAEA, 2009). Among the SMRs included is the barge-mounted plant with the two KLT-40S reactors. Safety features are presented in line with the defense-in-depth strategy defined by the IAEA Safety Standard NS-R-1 (IAEA, 2000). Since 2016, this standard is superseded by the IAEA Safety Standard SSR/2-1 (Rev. 1) titled safety of nuclear power plants: design requirements (IAEA, 2016a). The new standard introduces some updates to the terms; in particular, "design basis accidents" are replaced by "design basis events," and "beyond design bases accidents" are replaced by "design extension."

The IAEA-TECDOC-1785 titled design safety considerations for water-cooled small modular reactors incorporating lessons learned from the Fukushima Daiichi accident (IAEA, 2016b) is a follow-up to Nuclear Energy Series No. NP-T-2.2 (IAEA, 2009). It provides review of the engineering designs and performance of the engineered safety features of water-cooled small modular reactors in dealing with the design basis events and design extension. Included in the consideration are KLT-40S, RITM-200, and UNITHERM.

A publication by NEA-OECD titled current status, technical feasibility, and economics of small nuclear reactors, issued in 2011, provides analysis of economic characteristics of several SMRs being developed worldwide (NEA-OECD, 2011). Of the SMRs presented in this chapter, included are the KLT-40S, the ABV, and the SVBR-100 reactors. A key feature of reference (NEA-OECD, 2011) is that it summarizes the most reliable and detailed designers' economic data for SMRs and also provides independent economic evaluations for a number of SMRs, including the KLT-40S and the ABV reactors.

A more recent publication by NEA-OECD titled small modular reactors: nuclear energy market potential for near-term deployment includes a case study of Russia with a barge-mounted plant with two KLT-40S units (NEA-OECD, 2016). This study includes overall characterization of SMR market in Russia and analysis of market regulations and prices.

The IAEA Nuclear Energy Series No. NG-T-3.5 issued in 2013, entitled legal and institutional issues of transportable nuclear power plants: a preliminary study (IAEA, 2013), examines legal and institutional issues for export deployment of transportable NPPs. Addressed are land-based, barge-mounted, and sea bottom–based transportable NPPs. The conclusion is that issues different from those related to deployment of conventional NPPs may arise in the case when factory-fueled and tested reactors are transported through the territories or territorial waters of countries that are third parties to a transaction.

As mentioned in Section 19.1, several medium-sized reactors are also being developed in Russia to cater to a variety of the energy needs. Those include the downscaled VVER-1200 derivatives VVER-600 and VVER-300 and the VBER- 300 design merging the technologies of VVER reactors and the marine propulsion reactors. Further information on these designs can be found in NEA-OECD (2011), IAEA (2006, 2009, 2016b); ARIS, 2018, 2019). In the period since the first edition of this book (Elsevier, 2015), no major progress toward deployment of any of these designs has been observed.

References

AKME Engineering, 2019. http://www.akmeengineering.com/aboutus.html. (Accessed 30 December 2019).

ARIS, 2012. Status of Small and Medium Sized Reactor Designs, a Supplement to the IAEA Advanced Reactors Information System (ARIS), Vienna. Available from:https://aris.iaea.org/Publications/smr-status-sep-2012.pdf. (Accessed 30 December 2019).

ARIS, 2018. Advances in Small Modular Reactor Technology Developments, a Supplement to the IAEA Advanced Reactors Information System (ARIS), Vienna. Available from:https://aris.iaea.org/Publications/SMR-Book_2018.pdf. (Accessed 30 December 2019).

ARIS, 2019. Technical Data, KLT-40S. Available from:https://aris.iaea.org/PDF/KLT-40S.pdf. (Accessed 30 December 2019).

Elsevier, 2015. In: Carelli, M.D., Ingersoll, D.T. (Eds.), Handbook of Small Modular Reactors. In: Woodhead Publishing Series in Energy: Number 64, Elsevier, Cambridge.

IAEA, 2000. Safety of Nuclear Power Plants: Design Requirements, IAEA Safety Series No. NS-R-1, Superseded by SSR-2/1, Vienna. Available from:https://www.iaea.org/publications/6002/safety-of-nuclear-power-plants-design. (Accessed 30 December 2019).

IAEA, 2006. Status of Innovative Small and Medium Sized Reactor Designs 2005: Reactors With Conventional Refuelling Schemes, IAEA-TECDOC-1485, Vienna. Available from:http://www-pub.iaea.org/MTCD/publications/PDF/te_1485_web.pdf. (Accessed 30 December 2019).

IAEA, 2007. Status of Small Reactor Designs Without On-Site Refuelling, IAEA-TECDOC-1536, Vienna. Available from:https://www.iaea.org/publications/7772/status-of-small-reactor-designs-without-on-site-refuelling. (Accessed 30 December 2019).

IAEA, 2009. Design Features to Achieve Defence in Depth in Small and Medium Sized Reactors, IAEA Nuclear Energy Series No. NP-T-2.2, Vienna. Available from:https://www.iaea.org/publications/8094/design-features-to-achieve-defence-in-depth-in-small-and-medium-sized-reactors-smrs. (Accessed 30 December 2019).

IAEA, 2010. Small Reactors Without On-Site Refuelling: Neutronic Characteristics, Emergency Planning and Development Scenarios (Final Report of an IAEA Coordinated Research Project, IAEA-TECDOC-1652), Vienna. Available from:https://www.iaea.org/publications/8515/small-reactors-without-on-site-refuelling-neutronic-characteristics-emergency-planning-and-development-scenarios-final-report-of-an-iaea-coordinated-research-project. (Accessed 30 December 2019).

IAEA, 2012. Liquid Metal Coolants for Fast Reactors Cooled by Sodium, Lead, and Lead-Bismuth Eutectic, IAEA Nuclear Energy Series No. NP-T-1.6, Vienna., pp. 32–33. Available from:https://www.iaea.org/publications/8589/liquid-metal-coolants-for-fast-reactors-cooled-by-sodium-lead-and-lead-bismuth-eutectic. (Accessed 30 December 2019).

IAEA, 2013. Legal and Institutional Issues of Transportable Nuclear Power Plants: A Preliminary Study, IAEA Nuclear Energy Series No. NG-T-3.5, Vienna. Available from:https://www.iaea.org/publications/10516/legal-and-institutional-issues-of-transportable-nuclear-power-plants-a-preliminary-study. (Accessed 30 December 2019).

IAEA, 2016a. Safety of Nuclear Power Plants: Design, Specific Safety Requirements, IAEA Safety Standards Series SSR-2/1 (Rev. 1), Vienna. Available from: https://www.iaea.org/publications/8771/safety-of-nuclear-power-plants-design. (Accessed 30 December 2019).

IAEA, 2016b. Design Safety Considerations for Water Cooled Small Modular Reactors Incorporating Lessons Learned From the Fukushima Daiichi Accident, IAEA-TECDOC-1785, Vienna. Available from:https://www.iaea.org/publications/10981/design-safety-considerations-for-water-cooled-small-modular-reactors-incorporating-lessons-learned-from-the-fukushima-daiichi-accident. (Accessed 30 December 2019).

NEA-OECD, 2011. Current Status, Technical Feasibility and Economics of Small Nuclear Reactors. Nuclear Development, International Energy Agency/Nuclear Energy Agency, Organisation of Economic Cooperation and Development, Paris. Available from:http://www.oecd-nea.org/ndd/reports/2011/current-status-small-reactors.pdf. (Accessed 30 December 2019).

NEA-OECD, 2016. Small Modular Reactors: Nuclear Energy Market Potential for Near-Term Deployment, NEA No. 7213. Organisation for Economic Co-operation and Development NEA, Boulogne-Billancourt, France. Available from:https://www.oecd.org/publications/small-modular-reactors-9789264266865-en.htm. (Accessed 30 December 2019).

NIKIET, 2019a. https://www.nikiet.ru/. (Accessed 30 December 2019).

NIKIET, 2019b. https://www.nikiet.ru/index.php/2018-05-15-08-28-04/innovatsionnye-proekty/asmm. (Accessed 30 December 2019).

OKBM, 2019. Afrikantov. Available from:http://www.okbm.nnov.ru/en/. (Accessed 30 December 2019).

PRIS, 2019. IAEA Power Reactor Information System. Available from:https://pris.iaea.org/PRIS/home.aspx. (Accessed 30 December 2019).

RF Government Order No. 1451, 2018. RF Government Order No. 1451 from 30 November 2018 Regarding Early Termination of the Federal Program on Nuclear Power Technologies of New Generation for the Period 2010–2015 and for the Future up to 2020. Available from:http://docs.cntd.ru/document/551789487. (Accessed 30 December 2019).

RF Government Order No. 50, 2010. RF Government Order No. 50 from 3 February 2010 Regarding a Federal Programme on Nuclear Power Technologies of New Generation for the Period 2010–2015 and for the Future up to 2020. Available from:http://base1.gostedu.ru/57/57729/. (Accessed 30 December 2019).

RIA Novosti, 2017. Nuclear Technologies: The Government Has Postponed Construction of the SVBR-100 Power Unit. April 5, 2017. Available from:https://ria.ru/20170405/1491593190.html. (Accessed 30 December 2019).

Sozonyuk, A., 2011. Floating nuclear power plants, status and prospects (Technical Paper C03). In: Interregional Industrial Conference ASMM-Regionam 2010. IBRAE, Moscow, Russian Federation.

Toshinsky, G., Grigoriev, S., Dedul, A., et al., 2019. Safe controlled storage of SVBR-100 spent nuclear fuel in the extended-range future. World J. Nucl. Sci. Technol. 09 (03), 93552 Available from:http://www.scirp.org/Html/2-1090398_93552.htm. (Accessed 30 December 2019).

Velikhov, E.P., 2008. Evolution of Power Production in the XXI Century. IzdAt, Moscow, p. 159.

Veshnyakov, K., 2011. Results of development of detailed design of a reactor installation for multi-purpose nuclear icebreaker. In: Sudostroyenie, No. 3. Russian Federation, Moscow, pp. 32–37.

Small modular reactors (SMRs): The case of the United Kingdom 20

Kevin W. Hesketh[a] and Nicholas J. Barron[b]
[a]Fuel and Core, National Nuclear Laboratory, Preston, United Kingdom, [b]Reactor Core Technology, National Nuclear Laboratory, Sellafield, United Kingdom

20.1 Introduction

The purpose of this chapter is to provide a summary of UK research and development activities toward "SMRs." In the United Kingdom the term SMR is taken to mean small modular reactor that employ significant offsite (factory) manufacture and assembly. SMRs are exclusively water-cooled systems (PWRs and BWRs), and the term encompasses both integrated and nonintegrated designs. Modular reactors that do not use water as the coolant, that is, those cooled by gas, liquid metal, or molten salts, are deemed as advanced modular reactors (AMRs). It is also important to note that from an HM Government (HMG) perspective, AMRs encompass nuclear fusion and nuclear fission. In policy terms, AMRs and SMRs are grouped under the heading of advanced nuclear technologies (ANTs) (Fig. 20.1).

The United Kingdom has a strong history of nuclear research and development and a legacy of operating many systems that would now be classed as AMRs. This is briefly discussed in Section 20.2; Sections 20.3 and 20.4 detail the recent resurgence in UK interest in nuclear. Research programs supporting the advancement of UK capability and that align with the development needs of both SMRs are discussed in Section 20.5. Future UK trends and potential demands for SMRs and AMRs are covered in Section 20.6.

20.2 History of nuclear power development in the United Kingdom

The first nuclear power plants to operate in the United Kingdom were the four Magnox units at Calder Hall and the four largely identical units at Chapelcross. These were natural uranium metal-fueled, graphite-moderated, pressurized CO_2-cooled reactors, first connected to the grid in 1956. These plants were dual purpose units also used to supply nuclear materials for the UK's defense services. The UK Magnox design was developed for the first generation of commercial nuclear plants in the United Kingdom, with nine twin-unit power stations built between 1962 and 1971. Economies of scale led to a gradual evolution in the electrical power outputs, from less than 50 MWe per unit in Calder Hall to 600 MWe per unit at Wylfa, the last Magnox plant

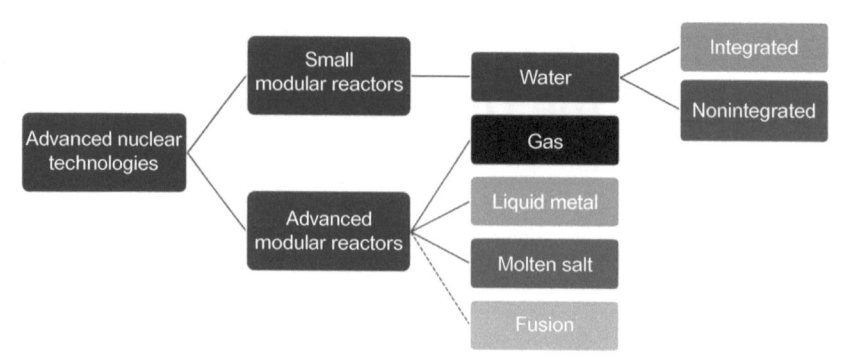

Fig. 20.1 Overview of UK headings for ANTs.

to be built. Wylfa was the last plant to remain in operation and shut down in 2015. Used fuel from the Magnox plants was all subject to reprocessing, and reprocessing operations are expected to continue until 2020 when all the used fuel is expected to have been processed.

The Calder Hall and Chapelcross power plants could arguably be regarded as the presaging modern SMRs. There were four modular units at each station, each with an electrical output in the SMR range. Safety of the reactors largely relied on them having low power densities, with decay heat removal by natural circulation. But they would fail to meet the economic requirements that current designs are expected to meet.

The Magnox fleet was superseded by the advanced gas reactors (AGR), of which seven twin-unit 600-MWe plants were built and remain in operation. The AGRs are graphite moderated and CO_2 cooled and use low-enriched UO_2 fuel, with stainless steel cladding. They are designed to operate with high gas outlet temperatures and have thermal efficiencies in excess of 40%. When they were being developed, AGRs were intended to be economically competitive with light water reactors (LWRs) and were selected in preference to LWRs for this reason. Their high thermal efficiencies and capability to refuel on-line were strong theoretical advantages in their favor. However, the AGRs are complex plants, and their early years of operation were affected by poor reliability. Subsequently, their reliability has improved with experience, and the plants have in recent years operated more successfully. Their theoretical economic advantage did not materialize, for various reasons. One factor is that though fuel burnups were comparable in the early years with LWRs of that time, LWRs have advanced considerably so that mean discharge burnups of 50 GWd/t are now routine. However, while AGR burnups have advanced, they have not been able to go much beyond 35 GWd/t burnup. The lower discharge burnup combined with the more complex fuel assembly design and fuel loading and discharge routes remains disadvantageous for AGRs. Another consideration is that LWRs have a large body of international experience that can be shared between designers and operators, which AGRs don't have the benefit of. For these reasons the AGR line of development will not be followed up when they are retired from operation.

The United Kingdom also invested heavily in fast reactor research up to the early 1990s and developed and operated the prototype fast reactor (PFR) at Dounreay, supporting research reactors. PFR was a pool-type sodium-cooled system during whose operation a considerable amount of valuable operating experience was accumulated. The United Kingdom also developed the steam-generating heavy-water reactor (SGHWR), which was a low-enriched heavy water-moderated, light water-cooled direct cycle system. SGHWR operated successfully as a demonstrator but was eventually abandoned in favor of PWRs to benefit from pooled international investment and experience.

In the late 1970s the United Kingdom decided to build a fleet of pressurized water reactors (PWRs) and chose the Westinghouse SNUPPs 4-loop design, adapted to meet UK licensing requirements. In the end political and market changes led to just one PWR, Sizewell B, being built. Sizewell B has demonstrated a reliable operating history from its inception and has been very successful. Subsequently, EDF Energy have started building a twin-unit EPR (Hinkley Point C) that will expand the PWR fleet to three units, with the possibility of additional plants to follow. Of the other planned LWR developments in the United Kingdom, two projects were recently cancelled, largely due to the difficulty of private investors financing large reactor projects in the current market conditions.

20.3 Strategic requirements and background to UK interest in modular reactors

In March 2016 the UK Government launched the first phase of the SMR competition, as an evidence-gathering phase to gauge market interest among technology developers, utilities, and potential investors (GOV.UK, 2016). Prior to the close of the competition in December 2017, 33 expressions of interest were received from eligible participants across technology vendors, research organizations, and utilities (which are listed in Appendix 20.1, Table 20.A1; Small Modular Reactors, n.d.). Phase 1 of the SMR competition was an information gathering process that did not involve technology downselection and provided eligible participants the opportunity to engage with HMG to inform policy development.

Following closure of the SMR competition, the UK Government sponsored the initial phase of its advanced modular reactor feasibility and development (AMR-F&D) project, which funded early feasibility studies for a number of advanced modular reactor (AMR) designs announced in the UK nuclear sector deal (HM Government, 2018). Small modular reactors (SMRs) with water moderation and cooling were specifically excluded from consideration.

UK interest in AMRs is driven by the possibility of achieving low generating costs, improved flexibility to respond to changes in grid demand, and their potential to support high-temperature heat for industry and/or hydrogen production and also provide low-temperature domestic heating. One of the main considerations for the UK Government was to facilitate designs that will require extensive research and development and that have the potential to generate intellectual property in the United Kingdom, with a significant component of UK manufacturing inputs eventually created. Nearer term SMRs based on existing light water reactor (LWR) designs were excluded because they were considered not to offer sufficient research and development and manufacturing opportunities.

Probably the single most important consideration underlying the interest in SMRs and AMRs has been the UK's experience of financing conventional new build reactors in the United Kingdom. All of the nuclear reactors currently operating were built as part of a nationalized electricity industry and were financed by government. The last government financed reactor build was Sizewell B that started up in 1994. Since then the UK electricity market has undergone a transition to a private sector enterprise and been subject to several major market reforms along the way. The initial capital investment required for new build units has proved very difficult to finance with private investors and has raised concerns over the future viability of nuclear new build and the management of financial risk. AMRs provide a possible means by which the initial capital investment can be de-risked and made more affordable.

Another driver has been the fundamental changes in the UK's generating profile that has occurred in recent years. The UK generating mix has transitioned from one dominated by coal, oil, and nuclear plants to one in which coal has been almost phased out and most of the generating capacity provided by gas, renewables, and nuclear. Throughout this transition, nuclear has continued to operate as baseload generation. However, with the continuing increase in variable capacity renewables, it seems increasingly likely that future nuclear plants will be required to contribute to load balancing. Therefore an increasing factor in the choice of future plants will be their ability to contribute to load balancing by operating in frequency control and load-follow modes. SMRs and AMRs may be better suited for this purpose than larger plants, though this has yet to be demonstrated. Another important consideration in a future grid will be maintaining sufficient inertia in the system to ensure stable operation. Phasing out gas generation in favor of renewables will reduce the number of synchronized generating sets that provide inertia in the system, and ANTs may be an attractive option to provide the required inertia.

In 2019 the United Kingdom became the first OECD nation to commit to net-zero emissions by 2050. This is based on recommendations by the Committee on Climate Change (Committee on Climate Change, 2019), who identified potential scenarios to support this. A pathway to decarbonization has been suggested that includes significant electrification, which could result in a doubling in UK electricity demand by 2050 (compared with 2018 levels). A diverse energy mix is expected to be required to support decarbonization and could require nuclear power to provide between 10 and 25 GWe "baseload" generating capacity and even more in the form of flexible supply. Based on current site analysis, the potential for circa 75-GWe capacity is thought to exist in the United Kingdom.

The UK Government's stated aim of net-zero emissions by 2050 will demand such fundamental changes in the UK's generating mix. Nuclear has historically provided safe, secure, and low-carbon electricity in the United Kingdom, and modular reactors could be a cost-efficient means to realize this. New requirements, such as the need to provide generation for electric or hydrogen transport, will add further impetus. Significant restructuring of the electricity supply industry will inevitably be required because of these changes. For example, there is currently no mechanism by which nuclear generators can be incentivized to provide load-balancing capability. All nuclear plants required to carry out load balancing will need a mechanism that values this important functionality. These requirements are discussed further in Section 20.5.

20.4 UK R&D activities to support modular reactor development

In 2013 the UK Government published the nuclear industry strategy (NIS) (HM Government, 2013) that highlighted the potential so-called small nuclear could play in the UK's energy mix. The identified potential led to the setting up of the Nuclear Innovation Research Advisory Board (NIRAB), to provide independent expert advice to government on the research and innovation needs for nuclear energy to play a significant role in the UK's future low-carbon energy mix. In 2015 NIRAB reported on the need for an ambitious program of fission research to support the development of UK capability and facilities in advanced nuclear (NIRAB, 2016). Funding was originally aligned to five key areas that are discussed in Section 20.4.1.6 encompassing advanced fuels, spent fuel recycle and waste management, strategic toolkit, reactor design, and advanced materials and manufacturing. This was subsequently expanded to include up to £44 million of funding for advanced modular reactors development (Section 20.4.1.7). The entire nuclear innovation program, including the AMR competition has received a funding commitment of £180 million to March 2021 from HMG (GOV.UK, 2019a). In addition, HMG has committed £18 million to the UKSMR consortium, led by Rolls Royce from the Industrial Strategy Challenge Fund; EU and HMG funding for fusion research will continue until at least the end of 2020 (GOV.UK, 2019a). Fig. 20.2 shows the research topics that are included in the UK nuclear innovation program from April 2018 to April 2021 and the approximate timescales.

20.4.1 Nuclear innovation program

The nuclear innovation program (NIP) involves HMG, academia, research organizations, and the supply chain. Its key objective is the development of UK capability, through enabling funded nuclear R&D.

20.4.1.1 Advanced manufacturing and materials

The use of advanced manufacturing techniques and novel-to-nuclear materials is potentially key to achieving significant cost savings during modular and offsite fabrication of SMRs and AMRs. To support realization of this, five areas were identified as part of the advanced manufacturing and materials nuclear innovation program (AMM-NIP) that encompassed materials, component manufacture, large-scale manufacturing and assembly, prefab module development, and codes and standards. Materials topics included investigation of component production techniques such as those using additive manufacturing or hot isostatic pressing and activities to enhance

	Research area	FY* 2017–2018	FY* 2018–2019	FY* 2019–2020	FY* 2020–2021
Advanced materials and manufacturing (AMM)	Materials				
	Component manufacture			(AMM) Phase 2 and 2b	
	Large scale manufacture and assembly				
	Pre-fabricated module development				
	Codes and standards				
Advanced fuels	Reactor physics				
	Accident tolerant fuels			The Advanced Fuels and Spent Fuel Recycle projects were expanded in scope and are managed as a single programme, the Advanced Fuel Cycle Programme	
	Coated particle fuels				
	Fast reactor fuels				
	Nuclear data				
Spent fuel recycle	Aqueous recycling of spent LWR fuel				
Advanced fuel cycle programme (AFCP)	Advanced fuels and recycle and waste management projects				
Reactor design	Modelling and simulation				
	Reactor safety and security engineering				
	Virtual engineering				
	Thermal hydraulic				
Nuclear facilities and strategic toolkit	Strategic assessments				
	Fast reactor knowledge capture				
	Regulatory engagement				
	Access to irradiations facilities				
Advanced modular reactors	Feasibility study				
	Design development				

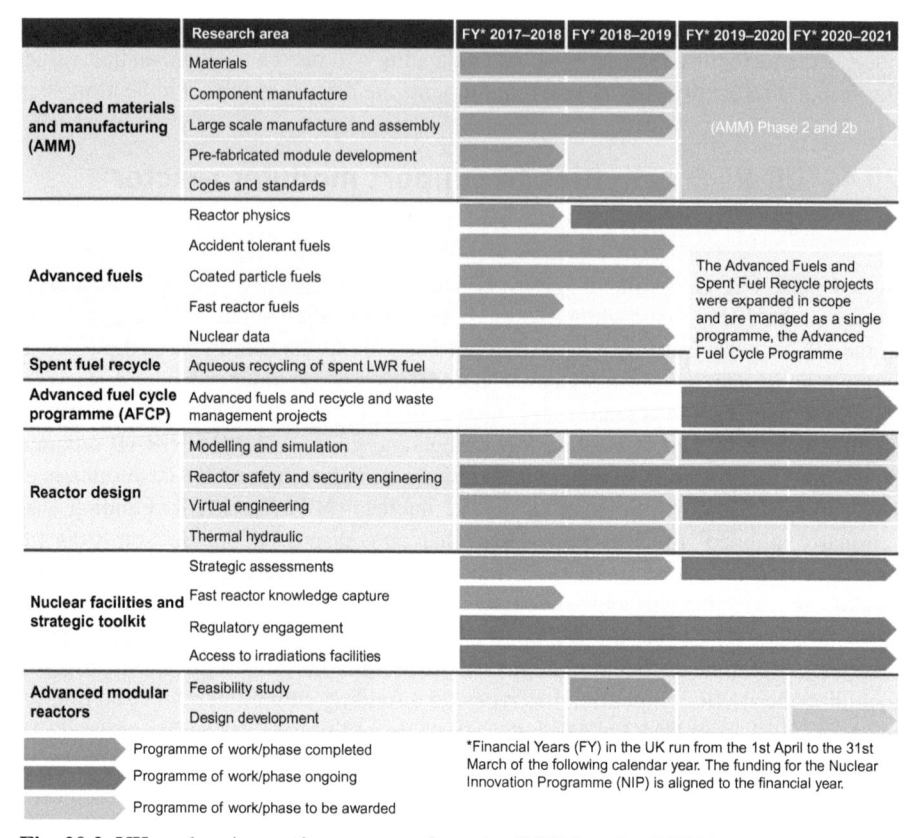

▬ Programme of work/phase completed
▬ Programme of work/phase ongoing
▬ Programme of work/phase to be awarded

*Financial Years (FY) in the UK run from the 1st April to the 31st March of the following calendar year. The funding for the Nuclear Innovation Programme (NIP) is aligned to the financial year.

Fig. 20.2 UK nuclear innovation program from April 2018 to April 2021.

understanding of advanced joining techniques. To reduce the number of component moves during manufacturing, the SIMPLE (single manufacturing platform environment) project investigated carrying out multiple processes on a single machine and increasing automation during manufacturing. INFORM (intelligent fixture for optimized and radical manufacture) is investigating the forging, machining, welding, and inspection of large components using advanced techniques throughout the manufacture. A key action for SMRs and AMRs is understanding how modular offsite builds can reduce the risk to nuclear new build, especially compared with traditional on-site construction. Initial work to understand this was included in a project on prefabricated module development.

Phase 1 part of the AMM-NIP also included funding to identify the nuclear design codes and standards needs for SMRs and Generation IV technologies (including AMRs). The initial work focused on a gap analysis and identified a roadmap to overcome these gaps for SMR and Gen-IV reactors/AMRs.

In summary, Phase 1 aimed to develop the fundamental technologies in the AMM theme. Phase 2 of the advanced manufacturing and materials component of the nuclear innovation program intends to build on the initial activities and move from lab scale/concept work to demonstration of the technology (i.e., toward commercialization). This includes a commitment to a further £20 million funding for innovation from government that requires a minimum of a further £12 million contribution from industry. Phase 2 activities are to complete by the end of March 2021, and as of December 2019, several proposals have been joint funded by industry and HMG, one of which is the development of nuclear design codes for high-temperature materials in reactors. The second announced project is looking to investigate coolant flow around fuel pins. A further funding round (Phase 2B) was announced in January 2020 to be awarded in April 2020 (GOV.UK, 2020). Many of the topic areas (large-scale nuclear component manufacturing and assembly, manufacturing modeling, prefabricated module development and verification, and advanced construction) could offer significant benefits for SMR and AMR development and ultimately deployment.

20.4.1.2 Advanced fuels

The initial phase of the advanced fuels NIP (AF-NIP) primarily focused on development of high-density and accident tolerant fuels (ATF) and cladding for Generation III/III+ water-cooled reactors. This included the investigation and development of novel production routes. Accident tolerant cladding development included R&D into coated zirconium alloys and advanced ceramics, such as silicon carbide (SiC), that could have applications across multiple SMR and AMR concepts. In addition, the advanced fuels NIP supported development of UK reactor physics capability, initially focusing on fuel behavior using multiphysics engines to link various models. Given the cost of constructing and operating experimental facilities, modeling is expected to play a greater role in supporting new fuel qualification. Phase 1 funding also supported development of facilities to enable the United Kingdom to carry out coated particle fuel research.

In November 2019 funding for a follow-on phase of the advanced fuels NIP was announced (GOV.UK, 2019b). This was via a single advanced fuel cycle program (AFCP) containing both the advanced fuels and recycle and waste management NIPs. The AFCP is funded to the end of March 2021 and significantly expands the scope. Advanced fuels research continues into high-density fuels and advanced claddings for LWRs. To support the development of UK facilities, researchers, and production expertise techniques for both coated and ceramic claddings, in the expanded AFCP, projects to develop UK capability and knowledge in coated particle and fast reactor fuels have also been included for gas-cooled and liquid metal-cooled reactors, respectively.

Funding to continue reactor physics modeling component is focused on coupling codes and on the development of an integrated model of fuel behavior in accident conditions across various advanced systems.

20.4.1.3 Recycle and waste management

Phase 1 of the recycle and waste management NIP focused exclusively on aqueous recycling of spent LWR fuel, with an overarching aim to provide credible technical options to policy makers by 2030, for the subsequent UK deployment of an advanced aqueous spent (LWR) fuel reprocessing facility. This has involved various technical programs involving entire baseline flowsheet development for dissolution process, separation processes, and product finishing for high-burnup oxide reference fuels. Flowsheet trials to experimentally underpin the advanced PUREX process have been carried out, as well as the development of an innovative photochemical conditioning process. In addition, the Sim-Plant model has been developed to visualize impacts of R&D on plant footprint and support the assessment of waste arising from spent fuel reprocessing. Combining the modeling and flowsheet activities has led to the advanced PUREX being assessed as able to process a more challenging range of fuels in a simpler plant and generating less wastes at source than the current commercially deployed PUREX process.

The follow-on phase of the recycle and waste management NIP is through the advanced fuel cycle program (AFCP), which encompasses both the advanced fuels and recycle and waste management NIPs (GOV.UK, 2019b). This has resulted in a significantly enlarged scope of work, encompassing aqueous recycle of spent LWR fuel (continuing Phase 1 work), aqueous fast reactor recycle, pyroprocessing, and multiple waste immobilization projects.

The aqueous recycle project aims to further develop the advanced PUREX flowsheet, with larger-scale rig trials, and enhance the capabilities of the Sim-Plant model. An industry engagement component to understand wider sector concerns and capture critical operational knowledge from the UK's thermal oxide reprocessing facility (THORP) is also a component, given the cessation of UK spent fuel reprocessing by 2020. Fast reactor reprocessing will include activities to capture the reprocessing experience from the UK's two fast reactors at Dounreay. In tandem an experimental program will include flowsheet development and initial validation to support this. Pyroprocessing in molten salts is viewed as the leading alternative to aqueous-based reprocessing of spent fuel. This is expected to focus upon the development of a baseline flowsheet for electrorefining and investigate innovative monitoring processes, as well as providing data across topics including salt handling and properties, materials corrosion, and materials accountancy. The waste immobilization technologies program area looks to further develop the UK's knowledge in this area, demonstrating effective immobilization technologies for wastes generated by advanced fuel cycles.

20.4.1.4 Reactor design

This topic aligns to a long-term ambition to include state-of-the-art digital engineering and design technology in the UK supply chain. Phase 1, to March 2019, focused upon modeling and simulation, reactor safety and security engineering, and virtual engineering. To support these activities an aspect of UK capability assessment was required to identify necessary development requirements. Various models, ranging from those to simulate radiation effects on components throughout their operational life, to security modeling were created as well as setting up of a nuclear virtual engineering center (Levers, 2017).

Capability development in the field of thermal hydraulics is also encompassed by the reactor design component of the nuclear innovation program, aiming to deliver the capability and facility requirements to support the design of advanced nuclear technology systems and their components. The nuclear sector deal provided UK Government commitment to the hosting of a thermal hydraulic facility, something the United Kingdom has not had since its fast reactor program. Work to March 2019 focused upon developing a specification for a UK thermal hydraulic facility, based on existing state-of-the-art facilities and identifying cross-cutting themes.

20.4.1.5 Strategic toolkit and facilities

To support decision-making on advanced nuclear technologies of the future, tools, methods, and data are required. The nuclear facilities and strategic toolkit component of the NIP is supporting this through developing necessary strategic assessments, knowledge capture of historic UK programs, regulatory engagement, and access to irradiation facilities and nuclear data. Each of these activities is scheduled for a varying length of time. Ongoing activities include developing frameworks to support the assessment of advanced technologies fuel cycles and energy scenarios and regulatory advancement. To date, only the knowledge capture task has been completed. This involved digitizing records from the UK's fast reactor program and creating network maps to aid future researchers in accessing information on topics covering fuels and core materials (SFR), reactor operations, core physics, and thermal hydraulics.

20.4.1.6 AMR competition

The UK Government's advanced modular reactor (AMR) competition invited designers to submit applications for inclusion in an initial technical feasibility study. Eight systems were successful at the initial stage and are described in this section.

Three of the successful applicants are high-temperature gas reactor (HTGR) systems. The fact that the United Kingdom has extensive experience of building and operating gas-cooled graphite reactors from the (former) Magnox and (current) AGR fleets, along with past R&D programs on HTGR, is likely to have been a factor in the HTGR systems having been selected.

The United Kingdom had a long-standing R&D program on sodium-cooled fast reactors, and therefore it is perhaps not surprising to see a sodium fast reactor selected. However, the use of metal alloy fuel is a major departure for the United Kingdom, given that its experience was almost exclusively with UO_2 fuels.

The UK Government decided that fast reactors with lead-bismuth coolant would be excluded from further consideration because of the issue of Po-210 generation. The two lead fast reactors chosen for the initial phase are therefore both systems that use lead rather than lead-bismuth eutectic, which has a lower melting point.

The inclusion of a very novel molten salt fast reactor and a compact spherical tokamak fusion system in the initial feasibility study fits with the stated intent of maximizing potential intellectual property rights for the United Kingdom.

Fig. 20.3 U-Battery schematic.
(Reproduced with the permission of Urenco.)

U-Battery
U-Battery is a micro modular reactor (MMR) for electricity and heat applications. It is being developed by a consortium of UK companies led by Urenco and partnered by Jacobs, Laing O'Rourke and Kinetrics, with support from Canada. U-Battery (Fig. 20.3) is a high-temperature gas reactor (HTGR), with a graphite-moderated prismatic core, cooled by helium gas and fueled with LEU TRISO fuel particles embedded in a graphite matrix (U-Battery Local Modular Energy, 2018). The nominal power output is 4 MWe (10 MWth) per unit with an outlet temperature of 750°C. U-Battery is designed to be deployed as single or multiple units. A secondary circuit uses nitrogen gas as the working fluid to power the generator.

USNC MMR
The micro modular reactor (MMR) is a US-designed reactor system being developed by Ultra Safe Nuclear Corporation. It is a prismatic core HTGR, with graphite moderation and helium coolant, and uses fully ceramic microencapsulated (FCM) LEU particle fuel (Jo et al., 2014). A molten salt secondary circuit is used for power generation. The nominal electrical power unit is 5 MWe, and up to 10 modules can be combined for 50 MWe total output. MMC has a single-batch core with a 20 year lifetime and no refueling.

DBD HTR-PM
HTR-PM is a pebble HTGR being developed in China which is being represented in the United Kingdom by the engineering consultancy DBD. HTR-PM is graphite moderated and helium cooled and uses coated particle fuel microspheres embedded in graphite that is partly based on the earlier HTR-10 prototype reactor (Zhang et al., 2009).

A steam secondary circuit provides power generation. HTR-PM is deployed as twin reactor modules sharing a single steam turbine rated at 210 MWe. A demonstration plant is under construction at Rongcheng in China, and commercial plants to follow will consist of three twin reactor modules.

Advanced reactor concept ARC-100 SFR

Advanced reactor concept is developing a 100-MWe sodium-cooled fast reactor (ARC-100) (Wade and Walters, 2011). ARC-100 is an entirely US design that features a pool-type configuration and has a sodium secondary circuit and a tertiary power-generating loop. ARC-100 builds on the US experience with operating the EBR-II prototype reactor and uses metal alloy fuel. It is designed for modular factory construction, and its small output allows a passive approach to decay heat removal.

LeadCold LFR (SEALER-UK)

SEALER LeadCold Reactors (Blykalla Reaktorer) is developing the SEALER-UK reactor. This is a 55-MWe lead-cooled fast reactor with a single-batch 25-year lifetime core using low-enriched uranium nitride fuel (Wallenius et al., 2018). SEALER-UK features a conservatively low output temperature to limit the potential for the lead coolant to corrode reactor components.

Westinghouse LFR

Westinghouse Electric Company UK Ltd. is undertaking the conceptual design of a 450-MWe lead-cooled fast reactor (LFR) (Ferroni et al., 2017). This uses a pool-type configuration for the primary circuit. LFR has a high operating temperature compatible with process heat applications and power production and load-follow capability.

Moltex stable salt reactor (SSR) MSR

The stable salt reactor (SSR) is a UK-origin design of molten salt reactor (MSR) being developed by Moltex Energy, with assistance from Canada (Moltex Energy, 2018). It is a novel design that separates the highly active fuel salt from a low activity coolant salt. The fuel salt is contained within cylindrical fuel rods which melts when the fuel is loaded in the reactor in contact with the coolant salt. The fuel salt is chloride based, while the coolant salt is fluoride based. SSR is very innovative and has no precedent in previous reactor designs.

Tokamak energy spherical tokamak

Tokamak Energy Ltd. is a UK startup that has been developing a compact spherical tokamak nuclear fusion reactor (Sykes et al., 2017). This is designed to achieve nuclear fusion in a much more compact space than is possible with conventional tokamaks. The spherical tokamak has become feasible following recent advances in high-temperature superconductors.

20.4.1.7 Additional activities

Nuclear innovation and advisory board (NIRAB)

The Nuclear Innovation and Research Advisory Board (NIRAB) was first convened to provide independent technical advice to HMG to underpin energy policy and the 2013 nuclear industry strategy (HM Government, 2013). The output in the form of a report to HMG in 2015 included the recommendation, for a program of publicly funded nuclear fission R&D that became the nuclear innovation program (NIP) (NIRAB, 2016). NIRAB was reconvened in 2018, with key asks being to monitor the NIP to date and provide any recommendations based on program outputs or the current nuclear landscape.

The NIRAB report for FY 2018–19 sets out a series of recommendations (listed in Appendix 20.2) for HMG (NIRAB, 2019). This included recommending that HMG carry out activities to de-risk future nuclear new build projects through both legislation and partnering with industry to support ANT (i.e., SMR and AMR) development and deployment. This included a recommendation for the United Kingdom to "facilitate an ANT build programme in the UK (operation of a mature commercial advanced nuclear technology by 2030 and a demonstrator of a lower maturity technology by mid 2020s)" and the continuation of publicly funded nuclear fission R&D funding through the nuclear innovation program.

UKSMR funding

A consortium led by Rolls Royce has been developing a small modular PWR with an electrical output in the range 220–440 MWe. In July 2019 the UK Government committed £18 m of initial funding to support design development for this system, subject to final confirmation in late 2019 (UK Research and Innovation, 2019). Rolls Royce's design is a conventional PWR with the reactor vessel connected to external steam generators, pressurizer, primary pumps, and pipework. However, Rolls Royce emphasizes the modular construction and factory build with the aim of significantly reducing construction times and costs.

Fusion

Fusion research in the United Kingdom is predominantly led by the UK Atomic Energy Authority (UKAEA) at the Culham Centre for Fusion Energy (CCFE) (UK Atomic Energy Authority, 2020). CCFE hosts the Mega Amp Spherical Tokamak (MAST) and Joint European Torus (JET), and the national fusion technology platform is expected to open in 2020 (GOV.UK, 2018). JET is currently the most powerful magnetic fusion device in the world and is hosted on behalf of the EUROfusion consortium funded as part of EURATOM 2020; additional funding is provided by UK Engineering and Physical Sciences Research Council (EPSRC). Experimental programs involve the participation of scientists from across Europe and around the globe. Continuation of research at the JET facility has been confirmed with funding announced until the end of 2020 (GOV.UK, 2019a).

Enabling regulation

As part of the 2017 Clean Growth Strategy, HMG announced funding for the Office for Nuclear Regulation (ONR) and Environment Agency (EA) to support the development of capability and capacity to support and regulate the development of ANTs (ONR, 2018). To meet the aims of HMG, ONR has been carrying out a program of work to

- develop ONR capability and expertise in ANTs,
- provide regulatory advice for HMG policy development,
- undertake industry engagement,
- review existing regulation (SAPs, SyAPs, and TAGs) to ensure it is fit for purpose to regulate modular reactor technology,
- advise HMG on the regulatory approaches of AMR vendors as part of the AMR feasibility and development program.

These upskilling activities are expected to complete in 2020 and will support the ONR with future Generic Design Assessments of both SMRs and AMRs.

Construction of the UK nuclear reactor fleet has previously been publicly funded. The sole exception (as of December 2019) is Hinkley Point C in Somerset, which began construction in late 2018. This is funded through a "contracts for difference" financing model that has supported other low-carbon technology deployment (renewables) in the United Kingdom. During 2019 a consultation was held into the use of an alternative funding model, the so-called "regulated asset base" model that has been used by other large infrastructure projects in the United Kingdom. This consultation is due to report back in 2020 and may aid the raising of finance for future SMR and AMR deployment in the United Kingdom.

20.5 Future role of SMRs/AMRs in low-carbon energy generation

20.5.1 Role in a low-carbon economy

The UK Government is committed to net-zero emissions by 2050, and there are demands for this target date to be brought forward. The likelihood is that increasing future demand will be met largely by renewables and this will further accentuate the need to manage intermittency of supply and still balance demand and output. Large-scale storage systems will be required, but it is not clear if installing sufficient storage capacity will be feasible. The availability of large and small nuclear plants to provide a mix of baseload generation and load balancing may well be a key factor in future. Realizing the load-balancing potential of nuclear will probably demand technical innovations (that SMRs may be well suited to) and/or market mechanisms to value load balancing as a specific role in secure generation.

20.5.2 Domestic heating

Space and water heating accounts for about 35% of UK energy consumption (ETI, 2018) and is a major contributor to carbon emissions. Decarbonizing the space and water heating therefore needs to be a central component of carbon-neutral future energy scenarios. The Energy Technologies Institute (ETI) has highlighted that nuclear plants sited within 100 km of cities would be attractive heat sources. SMRs would be suited to meet the demand profile and siting requirements, though questions remain over the management of seasonal variations and whether the market price of low-carbon heat would be sufficient.

20.5.3 Grid balancing frequency response and inertia

Up to the mid-1990s, UK generation was dominated by coal, oil, and nuclear, all of which are synchronous generators in which there is a turbine and generator set synchronized at 50 Hz with all the other generators in the system. Effectively the rotational inertias of the generators are coupled together with all the individual contributions combined. This allows transient imbalances of load caused by, for example, a generator disconnecting from the grid to be met, for a short period of time, by the stored inertial energy and gives time for frequency response and backup generators to take up the shortfall on longer timescales. The current grid has nuclear, biomass, gas generators, and a small amount of coal to provide synchronous inertia. Low inertia can lead to instability, and the stability of the UK grid was highlighted in August 2019 with a large-scale load shedding event that occurred following the loss of two generators within a few minutes (although low inertia itself may not have been a contributory factor in this instance).

A recent study (Peakman et al., 2020) has considered the role of inertia in future UK grid scenarios and helps clarify whether flexibility from large nuclear plants and SMRs would be beneficial. The study considered a base scenario with high demand, high gas and nuclear installed capacity and high interconnector capacity. The base case is not constrained by inertia, and there is little demand for nuclear flexibility. In a low-carbon scenario where gas plants are replaced by nuclear plants, again there is sufficient inertia that nuclear flexibility is not needed for grid stability, but some nuclear plants are required to operate in load-follow mode and therefore are deloaded for some of the time. A low-carbon scenario where gas generation is replaced by variable renewable energy sources (VRES) highlights the value of an enhanced frequency response (EFR) capability in the grid through the use of energy storage systems such as batteries. In this scenario, EFR reduces the requirement for primary response and decreases the requirement for flexibility from the nuclear plants. This scenario not only benefits from being able to keep the nuclear plants fully loaded for a greater proportion of the time but also shows the value of the nuclear plants being able to operate flexibly down to a low percentage of their nominal full power output [this is designated the minimum stable factor (MSF)].

20.5.4 Industrial heat applications

A UK study of the potential application of nuclear plants to industrial heat supply has recently been carried out. The study reviewed industrial processes in the United Kingdom that could potentially be met by future nuclear plants, concluding that nuclear plants could play a role in meeting up to 80% of total demand for nonelectric industrial heat demand. The study paid specific attention to the working temperatures required in UK industry and concluded that the potential demand could be met by a mix of plants capable of relatively modest operating temperatures, specifically LWRs and many of the Gen-IV systems capable of operating up to around 500°C. The study highlighted that there would be a limited role for nuclear to supply processes heat in the range 500–1000°C and indeed that there are some industrial processes where working temperatures in excess of 1000°C are needed and that cannot currently be directly met by any reactor technology, even by very high-temperature gas reactors (VHTRs).

20.6 Conclusions

To conclude, since the first edition of this book in 2015, multiple recommendations provided to Her Majesty's Government (HMG) by the independent Nuclear Innovation and Research Advisory Board (NIRAB) have been enacted. This included a publicly funded program of nuclear fission R&D that ultimately became the nuclear innovation program (NIP) and UK reengagement internationally. This step change in funding for nuclear fission R&D in the United Kingdom has not been seen in a generation. A key focus of this is supporting the development of advanced nuclear technologies (ANTs) for future UK deployment. These ANTs are expected to be smaller than conventional nuclear power reactors and employ significant offsite (factory) fabrication, encompassing both small modular (water-cooled) and advanced modular reactors (gas, liquid metal, or molten salt coolants).

The nuclear innovation program is primarily targeted toward four key areas (advanced fuel cycle program, advanced manufacturing and materials, advanced reactor design, and nuclear facilities and strategic toolkit) that are needed to advance the development state of both SMRs and AMRs. Further to funding R&D, HMG is working with industry and research organizations to support the development, UK demonstration, and ultimately deployment of modular fission and fusion reactor technologies. This includes support UK regulatory upskilling and engagement with technology developers, as well as potential funding routes.

Appendix 20.1

Table 20.A1 BEIS SMR competition Phase 1: list of eligible participants.

1. Algometrics Ltd., advanced hybrid SMR	17. Hydromine Nuclear Energy S.a.r.l., LFR-AS-200
2. Amec Foster Wheeler	18. LeadCold Reactors, SEALER-UK
3. Advanced Reactor Concepts LLC, ARC-100 advanced SMR	19. Moltex Energy, stable salt reactor
4. Atkins Ltd.	20. Nuclear Advanced Manufacturing Research Centre
5. Bechtel Management Company Ltd., Generation mPower SMR	21. Nuclear Cogeneration Industrial Initiative, GEMINI
6. China National Nuclear Corporation, ACP100 SMR	22. National Nuclear Laboratory Ltd.
7. China National Nuclear Corporation, ACP100+ SMR	23. NuScale Power LLC, NuScale Power—A UK-US SMR Collaboration
8. Costain	24. Nuvia Ltd.
9. Critical Path Energy Ltd.	25. Penultimate Power UK Ltd.
10. EDF Energy PLC, EDF Energy SMR	26. Rolls Royce PLC, UK SMR
11. Empresarios Agrupados Internacional S.A.	27. Sainc Energy Ltd., innovative SMR
12. Ernst & Young LLP	28. Sheffield Forgemasters International Ltd.
13. Frazer-Nash Consultancy	29. Terrestrial Energy Inc., integral molten salt reactor
14. GE-Hitachi Nuclear Energy International LLC, PRISM SMR	30. Tokamak Energy Ltd., small modular fusion power
15. GF Nuclear Limited/Korea Atomic Energy Research Institute, The SMART	31. TWI Ltd. UK
	32. URENCO Ltd., U-Battery
16. Holtec International, SMR-160	33. Westinghouse Electric Company UK Ltd., Westinghouse SMR

Appendix 20.2

NIRAB recommendations

Recommendation 1. Government should, as a matter of urgency, work with private industry to define a roadmap for future nuclear new build to meet the clean energy and growth challenge out to 2050.

Recommendation 2. Government should continue to develop and implement energy policy to foster technologies that deliver significant impact through clean growth. This policy development should include an enabling framework for the manufacture, testing and evaluation, and commercial deployment of advanced nuclear technologies, which deliver economic growth and energy system value in decarbonization.

Recommendation 3. Government should invest with private industry to facilitate an advanced nuclear technologies build program in the United Kingdom (operation of a mature commercial advanced nuclear technology by 2030 and a demonstrator of a lower maturity technology by mid-2020s).

Recommendation 4. Government should commission without delay the remainder of the prioritized program recommended previously by NIRAB and deliver on the commitment to spend £180 million on nuclear innovation over this spending review period to 2021.

Recommendation 5. Between 2021 and 2026, to meet ambitions for nuclear to play a broader decarbonization and clean growth role, government should consider investment in a nuclear innovation program in the region of £1 billion and include support for the construction of advanced nuclear technology demonstrators. In return, government should expect to attract significant private sector leverage as a direct result of this support.

Recommendation 6. Government should ensure value for money by assigning a strategically focused expert delivery body to actively manage and integrate public investment in civil nuclear innovation through a nuclear innovation program.

Recommendation 7. New build 30% cost reduction by 2030—government support for new build should be contingent on the application of cost and risk reduction best practice, with full transparency on how industry intends to deliver these strategies and where innovation will increase productivity and result in cost savings.

Recommendation 8. Decommissioning cost savings of 20% by 2030—government should ensure that the waste management and decommissioning sector baseline cost estimates from which the cost-reduction targets are to be measured are transparent and publicly available and that the sector's strategy of how targets are to be met is understood and articulated such that it can work with industry to deliver the requisite cost savings through targeted innovation and productivity increases.

Recommendation 9. Government should identify the role it needs to play in de-risking civil nuclear projects, including innovative finance models, such that they are investible to the private sector.

Recommendation 10. Government should establish an effective international collaboration strategy that balances goals relating to diplomatic relations, trade ambitions, and research and development programs.

Recommendation 11. Government should review the impact of BREXIT and BREXATOM on UK nuclear research and innovation programs once the new arrangements are clear.

References

Committee on Climate Change, May 2019. Net Zero—The UK's Contribution to Stopping Global Warming. https://www.theccc.org.uk/publication/net-zero-the-uks-contribution-to-stopping-global-warming/.

UK Atomic Energy Authority, 2020. Culham Centre for Fusion Energy (CCFE). http://www.ccfe.ac.uk/.

ETI, November 2018. District Heat Networks in the UK: Potential, Barriers and Opportunities. https://d2umxnkyjne36n.cloudfront.net/insightReports/District-Heat-Networks-in-the-UK-Final.pdf?mtime=20181105145836.

Ferroni, P., Franceschini, F., Stansbury, C., Harkness, A., Grasso, G., 2017. Westinghouse demonstration lead fast reactor in the context of the US DOE advanced demonstration and test reactor options study. In: Proceedings of 2017 International Congress on Advances in Nuclear Power Plants (ICAPP2017), Japanp. 2573.

GOV.UK, March 17, 2016. Small Modular Reactors Competition Phase 1, Guidance. (URN 16D/045)https://www.gov.uk/government/publications/small-modular-reactors-competition-phase-one.

GOV.UK, 2019a. Future of JET Secured With New European Contract. Department for Business, Energy and Industrial Strategy. Marchhttps://www.gov.uk/government/news/future-of-jet-secured-with-new-european-contract.

GOV.UK, 2019b. Policy Paper Advanced Nuclear Technologies. Department for Business, Energy and Industrial Strategy. November 5https://www.gov.uk/government/publications/advanced-nuclear-technologies/advanced-nuclear-technologies.

GOV.UK, January 2020. Call for Advanced Manufacturing and Materials Phase 2B (Advanced Materials and Construction) Guideance Notes. Department for Business, Energy and Industrial Strategy.https://www.gov.uk/government/publications/nuclear-innovation-programme-advanced-manufacturing-and-materials-phase-2.

GOV.UK, 2018. National Fusion Technology Platform. https://www.gov.uk/government/news/ukaea-launches-national-fusion-technology-platform.

HM Government, March 2013. The UK's Nuclear Future, Industrial Strategy: Government and Industry in Partnership. (BIS/13/627)https://assets.publishing.service.gov.uk/government/uploads/system/uploads/attachment_data/file/168048/bis-13-627-nuclear-industrial-strategy-the-uks-nuclear-future.pdf.

HM Government, 2018. Industrial Strategy, Nuclear Sector Deal. https://assets.publishing.service.gov.uk/government/uploads/system/uploads/attachment_data/file/720405/Final_Version_BEIS_Nuclear_SD.PDF.

Jo, C.K., Chang, J., Venneri, F., Hawari, A., 2014. Preliminary core analysis of a micro modular reactor. In: Transactions of the Korean Nuclear Society Spring Meeting Jeju, Korea, May 29–30.

Levers, A., 2017. The UK national program R&D on digital nuclear reactor design. In: Transactions of the American Nuclear Society, Washington, DC, October 29–November 2, vol. 117.

Moltex Energy, January 2018. An Introduction to the Moltex Energy Technology Portfolio.

NIRAB, March 2016. NIRAB Annual Report 2015. (NIRAB-88-3)https://www.gov.uk/guidance/funding-for-nuclear-innovation.

NIRAB, April 2019. NIRAB Annual Report 2018/19—Clean Growth Through Innovation—The Need for Urgent Action (NIRAB-213-3). Nuclear Innovation and Research Advisory Board.www.nirab.org.uk.

ONR, 2018. ONR Corporate Plan 2018–2019. http://www.onr.org.uk/documents/2018/onr-corporate-plan-2018-2019.pdf.

Peakman, A., Merk, B., Hesketh, K., 2020. The potential of pressurised water reactors to provide flexible response in future electricity grids. Energies 13, 941.

Small Modular Reactors, n.d. List of Eligible Participants in Phase One of the BEIS SMR Competition https://assets.publishing.service.gov.uk/government/uploads/system/uploads/attachment_data/file/665376/List_of_Eligible_Participants_in_Phase_One_of_the_SMR_Competition.pdf

Sykes, A., et al., 2017. Compact fusion energy based on the spherical tokamak. Nucl. Fusion. 58, 016039.

U-Battery Local Modular Energy, July 24, 2018. U-Battery—Presentation to YGN: Intro to Nuclear New Build. Nuclear Institute.https://www.u-battery.com/media.

UK Research and Innovation, December 2019. Low-Cost Nuclear. https://www.ukri.org/innovation/industrial-strategy-challenge-fund/low-cost-nuclear/?_ga=2.90262810.1465065549.1579446689-126580992.1579446689.

Wade, D.C., Walters, L., 2011. ARC-100, a modular nuclear plant for emerging markets: safety strategy. In: PHYTRA 2—The Second International Conference on Physics and Technology of Reactors and Applications. Fez, Morocco, September 26–28, 2011, on CD-ROM, GMTR, Rabat, Morocco.

Wallenius, J., Qvist, S., Mickus, I., Bortot, S., Szakalos, P., Ejenstam, J., 2018. Design of SEALER, a very small lead-cooled reactor for commercial power production in off-grid applications. Nucl. Eng. Des. 338, 23–33.

Zhang, Z., Wu, Z., Wang, D., Xu, Y., Sun, Y., Li, F., Dong, Y., 2009. Current status and technical description of Chinese 2×250 MWth HTR-PM demonstration plant. Nucl. Eng. Des. 239, 1212–1219.

Small modular reactors (SMRs): The case of the United States of America

Gary Mays*
Oak Ridge National Laboratory, Oak Ridge, TN, United States

21.1 Introduction

This chapter presents an overview of the research and development (R&D) underway in the United States on small modular reactors (SMRs) including all reactor technologies. The responsibility for conducting SMR R&D within the US government is that of the US Department of Energy's Office of Nuclear Energy (DOE-NE). DOE-NE's programs have continued to include a spectrum of activities spanning support for near-term deployment with emphasis on (1) licensing support to secure design certification approval from the US Nuclear Regulatory Commission (NRC); (2) first-of-a-kind design, engineering, and construction; and (3) generic issues such as source terms, staffing requirements, siting, and economics. Longer-term initiatives are aimed at developing advanced technologies and conceptual designs for advanced SMRs (A-SMRs) employing coolants other than water. For FY2019, DOE-NE added an "advanced SMR" programmatic element focusing on more specific A-SMR R&D needs. DOE-NE's Licensing Technical Support (LTS) program has achieved considerable success in support of the development of a near-term light water SMR (LW-SMR) design. In particular the LTS program has supported the submission of a design certification application for the NuScale design that is under current review by NRC.

More recently, DOE-NE's R&D approach has included a targeted and cooperative "public-private" approach with the A-SMR nuclear industry in (1) making capabilities and facilities available at its DOE national labs to developers via its Gateway for Accelerated Innovation in Nuclear (GAIN) program and (2) issuing Funding Opportunity Announcements (FOAs) aimed at accelerating the development and ultimate commercial deployment of A-SMRs that include cost sharing arrangements between DOE-NE and the A-SMR industry. Furthermore, DOE-NE announced in August 2019 its plans to create a new center for the demonstration of advanced nuclear technologies, the National Reactor Innovation Center (NRIC). Of potential interest for SMR developers, overall is that one of the NRIC's key objectives includes a "proof-of-concept" focus. This objective is to support

---

* Retired.

conducting R&D to address technical feasibility issues such as material and fuel testing. This chapter will highlight these recent developments and opportunities.

Also, briefly summarized in this chapter are two DOE initiatives that may have some relevance for the development of both LW-SMRs and A-SMRs. DOE through its Advanced Research Projects Agency-Energy (ARPA-E) announced awards in June 2018 for advanced reactor technology development. This new ARPA-E initiative is entitled "Modeling-Enhanced Innovations Trailblazing Nuclear Energy Reinvigoration" (MEITNER). The overall goal is to develop innovative technologies that enable designs for lower-cost, safer, advanced nuclear reactors. The second related initiative is in response to the FY2019 National Defense Authority Act. DOE-NE is in the early stages of establishing an R&D program to support the development and demonstration of a "microreactor" for commercial and/or defense applications. DOE-NE is to prepare a report in response to this legislation in 2019 discussing the requirements for and key elements of a pilot program to provide energy resilience to either DoD or DOE facilities with the operation of one microreactor by December 2027.

Finally, comments are offered as to the direction and content of future A-SMR R&D in the United States as might be indicated by R&D initiatives currently underway as was done also in the first edition.

21.2 Near-term SMR activities in United States

DOE-NE's overall objective for SMRs in general is to conduct R&D and activities that support the accelerated deployment of SMRs near term and development of advanced SMR technologies employing innovative designs that offer options for enhanced economic performance, safety, and security and for other applications such as process heat and hybrid energy in addition to producing electric power. Obviously the time horizons for ultimate deployment for these advanced SMR technologies are beyond the early to mid-2020 time frame currently projected for the nearer-term LW-SMRs.

DOE-NE's program for supporting the near-term deployment of the LW-SMRs has been the "SMR Licensing Technical Support" (LTS) program, which spanned the FY2012-FY2017 time frame and is now essentially complete. Recent accomplishments from the LTS program are described later. Also, summary design descriptions are provided in this section for two of the more mature LW-SMR designs.

21.2.1 DOE-NE LTS program

The NuScale LW-SMR design has emerged as the principal design moving forward commercially. The LTS program's focus has supported, in part, the development of the design certification document (DCD) for licensing by the US Nuclear Regulatory Commission (NRC). The submission of the DCD was completed in January 2017, and it is currently undergoing review by NRC. NRC's current schedule shows that their final technical review is to be completed by the end of CY2019 with approval of the DCD by September 2020.

The LTS program received its last year of planned funding in FY2017 and as noted has continued to support the NuScale design certification application and furtherance of its first-of-a-kind engineering design, as well as two, site permitting and licensing projects. Additionally, the program has supported projects to resolve technical, economic, and regulatory barriers to deployment of advanced SMRs (US Department of Energy Small Modular Nuclear Reactors, 2012). Of note are the current plans (agreement is under development) among DOE, the cooperative utility Utah Associated Municipal Power Systems (UAMPS), and Idaho National Lab (INL) to deploy the first-of-a-kind NuScale plant at INL. This first deployment will be NuScale's 12-module plant design and is projected for commercial operation in 2026. Through testing and advanced modeling efforts, NuScale announced in 2018 that they were able to optimize designs for the NuScale fuel and helical-coil steam generator to increase the power output from each module from the original 50 MWe to the current 60 MWe. Furthermore, it is expected that one of the 12 NuScale modules will be dedicated for R&D at INL. One of the primary objectives is to interface this module as part of nuclear hybrid energy systems studies (US Department of Energy Nation's First SMR, n.d.). Also, one additional module may be designated to provide power to INL as part of a potential power purchase agreement. Overall, DOE has awarded some $63 M in matching funds to conduct site selection activities, secure the site, and support the DCA combined license application to NRC and advance the site-specific design (Landgon, 2019).

21.2.2 Additional DOE-NE LW-SMR support

While not explicitly a part of this program but related to the NuScale design, DOE awarded grants in August 2019 to NuScale to construct reactor simulators at three universities—Oregon State University, Texas A&M University, and the University of Idaho. These simulators will model the operations and performance of the NuScale LW-SMR and will include a virtual control room displaying parameters for simulating plant responses to transients and allowing input from operators. These simulators will potentially support R&D into human factors engineering, human-system interface design, advanced diagnostics, cybersecurity, and plant control room automation.

21.2.3 NuScale design description

The following design summary information on the NuScale SMR was excerpted from an IAEA 2018 report (IAEA Advanced Reactors Information System (ARIS), 2018) and with the permission of NuScale. The NuScale design is a small, modularized pressurized-water reactor composed of up to 12 nuclear power modules. Each module has a dedicated turbine and power conversion system for independent operation and can produce up to 60 MWe. Thus, for a fully deployed 12-module plant, the gross electrical output totals 720 MWe with a net output of 684 MWe. The 12-module plant is the reference plant size for design and licensing purposes. All 12 modules are operated from a single control room. The NuScale plant's scalability allows a potential customer to meet increasing energy demands and is designed for both electric power and process heat applications.

The NuScale integral design relies on proven light-water reactor (LWR) technology while eliminating the need for reactor coolant pumps, large diameter-sized piping, and other systems/components that are included in large LWRs. Additional design features include natural circulation cooling for coolant flow during all operational phases and passive safety systems. Table 21.1 presents information for key design parameters and their values for one reactor module.

Overall reactor description. The NuScale nuclear steam supply system (NSSS) includes a reactor core, helical-coil steam generators, and pressurizer, all of which are integrated inside of its reactor pressure vessel (RPV). The NSSS is enclosed in a nearly cylindrical containment vessel that sits in the reactor pool structure. Each of the power modules connects to its own dedicated turbine generator and corresponding balance-of-plant systems. Figs. 21.1a and 21.1b present a cutaway view of the NuScale SMR plant and reactor module, respectively.

Reactor core. The NuScale core for each module is made up of 37 fuel assemblies with UO_2 fuel at less than 4.95% enrichment. The fuel assembly is approximately half the height of a standard 17 × 17 LWR assembly and is supported by 24 guide tubes for control rod fingers along with a central tube of instrumentation. A burnable absorber, Gd_2O_3, is mixed homogeneously in the fuel at certain rod locations. A three-batch refueling approach is performed on a 24-month cycle.

The primary method of reactivity control is accomplished principally via soluble boron in the primary coolant and 16 control assemblies. The control rod absorber material is B_4C. Four of the rods are positioned symmetrically in the core and serve as the control group employed during normal plant operations. The 12 other rods are shutdown rods used during scram events and shutdown.

Table 21.1 NuScale module design summary information.

Design parameter	Value
Thermal/electrical capacity (MWt/MWe)	160/60
Coolant/moderator	Light water/light water
Core inlet/exit temperature (°C)	258/314
Reactor system pressure (MPa)	12.8
Primary cooling	Natural circulation
Fuel enrichment (%)	<4.95
Fuel type/assembly array	UO_2 pellet/17 × 17 square
Number of assemblies	37
Fuel cycle length (months)	24
Fuel burnup (GWd/ton)	>30
Principal reactivity control method	Control rod drive—boron
RPV height/diameter (m)	17.8/3.0
Design life (years)	60
Reactor module weight (metric tons)	~700

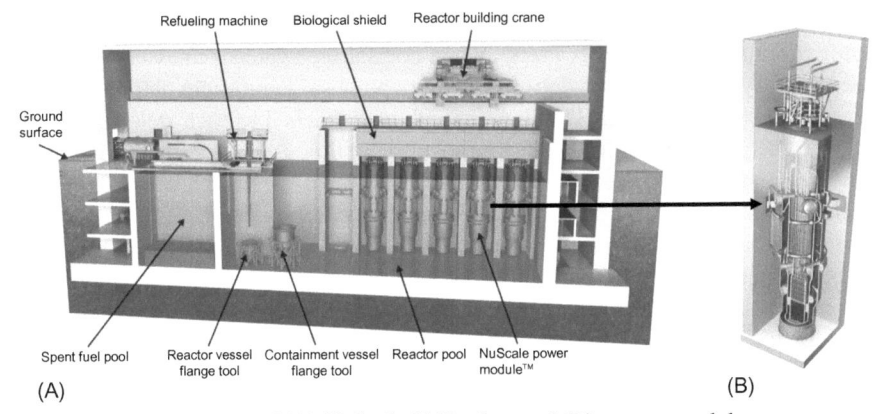

Fig. 21.1 Cut-away view of (A) NuScale SMR plant and (B) reactor module.

Major components. The NuScale RPV is a nearly cylindrical steel vessel designed for an operating pressure of 12.8 MPa and measuring ~17.8 m high and ~3 m in diameter. The lower part of the RPV has flanges located just above the core region providing access for refueling, while the RPV upper head supports the control rod drive mechanisms and has nozzles providing connections for reactor safety valves, reactor vent valves, and the secondary system steam piping.

For steam production, each module has two helical-coil steam generators located in the annular space between the hot leg riser and RPV inside diameter wall. Heat is transferred from the primary coolant to preheated feedwater as it circulates through the inside of the steam generator tubes. Fluid in the secondary side of the steam generator is then heated, converted to steam, and further superheated to generate dry steam for the turbine generator.

The pressurizer is located inside of the RPV and controls the reactor coolant system pressure at a constant pressure during reactor operations. To increase reactor coolant pressure, heaters located above the pressurizer baffle plate are turned on. Water spray from the reactor's chemical and volume control system is used to reduce pressure.

Safety systems. Key engineered safety features that provide inherent safety include the integral primary system configuration, high-pressure containment vessel, and passive heat removal systems. Each module includes redundant and independent features as described later.

The NuScale emergency core cooling system (ECCS) is made up of two independent reactor recirculation valves and three independent reactor vent valves. When a loss-of-coolant accident occurs inside containment, the ECCS returns coolant from the containment vessel to the reactor vessel, which ensures that the core remains covered with coolant and that decay heat is removed. Furthermore the ECCS system provides additional decay heat removal in the unlikely circumstances involving a loss of feedwater flow in conjunction with the loss of both trains of the decay heat removal system. The ECCS is also designed to remove heat and limit containment pressure via steam condensation on and convective heat transfer to the inside surface of the containment vessel.

The decay heat removal system (DHRS) is designed to provide secondary side reactor cooling for non-LOCA events when normal feedwater is unavailable. It is a closed-loop system removing decay heat by two-phase, natural circulation cooling. There are two trains of DHRS equipment with one attached to each steam generator loop with each train having 100% heat removal capability. As well, each train includes a passive condenser immersed in the reactor pool.

The containment vessel performs the following safety functions: (1) contains the release of radioactivity following postulated accidents, (2) protects the RPV and the equipment located inside from potential hazards, and (3) provides heat rejection to the reactor pool after an ECCS actuation. Each containment vessel is a steel cylinder with an outside diameter of ~4.5 m and overall height of ~23.1 m and contains the RPV, control rod drive mechanisms, and associated NSSS piping and components. The containment vessel also provides a passive heat sink for containment heat removal under LOCA conditions since it is immersed in the reactor pool.

Additional design information. Each module is operated independently of the other modules. The design employs a fully digital control system using field programmable gate arrays. The main control room is located below grade in the control building, which is located next to the reactor building. The projected plant footprint is 140,000 m^2.

Licensing status. The following key milestones are compiled from both NRC's and NuScale's websites as of March 2020:

• Design certification application (DCA) submitted (actual)	Dec 2016
• NRC acceptance of DCA (actual)	Mar 2017
• Phase 4—NRC issues advanced safety report—no open items (actual)	Dec 2019
• NRC completion of DCA process (target)	Jan 2021
• First module to go into operation (target)	2026

21.2.4 Holtec SMR-160 design description

The following design summary information for the Holtec SMR-160 is based upon an IAEA 2018 report (IAEA Advanced Reactors Information System (ARIS), 2018) as updated with information provided by Holtec and with the permission of Holtec. The Holtec design is an advanced PWR-type SMR rated at 525 MWt and 160 MWe. It employs both passive safety systems and natural circulation cooling for the primary loop. The design makes use of fewer major components including valves, pumps, heat exchangers, instrumentation, and control loops. Collectively, these design features serve to simplify the design and reduce capital costs. Holtec intends to implement a modular construction plan for fabricating those large components prior to their being shipped directly to the site. The plant is designed for electrical production with optional cogeneration equipment for such applications as district heating, seawater desalination, and hydrogen generation. Also the SMR-160 is configurable for siting at "dry locations" via use of Holtec's air-cooled condenser technology. Table 21.2 presents information for key design parameters and their values.

Table 21.2 Holtec SMR-160 summary information.

Design parameter	Value
Thermal/electrical capacity (MWt/MWe)	525/160
Coolant/moderator	Light water/light water
Core inlet/exit temperature (°C)	229/321
Reactor system pressure (MPa)	15.5
Primary cooling	Natural circulation
Fuel enrichment (%)	4.95 maximum
Fuel type/assembly array	UO_2 pellet/square array
Number of assemblies	57
Fuel cycle length (months)	24 (flexible)
Fuel burnup (GWd/ton)	50 (maximum and initial design)
Principal reactivity control method	Control rods and soluble boron
RPV height/diameter (m)	15/3
Design life (years)	80

Overall reactor description. The Holtec PWR nuclear steam supply system (NSSS) includes the reactor coolant system (RCS) that consists of the reactor pressure vessel (RPV) and steam generator (SG) in an offset configuration with an integrated pressurizer flanged to the top of the SG. The RPV and SG are connected by a single connection that houses both the hot and cold legs. Fig. 21.2 presents the major components of the SMR-160 RCS. This offset arrangement supports easy access to the core without having to move the RPV or SG during refueling. With high SG superheat, neither a moisture separator reheater nor multiple trains of feedwater heaters are needed. The secondary loop includes one stage of feedwater heating and eliminates high-pressure turbine stages.

Reactor core. The SMR-160 core makes use of currently licensed LWR technology and conventional reactor materials. It is designed to ensure a large margin relative to thermal-mechanical fuel limits. The core contains standard length 17×17 PWR fuel assemblies currently available from commercial suppliers along with typical control rod assemblies. The reactor vessel internals support the following: the reactor core, control rod assemblies, and control rod drive shafts. The heavy-walled RPV contains the fuel, reactor vessel internals, and coolant. A standard reload fuel reload shuffle is used, and the core is designed for a nominal 2-year cycle. Depending on the requirements, there is flexibility for shorter cycles. Burnable absorbers integral to the fuel provide long-term reactivity control. The burnable absorbers allow optimization of the 3D power distributions, cold shutdown margin, and hot excess reactivity. Short-term reactivity changes are controlled by adjusting soluble boron and movements of control rod assemblies (CRAs). CRAs are positioned by control rod drive mechanisms (CRDMs) based on current electromechanical technology. The CRDMs are located outside the reactor coolant system on the RPV upper head.

Major components. The SMR-160 has no reactor coolant pumps since natural circulation cooling is used. The three major components include the RPV, steam generator, and pressurizer.

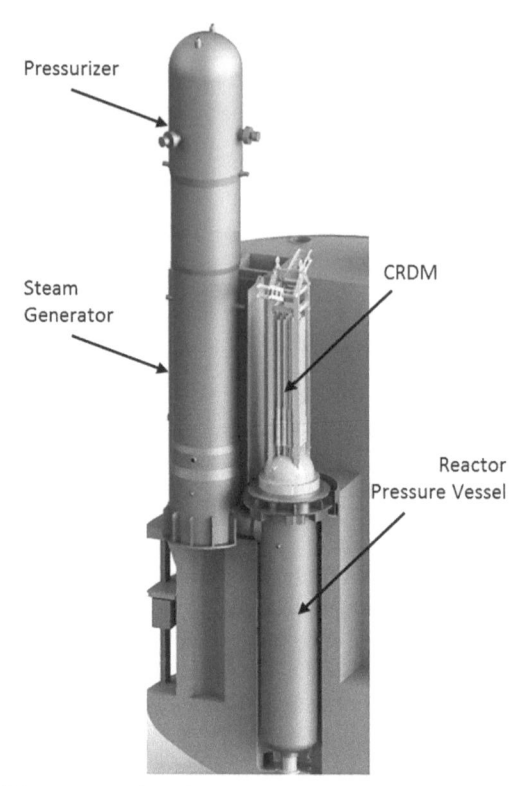

Fig. 21.2 SMR-160 Reactor Coolant System—Major Components.

The SMR-160 RPV is an ASME Section III Class 1 qualified thick-walled cylindrical vessel with an integrally welded bottom head and a removable top head. The upper part of the RPV shell has a tapered hub flange that is bolted to a similar flange welded to the top head. The aforementioned offset configuration of the SG and RPV allows the use of traditional external CRDMs and simplified refueling operations relative to typical integral PWR designs. The reactor internal structures are supported from the bottom of the vessel and are completely replaceable. There are no penetrations to the RPV below the elevation of the safety injection lines.

The SMR-160 SG is a single, vertically oriented, once-through straight tube design where the reactor coolant flows inside the thermally treated Inconel 690 tubes. Using straight tubes ensures easy access for in-service inspection. Subcooled feedwater is used to produce superheated steam on the shell side. The SG includes a large inventory of secondary water on the shell side that provides substantial margin to dry out.

The pressurizer is integral to the SG and employs heaters and cold water spray nozzles to perform the function of a typical pressurizer. Integration of the pressurizer with the SG eliminates significant primary piping and typical supporting structures normally connecting the primary external loop of a PWR to an external pressurizer and reactor coolant pumps. Also the large, relative size of the pressurizer eliminates the need for power-operated relief valves.

Safety systems. Overall the SMR-160 safety basis incorporates a defense in-depth approach using multiple and diverse pathways for rejecting decay heat. All safety systems are protected within a robust containment enclosure structure, thus protecting them from external threats. All makeup water needed for a postulated loss of coolant accident (LOCA) is inside containment making the containment isolable during a LOCA reducing possible dose to the public and effects on the environment. Another large inventory of water is contained within a reservoir between the containment enclosure structure and the containment vessel providing long-term postaccident coping times and then allows the structure to transition to air cooling for decay heat removal for essentially an unlimited coping time period after a design basis accident.

The SMR-160 engineered safety features rely on passive and redundant safety systems that also operate by natural circulation to ensure that the plant is maintained in a safe configuration. The passive safety systems ensure safe shutdown can be maintained and decay heat removal occurs for an unlimited period without the need for power, makeup water, or operator actions. If available, active nonsafety systems can be used by operators to mitigate events and preclude the need for using the engineered safety features.

The passive core cooling system (PCCS) provides emergency core cooling and makeup to the RCS during postulated accidents. It utilizes natural circulation for core cooling and compressed gas expansion for core makeup and cooling without using active components such as pumps. The PCCS consists of four major subsystems:

- Primary decay heat removal system—directly cools the primary coolant by re-routing the reactor coolant through a heat exchanger and rejecting heat to a second water loop that rejects heat to the large, annular reservoir around the containment.
- Secondary decay heat removal system—provides an alternative and diverse passive means to reject decay heat via a closed-loop system that relies on buoyancy-driven flow to route steam from the SG to a heat exchanger in the annular reservoir. Here the steam then condenses and rejects its latent heat with the condensate returned to the shell side of the SG.
- Automatic depressurization system—safely "lets down" the RCS pressure to allow staged safety injection by the primary core makeup system and permit long-term recirculation within the containment vessel.
- Passive core makeup water system—as just noted provides source of water for safety injection.

The containment system for the SMR-160 consists of a steel containment structure that is enclosed within a reinforced concrete containment enclosure structure (CES). The CES provides shielding and protection from external events. The CES walls are constructed with robust steel-concrete modules designed to protect against impacts from large commercial aircraft and other external hazards. Furthermore the containment system is partially embedded below grade (half of the total height) to maximize protection against external hazards and dampens seismic effects on critical components. The SMR-160 containment system also acts as a large passive heat exchanger.

The passive containment heat removal (PCHR) system passively cools the containment volume. During a postulated high energy release, heat from steam is rejected to the inner wall of the containment and condenses as the heat is transported to the water reservoir between the enclosure and the containment vessel. The large heat transfer area and high heat conductance of the metal containment wall results in near-instantaneous heat rejection to the annular reservoir, which in turns rejects heat to the environment. The large inventory of water in the annular reservoir is sufficient to extract energy from the containment for over 3 months without replenishment, thus enabling the PCHR to transition to air cooling as previously described. The containment houses the reactor and spent fuel pool, which is a large open volume of water designed to passively reject decay heat through surface evaporation to the containment volume in the event of the loss of active cooling. This design supports the safe and indefinite cooling of spent fuel, thus protecting against spent fuel-related events.

Additional design information. The SMR-160 uses the Mitsubishi Electric Total Advanced Controller (MELTAC) platform for its instrumentation and control/human system interface. MELTAC provides unique nuclear specific I/O and configuration flexibility to perform all nuclear safety and nonsafety functions using the same digital platform. The reactor auxiliary building, in addition to housing many of the plant auxiliary systems, is also designed to process spent fuel for interim on-site storage with Holtec's International HI-STORM UMAX modules (an underground dry cast storage technology). The plant's electric power system is designed to permit isolated operation in "island mode" and start-up operations independent of the grid, that is, "black start." The plant footprint is 20,500 m^2.

Development status. Holtec has entered into preapplication discussions with NRC and other international regulators. Key development milestones as of March 2020 include the following:

• Completed conceptual design of SMR-160	2015
• Mitsuibishi joins Holtec's SMR-160 development team	Jun 2017
• Holtec and SNC-Lavalin agree to develop and deploy SMR-160	Jul 2017
• Holtec and GE Hitachi to collaborate on advancing SMR-160	Feb 2018
• Energoatom and Holtec announce plans to build SMR-160 in Ukraine	Mar 2018
• Holtec, Energoatom, and SSTC establish consortium to advance deployment of SMR-160 across the Ukraine	Jun 2019
• Targeted deployment of SMR-160 in Ukraine	2026–30

21.3 Longer-term activities: US Department of Energy Office of Nuclear Energy (DOE-NE) small modular reactor (SMR) R&D program

R&D applicable for A-SMRs as conducted under DOE-NE now and moving forward covers both basic and applied R&D and technology development and demonstration.

The one new facet of DOE-NE support is providing support to potentially enable the testing and demonstration of these advanced SMR concepts. Work being supported as part of this more holistic approach for developing advanced SMRs is found in several DOE initiatives as described later.

The basic R&D and applied R&D aspects of DOE-NE's A-SMR program generally fall under its advanced reactor technology (ART) program. The intent of the ART R&D program is to conduct R&D on non-LWR designed SMRs to address their respective technology challenges leading ultimately to the design, licensing, and operation of enhanced SMR designs in the future. This class of reactors, A-SMRs, is defined within the context of the ART program as non-LW-SMR designs. The non-LWR coolants for which R&D is underway include such coolants as liquid metals, helium, and liquid salts including both fuel and coolant salts. DOE-NE's objective is to conduct impactful R&D to accelerate the development of the technologies for these innovative concepts that obviously are candidates for deployment beyond the more near-term concepts. ART program information follows in the next section.

21.3.1 DOE-NE ART R&D program

The current construct of DOE-NE's ART R&D program most relevant to A-SMRs includes three elements: (1) reactor technology-specific R&D projects, (2) cross-cutting technology development for classes of reactors, and (3) collaborative activities with A-SMR and microreactor developers.

The reactor technology-specific element of ART conducts R&D on the following reactor-specific technologies: (1) fast reactors, (2) gas-cooled reactors, (3) molten salt reactors, and (4) microreactors. For these first three reactor types, work is performed in the following subcategories:

- technology development and demonstration
- methods, modeling, and simulation
- fuel development
- advanced materials development
- chemistry (molten salt reactors only)

Given the special design aspects and potential applications for microreactors, the R&D structure for this reactor type is somewhat different. Its subcategories include the following:

- systems integration and analysis
- demonstration support capabilities
- technology maturation
- licensing-regulatory

The cross-cutting programmatic element is comparable with the various research pathways as described in the first edition of the handbook, that is, materials, instrumentation, and controls. The cross-cutting area of most interest for this the purposes of this handbook is entitled nuclear reactor technologies as organized by reactor classes including work for (1) advanced SMRs, (2) high-temperature reactors, and (3) advanced reactor concepts. For FY2019 the advanced SMRs was a new category.

The collaborative programmatic element represents the public-private part of DOE-NE's overall approach for development and deployment of advanced reactors including A-SMRs and microreactors. It includes the GAIN program, special FOAs for advanced reactor technology development, and the NRIC for demonstrating advanced technologies, all noted in the Introduction to this chapter and for which more detailed information is included in the sections that follow. All three of these involve some type of cost-share arrangement aimed at providing early-stage design-related technical assistance. The results from this support are expected to be widely applicable and employed by A-SMR vendors for the purpose of accelerating the development of their technologies.

Summary information is included later in this chapter on (1) the completion of DOE-NE's A-SMR concept evaluations that were highlighted in the first edition and have since been completed and documented, the Gateway for Accelerated Innovation in Nuclear (GAIN) program, and (3) those Nuclear Energy University Program awards related to A-SMRs.

21.3.2 A-SMR development related R&D program

As noted in the Introduction, one of DOE-NE's key approaches in developing A-SMR technologies and advancing A-SMR conceptual designs is focusing on public-private collaborations. In December 2017, DOE-NE issued "a multiyear cost-shared funding opportunity to support innovative, domestic nuclear industry-driven concepts that have high potential to improve the overall economic outlook for nuclear power in the United States. This funding opportunity is to enable the development of existing, new, and next-generation reactor designs, including SMR technologies. The scope of the funding opportunity is very broad and solicits activities involved in finalizing the most mature SMR designs; developing manufacturing capabilities and techniques to improve cost and efficiency of nuclear builds; developing plant structures, systems, components, and control systems; addressing regulatory issues; and other technical needs identified by industry. The funding opportunity is to provide awards sized and tailored to address a range of technical and regulatory issues impeding the progress of advanced reactor development (US Department of Energy Advanced Small Modular Reactors (SMRs), n.d.)." DOE-NE plans for FY18 were to issue some $30 M in awards while encouraging the US companies to partner with other US federal agencies, public and private laboratories, institutions of higher education, and other domestic entities to take advantage of expertise required to successfully develop A-SMR technologies.

The funding opportunity announcement (US Department of Energy, 2018) is entitled "U.S. Industry Opportunities for Advanced Nuclear Technology Development (Funding Opportunity Number: DE-FOA-0001817)" and describes the type of activities and projects in general that are anticipated to be funded. These include the following:

- advanced nuclear reactor designs, *including small modular reactors of various technology types*;
- engineering, analyses, and experimentation that would address first-of-a-kind reactor design certification, and licensing issues;
- sensors, instrumentation, and control systems;
- plant auxiliary and support systems;
- operational inspection and monitoring capabilities;
- modeling and simultion of various elements of plant life cycle;
- procedures, processes, and methodologies that can impact operational efficiencies;
- integration of nuclear energy into microgrid, nonelectric and/or hybrid applications;
- other components, systems, processes, or capabilities, including dynamic convection technologies, that could result in performance and economic improvements in advanced nuclear reactor designs;
- efforts to address regulatory and licensing issues with the NRC.

This FOA provides three separate funding pathways to support the aforementioned activities: (1) first of a kind nuclear demonstration readiness projects, (2) advanced reactor development projects, and (3) regulatory assistance grants and technology development opportunities.

The first funding pathway describes the type of activities and projects in general that are anticipated to be funded. These include the following:

- any new technology that has the ability to improve operations and extend the life of the existing fleet of domestic reactors;
- completion of certification and licensing activities for advanced reactor designs;
- design development, testing, analyses, first-of-a-kind engineering, and efforts leading to design finalization;
- development of fabrication capabilities, supply chains, procurement tasks, and other efforts that assure the ability to economically manufacture and construct advanced reactors;
- efforts involved in identifying, characterizing, permitting, and licensing sites associated with the proposed advanced reactor projects.

The second funding pathway is quite broad to support a range of proposals that offer the best opportunity to improve both the capabilities and commercialization of advanced reactor designs and technologies. Another principal objective under this pathway is to further accelerate advances in the innovation and competitiveness of a wide range of the US nuclear reactor technologies and designs.

There are two elements of the third funding pathway: (a) regulatory assistance grants and (b) technology development vouchers. DOE-NE will issue cost-shared grants for potential applicants to interact with NRC to address and resolve regulatory issues associated with designs, review topical reports and/or papers expected to support applications, and other activities aimed at obtaining license certifications and licensing approvals. In providing funding under this pathway, DOE-NE recognizes the importance of early interactions with the NRC is a key element in the ultimate deployment of A-SMRs. This pathway serves to facilitate both NRC's understanding and regulatory acceptance of these advanced technologies. The regulatory areas targeted in the FOA are the following:

- analysis, testing, computation, or engineering required to develop regulatory positions that will be presented to the NRC in reports, papers, meetings, applications, or other media;

- development of regulatory documentation, including topical reports and papers, permits, applications, and any other NRC deliverable;
- costs to support interactions with the NRC, including informational meetings, technical and topical report reviews, and pre- and postapplications reviews;
- other proposed activities that show promise for resolving regulatory issues.

The technology development voucher element of this pathway is structured to provide support comparable with that as included in small business voucher pilot program in 2016 within DOE-NE's Gateway for Accelerated Innovation in Nuclear (GAIN) as described in Section 21.5 in this chapter. The focus here is to expand what has been under GAIN to provide similarly-sized award vouchers to all types of US companies, regardless of size, and assist applicants seeking access to DOE national laboratories expertise and capabilities.

To date, DOE-NE has made six announcements that provide funding under this FOA. As indicated in the FOA, a broad range of technical areas will be funded including some advanced technology development awards for LWRs. Table 21.3 presents a partial listing of projects and brief technical descriptions for those successful applications that are considered most directly related to the A-SMRs and A-SMR technology development. Further information on additional awards not included in Table 21.3 can be found at DOE-NE's website by reviewing their news releases based upon the dates of award as indicated in Table 21.3.

21.4 A-SMR concept evaluations

DOE initiated preliminary work via its labs on concept studies for A-SMR technologies at the time of publication of the first edition of this handbook. These studies have since been completed and documented in a report published in January 2017 (Petti et al., 2017). This multi-DOE national lab supported effort described the advantages that advanced reactors provide overall, established objectives for conducting these evaluations, created metrics and scoring criteria, and developed point designs for the four concepts that were assessed. While not necessarily modular designs per se, the results from these studies certainly have relevance for addressing technology challenges for A-SMR designs. The four strategic objectives that served to focus these studies spanned a range of key advanced reactor nuclear energy missions and needs. The first three were focused on potential demonstration reactor options and the fourth on irradiation test reactor options. These four strategic objectives as noted in the report were the following:

- *Deploy a high-temperature process heat application (e.g., synfuels production) for industrial applications and electricity demonstration using an advanced reactor system to illustrate the potential that nuclear energy has in reducing the carbon footprint in the U.S. industrial sector.*
- *Demonstrate actinide management to extend natural resource utilization and reduce the burden of nuclear waste for future generations.*
- *Deploy an engineering demonstration reactor for a less-mature reactor technology with the goal of increasing the technology readiness level (TRL) of the overall system for the longer term.*

Table 21.3 A-SMR related R&D projects awarded under the US Industry Opportunities for Advanced Nuclear Technology Development FOA.

Issue date	Pathway	Title/award amount	Brief project description
September 10, 2019	Regulatory assistance	Advanced fuel qualification methodology report for TerraPower Traveling wave reactor (TWR) Total award: $0.98 M	Develop an advanced fuel qualification Methodology Report for TerraPower's TWR fuel and to deliver report to NRC for review and approval. The report will be developed for a metal-fueled sodium fast reactor. The processes and methodologies described will be generally applicable to other fuel types
March 27, 2019	FOAK demonstration	eVinci (TM) Microreactor Nuclear Demonstration Unit Readiness Project Total award: $28.5 M	Westinghouse and its team to prepare for the nuclear demonstration unit (NDU) of the eVinci microreactor through design, analysis, testing, and licensing to manufacture site and test the NDU by 2022
March 27, 2019	Advanced reactor development	Passive radio frequency tags and sensors for process monitoring in advanced reactors Total award: $1.25 M	Develop and commercialize next-generation specialized wireless sensing and monitoring passive and semipassive tags integrated with sensors for the remote process monitoring of advanced reactors
March 27, 2019	Advanced reactor development	Modeling and simulation development pathways to accelerate Kairos Power Fluoride-Cooled High-Temperature Reactor (KP-FHR) Licensing Total award: $10.1 M	Kairos Power, LLC will bring forward in the schedule critical advanced modeling and simulation capability through the DOE NEAMS program (Nuclear Energy Advanced Modeling and Simulation)
March 27, 2019	Regulatory assistance	Technology Preapplication Licensing Report on development of a mechanistic source term methodology for the KP-FHR Total award: $0.7 M	Develop a mechanistic source term for the KP-FHR design including consideration of radionuclides generated and transported in the fuel particle and the barriers to release for licensing basis event analyses

Continued

Table 21.3 Continued

Issue date	Pathway	Title/award amount	Brief project description
November 13, 2018	FOAK demonstration	Integral and separate effects test program for the investigation and validation of passive safety system performance of SMRs-Phase 1 Total award: $3.2 M	Develop a uniquely configurable set of testing platforms to demonstrate SMR passive safety system performance, accelerate the SMR-160 and other SMR designs to market, and help license these designs with NRC and international regulators
November 13, 2018	Regulatory assistance	TEUSA-USNRC Prelicensing activities for the integral molten salt reactor (IMSR) Total award: $0.6 M	Terrestrial Energy USA will conduct preapplication interactions with NRC to advance the progress of licensing the IMSR
July 10, 2018	FOAK demonstration	Calendar year 2018 activities for Phase 2 of NuScale small modular reactor project—NuScale Power Total award: $14.1 M	NuScale Power will build on Phase 1 project activities to advance licensing and design maturity to meet a commercial operation date of 2026 for first NuScale plant. Specific project scope funded represents CY2018 activities associated with the efforts (Phase 2) to bring NuScale design to market. Specific project activities include completion of the independent verification and validation licensing report, completion of the reactor building design optimization, and conduct of level sensor prototypic testing
July 10, 2018	Advanced reactor development	Conceptual engineering for a small modular reactor power plant based on lead-bismuth fast reactor (LBFR) technology Total award: $0.5 M	Columbia Basin Consulting Group is to develop a preconceptual design and preliminary cost estimate for a lead-bismuth small modular reactor

July 10, 2018	Advanced reactor development	Experimental verification of postaccident integral pressurized water reactor (iPWR) aerosol behavior, Phase 3 Total award: $1.4 M	Electric Power Research Institute will further improve models used to estimate postaccident radionuclide releases from iPWRs with a goal of reducing regulatory requirements for emergency planning zones
July 10, 2018	Advanced reactor development	Fluorination of lithium fluoride-beryllium fluoride (FLiBe) molten salt processing Total award: $2.6 M	Flibe Energy teamed with Pacific Northwest National Laboratory to examine the use of nitrogen trifluoride as an agent to remove uranium from a molten salt fuel mixture as a preliminary step for the removal of fission products
April 27, 2018	FOAK demonstration	Design and license application development for TRISO-X: a cross-cutting, high-assay low-enriched uranium fuel fabrication facility Total award: $8.98 M	X Energy, LLC is to develop the design and license application for a fuel fabrication facility capable of handling high-assay, low-enriched uranium and production of US developed uranium oxycarbide (UCO) TRIstructural ISOtropic (TRISO) particle-based fuel elements required for the future fleet of advanced reactors
April 27, 2018	FOAK demonstration	Phase 1 NuScale small modular reactor FOAK Nuclear Demonstration Readiness Project—NuScale Power Total award: $80.0 M	NuScale Power is to conduct design finalization activities and ensure supply chain readiness to meet a commercial operation date of 2026 for the first NuScale plant
April 27, 2018	Advanced reactor development	Combining multiscale modeling with microcapsule irradiation to expedite advanced	General Atomics is to combine advances made in microstructurally informed fuel performance modeling and simulation tools with a new

Continued

Table 21.3 Continued

Issue date	Pathway	Title/award amount	Brief project description
		fuels deployment Total award: $2.6 M	microcapsule irradiation capability that can substantially reduce the schedule and cost burden associated with qualifying new fuel systems for commercial deployment
April 27, 2018	Advanced reactor development	Modeling and optimization of flow and heat transfer in reactor components for Molten Chloride Salt Fast Reactor Application Total award: $3.2 M	Elysium Industries USA is to develop the computational fluid dynamics models needed to simulate and optimize the flows of chloride molten salt fuel in a reactor vessel and heat exchangers for their molten chloride salt fast reactor design
April 27, 2018	Advanced reactor development	Establishment of an integrated advanced manufacturing and data science-driven paradigm for advanced reactor systems Total award: $9.8 M	BWXT Nuclear Energy, Inc. is to develop the ability to implement additive materials manufacturing to the fabrication process for nuclear components and subcomponents that will yield acceptable material structure and strength that can be accepted by the national code organizations and the regulator
April 27, 2018	Regulatory assistance	Preapplication license review of silicon carbide composite clad uranium carbide fuel for long-life gas-cooled fast reactor cores Total award: $0.47 M	General Atomics will engage NRC to execute a prelicensing review of a silicon carbide composite clad uranium carbide fuel system for use in a gas-cooled fast reactor long-life core

- *Provide an irradiation test reactor to support development and qualification of fuels, materials, and other important components/items (e.g., control rods, instrumentation) of both thermal and fast neutron-based Generation-IV (Gen-IV) advanced reactor systems.*

The four concepts for which demonstration point designs were developed and that were subsequently assessed against the criteria included the following:

1. AREVA Steam Cycle-High Temperature Gas Reactor (SC-HTGR)
2. General Electric Hitachi (GEH) Sodium-Cooled Fast Reactor (PRISM SFR)
3. Oak Ridge National Laboratory Fluoride-Salt-Cooled High-Temperature Demonstration Reactor (FHR-DR)
4. Westinghouse lead-cooled fast reactor (DLFR)

The design parameters and features for each of these four reactor concepts are presented in Table 21.4.

Table 21.4 Design parameters and features for four advanced reactor concepts evaluated by DOE-NE.

Parameters/concepts	SC-HTGR	PRISM SFR	FHR DR	DLFR
Rated power	625 MWth / 272 MWe	471 MWth / 165 MWe	100 MWth / 42 MWe	500 MWth / 210 MWe
Thermal efficiency	43.5%	35%	42%	42%
Power conversion system	Conventional Rankine steam	Conventional Rankine steam	Open-air Brayton	Superheated Rankine steam
Core inlet/outlet temperature	325°C/750°C	352°C/500°C	660°C/700°C	390°C/510°C
Primary pressure	6 MPa	~0.1 MPa	~0.1 MPa	~0.1 MPa
Cycle length	18 months	18 months	18 months	18 months
Reactor vessel	SA-508/533	SS-316	Alloy 800H lined with alloy N	SS-316
Primary coolant	Helium	Sodium	FLiBE (99.995% Li-7)	Lead
Moderator	Graphite	N/A	Graphite	N/A
Fuel type	TRISO-coated UCO particles	U-Zr in HT-9 cladding	TRISO-coated UCO particles	UO2 in alumina-coated D-9 cladding
Fuel enrichment	15.5%	<20%	15.5%	17.5% (inner) 19.9% (outer)

The following high-level results for these concept evaluations are excerpted directly from Petti et al. (2017):

Strategic Objective 1: Process heat	The assessment for this objective identified two discriminators between point designs: suitability and prototypicality for a variety of high-temperature applications and commercially relevant scale of the demonstration. The higher maturity options were favored, with the most promising option being the HTGR because of its high outlet temperature ($>700°C$), flexibility for energy applications, and its state of development. Of the point designs in the study, the modular SC-HTGR design by AREVA best supports generating electricity using higher efficiency Rankine cycles in addition to process heat (e.g., synfuels) demonstration
Strategic Objective 2: Resource utilization and waste reduction	The assessment for this objective identified two discriminators between point designs: suitability and the prototypicality for demonstrating high natural resource utilization and commercially relevant scale of the demonstration. The higher maturity options were favored, with the most promising option for resource utilization and waste reduction being the SFR because of its ability to efficiently convert uranium and utilize recycled fuel and its state of development. Of the point designs in the study, the SFR proposed by GEH best supports the extension of natural resources and reduction of the nuclear waste burden, as well as fulfilling the fundamental mission of efficient and reliable electricity production
Strategic Objective 3: Engineering demonstration reactor—less-mature reactor technology	The ability of engineering demonstration reactors to advance the TRL of a less-mature reactor technology was assessed. Both the FHR and LFR are both low-maturity technologies that require significant research, development, and demonstration to significantly advance the TRL toward eventual commercialization. Significant discriminators between the FHR and LFR point designs related to readiness for an engineering demonstration

Strategic Objective 4: Irradiation test reactors to support development of advanced reactor systems	were not identified in the assessment. However, the distinct approaches that the two design teams took to advancing the TRL were used to illuminate potential pathways to reactor technology development. The FHR and LFR point designs, as engineering demonstration reactors, scored essentially the same in their ability to increase the technology readiness of these less-mature technologies Mature technologies are required for reliable operations of an irradiation test reactor. Thus only SFR and a HTGR were examined to determine their ability to provide neutron irradiation services to support nuclear fuels and materials testing for advanced reactor systems. The assessment for the irradiation test reactor objective identified two discriminators between point designs: fast flux levels and irradiation volumes. Overall the SFR irradiation test reactor design is the preferred option, because it can provide very high fast neutron flux and high-thermal neutron flux in moderated zones to meet many of the needs of both the fast and thermal reactor developers

Figs. 21.3–21.6 show the point designs for each of these four concepts as presented in the evaluation report (Petti et al., 2017).

21.5 DOE-NE GAIN program and A-SMRs

In November 2015, DOE announced the formation of its Gateway for Accelerated Innovation in Nuclear (GAIN) program. Given the considerable interest, investment, and number of companies developing A-SMR concepts in the private sector, DOE's objective in creating GAIN was to connect these innovators in the private sector with the research and development capabilities of its national laboratories. GAIN now serves as the single point of contact to a wide range of capabilities including facilities, data, and staff across the DOE system. In bringing the private sector developers together with these capabilities at the laboratories, DOE's expectation is that these collaborations will serve to facilitate the commercialization of these advanced concepts. Thus GAIN represents the process or framework for establishing these private-public partnerships. The overall capability areas that are accessible via GAIN include the following:

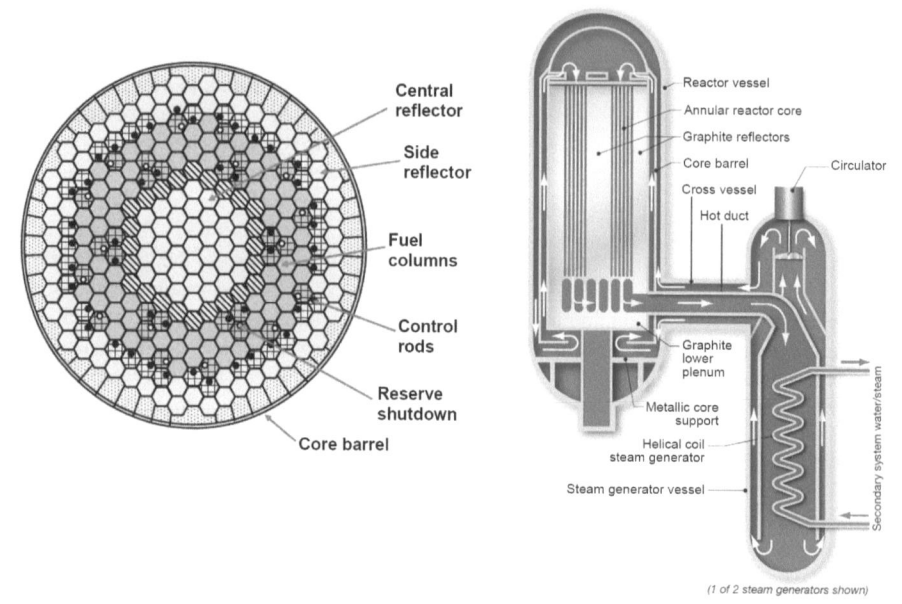

Fig. 21.3 SC-HTGR core cross section and primary circuit layout. Reproduced courtesy of Holtec.

- not only experimental capabilities with primary emphasis on nuclear and radiological facilities but also other testing capabilities (e.g., thermal-hydraulic loops and control systems testing),
- computational capabilities along with state-of-the-art modeling and simulation tools,
- information and data through knowledge and validation center,
- land use and site information for demonstration facilities.

GAIN identifies focused research opportunities and provides funding directly to the national laboratories as GAIN vouchers to collaborate with A-SMR developers. GAIN NE voucher recipients do not receive direct financial awards. The GAIN nuclear energy vouchers provide access to the national laboratory capabilities at no cost to the voucher recipients. All awardees are responsible for a minimum 20% cost share, which can be an in-kind contribution.

GAIN awards have been made on an ongoing basis since 2016. Summary level information on all the awards can be found at the GAIN website (Idaho National Laboratory GAIN Vouchers, n.d.). Table 21.5 provides a summary of the more recent awards from 2018 and 2019 that are related to the development and design of A-SMRs.

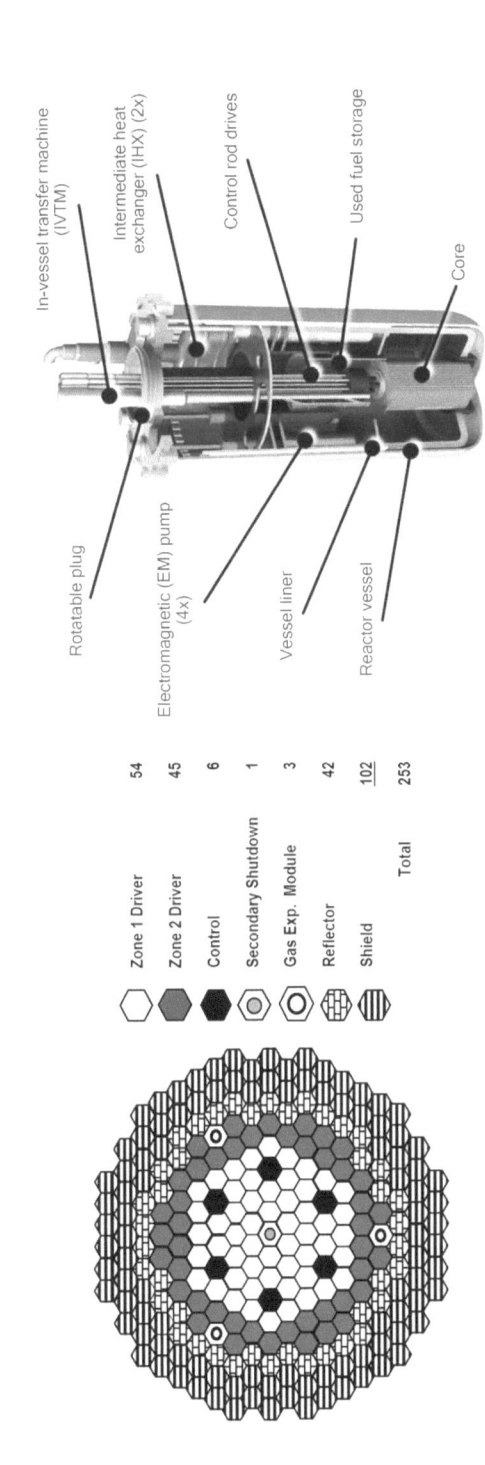

⬡	Zone 1 Driver	54
⬢	Zone 2 Driver	45
●	Control	6
◎	Secondary Shutdown	1
⊙	Gas Exp. Module	3
▦	Reflector	42
▥	Shield	102
	Total	253

Fig. 21.4 PRISM core cross section and reactor module isometric view.

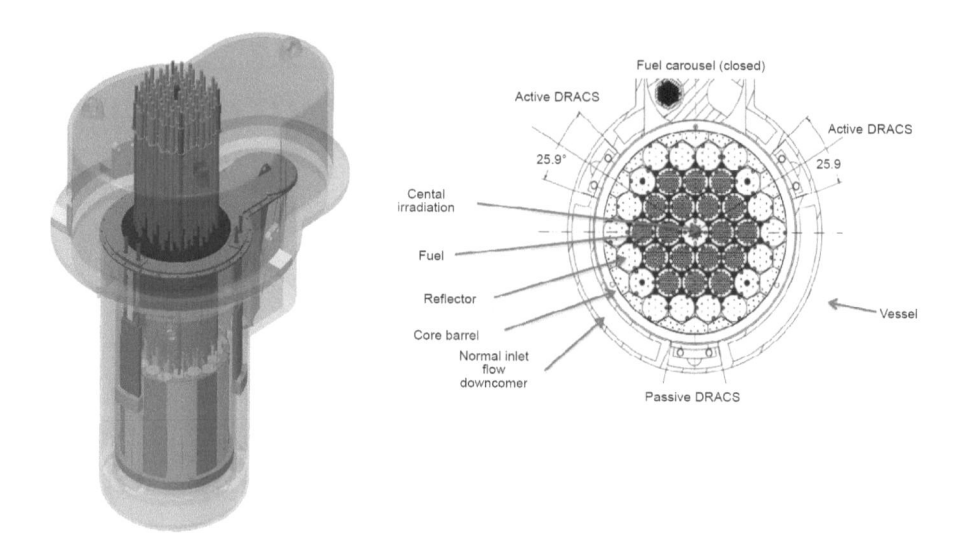

Fig. 21.5 FHR DR isometric view and horizontal cross section.

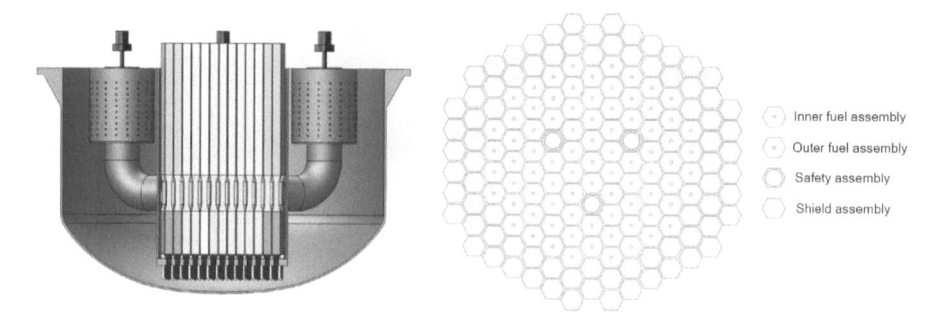

Fig. 21.6 DLFR primary system layout, vertical and horizontal cross sections.

Table 21.5 A-SMR related GAIN voucher awards.

Issue date	Award team	Brief project description
September 19, 2019	Analysis and Measurement Services Corporation—Oak Ridge National Lab	Testing of instrumentation and control sensors and cables for small modular reactors
September 19, 2019	HolosGen, LLC—Argonne National Lab	Advanced coolant and moderator enclosure solutions for microgas-cooled reactors with enhanced efficiency and safety

Table 21.5 Continued

Issue date	Award team	Brief project description
June 19, 2019	Westinghouse Columbia Fuel Fabrication Facility—Oak Ridge National Lab	Nuclear material holdup modeling and measurement campaign for the Columbia fuel fabrication facility
June 19, 2019	Flibe Energy, Inc.—Pacific Northwest National Lab	Metal organic frameworks for noble gas management in the liquid fluoride thorium reactor (LFTR)
June 19, 2019	TerraPower LLC—Argonne National Lab	Improvements to SAS4A severe accident modeling capabilities to support licensing and commercialization of TerraPower's traveling wave reactors
April 4, 2019	Flibe Energy, Inc.—Oak Ridge National Lab	LFTR preliminary safeguards assessment
April 4, 2019	Kairos Power—Argonne National Lab	Develop ASME Section III Division 5 design rules for elevated temperature cladded Class A Type 316 stainless steel components
January 24, 2019	Framatome—Argonne National Lab	Advanced fuel stability analysis using high-fidelity large-scale computational fluid dynamic simulations
January 24, 2019	Kairos Power—Argonne National Lab and Oak Ridge National Lab	Chemical method development for quantifying oxygen in beryllium salt
November 13, 2018	Eastman—Idaho National Lab and Oak Ridge National Lab	Integrated nuclear hybrid energy system
November 13, 2018	Elysium Industries USA—Argonne National Lab	Assessing fuel cycle options for elysium molten chloride salt fast reactor from spent nuclear fuel, plutonium, and depleted uranium
July 10, 2018	ThorCon—Argonne National Lab	Quantify sodium fluoride/beryllium fluoride salt properties for liquid-fueled fluoride molten salt reactors
July 10, 2018	Yellowstone Energy—Sandia National Lab	Characterization of the radiation stability of molten nitrate/nitrite salts for use as heat transfer fluids in nuclear reactor power plant
April 27, 2018	Oklo, Inc.—Argonne National Lab and Idaho National Lab	Accelerate development of industry-relevant features in modern simulation tools
April 27, 2018	Terrestrial Energy USA—Oak Ridge National Lab	Advancement of instrumentation to monitor IMSR® core temperature and power level
April 27, 2018	Thorcon—Argonne National Lab	Electroanalytical sensors for liquid-fueled fluoride molten salt reactor
April 27, 2018	Urbix Resources—Oak Ridge National Lab	Nuclear grade graphite powder feedstock development

21.6 DOE-NE Nuclear Energy University Program and A-SMRs

The DOE-NE created NEUP in 2009 to consolidate university support under one initiative and better integrate university research within DOE-NE's technical programs including SMRs and has continued to make awards on an ongoing basis. The objective of the program is to engage the US colleges and universities to conduct R&D, enhance infrastructure at these institutions, and support student education. One of the key means by which DOE-NE accomplishes its goals under NEUP of interest to A-SMR developers is by administering NEUP R&D awards to integrate R&D at universities, national laboratories, and industry. This serves to revitalize nuclear education and coordinate support of NE's R&D programs. The integrated research project (IRP) portion of NEUP represents the program directed component that is designed to address near-term, significant needs of DOE-NE R&D programs. The IRP awards are typically for 3-year periods and more significant funding amounts. NEUP awards are made in several different categories that include fuel cycle, modeling and simulation, and cyber security. The most relevant NEUP category for including in this handbook is "reactor concept research development and demonstration." Table 21.6 provides a summary of most recent awards from 2018 in this reactor concept category that involve advanced reactor concepts. Note that this R&D, while not necessarily aimed at A-SMRs, likely is relevant to the resolution of technical issues and development of technology to support A-SMRs. Summary level information on all NEUP awards can be found at the NEUP website (Idaho National Laboratory NEUP Awards, n.d.).

21.7 DOE-NE National Reactor Innovation Center

The DOE-NE announced the formation of its National Innovation Center (NRIC) in August 2019 (US Department of Energy National Reactor Innovation Center, n.d.). The mission of NRIC is to facilitate the development of advanced nuclear technologies and concepts by providing private sector technology developers the necessary support to test and demonstrate their reactor concepts and assess their performance. This support is anticipated to accelerate the licensing and commercialization of these advanced reactor concepts. The NRIC activities will be coordinated with the GAIN program. Funding for NRIC is expected in FY2020.

NRIC's objectives (US Department of Energy National Reactor Innovation Center Fact Sheet, n.d.) are as follows:

- enable testing and demonstration of reactor concepts by the private sector,
- validate advanced nuclear reactor concepts,
- resolve technical challenges of advanced nuclear reactor concepts,
- provide general research and development to improve innovative technologies.

Fig. 21.7 illustrates the overall philosophy and approach envisioned for NRIC as to how it will work in supporting the advancement of these concepts from proof of concept through proof of operations based on technology readiness levels.

Table 21.6 A-SMR related NEUP R&D awards.

Issue date	University	Award amount	Brief project description
2018	Georgia Institute of Technology	$800 K	Corrosion testing of new alloys and accompanying online redox measurements in ORNL FLiNaK and FLiBe molten salt flow loops
2018	Georgia Institute of Technology	$800 K	A novel high-fidelity continuous-energy transport tool for efficient fluoride salt-cooled high-temperature reactors (FHR) transient calculations
2018	Massachusetts Institute of Technology	$800 K	Determination of molecular structure and dynamics of molten salts by advanced neutron and X-ray scattering measurements and computer modeling
2018	University of Connecticut	$799 K	In situ measurement and validation of uranium molten salt properties at operationally relevant temperatures
2018	University of Michigan	$800 K	High-resolution experiments for extended LOFC and steam ingress accidents in high-temperature gas-cooled reactors (HTGRs)
2018	University of New Mexico	$800 K	Integral experimental investigation of radioisotope retention in stagnant and flowing lead for the mechanistic source term evaluation of lead-cooled fast reactor
2018	University of Tennessee	$800 K	Understanding molten salt chemistry relevant to advanced molten salt reactors through complementary synthesis, spectroscopy, and modeling
2018	University of Texas at San Antonio	$800 K	Oxidation of tristructured isotropic (TRISO) fuel forms in low oxygen and steam partial pressures and the role of matrix burn off in the oxidation rate at high temperature
2018	University of Wisconsin Madison	$800 K	Development of corrosion resistant coatings and liners for structural materials for liquid-fueled molten salts reactors
2018	University of Wisconsin Madison	$796 K	Advanced alloy innovations for structural components of molten salt reactors

In October 2019, DOE-NE and NRC signed a memorandum of understanding (MOU) (US Department of Energy, n.d.-a) "to share technical expertise and computing resources to speed up the deployment of advanced nuclear technologies." NRC is to have access to NRIC capabilities, as well as other DOE facilities, including high-performance computing and computer codes to assist in their licensing of advanced reactors. In the MOU, NRC is to provide information to DOE and the industry on

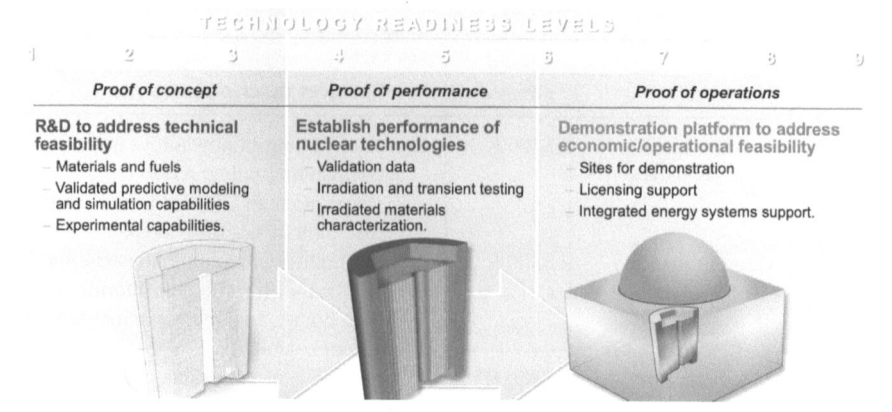

Fig. 21.7 NRIC support of private sector development of advanced reactor concepts

its licensing processes and other related guidance for both new and advanced nuclear technologies. DOE-NE will help new or advanced reactor applicants in understanding and interpreting these processes.

21.8 DOE-NE R&D efforts related to development of microreactors

As noted in the Introduction to this chapter, the FY2019 National Defense Authority Act (CONGRESS.GOV, 2018) requests that DOE "prepare a report describing the requirements for a pilot program to provide energy resilience for critical national security infrastructure at Department of Defense (DoD) and DOE facilities using at least one micro-reactor." Furthermore, this Act includes guidance that this pilot program should contract with a commercial company for the purposes of siting, constructing, and operating a licensed microreactor by December 31, 2027. In support for preparation of the report to Congress, DOE issued a request for information (RFI) (US Department of Energy Request for Information, 2018) to assist DOE-NE in obtaining information from the nuclear industry interested in supporting this pilot program on reactor technology options, fuel types, regulatory issues, financial support, etc. This RFI requested that responses be submitted by October 15, 2018. Additionally, DOE-NE is engaging with two DoD offices—Office of the Assistant Secretary of Defense for Sustainment and Strategic Capabilities Office—to coordinate efforts on the overall effort to develop and demonstrate microreactors.

The aforementioned RFI describes microreactors as a class of reactors up to 50 MWe, while other definitions characterize them in the 1–10-MWe range. The principal distinguishing features for microreactors include the following:

- components to be factory fabricated and assembled and then shipped to the site;
- reactor units transportable via truck, railcar, plane, etc.;

- designs to be self-regulating employing passive safety features;
- operations will not require a large number of specialized operators;
- operations for 10 years or more without refueling.

Fuel for most microreactor designs is expected to require higher uranium-235 concentrations than today's current reactors. Microreactors have the potential to provide both defense and commercial applications with a reliable and resilient source of power and/or process heat for remote and off-grid locations. They are also envisioned for emergency and disaster relief use.

The three principal objectives of DOE-NE's microreactor R&D program (Sowinski, 2019) are the following:

- Identify R&D requirements for microreactors.
- Conduct laboratory-led, early-stage R&D supporting cross-cutting microreactor technology development and licensing work.
- Coordinate activities with NRC, DoD, and industry to demonstrate microreactor technology on a DOE national laboratory site.

Note that DOE-NE earlier funded two microreactor projects—one under the "Advanced Reactor Technology Development" FOA described in Section 21.3.2 and one under its GAIN voucher program discussed in Section 21.5.

The Nuclear Energy Institute has prepared a roadmap (Nuclear Energy Institute Technical Report, 2018) outlining a path forward for the deployment of microreactors at a DoD installation within the United States only—not for forward operating bases. This document "identifies the timeline, major challenges and recommended actions to ensure successful deployment of the first micro-reactor at a DoD domestic installation." The deployment approach is noted as consistent with the FY2019 National Defense Authorization Act and assumes that (1) a power-purchase agreement with a commercial entity is employed for the purposes of siting, constructing, and operating a microreactor at a DoD installation and (2) the microreactor is licensed by the NRC.

21.9 DOE-ARPA-E R&D for modeling and simulation of innovative technologies for advanced reactors

In June 2018, DOE's Advanced Research Projects Agency-Energy (ARPA-E) announced awards for 10 projects totaling up to $24 M for advanced reactors technology development (US Department of Energy, n.d.-b). This new ARPA-E program is entitled "Modeling-Enhanced Innovations Trailblazing Nuclear Energy Reinvigoration" (MEITNER). The overall goal of these MEITNER awards is to develop innovative technologies that enable designs for lower-cost, safer, advanced nuclear reactors. More specifically, it is expected that these projects will take advantage of design features, advanced manufacturing processes that can potentially lead to decreased construction costs, autonomous operations, and enhanced safety. DOE ARPA-E coordinated the development of this funding call with DOE-NE. The MEITNER projects are to have access to DOE's modeling and simulation capabilities including the DOE national labs. Additionally, these innovative technologies are expected to assist in establishing the basis for a modern, domestic supply chain supporting nuclear technology. The list of MEITNER awards can be found at US Department of Energy (n.d.-b).

21.10 Future trends

The discussion included in this section in the first edition has been updated to reflect relevant developments as related to future or emerging trends. However, much of that discussion remains applicable today as it relates to the factors that will influence deployment of A-SMRs in the future and the R&D required to support and demonstrate these new technologies and is included again. The one, new emerging trend is the interest and support for the development and deployment of microreactors as discussed earlier in this chapter and noted again later.

The factors that will determine the ultimate deployment of SMRs and in particular A-SMRs will focus on the ability of these concepts to compete economically with large LWRs that are (1) offering lower costs in terms of both construction and operation, improved performance including conversion efficiencies perhaps with advanced power conversion system technologies, demonstrated enhancements in safety, and where possible generating less waste and (2) expanding the utility of nuclear energy for process heat applications that require higher temperatures in addition to producing electricity. The A-SMRs typically involve innovative designs where new fuels and materials are introduced. Several of these designs are for high-temperature applications. For new non-LWR reactor concepts, the emphasis of future R&D will likely focus on the development, demonstration, and qualification of these new materials and fuels.

R&D for new materials will include the development of materials that will need to be compatible with the several coolant types (liquid metals, gas, and liquid salts) at elevated operating temperatures and core structural materials such as graphite for HTGR and FHR applications. Advanced steels will be required for fast reactor concepts. R&D efforts are ongoing in the United States on the development and qualification of new materials for advanced reactors including A-SMRs under the ART program at DOE-NE.

Work in the United States is expected to continue on further development and qualification of TRISO fuels for use in HTGRs and FHRs and fast reactor fuels (oxide and metallic). These new fuels and cladding materials will need to be capable of withstanding irradiation at higher fuel burnup as well.

The demonstration of these new materials and fuels will require both modeling/ simulation and experimental capabilities. Given the significant resources required for new facilities, additional emphasis will be placed on modeling and simulation tools for the integration of reactor physics, thermal hydraulic, and structural mechanics models to evaluate the expected performance of these new concepts. Also the repurposing or refurbishing of experimental facilities used in the R&D performed earlier the United States will likely be evaluated as options to building new facilities to conduct verification and validation of the models as was done in November 2017 when operations at the Transient Reactor Test (TREAT) Facility resumed at INL. The TREAT facility had been shut down and maintained in standby status since 1994. TREAT is designed specifically to test nuclear reactor fuels and materials under extreme conditions.

Additional R&D will continue in looking at how to integrate digital instrumentation and control (I&C) systems and advanced control architectures to enable integration of control, diagnostics, and decision making for highly automated multiunit plant operations and devising alternate concepts of operation for multiunit SMR designs. Successfully integrating diagnostics capability into new A-SMR designs will assist in providing a sound technical basis for extended operation beyond the initial licensed time frame.

Integrating SMRs in general with renewable electric power sources as hybrid systems as a way to balance power on the electric grid due to the intermittency of power produced by such renewable sources as wind and solar and utilize the excess thermal power from the reactor for process heat applications continues to garner attention. With some large LWRs being shut down due to economic factors resulting from the lower cost of natural gas and decreased electrical demand, these hybrid systems potentially offer a new application for SMRs in general. To effectively evaluate such systems, R&D will be needed in the way of modeling to understand the interactions and control systems for coupling the renewable and nuclear plants. Also, in these coupled systems, the reactor may be used more in a load-following manner, which will require R&D for controls based upon process heat application and ensuring the safe operation when cycling the reactor through increasing and decreasing power operations.

For microreactor concepts the economic requirement may not be quite as important as for A-SMRs for baseload power on the grid depending on the application. For example, there may be overriding considerations in deploying microreactors at remote locations or forward military base operations. The economic requirement for deploying microreactors to support domestic military and other governmental needs is likely to be just as important as for A-SMRs. Depending on the reactor technology, there will continue to be R&D needs for fuels and materials for microreactors. Instrumentation and controls will likely be more challenging given the remote sensing and control requirements for microreactors requiring some additional R&D. Also, novel, smaller-sized PCS systems may require further development and demonstration. Given the emerging interest in microreactors as driven by legislation as described earlier for deployment of a microreactor at a DoD or DOE facility by the end of 2027, decisions will need to be made regarding the applicable licensing authority, that is, NRC, DOE, or DoD for this microreactor. Thus there may be more applied R&D needs related to supporting standards or other licensing basis requirements depending on the licensing authority.

As some of these A-SMR concepts mature and evolve as more realistic options for future deployment, one may see two trends emerge. Within the United States the first will be focused on potential industry-government partnerships somewhat different from to the cost-share arrangements between DOE and two of the LW-SMR vendors, m-Power and NuScale. These partnerships are focused on a cost-shared arrangement to support a design certification approval from NRC. After having demonstrated to a reasonable degree the favorable attributes of these advanced concepts in terms of economic attractiveness, efficient operations, and enhanced safety via government supported R&D, it will likely be necessary for an industry-government arrangement to support the final demonstration in the form of a test or prototype reactor. Such an arrangement would be more appealing to industry if the government were to identify scenarios for deployment to assist federal agencies including the Department of Defense in meeting clean power goals. It is straightforward to identify areas where there is significant power demand by groups of collocated federal agencies. Such a consideration would benefit the government agencies with a dedicated source of reliable power and provide the nuclear industry the opportunity to demonstrate the safe and reliable operation of these advanced concepts for further commercial deployment.

Since publication of the first edition, DOE-NE has undertaken two initiatives that reflect this trend. These two initiatives are (1) the advanced nuclear technology development FOA that includes A-SMRs and microreactors as discussed in Section 21.3.2 and (2) the development of the National Reactor Innovation Center (NRIC) for providing industry access to infrastructure for development, demonstration, and ultimate deployment of new nuclear technologies as noted in Section 21.7. It is anticipated that this trend, the industry-government partnership, will continue.

A second emerging trend may involve more cooperation between the United States and the international community in the development of new and/or shared use of existing experimental facilities, test loops, and perhaps test reactors. Such arrangements would support the government reaching the state of development described in the preceding paragraph at which point the nuclear industry could be engaged to complete final development. Presently, international cooperation is underway on advanced reactor concepts among some 13 countries under the Generation-IV International Forum (GIF) cooperative effort focusing on some six advanced concepts. The costs to construct and operate such test and experimental facilities may be viewed as prohibitive for any one country to undertake. Other opportunities for international collaboration exist via the International Atomic Energy Agency and the Nuclear Energy Agency with the Organization for Economic Cooperation and Development.

Overall in terms of trends, there continues to be serious interest in A-SMRs and now for microreactors as indicated by the number of industry companies developing various concepts and providing funding support. Many of the A-SMR developers have continued to pursue development of their respective concepts since publication of the first edition and are anticipated to continue to pursue demonstration and ultimate deployment in the future.

References

CONGRESS.GOV, 13 August 2018. H.R. 5515—John S. McCain National Defense Authorization Act for Fiscal Year 2019, Pub. L. 115-232. https://www.congress.gov/bill/115th-congress/house-bill/551/text.

IAEA Advanced Reactors Information System (ARIS), 2018. Advances in Small Modular Reactor Technology Developments—A Supplement to: IAEA Advanced Reactors

Information System (ARIS)—2018 Edition. International Atomic Energy Agency.http://aris.iaea.org.

Idaho National Laboratory GAIN Vouchers, n.d. https://gain.inl.gov/SitePages/Nuclear%20Energy%20Vouchers.aspx.

Idaho National Laboratory NEUP Awards, n.d. https://neup.inl.gov/SitePages/FY17_RandD_Awards.aspx

Landgon, K., 2019. NuScale small modular reactor (SMR) overview. In: INPRO Dialogue Forum on Opportunities and Challenges in Small Modular Reactors, Ulsan, Republic of Korea, July 2–5.

Nuclear Energy Institute Technical Report, October 4, 2018. Roadmap for the Deployment of Micro-Reactors for U.S. Department of Defense Domestic Installations. https://www.nei.org/CorporateSite/media/filefolder/resources/reports-and-briefs/Road-map-micro-reactors-department-defense-201810.pdf.

Petti, D., Hill, R., Gehin, J., et al., January 2017. Advanced Demonstration and Test Reactor Options Study, INL/EXT-16-37867, Revision 3.

Sowinski, T., June 2019. DOE-NE Micro-Reactor RD&D Program Mission and Objectives, GAIN Micro-Reactor Workshop. https://gain.inl.gov/SiteAssets/Micro-ReactorWorkshop Presentations/Presentations/02-Sowinski-MRProgramMission_June2019.pdf.

US Department of Energy, 2018. US Industry Opportunities for Advanced Nuclear Technology Development Funding Opportunity Number: DE_FOA-0001817, Issue date December 7, 2017, Amendment 0005 (May 1, 2018). https://www.grants.gov/web/grants/search-grants.html.

US Department of Energy, n.d.-a. DOE-NRC Agreement. https://www.energy.gov/ne/articles/new-doe-and-nrc-agreement-will-lead-faster-deployment-and-licensing-us-nuclear.

US Department of Energy, n.d.-b. Advanced Research Projects Agency-Energy MEITNER awards for advanced reactor technology, https://www.arpa-e.energy.gov/?q=news-item/department-energy-announces-10-projects-support-advanced-nuclear-reactor-power-plants.

US Department of Energy Advanced Small Modular Reactors (SMRs), n.d. https://www.energy.gov/ne/nuclear-reactor-technologies/small-modular-nuclear-reactors.

US Department of Energy Nation's First SMR, n.d. https://www.energy.gov/ne/articles/nations-first-small-modular-reactor-plant-power-nuclear-research-idaho-national.

US Department of Energy National Reactor Innovation Center, n.d. https://www.energy.gov/ne/articles/energy-department-launches-new-demonstration-center-advanced-nuclear-technologies.

US Department of Energy National Reactor Innovation Center Fact Sheet, n.d. https://www.energy.gov/ne/downloads/national-reactor-innovation-center-fact-sheet.

US Department of Energy Request for Information, September 31, 2018. Input on a Pilot Program for Micro-Reactor Demonstration (89243218NNE000001).

US Department of Energy Small Modular Nuclear Reactors, 2012. SMR Licensing Technical Support Program. https://www.energy.gov/ne/smr-licensing-technical-support-lts-program.

Part Five

Global perspectives

Small modular reactor (SMR) adoption: Opportunities and challenges for emerging markets

22

Geoffrey Black[a], David Shropshire[b], and Kathleen Araújo[c]
[a]Department of Economics, College of Business and Economics, Boise State University, Boise, ID, United States, [b]Nuclear Science and Technology Directorate, Idaho National Laboratory, Idaho Falls, ID, United States, [c]Energy Policy Institute, Boise State University, Boise, ID, United States

22.1 Introduction

The growth in global energy demand, coupled with priorities to address a changing climate and sustainable development, has brought attention to the need for informed development of low-carbon energy sources in both developed and emerging economies (Araujo, 2017). Fossil fuels are likely to continue playing a role in energy utilization; however, low carbon sources—including nuclear and renewable energy sources—require critical consideration for adapting the global energy system in line with current pressures. The need for changes in energy provision is particularly important given that the production of electricity is the largest single source of greenhouse gas emissions on a global level, particularly in emerging economies (European Environmental Agency, 2018).

In line with low-carbon energy priorities, nuclear energy has the potential to meet a critical share of related targets, presenting both challenges and opportunities for commercial adoption of nuclear power. Large nuclear power plants (NPPs) that have benefitted from improvements in safety and performance, including a new generation of advanced reactors, could serve as replacements for retiring nuclear plants and potentially increase nuclear power's contribution by mid-century (Massachusetts Institute of Technology, 2018). Additionally, an emerging share of future electricity and heat demand could be met by the deployment of small modular reactors (SMRs). Several characteristics of SMRs are expected to provide distinct advantages over large nuclear and potentially other baseload plants in certain markets. Such advantages stem from the reactors' smaller size, simplicity of design, and modularity. These attributes may result in lower capital and siting costs compared with large NPPs, both in total and on a per megawatt basis (Black et al., 2019; Boarin and Ricotti, 2014) as well as reduced construction time and lower financing risk and risk from cost overruns. Other advantages include operational flexibility and their ability to be deployed in areas where the infrastructure to support large power plants is insufficient. Further, SMRs are suited for meeting demand growth through incremental adoption of SMR modules as well as in the flexibility to adapt to changing energy market conditions. Other anticipated advantages may include new modes of inherent safety, passive cooling, the ability to integrate with renewable energy sources, and a range of nonelectrical cogeneration and hybrid process applications such as desalination, domestic and industrial heating, and hydrogen production. These SMR features make them a unique low-carbon energy technology that may be deployed in a diverse set of locations and applications. When coupled with increased demands to address sustainability, climate, and growth, such attributes make SMRs a contender for global markets dominated by fossil fuels, which can further the energy share from nuclear power over the coming decades.

Given the present conditions, assessments of the global market potential and suitability of SMRs across a variety of locations and applications are warranted. This chapter begins by discussing the market assessment and potential for SMR adoption in both developed and emerging economies. The methodology used in recent studies of this energy technology generally involves assessing specific characteristics of countries, including economic, technological, infrastructure, and other conditions, and then matching those characteristics to the capabilities and attributes of SMRs. For example, in emerging markets, the size and scalability of SMRs make them suited for grids with relatively low capacity and where the population may be more dispersed. In remote locations with relatively low populations, but one or more industrial facilities, the simplicity in on-site construction, high fuel economy, and planned safety systems could allow SMRs to address both electricity demand and industrial process heat applications and district heating. Thus, in Section 22.2 later, the potential is discussed for SMR utilization across countries and markets.

The remaining sections of this chapter address the implications of prospective changes in policies and goals on a global scale and changes in nuclear technology. Section 22.3 discusses SMR deployment in terms of goals and initiatives tied to climate change and energy access. These efforts include country carbon mitigation plans supporting the COP21 Paris Agreement's nationally determined contributions and the 2030 UN Sustainable Development Goals. The overarching theme of this section is the relevance of SMRs for sustainability goals, improved access to energy, economic development, and aspects of energy security. Section 22.4 focuses on the implications of disruptive changes in technology and carbon reduction efforts. Important topics, here, include the role of SMRs in a scenario of robust scaling of low-carbon technologies to replace high-carbon energy sources in developing economies and to support the growing energy needs of emerging megacities. Disruptive scenarios using advanced nuclear technologies, including advanced SMR designs and microreactors, are examined. Particular attention is paid to the integration of SMRs with renewables in distributed and hybrid energy systems. Finally, Section 22.5 discusses both the opportunities and challenges facing SMR deployment going into the future. Issues such as the transport of SMRs and their fuel, remote operations and security, the storage and handling of used fuel, and cost competitiveness are among the issues examined.

22.2 SMR market deployment potential

As noted earlier in this chapter and also reviewed thoroughly in Part 1 of the Handbook, certain advantageous characteristics to SMRs position them as a prospective low-carbon energy source to meet future energy demands in developed and emerging economies. As outlined by the US Department of Energy, unique features of SMRs, particularly relative to large NPPs, include their modularity, lower capital investment, siting flexibility, greater efficiency, and planned safety and security features (U.S. Dept of Energy (US DOE), 2019). Modularity enables much less on-site preparation and reduced construction times compared with larger nuclear or fossil fuel facilities with attendant cost advantages due to a much lower initial capital investment (U.S. Dept of Energy (US DOE), 2019). Further, as described in the previous section, the modularity and smaller size of SMRs enable more flexibility in siting, grid matching, and incremental growth in energy demand. For example, SMRs may be deployed in smaller electrical markets and grids, isolated areas, sites with limited water and acreage, and unique industrial applications (U.S. Dept of Energy (US DOE), 2019). In larger markets, SMRs may also replace baseload power from retiring fossil fuel plants and energy needs of megacities. With the increasing demands for low-carbon energy sources, SMRs ability to pair with variable renewable energy sources present a key attribute to explore.

22.2.1 Global market assessments

Based on the unique characteristics of SMRs across an array of locations, applications and economic, market, and infrastructure conditions, several market assessments of the deployment for SMRs have been conducted in recent years. One of the first to estimate global demand for SMRs was Solan et al. (2010a). This study was a top-down assessment of the potential for SMR deployment based on global projections for nuclear energy demand. The projections ranged from a high of 8.5 GWe of SMR deployment by 2030 to a low of 1 GWe across four different adoption scenarios. The focus of this study was the estimation of the economic impacts of developing a domestic SMR industry in the United States for the manufacture, domestic deployment, and export of SMRs, rather than market assessment of individual countries. Similarly, several studies of the potential global demand for SMRs were subsequently produced by public and private entities with a wide range of estimates for SMR market potential. Proprietary studies by UxC (Ux Consulting Company LLC, 2013) and Strategic Insights (Strategic Insights, 2015) projected estimates of global SMR potential. The UxC study estimated 22 GWe of new SMR additions by 2040 in a medium demand scenario and a low-adoption scenario with an estimated 9 GWe of new SMR additions. The Strategic Insights report (Strategic Insights, 2015) estimated SMR demand by geographic region under three different scenarios through 2035, with the most potential in China and the United States. In their medium scenario, for example, the report estimated more than 5 GWe of new SMR additions in China and between 1 and 5 GWe in the United States. Similarly a report by the UK's National Nuclear Laboratory (National Nuclear Laboratory, 2014) estimated global SMR demand and found the highest potential in China and the United States. The focus of this latter report was the economic impacts of UK companies producing SMRs and being part of the supply chain for SMR systems. The study estimated the market potential of SMRs in the United Kingdom to be approximately 7 GW of new capacity by 2035 and with the total market potential to be in the range of 65–85 GW globally in the same timeframe, with nearly half in China and the United States (Strategic Insights, 2015). Lower projected SMR deployment potential on a global basis was estimated in a study by the Nuclear Energy Agency (NEA) in 2016 (Nuclear Energy Agency/OECD, 2016). In this study a low-adoption scenario estimated less than 1 GWe of new SMR additions by 2035. A high-adoption scenario, based on the near-term development of SMR modular manufacturing facilities and SMR supply chain, estimated 21 GWe of SMR capacity by 2035.

22.2.2 Deployment potential with SMR indicators

The studies earlier project future SMR deployment potential over a wide range (1–85 GWe worldwide by 2035) by utilizing top-down projections based on energy demand, volumetric market forecasts from vendor country perspectives, or qualitative measures utilizing survey data or expert opinion. By contrast the first study to analyze the suitability of SMRs for individual markets, based on the specific characteristics of SMRs, was performed by the U.S. Department of Commerce (2011). This study investigated 27 countries identified as markets of interest for new nuclear builds, some of which have existing nuclear facilities and others were identified as nearing nuclear readiness. These counties were rank ordered in terms of their SMR market potential on the basis of six proxy indicators. The six indicators used in the study are (1) low population density, (2) growth in carbon emissions, (3) economic growth, (4) growth in energy consumption, (5) energy importer, and (6) existing nuclear power production. Countries with the lowest population density, the highest growth in carbon emissions, the highest rates of economic and energy consumption growth, and that were also net importers of energy and had some existing nuclear capacity generated the highest ranking for SMR deployment potential with the given set of proxy indicators.

To date the most comprehensive study using SMR-specific indicators to assess market potential in individual countries is the recently published International Atomic Energy Agency (IAEA) Technical Document (IAEA, 2018a). This study was the culmination of a previous work performed by the Energy Policy Institute with review by IAEA personnel, member states representatives, and subject matter experts (Solan et al., 2015). This work greatly expanded both the methodology and the number of countries evaluated by the US DOC (U.S. Department of Commerce, 2011) report. In terms of the scope of the countries analyzed, the US DOC study assessed SMR demand in a specific subset of 27 countries preselected as potential markets for US produced SMRs. By contrast the IAEA study incorporated a much broader scope of countries, assessing all 170 IAEA member states. Further the IAEA study broadened the scope and applicability of the criteria indicators used in its methodology, using 18 proxy indicators of SMR suitability compared with 8 indicators in the DOC study (U.S. Department of Commerce, 2011). The rationale for the indicators used in the IAEA study is twofold. First the indicators employed in the IAEA analyses were chosen to enable assessment using publicly available data from international agencies and governments and accessible to all member states. Second the indicators were chosen to be both indicative of certain conditions that may be amenable to the development of nuclear power in general and indicators pertinent to SMRs specifically. At the same time the indicators were to provide quantitative proxy measures of country conditions that could be used to assess SMR deployment potential. These indicators utilized a methodology that is quantifiable, objective, and applicable to a broad spectrum of conditions within countries across the globe. They include economic and financial characteristics, technology and infrastructure, government policy, energy demand, and carbon reduction.

These indicators fall into six categories: national energy demand, SMR energy demand, financial and economic sufficiency, physical infrastructure sufficiency, climate change motivation, and energy security motivation. Each of these, along with the indicators within each category, are described later and shown in Figs. 22.1–22.6.

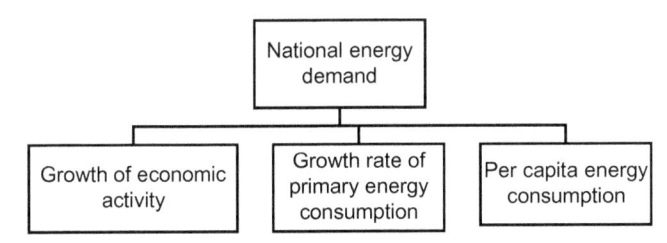

Fig. 22.1 National energy demand category and indicators.

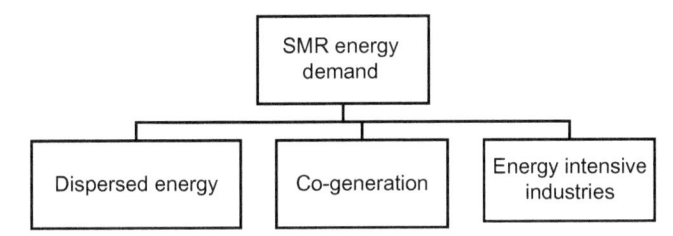

Fig. 22.2 SMR energy demand category and indicators.

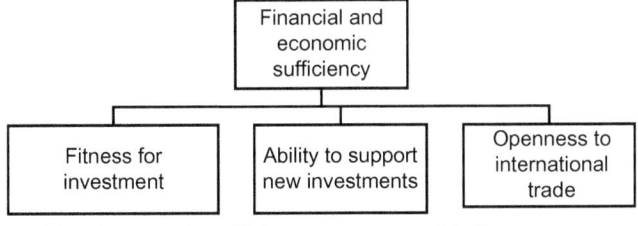

Fig. 22.3 Financial and economic sufficiency category and indicators.

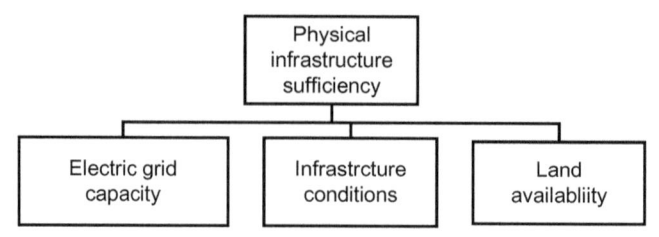

Fig. 22.4 Physical infrastructure sufficiency category and indicators.

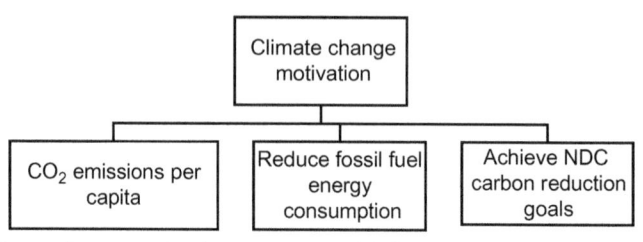

Fig. 22.5 Climate change motivation category and indicators.

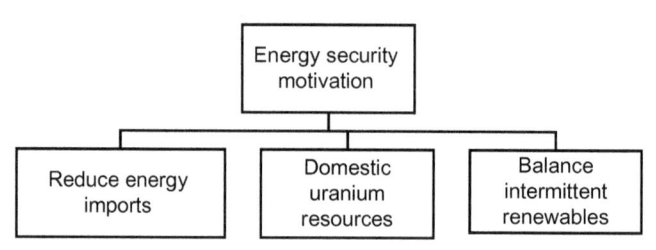

Fig. 22.6 Energy security motivation category and indicators.

The first category, national energy demand, assesses overall energy demand for a country or region. This category includes the indicators of the growth of economic activity, the growth rate of primary energy consumption, and the current per capita energy consumption, as shown in Fig. 22.1.[a]

This first category is aimed at assessing a country or region's overall need for new energy sources. Countries with more rapid economic growth, rapidly growing energy consumption, and high levels of per capita energy consumption are judged to have greater need for new energy production facilities. This could be met through the deployment of both small and large energy power generation technologies.

While the first category is applicable to both large and small power production facilities, the second category, SMR energy demand, is aimed at some of the differentiating features of SMRs specifically. These are shown in Fig. 22.2.

---

[a] This, like other deployment measures, does not account for all factors. Attributes, such as the introduction of increased efficiency, would be indirectly reflected. Institutional capacity, such as with regulatory capabilities, is not covered.

Within the SMR energy demand category, the dispersed energy indicator accounts for the distribution of a country's population. A population concentrated in large urban centers would be amenable to large power facilities, whereas a more dispersed population is likely more amenable to smaller, dispersed energy sources such as SMRs. The cogeneration indicator assesses the need for nonelectric applications such as desalination, oil recovery and refinery power, district heating, and process heat applications such as hydrogen production with high steam temperature electrolysis. As described in Section 22.4.2 later, such cogeneration applications present significant potential to use the design-specific features of SMRs to improve energy supply. Similarly the energy-intensive industries indicator measures the degree to which overall energy demand is composed of industrial demand in locations away from major population centers. In such locations, such as large-scale mining operations, energy demand could be satisfied with the deployment of one or more SMRs rather than long transmission lines from a large centrally located power facility.

The third and fourth categories address whether a country has the economic sufficiency and the infrastructure to support investments in SMRs. Fig. 22.3 shows the indicators within the category of financial/economic sufficiency.

Within this category the indicator fitness for investment is aimed to assess a country's overall credit worthiness, as measured by the amount of governmental debt relative to this size of a country's economy. High levels of domestic and foreign debt as a percentage of a country's GDP would indicate a reduced ability for a country to finance investments advanced energy technologies. Similarly the indicator ability to support new investments assesses the degree to which a county has the economic capability to support SMR deployment. The third indicator in this category, openness to international trade, is aimed at the likelihood that a country without a domestic SMR manufacturing industry will be open to importing this technology from other countries.

The category of physical infrastructure sufficiency contains three indicators that measure the sufficiency of the country's electric grid and infrastructure, as shown in Fig. 22.4.

While the third category described earlier addresses the ability of a country's economy to support new energy investments, the fourth category, physical infrastructure sufficiency, is aimed at assessing a country's physical capacity to support new power sources. A country's electric grid must be of sufficient size to accept new energy sources. In addition, larger electric grids have the capacity to incorporate more SMRs into their system. As a larger grid capacity is applicable to large and small power sources, this indicator along with the dispersed energy indicator described earlier gives higher rankings to larger and more decentralized electric grids that are better suited to SMRs deployment than more centralized grids. The infrastructure condition indicator measures a country's ability to support the deployment and operation of advanced energy technology on the basis of its transportation, communications, and electrical distribution networks. Finally the land availability indicator assesses the amount of land open for SMR development.

The two remaining categories address motivations for low-carbon energy production and the utilization of domestic resources. The incentive to utilize low-carbon energy sources is addressed in the climate change motivation category through the indictors shown in Fig. 22.5.

Overall, this category addresses the incentive for a country to increase electricity production through the use of low carbon sources. The first two indicators, CO_2 emissions per capita and fossil fuel energy consumption, assess a country's existing power production carbon footprint. Countries with high CO_2 emissions per capital and have a high percentage of their energy production through high carbon sources such as fossil fuels are more likely to utilize low-carbon energy sources including nuclear and renewable energy technologies, such as wind and solar. Unlike some large nuclear power plants, however, SMRs have the load-following capability to complement renewable sources and being a low-carbon energy source. Similarly the indicator achieve NDC carbon reduction goals assesses a county's commitment to reduce overall carbon emissions with policies concerning the need to address climate change. Finally the energy security motivation category contains the indicators shown in Fig. 22.6.[b]

Two of the indicators in this category assess the motivation for a country to use domestic resources. A country with a high reliance on nondomestic fuel or technologies will likely favor the use of domestic resources to enhance energy security. Similarly a country with sufficient domestic uranium resources is likely to be motivated to increase the use of nuclear power. The balance intermittent renewable indicator assesses a country's motivation to pair SMRs with renewable energy sources. Having a high percentage of domestic renewable energy facilities is likely to provide an incentive to deploy SMRs to address the need for nonintermittent baseload power domestically.

In the earlier IAEA study, each indicator is measured quantitatively using data from a range of international agencies, including the World Bank, the International Energy Agency, the US Energy Information Agency, the UN Framework Convention on Climate Change, and the Organization for Economic Cooperation and Development. These were chosen to ensure the validity of the data, the ability for any country to freely access all of the data used, and the inclusion of every IAEA member state in the data source. Some potential indicators were rejected because of limitations in data availability across countries, including supply chain information, the availability of qualified technical and regulatory personnel, and specific electric grid characteristics.

The analysis in the 2018 IAEA report (IAEA, 2018a) found that countries with strong economic growth, a high rate of energy consumption growth, a high reliance on fossil fuels, and high levels of greenhouse gas (GHG) emissions are potential adopters of SMRs. However, these conditions are also compatible with the development of large NPPs. Additional conditions that indicate prospective suitability for SMRs include a relatively dispersed population, site-specific demand such as desalination or process heat, and possibly relatively remote energy-intensive applications. Having economically viable deposits of uranium domestically improves the potential for nuclear energy sources in general but is not a requirement for SMR deployment

---

[b] In focusing on motivation, this does not account for localized security and safety capabilities, or other conditions.

within a country. Countries that were judged to be relatively less suited for SMR deployment are those with poor economies, insufficient energy demand, and inadequate physical infrastructure. Poor economic conditions, such as low levels of GDP and per capita GDP, may indicate a limited ability to finance the costs necessary for SMR deployment and operation. Slow economic growth and low rates of energy consumption growth indicate that a country's energy market size may be insufficient to support the deployment of SMRs. Finally the infrastructure within a country was approximated to be insufficient to support SMRs, if the total electric grid size is less than 1.5 gigawatt (GWe).

Two important contributions of the IAEA (IAEA, 2018a) report are the consideration of regional groupings of countries and illustrative case studies for SMR deployment. In terms of regional groupings, individual countries may want to enter into partnerships with other countries to increase their future energy options. For example, while an individual country may have insufficient energy demand or financial resources to develop SMRs domestically, these limitations might be overcome by engaging in power generation and sharing agreements with other countries. The IAEA examined the effects of regional groupings for three distinct areas. These are the Baltic states of Latvia, Estonia, and Lithuania; the Southeast Asian states of Brunei Darussalam, Indonesia, Malaysia, and Singapore; and seven Persian Gulf countries. Although not analyzed, the report identified other groupings such as the northern African states of Algeria, Libya, and Tunisia entering into sharing arrangements with Italy and similar arrangements with other Adriatic countries such as Croatia and Slovenia.[c] In addition to assessing regional groupings, the report also examined six illustrative deployment case studies that represent a variety of economic conditions, energy needs, and infrastructure. The results here are consistent with the findings from the analysis described earlier. Countries with high economic growth and energy demand, sufficient infrastructure, a large land area with a dispersed population, and growing sectors such as manufacturing or resource extraction may be suited for SMR deployment. Similar countries, but with a small land area and lower overall energy demand, could benefit by entering into regional groupings. In countries with highly developed economies with substantial infrastructure extant for a prolonged period, there are opportunities for SMRs to replace aging power plants and sites with energy-intensive industries. In addition, countries developing desalination capability may also be suited for SMR deployment. Finally, countries that are actively engaged in the development of a domestic SMR manufacturing industry to capitalize on the economic impacts from doing so may be incentivized to deploy SMRs domestically and for export.

The IAEA's report (IAEA, 2018a) on the deployment potential for SMRs can be seen as a broadening of its Milestones in the Development of a National Infrastructure for Nuclear Power program (IAEA, 2015a). In the report, several of the characteristics and conditions within countries that are relevant to the development potential of large NPPs are pertinent for the deployment of SMRs. However, many of the unique characteristics of SMRs make them amenable to additional markets and applications.

---

[c] As noted earlier, additional key attributes, such as institutional capabilities, were not covered.

As a result the market assessment for SMRs encompasses more indicators than those used for the analysis of conditions amenable for large NPPs. The smaller size, broader range of outputs, and other features provide a wider range of opportunities.

22.2.3 SMR deployment conditions and regional energy aims

Puerto Rico provides a useful point of reference for understanding of the potential for SMR adoption and regional energy aims.

In 2017 the Puerto Rican region was devastated by Hurricane Maria, where 3000 people were killed, more than $102 billion in damage was sustained, and the region was left with an unprecedented challenge to rebuild the electrical power system (Siemens Industry, 2019). One year after Hurricane Maria, portions of the electrical grid were still not restored (Siemens Industry, 2019). Anticipating that future storms are inevitable, the Puerto Rico power authority initiated a reassessment of the power supply and infrastructure to prepare for future weather events (Siemens Industry, 2019). An Integrated Resources Plan was drafted to meet the near-term needs and long-term goals of energy independence, zero emissions, and resilient energy. Subsequently the Puerto Rico House of Representatives requested a formal investigation (House Resolution 1189) to evaluate the feasibility for the use of microreactors and SMRs (Caribbean Business News, 2018). This study, led by the Nuclear Alternative Project, includes a look at the high-level objectives and specific factors associated with siting a nuclear reactor in Puerto Rico (The Nuclear Alternative Project, 2019).

These factors are used to support a needs assessment that accounts for aspects of an energy, economic, environmental, and societal profile for a given region. The high-level factors are correlated in Table 22.1 with the IAEA SMR deployment indicators, described in Section 22.2.2 (IAEA, 2018a):

These indicative factors are not comprehensive in outlining requisite conditions for SMR adoption, yet they provide a useful basis for understanding the needs profile of potential adopter countries, particularly island nations. Feasibility study results from Puerto Rico are expected in 2020 and may serve as a guide for others considering future use of SMRs (The Nuclear Alternative Project, 2019).

22.3 Recent climate goals and initiatives

22.3.1 Implications of the COP21 Paris agreement and 2030 UN sustainable development goals on nuclear energy utilization

The Paris climate accord, known as the "Paris Agreement," entered into force on November 4, 2016, to address international priorities on climate change. Since then, scientific studies on the climate threat continue to be reported by the Intergovernmental Panel on Climate Change (IPCC) (IPCC, 2018). In the 2019 IPCC special report on global warming of 1.5°C, scientists indicated that human activities have already caused approximately 1.0°C of global warming above preindustrial levels (IPCC, 2018).

Table 22.1 SMR siting factors and SMR deployment indicators.

Objectives	Specific factors	IAEA SMR deployment indicators
Reliable baseload	• Resilience with extreme weather events • Flexibility to integrate with renewable energy sources	*Physical infrastructure*: infrastructure conditions *Energy security*: balance variable renewable energy
Zero-carbon emissions	• Zero-carbon emission and clean air goals • Effects of climate change in Puerto Rico	*Climate change*: reduce CO_2 emissions, achieve carbon reduction goals
Economic boost	• Job growth • Research funds for the University of Puerto Rico and public education • Attracting private investment • Retaining college graduates and attracting young professionals • Reduce public debt	*National energy demand*: growth of economic activity, per capita energy consumption, growth of primary energy consumption *Financial/economic sufficiency*: ability to support new investments
Protection of natural resources	• Preservation of agricultural land and water resources • Grid decentralization	*Physical infrastructure*: land availability, electrical grid *SMR energy demand*: cogeneration (desalination)
Energy independence	• Energy security • Self-sustainability • Cost of electricity • Create global export markets • Develop local manufacturing industries	*Energy security*: reduce energy imports, balance variable renewable energy *Financial/economic sufficiency*: openness to international trade

Note: These do not represent all features of a given attribute.
Source: Adapted from IAEA, 2018a. Deployment Indicators for Small Modular Reactors, Methodology, Analysis of Key Factors and Case Studies (IAEA TECDOC No. 1854). International Atomic Energy Agency, Vienna, Austria; The Nuclear Alternative Project, 2019. Nuclear4PuertoRico. https://www.nuclearalternativeproject.org/nuclear-puertorico.

Pathways to limiting the global temperature rise to 1.5°C require deep emission reductions in all sectors and a wide portfolio of mitigation options (IPCC, 2018). Nuclear and hydropower are the lowest life cycle carbon emitters among energy technologies (IAEA, 2018b). Further, nuclear power has avoided production of more than 60 Gt of carbon dioxide over the past half century and continues to avoid 2 Gt of carbon each year (IAEA, 2018b).

According to the IAEA PRIS database, there are 449 nuclear power reactors supplying about 10% of the world's electricity and contributing to 33% of all low-carbon electricity (IAEA, 2018c). Around the world, 53 reactors are currently under construction, with hundreds more needed by 2050 to replace retiring reactors and to meet future energy demands, particularly in developing economies (IAEA, 2018c). IAEA's annual nuclear projections report a different story (IAEA, 2017). Trends in country's nuclear power projections have shown declines due to challenges from low natural gas prices, variable renewable energy (VRE) sources, financial uncertainty, and concerns about nuclear safety. However, longer-term outlooks are much more positive when country policies account for climate change. OECD/IPCC analysis indicates that around 90% of electricity will need to be low carbon, if the 2.0°C climate change targets are to be met by 2050 (IPCC, 2018). To further limit temperature rises to 1.5°C above preindustrial levels, significantly, greater shares of nuclear power can be leveraged. Demands to replace fossil energy, especially retiring coal plants during the next decade, could provide a strong motivator for countries to consider replacing this power with nuclear power. The World Nuclear Association's Harmony Program suggests that at least 25% of electricity be provided by nuclear by 2050 as part of a clean low-carbon mix (World Nuclear Association (WNA), 2019a). Achieving this means a tripling of nuclear generation by 2050. Achieving this goal would require a considerable rollout of large reactors and an expeditious introduction of micro-, small-, and medium-sized reactors to markets not currently served by nuclear power (e.g., hydrogen and process heat) to replace fossil sources. To achieve increased nuclear power deployment, more conducive market and policy conditions are needed.

22.3.2 *Country use of nuclear in carbon mitigation plans*

The United Nations Framework Convention on Climate Change (UNFCCC) organizes and supports countries to implement agreements under the 2015 Paris Agreement. The elements of the Agreement most relevant to nuclear power are Article 4 (mitigation), Article 9 (finance), and Article 10 (technology) (Bodle et al., 2016). Nuclear power contributes to mitigation by impacting the magnitude and timing of the peaking of global greenhouse gas emissions and subsequent reductions. Projections on future nuclear deployment and the associated contributions toward emissions reductions are reflected in a country's nationally determined contributions (NDCs). The NDCs are updated every 5 years (beginning 2020) and will require progressively greater contributions to meet climate goals. In 2023 the first "global stocktaking" will quantify progress in emission reductions in view of the long-term goals of the Paris Agreement. Light water reactor-based SMRs (LWR-SMRs) could be included in country NDC updates in 2025. By 2030, advanced SMR designs may be included, along with increasing rates of deployment of the LWR-SMRs by 2035. Continuous innovation of technologies and processes should boost global response to climate change in the longer term, at the same time promoting economic growth and sustainable development. The recurring nature of the Paris Agreement provides a mechanism to periodically reevaluate the role of SMR technologies and to develop appropriate policies for research, development, and demonstration programs.

Access to nuclear is impacted by the availability of finance to enhance implementation. In the past, nuclear power was restricted under the Clean Development Mechanism and Joint Implementation, in conjunction with the Kyoto Protocol (UNFCCC, 2020). Under the new Paris Agreement, countries are free to specify any energy technology, including nuclear power, to reduce emissions in future NDCs (IAEA, 2018b). International cooperation is a critical enabler for developing countries and vulnerable regions to strengthen their action for the implementation of 1.5°C consistent climate responses, including improved access to finance and technology, enhanced domestic capacities, and focus on national and local circumstances.

Out of 30 countries currently using nuclear power, eight countries including Argentina, China, India, Japan, Jordan, Niger, UAE, and Iran specifically identified nuclear energy in their Initial NDCs to reduce carbon emissions (IAEA, 2018b). The United States and several EU countries are projecting a continued level of nuclear power as part of an economy-wide target for emission reductions (IAEA, 2018b). Countries with developing economies, including China and India, are leading in new reactor builds to meet their increasing energy needs while limiting environmental impacts from pollution and climate change (IAEA, 2018b). Many countries' NDCs stress the potential cobenefits of climate actions for poverty reduction, afforestation, ecosystem conservation, and biodiversity central to countries' commitments (IAEA, 2018b).

In addition to renewable energy sources, SMRs can improve access to clean, reliable, and affordable energy (UN Sustainable Goal #7) by providing transportable and mobile energy sources, not requiring large electrical distribution systems (IAEA, 2016). Nuclear can be used to meet energy demands for development, while improving certain aspects of energy security, reducing environmental and health impacts from fossil fuels, and mitigating climate change (UN Sustainable Goal #13) (See Sections 22.5.1, 22.5.2, and 22.5.10 for more discussion of security and environmental considerations).

22.3.3 Relevance of SMRs in climate goals, access to energy, and economic development

Beyond the existing countries with nuclear power programs, there are another 30 countries that are considering, planning, or starting a nuclear power program and have sought support from the IAEA (IAEA, 2016). Of these countries, four have power reactors under construction (UAE, Belarus, Bangladesh, and Turkey) (IAEA, 2018c). For many countries the size of their grid system is too small to host a large nuclear power plant (>1000 MWe). The rule of thumb is to not build a power plant any larger than one-tenth of the capacity of the grid to maintain reliability should the plant be taken offline for refueling or maintenance, or due to unforeseen events (IAEA, 2015a). Upgrades to infrastructure including enlarging the capacity of transmission grids may result in a larger investment than the power plant. For many of these nuclear newcomer countries, a smaller reactor not requiring major upgrades to the grid may be a more strategic choice. Another challenge for nuclear newcomer countries is the licensing of reactor designs. Many countries use nuclear technologies for nonpower purposes (e.g., medical and industrial) but lack the technical capabilities for nuclear plant licensing (IAEA, 2012). Instead, they count on existing nuclear-powered countries to initially gain license certification (e.g., the United States, the United Kingdom, France, Russia, and China), while they develop their nuclear infrastructure, expertise, and institutional capabilities to license nuclear plants (IAEA, 2012). These nuclear newcomers' first nuclear power plant would use essentially the same design and safety features as a nuclear power plant that is already licensed by an experienced regulator (IAEA, 2012).

Nuclear energy can be considered a resilient energy resource by a number of counts, as it is able to withstand various forms of strong weather events due to reactor containment structures being composed of steel-reinforced concrete, has highly trained operators, and closes operation coordination with emergency services. A recent example is the continued operation of Florida Power and Light nuclear plants during Hurricane Irma (ANS Nuclear Café, 2017). However, most current reactors require emergency diesel generators to provide reactor cooling in case of the loss of external electrical supplies. The knockout of the emergency generators led to the loss of cooling accident at Fukushima Daiichi in March 2011. Since the accident the NRC and the nuclear industry reevaluated the risks from potential flooding and seismic effects on their sites. As a result, many plants were modified to add protection to plant structures, systems, and components, or the plant operators identified alternative strategies to facilitate the safety of the reactors in the event of a flooding or seismic event (IAEA, 2015b).

In the future, nuclear reactors might serve as a climate adaptation tool to support communities recovering from severe weather events (hurricanes, tornados, etc.) by providing energy, when other sources are disabled or cut off from fuel supply channels. Microreactors could 1 day replace diesel generators as an emergency energy source. The Puerto Rico study on the feasibility of microreactors and SMRs could inform future policy decisions on the use of nuclear energy.

22.4 Disruptive change: A closer look at global shifts and SMR options

For decades, small nuclear reactors have been at the forefront of technologies used for propulsion in submarines and large surface ships (CRS, 2019). With increases in global energy demand and pressures to reduce carbon, various options exist for new small reactor technologies, including microreactors.

22.4.1 The role of SMRs in connection to global energy demands

In today's global energy system, notable trends are evident as indicated earlier with a continued rise in energy demand, along with advances in urbanization and growth of megacities, plus more targeted prioritization of carbon reduction (Araujo, 2017; UN DESA/Population Group (UN DESA), 2018; Urban Gateway, 2018). Global energy demand, for instance, increased 60% between 1990 and 2018 (IEA, 2019a,b). Under current and planned policies, it is projected to grow by more than 25% to 2040, requiring an estimated $2 trillion a year in investment in the new energy supplies (IEA, 2018). In the period from 1990 to 2018, the number of megacities (e.g., cities with more than 10 million residents) also increased from 10 to 28, as global populations shifted from rural to urban areas (Urban Gateway, 2018). This trend is expected to continue, with over 1 billion additional people lacking access to electricity by 2030 (IAEA, 2016). Alongside the earlier trends the focus on carbon reduction has risen on public agendas, as fossil fuels represent about 85% of the global primary energy (Araujo, 2017; IEA, 2018, 2019b; BP, 2019; Araújo, 2014). Future mass migrations are also possible due to wars, natural and man-caused disasters, and climate change, creating demands for future energy systems that are more mobile and adaptive (Grafham, 2019).

SMRs could support such demands due to the reactors' flexible siting attributes, ability to integrate with VREs to balance energy demands on the grid, conduciveness for smaller electrical grids, shorter construction times, capability to match demand growth by incremental additions of reactor modules, and suitability for nonelectric applications (e.g., desalination) to replace fossil sources.

As governments and industry grapple with ways to meet low-carbon energy priorities, particularly in megacities, the development of SMR technologies offers a somewhat novel pathway for replacing high-carbon energy sources. Currently, 50 SMR designs are estimated as being under development globally (IAEA, 2019a).

In conjunction with the design and development of SMR technology, international institutions and partnerships are playing a role in the technology's evolution. The IAEA, for instance, convenes technical meetings and reports on globally accepted practices, among its activities. Since 2015 the IAEA has supported knowledge sharing on regulatory experience through the Small Modular Reactor Regulators' Forum (IAEA, 2019b). The Nuclear Energy Agency (NEA) initiative, Nuclear Innovation 2050, likewise aims to accelerate R&D and market deployment of innovative nuclear fission technologies (Nuclear Energy Agency/OECD, 2019). The Generation IV International Forum (GIF) is involved in international nuclear technology knowledge sharing. Established in 2000 the GIF represents the governments of 13 countries, of which most are committed to joint research and development for the next generation of nuclear technology (Generation IV International Forum (GIF), n.d.). The World Association of Nuclear Plant Operators is another international entity of relevance that fosters safety cultures and practices among its members (World Association of Nuclear Plant Operators, n.d.).

22.4.2 *Pathways with advanced nuclear technologies including SMRs and microreactors*

As small modular reactors are brought closer to market for commercial energy and industrial use, several pathways could be deployed at scale. Specific to energy utilization, SMRs may contribute to an increasingly distributed power system and sector. As a baseload source of power with load-following capabilities, nuclear energy sourced from the smallest SMRs (i.e., microreactors) have a profile that could meet needs of remote operations and provide flexible integration with VRE sources (see

Section 22.4.3). Sites with smaller grids or off-grid communities (e.g., islands, Arctic villages, remote research, and exploratory and extractive activities) are prospective candidates for "community scale" SMRs as electricity costs are typically high and based on diesel, oil, or gas-powered generation. Deploying SMRs to remote communities and operations could allow for capabilities in combined heat and power, as the offtake can provide higher temperatures used in process industries and lower temperatures for water desalination and district heating.

Microreactors, as a class of reactors (smaller than 20 MWe) within SMRs (typically ranging from 20 to 300 MWe), provide further potential for disruptive change, since they require little in terms of siting or infrastructure. In addition, microreactors offer options in terms of transport and mobility, can operate for years without refueling, may cogenerate electricity or heat, and can be operated independent from the grid. Among potential early users the US Department of Defense is a potential one for operations that are energy intensive and require portable, dense power sources (Idaho National Laboratory (INL), 2019).

Going further, SMRs could also be used to replace the electricity from retiring coal plants. Like large reactors, SMRs have a small greenhouse gas footprint. For the period between 2010 and the first quarter of 2019 in the United States, for instance, more than 546 coal-fired power plants, totaling approximately 102 GW of generation capacity, were announced as retiring, and another 17 GW is expected by 2025 (US EIA, 2019) (see also Fig. 22.7). US coal-fired plants that were taken offline from 2010 to 2012 averaged 97 MWe, and for the period from 2015 to 2025, the average coal plant size is expected to be 145 MWe (World Nuclear Association (WNA), 2019b)—all conducive to SMR technology, with capacities generally up to 300 MWe for which their structures, systems, and components can be shop fabricated and transported to a site as modules.

Another area to consider is industrial heat. Such heat provided by fossil sources currently equates to roughly one-fifth of global energy consumption (Bellverat, 2018). With increases in industrial heat use projected to rise from 2500 millions of

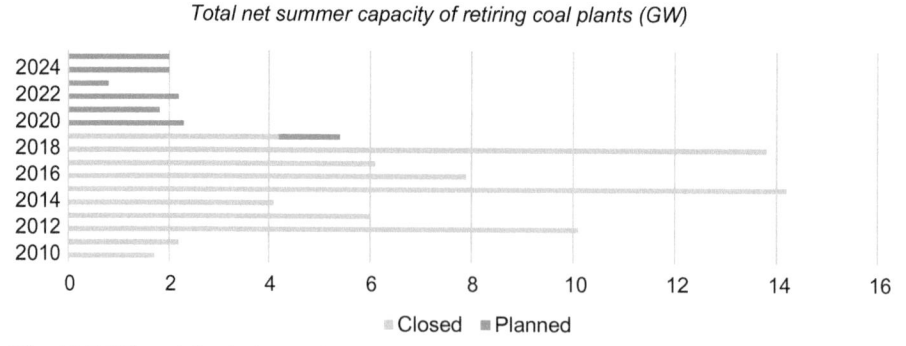

Fig. 22.7 US coal-fired plant retirements.
Source: Adapted from US EIA, July 26, 2019. More U.S. Coal-Fired Power Plants are Decommissioning as Retirements Continue. https://www.eia.gov/todayinenergy/detail.php?id=40212.

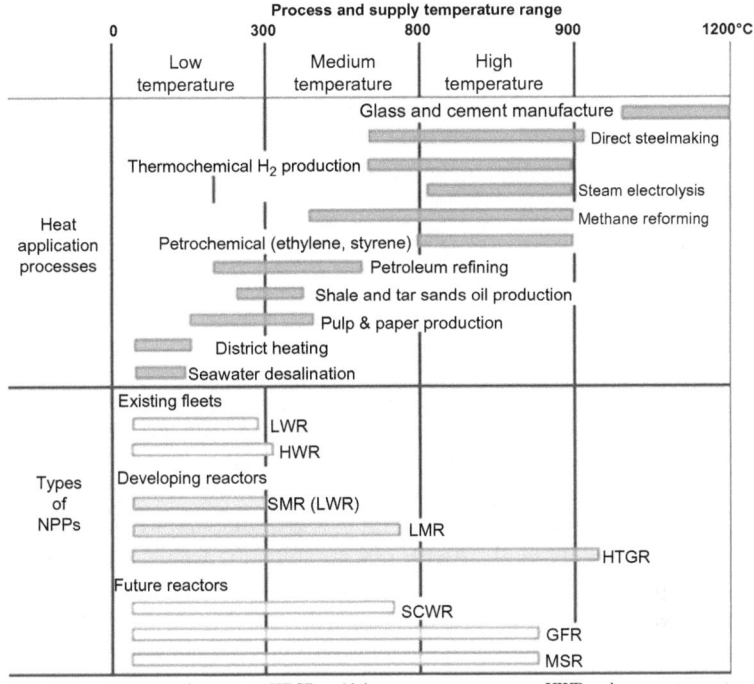

Fig. 22.8 SMR designs for nonelectric applications.
Source: From IAEA, 2017b. Opportunities for Cogeneration With Nuclear Energy (IAEA Nuclear Energy Series No. NP-T-4.1). IAEA, Vienna, Austria.

tons of oil equivalent (Mtoe) in 2016 to over 3000 Mtoe in 2040 (Bellverat, 2018), this area has potential for SMR nonelectric applications. Among such applications, desalination, district heating, and hydrogen production prominently figure in SMR designs (Fig. 22.8). NuScale's light water reactor design in the United States, for example, is exploring cogeneration for desalination, oil recovery, and refinery power, plus hydrogen production with high steam temperature electrolysis (World Nuclear Association (WNA), 2019b). High and very high-temperature reactors can produce "high value" process compatible heat from 500 to 700°C, which opens SMRs to additional uses. SMRs with smaller capacities (50–250 MWt) are the most compatible with process industries that are sited locally to minimize heat transmission losses.

22.4.3 SMR integration with renewables in distributed and hybrid energy systems including storage

In distributed systems, SMRs can provide baseload electricity while integrating with variable renewable energy (VRE) sources to help balance energy supply and demand on the power grid. SMRs can further integrate with variable sources by flexible operation, utilizing energy storage, enhanced maneuverability, and connection to cogeneration applications (e.g., heating and desalination) in a hybrid energy system (HES). HESs use multiple energy resources (e.g., nuclear, VREs, and natural gas) as inputs to physically coupled subsystems to produce one or more energy products (electricity, heat, and hydrogen). Integrated energy conversion processes can optimize energy management, reliability, security, and sustainability. HES can be used to provide reliable energy supplies and increase efficiency of conversion of available energy resources into infrastructure compatible products (IAEA, 2018d). SMRs with advanced engineered features are ideal for use in an HES due to their size and flexibility and use of enhanced safety features. They can support electric and heat needs of a wide range of users.

Integrated systems using SMRs can support the larger scaling of VRE sources on the grid. Specifically, SMRs may provide firming power generation to back up supplies from the wind and follow load. Integrated systems can also provide economic benefits by reducing power reserve requirements, which in turn reduces the backup capacity needed to maintain system reliability. In a study conducted in the EU, a 700-MWe SMR was integrated with a 1000-MWe offshore wind farm located in the North Sea (Shropshire et al., 2012). The combination resulted in 80% less wind power variation to the grid, while at the loss of 30% capacity utilization by the SMR. The research suggested that minimizing variability was best achieved with an SMR using multiple modules that could be turned off or on, versus power ramping of all modules. In demand-following mode the integrated system maneuvered output to improve synchronization with demand by 60%–70% over a wind-only system. Power variability was indifferent to the SMR module size. Balancing to eliminate all variation required additional balancing measures (energy storage, H_2 production, demand-side management, etc.) (Shropshire et al., 2012). Other EU studies expanded the integrated system (SMR and wind farm) to include biomass processing and hydrogen (H_2) production. Researchers found that the hybrid system reduced variability with demand by 32%. Notably, offshore wind has evidenced considerable improvements in performance in recent years, so the examples are illustrative (Araújo, 2019). Further, reductions could be made by adding H_2 generation (as energy storage during peak VRE production), however, at a very low utilization (15%) of the H_2 plant (Papaioannou et al., 2013).

A more comprehensive discussion on SMR integration is provided in Chapter 13.

22.5 Challenges and opportunities

The previous sections of this chapter have reviewed the features of SMRs, both in terms of design and in terms of their amenability to different types of markets and applications. A country considering the deployment of SMRs as part of its future energy supply, however, must contemplate a number of issues regarding not only a nuclear power but also a nascent technology. In this section, specific considerations influencing the deployment of nuclear technologies in general, and SMRs specifically, are addressed. The opportunities provided by SMRs to meet low-carbon energy development in a variety of market settings, as discussed earlier, must also take into consideration some of the issues particular to nuclear energy and to a new energy technology that requires both off-site manufacturing and on-site construction. Issues with regard to nuclear technology include attention to matters relating to risk, high capital expenditures, the transport of and storage of new and spent fuel, decommissioning, and the economic viability of a new nuclear technology. In addition to the considerations involving nuclear energy, SMRs pose both challenges and opportunities due to their use of modular construction that requires the development of new manufacturing sectors and supply chains. However, because of the desirability of having a diversified energy portfolio and of utilizing low-carbon energy technologies, the policies being employed in a variety of countries to address these concerns are also addressed in this section.

22.5.1 Fuel requirements and the transport of nuclear fuel and modules

As seen with other sections of this chapter, the fuel and transport needs of SMR technology have similarities and differences with larger nuclear reactors.

As with larger reactors, SMR's fuel requirements depend on design choices, among other factors. Specific to LWR-SMRs, UO_2 fuel may be expected with shorter assemblies for many of the designs reviewed to date (Bradford, 2014). Uranium enrichment could then be expected to follow the fuel cycle rules and requirements for <5% enrichment that also applies for larger nuclear power plants. In terms of waste, the composition of low-level waste can then also be expected to be similar to that for larger LWRs. In the United States, 10 CFR 61 regulation and guidance would apply (Bradford, 2014). Due to the differences in the size, vendors for transport and casks may be required to revise their designs (Bradford, 2014).

In contrast to LWR-SMRs, non-LWR SMRs typically require higher enrichments from 5% up to <20% and different fuel forms, such as ceramic coated particle fuel (CRS, 2019; Bradford, 2014). The design, licensing, and construction of new types of fuel fabrication facilities will then also likely be required (Bradford, 2014). Similarly, transportation and cask vendors may be asked to develop new designs. Here, new research and computer models may be necessary to support assumptions about fuel and spent fuel characteristics or behavior. NRC regulations may also require modification (Bradford, 2014).

Many SMRs require less frequent fueling, for example, every 3–7 years versus 1–2 years, and some could even operate without refueling for the life of the reactor. The KLT-40S refueling interval is 3–4 years on-site, and at the end of a 12-year operating cycle, the whole plant will be returned to a shipyard for a 2-year overhaul and storage of used fuel, before being returned to service (IAEA, 2018b). In essentially all cases, some amount of used fuel will require management during the life of the reactor.

Specific to the management of fuel and related infrastructure, processes for fuel from SMRs in existing nuclear countries could be similar to that for large reactors. New SMR reactors will likely first be deployed in a vendor country where fuel management infrastructure exists. Nuclear newcomers could then learn with initial deployments about management practices and related procedures. New reactor technologies, such as high-temperature gas-cooled reactors or fast reactor technologies, could also use existing infrastructures or adjust to new fuels.

In addition to enrichment and fuel management needs, consideration must also be given to the transport of radioactive waste and how it affects surrounding communities, the risk of water contamination, accident prone risk, and the ecology of the area. Transport of used or spent nuclear fuel generally follows established rules. However, the transport of a commercial nuclear plant loaded with fuel is a new concept, which can be expected to require new security norms and standards (Nuclear Energy Agency/OECD, 2016).

Some SMR and microreactor modules may be transported to a site with the fuel intact and then be returned for servicing or decommissioning. This activity may require transport and cask vendors to revise designs due to differences in size (Bradford, 2014). The transport of SMRs and SMR fuel is an area that will become more prominent in analysis as designs mature and are licensed for implementation.

Specific to methods for transporting nuclear fuel and refueling, these follow practices and guidelines developed by regulatory agencies and the IAEA. In the United States, for example, transportation of civilian used nuclear fuel is regulated by the US Department of Transportation (DOT) and the US Nuclear Regulatory Commission (NRC). DOT regulates the shipping routes and transportation conveyances. The NRC approves the design, fabrication, use, and maintenance of shipping containers and regulates security of the used fuel. Transportation containers are, in particular, designed to protect the public from releases of radioactive material in the unlikely event of an accident.

For international shipments the process is much more complex. There is transport within the country of origin, shipment across international space (air, water, or land of third countries), and final delivery in user country. Some reactor systems can be broken into parts for shipment. If a system can be fueled on-site, problems associated with shipping a fueled reactor can be eliminated.

Looking ahead, transportable SMRs could open markets to new customers with project energy needs of limited duration (e.g., 20 years). For example, the KLT-40S floating nuclear reactor is placed on a ship that is barged by sea to users at coastal locations. The onshore infrastructure requires less transportation links, power transmission lines, or preparatory infrastructure than land-based NPPs, and there is a higher degree of freedom in selecting the location for the SMR, as it technically may be moored in any coastal region (IAEA, 2018e).

22.5.2 Remote operations and security

Challenges and opportunities exist for SMRs in remote operations and with respect to security.

The smaller size and modularity attributes of the technology allow for additional installation possibilities including off-grid applications and the potential to install the reactors underground or on floating vessels. If installed underground, security might be improved by the natural isolation of the containment area from external disruptors and releases. However, such siting also has inherent natural risks associated with underground conditions, including flooding, ground shifts associated with ice melt or landslides, and seismicity (see Section 22.5.10). If deployed on a marine vessel, an SMR may be less susceptible to those hazards but will be required to withstand more marine-based conditions.

Siting of an SMR will also have certain inherent security needs for protection against terrorist interference. SMRs sited on an existing NPP site or a military base will have incumbent security infrastructure that will need to continue with ongoing updates and vigilance. SMRs sited at a greenfield location and especially remote operations will require additional security measures, such as semiautonomous operation, in person and remote monitoring systems, and related measures. The security and resiliency of some microreactors may be hardened through packaging in a secure canister, installation in a strong secure vault on-site, and connection to a microgrid that coordinates both heat and power.

Specific to SMR security risks and costs, such considerations may be reduced by eliminating the need for on-site fuel cycle operations by designing long-life cores, by sealing the reactor vessel, and physically locating the reactor underground. Extending the refueling intervals can be achieved through use of high-assay low-enriched uranium (HA LEU) with fuel enrichments from >5% up to 19.75%, achieved through high-enriched uranium (HEU) recovery from used fuel (US DOE, 2019). Reactor systems with HA LEU need to be designed to be hardened against cyber and electromagnetic pulse attacks.

Turning to remote Arctic locations, there are added considerations such as a limited time window when fuel and coolant can be delivered by secure transport on open sea or by ice road. More flexibility is offered if the primary vessel dimensions and weight are kept within limits for air transport. Air freighters capable of landing on nonpaved runways indicate that a maximum vessel weight of less than 30 tons and a diameter of less than 2.8 m are desired design objectives (i.e., transportable by a C-17 aircraft) (IAEA, 2018e).

22.5.3 Used fuel storage

Specific to storage of used fuel, characterizations relative to large reactors may also be seen.

For water-cooled SMRs using standard fuel and fuel assemblies, fuel management is similar to conventional large PWRs. However, fuel may not achieve high burnups (due to reduced reshuffling in small cores). The fuel throughput and the relative volume (per unit of energy produced) of spent fuel could also be higher. Wet (pool) storage capacity should support 20 years of storage, which can be followed by dry storage and/or reprocessing. High-temperature reactor (HTR) SMRs using TRISO fuels can retain fission products to provide high proliferation resistance, but these fuels are impractical to reprocess.

SMR procurement contracts may be designed to simplify the fuel cycle infrastructure requirements needed by nuclear newcomers by contracting of fresh fuel and refueling operations, arranging used fuel take back that minimizes localized considerations about the development of a permanent repository, contracted services for low-level waste treatment and disposal, and technical training provided by vendors or through agreement with nuclear countries.

22.5.4 Decommissioning and decontamination

The Nuclear Energy Agency notes that transparency around decommissioning and decontamination (D&D) costs and financing is fundamental to understand the actual costs of current D&D projects and to prepare better estimates of future projects (Nuclear Energy Agency, 2016).

Nuclear projects in the United States are required by national legislation to have adequate financial resources in place before the start of facility operations for D&D. Cost estimates are prepared at the onset of the nuclear project that include D&D planning, pre-D&D actions (e.g., radiological characterization), D&D actions (building dismantlement and demolition and refurbishment), waste removal, and site monitoring. D&D costs are historically viewed as a small sum compared with the cost of construction and operation of the reactor. However, recent decommissioning experience on large reactors has not produced reliable and comparable information on costs, largely due to vendor competitiveness concerns.

SMRs will have similar D&D requirements as large reactors. However, costs should generally scale with the smaller source term, with some exceptions based on the type of reactor. For integral PWR-type SMRs, the reactor vessel would be large to contain the primary system, so structural waste for decommissioning can be relatively larger. For a multimodule SMR plant, issues for plant decommissioning include Co-60 from reactor vessel, internals, and system piping, C-14, Ni, and Fe isotopes. For high-temperature reactor SMRs, key issues for plant decommissioning may include activated cylindrical steel core and graphite brick reflector C-14 and Cl-36 isotopes. Less conventional SMRs using a monolithic core with heat pipe technology (e.g., Westinghouse eVinci microreactor) lack a large bulk coolant and do not require the separation of fuel from the coolant that minimizes the need for separate fuel storage and minimizes the need for D&D. These designs allow for the reactor to be repacked for reuse or prepared for long-term storage at the factory (IAEA, 2018e).

22.5.5 Financing

Since the 1960s in the United States, the public sector was actively involved in the financing of the first commercial nuclear reactors. Financing often required both public (guaranteed funds) and private mechanisms to enable risk sharing across debt and equity funding. Higher-risk projects require a higher share of equity funding (e.g., 60%–80% equity/40%–20% debt) with higher loan rates than for debt financing (e.g., 10% vs 4%, respectively). The lowest project finance costs are for fully government-constructed facilities that are privately sold, due to government forgiveness of debt as sunk costs. In contrast the major cost contributor under a private ownership model is the interest on debt and equity for the initial capital investment. Any mechanism causing the overnight capital investment or loan rates to decrease will subsequently result in lower project costs (Shropshire and Chandler, 2006).

Nuclear projects share characteristics with large construction projects that carry similar risks (e.g., regulatory uncertainty, cost overruns, price risks, and technical performance). However, some aspects are unique to nuclear (i.e., NRC licensing, emergency planning zones, and radioactive waste disposition). Project financing requires a multifaceted approach that is managed by multiple mechanisms. The finance package for a nuclear project represents the summation of the valuation of the risks and equates them to the financial returns required by investors to accept these risks. The higher the perceived risk of the investment, the higher the rates negotiated. A new SMR, like building an advanced reactor, lacks prior construction and operational experience and therefore is subject to higher risks. However, many of the current SMR designs are based in part on current light water reactor designs to justify lower-risk premiums.

Due to changing electricity markets, SMRs and advanced reactors may require additional mechanisms for financing. Nuclear plants face more financial uncertainty in liberalized markets compared with regulated markets due to low-cost competitors (e.g., natural gas or renewable energy with policy support or declining costs). Risks can be managed through direct government financing in regulated markets, intergovernmental loans between governments, corporate and state-backed vendor financing (e.g., Russia and China), loan guarantees (e.g., DOE's Title XVII loan program), regulatory reforms, and negotiated risk sharing and credit support between industry and government agencies. Different types of risk (e.g., capital costs, market pricing, lack of penalty for emissions, or used fuel disposition) are weighted more heavily in assessment of premiums.

In several nuclear energy-powered countries (France, China, Russia, and India), energy resources are owned by the national government due to the necessity for government function. Plants in these countries enjoy less market restrictions and more reliable demand growth. Other countries (the United States) leave energy choices largely up to the market to decide. To increase nuclear adoption, legislative/policy tools (e.g., state zero emission credits) could compensate nuclear power for not emitting greenhouse gases to produce electricity. Also, new financing mechanisms could facilitate the financing of new plants, including SMRs. These include public-private partnership (PPP) structures and new ownership models (e.g., Mankala model, power purchase agreements, and contract for difference) that provide price guarantees to share and balance the risks of investing in a power plant. Other support mechanisms that could benefit SMRs include capacity remuneration, tax incentives, and carbon pricing and trading (IAEA, 2018f).

Compared with traditional nuclear reactors, SMRs require less capital to finance, are faster to construct, and have shorter payback periods. They also have facility footprints that are a fraction of the size of large NPPs and small exclusion zones (e.g., facility boundary). SMRs can earn additional revenues due to their flexibility for a wide range of users and production of nonelectric products (e.g., desalinated water and hydrogen). Due to modularity, SMRs can also benefit from self-financing of additional modules from revenues gained by initial modules (Mignacca and Locatelli, 2019). Financing may also be available from federal facilities to purchase power for up to 30 years. In some communities impacted by extreme weather events, related financing is also offered for installation of affordable resilient energy systems that may apply (Scully Capital, 2018).

22.5.6 Cost competitiveness

Loss of traditional supplies of baseload energy from retiring coal and nuclear plants along with increased shares of VRE technologies is changing the composition and behavior of the energy portfolios of utilities. Energy technologies built today need to be more flexible, resilient, safe, and consistent with a low-carbon energy transition. This new competitive environment, although not fully understood, is changing the way we think about nuclear power.

Researchers have conducted numerous studies to analyze the potential for SMRs to achieve cost parity with large reactors based on traditional economic measures including direct and indirect capital costs and the levelized cost of electricity (LCOE). Several posit the possibility that the loss of scale economies compared with large NPPs (e.g., Mignacca and Locatelli, 2019; Kuznetzov, 2008; Carelli et al., 2008; Ramana and Mian, 2014) are likely to result in higher capital costs for SMRs than large NPPs. Others note that SMR features such as design simplicity, reduced componentry, modularity, and faster construction are likely to overcome the loss of scale economies and make their initial capital costs lower, on a per megawatts electric basis, than traditional large NPPs (Kuznetzov, 2008; Locatelli et al., 2014; Solan et al., 2010b). A recent study by Black et al. (2019) supports the latter view by utilizing a general methodology and vendor data from the NuScale SMR design to demonstrate that the higher direct capital costs for SMRs are more than offset by the significantly lower indirect capital costs. The cost reductions in indirect capital costs, due to SMR design advantages of the NuScale design, were significant, with the result that significant overall cost savings on a per kilowatt-hour basis in the overnight capital costs for SMRs compared with large NPPs (Locatelli et al., 2014).

In addition to overnight capital costs, an important metric for evaluating the economic viability of SMRs relative to other energy sources is the LCOE. Estimation of an energy source's LCOE provides a basis for evaluating the cost per unit of electricity produced over the lifetime of a power-producing facility. While it considers construction and other capital costs involved in overnight cost estimations, it also accounts for the ongoing costs over the lifetime of a plant and the amount of energy produced. As a result the LCOE method allows power plant investors to compare energy producing technologies with different capital, operating, fuel, and other costs. The relatively high capital costs of traditional nuclear power plants have contributed to their historically higher LCOE estimates as compared with other energy sources (U.S. Energy Information Administration, 2016). However, the same features of SMRs that contribute to the possible mitigation of the loss of economies of scale in their capital costs also contribute to their relatively lower LCOE estimates. For example, design simplification, reduced componentry, smaller size, and other features of SMR contribute to their relatively lower operations and maintenance costs (Black et al., 2017). In addition, their relatively low fuel costs, high capacity factor, and long production life all serve to reduce the lifetime LCOE costs and are likely to make SMRs economically competitive, especially with respect to fossil fuel technologies. An additional advantage is that the operating costs and attendant LCOE estimates of SMRs are significantly less subject to risks associated fuel costs. For example, because fuel costs represent only about 15%–20% of total generating costs for nuclear power, the generating costs for nuclear plants are much less sensitive to fuel price volatility than for coal or natural gas (Nuclear Energy Institute, 2017). Estimates of the effects of fuel price increases are that a doubling of fuel cost will result in an increase of approximately 10% for nuclear, while a doubling of fuel costs will increase the generation costs by 32% for coal and 77% for natural gas (University of Utah Bureau of Economic and Business Research, 2012; IEA, Nuclear Energy Agency, 2015). The sensitivity of generating costs to changes in fuel costs is especially important when comparing LCOE estimates for SMRs versus natural gas, given the significant volatility in natural gas prices (Black et al., 2017).

It is important to note, however, that although comparisons of capital costs and LCOE are useful, they may well provide a misguided view of future SMR competitiveness. Electricity markets that value reliable energy sources that do not pollute, that can function flawlessly with high shares of variable renewable sources, where fossil fuels are no longer needed to produce process heat for industry, and where hydrogen is produced at the scales needed to run transportation systems will ultimately result in a new set of metrics for economic valuation, where SMRs are well positioned to be competitive. On the other hand, given that no commercial SMRs have yet been deployed and operational, commercial banks and electrical utilities are not generally inclined to take on first-of-a-kind (FOAK) risk of a new reactor without a government guarantee. Initial deployments of SMRs will be in selected markets that value "conservation and efficiency to address the long-term energy needs." Federal support of FOAK deployment of SMRs may use loan guarantees and federal facility power purchase agreements (PPA) to help FOAK reactors by guaranteeing a contract for at least a portion of the SMR output. The first SMR deployment in the United States is the Utah Associated Municipal Power Systems (UAMPS) Carbon-Free Power Project, consisting of the installation of 12 NuScale 60 MWe modules, for a net capacity of 685 MWe of power, at the Idaho National Laboratory site in eastern Idaho (NuScale Power, LLC, 2020; Black and Peterson, 2019). In this intermountain regional market, the FOAK SMR competes with alternative resource portfolios consisting of wind, solar, and natural gas as a replacement for retiring coal generation (Energy Strategies, 2019).

Competing factors from VRE depend on renewable energy capital costs, geographic specific capacity factors, energy storage costs, market purchases, and costs for additional dispatchable sources of supply to provide load balancing. NEA reports that system costs vary between less than $10/MWh of VRE for a share of 10% of wind and solar PV to more than $50/MWh of VRE for a share of 75% of wind and solar PV (NEA, 2019). Consideration of system costs can make SMRs more competitive in portfolios consisting of high shares of VREs.

The potential economic viability of SMRs in future European energy and cogeneration markets was studied by the European Commission Joint Research Centre

(Shropshire, 2011; Carlsson et al., 2012). Researchers developed "target costs" for SMRs deployed in a range of competitive market situations. SMRs were found to be most viable in small electrical distribution markets competing with natural gas with carbon capture and storage (CCS) (80–150€/MWh) (Shropshire, 2011). The greatest potential for SMRs producing combined heat and power (CHP) is in chemical/petroleum, paper, metal, and bioenergy markets with small capacities (50–250 MWth), competing with coal CHP (95–250€/MWh) and coal CHP with CCS (104–208€/MWh) (Carlsson et al., 2012). Sensitivity analysis showed that the variables with highest uncertainty were capital costs, financing, reactor capacity factors, and fuel costs. The studies reported that SMRs were most competitive when (1) they are well matched to a market niche, (2) financial risks are managed, (3) innovation is applied to reactor and fuel designs, (4) economies of scale are replaced by economies of replication, (5) FOAK deployments are successful, and (6) operational efficiencies are maintained for the life of the plant (Shropshire, 2011; Carlsson et al., 2012).

Initial military applications of FOAK SMRs can also reduce the anticipated risks of future SMR commercial deployment. Military deployment and practical experience gained from the reactor can enable countries building their first SMR to have greater confidence in their deployment. In the United States the 2019 National Defense Authorization Act (NDAA) directed the secretary of energy to report to congress within a year on the requirements and components needed for a pilot program to "site, construct and operate at least one licensed microreactor that provides resilience for national security infrastructure at a DOD or DOE facility" (Nuclear Energy Institute, 2019). Subsequently a solicitation for a mobile nuclear reactor was launched by the US DOD under the Pele Program. The microreactor (1–10 MWe) is required to meet technical requirements to produce energy while deployed in the field of operation under specific size/weight, fuel, power, safety, black start, cooling, health and safety, and other military readiness requirements (Nuclear Energy Institute, 2019).

22.5.7 Policies in the changing playing field

To level the playing field for nuclear power, some states in the United States are using legislative and policy enablers including clean energy standards (CES). For example, large nuclear plants in nonregulated markets in New York and Illinois were becoming uneconomic due to price pressure from low-cost natural gas and renewable energy with policy support or declining costs. CES were implemented to keep these low-carbon baseload plants running. New York has expanded CES so that 70% of New York's electricity comes from renewable energy sources such as solar and wind by 2030 (New York State, n.d.). SMRs could benefit from the CES as a backup to VREs. The policy is intended to bring investment, economic development, and jobs to New York State. Several other states are following this lead, including Washington, Colorado, New Mexico, New York, and Alaska (DSIRE, n.d.). Further, implementation of these policies at the state and federal level could improve the economics of SMRs.

Zero emission credits (ZEC) help ensure that low carbon sources are used to provide a share of the energy. ZEC are payments that electricity generators receive to compensate them for the favorable attribute of not emitting greenhouse gases in the production of electricity. Several states have implemented ZECs (New York, Illinois, Connecticut, New Jersey, and Ohio). They provide a source of revenue for clean energy that would benefit low-carbon energy sources including SMRs (Nuclear Energy Institute, 2018).

22.5.8 Nuclear plant construction

Trends in nuclear markets around the world show sovereign governments (e.g., Russia, China, France, and ROK) constructing reactors for domestic use while concurrently investing in export markets (Southern States Energy Board, 2019). Many of these countries do not leave nuclear competitiveness solely to market forces. They value nuclear beyond economics, including for energy supply security or foreign policy aims. The engaged export markets are often characterized by limited domestic energy sources (oil and natural gas). Here, nuclear suppliers offer financing, while retaining controls over suppliers, fuel, and electricity rates (Southern States Energy Board, 2019). For example, China plans to build 30 nuclear power plants in countries involved in its Belt and Road initiative by 2030 (GB Times, n.d.). This initiative, building trade partnerships along the ancient Silk Road across Asia, Europe, and Africa, is expected to help countries build a nuclear power industry chain (GB Times, n.d.). Similarly, Russia's Rosatom uses nuclear to create partnerships with nuclear newcomer countries and BRICS countries. Rosatom constructed six nuclear power reactors in India, Iran, and China and has nine reactors under construction in Turkey, Belarus, India, Bangladesh, and China (The Moscow Times, n.d.). Additional projects are in the planning stages within emerging markets around the world (The Moscow Times, n.d.).

To help market nuclear services including nuclear plant construction, countries create special economic zones to attract foreign capital for nuclear finance. In mainland China, special economic zones (SEZs) operate closer to a free market as compared with the Chinese-planned economy. SEZs are attractive for foreign and domestic firms to do business and offer lower taxes, reduced regulations, and business incentives to attract foreign investment and technology. The products produced are primarily export oriented and are driven by market forces (Wikipedia, n.d.-a).

22.5.9 Economies of production

Producing mass scale, standardized modules to reduce costs will be challenging. Achieving cost reductions from FOAK to NOAK requires an economy of production (vs scale). Factory fabrication of modules provides an environment where quality and safety may be closely controlled in the production line, workers gain proficiency through familiarity and repetition, and machine cells replace manual operations for less costly production (see Section 22.5.10 for discussion of regulation of factory fabrication). Standardization and producibility engineering are built into the design. Setting up a supply chain to support factory fabrication requires an order book of SMRs large enough to satisfy investors of cost recovery, which is not yet secured. A current study estimates that cost savings for modular over stick-built reactors may be about

15% with a schedule improvement of about 37% on average (Mignacca and Locatelli, 2019).

As a corollary to factory production of SMRs, we look to other highly regulated/low-medium production industries. Space X developed the Falcon 9, a two-stage rocket for reliable and safe transport of satellites and the Dragon spacecraft into orbit. Development was accelerated by NASA providing funding for development and purchase of several commercial flights. The SpaceX Hawthorne production facility produces about 70% of each Falcon launch vehicle to avoid single-source parts dependency and to improve quality, cost, and schedule control (Wikipedia, n.d.-b). The Falcon 9 was designed for ease of manufacturing, with commonality across subassemblies to facilitate rapid production. For example, different size tanks and engines share the same tooling, material, and manufacturing techniques. Starting December 2010, production rates started at four rocket cores/year increasing to eight/year over 2 years. In September 2013, manufacturing space was increased to nearly 1,000,000 ft^2 (93,000 m^2), to provide a capability to produce 40 rocket cores/year. To date, 77 Falcon 9 rockets (priced at $62 M each) have been launched over the past 9 years (Wikipedia, n.d.-b). The SpaceX experience provides some useful insights to how and what may be achieved for factory production of SMRs.

22.5.10 Sociopolitical and related environmental considerations

The licensing and regulation of SMRs will play a critical role in the technology's adoption. Proponents of small reactors indicate that such technology designs can pose less of a safety risk due to a reduced total source term of radioactive material on a site, lower risk of an atmospheric release of radionuclides especially when sited underground, and incorporation of passive safety measures (IAEA, 2012; Boborov et al., 2019).

New designs can include passive safety measures that rely on forces of physics such as convection, gravity, or conduction rather than operator intervention to cool a core during a serious event, high-temp resistant materials to minimize core meltdowns, forms of seismic protection, and simplified designs that can enhance access to vital valves and corridors (Boborov et al., 2019; Benedict, 2014).

According to some the technology should then have streamlined approval processes, based on these features, in conjunction with risk-informed regulation (CRS, 2019). Others, however, counter that safety from an accident or terrorist attack with underground siting of a reactor remains problematic as cleanup efforts may be impeded and releases may manifest in groundwater contamination (Benedict, 2014).

Existing regulatory frameworks may be able to address the safety features of SMRs, particularly for LWR-SMRs. However, the factory assembly of SMRs, particularly automation stages, will require adaptation (Nuclear Energy Agency/OECD, 2016). More broadly, reform is needed in the licensing process to harmonize across countries and regions, while complying with local requirements, including environmental impact assessments and stakeholder processes (IEA, 2019c).

In the United States the advanced features of SMRs are treated as exemption requests and for establishing Design-Specific Review Standards. SMR vendors then need to prove that their designs are compliant with Nuclear Regulatory Commission (NRC) requirements, including equipment fabrication, security, and operations rules. To advance SMR regulation and licensing, vendors and prospective adopters have been encouraged to work with nuclear regulators to facilitate early resolution of issues in SMR deployment and factory fabrication (IEA, 2019c). Going further with export of SMRs, nuclear regulators and export control agencies must be engaged.

The standardization of mass-produced reactors introduces potential gains and limits that should be factored by safety regulators and operators. Standardization allows for a more expeditious integration of safety designs. However, the process by nature can also replicate a technical issue. Here, work such as that for the Multinational Design Evaluation Program, established by the United States and France, aims to develop multinational standards for novel designs (Boborov et al., 2019).

Energy, environmental emergency planning, and industrial policies will continue to be a critical dimension in the adoption of SMRs, given the difficulties that have been evident, for instance, in the financing of large nuclear reactors (Nuclear Energy Agency/OECD, 2016). Specific to energy and environmental policy options, low carbon targets are potentially strong areas of support for SMRs. However, the specific definitional scope or "carve outs" must be noted, as forms of energy like nuclear power and large hydropower are excluded at times, based on legislative preferences. Carbon taxes and cap-and-trade systems represent another set of policy instruments that could favor SMR-generated energy. In this instance, calls for carbon pricing are voiced by certain stakeholder groups, including energy and other industrial executives (Market Watch, 2019; Total, 2015), yet there has been limited response to date.

Emergency planning is another policy area of relevance to SMRs. Reductions in the zonal size have been proposed in conjunction with the potential safety advantages of SMRs. For the Tennessee Valley Authority, for instance, this reflects a shift from the traditional 10-mile standard radius for current fleet in the United States to a 2-mile standard (Dillon and Swartz, 2019). The reduced zone size allows siting near population centers and in locations such as decommissioned plant brownfields and military bases. Here, critics argue that such proposals are based on models and assumptions that haven't yet been tested (Union of Concerned Scientists, 2013).

Turning to industrial policy, investment and research and development are primary means to advance SMR deployment. For first mover projects, investment support could include long-term contracts and capital grants (Nuclear Energy Agency/OECD, 2016). Yet over a longer term, markets will likely need to be redesigned to value system services and low carbon benefits of nuclear power. Additional forms of related policy could include sovereign guarantees for borrowing, joint ventures, direct investment by publicly owned entities, production tax credits, and investment tax credit set asides, as well as export-import support (Nuclear Energy Agency/OECD, 2016).

In terms of research and development, a number of key measures were passed in the United States in 2018: the Nuclear Energy Innovation Capabilities Act and the Department of Energy Research and Innovation Act. The former allows private and public institutions to conduct civilian R&D for advanced nuclear energy technologies by siting privately funded prototypes at Department of Energy (DOE) sites (World Nuclear

News, n.d.). The latter consolidated science bills provide direction to the DOE on nuclear energy R&D (World Nuclear News, n.d.). To support the development of new reactor technologies with higher efficiencies and longer core lifetimes, the US DOE further proposed to convert high-enriched uranium (HEU) in multiple physical forms (metallic, oxide, etc.) into high-assay low-enriched uranium (HA LEU) fuel with enrichment levels between 5% and 20% U-235, for R&D purposes on advanced reactor designs (IEA, 2019c).

Workforce issues are also likely to influence the deployment of SMRs on a global scale. An advantage with the factory fabrication of SMR components is in the retention of a skilled workforce in a fixed locale. This approach has potential for lower labor costs, localized learning and knowledge transfer, and more steady employment. Such build-and-ship models of SMR production or build-own-operate approaches may not be as advantageous to client regions, such as rapidly growing developing countries, which could miss opportunities to train their domestic workforce on plant construction and fuel fabrication (Benedict, 2014). Here, local content agreements may come into play in which the supplier or developer agrees to hire local employees to meet a certain percentage of project finance. This approach is not uncommon in international trade but can have problems with cultural differences and language in knowledge transfer.

In addition to policy and workforce issues, cooperation agreements reflect an area of international policy that can support the trade of SMRs. In the United States the "123 agreements" are based on a section of the Atomic Energy Act of 1954, as amended (P.L. 95-242), which requires a cooperative agreement as a precondition for nuclear deals between the United States and another country (Kerr and Nikitin, 2017). For a country to enter into such an agreement with the United States, it must abide by the nuclear nonproliferation guidelines set in the agreement. Once such an agreement is in place between the United States and its partner, cooperation may include the export of nuclear power plants and components, or the transfer of nuclear material. Currently the United States has 23 of these agreements in force with 48 countries, the IAEA, and the governing authority in Taiwan (NNSA, n.d.). Among the potential 30 newcomer countries, seven have 123 agreements with the United States: Australia, Indonesia, Kazakhstan, Morocco, Norway, Turkey, and the UAE (Kerr and Nikitin, 2017; NNSA, n.d.; World Nuclear Association, n.d.).

22.6 Conclusion

This chapter provides an overview of the potential for SMRs to meet certain energy growth needs without increasing greenhouse gas emissions. For countries interested in increasing their energy supply, a variety of options may be appropriate after the consideration of issues such as climate change, economics, energy security and resilience, and country-specific characteristics. For many countries, there is a role for nuclear power to meet future energy needs. Large NPPs are suited for markets where energy demand is high and concentrated where large electric grid capacity and infrastructure exists. In smaller electric markets, SMRs may be suited where the population is dispersed and the electric grid is decentralized, growing energy demand can be met with the sequential deployment of smaller energy sources, remote locations are in need of resilient energy supply, access to large capital requirements is more limited, renewable energy development would benefit from load-following capability of baseload energy sources, and nonelectric cogeneration applications are important. Among emerging economies the market assessments reviewed here suggest that countries with relatively high economic and energy demand growth, a robust infrastructure, reliance on energy production from carbon-intensive sources, and with government policies aimed at reducing greenhouse gas emissions are likely markets. SMRs may also be viable in developed economies where economic, financial, infrastructure, technological, regulatory factors, and others (such as those discussed in Section 22.2) can be met. In these markets, factors such as those addressed in the SMR energy demand, climate change motivation, and energy security categories are relevant to decisions regarding the deployment of SMRs or other energy sources.

Considerations that are relevant to the market characteristics and suitability of SMR deployment were reviewed in the first part of this chapter. There are, however, further issues to be evaluated by countries considering new energy sources and whether nuclear energy technologies in general, and SMRs specifically, are to play a part in their future energy development plans. A number of these issues were examined in the latter part of this chapter. Areas for continued evaluation for all nuclear technologies include the transport and storage of new and spent nuclear fuel, the degree to which nuclear energy involves higher capital costs and attendant higher-risk premiums relative to other energy sources, and the necessity of a robust regulatory infrastructure and a well-trained workforce. Further, while on-site construction issues are much reduced for SMRs relative to large NPPs or other energy sources, there are additional considerations involving the nascent nature of SMR technologies and their modular manufacturing capabilities. While some of these issues remain uncertain at this time, advantages of developing SMR energy sources and deploying them to meet a portion of the future energy supply while also addressing climate change are informing nuclear policies.

22.7 Sources of further information and advice

An extensive collection of academic research papers, government publications, reports from international agencies, and websites was utilized to present the information provided in this chapter. While several of these are technical or academic in nature, many of the resources used in this chapter are recommended for an overview of the market assessment and deployment potential for SMRs as well as the opportunities and challenges the industry faces moving forward. For information about future global energy demand and the role of nuclear power can be found in the following recommended sources:

> Energy Information Administration. International Energy Statistics. EIA, US Department of Energy, Washington DC. http://www.eia.gov/cfapps/ipdbproject/IEDIndex3.cfm.

International Atomic Energy Agency, International Energy Statistics, IEA Paris, France, http://www.iea.org/Sankey/#?c=World&s=Balance.

International Atomic Energy Agency (2019). IEA World Energy Balances 2019, https://webstore.iea.org/world-energy-balances-2019.

International Atomic Energy Agency (2017). Energy, Electricity and Nuclear Power Estimate for the Period up to 2050, Reference Data Series No. 1. International Atomic Energy Agency, Vienna, Austria.

International Atomic Energy Agency (2015). Milestones in the Development of a National Infrastructure for Nuclear Power, IAEA, Vienna. http://www-pub.iaea.org/books/IAEABooks/10873/Milestones-in-the-Development-of-a-National-Infrastructure-for-Nuclear-Power.

The following sources are recommended for information about market assessment and global deployment indicators for SMRs:

Black, G., Black, M., Solan, D., Shropshire, D. (2015). Carbon free energy development and the role of small modular reactors: a review and decision framework for deployment in developing countries. Renew. Sustain. Energy Rev. 43(1).

International Atomic Energy Agency (2018). Deployment Indicators for Small Modular Reactors, Methodology, Analysis of Key Factors and Case Studies, IAEA TECDOC No. 1854, International Atomic Energy Agency, Vienna, Austria.

Kessides, I., Kuznetsov, V. (2012). Small modular reactors for enhancing energy security in developing countries. Sustainability 4(18).

Kupitz, J., Mistra, B. (2007). The Role of Nuclear Power in Meeting the Potable Water Needs in Water Scarce Areas for the Next Decades, International Atomic Energy Agency, IAEA Vienna.

For further information about the challenges and opportunities facing the future of global SMR deployment going forward, please see the following:

Carless, T., Griffin, W., Fischbeck, P. (2016). The environmental competitiveness of small modular reactors: a life cycle study. Energy 114, 84–99.

Vegel, B., Quinn, J. (2017). Economic evaluation of small modular nuclear reactors and the complications of regulatory fee structures. Energy Policy 104, 395–403.

References

ANS Nuclear Café, 2017. Florida's Nuclear Plants Power Through Hurricane Irma. http://ansnuclearcafe.org/2017/09/12/floridas-nuclear-plants-power-through-hurricane-irma/#sthash.syxaWRkY.dpbs.

Araújo, K., 2014. The emerging field of energy transitions: progress, challenges, and opportunities. Energy Res. Soc. Sci. 1 (1), 112–121.

Araujo, K., 2017. Low Carbon Energy Transitions: Turning Points in National Policy and Innovation. Oxford University Press, New York, NY ISBN: 9780199362554.

Araújo, K., 2019. Offshore Wind Technology: Deployment, Cost-Value Trends, and Outlook. (Technical Report) Long Island Power Authority, Uniondale, NY.

Bellverat, E., 2018. Clean and Efficient Heat for Industry. IEA. January 23, 2018. https://www.iea.org/newsroom/news/2018/january/commentary-clean-and-efficient-heat-for-industry.html.

Benedict, K., January 29, 2014. Are Small Nuclear Reactors the Answer? The Bulletin of Atomic Scientists.

Black, G., Peterson, S., 2019. Economic Impact Report: Construction and Operation of a Small Modular Electric Power Generation Facility at the Idaho National Laboratory Site. Report for the Regional Economic Development Idaho, Butte County, Idaho.

Black, G., Aydogan, F., Koerner, C., 2017. Levelised Costs of Electricity: Small Modular Reactors and Comparisons with Other Power Generation Sources. Economic Evaluation Report for NuScale Power, LLC. Economic Policy SeriesCenter for Advanced Energy Studies.

Black, G., Adoyagan, F., Koerner, C., 2019. Economic viability of light water small modular nuclear reactors: general methodology and vendor data. Renew. Sustain. Energy Rev. 103, 248–258.

Boarin, S., Ricotti, M., 2014. An evaluation of SMR economic attractiveness. Sci. Technol. Nucl. Install. 2014, 803698.

Boborov, R., Diamond, J., Shemshedinov, K., June 20, 2019. Resolving Safety and Security Concerns about Nuclear Power. The Bulletin of Atomic Scientists. https://thebulletin.org/2019/06/resolving-safety-and-security-concerns-about-nuclear-power/.

Bodle, R., Donat, L., Duwe, M., 2016. The Paris Agreement: Analysis, Assessment and Outlook. Ecologic Institute. https://www.ecologic.eu/13321.

BP, 2019. BP Statistical Outlook. http://bp.com/.

Bradford, A., June 11, 2014. Fuel and Waste Considerations for SMRs and Adv Reactors, Presentation, ML1417. FCIX.

Carelli, M.D., Mycoff, C.W., Garrone, P., Locatelli, G., Mancini, M., Ricotti, M.E., 2008. Competitiveness of small-medium, new generation reactors: a comparative study on capital and O&M costs. In: 16th International Conference On Nuclear Engineering, ASME ICONE16, Orlando, Florida, USA.

Caribbean Business News, 2018. Resolution Introduced in Puerto Rico House to Investigate Nuclear Power Production. https://caribbeanbusiness.com/lawmaker-introduces-resolution-to-investigate-nuclear-power-in-puerto-rico/.

Carlsson, J., Shropshire, D., van Heek, A., Fütterer, M., 2012. Economic viability of small nuclear reactors in future European cogeneration markets. Energy Policy 43, 396–406.

CRS, 2019. Advanced Nuclear Reactors: Technology Overview ad Current Issues. (R45706) Congressional Research Service, Washington, DC.

Dillon, J., Swartz, K., August 15, 2019. NRC Holds First Hearing for SMR Site, Energy Wire. https://www.eenews.net/energywire/stories/1060954519.

DSIRE, n.d. DSIRE Database. https://www.dsireusa.org/.

Energy Strategies, 2019. Analyzing the Cost of Small Modular Reactors and Alternative Power Portfolios. Health Environment Alliance of Utah. Energy Strategies, LLC May 2019.

European Environmental Agency, 2018. Total Greenhouse Gas Emission Trends and Projections. (EEA Prod-ID: IND-37-en). https://www.eea.europa.eu/data-and-maps/indicators/greenhouse-gas-emission-trends-6/assessment-2.

GB Times, n.d. China to Build 30 Nuclear Plants in Silk Road Countries. https://gbtimes.com/china-build-30-nuclear-plants-silk-road-countries

Generation IV International Forum (GIF), n.d. https://www.gen-4.org/gif/

Grafham, O. (Ed.), 2019. Energy Access and Forced Migration. In: IEA World Energy Balances 2019, IEA. https://webstore.iea.org/world-energy-balances-2019.

IAEA, 2012. Licensing the First Nuclear Power Plant. (INSAG Report 26). https://www-pub.iaea.org/MTCD/Publications/PDF/Pub1573_web.pdf.

IAEA, 2015a. Milestones in the Development of a National Infrastructure for Nuclear Power. Nuclear Energy Series No. NG-G-3.1 (Rev 1)IAEA.

IAEA, 2015b. The Fukushima Daiichi Accident. International Atomic Energy Agency, Vienna, Austria.
IAEA, 2016. Nuclear Power and Sustainable Development. (STI/PUB/1754) International Atomic Energy Agency, Vienna, Austria.
IAEA, 2017. Energy, Electricity and Nuclear Power Estimate for the Period up to 2050. Reference Data Series No. 1International Atomic Energy Agency, Vienna, Austria.
IAEA, 2018a. Deployment Indicators for Small Modular Reactors, Methodology, Analysis of Key Factors and Case Studies. (IAEA TECDOC No. 1854) International Atomic Energy Agency, Vienna, Austria.
IAEA, 2018b. Climate Change and Nuclear Power 2018. International Atomic Energy Agency, Vienna, Austria.
IAEA, 2018c. PRIS Database on Nuclear Power Reactors. https://pris.iaea.org/pris/ (Accessed November 12, 2019).
IAEA, 2018d. Nuclear-Renewable Hybrid Energy Systems for Decarbonized Energy Production and Cogeneration. (IAEA TECDOC No. 1885) International Atomic Energy Agency, Vienna, Austria.
IAEA, 2018e. Advances in Small Modular Reactor Technology Developments. A Supplement to IAEA Advanced Reactors Information System. ARIS. https://aris.iaea.org/Publications/SMR-Book_2018.pdf.
IAEA, 2018f. Financing Nuclear Power in Evolving Electricity Markets. International Atomic Energy Agency, Vienna, Austria.
IAEA, 2019a. Small Modular Reactors. https://www.iaea.org/topics/small-modular-reactors.
IAEA, 2019b. Small Modular Reactor (SMR) Regulators' Forum. https://www.iaea.org/topics/small-modular-reactors/smr-regulators-forum.
Idaho National Laboratory (INL), 2019. https://inl.gov/trending-topic/microreactors/.
IEA, 2018. World Energy Outlook. https://www.iea.org/.
IEA, 2019a. IEA World Energy Balances 2019. https://webstore.iea.org/world-energy-balances-2019.
IEA, 2019b. Global Energy and CO_2 Status Report. https://www.iea.org/geco/data/.
IEA, 2019c. Nuclear Power in a Clean Energy System. (Report), May 2019.
IEA, Nuclear Energy Agency, 2015. Projected Costs of Generating Electricity. https://www.oecd-nea.org/ndd/pubs/2015/7057-proj-costs-electricity-2015.pdf.
IPCC, 2018. Summary for policymakers. In: Masson-Delmotte, V., Zhai, P., Pörtner, H.-O., Roberts, D., Skea, J., Shukla, P.R., ... Waterfield, T. (Eds.), Global Warming of 1.5°C. An IPCC Special Report on the Impacts of Global Warming of 1.5°C Above Pre-Industrial Levels and Related Global Greenhouse Gas Emission Pathways, In the Context of Strengthening the Global Response to the Threat of Climate Change, Sustainable Development, and Efforts to Eradicate Poverty. World Meteorological Organization, Geneva, Switzerland, 32 p.
Kerr, P., Nikitin, M., December 27, 2017. Nuclear Cooperation With Other Countries. Congressional Research Service, Washington, DC.
Kuznetzov, V., 2008. Options for small and medium sized reactors (SMRs) to overcome loss of economies of scale and incorporate increased proliferation resistance and energy security. Prog. Nucl. Energy 50, 242–250.
Locatelli, G., Bingham, C., Mancini, M., 2014. Small modular reactors: a comprehensive overview of their economics and strategic aspects. Prog. Nucl. Energy 73, 75–85.
Market Watch, May 22, 2019. CEOs of Major Companies Call on U.S. Congress to Set a Natioanl Price on Carbon. https://www.marketwatch.com/press-release/ceos-of-major-companies-call-on-us-congress-to-set-a-national-price-on-carbon-2019-05-22.

Massachusetts Institute of Technology, 2018. The Future of Nuclear Energy in a Carbon-Constrained World. https://energy.mit.edu/wp-content/uploads/2018/09/The-Future-of-Nuclear-Energy-in-a-Carbon-Constrained-World.pdf.

Mignacca, B., Locatelli, G., 2019. Economics and finance of small modular reactors: a systematic review and research agenda. Renew. Sustain. Energy Rev. 118(2020), 109519.

National Nuclear Laboratory, 2014. Small Modular Reactor Technology. United Kingdom National Nuclear Laboratory.

NEA, 2019. The Costs of Decarbonisation: System Costs with High Shares of Nuclear and Renewables. Nuclear Energy Agency OECD (NEA No. 7335).

New York State Clean Energy Standard, n.d. https://www.nyserda.ny.gov/All-Programs/Programs/Clean-Energy-Standard.

NNSA, n.d. 123 Agreements for Peaceful Cooperation. https://www.energy.gov/nnsa/123-agreements-peaceful-cooperation

Nuclear Energy Agency, 2016. Costs of Decommissioning Nuclear Power Plants. (OECD 2016, NEA No. 7201).

Nuclear Energy Agency/OECD, 2016. Small Modular Reactors: Nuclear Energy Market Potential for Near-Term Deployment, Paris, France.

Nuclear Energy Agency/OECD, 2019. Nuclear Innovation 2050 (NI2050). https://www.oecd-nea.org/ndd/ni2050/.

Nuclear Energy Institute, 2017. Nuclear Costs in Context. https://www.nei.org/CorporateSite/media/filefolder/Policy/Papers/Nuclear-Costs-in-Context.pdf?ext=.pdf.

Nuclear Energy Institute, 2018. Zero-Emission Credits. https://www.nei.org/CorporateSite/media/filefolder/resources/reports-and-briefs/zero-emission-credits-201804.pdf.

Nuclear Energy Institute, 2019. Microreactors Could Power Remote Military Bases Within a Decade. https://www.nei.org/news/2018/microreactors-power-remote-military-bases.

NuScale Power, LLC, 2020. Utah Associated Municipal Power Systems (UAMPS) Carbon Free Power Project. https://www.nuscalepower.com/projects/carbon-free-power-project.

Papaioannou, I., Purvins, A., Shropshire, D., Carlsson, J., 2013. Role of a hybrid energy system comprising a small/medium-sized nuclear reactor and a biomass processing plant in a scenario with a high deployment of onshore wind farms. J. Energy Eng. 140, 1–10.

Ramana, M., Mian, Z., 2014. One size doesn't fit all: social priorities and technical conflicts for small modular reactors. Energy Res. Soc. Sci. 2, 115–124.

Scully Capital, 2018. Examination of Federal Financial Assistance in the Renewable Energy Market, Implications and Opportunities for Commercial Deployment of SMRs.

Shropshire, D., 2011. Economic viability of small to medium-sized reactors deployed in future European energy markets. Prog. Nucl. Energy. https://doi.org/10.1016/j.pnucene.2010.12.004.

Shropshire, D., Chandler, J., 2006. Financing strategies for a nuclear fuel cycle facility. In: 14th International Conference on Nuclear Engineering, ICONE14-89255, July 17–20, Miami, FL.

Shropshire, D., et al., 2012. Benefits and cost implications from integrating small flexible nuclear reactors with off-shore wind farms in a virtual power plant. Energy Policy. https://doi.org/10.1016/j.enpol.2012.04.037.

Siemens Industry, 2019. Puerto Rico Integrated Resource Plan 2018–2019, Draft for the Review of the Puerto Rico Energy Bureau. (Siemens PTI Report Number: RPT-015-19, Rev. [1]), 2/12/2019.

Solan, D., et al., 2010a. Economic and Employment Impacts of Small Modular Reactors. Economic Policy SeriesEnergy Policy Institute—Center for Advanced Energy Studies.

Solan, D., Black, G., Louis, M., Peterson, S., Carter, L., Peterson, S., Bills, R., Morton, B., Arthur, E., 2010b. Economic and Employment Impacts of Small Modular Reactors. Center for Advanced Energy Studies Policy Series.

Solan, D., Black, G., Black, M., 2015. Assessing the Potential Global Demand for Small and Medium-Sized Reactors: 2020-2050. Economic Policy SeriesEnergy Policy Institute—Center for Advanced Energy Studies.

Southern States Energy Board, 2019. U.S. Global Competitiveness in Nuclear Energy. www.sseb.org/wp-content/uploads/2019/02/Andy-Paterson.pdf.

Strategic Insights, 2015. Small Modular Reactor Market (SMR) Outlook. https://www.bigmarker.com/strategic-insights-inc-/Strategic-Insights-2015-Small-Modular-Reactor-Market-Outlook.

The Moscow Times, n.d. Russia's Nuclear Power Exports are Booming. https://www.themoscowtimes.com/2019/05/09/russias-nuclear-power-exports-are-booming-a65533

The Nuclear Alternative Project, 2019. Nuclear4PuertoRico. https://www.nuclearalternativeproject.org/nuclear-puertorico.

Total, May 29, 2015. Letter to C. Figuers From BP, Total, BG Group, ENI, Shell, and Statoil. https://www.total.com/sites/default/files/atoms/files/letter_to_christiana_figueres.pdf.

U.S. Department of Commerce, 2011. The Commercial Outlook for U.S. Small Modular Nuclear Reactors. International Trade Administration. Available at:http://trade.gov/publications/abstracts/the-commercial-outlook-for-us-small-modular-nuclear-reactors.asp.

U.S. Dept of Energy (US DOE), 2019. Benefits of Small Modular Reactors. https://www.energy.gov/ne/benefits-small-modular-reactors-smrs.

U.S. Energy Information Administration, 2016. Capital Cost Estimates for Utility Scale Electricity Generating Plants. https://www.eia.gov/analysis/studies/powerplants/capitalcost/pdf/capcost_assumption.pdf.

UN DESA/Population Group (UN DESA), 2018. World Urbanization Prospects. https://population.un.org/wup/Publications/.

UNFCCC, 2020. Mechanisms Under the Kyoto Protocol. https://unfccc.int/process/the-kyoto-protocol/mechanisms.

Union of Concerned Scientists, September 2013. Small Isn't Always Beautiful, Edwin Lyman. https://www.cleanenergy.org/wp-content/uploads/UCS-SMR-Rpt-Sept-2013.pdf.

University of Utah Bureau of Economic and Business Research, 2012. A Review of the Costs of Nuclear Power Generation. http://gardner.utah.edu/bebr/Documents/studies/Nuclear_Report_Final_Web_7Mar2012.pdf.

Urban Gateway, 2018. The Number of Megacities has Tripled Since 1990. https://www.urbangateway.org/news/number-megacities-has-tripled-1990.

US DOE, 2019. Nuclear Energy Advisory Committee Meeting. March 28, 2019. https://www.energy.gov/sites/prod/files/2019/04/f61/HALEU%20Report%20to%20NEAC%20Committee%203-28-19%20%28FINAL%29.pdf.

US EIA, July 26, 2019. More U.S. Coal-Fired Power Plants are Decommissioning as Retirements Continue. https://www.eia.gov/todayinenergy/detail.php?id=40212.

Ux Consulting Company LLC, 2013. SMR Market Outlook. UxC, GA, USA.www.uxc.com.

Wikipedia, n.d.-a. Special Economic Zones in China. https://en.wikipedia.org/wiki/Special_economic_zones_of_China

Wikipedia, n.d.-b. Falcon 9. https://en.wikipedia.org/wiki/Falcon_9#Development_history

World Association of Nuclear Plant Operators, n.d. https://www.wano.info/

World Nuclear Association (WNA), 2019a. The Harmony Programme. https://world-nuclear.org/our-association/what-we-do/the-harmony-programme.aspx.

World Nuclear Association (WNA), 2019b. Nuclear Power in the US. https://www.world-nuclear.org/information-library/country-profiles/countries-t-z/usa-nuclear-power.aspx.

World Nuclear Association, n.d. Emerging Nuclear Energy Countries. https://www.world-nuclear.org/information-library/country-profiles/others/emerging-nuclear-energy-countries.aspx

World Nuclear News, n.d. U.S. Launches Advanced Nuclear Technologies Demonstration Centre. https://www.world-nuclear-news.org/Articles/US-launches-advanced-nuclear-technologies-demonstr

Small modular reactors (SMRs): The case of developing countries

23

D. Goodman
Consultant, USA

23.1 Introduction

The planet is experiencing a combination of increased economic output from the global South with acute and cumulative effects of climate change that affect poorer countries the most, a heightened general awareness of the disparities of quality of life, and a deficit of social trust (UNDP, 2013). The current global predicament demands a revaluation.

Developing countries present distilled instances of systemic problems and solutions. Similarly, small modular reactors (SMRs) distil the operational experience and lessons learned, from submarines and large reactors and hazards; and incorporate cooperation with natural processes, such as gravity, evaporation and condensation, in their concepts and designs. SMRs are to nuclear as 'developing countries' are to the world: a concentrated instance of the whole; and they belong together.

Generic development issues are intensified in the case of nuclear. SMR deployment brings into focus poignant, generic issues of development, such as debt, national priority, natural resource endowments and industrial planning, climate change, technological competence, the need for durable institutions, educational and human resources, and in some cases limited capital. Nevertheless, SMRs' intrinsic technical features, such as their size, modularity, simplified operation and inherent safety features, make them particularly well suited to the circumstances of developing countries.

As discussed in Chapter 3 of this Handbook, SMRs are a robust, 'forgiving' nuclear power technology, designed for simplicity and plant resilience, and to incorporate inherently or 'passively' safe features such as natural circulation for core cooling, incorporating most primary components in a single vessel, long coping periods to handle interrupted backup power supply, and generally designed for below-grade or barge deployment with the intended result of much reduced risk of accident, and fewer and less acute consequences in case of accident (Carelli, 2014; Ingersoll, 2011).

A promising feature for the economics of wide SMR deployment in developing countries is the modularity of some designs, which refers to the incorporation of all major safety-significant systems within one module, to the potential of modules to be standardized and factory-built in series, with minimum site-specific design and construction. It also refers to the possibility of scaling-up power production in increments, building and putting additional modules into service over time, which would reduce the initial capital outlay and overall investment risk (Barkatullah, 2011; Kessides and Kuznetsov, 2012). There is also potential for SMR components to be manufactured indigenously, with a corresponding positive impact on industry.

In themselves and by analogy, SMRs can become the catalyst for development and capability: 'There is no country in the world that has made significant development without embracing nuclear technology. Not just because nuclear generates electricity in an affordable way, but if you are able to demystify the mysteries and the myths surrounding nuclear technology, then you can manufacture and process anything.' (Ayacko, 2012). Development and SMRs could have a synergistic relationship, with potential for these small reactors to be as disproportionately important to developing countries as are mobile telephones and air travel.

This chapter will mention the 'capabilities approach' to human development, as a foundational concept; look at various characteristics of developing countries for which SMR deployment is suitable; and discuss development and some associated trade-offs that affect SMR deployment. Although this chapter affirms SMRs as a 'counterfactual choice' – what one would have chosen if one had the choice – for energy in developing countries, some alternative choices and potential consequences will be discussed.

The fulcrum of the argument is the opportunity cost incurred by the continued use of greenhouse gas (GHG)-emitting and costly fuels, once an SMR is viable for deployment in countries that disproportionately suffer the effects of climate change and lack energy security in the face of high and volatile fuel costs, while hampered by low income or high debt.

Looking ahead, some obstacles to SMR deployment, and innovations that could address them, will be identified. The chapter concludes on the need for a revised understanding of the wide context of SMR deployment and developing countries.

23.2 Measuring development

This chapter does not, like others in this Handbook, concern technology development, but rather human development, in the global context. Our understanding of development tends to be one-dimensional, concentrating on economic indices. Development is regarded by some as a measure of economic growth, as measured by GDP (gross domestic product), denoting the total value of a country's economic production or GNI (gross national income), the World Bank economic indicator formerly known as GDP.

Even this classification is problematic: for example, income is not coterminous with development (World Bank, 2013) and GDP by itself is not necessarily a good indicator of economic sustainability (SIDS Outcome, 2013). Development is also denoted by the presence of other human development indicators (HDI), which include life expectancy at birth. One catalogue of development desiderata is represented by the Millennium Development Goals, which were agreed by the UN in 2000 as the most pressing goals to achieve, namely to eradicate extreme poverty and hunger; achieve universal primary education; promote gender equality and empower women; reduce child mortality;

improve maternal health; combat HIV/AIDS, malaria and other diseases; ensure environmental sustainability; and enable global partnership for development.

Finally, development can be measured by the ability of citizens to realize their innate capabilities. Following Sen (1995) and Nussbaum (2000), one might sum up development in this way: once anyone anywhere on the planet has the basic human capabilities, they can exert their creativity to even higher reaches of capability and functioning. These are defined as: '1. Life, 2. Bodily health, 3. Bodily integrity, 4. Senses, Imagination, and Thought, 5. Emotions, 6. Practical reason, 7. Affiliation, 8. Other Species, 9. Play, 10. Control over One's Environment a. political. b. material' (Nussbaum, 2000).

23.3 Trade-offs of small modular reactors (SMRs) in developing countries

When the basic human capabilities are not present, trade-offs have to be made between competing goods and harms. The effects of these trade-offs are not confined to the person or country: we live on a planet of increasingly shared benefits and detriments. If the avoidance of one risk means running another, how can their equivalence or difference be discerned?

In 2011, the South African Planning Commission's diagnostic report asked some pointed questions that sum up the generic development issues and make explicit some embedded trade-offs to which SMRs have unique potential to respond:

> *Is it possible to reduce carbon emissions and environmental impacts and still remain a competitive commodity exporter? How quickly can the economy shift from being a high resource-intensive one to a more knowledge-intensive or labour-intensive one? How does the country balance the need for infrastructure to suit today's economy, without locking in the present resource-intensive development path? (SAPC, 2011)*

In a powerful 2009 appeal for a radical reconsideration of the precautionary principle that informs radiation regulation, Abdel-Aziz et al. (2009) argued that '[i]n the social and economic contexts of the developing world, lack of energy poses its own considerable risks to health and well-being [...] where access to electrical energy is a direct covariant of health, education, life expectancy and child mortality.' The instrumental quality of electricity in supporting the basis for human development should be acknowledged, as shown in the correlation between electricity usage and human development indicators: as one rises, so does the other (Pasternak, 2000).

All countries in the global economy are interdependent in that they all produce, suffer from, and can mitigate the effects of economic paths, social factors and climate change. This implies a corresponding duty to consciously take responsibility for the widest possible effects of their decisions and actions (UNDP, 2013). The advocacy of SMRs in this chapter is grounded in ethical concerns. Recent work by Kharecha and Hansen (2013) on the net reduction in harm to human health and the environment due to the use of nuclear power, as well as the remark in the proceedings of the INPRO 5th

Dialogue Forum on Long-term Prospects for Nuclear Power that 'eliminating nuclear power from the energy mix [...] might have greater societal effects than the added risk' reinforce the validity of this approach.

Some advantages of SMRs for developing countries, as compared to fossil generation technologies, include the comparative long-term stability of nuclear fuel prices versus volatile market prices of potentially scarce fossil fuels; and the lack of carbon emissions from nuclear electricity production, and in comparison to large-scale nuclear plant, its comparative flexibility and application to smaller grids, and its size compatibility with variable technologies such as solar or wind power. In addition, SMRs could be used for desalination purposes.

The energy-security benefits of SMRs, therefore, are considerable for developing-country fossil-fuel importers, when compared with reliance on costly fossil fuels from abroad. For example, in Jamaica, whose 95% dependency on petroleum makes it highly vulnerable to oil-price volatility, like the rest of the Caribbean and Central America (Yepéz-Garcia et al., 2012), SMRs are explicitly considered as a future option in its 30-year national energy policy (Jamaica Ministry of Mining and Energy, 2009; Mian, 2011).

As another example, the increased need for power and therefore gas in Nigeria means that less gas is made available to export to neighbouring Ghana. As a result of the gas shortfall, the Ghanaian utility is forced to switch fuels from gas to crude oil or diesel and back, which damages plant meant to use a single fuel type. The utility resorts to load-shedding and power cuts (Osabutey, 2012). There, SMRs are also being considered as an option for energy security reasons, among others.

Where economic considerations might impinge on fossil-fuel supply, geopolitical ones could affect developing countries' choice of SMR technology and nuclear fuel supply and disposal (IAEA INPRO DF 3, 2011).

23.4 Characteristics of developing countries that make deployment of SMRs viable

23.4.1 The increasing importance of the information economy

Feedbacks arise from the growth of connectivity and information economies in the developing world. The immense growth in mobile-phone uptake and information and communications technology (ICT) in the developing world has knock-on effects in increasing the demand for electricity. However, the ICT aspect could make energy demand planning problematic, reduce the effectiveness of conservation strategies, pit GHG-reducing plans against poverty reduction, and ultimately lead to a shortfall in supply and unplanned electricity shortages (Sadorsky, 2012).

With their comparatively small output and ability to be deployed in series, SMRs can help to ramp up to meet demand growth, mirroring the expansion of ICT – which itself has been a driver of development and social progress throughout the developing world.

23.4.2 Water precarity or scarcity

Water precarity or scarcity is exacerbated by the effect of climate change on rainfall patterns. In Ghana, for example, this has affected the reliability of hydropower resources with a resulting shortfall in electricity supply (IBRD, 2010).

The use of coal plant requires large quantities of water, which may affect its availability for other needs, and additionally creates downstream pollution. The water component in energy planning has become increasingly important in this perspective, with more stringent requirements for water usage being put in place when funding and planning for large plants is done. An example is the Medupi coal plant in South Africa, discussed in Section 23.5.4. Some SMR designs are air-cooled, obviating the need for a great deal of water usage for electrical power production. This would greatly expand the range of suitable sites.

In addition, the desalination capability of SMRs could provide water for desert or polar sites. Indeed, as after the Haiti earthquake of 2010, SMR desalination could be applied temporarily in disaster situations where sites are accessible to the sea and nuclear vessels like the USS *Carl Vinson* could share their excess distilled water (Padgett, 2010).

23.4.3 The high cost of grid power compared to the developed world

The predicament of many developing countries is difficult: not only is electricity scarce and intermittent, it is also dear. In countries such as Ghana, electricity tariffs do not cover actual generation costs, resulting in major financial losses to the utility.

> *Ghana's power tariffs are based on the costs of baseload hydropower priced at $0.05 per kilowatt-hour. However, the oil-based generation used to meet incremental demand is priced at more than $0.20. Since there is no mechanism for automatically adjusting tariffs, this situation generates annual financial losses for the Volta River Authority (VRA) of $400 million – 3 percent of GDP. (IBRD, 2010)*

The handicap to economic activity is considerable: in 2007, it was reported that more than 2% of GDP was sacrificed to power shortfalls in vulnerable African countries (Wines, 2007). Nevertheless, although energy consumption is rising in developing countries throughout the world, it is worth remembering that the overall quantity is low in comparison with high-income countries. '[T]he annual average per capita consumption of electricity in the developing world is 1155 KWh and 10,198 kWh in high-income countries' (IBRD, 2009).

23.4.4 Energy infrastructure weakness

Diesel-generator backup systems are prevalent in the developing world, with high-cost, high-polluting fuel. Insufficient grid power leads to extra capital stress on private firms and families who can invest in backup generators, in order to maintain operations during routine brownouts and blackouts.

Average figures from enterprise surveys in (admittedly vast) ranges of countries in the Latin American-Caribbean region and Sub-Saharan African region show that percentages of manufacturing firms identifying electricity as a major constraint were, respectively, 37.6% and 50.3% (World Bank Enterprise Survey, 2012). From the perspective of domestic electricity users, the most vocal figures might be the average number of electrical outages in a typical month (3.7 in the Latin American-Caribbean region and 10.7 in the Sub-Saharan African region) and the duration of a typical electrical outage (2.1 hours and 6.6 hours, respectively).

This results in significant generator use for business, with 28.1% of Latin American-Caribbean firms owning a generator and 43.6% of Sub-Saharan African firms owning a generator – with concomitant fuel and operations and maintenance (O&M) costs (World Bank Enterprise Survey, 2012). State electricity firms also commonly resort to mobile generators for emergency power (*New York Times*, 2013).

23.4.5 The growth of megacities

Half of the world's population lives in cities, and this will grow as increased urbanization occurs in developing countries. Lagos, Dhaka, Manila and Cairo are examples of megacities (those with over 10 million inhabitants) in the developing world.

Air pollution and climate impacts of the world's megacities are worsening (WMO/IGAC, 2012; Cossardeux 2013). SMRs could be a source of emissions-free electricity to help power megacities and mitigate their effect on the air and the climate. To serve economic and city planning, they could be deployed as scalable multi-module plants ramping up to meet demand. In addition, they could be sited closer to load centres if a reduced surrounding safety zone due to their designs' safety benefits (IEAE INPRO DF 5, 2012) could be envisaged.

23.4.6 Sociological public-acceptance factors

These might be markedly different from those in developed countries (Hecht, 2012). This has to do with differing value systems and weight given to different socioeconomic aspects, for example, antipathy to coal mining or conversely national reliance on coal mining; environmental desiderata; water scarcity; the existence of mining operations; the relative openness and transparency of energy policymaking and bid procedures; the potential for participation in the supply chain and for technology transfer. The pressing need for clean, sustainable power and water supplies in deprived and burgeoning countries alike might create greater openness for small-scale nuclear power in the form of SMRs. Moreover, public-acceptance factors are as potent or determinant for developed countries as for developing countries when the prospect of wide deployment of nuclear is raised, whether it be in the shape of SMRs or not. Fear of increased access of developing countries to nuclear power would contribute to a policy that locks countries with highest growth prospects and most acute need for low-carbon electricity into unsustainable, high-emitting, or costly renewable technologies (Ropeik, 2012). Nevertheless, how nuclear is construed is historically or culturally determined, and can change (Weart, 2012).

SMR provides an opportunity and vehicle for development and should not just be considered a tool, or even a goal in itself, to demonstrate technical advancement.

23.5 SMR choices in developing countries

This chapter posits SMR as a counterfactual choice – 'what one would have chosen if one had the choice' – for energy for developing countries (Sen, 1995). At an historical moment when developing countries disproportionately suffer the effects of climate change and, for some, the energy-security challenge of importing fuels with high price volatility, there is considerable value in the potential for an emissions-free energy source with comparatively stable fuel prices.

Embedded risk biases may currently prevent recognition of it; however, in the looming crisis of climate change, the true costs of business as usual will fall disproportionately on developing countries. That they may prove higher than going along the SMR route is a matter of opinion. However, it is not an outlandish conclusion to draw: 'The biggest factor in the decision to construct a plant, may shift from the customer's ability to finance the project to a careful consideration of opportunity cost,' observe Abdulla *et al.* (2013), in the context of the possibility of more stringent carbon regulation.

As a developing country, if SMR were *not* available, what would be the alternatives? What are the potential consequences (including for the climate, societal development, the economy, and thus the ethical dimension), for developing and developed countries, viewed as a dyad?

23.5.1 Technology lock-in and decarbonization

If fossil technology is locked in for the developing world, with its growing energy needs (IEA, 2011), then even if high-income countries achieve more rapid decarbonization, the deleterious environmental effects will nevertheless be shared by all for longer. Conversely, once carbon is penalized, however, existing investment in CO_2-emitting energy types may need to be trashed (Stern, 2012), thus wasting the scarce capital of developing countries.

In a global perspective, where energy demand goes unsatisfied in poorer or energy-hungry regions, or the water supply (often intertwined with energy) is imperilled (G-Science Academies, 2012), social unrest and political instability could result (Lee *et al.*, 2012).

Energy planning decisions made now will threaten or support the economic future of developing countries; oil-price volatility, for example, will affect them in proportion to their reliance on diesel and heavy fuel oil (Yepéz-García and Dana, 2012). Conversely, even the shortest refuelling interval of proposed SMR designs (14 months, with an upper range of 10 to 30 years for sealed-unit designs), would protect them from fuel-price volatility (IAEA, 2012). Nevertheless, long-term fuel-supply contracts, with a limited number of suppliers globally, could present multifaceted challenges for developing countries (IAEA INPRO DF3, 2011).

Finally, timing is a crucial factor because of the overlap of climate, economic and social pressures (IEA, 2011). The less able, because of polluting energy choices, developing countries are to meet CO_2-reduction targets, the more responsibility they will have to assume for increased climate change, and the greater burden will fall on them and on developed countries to curtail emissions in order to compensate.

23.5.2 Sustainable energy choices and the role of debt

The trade-offs can be heavy. Developing countries would be penalized and their access to international development aid and even infrastructural finance could be imperilled because in the increasingly carbon-constrained environment, under new standards, the use of polluting fossil fuel and concomitant water usage would prevent them from meeting carbon-reduction targets and harm the environment.

Moreover, in the perspective of development itself, if a disproportionate amount of the national budget is allocated to purchase fossil fuels, these sums are effectively sequestered from other high development and national priorities such as potable water distribution, adequate schooling, health care, housing, judiciary, and so forth. A switch to small-scale nuclear power could therefore release these funds for social services and other government functions, strengthening the country, and help to free it from debt.

In the Jamaican example, the causal link between energy cost, industrial activity and national wealth is not opaque, but a matter of discussion for the daily papers: '[a proposed fossil] plant will lower the high cost of energy that currently threatens the viability of the bauxite/alumina sector, which earns the third-highest levels of foreign exchange for Jamaica,' (Jackson, 2013).

The potential still remains for an SMR program itself and the attendant infrastructure to saddle countries, especially nuclear newcomer countries, with increased debt. Therefore appropriate financing and institutional mechanisms are needed to support timely SMR access in newcomer countries.

SMRs' comparatively small project size could make them apt for innovative fleetwide project-finance mechanisms; alternatively, BOO (Build–Own–Operate by the vendor) or BOOT (Build–Own–Operate–Transfer to the user) mechanisms could buy countries time to build up the human-resources and operational contingent without in the meantime sacrificing the benefits of SMR power.

There is increasing political, social and market scrutiny of the environmental and economic sustainability of investments and infrastructure. On the one hand, sustainability is becoming a higher priority for energy and infrastructure planners, as well as commercial utilities, and funds that finance these ventures, in order to protect their infrastructure investments. On the other, since vulnerable populations in developing countries are disproportionately affected by climate change (Dell *et al.*, 2012), sustainability is not abstract, but is a survival issue (Hinshaw, 2010; Chonghaile, 2012). SMRs would be able to fulfil sustainability criteria to the satisfaction both of the funders and of the overarching purpose the criteria represent.

23.5.3 Energy resource-rich countries

To offset the growing internal demand for fuels they produce and still maintain export volumes (Lee *et al.*, 2012; Osabutey, 2012), countries with coal, oil, or gas resources may find small-scale nuclear useful as replacement.

SMR generation could provide power for industry outside the resource realm, contributing to diversified economic activity. The caveat exists that where countries are highly dependent on coal and petroleum industries, as for example Indonesia or Venezuela, an SMR program would need to be integrated into the existing economic structure. Their comparatively small size might more easily permit this.

23.5.4 Financing and the effect of external policy preferences

For developing countries, as with high-income countries, financing is central. Here, the trade-offs are acute. Countries that rely on external aid or multilateral aid to help in financing major infrastructure will face those institutions' policy preferences as to the types of energy and institutional programs they will support (World Bank, 2010). This could be an obstacle to SMR programs in the developing world. For example, in 2012, Kenya's nascent nuclear program, one of the key developmental priorities in its country's Vision 2030 plan, was criticized by the head of UNEP (which institutionally favours renewable energy), who urged exploration of other options first (Orengo, 2012; *The Standard*, 2012).

As a parallel example of the impact of aid issues, this time for a thermal plant, the Medupi coal-fired plant funded by the World Bank in South Africa has come under scrutiny – and resulted in additional costs – because of its environmental impacts on water usage and sulfur dioxide emissions. Alternatively, a revision of multilateral lending criteria for energy projects could take greater account of externalities (World Bank, 2010). This could facilitate SMR fleet deployment in developing countries.

23.6 Obstacles and innovations

SMR deployment in developing countries, correspondingly, will not work out optimally if it is done in the same way that nuclear power deployment has been done in the past in high-income countries. The institutional and human-resources challenge is just too difficult to overcome in the window of opportunity available; and in the case of SMRs, the degree of regulatory intensity may not be appropriate.

SMRs, with their simple and 'forgiving' technology, work for developing countries on that very basis (Carelli, 2014). They do not need the scale of supporting infrastructure and expertise that is currently demanded by existing large-scale nuclear power plants in rich countries, and thus could provide the starting point for a new nuclear country of more limited means (Mian, 2011).

What are the new basic structures needed to ensure that this matchup of a situation and a technology that are similar in kind can take place responsibly?

23.6.1 The role of standardization of technology and licensing

The first potential approach arises from the technology: for deployment in developing countries, the highest degree of standardization of plant design, manufacturing and construction methods would be best. This would overcome some temporal or material limitations of expertise and resources, and permit greater ease of technology transfer and capacity-building.

This would not just affect technology deployment but also licensing. Currently, the multiplicity of incompatible licensing approaches (Söderholm, 2013) is an obstacle to widespread international SMR deployment. Because of their scale and comparative simplicity, SMRs present the opportunity to rationalize and innovate in the areas of policy and regulation. Reciprocally, a technology-neutral regulatory approach to SMRs eventually could be the template for new countries initiating a civil nuclear program and a new nuclear regulatory framework, especially if finances or grid size preclude embarking on a large-scale nuclear power program. Safety and security measures that are designed to enable remote operation and monitoring are also applicable and apt in developing-country situations.

23.6.2 Utilization of regional mechanisms

Second, a template for education, training, skills- and institution-building to support specifically SMR nuclear programs in multiple countries would shorten an indispensable rate-determining step (Goodman and Storey, 2012). A new country aiming to develop a nuclear regulatory body from scratch is presented with a daunting prospect, especially if the achievement of regulatory competence is tied to delivery of a national project to build a nuclear power plant for the first time.

Developing countries, grouped by region, could benefit from pooling of resources and sharing of processes. To lighten the institutional and infrastructural burden on individual developing countries, especially those that are poor or small, regionalizing SMR mechanisms would be useful.

This could mirror regional electricity grid interconnections such as the WAPP (West African Power Pool) or the SAPP (Southern African Power Pool), or regional economic communities such as ASEAN or that comprise mainly SIDS (small island developing states), such as CARICOM (the Caribbean Economic Community area).

Financing, grid installation, modernization and networking, nuclear regulation, operation, training, and monitoring are all good candidates for regionalization. Regional fleet mechanisms for a particular SMR type could be developed to handle all stages of project deployment, operation and decommissioning. This would be advantageous financially and also from the point of view of safety and security.

Regional regulators, especially or even specifically for pre-licensing of designs, would reduce the number of steps for an individual country regulator, and additionally improve financing potential, because of the shorter gap before the site-specific licensing and build process begins. Regional TSOs (technical safety organizations) could provide pooled technical resources, while maintaining the sovereignty and independence of the national regulator. A precursor could be ETSON (European Technical

Safety Organisations Network), a European network of TSOs that support separate national regulators (ETSON, 2013).

To ease the burden on individual regulatory resources, the regional regulators would as a body or jointly perform a design verification delivering some form of 'acceptable-design' instrument, which could be conferred by an internationally recognized body. For example, the MDEP (Multinational Design Evaluation Program) could be extended into a group examination process towards common international design certification of SMRs (Goodman and Storey, 2012).

In the interim, innovative business models such as BOO; bilateral and regional tutelary programs; and international SMR-specific TSOs could help to bridge the human-resources gap for newcomer developing countries (Goodman and Raetzke, 2013). Developing countries will be able to operate and maintain SMR plants, by themselves, to global standards of safety.

23.6.3 Inclusion rather than 'exceptionalism'

Third, these standards themselves, and regulations and international rules governing the use of nuclear power, need to be reconsidered in a more inclusive perspective. Instead of persisting in a 'nuclear exceptionalism' that no longer adequately serves the needs of the whole global community, a wider perspective should be taken (Abdel-Aziz *et al.*, 2009; Hecht, 2012).

Regulatory innovators or change agents, would consider more, and perhaps different, factors than have hitherto influenced the formation of rules (Alexandre, 2011). A much broader spectrum of needs, capabilities, harms and risks – not just to humans, but to the environment and the atmosphere, and over time – will need to inform the design of more equitable rules for a more equitable and sustainable result (Sunstein, 2005; Mossman, 2006; Dell *et al.*, 2012; Rowell, 2012; Kharecha and Hansen, 2013).

For example, a suite of goal-based SMR standards, set against the background of a more inclusive understanding of commensurable societal risks, costs and harms, could (directly or obliquely) take into consideration (among other factors) the radiological consequences of Fukushima, the higher cost of climate change to developing countries, the inherent and overlapping safety characteristics of the technology, and the social and economic detriment of lack of electricity (Wiener, 2004; Abdel-Aziz *et al.*, 2009; IBRD, 2009; IAEA INPRO DF5, 2012; MacKay, 2009; UNDP, 2013).

23.6.3.1 A proposed approach

I would suggest a consensus for a multinational framework for SMRs, as a formal statement of principle. This would be a tier below an internationally binding convention and could draw from models of the International Maritime Organization, International Civil Aviation Organization and the Naval Ship Code.

There are templates outside of the nuclear sector, for global fleet design and safety licensing, design change management and design authority, and requirements management, that have proven to be successful and could be employed (Söderholm, 2013). These include maritime regulation mechanisms for nuclear transport, and aircraft licensing. Aircraft licensing in particular could be a useful template to regard for standard SMR plants deployed in multiple countries (Goodman and Raetzke, 2013).

23.7 Conclusion

Keeping in mind the global consequences and effects of both disasters and advances, we should look at the risks of a lack of shared cooperative mechanisms for nuclear licensing, safety, monitoring and fuel. Developing countries and the rest of the world share so many systemic risks (economic, health, political, environmental) that new principles of sharing and cooperation, responsibility and mutual obligation will have to supersede narrower concerns such as sovereignty. Arguably, this is already taking place within the nuclear realm, under the NPT (nuclear non-proliferation treaty), and the concept would simply be extended, as far as SMRs are concerned, to the arenas of climate change and technology transfer. This idea deserves further study. SMRs, standardized, safety-enhanced, small-scale and flexible nuclear power technology, would be suitable for such new cooperative arrangements.

There will be a change of global understandings of risk, due to the raised awareness about climate change (because of acute observable effects and incidents). This will affect the economics of SMRs. The trade-offs and embedded risk biases will become more transparent in crisis. As a result, it is anticipated that emissions-reduction and climate-change strategies and policies will gain importance; and that nuclear safety monitoring, regulatory activity, and supply chain will be increasingly internationalized.

Above all, in the very idea of the 'developing' country, there is a continuum, the idea of progress. What is development for? We are aiming for the attainment of basic human functional capabilities by all, no matter where, with no exceptions (Nussbaum, 2000), for the 'unfolding of powers that human beings bring into the world' (Nussbaum, 2011).

To reiterate, SMRs provide the opportunity to go back to basic first principles (just as technology designers have) and consider what supporting institutions, infrastructure, and rationales for their use, are actually fit for purpose in a world of increasingly shared benefits and detriments.

We have to start somewhere.

Acknowledgments

I am grateful to Peter Storey for illuminating discussions about the subject of this chapter.

References

Abdel-Aziz, A., Rao, S., Lee-Shanok, B., 2009. Before the Renaissance: A Reformative Challenge to Precautionary Dogma in Nuclear Safety Regulation. In: Presentation at International Nuclear Law Association (INLA) Nuclear Inter Jura Congress, October 5–9.

Abdulla, A., Ines Lima, A., Granger Morgan, M., 2013. Expert assessments of the cost of light water small modular reactors. PNAS. 100, (24). https://doi.org/10.1073/pnas.1300195110.

Alexandre, C., 2011. Regulators as change agents. Innovations: Technology, Governance, Globalization. 6(4).
Ayacko, O., 2012. Solving Kenya's Power Crisis. Kenya, posted May 28, 2012. Accessed at: http://www.ntv.co.ke/news2/ntv-business-agenda-solving-kenyas-power-crisis/.
Barkatullah, N., 2011. Possible Financing Schemes for Current and Near Term Nuclear Power Projects. In: Workshop on Technology Assessment of Small and Medium Reactors for Near Term Deployment. IAEA, Vienna, Austria 8 December 2011. Accessed at: www.iaea.org/…/23_IAEA_Barkatullah-Economics_SMRDec2011.pdf.
Carelli, M.D., 2014. Chapter 3 in Handbook of Small Modular Reactors. Woodhead Publishing.
Chonghaile, C.N., 2012. Kenya's bid to become the first African nation to set up a climate authority. The Guardian.May 25, 2012. Accessed at: http://www.guardian.co.uk/environment/2012/may/25/kenya-climate-change-authority.
Cossardeux, J., 2013. Les mégapoles du monde gagnées par l'étouffement. Les Echos.February 5, 2013. Accessed at: http://www.lesechos.fr/vg/3ca45bb1eb/journal20130205/lec1_monde/0202543980693-les-megapoles-emergentes-gagnees-par-l-etouffement-535240.htm.
Dell, M., Jones, B.F., Olken, B.A., 2012. Temperature shocks and economic growth: Evidence from the last half century. American Economic Journal: Macroeconomics 4 (3), 66–95. Web accessed at: https://doi.org/10.1257/mac.4.3.66.
ETSON, 2013. http://www.etson.eu/About/missionvision/Pages/TSO-Requirements.aspx also http://www.etson.eu/Join/Pages/default.aspx.
Goodman, D., Raetzke, C., 2013. SMR licensing: Looking to the Skies. Nuclear Future. 9(6).
Goodman, D., Storey, P., 2012. How SMRs Can Support Developing Countries. In: Presentation at Platts SMR Conference, Arlington, Virginia, May 21–22.
G-Science Academies, 'Energy and Water Linkage: Challenge to a Sustainable Future', in G-Science Academies Statements. Accessed at: http://www.g20mexico.org/index.php/en/mexican-academy-of-sciences.
Hecht, G., 2012. Being Nuclear: Africans and the Global Uranium Trade. The MIT Press, Cambridge, MA.
Hinshaw, D., 2010. 'Africa Looks to Nuclear Power'. Christian Science Monitor.April 2, 2010. Accessed at: csmonitor.com/World/Africa/2010/0402/Africa-looks-to-nuclear-power.
IAEA, 2012. Status of Small and Medium Sized Reactor Designs; A Supplement to the IAEA Advanced Reactors Information System (ARIS), September 2012.Accessed at: http://aris.iaea.org.
IAEA INPRO DF3, 2011. Common User Considerations for Small and Medium-sized Nuclear Power Reactors. In: Proceedings of 3rd INPRO Dialogue Forum, Vienna, Austria, 14–18 November 2011 Accessed at: www.iaea.org/INPRO/3rd_Dialogue_Forum/index.html.
IAEA INPRO DF 5, 2012. Long term Prospects for Nuclear Energy post-Fukushima. In: Proceedings of 5th INPRO Dialogue Forum, 27-31 August 2012 Accessed at: www.iaea.org/INPRO/5th_Dialogue_Forum/index.html.
IBRD, 2009. Africa Infrastructure Country Diagnostic. Background Paper 6. Underpowered: The State of the Power Sector in Sub-Saharan Africa. 2009, The International Bank for Reconstruction and Development/The World Bank, Washington, DC. Accessed at: http://web.worldbank.org/WBSITE/EXTERNAL/COUNTRIES/AFRICAEXT/EXTAF_RREGTOPENERGY/0,contentMDK:22679058~pagePK:34004173~piPK:34003707~theSitePK:717306,00.html.
IBRD, 2010. Africa Infrastructure Country Diagnostic Country Report. Ghana's Infrastructure: A Continental Perspective. The International Bank for Reconstruction and Development/The World Bank, Washington, DC March 2010.
IEA, 2011. World Energy Outlook 2011. International Energy Agency.

Ingersoll, D.T., 2011. An Overview of the Safety Case for Small Modular Reactors. In: Proceedings of the ASME 2011 Small Reactors Symposium, September 28–30, 2011, Washington, DC.

Jackson, S., 2013. Jamalca To Press Ahead With Coal Plant. Daily Gleaner.April 26, Accessed at: http://jamaica-gleaner.com/gleaner/20130426/business/business2.html.

Jamaica Ministry of Mining and Energy, 2009. Jamaica's National Energy Policy 2009–2030. accessed at http://www.men.gov.jm/PDF_Files/Energy_Policy/Jamaica's National Energy Policy 2009–2030.pdf.

Kessides, I.N., Kuznetsov, V., 2012. Small modular reactors for enhancing energy security in developing countries. Sustainability 4, 1806–1832. https://doi.org/10.3390/su4081806.

Kharecha, P.A., Hansen, J.E., 2013. Prevented mortality and greenhouse gas emissions from historical and projected nuclear power. Environ. Sci. Technol. https://doi.org/10.1021/es3051197 Accessed at: http://pubs.giss.nasa.gov/abs/kh05000e.html.

Lee, B., Prestor, F., Kooroshy, J., Bailey, R., Lahn, G., 2012. Resources Futures. Chatham House Report. The Royal Institute of International Affairs December.

MacKay, D.J.C., 2009. Sustainable Energy – Without the Hot Air. UIT, Cambridge.

Mian, Z., 2011. Jamaica: Energy Future (Part 2). Daily Gleaner.February 20. Accessed at: jamaica-gleaner.com/gleaner/20110220/focus/focus6.html.

Mossman, K., 2006. Radiation Risks in Perspective. Taylor and Francis Publishing Group, Boca Raton, FL.

New York Times, The (Reuters), 2013. Guinea to Buy Electricity from UK's Aggreko to Stem Power Cuts. August 7, www.nytimes.com/reuters/2013/08/07/business/07reuters-guinea or www.reuters.com/article/2013/08/07/guinea-power-aggreko-idUSL6N0G83A520130807.

Nussbaum, M.C., 2000. Women and Human Development: The Capabilities Approach. Cambridge University Press, Cambridge, pp. 78–80.

Nussbaum, M.C., 2011. Creating Capabilities; The Human Development Approach. Belknap Press of Harvard University Press, Cambridge, MA p. 23.

Orengo, P., 2012. Abandon nuclear energy programme, Unep boss urges Kenya. The Standard. February 14. Accessed at: www.standardmedia.co.ke/?articleID=2000052072&pageNo=1.

Osabutey, P.D., 2012. Nigeria's Desire for More Power, Source of in Ghana. Ghanaian Chronicle.April 24. Accessed at: the chronicle com.gh/nigeria's-desire-for-more-power-source-of-in-ghana/.

Padgett, T., 2010. The Postquake Water Crisis: Getting Seawater to the Haitians. Time.January 18, Accessed at: http://www.time.com/time/specials/packages/article/0,28804,1953379_1952494_195484,00.html.

Pasternak, A.D., 2000. Global Energy Futures and Human Development: A Framework for Analysis. US Department of Energy Report UCRL-ID-140773, Lawrence Livermore National Laboratory, Livermore, CA.

Ropeik, D., 2012. The Wages of Eco-Angst. The New York Times.February 26. Accessed at http://opinionator.blogs.nytimes.com/2012/02/26/the-wages-of-eco-angst.

Rowell, A., 2012. Allocating pollution. The, University of Chicago Law Review 79 (985), 1044.

Sadorsky, P., 2012. Information communication technology and electricity consumption in emerging economies. Energy Policy 48, 130–136.

SAPC, 2011. National Planning Commission, Diagnostic Overview, 2011. South Africa. Accessed at: http://npconline.co.za/pebble.asp?relid=33.

Sen, A., 1995. Inequality Reexamined. Russell Sage Foundation, NY, p. 67 Harvard University Press, Cambridge.

SIDS Outcome, 2013. par 4. Accessed at: Kingston Outcome of the Caribbean Regional Preparatory Meeting for the Third International Conference on Small Island Developing States (SIDS)', Kingston, Jamaica, Julyp. 7. http://www.mwh.gov.jm/new/index.php/misc/sids-conference.

Söderholm, K., 2013. Licensing Model Development for Small Modular Reactors (SMRs) – Focusing on the Finnish Regulatory Framework. PhD dissertationIn: Acta Universitatis Lappeenrantaesis research series, Number 528, Lappeenranta University of Technology Accessible at: http://urn.fi/URN:ISBN:978-952-265-452-6.

Stern, N., 2012. What we risk and how we should cast the economics and ethics. In: Lionel Robbins Memorial Lecture Series, Lecture 1, 21 February, London School of Economics and Political Science Accessed at: http://www2.lse.ac.uk/publicEvents/events/2012/02/20120221t1830vOT.aspx.

Sunstein, C.R., 2005. Laws of Fear: Beyond the Precautionary Principle. Cambridge University Press, New York.

The Standard, 2012. Nuclear energy source versus safe renewable energy sources: Which way for Kenya. February 9, 2012. Accessed at: www.standardmedia.co.ke/?articleID=2000052418&pageNo=1.

UNDP, 2013. Human Development Report 2013 summary. The Rise of the South: Human Progress in a Diverse World. United Nations Development Programe, New York, NY. Accessed at: undp.org/content/dam/undp/library/corporate/HDR/2013GlobalHGR/English/HDR2013%20Summary%20English.pdf.

Weart, S.R., 2012. The Rise of Nuclear Fear. Harvard University Press, Cambridge, MA.

Wiener, J.B., 2004. Hormesis, hotspots, and emissions trading. Human and Experimental Toxicology 23, 289–301.

Wines, M., 2007. Toiling in the Dark: Africa's Power Crisis. The New York Times.July 29.

WMO/IGAC, 2012. Impacts of Megacities on Air Pollution and Climate.September. Accessed at: http://www.igacproject.org/Megacities.

World Bank, 2010. Operational Guidance for World Bank Group Staff. Criteria for Screening Coal Projects under the Strategic Framework for Development and Climate Change. MarchIn: The International Bank for Reconstruction and Development/The World Bank, Washington, DC Accessed at: http://documents.worldbank.org/curated/en/2010/03/16286802/operational-guidance-world-bank-group-staff-criteria-screening-coal-projects-under-strategic-framework-development-climate-change.

World Bank, 2013. How We Classify Countries. Accessed at http://data.worldbank.org/about/country-classifications.

World Bank Enterprise Survey, 2012. International Finance Corporation, 'World Bank Enterprise Surveys: What Businesses Experience'. Topic: Infrastructure Accessed at: http://www.enterprisesurveys.org/Data/ExploreTopics/infrastructure.

Yepéz-García, R.A., Dana, J., 2012. Mitigating Vulnerability to High and Volatile Oil Prices: Power Sector Experience in Latin America and the Caribbean, World Bank, Washington, DC. License: Creative Commons attribution CCBY 3.0.

Index

Note: Page numbers followed by *f* indicate figures and *t* indicate tables.

A

ABV-6E reactor, 470. *See also* OKBM Afrikantov small modular reactor projects
   chromium-nickel alloy, 477
   construction period, 476–477
   core and fuel design characteristics, 473*t*
   deployment in Russia, 495
   design and operating characteristics, 470, 471–472*t*
   design layout, 470–474, 475*f*
   factory refueling, 477
   liquid boron system, 477
Accident source terms, 285
ACP-100, 33
   deployment
     licensing, 407
     site selection, 407
   design parameters, 400
   engineered safety feature plan, 403–404
   licensing, 407
   nuclear steam supply system, 401–403
   passive safety design features, 404–405
   plant layout, 400
   post-Fukushima actions, 405–406
   site selection, 407
   technical aspects, 399–400
   testing and verification, 406
Active safety systems, 188–189
Adaptive HSIs, 173
Additive layer manufacturing (ALM), 309–311
   benefits, 310–311
Advanced gas reactors (AGRs), 504
Advanced high-temperature reactor (AHTR), 46
Advanced light water reactor (ALWR), 9
Advanced modular reactor feasibility and development (AMR-F&D) project, 505
Advanced modular reactors (AMRs), 503
   role in low-carbon energy generation, 515–517
   United Kingdom
     DBD HTR-PM, 512–513
     grid balancing frequency response and inertia, 516–517
     high-temperature gas reactor (HTGR) systems, 511
     industrial heat applications, 517
     low-carbon economy, 515
     micro modular reactor (MMR), 512
     Moltex stable salt reactor (SSR), 513
     SEALER-UK reactor, 513
     space and water heating, 516
     Tokamak energy spherical tokamak, 513
     U-Battery, 512, 512*f*
     Westinghouse lead-cooled fast reactor (LFR), 513
Advanced nuclear technologies (ANTs), 503, 504*f*
Advanced Research Projects Agency-Energy (ARPA-E) program, 522, 549–550
Advanced SMRs (A-SMRs), 521, 532
   concept evaluations, 534–541
   development related R&D program, 532–534
   funding opportunity announcement, 532–533
     awarded R&D projects, 535–538*t*
     funding pathways, 533–534
     regulatory areas, 533–534
   and Nuclear Energy University Program, 546
Aircraft licensing, 605
Aircraft Nuclear Propulsion (ANP) program, 5–6
Aircraft Reactor Experiment (ARE), 5–6
Aircraft shield test reactor (ASTR), 5–6
Ammonia production, 347
AREVA Steam Cycle-High Temperature Gas Reactor (SC-HTGR), 539, 543*f*

Argentina SMRs. *See also* Central Argentina de Elementos Modulares (CAREM) project
  heavy water reactors, 360–362
  research reactors, 359–360
  water-cooled reactors, 32–33
Argonne National Laboratory (ANL), 453–454
ATGOR transportable NPP, 488, 488*f*
Atomic Energy Commission of Argentina (CNEA), 360–362, 370
Atomic Energy of Canada Limited (AECL), 375
Atucha-1 reactor, 360–362
Atucha-2 reactor, 361
Auditory interfaces, 170–171
Automatic depressurization system (ADS) valves, 107
Auxiliary feedwater (AFW) system, 108–109

B

Baffle plate, 101–103
Balance of Plant (BOP) instrumentation, 131
  nonsafety systems, 141–142
  wireless communication, 142
BANDI-60S
  design features, 446, 446*t*
  design goals, 444
  future plan, 445
  in-vessel control element drive mechanism, 447
  KEPCO E&C, 444–445
  passive safety systems, 447–449
  reactor coolant system, 446
  soluble boron-free design, 447
  technical data, 446–449
Barge-mounted nuclear power plants, 470, 479, 496
  deployment, 494
  external events, 479
  with KLT-40S reactors, 479
  plant mooring devices, 479
  with RITM-200M reactors, 475
BEIS SMR competition Phase 1, 518*t*
Beyond design basis events (BDBE), 204–206
BOO (Build – Own – Operate by the vendor) mechanisms, 602
BOOT (Build – Own – Operate – Transfer to the user) mechanisms, 602

Bottom-up cost estimation, 244, 273
Brain-machine interaction, 172–173
Bupp–Derian–Komanoff–Taylor hypothesis, 251–252
Busy-hands busy-eyes tasks, 170–171
B&W mPower integrated system test (IST) facility, 204
BWRX-300 design, 36

C

Cabling, 135–136
Calder Hall, 503–504
Canada Deuterium Uranium (CANDU) reactor, 375
Canadian Nuclear Laboratories (CNL), 386–387
Canadian Nuclear Safety Commission (CNSC), 45–46, 289, 375
Canadian SMRs, 45–46
  case study
    province of New Brunswick, 377–378
    province of Ontario, 377
  CNL's demonstration siting initiative, 386–387
  development and deployment, support for, 386–387
  electricity applications, 378–379
  greenhouse gas emissions, 388–389
  heavy industry applications, 380–381
  markets and applications, 378–383
  power generation industry, 389–390
  R&D support, 387
  regulatory framework, 384–385
  remote communities, 381–382
  roadmap, 376–377
  supply chain readiness, 386
Cap-and-trade systems, 585
Capital costs
  economy of scale, 252–254
  multiple units, 255–260
  size-specific factors
    design factor, 261–263
    modularization, 260–261
Carbon mitigation plans, 568–569
Carbon taxes, 585
Central Argentina de Elementos Modulares (CAREM) project, 32–33, 359
  commercial modules' development, 370–372

Index 613

control rod drive mechanisms, 106–107
developments, 366–369
post-Fukushima actions, 369
prototype design, 362–366
pumps, 103–104
steam generators, 105
Chapelcross power plants, 504
Chemical and volume control system (CVCS), 108
Chernobyl Unit 4 accident, 52, 410–411
China General Nuclear Power Corporation, 408
China National Nuclear Corporation (CNNC), 33, 407–408
Cladding, 312–314
Class 1E safety qualification, 140
Clean energy standards (CES), 582
Climate goals and initiatives
  carbon mitigation plans, 568–569
  Paris Agreement, 566–569
Coal and combined cycle gas turbine (CCGT), 270
Coal-to-gasoline production, 346
Coatings systems, 318
Cogeneration systems, 323
Combined construction and operating license (COL), 280
Combustible Avanzado para Reactores Argentinos (CARA) project, 361–362
Commission peer reviews, 228
Common cause failure (CCF), 134, 138–139
Compact containment boiling water reactor (CCR), 414–416
Component fabrication
  additive manufacture, 309–311
  cladding, 312–314
  electron beam melting, 311
  hot isostatic pressing, 314–316
  shaped metal deposition, 314–316
Components performance test, 442
Construction costs, 244–245
Construction methods
  component fabrication, 309–316
  economic development, 299
  production line approach
    automotive applications, 302
    component sizing, 308
    flowline, 303–308
    sales, volume and profile of, 302–303

standardisation, 308
supply chain implications, 318–321
Control rod drive mechanisms (CRDMs), 106–107
Control rod drives (CRDs), 364–365
Control rods, 105–106
Control room habitability equipment, 111
Control room staffing, 283–284
Cooperation in Reactor Design Evaluation and Licensing (CORDEL) Working Group, 294
Core baffle, 107–108
Core barrel, 107–108
Core basket, 107–108
Core damage frequency (CDF), 53–54, 206–208
Core makeup tank (CMT), 447–449
Co-siting economies, 260
Cost competitiveness, 580–582
Critical heat flux (CHF) tests, 441–442
Cybersecurity, integral pressurized water reactor, 139–140

D

DBD HTR-PM, United Kingdom, 512–513
Decay heat, 187
Decay heat removal system (DHRS), 455–456
Decommissioning cost, 242
Decontamination and decommissioning (D&D) costs, 268, 578
Defense in depth (DiD), 206, 293
Department of Energy Research and Innovation Act, 585–586
Design basis accident (DBA), 204–206, 285
Design basis events (DBEs), 56
Design basis threat (DBT), 233–234
Design certification rule (DCR), 281–282
Design-Specific Review Standards, 584–585
Developing countries
  energy planning decisions, 601
  energy resource-rich countries, 603
  financing and external policy preferences, 603
  SMR deployment
    cost of grid power, 599
    electricity tariffs, 599
    energy infrastructure weakness, 599–600

Developing countries *(Continued)*
  growth of megacities, 600
  information and communications technology (ICT), 598
  regional regulators, 604–605
  regulatory innovators/change agents, 605
  sociological public-acceptance factors, 600–601
  technology and licensing, standardization of, 604
  technology-neutral regulatory approach, 604
  trade-offs, 597–598
  utilization of regional mechanisms, 604–605
  water precarity or scarcity, 599
  sustainable energy choices and debt, 602
  technology lock-in and decarbonization, 601–602
Diagnostic technology, 131–132
Diamond-like-coatings (DLC), 318
Diesel generators, 111–112
Digital technology, 133–134, 141
Diode laser powder deposition (DLPD), 313, 313–314f, 314t
Direct metal laser sintering, 309
Discounted cash flow (DCF) analysis, 244–245
Distributed control system (DCS), 150
Doppler broadening, 72–73
Dynamic system modeling, 350

E
Early site permit (ESP) process, 281–282
Economic competitiveness, 52, 61–63, 252
Economic development, 299
Economic viability, 581–582
Economy of multiples, 269
Economy of scale, 252–254
Electric power, 4, 7–8
Electron beam melting (EBM), 311
Electron beam welding, 317–318
Electronic advanced reactor information system (ARIS)
  KLT-40S reactor design, 498
  periodically updated brochure, 498
Emergency cool-down tanks (ECTs), 449

Emergency core cooling system (ECCS), 109–112
Emergency diesel generator (EDG), 111–112
Emergency operating facility (EOF), 166
Emergency planning, 285–286, 585
Emergency planning zones (EPZs), 293
Empresa Nuclear Argentina de Centrales Electricas (ENACE), 361
Energy Multiplier Module (EM^2) reactor, 40–41
Energy resource-rich countries, 603
Engineered safety feature plan, 398–399, 403–404
Engineering development tests, 198
Example sodium fast reactor (ESFR) case study, 229–230
External factors, 271–273
External pool, 111

F
Factory fabrication of modules, 583–584
Falcon 9, 584
Field-programmable gate array (FPGA) technology, 134–135, 138, 141
Financing, 578–580
First of a kind (FOAK) reactor technology, 152, 263–264, 280, 288, 581–582
First shutdown system (FSS), 364
Fischer-Tropsch reaction, 346–347
Flexblue SMR, 33–34
Flibe Energy, 46–47
Flowline approach, 303–308
Flow transmitters
  nuclear steam supply system (NSSS) control systems instrumentation, 130–131
  safety system instrumentation and controls, 125–126
Fluoride salt-cooled high-temperature reactor (FHR), 6–7, 44–45
Fossil fuels, 557
French NPP
  fleet deployment costs, 254, 256–257f
  water-cooled reactors, 33–34
Fuel burnup, 71–72
Fuel cost, 242
Fuel requirements, 575–576
Fuel temperature coefficient, 72–73

Fukushima Daiichi nuclear power plant accident, 52, 422, 425–426
FY2019 National Defense Authority Act, 548–549

G

Gamma thermometers, 127
Gas-cooled reactors
  commercial designs, 38, 39t
  EM^2 reactors, 40–41
  GT-MHR design, 39–40
  helium-cooled reactors, 38
  HTR-PM design, 39
  Xe-100 design, 41
Gas turbine high temperature reactor 300 (GTHTR300), 417–420
Gas turbine modular high-temperature reactor (GT-MHR) reactor, 39–40
Gateway for Accelerated Innovation in Nuclear (GAIN) program, 541–545
General design criteria (GDC), 289
General Electric Hitachi (GEH) Sodium-Cooled Fast Reactor (PRISM SFR), 539, 544f
Generation IV International Forum (GIF), 217, 226–230, 337, 571
Gesture interaction, 171–172
Global energy demands, 570–571
Global market assessments, 559–560
Global warming, 566–567
Graded approach, licensing, 293
Greenhouse gas (GHG) emissions, 378–379, 379f, 388–389

H

Haptic interaction, 172
Heaters, 101–103
Heating, ventilation and air conditioning (HVAC) system, 131
Heat Transfer Reactor Experiment-1 (HTRE-1), 5–6
Heavy water reactors (HWRs), 359–362
Helium-cooled test reactors, 38
High-temperature electrolysis (HTE), 346
High-temperature gas-cooled reactor (HTR)-200
  deployment, 406–407
  design parameters, 397
  engineered safety feature plan, 398–399
  technical aspects, 396
  testing and verification, 399
High-temperature gas reactor (HTGR) systems, 511
High-temperature reactor pebble bed module (HTR-PM) reactor, 39, 512–513
High-temperature reactors, 48
Holographic display technology, 170
Holtec SMR-160, 98–99, 103–105, 112
Hot isostatic pressing (HIP), 314–316
HSI. *See* Human-system interfaces (HSI)
Human development indicators (HDI), 596–597
Human factors engineering (HFE) analysis, 284
Human-system interfaces (HSI)
  architecture and functions, 173–175
  auditory interfaces, 170–171
  classification, 167–173
  context of use, 159–160
  control and monitoring centers, 163–165
  control devices and mechanical interaction, 171
  design considerations, 176–178
  emergency operating facility, 166
  functional criteria, 150–151
  future trends, 178–182
  hardware features, 150
  human factors considerations, 157–158
  hybrid interfaces, for multimodal interaction, 171–173
  implementation and design strategies, 175–178
  instrumentation and control engineers, 181–182
  interaction modalities, 167–168
  materials and waste fuel handling, 165–166
  NUREG-0700 definition, 149–150
  operational domains, 162–167, 163t
  operational requirements, 160
  organizational requirements, 160
  outage control center, 166
  regulations, 161–162
  regulatory requirements, 176
  software criteria, 150
  standards and design guidance, 176
  technical characteristics, 159
  technical support center, 167

616 Index

Human-system interfaces (HSI) *(Continued)*
  technology forecast, 178, 179–181*t*
  visual interfaces, 168–170
    large screen displays, 169
    3D displays, 170
    wearable displays, 169–170
Hybrid energy system (HES), 573–574
  configuration selection and optimization, 349
  definition, 325–326
  electricity markets, 336–337
  integrated system testing, 351–353
  modeling and simulation, 350–351
  nuclear-renewable integration, 340–342
  performance assessment
    environmental impacts, 332–333
    overall public or political attractiveness, 335
    overall system economics, 332
    production reliability/on-demand availability, 333
    system resiliency, 333–334
    system security, 334–335
    technical feasibility, 331
  principles, 330
  process heat applications
    ammonia production, 347
    coal-to-gasoline production, 346
    general considerations, 342–344
    hydrogen production, 345–346
    natural gas-to-diesel production, 346–347
    oil shale, 348–349
    olefins via methanol production, 349
    overview, 342, 343*t*, 344–349
    steam-assisted gravity drainage, 348
    water desalination, 347–348
  reactor classes, 337, 338*t*
  system siting and resource integration, 336–337
Hydraulic control rod drive (HCRD) mechanism, 368–369
Hydrogen production, 345–346

I

In-core fuel management, 79–82
Independent peer review, 228
Industrial heat, 572–573

Inherently safe and economical reactor (ISER), 410
Inherent safety, 189, 191
In-process peer review, 228
In-service deployment model, 321
Instrumentation, 108
Instrumentation and controls (I&C) design
  BOP instrumentation, 131
  cabling, 135–136
  components, 118–119
  diagnostics and prognostics, 131–132
  future trends and challenges, 136–143
  general requirements, 119–127
  licensing, 136–140
  nonsafety systems, 141–142
  nuclear steam supply system, 127–131
  processing electronics, 132–135
  safety systems, 119
  wireless *versus* wired solutions, 142–143
Integral effect test, 199, 442
Integral pressurized water reactor (iPWR), 54–56, 69, 359, 362–369
  burnable poison effects, 77–79, 88–89
  connected system components
    auxiliary feedwater system, 108–109
    chemical and volume control system, 108
    control room habitability equipment, 111
    diesel generators and electrical distribution, 111–112
    emergency core cooling system, 109–111
    external pool, 111
    refueling water storage tank, 109–111
    residual heat removal system, 108–109
  cybersecurity, 139–140
  economic competitiveness, 61–63
  evolution, 54–56
  fissile material enrichment, 75–77
  generic components, 95, 97*f*
  in-core fuel management, 79–82
  instrumentation and controls
    (*see* Instrumentation and controls (I&C) design)
  integral components
    automatic depressurization system valves, 107
    baffle plate, 101–103
    control rod drive mechanisms, 106–107

Index   617

control rods and reactivity control, 105–106
core baffle, 107–108
core barrel, 107–108
core basket, 107–108
instrumentation, 108
pressure vessel and flange, 96–101, 98f
pressurizer and heaters, 101–103
pressurizer relief tank, 101–103
pumps, 103–104
reactor coolant system, 101
relief valves, 107
riser, 104
spray valve, 101–103
steam generators, 104–105
nuclear design
control rods, 88–89
core loading, 90–91
design features, 69–70, 75–82
objectives, 69
process overview, 82, 83f
safety design criteria, 70–75
smaller cores, 82–88
safety, 56–61 (see Safety, iPWRs)
safety system end-state architecture, 137–138
Integral primary system, 188
Integrated modular water reactor (IMR), 103–104, 412–413
Intelligent HSIs, 173
Interaction modalities, 167–168
Intermediate reactor auxiliary cooling system (IRACS), 422
International Atomic Energy Agency (IAEA), 49
International Atomic Energy Agency (IAEA), 219, 292–293, 296, 409, 477–478, 498–499, 565
  international certification, 296
  SMR deployment indicators, 566, 567t
International Atomic Energy Agency (IAEA) Technical Document, 560–561
International certification, 295–296
International Reactor Innovative and Secure (IRIS) project, 55–56, 59–60
  pressure vessel, 99
  pressurizer, 102
  pumps, 103–104
INVAP, 360

Investment and risk factors, 245–252
Ion chambers, 127
IRIS SPES3 facility, 200

J

Japan Atomic Energy Agency (JAEA), 417
Japan Atomic Power Company (JAPC), 412–413
Japanese SMR
  BWRX-300, 36
  CCR, 414–416
  deployment, 422
  DMS, 416–417
  GTHTR300, 417–420
  IMR, 412–413
  R&D activity, 410–411
  4S reactor, 42–43, 420–422
Johnson noise thermometer (JNT), 124
Joint cognitive system, 153
Joint Stock Company (JSC) NIKIET SMR projects
  ATGOR, 481
  ATGOR transportable NPP, 488
  core and fuel design characteristics, 484t
  design and operating characteristics, 482–483t
  SHELF, 481, 486–488
  UNITHERM, 481, 485–488, 485f

K

Kairos Power, 46
KAIST micromodular reactor (KAIST-MMR), 428
  air-cooling layout, 459–460, 462f
  concept, 458–459
  core layout, 459, 461f
  future plan, 459
  internal layout, 459, 460f
  passive decay heat removal, 461, 463f
  technical data, 459–462
KEPCO E&C, 444–449
KLT-40S reactor, 34–35, 470, 477.
  See also OKBM Afrikantov small modular reactor projects
  core and fuel design characteristics, 473t
  design and operating characteristics, 470
  layout, 470, 474f

KLT-40S reactor *(Continued)*
    on-site and off-site emergency planning, 479
Korea Atomic Energy Promotion Council, 453–454
Korea Atomic Energy Research Institute (KAERI), 425–426, 453–458
Korean Atomic Energy Commission, 453–454
Korean NPP fleet deployment costs, 257, 258–259*f*

L

Large and early release frequency (LERF), 206–207
Large-scale nuclear reactors, 323
    cumulated cash flows, 264, 266*f*
    deterministic scenarios, 263–264
    economy of multiples, 269
    external factors, 271–273, 272*f*
    operating costs, 267–269
    profitability index, 265–267, 267*f*
    uncertainty, 264–267
Large screen displays, 169
Laser sintering, 309
Levelized cost of electricity (LCOE), 242, 242*t*, 264–265, 580–581
Levelized unit electricity cost (LUEC), 7, 9–10, 242
Level transmitters
    nuclear steam supply system (NSSS) control systems instrumentation, 129
    safety system instrumentation and controls, 122–123
Licensing
    ACP100, 407
    control room staffing, 283–284
    deterministic or risk informed approaches, 282–283
    industry codes and standards, 290–291
    instrumentation and controls (I&C) design, 136–140
    international certification, 295–296
    international strategy and framework, 291–297
    non-LWR SMR designs, 288–290
    NRC requirements, 280–288
    risk mitigation, 288
    security requirements, 284–285
    timeliness, 287–288
    worldwide informational data, 296–297
Licensing Technical Support (LTS) program, 521–523
Light-water reactor (LWR), 61, 95, 326, 407, 410
    vs. advanced gas reactors (AGRs), 504
    worldwide design, 31–32, 32*t*
Liquid fluoride thorium reactor (LFTR) design, 46–47
Liquid metal-cooled reactors
    advantages, 41–42
    commercial designs, 42, 42*t*
    PRISM design, 43–44
    4S design, 42–43
    SVBR-100 design, 43
Local control station (LCS), 165
Loss-of-coolant accidents (LOCAs), 54, 59–60, 100–101, 220, 262–263
Low-temperature electrolysis (LTE), 345–346
LUEC. *See* Levelized unit electricity cost (LUEC)
Luna Innovations' fiber optic distributed temperature sensor, 124–125, 124*f*

M

Magnox fleet, 504
Main control room (MCR), 164
Man Machine Interface Systems (MMIS), 439, 440*f*
Manufacturing license, 287
Manufacturing processes, 309–316.
    See also Production line approach
Marine Reactor X (MRX), 410
Master Facility License, 294–295
Maximum reactivity insertion rate, 74
Micro modular reactor (MMR), United Kingdom, 512
Micro Nuclear Energy Research and Verification Arena (MINERVA), 463
Micropocket fission detectors, 127
Microreactor R&D program, 548–549
    features, 548–549
    objectives, 549
    request for information (RFI), 548
Microreactors, 570, 572

Index 619

Military reactors, for terrestrial application, 4
Minimum Attention Plant (MAP), 55
Mitsubishi intrinsic safe integrated reactor (MISIR), 410
Mobile reactors, 48
Modeling-Enhanced Innovations Trailblazing Nuclear Energy Reinvigoration (MEITNER), 522, 549–550
Modular build approach, 319–320
Modular simplified and medium small reactor (DMS), 416–417
Molten-salt-cooled reactors
 applications, 44
 commercial designs, 45, 45t
 IMSR design, 45–46
 KP-FHR design, 46
 LFTR design, 46–47
Moltex stable salt reactor (SSR), United Kingdom, 513
Moody's rating methodology for electric utilities, 250, 251t
mPower reactor, 105, 199
Multifunction LCS, 165
Multimodule control rooms, 164–165
Multinational Design Evaluation Program (MDEP), 292, 585
Multiple-module licensing, 286–287

N

National Atomic Energy Commission (CNEA), 32
Nationally determined contributions (NDCs), 568
National Reactor Innovation Center (NRIC), 521–522, 546–548
National Renewable Energy Laboratory (NREL), 339
Natural gas-to-diesel production, 346–347
Natural Resources Canada, 376
Next-generation nuclear plant (NGNP) program, 329–330, 342–345
Nonelectric applications, SMR designs for, 572–573, 573f
Non-LWR SMRs licensing, 288–290
Nonsafety instrumentation systems, 141–142
Nuclear design, 69

integral pressurized water reactor (iPWR)
 control rods, 88–89
 core loading, 90–91
 design features, 69–70, 75–82
 objectives, 69
 process overview, 82, 83f
 safety design criteria, 70–75
 smaller cores, 82–88
Nuclear energy, 570
 countries using, 569
 2030 UN sustainable development goals on, 566–568
Nuclear Energy Innovation Capabilities Act, 585–586
Nuclear Energy Institute (NEI), 284–285, 289–290
Nuclear Energy University Program and A-SMRs, 546
Nuclear fuel and modules transport, 575–576
Nuclear hybrid energy systems (NHESs), 323
Nuclear hydrogen development and demonstration (NHDD) project, 456, 457f
Nuclear industry strategy (NIS), 507
Nuclear Innovation 2050, 571
Nuclear innovation program (NIP), United Kingdom
 advanced fuels, 509
 advanced manufacturing and materials, 507–509
 nuclear facilities and strategic toolkit component, 511
 reactor design, 510–511
 recycle and waste management, 509–510
Nuclear Innovation Research Advisory Board (NIRAB), 507, 518–519
Nuclear newcomer countries, challenges for, 569–570
Nuclear plant construction, 583
Nuclear power, 51
 components, 63
 imperatives, 52–54
Nuclear power development, United Kingdom, 503–505
Nuclear Power Institute of China (NPIC), 33
Nuclear power plants (NPPs), 359
 design strategies, 147
 life-cycle costs, 241–242
Nuclear propulsion system, 5–6

Nuclear Regulatory Commission (NRC)
 Regulatory Guide, 113, 132–133,
 139–140
Nuclear steam supply system (NSSS), 3
 ACP100, 401–403
 control systems instrumentation
  flow transmitters, 130–131
  general requirements, 127–128
  level transmitters, 129
  pressure transmitters, 128
  temperature devices, 129–130
Nuclear technologies
 cost competitiveness, 580–582
 decommissioning and decontamination, 578
 financing, 578–580
 fuel requirements, 575–576
 nuclear fuel and modules transport, 575–576
 remote operations and security, 576–577
 used fuel storage, 577–578
Nucleoelectrica Argentina SA (NA-SA), 361
NuScale integral system test (NIST), 200–201
NuScale LW-SMR design, 522
NuScale reactor, 37, 199
 control rod drive mechanisms, 106
 external pool, 111
 flow transmitters, 126
 NRC's final safety evaluation report, 280–281
 pressurizer, 102–103
 pumps, 103–104
 steam generators, 105

O

Oak Ridge National Laboratory (ORNL), 339–340
Oak Ridge National Laboratory Fluoride-Salt-Cooled High-Temperature Demonstration Reactor (FHR-DR), 539, 544f
Oak Ridge Siting Analysis for power Generation Expansion (OR-SAGE), 339–340
Oil shale, 348–349
OKBM Afrikantov small modular reactor projects
 ABV-6E reactor, 470–474, 475f
 construction period, 476–477

core and fuel design characteristics, 473t
design and operating characteristics, 470, 471–472t
design features, 476
fourth level of defence, 478–479
in-vessel corium retention without core catcher, 478–479
KLT-40S reactor, 470, 474f
on-site and off-site emergency planning, 479
probabilistic safety analysis, 479–480
protection against external event impacts, 479
RITM-200 reactor, 470, 474–475, 476f
second level of defence, 478
third level of defence, 478
Olefins production, 349
Operation and maintenance (O&M) costs, 241, 267–268
Optical fiber technology, 124–125, 124f
Organization for Economic Co-operation and Development Nuclear Energy Agency (OECD NEA), 324
Organization of Canadian Nuclear Industries (OCNI), 386
Original equipment manufacturer (OEM), 310
Otto Hahn reactors, 5, 54
Outage control center (OCC), 166

P

Paris Agreement, 566–569
Passive decay heat removal (PDHR) system, 461
Passive nuclear power plant electrical systems, 139
Passive safety design features, 188–189
 ACP100, 404–405
 BANDI-60S, 447–449
 4S reactor, 422
People's Republic of China
 ACP100, 399–407
 gas-cooled reactors, 39
 HTR-200, 396–399, 406–407
 water-cooled reactors, 33
Periodic cable testing, 135
Physical security (PS), 232–234
 integral pressurized water reactor, 140

Plant cost *vs.* power output, 299, 300*f*
Polarized 3D system, 170
Power distribution, 73–74
Power operated relief valve (PORV), 102–103
Power plant critical path, 320–321
Power Reactor Inherently Safe Module (PRISM) sodium-cooled reactor design, 43–44
Power stability, 74–75
Preliminary safety information document (PSID), 454
Pressure transmitters
  nuclear steam supply system (NSSS) control systems instrumentation, 128
  safety system instrumentation and controls, 121–122
Pressure vessel, 96–101
Pressurized heavy-water reactors (PHWR), 375. *See also* Canadian SMRs
Pressurized water reactor (PWR), 69
  safety design criteria, 70–75
  system and component design features, 95, 96*t*
Pressurizer, 101–103
Pressurizer relief tank, 101–103
Principle design criteria (PDC), 289
Probabilistic risk assessment (PRA), 59, 59*f*, 137, 283
  defense in depth, 206
  design basis accident, 204–206
  off-site emergency planning zone, 209–210
  PRA-guided design approach, 207–208
  safety indicators, 206
  seismic isolators, 210
  WANO's performance assessments, 296
Probabilistic safety assessment (PSA), 204–210
Process heat applications
  ammonia production, 347
  coal-to-gasoline production, 346
  general considerations, 342–344
  hydrogen production, 345–346
  natural gas-to-diesel production, 346–347
  oil shale, 348–349
  olefins via methanol production, 349
  overview, 342, 343*t*, 344–349
  steam-assisted gravity drainage, 348
  water desalination, 347–348

Processing electronics, 132–135
Process Inherent Ultimate Safety (PIUS) reactor, 54–55, 410
Production line approach
  automotive applications, 302
  component sizing, 308
  flowline, 303–308
  sales, volume and profile of, 302–303
  standardisation, 308
Projected LUEC, 7, 9–10
Proliferation resistance and physical protection (PR&PP)
  basic framework, 217, 218*f*, 222, 223*f*
  case study, 229–230
  definitions, 217–218
  future trends, 234–236
  general considerations, 221–222
  Gen IV International Forum (GIF) webinar, 217, 226–229
  importance, 218–222
  physical security, 232–234
  system response
    element identification, 224
    estimation of measures, 225–226
    pathway identification, 224–225
    steps, 223, 224*f*
    target identification, 224
Prototype fast reactor (PFR), United Kingdom, 504–505
Prototype Gen IV sodium-cooled fast reactor (PGSFR), 426
  design parameters, 455, 455*t*
  design phases, 454
  future plan, 454–455
  safety design features, 456
  safety performance, 454
  spent nuclear fuel (SNF) problem, 453
  technical data, 455–456
Pumps, 103–104

R

Radioisotope, 359–360
RA-1 research reactor, 359–360
RA-3 research reactor, 360
RA-6 research reactor, 360
RA-10 research reactor, 360
Reactivity coefficients, 72–73
Reactivity control, 105–106

Reactor coolant system piping, 101
Reactor footprint, 188
Reactor module certification, international transfer of, 294
Reactor pressure vessel (RPV), 220, 363
Reactor vessel auxiliary cooling system (RVACS), 422
Real options, 274
Refueling water storage tank (RWST), 109–111
Regional Energy Reactor 10 MWt (REX-10), 427–428
  design parameters, 451, 452t
  divisions, 450
  engineered safety systems, 452, 453f
  future plans, 450
  goals, 449–450
  schematic diagram, 450, 451f
  technical data, 450–452
Regulated asset base model, 515
Regulatory Guide (RG) 1.232, 113
Relief valves, 107
Remote operations and security, 576–577
Republic of Korea SMRs
  BANDI-60S, 444–449
  MINERVA, 463
  MMR, 458–462
  PGSFR, 453–456
  REX-10, 449–452
  SMART, 428–442
  VHTR, 456–458
  water-cooled reactors, 34
Research and Development Institute of Power Engineering (RDIPE), 480–481
Research reactors, 359–360
Residual heat removal (RHR) system, 108–109
Resistance temperature devices (RTDs), 123–124
Riser, 104
Risk-informed licensing, 212
RITM-200 reactor, 35–36, 470, 494, 496–497. *See also* OKBM Afrikantov small modular reactor projects
  construction period, 476–477
  core and fuel design characteristics, 473t
  core design, 477
  deployment in Russia, 494–495
  design and operating characteristics, 471–472t
  factory refueling, 477
  layout, 474–475, 476f
2020 Roadmap for Integrated Energy Systems, 349–350
Russian Federation
  GT-MHR, 39–40
  KLT-40S, 34–35
  RITM-200, 35–36
  SVBR-100, 43
  VK-300, 36
Russian Federation small modular reactors
  deployment, 494–495
  design development
    activities, 468
    vs. other countries, 468–469
  future trends, 496–497
  Joint Stock Company (JSC) NIKIET, 480–488
  JSC AKME Engineering projects, 489–494
  OKBM Afrikantov projects, 469–480
Russian PAMIR reactor, 4
Russian territory
  characteristics, 467–468
  permafrost, 467

S

Safeguards by design (SBD), 235
Safe Integral Reactor (SIR), 55
Safety-by-Design approach, 56, 57–58t, 61, 189, 191, 192–193t
Safety design criteria
  fuel burnup, 71–72
  maximum reactivity insertion rate, 74
  power distribution, 73–74
  power stability, 74–75
  reactivity coefficients, 72–73
  shutdown margin, 74
Safety evaluation report (SER), 280–281
Safety, iPWRs
  analytic capabilities and licensing, 212
  challenges, 187–188, 210
  design trends, 211–212
  low-power level systems, 212
  PRA/PSA (*see* Probabilistic risk assessment (PRA))
  research needs, 212–213

Index   623

safety indicators, 206
security characteristics, 210–211, 211*t*
simplicity, 188
testing and validation, 212
Safety, nuclear power plants, 52, 56–61
Safety system instrumentation and controls
  flow transmitters, 125–126
  general requirements, 119–120
  level transmitters, 122–123
  old *vs*. new, 140–141
  power/flux devices, 127
  pressure transmitters, 121–122
  temperature devices, 123–125
SEALER-UK reactor, United Kingdom, 513
Security-by-Design, 60
Seismic isolators, 210
Seoul National University (SNU), 449–452
Separate effects component tests, 198–199
Shape deposition manufacturing (SDM), 132
Shaped metal deposition (SMD), 314–316
SHELF, 488
  core and fuel design characteristics, 484*t*
  deployment in Russia, 495
  description, 486
  design and operating characteristics, 482–483*t*
  safety features, 487
  schematic illustration, 486*f*
  seabed-based variant, 486–487
  self-spaced cylindrical fuel elements, 487
  for unmanned operation, 487
Shutdown margin, 74
Single-function LCS, 165
Small modular advanced reactor training (SMART), 387
Small modular reactor (SMR)
  advantages, 598
  Canada (*see* Canadian SMRs)
  commercial deployment, 279
  deployment, 319–320, 595–596, 598–601
  in developing countries, 597–601
  energy-security benefits, 598
  intrinsic technical features, 595
  Japan (*see* Japanese SMR)
  license (*see* licensing)
Small Modular Reactor Ad hoc Group (SMRAG) report, 294–295
Small modular reactors
  (SMRs), 52–53, 69

capital cost, 62, 62*t*
challenges, 8–10
characteristics, 557–558
commercial industries, 6, 6*t*
cost competitiveness, 580–582
current status, 23–24
decay heat removal, 22–23
decommissioning and decontamination, 578
definition, 3–4
deployment conditions and regional energy aims, 566
deployment potential with indicators, 560–566
development challenges, 48
disruptive change, 570–574
economic implications, 18–19
energy sources, 47
evolution, 4–7
financing, 578–580
fission products, mitigation of, 22
fuel requirements, 575–576
future trends, 24
global energy demands, 570–571
global market assessments, 559–560
HSI (*see* Human-system interfaces (HSI))
incentives, 7–8
indicators
  climate change motivation category, 561, 562*f*, 563–564
  energy security motivation category, 561, 562*f*, 564
  financial and economic sufficiency category, 561, 561*f*, 563
  national energy demand category, 561–562, 561*f*
  physical infrastructure sufficiency category, 561, 562*f*, 563
instrumentation and controls
  (*see* Instrumentation and controls (I&C) design)
vs. large reactors
  cumulated cash flows, 264, 266*f*
  deterministic scenarios, 263–264
  economy of multiples, 269
  external factors, 271–273, 272*f*
  operating costs, 267–269
  profitability index, 265–267, 267*f*

Small modular reactors
  (SMRs) *(Continued)*
    uncertainty, 264–267
  LOCA, 22–23
  nuclear fuel and modules transport,
    575–576
  operational reliability, 14–17
  potential energy release, 19–22
  public health and safety, 19–23
  reactor mission, 13–14
  recommended sources, 587–588
  remote operations and security, 576–577
  Republic of Korea
    BANDI-60S, 444–449
    MINERVA, 463
    MMR, 458–462
    PGSFR, 453–456
    REX-10, 449–452
    SMART, 428–442
    VHTR, 456–458
  role in low-carbon energy generation,
    515–517
  sociopolitical and environmental
    considerations, 584–586
  sources of financing, 264, 265f
  strategy for development, 4
  types, 10–19
  used fuel storage, 577–578
  workforce issues, 586
  worldwide design, 29–31, 30t
SMART integral test loop (SMART-ITL)
  facility, 201–203
SMART technology. *See* System-integrated
  modular advanced reactor (SMART)
  technology
SMR-160 design, 37–38
SMR Regulators' Forum, 293
Sodium Cooled Fast Reactor Development
  Agency, 453–454
Sodium thermohydraulic test program
  (STELLA), 454
Software controlled nonsafety digital
  systems, 141
Source term, 188, 285
SpaceX Hawthorne production facility, 584
Special economic zones (SEZs), 583
Specific design safety analysis report
  (SDSAR), 454
Speech recognition, 170–171

Spent nuclear fuel (SNF), 453
Sporian temperature sensor, 125, 125f
Spray valve, 101–103
4S (supersafe, small, and simple) reactor,
  420–422
Standardization, of mass-produced reactors,
  585
Standard review plan (SRP), 280
Standards development organizations
  (SDOs), 290–291
Steam-assisted gravity drainage (SAGD), 348
Steam-generating heavy-water reactor
  (SGHWR)
  United Kingdom, 504–505
Steam generator(s), 104–105
Structures, systems and components (SSCs),
  296
SVBR-10 reactor, 494
SVBR-100 reactor, 43, 489
  baseload electricity generation, 489–490
  core and fuel design characteristics, 491t
  deployment in Russia, 495
  design and operating characteristics, 489,
    490t
  equipment layout schematics, 490, 492f
  reactor module layout, 491, 493f
  safety features, 491–492
  seismic design, 492–493
System-integrated modular advanced reactor
  (SMART) technology, 34
  aim, 425
  applications, 426
  chronicles, 428–429
  configuration, 430, 431f
  control rod drive mechanisms, 106
  design characteristics, 430–439
  design parameters, 430, 432–433t
  engineered safety features, 435
  fuel assembly and core, 435
  nuclear safety, 436
  pumps, 103–104
  reactor coolant pump, 435
  reactor vessel assembly, 434
  safety design principles, 436–438
  Standard Design Approval (SDA), 428
  steam generator cassette, 435
  steam generators, 105
  thermohydraulic test, 441–442
  verification, 439–442

Index 625

System-integrated PWR (SPWR), 410
Systems engineering (SE), 175–176
Systems engineering process (SEP), 175–176
System Steering Committees (SSCs), 230–232

T

Target costs, for SMRs, 581–582
Technical support center (TSC), 167
Technology development, 596
  measuring, 596–597
Temperature devices
  nuclear steam supply system (NSSS) control systems instrumentation, 129–130
  safety system instrumentation and controls, 123–125
Testing, of SMRs, 196–204
Thermocouples, 123–124
Thermohydraulic test, SMART
  components performance test, 442
  critical heat flux tests, 441–442
  integral effect test, 442
  purpose, 441
  two-phase critical flow test, 442
3D displays, 170
3D-printing, 309
Three Mile Island Unit 2 (TMI-2) accident, 51–52, 59, 153, 409–410
Tokamak energy spherical tokamak, United Kingdom, 513
Top-down estimation cost, 244
Toshiba Super Safe Small and Simple (4S) reactor design, 42–43
Total capital investment cost, 241
Transient Reactor Test (TREAT) Facility, 550–551
TRISO-coated uranium particle fuel, 40–41, 46
Tube sheets, 104–105
Two-phase critical flow test, 442

U

U-Battery, United Kingdom, 512, 512f
Ulsan Institute of Science and Technology (UNIST), 463
United Kingdom
  enabling regulation, 515
  Fusion research, 514
  interest in modular reactors, 505–507
Nuclear Innovation and Research Advisory Board (NIRAB), 514
nuclear innovation program (NIP)
  advanced fuels, 509
  advanced manufacturing and materials, 507–509
  nuclear facilities and strategic toolkit component, 511
  reactor design, 510–511
  recycle and waste management, 509–510
nuclear power development, history of, 503–505
research and development, SMR, 503
UK SMR funding, 514
United Kingdom Atomic Energy Authority (UKAEA), 55
United Nations Framework Convention on Climate Change (UNFCCC), 568
United States
  BWRX-300, 36
  EM^2 reactors, 40–41
  FOAK reactor technology, 288
  KP-FHR, 46
  LFTR design, 46–47
  longer-term SMR activities, 530–534
  near-term SMR activities, 522–530
  NuScale, 37
  PRISM design, 43–44
  SMR-160, 37–38
  Xe-100, 41
UNITHERM, 488
  autonomous operation, 485
  core and fuel design characteristics, 484t
  deployment in Russia, 495
  design and operating characteristics, 481, 482–483t
  intermediate circuit, 481–485
  safety features, 487
  self-spaced cylindrical fuel elements, 487
  three-circuit scheme, 481, 485f
US Army Nuclear Power Program, 4
US coal-fired plant retirements, 572, 572f
US Department of Energy (DOE), 288
US Department of Energy's Office of Nuclear Energy (DOE-NE) programs, 521
  additional LW-SMR support, 523
  Advanced Research Projects Agency-Energy (ARPA-E) program, 549–550

US Department of Energy's Office of Nuclear Energy (DOE-NE) programs *(Continued)*
  ART R&D program, 531–532
  future trends, 550–552
  Gateway for Accelerated Innovation in Nuclear (GAIN) program, 541–545
  Holtec SMR-160 design description, 526–530
  Licensing Technical Support (LTS) program, 521–523
  microreactor R&D program, 548–549
  modeling/simulation and experimental capabilities, 550–551
  National Reactor Innovation Center, 546–548
  NuScale design description, 523–526
  R&D approach, 521–522
US Department of Transportation (DOT), 576
Used fuel storage, 577–578
US National Weather Service, 296
US Nuclear Energy for the Propulsion of Aircraft (NEPA) project, 5–6
US Nuclear Regulatory Commission (NRC), 232, 234, 576
  licensing, 280–288
US Secretary of Energy, 241

V

Variable renewable energy (VRE) sources, 573–574
VBER- 300 design, 499
Very small reactors (VSR), 262–263
VHTR
  future plan, 458
  key technologies development project, 456
  nuclear hydrogen production, 456
Vibrating fork technology, 129
VISTA facility, 442, 443f
Visual interfaces, 168–170
  large screen displays, 169

3D displays, 170
wearable displays, 169–170
VK-300 design, 36
VVER-600 and VVER-300 design, 499

W

Waste disposal, of nuclear power plants, 52
Water-cooled reactors
  ACP-100, 33
  BWRX-300, 36
  CAREM, 32–33
  commercial designs, 31–32, 32t
  Flexblue design, 33–34
  KLT-40S, 34–35
  NuScale design, 37
  RITM-200, 35–36
  SMART design, 34
  SMR-160, 37–38
  VK-300, 36
Water desalination, 347–348
Wearable displays, 169–170
Westinghouse lead-cooled fast reactor (LFR), United Kingdom, 199, 513, 539, 548f
  control rod drive mechanisms, 106–107
  steam generators, 105
Westinghouse SNUPPs 4-loop design, 505
Wireless safety systems, 142
World Association of Nuclear Operators (WANO), 296
World Association of Nuclear Plant Operators, 571
World Bank economic indicator, 596
World Nuclear Association, 294, 568

X

Xe-100 reactor, 41

Z

Zero emission credits (ZEC), 582–583